Volcanism and Tectonism Across the Inner Solar System

Geological Society books refereeing procedures

The Society makes every effort to ensure that the scientific and production quality of its books matches that of its journals. Since 1997, all book proposals have been refereed by specialist reviewers as well as by the Society's Books Editorial Committee. If the referees identify weaknesses in the proposal, these must be addressed before the proposal is accepted.

Once the book is accepted, the Society Book Editors ensure that the volume editors follow strict guidelines on refereeing and quality control. We insist that individual papers can only be accepted after satisfactory review by two independent referees. The questions on the review forms are similar to those for *Journal of the Geological Society*. The referees' forms and comments must be available to the Society's Book Editors on request.

Although many of the books result from meetings, the editors are expected to commission papers that were not presented at the meeting to ensure that the book provides a balanced coverage of the subject. Being accepted for presentation at the meeting does not guarantee inclusion in the book.

More information about submitting a proposal and producing a book for the Society can be found on its website: www.geolsoc.org.uk.

It is recommended that reference to all or part of this book should be made in one of the following ways:

PLATZ, T., MASSIRONI, M., BYRNE, P. K. & HIESINGER, H. (eds) 2015. *Volcanism and Tectonism Across the Inner Solar System*. Geological Society, London, Special Publications, **401**.

CARLI, C., SERVENTI, G. & SGAVETTI, M. 2015. VNIR spectral characteristics of terrestrial igneous effusive rocks: mineralogical composition and the influence of texture. *In*: PLATZ, T., MASSIRONI, M., BYRNE, P. K. & HIESINGER, H. (eds) *Volcanism and Tectonism Across the Inner Solar System*. Geological Society, London, Special Publications, **401**, 139–158. First published online June 17, 2014, http://dx.doi.org/10.1144/SP401.19

GEOLOGICAL SOCIETY SPECIAL PUBLICATION NO. 401

Volcanism and Tectonism Across the Inner Solar System

EDITED BY

T. PLATZ
Planetary Science Institute, USA

M. MASSIRONI
Università degli Studi di Padova, Italy

P. K. BYRNE
Lunar and Planetary Institute, USA

and

H. HIESINGER
Westfälische Wilhelms-Universität Münster, Germany

2015
Published by
The Geological Society
London

THE GEOLOGICAL SOCIETY

The Geological Society of London (GSL) was founded in 1807. It is the oldest national geological society in the world and the largest in Europe. It was incorporated under Royal Charter in 1825 and is Registered Charity 210161.

The Society is the UK national learned and professional society for geology with a worldwide Fellowship (FGS) of over 10 000. The Society has the power to confer Chartered status on suitably qualified Fellows, and about 2000 of the Fellowship carry the title (CGeol). Chartered Geologists may also obtain the equivalent European title, European Geologist (EurGeol). One fifth of the Society's fellowship resides outside the UK. To find out more about the Society, log on to www.geolsoc.org.uk.

The Geological Society Publishing House (Bath, UK) produces the Society's international journals and books, and acts as European distributor for selected publications of the American Association of Petroleum Geologists (AAPG), the Indonesian Petroleum Association (IPA), the Geological Society of America (GSA), the Society for Sedimentary Geology (SEPM) and the Geologists' Association (GA). Joint marketing agreements ensure that GSL Fellows may purchase these societies' publications at a discount. The Society's online bookshop (accessible from www.geolsoc.org.uk) offers secure book purchasing with your credit or debit card.

To find out about joining the Society and benefiting from substantial discounts on publications of GSL and other societies worldwide, consult www.geolsoc.org.uk, or contact the Fellowship Department at: The Geological Society, Burlington House, Piccadilly, London W1J 0BG: Tel. +44 (0)20 7434 9944; Fax +44 (0)20 7439 8975; E-mail: enquiries@geolsoc.org.uk.

For information about the Society's meetings, consult *Events* on www.geolsoc.org.uk. To find out more about the Society's Corporate Affiliates Scheme, write to enquiries@geolsoc.org.uk.

Published by The Geological Society from:
The Geological Society Publishing House, Unit 7, Brassmill Enterprise Centre, Brassmill Lane, Bath BA1 3JN, UK

The Lyell Collection: www.lyellcollection.org
Online bookshop: www.geolsoc.org.uk/bookshop
Orders: Tel. +44 (0)1225 445046, Fax +44 (0)1225 442836

The publishers make no representation, express or implied, with regard to the accuracy of the information contained in this book and cannot accept any legal responsibility for any errors or omissions that may be made.

British Library Cataloguing in Publication Data

A catalogue record for this book is available from the British Library.
ISBN 978-1-86239-632-6
ISSN 0305-8719

Distributors
For details of international agents and distributors see:
www.geolsoc.org.uk/agentsdistributors

Typeset by Techset Composition India (P) Ltd., Bangalore and Chennai, India.
Printed by Berforts Information Press Ltd, Oxford, UK

Contents

Preface

Our understanding of the inner solar system has dramatically expanded over the past three decades, with unmanned spacecraft visiting worlds as diverse as enigmatic Mercury, Earth's hellish sister planet Venus, and far flung asteroids. Advances in spacecraft navigation, longevity and instrument design mean we now have more information about our neighbouring planetary bodies than we do for parts of Earth's oceans. And with these advances comes a far greater comprehension of the ways in which planetary surfaces are shaped.

Volcanic and tectonic processes are the dominant means by which a planet's surface is modified from within, and it is the various aspects of these processes that form the basis for the book you now hold. This Special Publication arose from a series of planetary volcanism and tectonism sessions at the European Geosciences Union General Assembly held between 2010 and 2014. In its development we have sought not to exhaustively document the myriad ways in which volcanism and tectonism are manifest on planetary bodies, but instead to provide a basis from which a non-expert audience can explore these facets of planetary science, and with which the specialist reader can expand her or his interests.

Of course, this book would not be possible without the reviewers who played a vital role in ensuring its scientific content was robust and thematically appropriate. We are therefore very grateful to Sebastian Besse, David Blair, Marco Bonini, Giacomo Corti, Brett Denevi, Caleb Fassett, Taras Gerya, Agust Gudmundsson, Christopher Hamilton, Karen Harpp, Robert Herrick, Mikhail Ivanov, Christian Klimczak, Nicholas Lang, Lucia Marinangeli, Tamsin Mather, Lucie Mathieu, Francesco Mazzarini, Patrick McGovern, Chris Okubo, Mark Robinson, Stefano Tavani, Stephanie Werner and Nigel Woodcock, and to 11 reviewers who wished to remain anonymous. We are also indebted to Angharad Hills and the staff at the Geological Society of London, who together patiently and expertly guided us through the entire development of this book.

THOMAS PLATZ
MATTEO MASSIRONI
PAUL K. BYRNE
HARALD HIESINGER

Volcanism and tectonism across the inner solar system: an overview

T. PLATZ[1,2]*, P. K. BYRNE[3,4], M. MASSIRONI[5] & H. HIESINGER[6]

[1]*Planetary Science Institute, 1700 East Fort Lowell Road, Tucson, AZ 85719-2395, USA*

[2]*Freie Universität Berlin, Institute of Geological Sciences, Planetary Sciences & Remote Sensing, Malteserstrasse 74-100, 12249 Berlin, Germany*

[3]*Lunar and Planetary Institute, Universities Space Research Association, 3600 Bay Area Boulevard, Houston, TX 77058, USA*

[4]*Department of Terrestrial Magnetism, Carnegie Institution of Washington, 5241 Broad Branch Road NW, Washington, DC 20015-1305, USA*

[5]*Dipartimento di Geoscienze, Universita' degli Studi di Padova, via G. Gradenigo 6, 35131 Padova, Italy*

[6]*Institut für Planetologie, Westfälische Wilhelms-Universität Münster, Wilhelm-Klemm-Strasse 10, 48149 Münster, Germany*

*Corresponding author (e-mail: platz@psi.edu)

Abstract: Volcanism and tectonism are the dominant endogenic means by which planetary surfaces change. This book, in general, and this overview, in particular, aim to encompass the broad range in character of volcanism, tectonism, faulting and associated interactions observed on planetary bodies across the inner solar system – a region that includes Mercury, Venus, Earth, the Moon, Mars and asteroids. The diversity and breadth of landforms produced by volcanic and tectonic processes are enormous, and vary across the inventory of inner solar system bodies. As a result, the selection of prevailing landforms and their underlying formational processes that are described and highlighted in this review are but a primer to the expansive field of planetary volcanism and tectonism. In addition to this extended introductory contribution, this Special Publication features 21 dedicated research articles about volcanic and tectonic processes manifest across the inner solar system. Those articles are summarized at the end of this review.

Volcanic and tectonic processes have profoundly shaped the surfaces of terrestrial planets in the inner solar system. Even minor bodies such as asteroids and small moons, where volcanism and tectonism have not played dominant roles, are still affected by fracturing and faulting as a result of other processes like dynamic loading and gravitational collapse. This Special Publication aims to encompass the broad range in character of volcanism, tectonism, faulting and associated interactions observed on planetary bodies across the inner solar system. By collating observations of the Earth and other planetary bodies, the interpretations of extraterrestrial landforms and their formational processes are appraised in the light of our current understanding of comparable processes on Earth.

The inner solar system comprises our star, the Sun, and the four terrestrial planets, Mercury, Venus, Earth and Mars, as well as Mars' moons Phobos and Deimos, and Earth's companion, the Moon (Fig. 1). Although the main asteroid belt, located between the orbits of Mars and Jupiter, divides our solar system into inner and outer portions, it itself is composed of asteroidal and cometary objects of which a large number enter the inner solar system. Some asteroids have received attention as the result of spacecraft flybys or orbital operations, and for that reason are included briefly in this volume.

In this Special Publication, the journey across the inner solar system begins at the planet closest to the Sun. From Mercury we move to Venus; Earth and its Moon are next, before we move yet further out, to Mars. This celestial journey terminates at the main asteroid belt (Fig. 1).

The first part of this introductory chapter highlights the current knowledge of, and recent discoveries regarding, volcanic and tectonic features and their formational processes on the Moon, Mars, Mercury and Venus. The second part is dedicated to summarizing the major conclusions of articles presented in this volume. In its writing, we have sought not to compose a comprehensive review

From: PLATZ, T., MASSIRONI, M., BYRNE, P. K. & HIESINGER, H. (eds) 2015. *Volcanism and Tectonism Across the Inner Solar System*. Geological Society, London, Special Publications, **401**, 1–56.
First published online September 17, 2014, http://dx.doi.org/10.1144/SP401.22

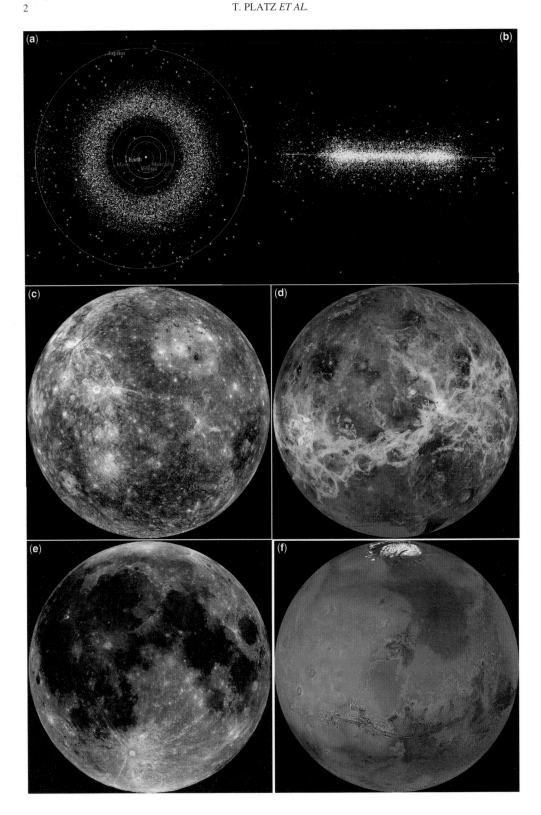

per se but, rather, to provide a detailed introduction to the diversity of observed volcanic and tectonic processes present throughout the inner solar system, from which the interested reader may explore further – and farther.

Mercury

Until very recently, Mercury (Fig. 1c) was the most enigmatic of the inner solar system's planets. Its proximity to the Sun rendered telescopic observations of Mercury from Earth difficult, and the planet's location in the Sun's gravity well challenged mission designers. It was not until NASA's Mariner 10 spacecraft flew past the planet in the 1970s that the surface of Mercury was imaged directly for the first time and, even then, only a single hemisphere was observed. Those early data showed the planet's surface to resemble superficially that of the Moon, with ancient, cratered plains interspersed with expanses of younger smooth plains. Yet, unlike its larger terrestrial counterparts, Mercury does not have primary volcanic features, such as the giant shield volcanoes that dominate the Tharsis province on Mars. The volcanic character of Mercury, therefore, remained an open question until the planet was visited by the MErcury Surface, Space ENvironment, GEochemistry, and Ranging (MESSENGER) mission. However, the tectonics of Mercury were readily visible from the outset of its exploration by spacecraft. Long, cliff-like escarpments were observed across the Mariner 10 hemisphere, with wrinkle ridges akin to those in lunar maria populating the planet's smooth plains units. Even so, the spatial extent, styles and amount of tectonic deformation of Mercury are questions that could only be fully explored from orbit. This section describes the current state of knowledge of Mercury's volcanic and tectonic character, places these findings in the context of how our understanding of the innermost planet has evolved, and highlights key aspects of the geological development of Mercury that have yet to be answered.

Volcanism

The three flybys of Mercury by the Mariner 10 spacecraft in 1974–1975 returned images that raised the prospect of volcanism on the innermost planet. Smooth plains deposits were identified across the approximately 45% of the planet observed during that mission; some workers interpreted their large volumes, together with their embayment relationships with, and spectral distinctiveness from, surrounding terrain, as evidence for a volcanic origin for these deposits (Murray *et al.* 1975; Strom *et al.* 1975; Dzurisin 1978; Kiefer & Murray 1987; Robinson & Lucey 1997). Yet, others argued that Mercury's smooth plains units were morphologically similar to lunar highland plains, which were shown to have been emplaced as fluidized ejecta (Wilhelms 1976; Oberbeck *et al.* 1977). The provenance of smooth plains on Mercury therefore remained unresolved until the three flybys of the MESSENGER spacecraft in 2008–2009 (Fig. 1c).

Smooth plains. MESSENGER imaged almost the entire surface of Mercury during its flybys, and showed the smooth plains to be a globally present unit, the majority of which is volcanic in nature (Fig. 2). This inference is based on superposition relations indicative of the sequential embayment of impact basins and ejecta, as well as spectral homogeneity but colour variation, partially buried impact structures, and thicknesses of hundreds to thousands of metres (Head *et al.* 2008, 2009; Denevi *et al.* 2009). Observations made after MESSENGER entered orbit about the planet in March 2011 have allowed for the spatial extent of Mercury's smooth plains to be quantified (Denevi *et al.* 2013): these plains are now known to occupy some 27% of the surface of Mercury (Fig. 2). Notably, the single largest contiguous smooth plains unit on the planet has been identified at high northern latitudes (Head *et al.* 2011). Occupying around 6% of the total planet surface, this region has been termed the northern volcanic plains (NVP) (Fig. 3a).

Fig. 1. The principal components of the inner solar system include the four terrestrial planets, Mercury, Venus, Earth and Mars, as well as Earth's moon and the two moons of Mars. The main asteroid belt separates the inner and outer portions of the solar system. On a yet smaller scale are objects that come close to Earth's neighbourhood (on astronomical scales) or cross Earth's orbit; these are collectively termed Near Earth Objects (NEO). At time of writing, there are 11 057 NEAs, of which 861 are larger than 1 km in diameter. (**a**) View of the inner solar system from above the ecliptic plane. The yellow dots denote Near Earth Asteroids; white triangles denote Near Earth Comets (courtesy of P. Chodas; 1 April 2014; NASA/JPL; http://neo.jpl.nasa.gov). (**b**) View of the inner solar system from the edge of the ecliptic plane. The orange line represents Jupiter's orbit (courtesy of P. Chodas; 1 April 2014; NASA/JPL; http://neo.jpl.nasa.gov). (**c**) Enhanced colour mosaic of Mercury in orthographic projection centred at 0° (wide-angle camera of the Mercury Dual Imaging System; NASA/John Hopkins University Applied Physics Laboratory/Carnegie Institution of Washington). (**d**) Global view of Venus centred at 180°E (Magellan Synthetic Aperture Radar Mosaic; NASA/JPL). (**e**) Nearside view of the Moon (Lunar Reconnaissance Orbiter wide-angle camera mosaic; NASA/GSFC/Arizona State University). (**f**) Global view of Mars centred at 20°N, 300°E (Viking Orbiter 1 mosaic; NASA/JPL/USGS).

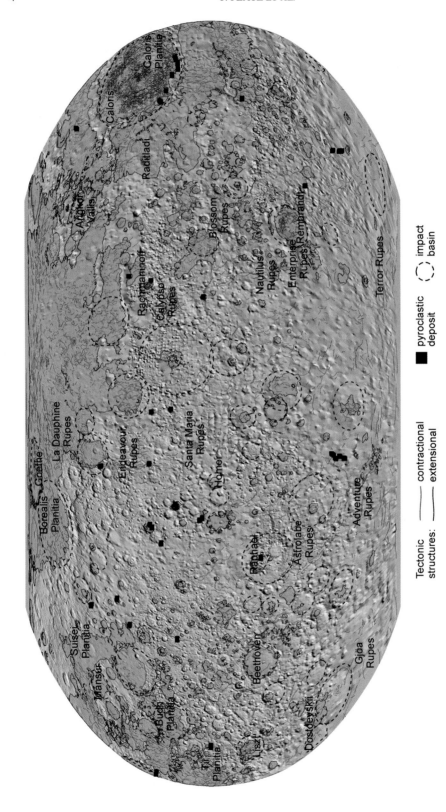

Fig. 2. All mapped tectonic structures on Mercury. Contractional (red lines: Byrne *et al.* 2014*b*) and extensional (blue lines: Byrne *et al.* 2013*d*) are shown, as well as the smooth plains units mapped by Denevi *et al.* (2013). Black squares represent the locations of pyroclastic deposits (Kerber *et al.* 2011*a*). Only impact basins >300 km in diameter (dashed ellipses) are highlighted for clarity (Fassett *et al.* 2011). Topography is from the controlled Mercury Dual Imaging System (MDIS) wide-angle camera global base map (2.7 km/px; Becker *et al.* 2012). The map is shown in a Robinson projection, centred at 0°E.

The NVP is not obviously related to one or more large impact basins, although an origin due to impact cannot be discounted (Byrne *et al.* 2013*b*). Nevertheless, these plains, emplaced near the end of the Late Heavy Bombardment (LHB) of the inner solar system at around 3.7–3.8 Ga, were likely to have formed in a single, voluminous event associated with extensive partial melting of Mercury's mantle (Head *et al.* 2011). The NVP hosts a variety of landforms characteristic of the planet's physical volcanological character in general. Numerous 'ghost craters', impact structures that predate the emplacement of the plains and that are partially to almost entirely filled with lava, are widespread throughout the NVP. Features interpreted to be lava flow fronts have also been identified throughout the region, often as linear, lobate contacts that embay older, more heavily textured terrain (Head *et al.* 2011).

Valles. One of the most notable types of landform spatially associated with the NVP is a set of five broad channels (or 'valles') located to the SE of the region. These channels, morphologically distinct from any other trough-like depression identified on Mercury, connect regions of smooth plains through surrounding older terrain, and are typified by steep, linear edges and smooth floors that feature rounded islands aligned with the long axes of the channels (Byrne *et al.* 2013*c*). In one case, at the end of Angkor Vallis, a group of such islands forms a 'splay'-like pattern as the channel opens into the adjoining Kofi basin (Fig. 3b). Four of the valles are orientated approximately radial to the 1640 km-diameter Caloris basin, and may have originally been impact-sculpted furrows carved by ballistically emplaced ejecta during the formation of that basin (Byrne *et al.* 2013*c*). Some combination of thermal and mechanical lava erosion, probably by voluminous, high-temperature, low-viscosity flows, then probably shaped the troughs to the forms the channels have today (Byrne *et al.* 2013*c*; Hurwitz *et al.* 2013*b*).

Taken together, these observations attest to a predominance of flood-mode lava emplacement for the majority of Mercury's smooth plains units. This inference is consistent with geochemical data returned by MESSENGER's X-ray spectrometer (XRS) instrument, which indicate that the planet's surface is relatively rich in Mg but poor in Al, Ca and Fe, relative to terrestrial and lunar basalts (Nittler *et al.* 2011). Mercury's surface therefore probably has a bulk composition intermediate between low-Fe basalt and high-Mg ultramafic lithologies. Moreover, on-going XRS observations suggest a compositional difference between the NVP and the surrounding plains (Stockstill-Cahill *et al.* 2012; Weider *et al.* 2012). The last widespread

volcanism on Mercury was therefore effusive, with MESSENGER observations indicating that some such activity, albeit highly spatially localized, may have occurred as recently as around 1 Ga (Prockter *et al.* 2010; Marchi *et al.* 2011).

Pyroclastic volcanism. Mercury is not without explosive volcanism. Evidence for pyroclastic activity has been identified at numerous sites across the planet, in the form of irregularly shaped depressions without raised rims that are often located atop low, broad rises. These features frequently appear to be coalesced from smaller, overlapping depressions and are typically encircled by haloes of high-reflectance material with a steeper spectral slope over visible to near-infrared wavelengths (and so they appear redder) than most of Mercury's surface. These haloes have been interpreted as proximal deposits of fine-grained pyroclastic material (Head *et al.* 2009; Kerber *et al.* 2009, 2011*a*) (Figs 2 & 3c). The Caloris basin hosts a number of such 'red'-haloed depressions, particularly along its southern margin (Head *et al.* 2009). Some of these depressions may, in fact, be long-lived loci of explosive eruptive activity, whose locations have been influenced by the underlying structural fabric of the Caloris basin (Rothery *et al.* 2014).

It should be noted that irregular, coalesced depressions occur across Mercury without red halos. For example, numerous impact craters across the planet host central 'pits' on their floors (and so are termed pit-floor craters), which appear unrelated to the impact process itself. Gillis-Davis *et al.* (2009) investigated the morphology, structural association, relative age and proximity to smooth plains units of seven such pit craters, and concluded that they formed through collapse into underlying magma chambers. A similar mechanism has been inferred for the origin of several non-impact crater-hosted depressions within the broad channel network proximal to the NVP (Byrne *et al.* 2013*c*). Such pits therefore represent a third form of surficial igneous activity on Mercury, in addition to pyroclastic and effusive eruptions (Gillis-Davis *et al.* 2009).

Cratered plains. What of Mercury's older terrain? Trask & Guest (1975) classified Mercury's non-smooth plains surface portions as either 'intercrater plains' or 'heavily cratered terrain'. Intercrater plains were interpreted to be the oldest surviving surface unit, its emplacement even predating the end of the LHB, whereas heavily cratered terrain described large craters, basins and their deposits. Despite the lack of primary volcanic landforms on Mercury (e.g. large shield volcanoes), Strom *et al.* (1975) considered their great volumes as evidence that these plains were formed by volcanism.

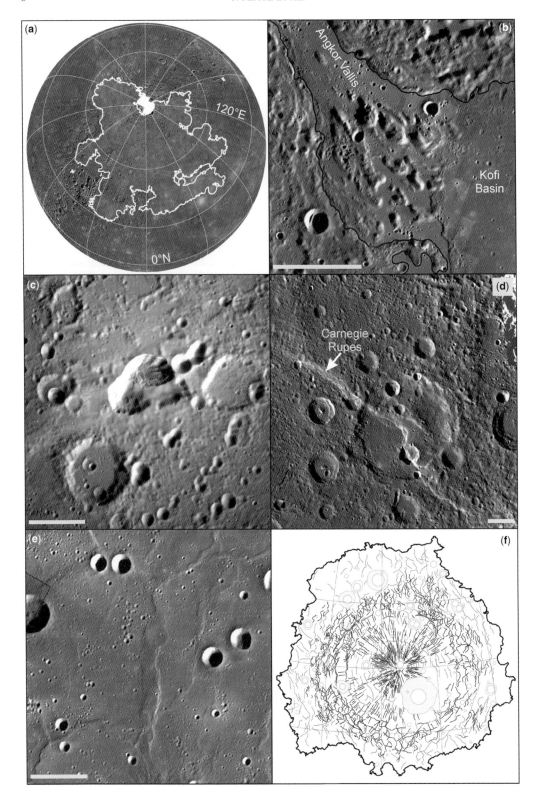

Indeed, there is a lower density of 100 km-diameter craters on Mercury than on the Moon (Strom & Neukum 1988; Fassett et al. 2011), and the oldest surviving terrain on Mercury is no older than 4.1 Ga (Marchi et al. 2013). These findings reinforce a geological history for the innermost planet that features global resurfacing due to volcanism before the end of the LHB. If so, then the majority of Mercury's surface – that portion consisting of smooth plains, intercrater plains and heavily cratered terrain – is dominantly volcanic.

A further observation is that the majority of smooth plains deposits are collocated with impact structures (Byrne et al. 2013b). This is best expressed by the expansive interior Caloris smooth plains deposits (Strom et al. 1975), and by smooth plains within other large impact basins such as Rembrandt, Beethoven and Tolstoj (Fig. 2). Although there is no consensus as to the existence of a causal link between the impact process and volcanism (Ivanov & Melosh 2003; Elkins-Tanton & Hager 2005), the impact process removes overburden, introducing thermal energy to the lithosphere, and resets pervasive regional- or global-scale stresses (see the next section) (Byrne et al. 2013b). This last point is particularly relevant to Mercury, since the global contraction of the planet (see the next section) will put the lithosphere into a pervasive state of net compression, inhibiting voluminous surface volcanism (e.g. Solomon 1977; Klimczak 2013). Whatever the mechanism responsible for the generation of large volumes of partial melt that then pools within impact structures – even if such structures penetrate to existing magma chambers or asthenospheric melts, or simply represent the shortest path for ascending magma to reach the surface – the presence of large craters and basins may have played a key role in the last phase of extensive volcanic resurfacing on Mercury.

Tectonism

A key observation made of Mercury with Mariner 10 data was the extent to which the planet's surface has been tectonized. Evidence for both crustal shortening and lengthening was identified, with the former process the overwhelmingly dominant style of deformation (Strom et al. 1975). Although Mariner 10 viewed less than half of Mercury's surface, the inference that such deformation was global in nature was confirmed when MESSENGER imaged the entire planet (Figs 1c & 2). To first order, crustal shortening has occurred universally across all surface units of the innermost planet and is manifest as one of two primary classes of structure (lobate scarp or wrinkle ridge) (e.g. Strom et al. 1975; Watters et al. 2009), whereas extension, in the form of graben, is restricted to volcanically flooded impact structures (e.g. Strom et al. 1975; Byrne et al. 2013d).

Lobate scarps. Lobate scarps are, in terms of overall length and accumulated relief, the larger of the two principal expressions of crustal shortening on Mercury. Like their counterparts on Mars and the Moon (e.g. Mueller & Golombek 2004), they are characterized by a steeply sloping scarp face and a gently sloping back limb (Fig. 3d), and probably represent a monocline or asymmetrical hanging-wall anticline atop a blind or surface-breaking thrust fault. Mercury's lobate scarps range in length from 9 to 900 km, and in places have accumulated some 3 km of relief (Byrne et al. 2014b). Although lobate scarps are superficially similar in morphology to inferred lava flow fronts (Head et al. 2011), their generally linear plan form, transection relationships with craters and surrounding terrain, and their cross-cutting of craters with orthogonal horizontal offsets support a tectonic origin for the overwhelming majority of such scarps (e.g. Strom et al. 1975). A less commonly observed form of scarp, termed a high-relief ridge, also occurs on Mercury (Dzurisin 1978; Watters et al. 2009). High-relief ridges are generally narrower than, but often transition into, lobate scarps, such that delineating these ridges from other structures is difficult.

Wrinkle ridges. Wrinkle ridges, by comparison, are substantially smaller landforms. Like on other terrestrial worlds, they are typically manifest on Mercury as broad, low-relief arches with opposite-facing leading edges, often superposed by a

---

Fig. 3. Examples of key volcanic and tectonic landforms on Mercury. (**a**) The northern volcanic plains (NVP) on Mercury (outlined in white) occupy some 6% of the planet's surface (Head et al. 2011). (**b**) Several valles occur proximal to the NVP, shaped by high-temperature, low-viscosity lavas. Here, lava flowing through Angkor Vallis into the Kofi basin has formed a 'splay'-like pattern of erosional remnants (Byrne et al. 2013c). (**c**) An example of an irregular depression, surrounded by a halo of high-albedo, fine-grained material interpreted as a pyroclastic vent (Kerber et al. 2009). This example lies between the Rachmaninoff and Copland basins. (**d**) A prominent lobate scarp, Carnegie Rupes, in Mercury's northern hemisphere; contrast its size and morphology with that of the wrinkle ridge, situated in the NVP, shown in (**e**). (**f**) Structural sketch of the tectonic structures within the Caloris basin (Byrne et al. 2013d). Ridges and scarps are shown in black, whereas graben are shown in grey. Superposed craters and their ejecta deposits are also shown. Azimuthal equidistant projections centred as follows: (a) 0°N, 70°E; (b) 57°N, 124.2°E; (c) 35.7°N, 64.1°E; (d) 59.1°N, 304.5°E; (e) 61.4°N, 49.9°E; and (f) 30°N, 161°E. Scale bar in (b)–(e) is 30 km.

narrow ridge (Fig. 3e). Although the dimensions of wrinkle ridges across the planet vary, they are usually tens to hundreds of metres in height and several kilometres in length, and occur in long, subparallel groups or in complex map patterns (Byrne *et al.* 2014*b*). Wrinkle ridges are likely to form due to some combination of faulting and folding but the orientations and depths of the causative faults, and the contribution to ridge formation played by folding, remain open questions (e.g. Golombek *et al.* 1991; Plescia 1993; Zuber 1995; Schultz 2000). Moreover, although wrinkle ridges appear to accommodate small amounts of shortening relative to lobate scarps (e.g. Plescia 1993) on all planets, debate continues regarding their subsurface structure and whether their faults penetrate to tens of kilometres (i.e. 'thick-skinned deformation': Zuber 1995; Golombek *et al.* 2001; Montési & Zuber 2003) or only the upper few kilometres of the lithosphere (i.e. 'thin-skinned deformation': Watters 1991; Mangold *et al.* 1998).

Graben. Where it has occurred, crustal lengthening on Mercury has been accommodated by linear, topographical troughs that are interpreted as graben (i.e. sets of antithetic normal faults) (e.g. Strom *et al.* 1975). Graben are typically 5–10 km long and up to 1 km wide, although some examples are larger (Klimczak *et al.* 2012; Watters *et al.* 2012). Some graben appear isolated and show no evidence of having interacted with other such structures, and display a characteristic continuous deepening of their floor towards the centre, whereas others are segmented or linked (Klimczak *et al.* 2012). Most graben show a generally constant displacement along their lengths, implying that the development of their component-normal faults was confined to a mechanical layer of limited thickness (Klimczak *et al.* 2012, 2013).

Distribution of tectonics. The tectonic deformation of Mercury is global but the distribution of different classes of structure is far from uniform (Fig. 2). In their global survey of contraction on the innermost planet, Byrne *et al.* (2014*b*) characterized shortening structures according to the primary terrain type in which they occur. Adopting the nomenclature of Trask & Guest (1975), the surface of Mercury can be described as consisting of smooth plains and cratered plains (the latter term incorporating both the intercrater plains and heavily cratered terrain units described from images returned by Mariner 10). Smooth plains structures overwhelmingly consist of wrinkle ridges, and represent about two-thirds of the almost 6000 shortening structures mapped on the planet. Over 1500 ridges were identified in the NVP alone, with the remainder situated within the circum-Caloris

plains or in smaller smooth plains deposits that occur in more heavily cratered terrain. In contrast, cratered plains structures are almost entirely lobate scarps, and represent about one-third of all mapped shortening structures.

In some cases, shortening structures demarcate volcanically filled or buried impact features. For example, wrinkle ridges can delineate the rims of buried craters (i.e. the Type-1 'ghost craters' described by Klimczak *et al.* 2012), particularly in the NVP, whereas lobate scarps situated within impact basins can follow, and verge outwards in the direction of, the basin perimeter (Byrne *et al.* 2014*b*). Moreover, approximately 100 lobate scarps border areas of high-standing terrain on Mercury and verge onto surrounding lows; these structures have some of the greatest accumulated relief of any tectonic landform on Mercury (Byrne *et al.* 2014*b*).

Earlier studies of Mercury's tectonics suggested that Mercury's lithospheric fracture pattern might contain evidence for ancient global stress states resulting from, for example, tidal despinning (Melosh & Dzurisin 1978; Melosh & McKinnon 1988). Byrne *et al.* (2014*b*) did not identify such a *globally* coherent pattern but did note that in places their structural survey was probably influenced, in part, by lighting geometry (Mercury has a remarkably low obliquity of *c.* 2 arcmin). Even so, in places there are systematic patterns of *regional*-scale deformation, where ridges and scarps form laterally contiguous, narrow bands of substantial length. These narrow zones of concentrated crustal shortening are Mercury's equivalent to the fold-and-thrust belts of Earth (Byrne *et al.* 2014*b*).

Planetary radius change. The widespread distribution of lobate scarps on Mercury implies that their formation is linked to a process that is global in scale. Thermal models for the innermost planet require a substantial contraction in response to secular cooling of the planet's interior (e.g. Solomon 1977; Schubert *et al.* 1988). It is this contraction, and the resultant horizontal compression of the planet's lithosphere, that probably formed the lobate scarp population observed today. Importantly, the morphology of lobate scarps can be used to estimate their contribution to planetary radius change (a direct measure of planetary contraction), wherein their relief is related to their horizontal component of shortening using an assumed fault dip.

There has been a long-standing disagreement between estimates of radius change made from photogeological studies of Mercury's brittle structures (e.g. Strom *et al.* 1975; Watters *et al.* 1998, 2009; Di Achille *et al.* 2012) and those predicted by thermal evolution models (e.g. Solomon 1977; Dombard & Hauck 2008), with those from the first approach typically about 1–3 km but those from modelling of the order of 5–10 km. The availability of global imaging and topographical data for Mercury from the MESSENGER mission, however, has enabled this issue to be resolved: Byrne *et al.* (2014*b*) have shown that the portion of radius change accommodated by tectonic structures is at least 5–7 km, bringing into accord photogeological observations and thermal history models. This finding is of importance to on-going studies of Mercury's bulk silicate abundances of heat-producing elements, mantle convection, and the cooling and present-day structure of the planet's large metallic core.

The provenance of Mercury's wrinkle ridges is not as obviously linked to global contraction as are its lobate scarps. Because they are generally associated with volcanic units within impact structures, subsidence of volcanic infill may play a dominant role in their formation (e.g. Watters *et al.* 2009; Byrne *et al.* 2013*d*), as has been suggested for wrinkle ridges within volcanic units on the Moon (e.g. Melosh 1978) and on Mars (e.g. Zuber & Mouginis-Mark 1992). However, given their small size relative to lobate scarps, removing the contribution of wrinkle ridges to Mercury's global contraction results in a change in radius only around 10% lower than that calculated for all structures.

Tectonics within impact structures. Whereas the widespread smooth and cratered plains structures attest to global crustal shortening, extension on Mercury is almost exclusive to volcanically in-filled craters and basins (Murchie *et al.* 2008; Watters *et al.* 2009; Byrne *et al.* 2013*d*). However, the complexity of extensional deformation varies considerably, and increases within progressively larger impact structures. For example, many ghost craters on Mercury, tens of kilometres across, contain sets of graben in their interior that have no preferred orientation and so form polygonal block patterns; these ghost craters may also feature graben superposed on (and following the strike of) the wrinkle ridges that outline the ghost crater rim (Klimczak *et al.* 2012). Several medium-sized basins on the planet also feature interior graben that show no preferred orientations, such as the 230 km-diameter Mozart basin, although some graben in this basin appear concentric to its perimeter (Blair *et al.* 2013). Yet, with an increase in basin diameter, the pattern of interior extension becomes even more complex: the 750 km-diameter Rembrandt basin features collocated basin-radial graben and wrinkle ridges, which are bound to the north by circumferential graben (and ridges) (Byrne *et al.* 2013*d*).

However, the greatest structural complexity of all occurs within the Caloris basin, where a prominent set of radial graben (termed Pantheon Fossae) dominates the basin interior, is superposed by basin-circumferential ridges and is bound by a near-complete annulus of circumferential graben located at about half a basin radius from the centre of Caloris (Byrne *et al.* 2013*d*). Beyond this annulus, graben once more lack a preferred orientation, and, forming a polygonal map pattern, steadily decrease in width, depth and length towards the basin rim (Fig. 3f).

Finite-element models show that the thermal contraction of thick, rapidly emplaced lava flows produces quasi-isotropic horizontal stresses that promote the formation of graben with mixed orientations, such as those observed within ghost craters and medium-sized basins on Mercury (Freed *et al.* 2012; Blair *et al.* 2013). This process may also account for the mixed orientations of graben within Caloris. Moreover, such models also show that graben can nucleate atop ghost-crater-delineating wrinkle ridges (Freed *et al.* 2012) and over buried basin rings (Blair *et al.* 2013), which may account for the circumferential graben within Caloris. As yet, however, there is no consensus for the origin of the remarkable Pantheon Fossae (Klimczak *et al.* 2013), although previous studies suggested that they may have formed due to dyke propagation (Head *et al.* 2008) or to flexural uplift of the basin centre, in response either to the volcanic emplacement of the circum-Caloris smooth plains (Freed *et al.* 2009) or to the inward flow of the lower crust (Watters *et al.* 2005).

Venus

The major findings on Venus' composition, volcanic forms and tectonic structures have been obtained mainly by analysing data from the late 1970s and 1980s, together with Magellan radar images (Figs 1d & 4). Nevertheless, Venus has always been of particular interest due to its Earth-like size, mass and internal structure, despite great differences between the physiography of the two planets. The main reason for this discrepancy is probably the apparent lack of plate tectonics on Venus, which is interpreted to be due to the planet's water-depleted bulk composition. For that reason, the following subsection starts by addressing Venus' general geodynamics, which serve as a basis from which to cover specific aspects of its tectonic and volcanic character.

Tectonics

Venus as a one-plate planet. Water, and its capacity to weaken rock, is essential in governing lithospheric rheology on Earth. Together with temperature, water is the major actor responsible for defining the thickness and strength of the elastic lithosphere, its decoupling from the deeper asthenosphere and, hence, plate tectonic onset and development. The absence of water on Venus' surface due to the high surface temperatures (460 °C on average: e.g. Lewis 2004), coupled with the dehydrated character of its interior as a result of extensive volcanic degassing (Kaula 1990; Grinspoon 1993; Smrekar & Sotin 2012), precludes plate tectonics on the planet (e.g. Smrekar *et al.* 2007; McGill *et al.* 2010).

The lack of Venusian plate tectonics is also implied by the global dominance of a basaltic composition of the crust, with limited possible exceptions on the highlands (Terrae) (Surkov 1983; Hashimoto *et al.* 2008; Helbert *et al.* 2008). A dehydrated, stiff diabasic crust prevents the onset of lithospheric break-up and ensuing subduction; moreover, dry systems possess a highly viscous and water-depleted mantle, which precludes any asthenosphere–lithosphere decoupling (Kohlstedt & Mackwell 2010; Huang *et al.* 2013), a process that modulates relative plate motions on Earth (e.g. McKenzie 1967; Isacks *et al.* 1968; Morgan 1968; Ranalli 1995, 1997; Gung *et al.* 2003; Scoppola *et al.* 2006; Anderson 2007; Doglioni *et al.* 2007, 2014; Fischer *et al.* 2010). The main consequence of a globally continuous lithosphere are a conductive stagnant lid on top of a convective and viscous deep mantle (Moresi & Solomatov 1998; Reese *et al.* 1999; Solomatov & Moresi 2000; O'Rourke & Korenaga 2012; Smrekar & Sotin 2012), and planetary tectonics that directly reflect underlying mantle processes (e.g. Smrekar & Phillips 1991; Herrick & Phillips 1992; Bindschadler *et al.* 1992; Kohlstedt & Mackwell 2010; Huang *et al.* 2013).

On Venus then, planetary cooling is achieved through both conduction across a stagnant lid and advection by mantle plumes, whose surface expressions are well represented by a great variety of volcanic forms (e.g. volcanic rises, coranae and shield volcanoes) (Solomon & Head 1982; Morgan & Phillips 1983; Smrekar & Parmentier 1996; Schubert *et al.* 1997; Hansen & Olive 2010). Compared with plate tectonics, however, near-surface conduction and deep advection are regarded as less efficient than mantle convection in cooling a terrestrial planet whose mass and heat production is thought to be more or less equal to that of Earth. The result, therefore, is the heating up of the mantle, with a consequent increase in mantle dynamics and volcanism that may ultimately lead to periodical global resurfacing and overturn (e.g. Parmentier & Hess 1992; Turcotte 1993; Nimmo & McKenzie 1998; Reese *et al.* 1999; Turcotte *et al.* 1999), as suggested by the low-density and nearly uniform distribution of craters on the surface of Venus.

These craters give an average age of between 500 and 800 Ma for almost the entire surface of Venus (Schaber *et al.* 1992; Strom *et al.* 1994; McKinnon *et al.* 1997; Campbell 1999), far younger than the scarred surfaces of Mercury or the Moon.

The topography of Venus is one of the principal pieces of evidence for the absence of modern plate tectonics on that planet. Unlike Earth, which is characterized by the classic ocean- continent bimodal hypsometry, Venus shows a unimodal hypsometric function, with over 80% of the surface covered by plains (Planitiae) and only 8% of the surface at elevations greater than 2 km above the datum. These uplands are in turn subdivided into Terrae and Regiones, with Regiones having more moderate relief than the Terrae (Fig. 4) (Banerdt *et al.* 1997; Tanaka *et al.* 1997).

Also, unlike Earth, the strong correlation between free-air gravity and topography, and the high gravity anomaly/relief ratios of some positive topographical features, indicate that topographical variations on Venus are deeply compensated through upwelling and downwelling processes within the mantle (e.g. Smrekar & Phillips 1991; Solomon 1993; McGill *et al.* 2010). This implies the absence of an Earth-like asthenosphere–lithosphere decoupling, as predicted for a dehydrated planetary lithosphere, and points to a direct correlation between crustal deformation and the upwelling–downwelling convective motions of the mantle (e.g. Kiefer & Hager 1991; Smrekar & Phillips 1991; Solomon 1993). Hence, dome-shaped Regiones (i.e. elevated expanses) associated with broad free-air gravity anomalies (e.g. Beta, Atla and Eislta Regiones), together with great compensation depths, have been interpreted as volcanic rises directly related to underlying mantle plumes. Moreover, the enigmatic Artemis, a feature 2400 km in diameter, may be the largest mantle-plume-derived landform in the solar system (Hansen & Olive 2010). Therefore, Venus is thought to be essentially dominated by vertical tectonism (e.g. Phillips & Malin 1983; Campbell *et al.* 1984; Kiefer & Hager 1991; Senske *et al.* 1992; Smrekar 1994; Solomon 1993; Smrekar & Parmentier 1996; Smrekar *et al.* 1997).

On the other hand, some extensive and highly deformed highlands on Venus, termed crustal plateaux (e.g. Ishtar and Aphrodite Terrae and Alpha, Ovda, Thetis, Phoebe and Tellus Regiones), are associated with low gravity anomalies and a low gravity/topography ratio, which are indicative of a shallow depth of compensation and a thickened, low-density crust (Bindschadler *et al.* 1992; Smrekar & Phillips 1991; Kucinskas *et al.* 1996; Simons *et al.* 1997). The two main hypotheses proposed to explain such peculiar geophysical signatures invoke either downwelling or upwelling mantle flows. According to the downwelling model,

Fig. 4. Simplified geological map of Venus, modified after Ivanov & Head (2011). From that original source, the following units were merged to represent: (1) tectonized terrains – units *t* (tessera, Fortuna Formation), *pdl* (densely lineated plains, Atropos Formation), *pr* (ridged plains, Lavinia Formation), *mb* (mountain belts, Akna Formation) and *gb* (groove belts, Agrona Formation); (2) global volcanic terrains – units *psh* (shield plains, Accruva Formation), *rp1* (regional plains 1, Rusalka Formation) and *rp2* (regional plains 2, Ituana Formation); (3) rifting-related terrains – units *pl* (lobate plains, Bell Formation), *ps* (smooth plains, Gunda Formation) and *sc* (shield clusters, Boala Formation); and (4) tectonic components related to (3) – units *rz* (rift zones, Devana Formation) and *ac* (Artemis Canyon materials). Impact craters and crater outflows are shown as units *c* and *cf*, respectively. Individual tectonic structures are omitted for clarity. Black dots represent the locations of volcanic edifices (Head *et al.* 1992). Grey areas correspond to the radar-based hillshade (i.e. no mapping data available). The geological map is superimposed on a Magellan Global Topography Data Record shaded relief map (4.6 km/px) shown here in a Robinson projection, centred at 0°E.

crustal thickening and plateau formation is achieved through accretion of a thin primordial lithosphere by downwelling of a cold mantle diapir (Bindschadler & Parmentier 1990; Bindschadler & Head 1991; Bindschadler *et al.* 1992; Gilmore & Head 2000; Marinangeli & Gilmore 2000). In contrast, upwelling may accomplish crustal thickening and plateau formation through magmatic underplating above a laterally spreading hot plume (Grimm & Phillips 1991; Phillips *et al.* 1991; Hansen & Willis 1998; Phillips & Hansen 1998; Ghent & Hansen 1999). The latter hypothesis is consistent with the evidence of low-density felsic rocks within crustal plateaux (Basilevsky *et al.* 1986, 2012; Nikolaeva *et al.* 1992; Jull & Arkani-Hamed 1995; Arkani-Hamed 1996; Hashimoto *et al.* 2008; Helbert *et al.* 2008; Mueller *et al.* 2008; Harris & Bédard 2014*a*), which could have been generated by high-temperature plumes in an earlier, wetter environment on Venus (McGill *et al.* 2010; Shellnutt 2013). Upwelling in response to bouyant melt formed by the impact of large bolides (*c.* 20–30 km in diameter) may have contributed to the formation of crustal plateaux and tesserated terrain (Hansen 2006).

However, the upwelling model cannot explain the widespread contractional tectonics observed at some plateau margins nor the complex compressional–extensional deformation recorded on crustal plateaux within so-called tesserae terrains (Ivanov & Head 1996; Gilmore *et al.* 1998; Romeo *et al.* 2005; Hansen & López 2010). On the other hand, the downwelling hypothesis fails to explain the flat-topped topography of plateaux, and, further, it requires an excessive amount of time for crustal thickening (Lenardic *et al.* 1995; Kidder & Phillips 1996). Interestingly, a recent model suggests that crustal plateaux (and tesserae terrains across Venus' surface) might represent buoyant continental crust remnants that were spared from catastrophic overturns due to their inherent buoyancy, and that thereafter were involved in cycles of compressional and extensional tectonism as a function of the evolving continental crust/lithosphere mantle thickness ratio (Romeo & Turcotte 2008). This last model is not strictly an alternative to the upwelling hypothesis as it still allows that the continental crust on Venus could have originated, and then be developed, through felsic magma production atop hot mantle plumes.

Rift systems and extensional deformation. Despite the absence of plate tectonics on Venus, the planet still hosts Earth-like rift systems called chasmata that extend for several thousand kilometres across its surface (e.g. Devana, Ganis, Daiana/Dali, Hecate and Parga Chasmata) (Fig. 4). These structures have been interpreted as linear zones of mantle upwelling and lithospheric extension, in several cases punctuated by single plumes or plume clusters manifest as volcanic rises, shield volcanoes, coronae and radiating graben–fissure systems (e.g. Schaber 1982; Head & Crumpler 1987; Hansen & Phillips 1993; Baer *et al.* 1994; Hamilton & Stofan 1996; Aittola & Kostama 2000; Magee & Head 2001; Stofan *et al.* 2001; Krassilnikov & Head 2003; Harris & Bédard 2014*a*). On the basis of structural observations and gravitational data, further major extensional rifts have been proposed to follow plains' depressions that were later covered by volcanic materials (Sullivan & Head 1984; Harris & Bédard 2014*b*).

Extensional deformation is widespread on planitiae, and commonly manifest as graben that appear generally associated with plume diapirism and volcanic centres (McGill *et al.* 2010). Graben that form polygonal patterns are also present on Venus, with that pattern being attributed variously to cooling–heating cycles induced by subsurface dynamics (Johnson & Sandwell 1992) or to climate changes (Anderson & Smrekar 1999; Smrekar *et al.* 2002).

Mountain belts. The large number of Venusian rift systems, and their corresponding lithospheric extension, is not compensated by a comparable cumulative length of compressive mountain belts (montes). Mountains are, instead, limited to crustal plateau boundaries, such as the Montes encircling Laksumi Planum at Isthar Terra (Danu, Akna, Freya and Maxwell Montes: Fig. 5a). This has been taken as further evidence for the lack of plate tectonics on Venus, where contractional deformation is widely distributed throughout the entire lithosphere rather than at discrete sites (Solomon *et al.* 1992; Solomon 1993). Indeed, folds and faults are particularly pervasive in the tesserated terrains, which are expressions of multiple deformations on high-relief crustal plateaux and crustal remnants within plains (Bindschadler & Head 1991; Hansen & Willis 1996; Hansen *et al.* 1999, 2000; Romeo & Turcotte 2008). Folds and faults are also common in the planitiae, either concentrated in narrow contractional ridge belts or in widely distributed wrinkle ridges (Tanaka *et al.* 1997; McGill *et al.* 2010).

Contractional deformation. The ridge belts are up to 20 km wide and several hundreds to thousands of kilometres long (Ivanov & Head 2001; Rosenberg & McGill 2001). As for terrestrial fold-and-thrust systems, these Venusian features are also thought to have formed due to long-lived tectonic stress fields. The most widely distributed contractional landforms are wrinkle ridges (e.g. Tanaka *et al.* 1997; Ivanov & Head 2008). The origin of these structures has been variously ascribed to gravitational spreading on shallow slopes (McGill 1993), rock thermal expansion induced by a hotter atmosphere than at present after global resurfacing (Solomon *et al.* 1991) or regional tectonics controlled by a stress state centred on the planetary geoid, with compression at the geoid's low-standing regions and extension at the geoid's high-standing regions, respectively (Sandwell *et al.* 1997; Bilotti & Suppe 1999).

Strike-slip tectonics. On a planet dominated by vertical tectonics and characterized by widely distributed deformations, there is no space for strike-slip shear belts that have traditionally been highly underconsidered on Venus (e.g. Solomon 1993; McGill *et al.* 2010) despite the apparently straightforward evidence collected so far (Raitala 1994; Brown & Grimm 1995; Ansan *et al.* 1996; Koenig & Aydin 1998; Tuckwell & Ghail 2003; Kumar 2005; Romeo *et al.* 2005; Chetty *et al.* 2010; Fernàndez *et al.* 2010; Harris & Bédard 2014*a*, *b*). For example, Brown & Grimm (1995) highlighted en echelon fractures and folds along and within Artemis Chasma. Koenig & Aydin (1998) and Fernàndez *et al.* (2010) showed horse-tail terminations, en echelon folds, contractional bends and strike-slip offsets in Lavinia Planitia. Finally, Romeo *et al.* (2005) and Chetty *et al.* (2010) identified conjugate shear zones in Ovda Regio associated with tear-folds and imbricate duplexes. Continental drift may account for the substantial horizontal tectonism required by prominent strike-slip belts (Harris & Bédard 2014*b*).

Volcanism

On a geodynamically active planet that lacks modern-day plate tectonics, upwelling plumes and their related surface expressions are its dominant volcanic traits. The volcanic landforms on Venus largely depend on the dimensions of the blooming head of such mantle plumes, which are mostly responsible for the development of volcanic rises ranging from 1400 to 2500 km in diameter and up to 2.5 km in elevation (e.g. Beta, Atla, Bell; Eistla, Luafey, Imdr, Themis and Dione Regiones). The enigmatic coronae, also attributed to volcanism, display variable diameters (with 250 km being the average). Other types of volcanic landform include shield volcanoes, which normally range between 100 and 600 km in diameter (and which rarely exceed 700 km in breadth: see Stofan *et al.* 1995; Glaze *et al.* 2002; Crumpler & Aubele 2000; Ivanov & Head 2013 for size distributions). These landforms are typically accompanied by smaller volcanic features (<100 km in diameter), such as small and intermediate volcanoes known as tholi and steep-sided domes ('pancakes': Fig. 5f).

Volcanic rises. Volcanic rises, corresponding to Venus' regions, have been classified either as volcano-dominated, coronae-dominated or rift-dominated (Stofan *et al.* 1997). Whereas in the first case volcanic rises are representative of major plume sites along rift zones, in the latter two cases the plume heads could have been broken up into a swarm of smaller upwelling diapirs (although not necessarily simultaneously) (Stofan *et al.* 1995). Smrekar & Stofan (1999) noted that the volcanic construction typical of volcano-dominated rises requires large volumes of pressure-release melting due to vigorous hot plumes generated at the core–mantle boundary, whereas corona-dominated rises are more likely to be associated with smaller plumes from shallower depths.

Coronae, novae and arachnoids. Coronae (Fig. 5d & e) are annular forms dominated by either positive or negative topography, and which are encircled by concentric systems of fractures and ridges (Type 1 coronae) or by a rim alone (Type 2) (Basilevsky *et al.* 1986; Head *et al.* 1992; Squyres *et al.* 1992; Stofan *et al.* 1992, 2001). Coronae are often associated with radiating, graben–fissure systems (Fig. 5d) that converge towards the coronas' rims and that in some cases reach their centres (with such structures being termed novae) (Barsukov *et al.* 1986; Crumpler & Aubele 2000; Aittola & Kostama 2000, 2002; Krassilnikov & Head 2003; Basilevsky *et al.* 2009; Studd *et al.* 2011). Indeed, radiating graben–fissure systems can also be associated with large volcanoes (e.g. Keddie & Head 1994; Galgana *et al.* 2013) or can occur individually, following rift zones (Grosfils & Head 1994; Aittola & Kostama 2000). The structures within radiating graben–fissure systems can extend for up to 2000 km. In places, they may become subparallel to each other, or they may form systems perpendicularly orientated to the prevailing maximum regional horizontal stress (Grosfils & Head 1994). Based on their appearance and lateral extent, radiating graben–fissure systems have been compared with giant radiating dyke swarms on Earth, such as the Proterozoic Mackenzie and Matachewan swarms in Canada, the Tertiary Sky Isle swarm in Scotland, and the Mesozoic central Atlantic reconstructed swarm (Ernst *et al.* 1995). The Venusian systems also resemble radiating extensional systems on Mars (Ernst *et al.* 2001; Wilson & Head 2002).

Arachnoids are characterized by ridges that converge towards annular depressions (Aittola & Kostama 2000; Crumpler & Aubele 2000; Frankel 2005). These landforms have often been considered as a subtype of coronae (Head *et al.* 1992; Price & Suppe 1995; Hamilton & Stofan 1996). Indeed, coronae, radiating graben–fissure systems (novae) and arachnoids may represent different

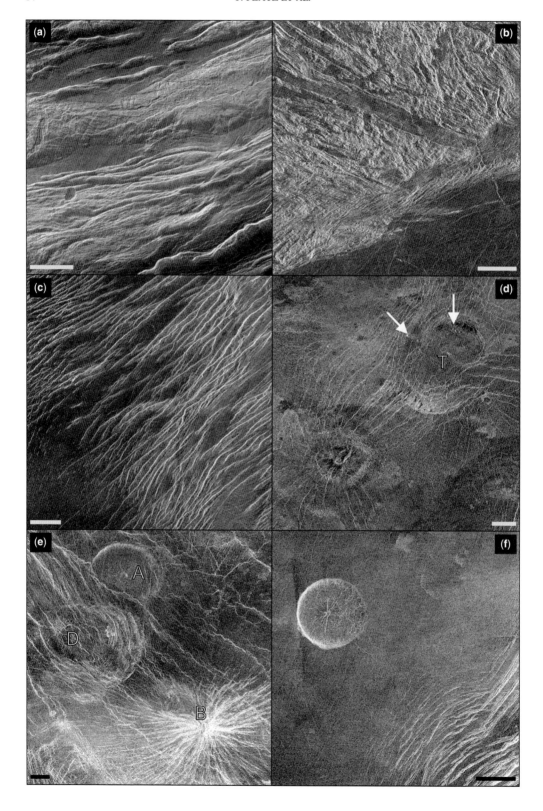

evolutionary stages of mantle plume upwellings (Hamilton & Stofan 1996). Under such a scenario, corona evolution begins with lithospheric up-doming and radial fracturing due to mantle plume (or diapir) impingement, followed by the gravitational collapse of the central dome due to radial spreading and thinning of the plume (Squyres *et al.* 1992; Stofan *et al.* 1991, 1992; Hamilton & Stofan 1996). Recent finite-element modelling indicates that radial fractures form during early mantle diapir impingement, with an outward propagation of dykes occurring at a later stage due to loading-induced downward lithospheric flexure (Galgana *et al.* 2013). Together with novae, then, coronae with a topographically positive central dome and radial fracture system may represent the first stage of mantle plume upwelling, and coronae with a central annular depression might be typical of the final stage (Stofan *et al.* 1992; DeLaughter & Jurdy 1999). In another analysis, however, Gerya (2014) proposed a variation of this general process. Gerya suggested a scenario in which novae and coronae structures are not directly related to astenospheric plume impingement but, instead, are the result of magma-assisted convection of a weak ductile crust, induced by decompressional melting of hot rising mantle plumes. Under this scenario, novae represent the initial stage of this convection, with coronae corresponding to the intermediate and final stages.

Whether arachnoid structures are a somewhat intermediate phase between novae and coronae (Hamilton & Stofan 1996) is still the subject of debate, however, given the different geological contexts in which arachnoids have been documented. In particular, these structures are largely located in volcanic plains, and are rarely situated along the equatorial deformation zones where most novae and corona are situated (Aittola & Kostama 2000).

Volcanic products. The primary volcanic products on Venus are extensive tholeitic–alkali basalt plains that cover about 70% of the planet. These plains differ in terms of the styles of deformation to which they have been subjected or their connection with small shield volcanoes (ridged, lineated, regional plains, lobate and shield plains) (e.g. Ivanov & Head 2011).

Typical volcanic landforms include lava flow fields, lava channels and steep-sided domes (Crumpler & Aubele 2000; Stofan *et al.* 2000; Magee & Head 2001). Lava flow fields are composed of very long (up to 1000 km) digitated flows, both with radar bright (interpreted to be aa-type lavas) or dark (pahoehoe-type lavas) surfaces, which erupted from shield volcanoes and fissures (including those of coronae). Their substantial lengths are attributed to endogenous spreading processes, such as inflation and lava tubes, which should be facilitated by an efficient cooling of the flow surface (due to convective processes within the dense atmosphere, whose pressure is *c.* 9.2 MPa), aided by a low cooling rate of the lava flow's inner core (due to the high surface temperature environment of *c.* 700°K) (Grosfils *et al.* 1999; Crumpler & Aubele 2000).

Remarkably long individual lava channels, of typical lengths of between 100 and 1000 km in general but in some cases extending to more than 5000 km, are explained by invoking very low-viscosity fluxes such as these expected for exotic lava types such as komatites, carbonatites and sulphur flows (Kargel *et al.* 1994; Baker *et al.* 1997; Williams-Jones *et al.* 1998; Komatsu *et al.* 2001; Lang & Hansen 2006).

There is evidence for highly viscous lavas on Venus, too. Such evidence includes steep-sided domes, which are circular, positive-relief landforms with diameters ranging between 20 and 100 km (e.g. Pavri *et al.* 1992), and flows with pronounced margins. The higher lava viscosities implied by these features could be due to a high silica content (i.e. evolved magmas), high crystal content or high

Fig. 5. Examples of volcanic and tectonic features on Venus. (**a**) Fold-and-thrust belt sequence of Akna Montes; note the presence of intra-mountain basins (image centre: 69°N, 318°E; Magellan full resolution radar mosaic archive MGN-RDRS FMAP: FL69n318). (**b**) Complex deformation in Sudice Tessera; at the lower right are the regional plains of Aino Planitia (37°S, 113°E; MGN-RDRS FMAP: FR37s113). (**c**) Tectonized ridge belt east of Aino Planitia (45°S, 116.5°E; MGN-RDRS FMAP: FR45s117). (**d**) Structural relations between Tituba corona (T), centred at 42.5°N, 214.5°E (termed Type 1, after Stofan *et al.* 2001) and the corona at 40.4°N, 212.4°E (Type 2, after Stofan *et al.* 2001). Note the concentric and radial features related to both the coronae and the small volcanic edifices at the NW margin of Tituba corona (white arrows). The radial fissures and graben between the coronae become parallel at a certain distance from the coronae centres, and are common to both the two main volcanic landforms (MGN-RDRS FMAP: Fl39n211, FL39n213, FL39n215, FL41n211, FL41n213, FL41n215, FL43n211, FL43n213, FL43n215). (**e**) Becuma radiating graben system (nova) (B) (34.13°N, 21.9°E), Dzudzdi corona (D) (35°N, 20.7°E) and the Aegina Farrum steep-sided dome (A) (35.54°N, 21.1°E) in the eastern Sedna Planitia. Note that Aegina Farrum is transected by a 220 km-long extensional fault. (MGN-RDRS FMAP: Fl35n021, FL35n023, FL33n021, FL33n023). (**f**) Steep-sided dome at 2.8°S, 150.9°E, to the west of Sella Corona (not shown) and its related graben system (lower right corner) (MGN-RDRS FMAP: FL03s151). The scale bar in all images is 20 km, except in (d) where it is 50 km. All images are shown with simple cylindrical projections, except (a), which has a stereographic projection.

vesicularity (Head *et al.* 1992; Pavri *et al.* 1992). It has been proposed that the considerable atmospheric pressure on Venus' surface prevents the formation of eruptive plumes due to explosive magmatic fragmentation and, instead, is conducive to outpourings of frothy lava domes.

It is widely accepted that the planet's high atmospheric pressure would strongly inhibit gas exsolution in magmas and, hence, would inhibit or preclude entirely explosive eruptions (e.g. Head & Wilson 1986). However, geomorphological evidence exists for an extensive pyroclastic flow deposit near Diana Chasma, which, if found to be a common form of volcanic deposit on Venus, would have implications for the mantle volatile content and the volcanic origin of atmospheric SO_2 on the planet (Ghail & Wilson 2013).

Global volcanic resurfacing. Global resurfacing models of Venus (e.g. Parmentier & Hess 1992; Turcotte 1993; Nimmo & McKenzie 1998; Reese *et al.* 1999) call for volcanic and tectonic events that are globally correlated throughout the planet's geological record. Indeed, according to several studies (e.g. Basilevsky & Head 1995, 1998, 2000; Basilevsky *et al.* 1997; Ivanov & Head 2011), all of the physiographical units recognized on the Venusian surface are expressions of a common sequence of events across the globe. The geological map of Ivanov & Head (2011) is currently the most complete and updated synthesis of this view, where each morphological unit is regarded as a stratigraphic unit of global significance and, when synthesized as such, describes a coherent Venusian geological history.

Under this scheme, all deformed terrains (tessera, densely lineated plains, ridged plains, mountain belts and groove belts) should belong to the earliest phase of tectonism, which was followed by extensive volcanic effusion manifest as regional and shield plains, and which was followed in turn by more spatially concentrated, long-lived rift volcanism, represented by morphological units such as lobate plains, smooth plains and shield clusters. The first phase of volcanism probably began with the construction of shield plains associated with small shield structures and steep-sided domes, the likely expression of shallow crustal melting. This form of volcanism was followed by the effusion and emplacement of huge volumes of lava, essentially mantle-derived, pressure-release melts as attested to by regional plains and extended lava channels. This second volcanic phase, well represented by lobate plains, is supposed to be dominated by voluminous volcanic effusions punctuated by long temporal gaps (Ivanov & Head 2013).

Guest & Stofan (1999) defined as a 'directional model' the view of a globally correlated geological history, which contrasts with a 'non-directional model' in which the recognition of similar sequences of events in different regions of Venus does not imply their coeval occurrence. In favour of the first model are the globally distributed regional plains that were utilized as the main stratigraphic reference for the entire Venusian time system by Ivanov & Head (2011). These authors argued that the global presence of these plains supports the general 'directionality' of Venus' geological history but some authors have highlighted several exceptions to this hypothesized commonality in the geological record across the entire surface (Guest & Stofan 1999; Rosenberg & McGill 2001; McGill 2004; McGill *et al.* 2010). Of note, and supported by observations of where in terrestrial geology equivalent morphologies and geneses do not necessarily mean a coeval origin, McGill *et al.* (2010) proposed an intermediate view within which a general 'directional' framework predominated but in which tectonic and volcanic events might still have developed at different times and in diverse locations.

Nevertheless, the resurfacing of Venus continues to be viewed within the narrative of either 'catastrophic' (e.g. Schaber *et al.* 1992) or 'equilibrium' (e.g. Phillips *et al.* 1992) scenarios, both of which must satisfy the near-random spatial distribution of fewer than 1000 impact craters across the planet and the lack of many obviously modified such craters. Although catastrophic models more readily satisfy these constraints, recent work has also shown that equilibrium models, within a select parameter space, also meet these requirements (Bjonnes *et al.* 2012). Moreover, numerous further constraints introduced by careful geological mapping question the underlying assumptions of the catastrophic scenarios. It may be that a protracted period of resurfacing set against the backdrop of a globally thin lithosphere (e.g. the 'SPITTER' hypothesis: Hansen & Young 2007) accounts more fully for these observational constraints. This viewpoint is bolstered by recent mapping and modelling, which indicates that Venus' tesserae terrain probably predates extensive resurfacing of the planet (Romeo & Turcotte 2008; Hansen & López 2010).

The Moon

The Moon is a key planetary body with which to study diverse volcanic activity, as it has produced a variety of volcanic landforms over an extended period of time (Figs 6 & 7). As the Moon is lacking plate tectonics, an atmosphere, water and life, it allows us to study volcanic processes in an unobscured form (e.g. Hiesinger & Head 2006; Jaumann *et al.* 2012; Hiesinger & Jaumann 2014). Analyses

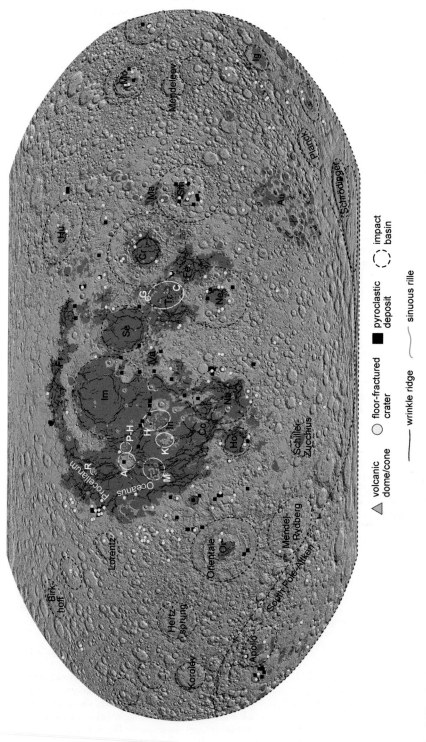

Fig. 6. The distribution of maria, volcanic edifices, sinuous rilles and selected tectonic structures on the Moon. The lunar maria are highlighted in red. Impact basins larger than 300 km in diameter are shown as dashed black ellipses (Kaddish et al. 2011). The locations of floor-fractured craters (yellow dots) are taken from Jozwiak et al. (2012). Identified pyroclastic deposits (black squares: Gaddis et al. 1998) and sinuous rilles (green lines: Gustafson et al. 2012; Hurwitz et al. 2012, 2013a) are shown. The approximate locations of the broad shield volcanoes identified by Spudis et al. (2013) are highlighted as white ellipses with bold white letters representing: A, Aristarchus; C, Cauchy; G, Gardner; H, Hortensius; K, Kepler; M, Marius; P–H, Prinz–Harbinger; and R, Rümker. The locations of the 118 domes and cones identified on the Moon are shown as green triangles (data from http://digilander.libero.it/glrgroup/consolidatedlunardomecatalogue.htm). Maria labels are abbreviated as follows: Au, Australe; Co, Cognitum; Cr, Crisium; Fe, Fecundidatis; Fr, Frigoris; Ho, Humorum; Hu, Humboldtianum; Ig, Ingenii; Im, Imbrium; In, Insularum; Ma, Marginis; Mo, Moscoviense; Ne, Nectaris; Nu, Nubium; Or, Orientale; Se, Serenitatis; Sm, Smythii; Tr, Tranquillitatis; Va, Vaporum. The background image is a Lunar Orbiter Laser Altimeter (LOLA)-based shaded-relief map (0.23 km/px) with a Robinson projection, centred at 0°E.

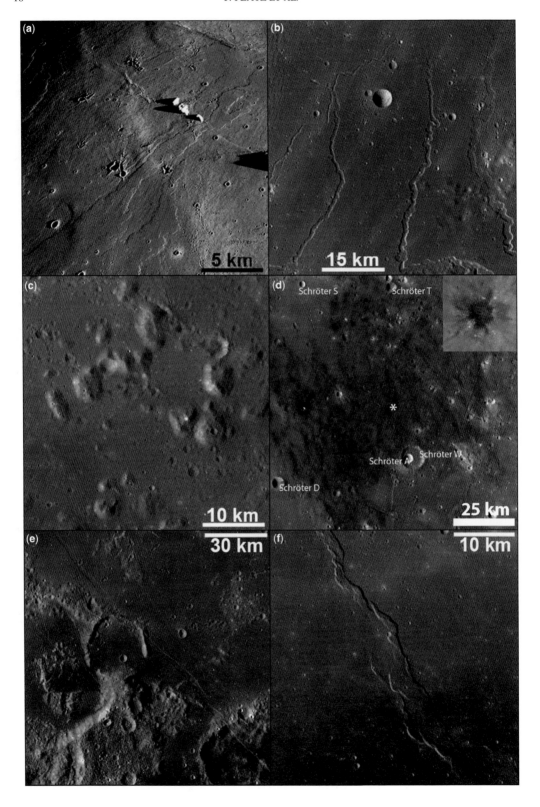

of lunar samples, coupled with remote-sensing investigations, have yielded detailed information on many aspects of lunar volcanism (i.e. extrusion) and, to a lesser amount, on lunar plutonism (i.e. intrusion). From those studies, it became apparent that magmatism has been a major crust-building and resurfacing process throughout the Moon's geological history (e.g. Lucey *et al.* 2006; Shearer *et al.* 2006). The available data also allow us to study the role of magmatic activity (both intrusive and extrusive) during the heavy bombardment, as well as during more recent lunar history (i.e. the mare stratigraphic record). Furthermore, the distribution of basalt types, and the implied spatial and temporal distribution of mantle melting, as well as volcanic volumes and fluxes, can be understood. Finally, the Moon is unique in that it allows us to assess a wide range of eruption styles, including pyroclastic activity, and their petrogenetic significance owing to the existence and preservation of an assorted suite of volcanic deposits (e.g. Hiesinger & Head 2006; Jaumann *et al.* 2012; Hiesinger & Jaumann 2014).

Igneous geochemistry

From the lunar sample collection it became apparent that lunar rocks can be classified, on the basis of texture and composition, into four distinct groups: (1) pristine highland rocks that are primordial igneous rocks, uncontaminated by impact mixing; (2) pristine basaltic volcanic rocks, including lava flows and pyroclastic deposits; (3) polymict clastic breccias, impact melt rocks and thermally metamorphosed granulitic breccias; and (4) the lunar regolith. In the following paragraphs we will only focus on highland rocks and mare basalts. Although breccias may contain igneous fragments, they are primarily produced by impacts and not by magmatic/volcanic processes.

On the basis of their molar $Ca/(Ca + Na + K)$ content v. the molar $Mg/(Mg + Fe)$ content of their bulk rock compositions, pristine highland rocks fall into two major chemical groups: ferroan anorthosites and magnesium-suite rocks (e.g. Warner *et al.* 1976; Papike 1998). These rocks appear to have different ages, with the ferroan anorthosites

being somewhat older (c. 4.56–4.29 Ga) than the magnesian-suite rocks (high Mg/Fe) (4.46–4.18 Ga) (Shearer *et al.* 2006). This latter group contains dunites, troctolites, norites and gabbronorites. Compared with these two rock types, the alkali suite is less abundant. Although this group contains similar rock types, they are enriched in alkali and other trace elements relative to, and are somewhat younger than (4.37–3.80 Ga), the ferroan anorthosites and the Mg-suite rocks (Shearer *et al.* 2006). This implies that the earliest Mg-suite rocks were formed contemporaneously with at least some ferroan anorthosite, which is not consistent with the idealized magma-ocean model, in which ferroan anorthosites form the oldest crust and are later intruded by younger, Mg-suite plutonic rocks. An alternative model to the magma-ocean scenario, which proposes a genesis of the lunar crust by intrusion of multiple magma bodies (i.e. serial magmatism) (e.g. Walker 1983; Longhi & Ashwal 1985; Longhi 2003; Shearer *et al.* 2006; Borg *et al.* 2011), appears to be more consistent with the observed age relationships of lunar pristine rocks.

From the Apollo samples, it is known that anorthosite is common in the lunar highlands. Compared with terrestrial rocks, the anorthosite abundances (An_{96}) of plagioclase in these rocks are much higher, ultimately reflecting the Moon's depletion in volatile elements such as sodium (Lucey *et al.* 2006 and references therein). The Mg# (i.e. Mg/Mg + Fe) of pyroxene and olivine in lunar anorthosite is much more ferroan than in terrestrial rocks of such high Ca/Na ratios and any other non-mare lunar rocks (e.g. Lucey *et al.* 2006 and references therein). Thus, ferroan anorthosite refers to lunar anorthosite with plutonic or relict plutonic textures (e.g. Dowty *et al.* 1974a, b). The ferroan-anorthositic suite (Warren 1993) consists of ferroan anorthosites (>90% plagioclase), as well as their more mafic but less common variants, ferroan noritic anorthosite and ferroan anorthositic norite. Pyroxene usually predominates in the ferroan anorthosites, although some samples also contain olivine. Lunar ferroan anorthosites are coarse-grained intrusive igneous rocks, formed during slow cooling at some depth below the surface (Lucey *et al.* 2006). Because of the high concentration of plagioclase feldspar in

---

Fig. 7. Examples of volcanic and tectonic features on the Moon. (**a**) Basaltic lava flow in Mare Imbrium. Note also wrinkle ridges and secondary impact craters (Apollo 15 orbital photograph; AS15-M-1556). (**b**) Sinuous rille (Rima Prinz) in Mare Imbrium (Lunar Reconnaissance Orbiter Camera (LROC) wide-angle camera (WAC)). (**c**) Basaltic lava dome complex, situated within Mare Imbrium (LROC WAC). (**d**) Pyroclastic deposits located between Sinus Aestum and Schröter crater. Asterisk marks the location of the inset showing a 170 m-diameter crater, which excavated fresh pyroclastic material that is also termed dark mantle deposit (LROC WAC and NAC). (**e**) Rimae Goclenius graben system in Mare Fecunditatis. Faults cut the volcanic plains and pre-existing craters (LROC WAC). (**f**) Dorsa Whiston in Oceanus Procellarum represents a typical mare ridge (LROC WAC). Figures are adopted from Hiesinger & Jaumann (2014).

ferroan anorthosites, they are interpreted as cumulate rocks, produced by the separation and accumulation of crystals from the remaining melt. Compared with other lunar rocks, ferroan anorthosites are low in FeO and incompatible trace elements (e.g. Th) (e.g. Lucey *et al.* 2006 and references therein). The Apollo sample collection contains several large, nearly monomineralic plagioclase rocks (e.g. Warren 1990), and outcrops of 'pure' anorthosite have also been identified from Earth-based and spacecraft observations (Hawke *et al.* 2003; Ohtake *et al.* 2009). However, its perceived importance to lunar crustal formation has been challenged by Lucey *et al.* (2006) because, although ferroan anorthosite is the most common pristine rock type at the Apollo 16 site, it is uncommon or rare at other sites. On the basis of early inspection of the Apollo samples, it was concluded that the lunar highlands are dominated by ferroan anorthosite, and that the ferroan-anorthositic suite component of the crust is highly feldspathic. However, Lucey *et al.* (2006 and references therein) pointed out that the feldspathic lunar meteorites, geochemical observations taken from orbit, and regolith samples from Apollo 16 and Luna 20 suggest that highly feldspathic ferroan anorthosite is not necessarily typical of the highlands surface, at least not on the Moon's nearside. In fact, the majority of feldspathic lunar meteorites are more mafic than the feldspathic material of the Apollo 16 regolith (Korotev 1996, 1997; Korotev *et al.* 2003), and the high Mg# of the upper feldspathic crust (70 ± 3) is at the high end for ferroan anorthosite. Together, this implies that magnesian feldspathic rocks contributed substantially to the make-up of the lunar highlands (e.g. Lucey *et al.* 2006).

The magnesian suite possibly represents the transition between magmatism associated with the magma ocean and serial magmatism that occurred between 30 and 200 Ma after the formation of that ocean (Solomon & Longhi 1977; Longhi 1980; Shearer & Newsom 2000). However, the exact duration of emplacement of the magnesian suite into the lunar crust is still debated due to the lack of deep sampling by impacts after 3.9 Ga, a function of a decreasing impact flux. Taylor *et al.* (1993) estimated that the magnesian rock suite constitutes approximately 20% of the uppermost 60 km of the crust, with the rest being composed of ferroan anorthosite. However, other studies suggest smaller amounts of magnesian- and alkali-suite rocks are present (e.g. Jolliff *et al.* 2000; Korotev 2000).

Probably the most prominent evidence of volcanic activity on the Moon is the emplacement of mare basalts that constitute the dark basin fills. In comparison to highland rocks, they are enriched in FeO and TiO_2, depleted in Al_2O_3, have higher CaO/Al_2O_3 ratios, contain more olivine and/or pyroxene, especially clinopyroxene, but contain less plagioclase (e.g. Taylor *et al.* 1991). Mare basalts most probably formed from remelting of mantle cumulates produced by the early differentiation of the Moon. The origin of KREEP basalts (rich in K, Rare Earth Elements and P), however, is likely to be related to remelting or assimilation by mantle melts of a late-stage magma-ocean residuum, the so-called ur-KREEP (Warren & Wasson 1979). There are numerous ways to distinguish different basalt types, including through petrography, mineralogy and chemistry (e.g. Neal & Taylor 1992; Papike *et al.* 1998). Taylor *et al.* (1991) utilized their TiO_2 abundances to define three basalt types: very low-Ti (VLT) basalts (<1.5 wt% TiO_2), low-Ti basalts ($1.5-9$ wt% TiO_2) and high-Ti basalts (>9 wt% TiO_2). On the basis of extensive laboratory studies of TiO_2 abundances of the lunar samples and spectroscopy (e.g. Papike *et al.* 1976; Papike & Vaniman 1978; Neal & Taylor 1992; Papike *et al.* 1998), global maps of the major mineralogy–chemistry of the Moon with remote-sensing techniques have been derived (e.g. Charette *et al.* 1974; Pieters 1978; Johnson *et al.* 1991; Melendrez *et al.* 1994; Shkuratov *et al.* 1999; Giguere *et al.* 2000; Lucey *et al.* 2000). Because early interpretations of the Apollo and Luna data suggested that lunar mare volcanism began with high-TiO_2 basalts that are older than later Ti-poor basalts, models were proposed in which this perceived correlation was coupled to the depth of melting (e.g. Taylor 1982). However, remote-sensing data indicate that young basalts exist with high TiO_2 concentrations (e.g. Pieters *et al.* 1980); moreover, there are also old (mostly >3 Ga) lunar basaltic meteorites that are very low in Ti content (Cohen *et al.* 2000; Terada *et al.* 2007). Finally, a combination of iron and titanium maps (e.g. Lucey *et al.* 2000) and crater size–frequency distribution measurements across the nearside and farside did not reveal a distinct correlation between mare ages and composition (Hiesinger *et al.* 2001; Pasckert *et al.* 2014). Thus, FeO and TiO_2 concentrations varied independently with time, whereas TiO_2 (FeO)-rich and TiO_2 (FeO)-poor basalts erupted contemporaneously.

The geochemistry of returned lunar samples suggests that the ultramafic sources of mare basalts are complementary to the anorthositic crust (e.g. Wieczorek *et al.* 2006) and that the Moon is, therefore, differentiated. To explain these observations, the magma-ocean model was developed (e.g. Smith *et al.* 1970; Wood *et al.* 1970; Warren & Wasson 1979; Warren 1990). The model assumes that large parts of the Moon were initially molten, such that a global, several-hundred-metres-thick magma ocean formed. Although the details of the crystallization of such a magma ocean are not fully understood, it is likely that the early crystallization of olivine and orthopyroxene produced cumulates in the deeper parts of the magma ocean because they were denser than coexisting melt (Shearer *et al.* 2006 and references therein). With time, early magnesium-rich cumulates became increasingly iron-rich. As a consequence of the early crystalliza- tion of Fe- and Mg-rich minerals, the melt became richer in Al and Ca, which in turn resulted in the crystallization of plagioclase after about 75–80% magma-ocean solidification. As plagioclase was less dense than its source melt, it eventually formed the anorthositic highland crust. Finally, continued crystallization of the magma ocean yielded a glob- ally asymmetric distribution of KREEP elements. On the basis of lunar samples, it became apparent that the main phases of the magma ocean were com- pletely crystallized by about 4.4 Ga. The KREEP residue was solid at about 4.36 Ga. In such a crystal- lization sequence, it is plausible that the later- formed, denser, iron-rich cumulate mantle overly- ing the earlier, less dense magnesium-rich mantle cumulates was gravitationally unstable, resulting in the development of an overturn of the cumulate mantle. Such an overturn might have delivered cold, dense, incompatible-element-rich material to the core–mantle boundary. Simultaneously, hot rising mantle plumes may have melted adiabati- cally to produce the first basaltic resurfacing crust of the Moon (i.e. the early mare basalts).

Volcanic landforms

The Moon hosts a wide variety of volcanic land- forms, including lava flows, cones, domes, shield volcanoes, sinuous rilles, pyroclastic deposits and cryptomaria (e.g. Hiesinger & Head 2006; Spudis *et al.* 2013) (Fig. 7). Since the Russian Luna 3 mis- sion in 1959, it has been known that lunar basalts are concentrated on the nearside. Many basalt deposits are located in the interiors of low-lying impact basins. Globally, mare basalts cover 7×10^6 km^2, or 17%, of the total lunar surface – amounting to 1% of the lunar crustal volume (Head 1976; Wilhelms 1987; Hiesinger & Head 2006) (Fig. 6). Although the specific details of mare basalt petrogenesis are still not fully under- stood, the presence of radioactive elements (e.g. K, U and Th) most probably resulted in the for- mation of partial melts of ultramafic mantle material at depths of between about 60 and 500 km (e.g. Wil- helms 1987; Hiesinger & Head 2006).

Lava flows. Mare basalts were formed by large volumes of low-viscosity, high-temperature basal- tic lava, which resurfaced vast areas (i.e. *c.* 30% of the lunar nearside) (Head 1976; Wilhelms 1987). In fact, laboratory measurements of molten lunar basalts indicate that their viscosity is only a few

tens of poise at 1200 °C, allowing them to flow for long distances across the surface before solidifying (Hörz *et al.* 1991). Lava flows several tens of metres thick extending for hundreds to thousands of kilometres across Mare Imbrium have been docu- mented (Schaber 1973; Schaber *et al.* 1976) (Fig. 7a). Thin lava flows similar to those in Mare Imbrium have been observed elsewhere on the Moon: within the Hadley Rille at the Apollo 15 landing site, for example. However, many lunar lava flows lack distinctive flow fronts due to their very low viscosities, high eruption rates, ponding of lava in shallow depressions, subsequent destruc- tion by impact processes and/or burial by youn- ger flows. Similarly, volcanic vents in the mare regions are rare because they were probably covered by the erupting basalts or were degraded by subsequent impacts.

Sinuous rilles. Sinuous rilles (Figs 6 & 7b) are meandering channels that often start at a crater-like depression and end by grading downslope into the smooth mare surface (Greeley 1971). Most such rilles originate along the margins of the basins and trend towards the basin centre. These channels range in width from a few tens of metres to approxi- mately 3 km, from a few kilometres to up to 300 km in length and are, on average, 100 m deep (Schubert *et al.* 1970; Hurwitz *et al.* 2012, 2013*a*). The Apollo 15 mission investigated one of these sinuous rilles, Rima Hadley, in detail. Despite early interpreta- tions of sinuous rilles as lunar rivers (e.g. Lingenfel- ter *et al.* 1968; Peale *et al.* 1968), no evidence for water or pyroclastic flows was found and so it was concluded that sinuous rilles formed by thermal erosion, resulting in widening and deepening of channelled lava flows due to the melting of the underlying rock by very hot lavas (Hulme 1973; Coombs *et al.* 1987; Williams *et al.* 2000; Hurwitz *et al.* 2012, 2013*a*). Lower gravity, higher melt temperatures, lower viscosities and higher extru- sion rates might be responsible for the much larger sizes of lunar sinuous rilles compared with lava tubes and channels on Earth. Bussey *et al.* (1997) and Fagents & Greeley (2001) showed that the process of thermal erosion is very sensitive to the physical conditions in the boundary layer between lava and solid substrate. They found that thermal erosion rates depend on the slopes, effusion rates and thermal conductivities of the liquid substrate boundary layer.

Cryptomaria. Cryptomaria are mare-like volcanic deposits that later were covered with lighter- coloured material (e.g. ejecta from craters and basins) (Head & Wilson 1992). These deposits can be studied with several techniques, including inves- tigations of dark halo craters (e.g. Schultz & Spudis

1979, 1983; Hawke & Bell 1981), multispectral images (e.g. Head et al. 1993; Greeley et al. 1993; Blewett et al. 1995; Mustard & Head 1996) and orbital geochemical observations (e.g. Hawke & Spudis 1980; Hawke et al. 1985). These studies have shown that if cryptomaria are included, the total area covered by mare deposits exceeds 20% of the lunar surface, compared with about 17% of typical mare deposits alone (Head 1976; Antonenko et al. 1995). Thus, cryptomaria not only indicate a wider spatial distribution of ancient volcanic products but also reveal that mare volcanism was already active prior to the formation of the Orientale basin (i.e. the youngest basin on the Moon) and the emplacement of its ejecta, which is an important stratigraphic marker horizon. Although Giguere et al. (2003) and Hawke et al. (2005) reported that the buried basalts in the Lomonosov–Fleming and the Balmer–Kapteyn regions are very-low- to intermediate-Ti basalts, sampling and subsequent analyses are required to understand the true nature of cryptomaria.

Domes, cones and shields. A small fraction of the vast mare regions is covered by positive topographical surface features, such as domes, cones and shields, that measure up to several tens of kilometres across, are up to several hundred metres high and are basaltic in composition (Head & Gifford 1980). Lunar cinder cones are often associated with lunar sinuous rilles (e.g. in Alphonsus crater; Head & Wilson 1979), are less than 100 m high, 2–3 km wide, have summit craters of less than 1 km diameter and have very low albedos (Guest & Murray 1976).

Lunar mare domes are generally broad, convex, semi-circular landforms with relatively low topographical relief (Fig. 7c). For example, Guest & Murray (1976) mapped 80 mare domes with diameters of 2.5–24 km, 100–250 m heights and 2°–3° slopes; most of them occur in the Marius Hills complex. Some of the Marius Hills domes are characterized by steeper slopes (7°–20°) and some have summit craters or fissures. Most probably, the mare domes were formed by eruptions of more viscous (i.e. more silicic) lavas, intrusions of shallow laccoliths or mantling of large blocks of older rocks with younger lavas (e.g. Heather et al. 2003; Lawrence et al. 2005). Large shield volcanoes (>50 km) are common on Earth, Venus and Mars, and are constructive features: that is, they consist of a large number of small flows derived from a shallow magma reservoir where the magma reaches a neutral buoyancy zone (e.g. Ryan 1987; Wilson & Head 1990; Head & Wilson 1992). Thus, the presence of shield volcanoes and calderas implies shallow buoyancy zones, the stalling and evolution of magma there, leading to numerous eruptions

of small volumes and durations, and shallow magma migration that causes caldera collapse. However, no shield volcanoes larger than about 20 km in diameter have been identified on the Moon (Guest & Murray 1976). This observation indicates that shallow buoyancy zones do not occur on the Moon, and that lavas did not extrude in continuing sequences of short-duration, low-volume eruptions from shallow reservoirs. However, there is evidence that in a few locations magma may have stalled near the surface to form shallow sills or laccoliths, as possibly indicated by the formation of floor-fractured craters (Schultz 1976; Wichman & Schultz 1995, 1996) (see below).

Apart from the basaltic domes in the mare regions, there are also domes that are plausibly linked to non-mare volcanism. These landforms are much less abundant than those associated with basaltic volcanism, are characterized by slopes steeper than those of mare domes, and exhibit a high albedo and a strong absorption in the ultraviolet (Malin 1974; Wood & Head 1975; Head et al. 1978; Chevrel et al. 1999; Hawke et al. 2003). Braden et al. (2010) and Tran et al. (2011) used Lunar Reconnaissance Orbiter Camera (LROC) stereo images to study the topography of mare and non-mare domes. Volcanic constructs with shallow flank slopes (<10°) are associated with preferentially low-viscosity eruptions, whereas steeper slopes (>20°) tend to indicate high-viscosity, silica-rich eruptions of lava that are more viscous than mare basalts (Hawke et al. 2003; Wilson & Head 2003; Glotch et al. 2011). The Marius Hills complex is a key locality for the first type of dome; the Gruithuisen domes are excellent examples of the second type. However, some domes in the Marius Hills complex display relatively steep slopes (Tran et al. 2011), indicating that localized eruptions of higher-silicic lavas occurred, confirming the LROC and Chandrayaan-1 Terrain Mapping Camera (TMC) results of Lawrence et al. (2010) and Arya et al. (2011), respectively. Detailed morphological and spectral studies of non-mare domes (e.g. Gruithuisen and Hansteen Alpha), as well as domes associated with the Compton–Belkovich thorium anomaly, using LROC and LRO Diviner data are consistent with high-viscosity, silicic, non-mare volcanism (Braden et al. 2010; Glotch et al. 2011; Hawke et al. 2011; Jolliff et al. 2011). Lunar Prospector gamma-ray data revealed that several domes exhibit high Th concentrations (Lawrence et al. 2003; Hagerty et al. 2006), and in LRO Diviner data the domes are characterized by elevated silica contents (Glotch et al. 2011). Some of the non-mare domes show a very specific spectral behaviour in the ultraviolet spectrum and are known as 'red spots'. Red spots show a much wider range in morphology, and also include bright shields and bright smooth plains with little topographical expression (Bruno *et al.* 1991; Hawke *et al.* 2003). Because of their unique characteristics, it is thought that red spots have been formed by more viscous lava, comparable to, for example, terrestrial dacites or rhyolites (Hawke *et al.* 2003; Wilson & Head 2003).

Pyroclastic volcanism. There are two types of pyroclastic deposit that have been identified on the surface of the Moon: extensive regional pyroclastic deposits (>1000 km) located on the uplands adjacent to younger maria (e.g. Gaddis *et al.* 1985; Weitz *et al.* 1998); and smaller pyroclastic deposits that are more widely dispersed across the lunar surface (Head 1976; Hawke *et al.* 1989; Coombs *et al.* 1990). Most pyroclastic deposits are concentrated in localized areas but some may cover larger areas, in excess of 2500 km^2 (e.g. Head *et al.* 2002; Hiesinger & Head 2006) (Figs 6 & 7d). It has been proposed that regional dark-mantle deposits were formed by eruptions in which continuous gas exsolution in the lunar environment caused Hawaiian-style fire fountaining that distributed pyroclastic material over tens to hundreds of kilometres (Wilson & Head 1981, 1983). Substantial progress in understanding the ascent and eruption conditions of pyroclastic deposits has been made (e.g. Head *et al.* 2002; Shearer *et al.* 2006). The Apollo 17 orange glasses and black vitrophyric beads, for example, formed during lava fountaining of gas-rich, low-viscosity, Fe–Ti-rich basaltic magmas (Heiken *et al.* 1974). Crystallized black beads from the Apollo 17 landing site had cooling rates of 100 °C s^{-1}, which is much slower than expected from black-body cooling in a vacuum (Arndt & von Engelhardt 1987). Apollo 17 is not the only landing site where pyroclastic glass beads have been identified. Rather, the Apollo 15 green glasses are also volcanic in origin, as are other pyroclastic glasses found in the regolith of other landing sites (Delano 1986). In comparison to mare basalts, the volumes of pyroclastic deposits are trivial. However, they demonstrate that lava fountaining occurred on the Moon. In addition, because of the presumably fast ascent and cooling history of pyroclastic materials, they are thought to be unmodified by crystal fractionation. Thus, volcanic glass beads are plausibly the best samples for studying the lunar mantle. Pyroclastic eruptions resulted in the emplacement of dark mantled deposits that cover areas of the lunar surface large enough to be visible in remotely sensed data (e.g. Hawke *et al.* 1979, 1989; Head & Wilson 1980; Gaddis *et al.* 1985; Coombs *et al.* 1990; Greeley *et al.* 1993; Weitz *et al.* 1998; Weitz & Head 1999; Head *et al.* 2002). In remote-sensing data, it is apparent that pyroclastic deposits often tend to occur along the margins of impact basins, and in association with vents and sinuous rilles, implying that they were formed by sustained, large-volume eruptions. In addition, at least some of the observed dark-halo craters are volcanic (i.e. pyroclastic) in origin, whereas others are impact craters that excavated darker material from the subsurface. Commonly, the volcanic dark-halo craters preferentially occur along fractures and on the floors of larger craters. From studies of dark-halo craters within Alphonsus crater, it appears likely that they were formed by vulcanian-style eruptions (Head & Wilson 1979; Coombs *et al.* 1990).

Ages of volcanic deposits. On planetary surfaces, the number of impacts on a specific surface unit can be correlated with the time that this unit was exposed to bombardment by asteroids and comets: the higher the crater frequency, the greater the age of the unit (e.g. Öpik 1960; Baldwin 1964; Neukum *et al.* 1975*a, b*, 2001; Basaltic Volcanism Study Project 1981). Thus, the frequency of craters superimposed on a specific surface unit at a given diameter, or range of diameters, is a direct measure of the relative age of the unit (e.g. Arvidson *et al.* 1979; Basaltic Volcanism Study Project 1981; Neukum & Ivanov 1994; Neukum *et al.* 2001). The stratigraphic record of geological events on the Moon was studied in detail in preparation for, and following, the US and Russian lunar missions (e.g. Shoemaker & Hackman 1962; Wilhelms & McCauley 1971; Wilhelms 1979, 1987). Those studies, in concert with radiometric dating of lunar samples, revealed that the lunar highlands are generally older than the mare regions (e.g. Wilhelms 1987), that mare volcanism occurred over an extended period of time (e.g. Shoemaker & Hackman 1962; Carr 1966; Hiesinger *et al.* 2000, 2003, 2011) and that there is considerable variation in the mineralogy of basalts of different ages (e.g. Soderblom *et al.* 1977; Pieters *et al.* 1980; Hiesinger *et al.* 2000). Accurate mare basalt ages are important data as they help characterize the duration and the flux of lunar volcanism, the petrogenesis of lunar basalt and the relationship of volcanic activity to the thermal evolution of the Moon.

Hiesinger *et al.* (2000, 2003, 2011, 2012) dated basalts in Oceanus Procellarum, Imbrium, Serenitatis, Tranquillitatis, Humboldtianum, Australe, Humorum, Nubium, Cognitum, Nectaris, Frigoris and numerous smaller occurrences. They found that: (1) in the studied locations, lunar volcanism was active for almost 3 Ga, starting at about 4.0–3.9 Ga and ceasing at around 1.2 Ga; (2) most basalts were erupted during the late Imbrian Period, at about 3.8–3.6 Ga; (3) substantially fewer basalts were emplaced during the Eratosthenian Period (3.2–1.1 Ga); and (4) basalts of possible

Copernican age (<1.1 Ga) are found only in limited areas within Oceanus Procellarum (Hiesinger et al. 2000, 2003, 2011, 2012). From these results it is also apparent that older mare basalts preferentially occur in the eastern and southern lunar nearside, and in patches of maria peripheral to the larger maria, in contrast to the younger basalt ages on the western nearside, for example, those in Oceanus Procellarum. Although older basalts certainly also erupted in the western mare areas prior to the emplacement of younger flows, young volcanism only occurred in the western hemisphere, and has been related to the concentration of heat-producing elements there. Mare basalts on the central lunar farside erupted between 3.5 and 2.7 Ga, which is well within the range of ages found for the nearside mare basalts (Morota et al. 2011; Passckert et al. 2014). However, farside mare volcanism ceased earlier than on the nearside, which might be a consequence of either a thicker crust or reduced abundances of radioactive elements in the farside mantle.

On the Moon, volcanism apparently resulted from the partial melting of mantle rocks. The decay of naturally radioactive elements resulted in the production of partial melts of mostly basaltic composition (45–55% SiO_2, and relatively high MgO and FeO content), requiring temperatures of >1100 °C and depths of >150–200 km (Hörz et al. 1991). Radiometric ages of returned lunar samples reveal that most volcanic eruptions stopped at approximately 3 Ga (Hörz et al. 1991). This finding was interpreted as evidence for an early cooling of the mantle below the temperature necessary to produce partial melts. However, this interpretation is inconsistent with crater counts on mare basalt surfaces, which indicate that some basalts erupted as 'recently' as about 1–2 Gyr ago (Hiesinger et al. 2003, 2011). Some geophysical models of the thermal evolution of the Moon suggest that the zone of partial melting necessary for the production of basaltic magmas migrated to depths too great for melts to reach the surface approximately 3.4–2.2 Gyr ago (e.g. Spohn et al. 2001). However, taking into account the insulating effects of a porous megaregolith, the interior can be kept warm enough to explain late-stage volcanic eruptions until about 2 Ga (Ziethe et al. 2009). Moreover, a non-uniform distribution of heat-producing elements in the mantle, as indicated by Lunar Prospector data, could extend the potential for melting to even more recent times.

Detailed crater size–frequency distribution measurements revealed that the two Gruithuisen domes in the northern Oceanus Procellarum region appear to be contemporaneous with the emplacement of the surrounding mare basalts, but post-date the formation of post-Imbrium crater Iridum (Wagner et al. 1996, 2002). Head et al. (2000) interpreted this contemporaneity with the maria as evidence for a petrogenetic link; one possibility is that mare diapirs stalled at the base of, and partially remelted, the crust, which produced the more silicic viscous magmas of the domes. Crater size–frequency distribution measurements were also performed for red spots in southern Oceanus Procellarum and Mare Humorum, and indicate a wide range of ages. For example, Hansteen Alpha is about 3.74–3.56 Ga old, and so is slightly younger than the Gruithuisen domes but post-dates craters Billy (3.88 Ga) and Hansteen (3.87 Ga). However, Hansteen Alpha is older than the surrounding mare materials (3.51 Ga) (Wagner et al. 2010). NE of Mare Humorum, red-spot light plains associated with a feature named 'The Helmet' (investigated in detail by Bruno et al. 1991) range in age from 3.94 Ga (Darney χ) to 2.08 Ga (Wagner et al. 2010). The ages of the Gruithuisen domes and of Hansteen Alpha show that high-silica, more viscous non-mare volcanism was active in a shorter time interval than mare volcanism activity, and appears to be restricted to more or less the Late Imbrian Epoch, at least in these two investigated areas (Wagner et al. 2002, 2010). However, a non-volcanic origin for the red-spot light plains cannot be excluded on the basis of the data currently available (Wagner et al. 2010 and references therein).

Tectonism

The Moon is a so-called one-plate or stagnant-lid planetary body (e.g. Solomon & Head 1979, 1980; Spohn et al. 2001; Hiesinger & Head 2006). Crystallization of the magma ocean resulted in a globally continuous, low-density crust that may have hindered the development of plate tectonics early in lunar history. Once such a nearly continuous, low-density crust or stagnant lid was established, conductive cooling dominated the transfer of heat from the Moon's interior to its surface. This resulted in the production of a globally continuous lithosphere rather than multiple, moving and subducting plates as on Earth. The heat flows measured at some of the Apollo landing sites are much lower than those on Earth (e.g. Langseth et al. 1976), and are consistent with heat loss predominantly by conduction. The large ratio of surface area to volume has been very effective in cooling the Moon by conduction (i.e. by radiating heat into space). Thus, it is thought that the lithosphere of the Moon thickened rapidly, and so the Moon became a one-plate planet quickly, losing most of its heat through conduction (Solomon 1978). Support for this model comes from nearside seismic data that indicate the presence of a relative rigid, 800–1000 km-thick lithosphere (Nakamura et al. 1973; Spohn et al. 2001;

Wieczorek et al. 2006). In summary, the crust of the Moon appears to be thick, rigid, immobile and cool, inhibiting large-scale motion. The lack of plate tectonics-style crustal deformation is consistent with the returned samples, which show virtually no textures typical of plastic deformation.

On the Moon, tectonic deformation is caused by: (1) impact-induced stress; (2) stress induced by the load of basaltic materials within impact basins; (3) thermal effects; and (4) tidal forces (e.g. Hiesinger & Head 2006), producing mostly extensional and contractional features (e.g. faults, graben, dykes and wrinkle ridges).

Models of lunar thermal evolution indicate that, during the first billion years, the lunar crust was subject to extensional stresses produced by thermal expansion, and that, during the following 3.5 Ga until the present, compressional stresses have dominated due to cooling and contraction (e.g. Solomon & Chaiken 1976). However, these thermal models have numerous unconstrained parameters that have influence on the model predictions for the evolution of the planetary radius and the resulting stress field (e.g. Pritchard & Stevenson 2000). Although the early stresses in the lunar crust are not accurately known, Solomon & Head (1979) proposed that graben that are related to the loading by basaltic flows of basin centres could only form in the presence of global, mildly tensile stresses. Pritchard & Stevenson (2000) pointed out that the end of graben formation at 3.6 Ga cannot be used to decipher the ancient global lunar stress field because local effects including flexure and magmatic activity (i.e. diapirism) could mask the signal. However, after about 3.6 Ga, the lunar stress field became compressional, and internally driven tectonic activity may have ceased for 2.5–3 Gyr, until sufficient stress (>1 kbar) had accumulated to produce small-scale thrust faults.

Tidal effects due to the Earth–Moon gravitational interaction would have been more intense early in lunar history, when the Moon was closer to Earth. In fact, early lunar tectonic activity may have been dominated by tidal stresses, and internally generated stresses may have been less important. However, the spatial distribution of lineaments mapped on the lunar surface is similar on the near- and farside, and thus is probably independent of tidal forces (e.g. the collapse of a tidal bulge) (Chabot et al. 2000). In particular, Watters et al. (2010) argued that the spatial distribution of small young lobate scarps is inconsistent with an origin related to tidal deformation. They concluded that a global contraction of the lunar radius by 100 m over the last 1 Gyr could explain the observed pattern and calculated stresses.

In the following paragraphs, we describe briefly some of the tectonic landforms on the Moon, including graben, wrinkle ridges, lobate scarps and floor-fractured craters.

Graben. Impacts are capable of creating radial and/ or concentric extensional troughs or graben (e.g. Ahrens & Rubin 1993), and impact-induced faults may be reactivated by seismic energy (e.g. Schultz & Gault 1975). Loading of impact basins with basaltic infill causes the development of extensional stresses at the edges of the basins and the formation of arcuate troughs or rilles there (e.g. Solomon & Head 1979, 1980; Wilhelms 1987) (Fig. 7e). Towards the basin interior, compressional stress due to downwarping of the basin centre leads to the formation of subradial and concentric ridges (Solomon & Head 1979, 1980; Freed et al. 2001) (Fig. 7f). Concentric graben around the Humorum basin are a few hundred kilometres long and are filled with basaltic lavas. Thus, they are evidence of early extensional forces in this area. Because these graben extend virtually unobstructed from the mare into the adjacent highlands and cut across pre-existing craters, their formation was possibly related to a substantial, deep-seated, basin-wide stress field.

Wrinkle ridges. Wrinkle ridges are common landforms on terrestrial planets (Figs 6 & 7f). On the Moon, the spatial dimensions of wrinkle ridges range from several kilometres up to 10 km in width, tens to hundreds of km in length and their average heights are of the order of 100 m (Wilhelms 1987). Lucchitta (1976) proposed that wrinkle ridges are caused by thrust faulting and folding, a model also favoured by Golombek (1999) and Golombek et al. (2000) on the basis of high-resolution topographical data of Martian wrinkle ridges. Although wrinkle ridges on the Moon are commonly interpreted as thrust faults formed by compressive stresses (e.g. Hodges 1973; Lucchitta 1976; Solomon & Head 1979, 1980; Plescia & Golombek 1986; Schultz & Zuber 1994; Golombek 1999; Golombek et al. 2000; Hiesinger & Head 2006), some may have an origin linked to the emplacement of magma in the subsurface (e.g. Strom 1964; Hartmann & Wood 1971). Observed en echelon offsets of wrinkle ridges in Mare Serenitatis were interpreted as evidence of compressional stresses. The wrinkle ridges in Mare Serenitatis are concentric to the basin, and Muehlberger (1974) and Maxwell (1978) estimated a centrosymmetric foreshortening of approximately 0.5–0.8% in order to produce the ridges. Ground-penetrating radar (GPR) data from the Apollo Lunar Sounding Experiment (ALSE) revealed substantial upwarping, and possibly folding and faulting of the basaltic surface down to about 2 km below the wrinkle ridges. Recent work has shown that deep-seated faults that penetrate approximately 20 km into the lunar

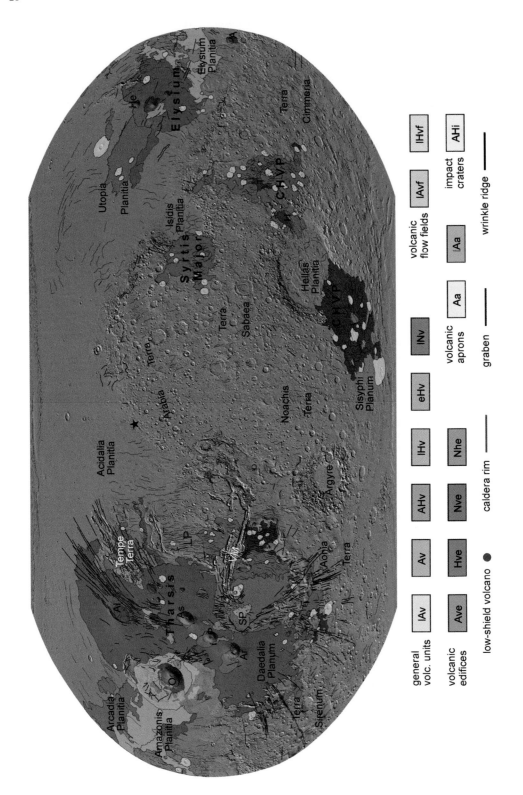

lithosphere underlie wrinkle ridges in Mare Crisium (Byrne *et al.* 2014*c*).

Lobate scarps. Although wrinkle ridges and graben dominate the tectonics of the nearside maria, lobate scarps are the dominant tectonic landform on the farside (Watters & Johnson 2010; Watters *et al.* 2010). Lunar lobate scarps are characterized by steep scarp faces and by linear or curvilinear asymmetric forms with arcuate fault surfaces, and consist of a series of smaller, connected structures that form complexes. Scarp complexes have lengths of up to about 10 km, occur in sets of up to 10 individual structures and commonly are less than approximately 100 m high (e.g. Binder 1982, 1986; Binder & Gunga 1985; Watters *et al.* 2010; Banks *et al.* 2012; Williams *et al.* 2013; Clark *et al.* 2014). Schultz (1972) realized that the scarps were very young and based on crater degradation measurements, Binder & Gunga (1985) estimated the ages of lobate scarps found in Apollo Panoramic Camera images to be <1 Ga. Such young ages are generally consistent with crater size–frequency measurements (van der Bogert *et al.* 2012), their relatively undegraded appearance and their cross-cutting relationships with small craters (Watters & Johnson 2010; Watters *et al.* 2010). Thus, Watters *et al.* (2010) proposed that the lobate scarps are evidence of late-stage contraction of the Moon, and that the estimated strains accommodated by these small scarps are consistent with thermal history models that predict low-level compressional stresses and relatively small changes (100–1000 m) in lunar radius. Although thermal history models for either a nearly or totally molten early Moon, or an early Moon with an initially hot exterior (i.e. a magma ocean) and a cool interior, predict late-stage compressional stresses in the upper crust and lithosphere, the timing of scarp formation and the estimated stress levels are more consistent with the magma-ocean hypothesis (e.g. Watters *et al.* 2010).

Floor-fractured craters. Numerous impact craters on the Moon exhibit an extensive system of fractures and graben on their floors (Fig. 6). Schultz (1976) proposed that fracturing of lava-filled crater floors (e.g. crater Gassendi) might be related to isostatic uplift of the floor materials and expansion due to the emplacement of sills below the crater floor. Support for this interpretation comes from the geophysical models of Wichman & Schultz (1995, 1996) and Dombard & Gillis (2001). Wichman & Schultz (1996) estimated the minimum depth of a 30 km-wide and 1900 m-thick intrusion beneath crater Tauruntius to be of the order of 1–5 km, resulting in an excess pressure of around 9 MPa. Similarly, on the basis of their model, Dombard & Gillis (2001) concluded that, compared with topographical relaxation, laccolith emplacement is the more viable formation process. More recent studies utilizing gravity data acquired by the Gravity Recovery and Interior Laboratory (GRAIL) mission demonstrated the presence and dynamics of magmatic intrusive bodies beneath floor-fractured craters (Jozwiak *et al.* 2014; Thorey & Michaut 2014).

Mars

Early Mars exploration by flybys of the Mariner 4, 6 and 7 spacecraft, and orbital observations by the Mariner 9 and Viking Orbiter I and II missions, revealed extensive volcanic surfaces and large volcanic edifices on the Red Planet (McCauley *et al.* 1972; Greeley & Spudis 1981). In the seminal paper of Greeley & Spudis (1981), most of the volcanic landforms were categorized into large shield volcanoes (e.g. Olympus Mons and the three Tharsis Montes, Arsia, Pavonis and Ascraeus), steep-sided domes (e.g. Tharsis Tholus) and highland paterae (e.g. Tyrrhenus and Hadriacus Montes) (Fig. 8) – the latter being the only large-scale volcanic

Fig. 8. The distribution of volcanic units, edifices and tectonic structures on Mars (adopted from Tanaka *et al.* 2014). Unit names consist of (1) age, (2) unit group and (3) unit subtype; (1) stratigraphic periods include: A, Amazonian; H, Hesperian; N, Noachian; with epochs shown in lower case (i.e. e, Early; l, Late); (2) a, apron; h, highlands; i, impact; v, volcanic; (3) e, edifice; f, field. Note that only Amazonian–Hesperian impact craters are shown, which may superpose volcanic units. Units Aa and lAa located around Olympus Mons, and west of the Tharsis Montes, respectively, represent edifice-derived volcaniclastic material. Mapped Noachian highland edifices may be volcanic constructs; those located in Sisyphi Planum are not shown at this map scale. All unit contacts are displayed as certain contacts for clarity. Symbols for graben and wrinkle ridges have been modified from the source (Tanaka *et al.* 2014). The locations of low-shield volcanoes are from Hauber *et al.* (2011), Platz & Michael (2011), and Manfredi *et al.* (2012). Note there exist more low-shield volcanoes in Tharsis; those present in Elysium Planitia are not included. Major volcanic provinces are abbreviated as CHVP, Tharsis, Elysium and Syrtis Major. A, Apollinaris Mons; Al, Alba Mons; Ar, Arsia Mons; As, Ascraeus Mons; CHVP, Circum-Hellas Volcanic Province; H, Hadriacus Mons; He, Hecates Tholus; LP, Lunae Planum; O, Olympus Mons; P, Pavonis Mons; SP, Syria Planum; T, Tyrrhenus Mons; VM, Valles Marineris. The star shows the approximate location of Eden Patera (Michalski & Bleacher 2013). The map has a Robinson projection, centred at 0°E; the background image is a Mars Orbiter Laser Altimeter (MOLA)-derived shaded-relief image (0.46 km/px).

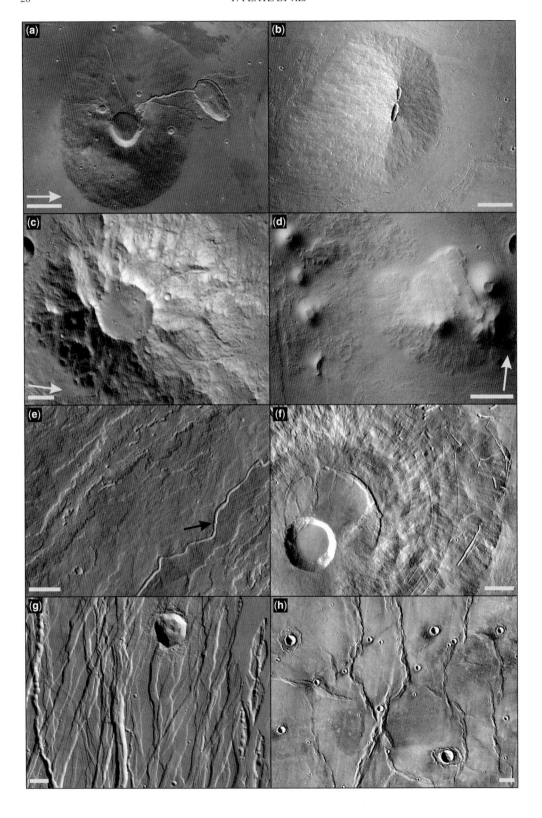

landforms attesting to explosive volcanism in early Mars history (Fig. 8). Extensive volcanic plains associated with large edifices were recognized and grouped into volcanic provinces, with Tharsis and Elysium being the largest, most active and longest-lived (e.g. Neukum *et al.* 2004, 2010; Hauber *et al.* 2011; Platz & Michael 2011). Greeley & Spudis (1981) found that volcanic activity on Mars spans the entire planet's history, and is recorded over large portions of its surface (Fig. 8). Major volcanic landforms, including large volcanic provinces and associated shield volcanoes, pyroclastic cones, lava flows, and fall-out deposits, are reviewed in the following subsections.

Volcanism

Volcanic provinces. The Tharsis region is the most dominant locus of volcanic activity on Mars, extending more than 6000 km (NNE–SSW) by 3500 km (east–west). This region has been mainly formed by five large volcanoes, with summit elevations of up to 21 km (Olympus Mons, Alba Mons and the Tharsis Montes). Several smaller volcanoes, known as paterae and tholi (Uraunius, Biblis, Ulysses Paterae and Tharsis, Ceraunius (Fig. 9a), Uraunius, and Jovis Tholi), are also present (Fig. 8). More than 700 small, low-shield volcanoes (Fig. 9b), vents and fissures located in summit calderas, on edifice flanks, at the periphery of large shield volcanoes or within fractured terrain have also been identified so far. Particularly noteworthy are two low-shield fields in Sinai Planum (Baptista *et al.* 2008; Hauber *et al.* 2011; Richardson *et al.* 2013) and Tempe Terra (Davis & Tanaka 1993; Hauber *et al.* 2011) (Fig. 8).

The majority of volcanic material in the Tharsis volcanic province was erupted in the Noachian Period (>3.71 Ga) to Early Hesperian Epoch (3.71–3.61 Ga), with infrequent eruptions thereafter scattered throughout the Late Hesperian Epoch (3.61–3.37 Ga) and Amazonian Period (3.37 Ga–present) (Neukum & Hiller 1981; Greeley & Schneid 1991; Neukum *et al.* 2004, 2010; Werner 2009) (Fig. 8). More recently, lava flows as young as 2 Ma were discovered, which are mostly related to low-shield volcanoes or late-stage effusive events (Neukum *et al.* 2004; Vaucher *et al.* 2009; Hauber *et al.* 2011; Platz & Michael 2011). Volcanic activity associated with partial caldera collapse at the summit of most large shield volcanoes has also occurred in the Middle–Late Amazonian (Werner 2009; Robbins *et al.* 2011). The total production of magma (i.e. intrusive and extrusive) within the Tharsis region is estimated to be about 3×10^8 km^3 (Phillips *et al.* 2001). However, to date, no data exist on what percentage of that volume was erupted during each of the three Martian periods (i.e. the Noachian, Hesperian and Amazonian).

The Elysium volcanic province (Fig. 8) is located in the eastern hemisphere of Mars, and comprises the 14 km-tall Elysium Mons and the smaller Albor and Hecates Tholi, all of which are situated on the broad, >1200 km-wide Elysium rise. The Elysium rise itself developed on the SE rim of the approximately 3300 km-diameter Utopia impact basin, which is thought to have formed at approximately 4.1 Ga (Frey *et al.* 2007; Frey 2008). This province comprises lava flow plains and reworked volcaniclastic material, which together extend over an area of approximately 3.4×10^6 km^2 (Tanaka *et al.* 2005). The total minimum

---

Fig. 9. Examples of volcanic and tectonic features on Mars. (**a**) Ceraunius Tholus, a large shield volcano, located in the Tharsis region between Ascraeus Mons and Tempe Terra. It is a large shield volcano partially buried by the surrounding lava plains. From its central caldera a large valley emanates towards the north which formed a delta within the Rahe impact crater. Rahe's butterfly-shaped ejecta blanket is clearly visible. Scale bar is 25 km (HRSC image mosaic; ESA/ DLR/FU Berlin (Gerhard Neukum)). (**b**) Low-shield volcano located south of Ascraeus Mons at 3.1°N, 106.8°W. The edifice is partially buried by external lava flows; channel and levée morphologies are partially visible. The edifice's summit is marked by two aligned and elongated craters. Lava flows emanated radially from the vents. Scale bar is 2 km (context camera [CTX] image P06_003185_1824). (**c**) Zephyria Tholus, an ancient, degraded highland volcano located at 19.7°S, 172.9°E that hosts a filled central, circular caldera (scale bar: 4 km; CTX image B18_016743_1597). (**d**) Pyroclastic cones located in the Ulysses Fossae area north of Biblis Tholus. This image is centred at 5.8°N, 122.8°W. From two cones a thick lava flow emanates. At others an extended surrounding lava field is observed (scale bar: 4 km; CTX image P21_009198_1858). (**e**) Examples of different types of lava flow morphology, approximately 30 km NNE of Olympus Mons (22.1°N, 129.8°W). Bottom right: a sinuous channel atop a lava tube (black arrow). Top left: lava flows with either channel-levée or sheet-flow morphologies (scale bar: 1 km; CTX image P20_008684_2011). (**f**) Concentric graben at the lower flanks of Pavonis Mons, whose summit region is marked by a large filled caldera and a younger, partially filled caldera. Note the radially oriented wrinkle ridges on the summit (THEMIS IR daytime mosaic centred at 1.3°N, 248.6°E; scale bar is 25 km). (**g**) A heavily fractured surface in Tharsis, located between Ceraunius Fossae and Tractus Catena. In some graben segments, classic pit chains developed; low-lying fractures have been flooded by Tharsis-sourced lava flows (upper left; HRSC image h9443_0000 centred at 26.0°N, 254.2°E; scale bar is 5 km). (**h**) Typical wrinkle ridges, here shown in Lunae Planum, with linear to concentric orientations. This image is centred at 18.9°N, 295.5°E (THEMIS IR daytime mosaic). Scale bar is 10 km. North is up in all images unless otherwise indicated by white arrow.

volume of erupted and reworked volcanic material is about 3.5×10^6 km^3 (Platz *et al.* 2010). The eruption frequency for the province has been recently studied by dating lava flows and caldera segments across Elysium (Platz & Michael 2011). Main volcanic activity occurred between 2.5 and 1 Ga but decreased thereafter, however, with 12 less frequent eruptions recorded in the past 500 myr (Platz & Michael 2011).

To the south and SE of Elysium there is yet another volcanically active region known as Elysium Planitia or the Cerberus volcanic plains (Fig. 8), which contains one of the youngest low-shield volcanoes and lava flows known on Mars (Plescia 2003; Werner *et al.* 2003; Baratoux *et al.* 2009; Vaucher *et al.* 2009). Because these lava flows and low-shield volcanoes are mostly associated with or originate at the Cerberus Fossae fracture system, which also extends into the Elysium volcanic province, Platz & Michael (2011) argued that both regions share the same magma source at depth. Hence, Elysium Planitia is considered as part of the broader Elysium volcanic province.

The Syrtis Major volcanic province is located SW of the 1350 km-diameter Isidis impact basin, near the Martian highland–lowland boundary. It extends over 7.4×10^6 km^2, and includes Syrtis Major Planum and the two edifices of Nili and Meroe Paterae (Fig. 8). The estimated thickness of erupted volcanic material in the province ranges from 0.5 to 1 km, with a total volume of about $1.6 \times 10^5 – 3.2 \times 10^5$ km^3 (Hiesinger & Head 2004). The low-relief volcanoes each contain a north–south-orientated caldera, whose formation ages are given as 3.73–2.33 Ga and 3.77–0.23 Ga for Meroe Patera caldera segments, and 3.55–1.61 Ga for Nili Patera (Werner 2009; Robbins *et al.* 2011). Based on geological mapping and observed areal densities for craters larger than 5 km in diameter, an Early–Late Hesperian age (i.e. 3.71–3.37 Ga) for Syrtis Major volcanism is suggested (Greeley & Guest 1987; Hiesinger & Head 2001; Tanaka *et al.* 2005, 2014). However, a recent study of mapping and dating of individual lava flows and volcanic crater infill clearly points to volcanic activity extending into the Early and Middle Amazonian Epochs across the province (Platz *et al.* 2014).

The Circum-Hellas Volcanic Province (CHVP) is located at the periphery of the large, approximately 2100 km-diameter Hellas impact basin (Fig. 8). The province consists of the two well-studied highland volcanoes Tyrrhenus and Hadriacus Montes, as well as the extensive volcanic plains of Hesperia Planum to the NE, and Malea Planum (including the putative volcanoes Amphitrites, Peneus, Malea and Pityusa Paterae) to the SW of Hellas (Greeley & Spudis 1981; Williams *et al.* 2007, 2008, 2009). The entire province covers more than 2.1×10^6 km^2 and formed after the Hellas basin-forming event at about 4.0–3.8 Ga (Williams *et al.* 2009). Major volcanic activity throughout the CVHP appears to be restricted to the Noachian and Hesperian, between 3.9 and 3.6 Ga (Williams *et al.* 2009), although isolated activity at Hesperia Planum did occur during the Early Amazonian (Lehmann *et al.* 2012).

Apollinaris Mons and other ancient highland volcanoes. Apollinaris Mons is a low-relief, stand-alone volcanic edifice located near the highland–lowland transitional zone SE of the Elysium volcanic province, at 9.2°S, 174.8°E (Fig. 8). The edifice is about 180 km in diameter and rises up to 3.2 km. Based on its relief, surface texture and degraded friable materials, it is likely to be composed of pyroclastic material (Robinson *et al.* 1993; Crumpler *et al.* 2007). Apollinaris Mons appears to have been active during the Noachian and Hesperian periods (Werner 2009; Tanaka *et al.* 2014), which may have included long-lived hydrothermal activity (El Maarry *et al.* 2012).

Recently, some large, irregularly shaped depressions in Arabia Terra were interpreted to represent ancient volcanic constructs that together could form the putative Arabia Terra volcanic province (Michalski & Bleacher 2013). Of these possible volcanoes, Eden Patera (Fig. 8) is the best candidate, a large depression 55×85 km in size, located at 33.6°N, 348.9°E. Fine-grained, layered and often sulphate-bearing material exposed throughout Arabia Terra (Malin & Edgett 2000; Edgett & Malin 2002) may represent volcanic fall-out deposits sourced from Eden Patera and nearby presumptive volcanic constructs (Michalski & Bleacher 2013).

In the cratered highlands of Mars, further edifices have been identified that potentially resemble ancient Noachian, partially highly degraded or deformed volcanoes (e.g. Dohm & Tanaka 1999; Stewart & Head 2001; Ghatan & Head 2002; Xiao *et al.* 2012). Xiao *et al.* (2012) mapped 75 such edifices; most of them are located at the southern periphery of Daedalia Planum and the Thaumasia highlands, in Terra Sirenum, Sisyphi Planum (including Sisyphi Montes) and Terra Sabea (Fig. 8). However, the best-preserved potential volcanic edifice is Zephyria Tholus (Fig. 9c), located at 19.8°S, 172.9°E (Stewart & Head 2001; Ghatan & Head 2002) (Fig. 8).

Volcanic landforms

Volcano morphometry. Volcano morphology on Mars differs substantially in terms of edifice diameter, height and flank slopes relative to volcanoes on Earth. Martian volcanoes consist primarily of large and low-angle shield volcanoes, domes, pyroclastic cones (scoria/cinder cones and tuff rings) and putative stratovolcanoes. Edifice formation and evolution essentially depends on whether these volcanic constructs were formed during a single eruptive phase (i.e. they are monogenetic) or during repeated eruptions (i.e. polygenetic), and/or on the longevity of volcanism at certain sites, leading to the large shield volcanoes and stratovolcanoes. The large volcanoes known as Montes, Tholi and Paterae in the major volcanic provinces developed over millions to billions of years, punctuated with periods of quiescence of currently unknown durations (e.g. Werner 2009; Williams *et al.* 2010; Platz & Michael 2011).

Owing to the differing geological settings in which major volcanoes formed (e.g. on slopes along the highland–lowland transitional zone (Olympus, Alba, Apollinaris Montes), in tectonically active regions (Alba and Tharsis Montes), or near or on large impact basin rims (Nili and Mereo Paterae or Elysium and Hadriacus Montes, respectively), these edifices are rarely symmetrical in shape and structure. As a result, basal diameters and slopes vary across individual edifices. The giants among the Martian volcanoes, with respect to height and basal diameter, are Olympus Mons (21.2 km in height) and Alba Mons (*c.* 1100 km in diameter) (Plescia 2004), respectively. Generally, the large shield volcanoes in the Tharsis and Elysium volcanic provinces have flank slopes of about $1°–6°$, although a few Tholi exhibit flank slopes of up to $27°$ (Plescia 2004; Platz *et al.* 2011). Syrtis Major and the circum-Hellas shield volcanoes are characterized by flank slopes of less than $1°$ (Plescia 2004).

Small shield volcanoes are observed throughout the Tharsis region and in the Elysium volcanic province, and commonly occur in clusters or chains (Figs 8 & 9b). Their basal diameters range from a few kilometres to a few tens of kilometres, and they frequently have summit elevations of up to several hundred metres and flank slopes of less than $5°$ (Baratoux *et al.* 2009; Bleacher *et al.* 2009; Hauber *et al.* 2009; Richardson *et al.* 2013). These shield volcanoes are typical of plains-style volcanism (Greeley 1982). It is likely that the formation of low-shield volcanoes was common throughout the evolution of the Tharsis volcanic province, although most of the older edifices were probably buried by successive, subsequent volcanic activity (Hauber *et al.* 2011).

Pyroclastic cones (Fig. 9d), defined here as scoria/cinder and tephra cones and tuff rings, have only been observed on Mars since the availability of high-resolution imagery in the following areas: Pavonis Mons (Keszthelyi *et al.* 2008), Ascraeus Mons (Mouginis-Mark & Christensen 2005), Syria Planum (Hauber *et al.* 2009), Ulysses Fossae (Brož

& Hauber 2012), Utopia Planitia (Lanz *et al.* 2010), Nili Patera (Skok *et al.* 2010) and the Nephentes– Amenthes region (Skinner & Tanaka 2007; Brož & Hauber 2013). Nevertheless, their possible existence in some locations had been reported earlier (e.g. Carr *et al.* 1977; Frey & Jarosewich 1982; Edgett 1990; Hodges & Moore 1994; Plescia 1994). Morphometric analyses of Martian pyroclastic cones are sparse. However, Brož & Hauber (2012) studied 29 pyroclastic cones in the Ulysses Fossae area in detail (see the cover picture of this Special Publication); cone basal diameter, height and slope can be up to 3.9 km, 650 m and $27.5°$, respectively.

Lava flows. Lava flows on Mars can be generally classified as channel fed and tube fed (Fig. 9e). Channel-fed lava flows are often narrow at proximal reaches, forming levées on either side of the flow with a confined inner channel. Further away, the levée-channel morphology can transition into a sheet flow, which is characterized by flat-topped, rough or smooth textured surfaces, steep flow margins and lobate escarpments. Sheet-flow width at distal reaches is often several times larger than the channelled lava flow near its source region. Tube-fed lava flows form curvilinear ridges up to several kilometres across. These long ridges exhibit either single, partially recognizable channels at their crests, or aligned isolated or coalesced pit-craters that formed after the encapsulated tubes had been drained (i.e. syn- or post-eruptive collapse). The tube morphology can transition into multiple overlapping, narrow, channelled to sheet-type flows that in turn form a lava field or fan at the base of the tube. This type of tube-fed flow is best observed on the western flanks of Olympus Mons (Fig. 9e). Channel-fed sheet flows are probably best preserved on the relatively young, Middle–Late Amazonian lava plains in the Tharsis province (Fig. 9e). In each of the Martian volcanic provinces, channel-fed and tube-fed lava flows were formed, although their state of preservation differs. Moreover, pre-existing surface textures, channel confinement and slope, lava composition, and effusion rate result in differences in lava-flow morphology (e.g. ridged or platy texture, and channel sinuosity) and morphometry (e.g. width, height, runout distance and volume).

Fallout deposits. Volcanic fallout or tephra deposits generally consist of particles that have been transported ballistically from a source vent or settled from an eruption column. In the latter case, particles may have been transported over large distances before settling down from the atmosphere, forming air-fall deposits. Clast sizes within fallout deposits vary considerably. In proximal reaches, tephra deposits can be poorly sorted, and are comprised of ash (<2 mm in width), lapilli (2–64 mm) and blocks/bombs fragments (i.e. wall rock/juvenile material; >64 mm). Further from the source, particles become increasingly better sorted, with tephra deposits continuously thinning out downrange of the volcanic plume and away from its apex.

On Mars, several fine-grained and friable deposits have been observed that probably have a volcanic origin. The largest deposits, collectively known as the Medusae Fossae Formation (Scott & Tanaka 1982; Mandt *et al.* 2008), are located along the highland–lowland transitional zone between 140°E and 230°E (i.e. the longitudinal range) and to the south of the Elysium rise through to Olympus Mons. The source for these large (potentially tephra) deposits has been attributed to either Apollinaris Mons (Kerber *et al.* 2011*b*) or to the Tharsis volcanic province (Bradley *et al.* 2002; Hynek *et al.* 2003). Interestingly, ground observations by the Mars Exploration Rover 'Spirit' in Gusev crater confirmed the presence of layered coarse and fine-grained tephra at 'Home Plate' (Squyres *et al.* 2007).

Volcaniclastic deposits. The collective term 'volcaniclastic deposit' includes lahar, debris-flow and debri-avalanche (landslide) deposits, dunes, and rootless cones, which together resemble reworked primary volcanic deposits such as lava flows, as well as pyroclastic flow and tephra deposits. On Mars, debris avalanche deposits caused by flank edifice failure are observed in association with Olympus Mons and Tharsis Tholus. Olympus Mons is surrounded by aureole deposits known as Lycus, Cyane, and Gigas Sulci and Sulci Gordii, which represent large-scale landslide deposits that were probably sourced from the circumferential scarp (Lopes *et al.* 1982; McGovern *et al.* 2004; Byrne *et al.* 2013*a*). Tharsis Tholus also experienced several flank failure events, of which only remnants of the western flank collapse are preserved as mounds and rotated blocks (Platz *et al.* 2011). The best examples of lahar (and debris-flow) deposits are those associated with the sudden release of groundwater at the western flank of the Elysium rise (Christiansen 1989). During the outburst(s) of water, a substantial portion of the lower western Elysium flank was eroded, transported and deposited onto the plains of Utopia Planitia. Since the observed channels and channel networks of Granicus Valles and Hrad Vallis are confined by debris levées at proximal to medial reaches, a 'pure' fluvial origin with subsequent valley formation can be excluded.

Dark sediments and dunes are frequently observed in the highlands of Mars. They occur mostly within craters, calderas and intercrater plains. Their aeolian origin was suggested in earlier studies (e.g. Thomas 1984; Edgett & Blumberg 1994) and they feature particle sizes ranging from medium to coarse sand, which appear coarser than their analogues on Earth (Edgett & Christensen 1991, 1994). Remotely sensed spectroscopic studies revealed that most of these dark dunes and sediments are composed of olivine and pyroxene, which suggests a volcanic origin (e.g. Poulet *et al.* 2007; Tirsch *et al.* 2011). Dark basaltic material is either deposited within craters by airfall or tephra layers are exposed by impact craters from which intracrater dunes formed (Tirsch *et al.* 2011). Probably the best examples of active, dark basaltic dunes are those exposed in the caldera Nili Patera (Silvestro *et al.* 2010), which are composed of abraded lava material and reworked tephra.

Rootless cones (or pseudocraters) form by explosive interactions between lava and external water (Thorarinsson 1953), either while lava is flowing over water-saturated strata or when it physically mingles with external water/ice. On Mars, there are abundant locations of so-called rootless cones – conical edifices with a summit depression – that have also been attributed to periglacial processes (i.e. pingos: e.g. Burr *et al.* 2005; Page 2007) and mud volcanism (Farrand *et al.* 2005). Recent detailed surveys have shown that small cone groups in Tartarus Colles (Hamilton *et al.* 2011), and in Athabasca Valles and Cerberus Palus (Keszthelyi *et al.* 2010), probably formed by rootless eruptions.

Composition of volcanic rocks

The morphology and morphometry of lava flows (e.g. Greeley 1974; Greeley & Spudis 1981; Keszthelyi *et al.* 2004), and their derived rheological parameters (e.g. Zimbelman 1985; Baloga *et al.* 2003; Garry *et al.* 2007; Baratoux *et al.* 2009; Hauber *et al.* 2011; Pasckert *et al.* 2012), together with the similarities between terrestrial (e.g. islands of Hawaii and Galapagos) and Martian shield volcanoes (e.g. Greeley & Spudis 1981; Hauber *et al.* 2009), suggest that volcanism on Mars is predominantly basaltic in nature. Geochemical analyses of Mars-sourced SNC (Shergottite, Nakhlite Chassignite) meteorites (e.g. McSween 1994, 2002) and *in situ* rover and lander investigations (e.g. Larsen *et al.* 2000; McSween *et al.* 2004) have also confirmed the dominant presence of basaltic rocks on Mars. Although andesitic rock compositions were also suggested to be present at rover/lander sites and across the northern plains (Bandfield *et al.* 2000; Larsen *et al.* 2000), Wyatt & McSween (2002) reanalysed mineral abundances from published work and attributed the 'andesite'-like signature to low-temperature, aqueous alteration of basalts. Igneous minerals such as olivine, pyroxene and feldspar, as well as volcanic glass present in the Martian regolith, have been detected by spectral analyses from orbital spacecraft (e.g. Bandfield *et al.* 2000; Christensen *et al.* 2003; Bibring *et al.* 2005). Rock compositional analyses at the Mars Exploration Rover 'Spirit' and Mars Science Laboratory 'Curiosity' landing sites provided evidence for a bulk-chemical, mineralogical and textural diversity of igneous samples (e.g. Squyres *et al.* 2006; Sautter *et al.* 2014).

Volcano-tectonics

Calderas. On all large shield and highland volcanoes across the main volcanic provinces, summit caldera(s) are observed, which attest to single or multiple cycles of large-scale magma storage, growth and replenishment within the edifices and subsequent summit collapse(s). Crumpler *et al.* (1996) studied Martian calderas and defined two distinct caldera types: (1) Olympus type and (2) Arsia type (Fig. 9a, f), which are characterized by complex/nested or by single summit calderas, respectively. Caldera dimensions vary substantially in diameter and depth. The largest caldera of the biggest shield volcanoes is hosted by Arsia Mons and measures 115 km in diameter, whereas, at highland volcanoes in the CHVP, calderas have formed with diameters up to 145 km (Crumpler *et al.* 1996). Caldera depths range from a few hundred metres up to 5 km (for Pavonis Mons: Fig. 9f). It has been noted that most of the 'smaller' volcanoes (Uraunius Mons and Tharsis, Ceraunius, Jovis, Biblis, and Ulysses Tholi) at the periphery of the Tharsis Montes exhibit greater caldera depths (i.e. between 1 and 3 km) than the calderas of most of the large shield volcanoes in Tharsis (Crumpler *et al.* 1996). Martian calderas show similar features to terrestrial calderas, including one or more of the following characteristics: steep, circular to elliptical caldera walls, terraced caldera margins, circumferential scarps, faults and graben, radially orientated ridges and faults, pit crater chains, linear arrangement of small vents, and volcanic flooding that has levelled caldera floors (e.g. Crumpler *et al.* 1996; Mouginis-Mark & Rowland 2001; Platz *et al.* 2011; Byrne *et al.* 2012).

Flank deformation. Volcanoes can experience tectonic deformation due to a number of exogenic processes (e.g. rifting) but gravitationally driven tectonism is one of the primary endogenic processes responsible for volcano flank deformation (Fig. 9f). Edifice spreading and sagging represent end members of a structural continuum along which a given volcano, subject to gravitational deformation, will lie; this continuum probably applies to Mars as equally as it does to Earth (Byrne *et al.* 2013a). Volcano spreading is characterized by the formation of a system of radial normal faults,

often forming 'leaf graben', on the flanks of the edifice and a concentric thrust belt at its base (Borgia *et al.* 2000). In contrast, volcano sagging will result in the development of concentric flank thrusts or 'terraces' (Byrne *et al.* 2009, 2013a), accompanied by the formation of a flexural moat and bulge (e.g. Comer *et al.* 1985).

There is little direct evidence for volcano spreading on Mars. This is because spreading requires that an edifice be mechanically detached from its underlying basement, such that the response to loading is accommodated in the main by the volcano itself (e.g. McGovern & Solomon 1993; Borgia 1994), and the conditions necessary for such decoupling (e.g. low-competency strata such as clays) are not widely observed on Mars. Nevertheless, the Tharsis Tholus edifice, in eastern Tharsis, does appear to have experienced sector collapse in a manner similar to volcanoes known to have spread on Earth (Platz *et al.* 2011). Spreading along phyllosilicates proximal to and beneath Olympus Mons has also probably played a role in shaping that volcano (Morgan & McGovern 2005; McGovern & Morgan 2009; Byrne *et al.* 2013a).

The effects of volcano sagging are seen much more widely across Mars. At least nine volcanoes (including Olympus Mons) show evidence of flank terraces, topographically subtle landforms that are difficult to see without the aid of topographical data (Byrne *et al.* 2009). These structures were observed on Olympus and on the Tharsis Montes initially (e.g. Thomas *et al.* 1990) but their prevalence on shields of a range of shapes and sizes indicates that their formation is likely to be tied to a process commonly experienced by volcanoes. Flank terraces were tied to lithospheric flexure by McGovern & Solomon (1993), an interpretation reinforced by more recent analogue modelling studies (e.g. Byrne *et al.* 2013a). Importantly, volcano sagging will serve to place an edifice into a state of net compression, which will impede or even inhibit magma ascent to its summit, and, in turn, will alter its eruptive behaviour and development (e.g. Byrne *et al.* 2012; McGovern *et al.* 2014).

Tectonic structures

It is widely accepted that Mars is a one-plate planet, although the prospect of plate tectonics having at some point operated on that planet has yet to be fully resolved (Sleep 1994; Yin 2012). This possibility was proposed because of the apparent hemispheric crustal dichotomy dividing Mars into the southern high-standing, cratered highlands and the northern lowlands, with the latter appearing, with Viking-based imagery, to be less cratered than the southern highlands. With new, high-resolution topographical data, however, large subdued basin structures were discovered, which most probably makes the lowlands as old as the highlands (Frey 2008). There are currently two theories for how the dichotomy on Mars developed: (1) by a giant impact (e.g. Wilhelms & Squyres 1984; Andrews-Hanna *et al.* 2008; Marinova *et al.* 2008); or (2) by endogenic processes (convective overturn of the interior (Wise *et al.* 1979) or by degree-1 convection with north–south asymmetry (Zhong & Zuber 2001; Roberts & Zhong 2006)).

In the Martian lithosphere, a suite of faults has been identified with normal (e.g. Plescia & Saunders 1982; Schultz *et al.* 2007), reverse (e.g. Schultz & Tanaka 1994) and strike-slip senses of movement (Andrews-Hanna *et al.* 2008; Yin 2012) (Fig. 8). Extensional features include normal faults, half-graben, graben and rift-like structures such as Acheron Fossae (Kronberg *et al.* 2007) (Figs 8 & 9g). As for other terrestrial worlds, most graben likely represent an hourglass-shaped subsurface pattern (Schultz *et al.* 2007). And as for Mercury and the Moon, wrinkle ridges (Figs 7f & 9h) are very common landforms on the Martian surface once more interpreted as arcuate, asymmetric ridges that have formed above an underlying, low-angle thrust fault. This type of faulting is thought to occur in layered rocks such as sedimentary sequences or successions of lava flows. Well-developed wrinkle-ridge systems are probably best preserved in Lunae, Solis and Syrtis Major Plana (Fig. 9h).

The global tectonic map of Mars (Knapmeyer *et al.* 2006) shows that the focus of activity is associated with the large Tharsis volcanic province. Here, large sets of normal faults, half-graben and graben radiate outwards (with minor occurrences of concentric faults) from the central Tharsis rise (Fig. 8). Plescia & Saunders (1982) studied in detail the tectonic evolution of Tharsis, and proposed four discrete centres of faulting that, from oldest to youngest, include the Thaumasia highlands, northern Syria Planum and two centres near Pavonis Mons.

The timing of the main tectonic activity on Mars was later determined by Anderson *et al.* (2001) to have peaked in five main phases. The oldest identified stage of activity occurred during the Noachian, when most of the graben in Syria Planum, Tempe Terra and Thaumasia formed. The Late Noachian–Early Hesperian (stage 2) and Early Hesperian (stage 3) tectonic phases formed extensional structures along the central Valles Marineris, and in Pavonis, Syria, Ulysses and Tempe Terra, respectively. Stage 3 tectonic activity is also associated with wrinkle-ridge formation in Lunae and Solis Plana, as well as in Thaumasia, Sirenum, Memnonia and Amazonis Planitia (Anderson *et al.* 2001). The tectonic structures formed in stage 4 (Late Hesperian–Early Amazonian)

developed around Alba Mons and the Tharsis Montes, whereas the latest activity (stage 5) occurred during the Middle–Late Amazonian with associated faults located around the large shield volcanoes (Anderson *et al.* 2001).

The main cause of extensional deformation within Tharsis and its periphery is its loading-induced stress on the lithosphere. It is thought that most graben are the surface expressions of giant dyke swarm intrusions (Ernst *et al.* 2001; Wilson & Head 2002; Schultz *et al.* 2004). Similar concentric and radial fault patterns (e.g. Cerberus Fossae), although far less numerous, are also observed in the Elysium volcanic province, where the mass and volume of the Elysium rise has also induced faulting.

Valles Marineris constitutes the largest, most spectacular and, perhaps, the most puzzling set of canyons in the solar system (Lucchitta *et al.* 1992). Although the linearity of canyon walls suggests a tectonic origin, differing driving mechanisms have been proposed for the canyons, including tectonic rifting associated with large-scale magmatism and/or extensive dyke emplacement (e.g. Blasius *et al.* 1977; Mège & Masson 1996; McKenzie & Nimmo 1999; Schultz & Lin 2001; Dohm *et al.* 2009), collapse along tectonic zones and subsequent catastrophic discharges (Sharp 1973; Tanaka & Golombek 1989; Spencer & Fanale 1990; Rodriguez *et al.* 2006), salt tectonics (Montgomery & Gillespie 2005; Adams *et al.* 2009), composite origins involving erosion (Lucchitta *et al.* 1994) and/or major distinct stages of collapse and normal faulting (Schultz 1998), and volcano-erosion where Tharsis-sourced lava tubes form pit chains due to roof collapse, which later evolve into fossae and chasmata (Leone 2014). A recent study by Andrews-Hanna (2012) showed that canyon formation occurred through displacement along steeply dipping faults, coupled with vertical subsidence.

Article summaries

P. Mancinelli, F. Minelli, A. Mondini, C. Pauselli & C. Federico

A downscaling approach for geological characterization of the Raditladi basin of Mercury

Through combining newly available photogeological, compositional and topographical data for Mercury, **Mancinelli *et al.* (2014)** first present a new synthesis of the surface units on the innermost planet. These authors then investigate an area of particularly diverse units in greater detail, an area that includes the 260 km-diameter Raditladi impact basin. In investigating the geological history of the basin, they construct a geological cross-section that shows how the volcanic units inside the interior of Raditladi were emplaced upon impact-related units. The chapter ends with a call for further regional- and local-scale mapping of Mercury, to elucidate the origin of units observed globally but whose nature is currently unclear.

N. P. Lang & I. López

The magmatic evolution of three Venusian coronae

Lang & López (2013) argue that the volcanic products and forms associated with three case-study coronae, Zemire, Bhumidevi and Aramiti, may not be consistent with widely accepted models of corona formation. Instead, their evolution can be explained by the mass evacuation of a stratified, shallow magma chamber. This evacuation and collapse would account for the observed extensive lava flows emanating from the annular fractures surrounding these coronae, as well as the steep-sided domes and tholi that formed along these fractures at the latest stage of the corona evolution, when crystal-rich magmas (or basaltic foams) were squeezed up and forced to the surface.

R. C. Ghail & L. Wilson

A pyroclastic flow deposit on Venus

Ghail & Wilson (2013) describe the morphological characteristics of a semi-circular, doughnut-shaped deposit on Venus that is morphologically consistent with pyroclastic flow deposits on Earth. The hydrodynamic interaction of this deposit, named Scathach Fluctus, with a volcanic cone indicate flow velocities of up to 48 m s^{-1}. Estimated volatile abundances associated with the explosive eruption imply high CO_2 and SO_2 concentrations in the mantle. Because the radar characteristics of Scathach Fluctus are similar to many parts of the Venusian surface, these authors suggest that pyroclastic flow deposits are more widespread on the second planet than previously thought.

C. M. Meyzen, M. Massironi, R. Pozzobon & L. Dal Zilio

Are terrestrial plumes from motionless plates analogues to Martian plumes feeding the giant shield volcanoes?

Hawaiian intraplate volcanism has long been thought an apt analogue to the giant, long-lived

volcanoes on Mars. However, **Meyzen et al. (2014)** argue for a revision of that view: that, instead, volcanoes on the slow-moving Nubian and Antarctic plates provide a better comparison. By comparing and contrasting the properties of volcanoes located on slow-moving plates on Earth with the large volcanoes in Mars' Tharsis region, these authors seek to understand more fully the nature and significance of the large-scale melting and differentiation processes of volcanoes on Mars.

T. Morota, Y. Ishihara, S Sasaki, S. Goossens, K. Matsumoto, H. Noda, H. Araki, H. Hanada, S. Tazawa, F. Kikuchi, T. Ishikawa, S. Tsuruta, S. Kamata, H. Otake, J. Haruyama & M. Ohtake

Lunar mare volcanism: lateral heterogeneities in volcanic activity and relationship with crustal structure

The asymmetry of lunar near- and farside maria is still under investigation. Here, **Morota et al. (2014)** study the relationship between mare distribution and crustal thickness on the Moon using remotely sensed geological and geophysical data. Their results show that magma extrusion is dominant in regions of relatively thin crust, which is consistent with previous studies. However, these authors also find lateral heterogeneities in the upper limits of crustal thickness, which would allow magma ascent to, and lava extrusion onto, the lunar surface. These heterogeneities may be due to lateral variations in melt/magma generation within the mantle and/or changes in crustal density.

C. Carli, G. Serventi & M. Sgavetti

VNIR spectral characteristics of terrestrial igneous effusive rocks: mineralogical composition and the influence of texture

Carli et al. (2014) discuss the utility of visible and near-infrared (VNIR) spectroscopy to map mineralogical variations across planetary surfaces. In particular, igneous rocks emplaced effusively have distinct crystal field absorption bands in the VNIR spectral range, bands that correspond to the rocks' constituent mineralogy. These authors review how petrological properties influence the interpretation of rock mineralogy using spectroscopy. Among other results, they show how grain and crystal size can influence the spectra of effusive rocks, and how glassy components in rock groundmass reduce or hide absorption bands of mafic minerals or feldspars. They also suggest that combining geomorphic and spectral data is the most reliable method of mapping of volcanic material on planetary surfaces.

S. Ferrari, M. Massironi, S. Marchi, P. K. Byrne, C. Klimczak, E. Martellato & G. Cremonese

Age relationships of the Rembrandt basin and Enterprise Rupes, Mercury

The time–stratigraphic relationship between the 715 km-diameter Rembrandt impact basin and the Enterprise Rupes scarp system, which extends for over 800 km across the surface of Mercury, is the focus of the work by **Ferrari *et al.* (2014)**. These authors find that the Rembrandt basin formed at about 3.8 Ga, with resurfacing of its interior by volcanic smooth plains occurring within 100–300 myr after basin formation. The most recent activity along Enterprise Rupes took place at about 3.6 Ga, cross-cutting (and therefore post-dating) the basin's volcanic infilling event(s). It is currently unclear whether the initiation of the Enterprise Rupes fault system pre- or post-dates the Rembrandt basin-forming event.

F. C. Lopes, A. T. Caselli, A. Machado & M. T. Barata

The development of the Deception Island volcano caldera under control of the Bransfield Basin sinistral strike-slip tectonic regime (NW Antarctica)

Deception Island is a small, volcanically active caldera volcano located in the Bransfield Strait, off the Antarctic Peninsula. **Lopes *et al.* (2014)** present evidence that the fractures that have shaped the edifice, and its elongate caldera, are the result of pervasive left-lateral simple shearing within the Bransfield Basin. They also review the formational history of the caldera, proposing that at least two phases of collapse have occurred: first in a small-volume event and, later, in a larger event that affected the flanks of the volcano itself.

P. K. Byrne, E. P. Holohan, M. Kervyn, B. van Wyk de Vries & V. R. Troll

Analogue modelling of volcano flank terrace formation on Mars

Flank terraces are laterally extensive, topographically subtle landforms on the slopes of large Martian shield volcanoes. In this chapter, **Byrne *et al.* (2014*a*)** use a series of scaled analogue models to test the hypothesis that flank terraces result from constriction of a volcano as it down-flexes its underlying lithospheric basement. They show that terrace formation on sagging edifices is largely independent of volcano slope, size or aspect ratio, but increasing lithospheric thickness will ultimately inhibit terrace development entirely. These authors conclude that understanding the structural evolution of large shields on Mars requires that these volcanoes be appraised within the context of lithospheric flexure.

R. Pozzobon, F. Mazzarini, M. Massironi & L. Marinangeli

Self-similar clustering distribution of structural features on Ascraeus Mons (Mars): implications for magma chamber depth

Pozzobon *et al.* (2014) use self-similar fractal clustering techniques to examine the distribution of pit craters on the Ascraeus Mons volcano on Mars. These pits are probably related to feeder dykes and, by understanding how the pits are distributed, the subsurface architecture of the magma system below Ascraeus can be understood. The authors find evidence for two discrete pit populations, indicative of two magma sources – one at shallow depths and the other deep below the volcano – and appraise this finding within the context of earlier studies of the volcano, suggesting that this analysis may provide insight into the deep structure of other large volcanoes on Mars.

P. J. McGovern, E. B. Grosfils, G. A. Galgana, J. K. Morgan, M. E. Rumpf, J. R. Smith & J. R. Zimbelman

Lithospheric flexure and volcano basal boundary conditions: keys to the structural evolution of large volcanic edifices on the terrestrial planets

McGovern *et al.* (2014) study the interplay between large volcanic edifices and the underlying lithosphere, which flexes in response to the exerted volcanic load. Lithospheric thickness influences the shape of the flexural response, and the associated stress states in turn can influence the structure and evolution of the overlying edifice – which in turn affects lithosphere response. The edifice–basement basal boundary condition (i.e. welded base or gliding basal plane) determines whether compression is transferred into the edifice, which can potentially inhibit magma ascent into the edifice. Volcanoes situated on a thick lithosphere and a clay-based décollement can grow to enormous sizes, whereas the growth of an edifice welded to a thin lithosphere is likely to be limited.

E. B. Grosfils, P. J. McGovern, P. M. Gregg, G. A. Galgana, D. M. Hurwitz, S. M. Long & S. R. Chestler

Elastic models of magma reservoir mechanics: a key tool for investigating planetary volcanism

Exploring the mechanics of magma storage, its ascent to the surface, and the interplay of subsurface and surface volcano-tectonic processes is the main objective of this contribution by **Grosfils *et al.* (2013)**. These authors use bespoke elastic numerical models that leverage field, laboratory and remote-sensing observations to study volcanic processes on terrestrial worlds. Their models provide renewed insights into how subsurface magma reservoirs inflate and rupture, and how these processes relate to volcano growth, caldera formation, and the associated emplacement of circumferential and radial dykes.

M. Massironi, G. Di Achille, D. A. Rothery, V. Galluzzi, L. Giacomini, S. Ferrari, M. Zusi, G. Cremonese & P. Palumbo

Lateral ramps and strike-slip kinematics on Mercury

Massironi *et al.* (2014) investigate contractional features on Mercury for evidence of strike-slip deformation. Such evidence includes en echelon fold arrays, restraining bends, positive flower structures, stike-slip duplexes and crater rims that have been displaced by lobate scarps and high-relief ridges. These authors find that the strike-slip to transpressional motion along faults they observe is inconsistent with a globally homogenous stress field predicted to result from secular cooling-induced global contraction alone. They conclude that other processes, such as mantle convection, may have played a contributory role during the tectonic evolution of Mercury.

L. Giacomini, M. Massironi, S. Marchi, C. I. Fassett, G. Di Achille & G. Cremonese

Age dating of an extensive thrust system on Mercury: implications for the planet's thermal evolution

Mercury's surface is characterized by abundant contractional features such as lobate scarps and

wrinkle ridges, which are principally attributed to the planet's secular cooling and resultant global contraction. **Giacomini *et al.* (2014)** study the formation age of an extensive fold and thrust belt of which Blossom Rupes is part, using different age determination techniques, including buffered crater counting. They find that thrust activity along this system terminated between 3.7 and 3.5 Ga. Should these techniques indicate that other large-scale contractional features on Mercury have similar ages, a revision of current thermal evolution models for Mercury, including an earlier onset of planetary contraction, is required.

V. Galluzzi, G. Di Achille, L. Ferranti, C. Popa & P. Palumbo

Faulted craters as indicators for thrust motions on Mercury

Is it possible to directly determine true dip angles and slip vectors for faults on other planets? **Galluzzi *et al.* (2014)** show that this can be accomplished by using digital terrain models of deformed craters on Mercury. In so doing, these authors demonstrate the broad range of dip angles and kinematics of Mercurian faults. This methodology, which allows for the quantitative structural characterization of remotely sensed faults, can be used to enhance our understanding of planetary geodynamics.

L. B. Harris & J. H. Bédard

Interactions between continent-like 'drift', rifting and mantle flow on Venus: gravity interpretations and Earth analogues

Harris & Bédard (2014*b*) identify major strike-slip shear zones at Ishtar and Afrodite Terrae and at Sedna Planitia, using offsets of Bouger gravity anomalies and gravity gradient edges. Their observations call for a new conceptual model capable of satisfying Venusian subduction-free geodynamics, dominant convective upwellings and the substantial horizontal tectonism required by the observed strike-slip belts. These authors suggest that mantle traction, generated and controlled by linear upwellings along rifts, has resulted in the substantial lateral motion of areas of continent-like crust on Venus, such as Lakshumi planum (in western Ishtar Terra). This process accounts for the fold-and-thrust belt that bounds Lakshumi planum to the north, as well as the transpressive regimes recognized at its eastern and western margins. Harris & Bédard (2014*b*) conclude by proposing that this new perspective of Venus may provide insight into the tectonics of the Archaean Earth.

M. T. Barata, F. C. Lopes, P. Pina, E. I. Alves &
J. Saraiva

Automatic detection of wrinkle ridges in Venusian Magellan imagery

Wrinkle ridges are common and widespread tectonic features on Venus. **Barata *et al.* (2014)** present an automated algorithm to detect wrinkle ridges using Magellan Synthetic Aperture Radar imagery, with which they characterize ridge morphology, including orientation, length and spacing. This procedure greatly enhances wrinkle ridge mapping and analysis. In addition, these authors also test an automated procedure to identify and characterize impact craters and their ejecta.

A. L. Nahm & R. A. Schultz

Rupes Recta and the geological history of the Mare Nubium region of the Moon: insights from forward mechanical modelling of the 'Straight Wall'

The Moon's famous Rupes Recta, or 'Straight Wall', situated in Mare Nubium on the lunar nearside has been known for more than three centuries. **Nahm & Schultz (2013)** investigate its fault characteristics. Detailed structural mapping and throw distribution measurements show that this structure has experienced bi-directional growth. Forward mechanical modelling of its topography indicates that the fault has a dip angle of 85°, almost 0.5 km of maximum displacement and penetrates over 40 km into the lunar lithosphere. These authors show that the development of Rupes Recta could have been strongly influenced by columnar cooling joints activated as shear planes during subsidence of Mare Nubium.

D. Y. Wyrick, A. P. Morris, M. K. Todt &
M. J. Watson-Morris

Physical analogue modelling of Martian dyke-induced deformation

Dykes are commonly thought to form, and thus underlie, laterally extensive graben on planetary surfaces. Using analogue modelling techniques, **Wyrick *et al.* (2014)** demonstrate that dykes injected into an undisturbed crust cause ridges and related contractional features to develop at the surface, instead of extensional structures. This finding has important implications for graben sets on numerous worlds, including our understanding of Tharsis-radial graben, which should predate dyke emplacement, the evolution of Venusian radiating fissure systems and, supposedly, dyke-induced graben on the Moon.

L. Guallini, C. Pauselli, F. Brozzetti & L.
Marinangeli

Physical modelling of large-scale deformational systems in the South Polar Layered Deposits (Promethei Lingula, Mars): new geological constraints and climatic implications

In a follow-on study of the Promethei Lingula ice sheet on Mars, **Guallini *et al.* (2014)** integrate structural analysis with thermal and mechanical models to quantify the deformation of part of the ice sheet's South Polar Layered Deposits. They show that parts of these deposits feature soft-sediment deformation and that internal compositions are dominated by CO_2 ice. Moreover, these authors determine that deformation of the layered deposits is unlikely to have occurred under present-day climatic conditions. Instead, warmer temperatures in the past were likely to have been responsible for soft-sediment deformation, and may even have triggered gravitational sliding of the entire ice sheet.

D. L. Buczkowski & D. Y. Wyrick

Tectonism and magmatism identified on asteroids

This contribution provides a review of linear features observed on a range of asteroids, including Gaspra, Eros and Itokawa. **Buczkowski & Wyrick (2014)** primarily focus on previous observations of tectonic structures, current models to explain linear feature formation and the implications for the internal structure of these small bodies. Even though Vesta is a unique and differentiated proto-planetary body, it hosts fractures and grooves that are morphologically similar to those observed on smaller asteroids, and is therefore also included in this review chapter. To date, no volcanic features have been identified on Vesta's surface, but these authors discuss the prospect that the geological history of Vesta may have included endogenic magmatism.

This volume would not have been possible without the assistance of all who willingly agreed to review these chapters. Thanks also go to the reviewer, D. A. Williams, of this chapter. The authors also wish to thank T. M. Hare (USGS Astrogeology Science Center, Flagstaff, AZ) for compiling global datasets in GIS-ready formats for Venus, the Moon and Mars. M. A. Ivanov (Russian Academy of Science, Moscow) kindly provided his geological map of Venus and assisted in simplifying the map units. P. K.

Byrne acknowledges support from the MESSENGER project, under contracts NASW-00002 to the Carnegie Institution of Washington and NAS5-97271 to The Johns Hopkins University Applied Physics Laboratory, and from the Department of Terrestrial Magnetism, Carnegie Institution of Washington. M. Massironi acknowledges the support from the Italian Space Agency (ASI) within the SIMBIOSYS Project (ASI-INAF agreement no. I/022/10/0) and from the University of Padua within the Ateneo Project CPDA112213/11. This research made use of NASA's Planetary Data System and Astrophysics Data System.

References

ADAMS, J. B., GILLESPIE, A. R. *ET AL.* 2009. Salt tectonics and collapse of Hebes Chasma, Valles Marineris, Mars. *Geology*, **37**, 691–694.

AHRENS, T. J. & RUBIN, A. M. 1993. Impact-induced tensional failure in rock. *Journal of Geophysical Research*, **98**, 1185–1203.

AITTOLA, M. & KOSTAMA, V. 2000. Venusian novae and arachnoids: characteristics, differences and the effect of the geological environment. *Planetary Space Science*, **48**, 1479–1489.

AITTOLA, M. & KOSTAMA, V. P. 2002. Chronology of the formation process of Venusian novae and the associated coronae. *Journal of Geophysical Research*, **107**, 5112.

ANDERSON, D. L. 2007. *New Theory of the Earth*. Cambridge University Press, Cambridge.

ANDERSON, F. S. & SMREKAR, S. E. 1999. Tectonic effects of climate change on Venus. *Journal of Geophysical Research*, **104**, 30 743–30 756.

ANDERSON, R. C., DOHM, J. M. *ET AL.* 2001. Primary centers and secondary concentrations of tectonic activity through time in the western hemisphere of Mars. *Journal of Geophysical Research*, **106**, 20 563–20 585.

ANDREWS-HANNA, J. C. 2012. The formation of Valles Marineris: 1. Tectonic architecture and the relative roles of extension and subsidence. *Journal of Geophysical Research*, **117**, E03006.

ANDREWS-HANNA, J. C., ZUBER, M. T. & BANERDT, W. B. 2008. The Borealis basin and the origin of the Martian crustal dichotomy. *Nature*, **453**, 1212–1215.

ANSAN, V., VERGELY, P. & MASSON, P. 1996. Model of formation of Ishtar Terra, Venus. *Planetary and Space Science*, **44**, 817–831.

ANTONENKO, I., HEAD, J. W., MUSTARD, J. F. & HAWKE, B. R. 1995. Criteria for the detection of lunar cryptomaria. *Earth, Moon, and Planets*, **69**, 141–172.

ARKANI-HAMED, J. 1996. Analysis and interpretation of high-resolution topography and gravity of Ishtar Terra, Venus. *Journal of Geophysical Research*, **101**, 4691–4710.

ARNDT, J. & VON ENGELHARDT, W. 1987. Formation of Apollo 17 orange and black glass beads. *Journal of Geophysical Research*, **92**, 372–376.

ARVIDSON, R. E., BOYCE, J. *ET AL.* 1979. Standard techniques for the presentation and analysis of crater size-frequency data. *Icarus*, **37**, 467–474.

ARYA, A. S., THANGJAM, G. & RAJASEKHAR, R. P. 2011. Analysis of mineralogy of an effusive volcanic lunar dome in Marius Hills, Oceanus Procellarum. Abstract 1845 presented at the EPSC-DPS Joint Meeting 2011, 2–7 October 2011, Nantes, France.

BAER, G., TERRA, L., BINDSCHADLER, D. L. & STOFAN, E. R. 1994. Spatial and temporal relations between coronae and extensional belts, northern Lada Terra, Venus. *Journal of Geophysical Research*, **99**, 8355–8369.

BAKER, V. R., KOMATSU, G., GULICK, V. C. & PARKER, T. J. 1997. Channels and valleys. *In*: BOUGHER, S. W., HUNTEN, D. M. & PHILLIPS, R. J. (eds) *Venus II: Geology, Geophysics, Atmosphere, and Solar Wind Environment*. University of Arizona Press, Tucson, AZ, 757–793.

BALDWIN, R. B. 1964. Lunar crater counts. *The Astronomical Journal*, **69**, 377–392.

BALOGA, S. M., MOUGINIS-MARK, P. J. & GLAZE, L. S. 2003. Rheology of a long lava flow at Pavonis Mons, Mars. *Journal of Geophysical Research*, **108**, E75066.

BANDFIELD, J. L., HAMILTON, V. E. & CHRISTENSEN, P. R. 2000. A global view of Martian surface compositions from MGS-TES. *Science*, **287**, 1626–1630.

BANERDT, W. B., McGILL, G. E. & ZUBER, M. T. 1997. Plains tectonics on Venus. *In*: GOUGHER, S. W., HUNTEN, D. M. & PHILLIPS, R. J. (eds) *Venus II: Geology, Geophysics, Atmosphere, and Solar Wind Environment*. University of Arizona Press, Tucson, AZ, 901–930.

BANKS, M. E., WATTERS, T. R., ROBINSON, M. S., TORNABENE, L. L., TRAN, T., OJHA, L. & WILLIAMS, N. R. 2012. Morphometric analysis of small-scale lobate scarps on the Moon using data from the Lunar Reconnaissance Orbiter. *Journal of Geophysical Research*, **117**, E00H11, http://dx.doi.org/10.1029/2011JE003907

BAPTISTA, A. R., MANGOLD, N. *ET AL.* 2008. A swarm of small shield volcanoes on Syria Planum, Mars. *Journal of Geophysical Research*, **113**, E9010.

BARATA, M. T., LOPES, F. C., PINA, P., ALVES, E. I. & SARAIVA, J. 2014. Automatic detection of wrinkle ridges in Venusian Magellan imagery. *In*: PLATZ, T., MASSIRONI, M., BYRNE, P. K. & HIESINGER, H. (eds) *Volcanism and Tectonism Across the Inner Solar System*. Geological Society, London, Special Publications, **401**. First published online January 9, 2014, http://dx.doi.org/10.1144/SP401.5

BARATOUX, D., PINET, P., TOPLIS, M. J., MANGOLD, N., GREELEY, R. & BAPTISTA, A. R. 2009. Shape, rheology and emplacement times of small Martian shield volcanoes. *Journal of Volcanology and Geothermal Research*, **185**, 47–68.

BARSUKOV, V. L., BASILEVSKY, A. T. *ET AL.* 1986. The geology and geomorphology of Venus surface as revealed by the radar images obtained by Venera 15 and 16. *Journal of Geophysical Research*, **91**, 378–398.

BASALTIC VOLCANISM STUDY PROJECT 1981. *Basaltic Volcanism on the Terrestrial Planets*. Pergamon, New York.

BASILEVSKY, A. T. & HEAD, J. W. 1995. Regional and global stratigraphy of Venus: a preliminary assessment and implications for the geological history of Venus. *Planetary Space Science*, **43**, 1523–1553.

BASILEVSKY, A. T. & HEAD, J. W. 1998. The geologic history of Venus: a stratigraphic view. *Journal of Geophysical Research*, **103**, 8531–8544.

BASILEVSKY, A. T. & HEAD, J. W. 2000. Geologic units on Venus: evidence for their global correlation. *Planetary and Space Science*, **48**, 75–111.

BASILEVSKY, A. T., PRONIN, A. A., RONCA, L. B., KRYUCHKOV, V. P., SUKHANOV, A. L. & MARKOV, M. S. 1986. Styles of tectonic deformation on Venus – Analysis of Venera-15 and Venera-16 data. *Journal of Geophysical Research*, **91**, D399–D411.

BASILEVSKY, A. T., HEAD, J. W., SCHABER, G. G. & STROM, R. G. 1997. The resurfacing history of Venus. *In*: BOUGHER, S. W., HUNTEN, D. M. & PHILLIPS, R. J. (eds) *Venus II: Geology, Geophysics, Atmosphere, and Solar Wind Environment*. University of Arizona Press, Tucson, AZ, 1047–1086.

BASILEVSKY, A. T., AITTOLA, M., RAITALA, J. & HEAD, J. W. 2009. Venus astra/novae: estimates of the absolute time duration of their activity. *Icarus*, **203**, 337–351.

BASILEVSKY, A. T., SHALYGIN, E. V. *ET AL.* 2012. Geologic interpretation of the near-infrared images of the surface taken by the Venus Monitoring Camera, Venus Express. *Icarus*, **217**, 434–450.

BECKER, K. J., WELLER, L. A., EDMUNDSON, K. L., BECKER, T. L., ROBINSON, M. S., ENNS, A. C. & SOLOMON, S. C. 2012. Global controlled mosaic of Mercury from MESSENGER orbital images. Abstract 2654 presented at the 43rd Lunar & Planetary Science Conference, March 19–23, 2012, The Woodlands, Texas.

BIBRING, J. P., LANGEVIN, Y. *ET AL.* 2005. Mars surface diversity as revealed by the OMEGA/Mars Express observations. *Science*, **307**, 1576–1581.

BILOTTI, F. & SUPPE, J. 1999. The global distribution of wrinkle ridges on Venus. *Icarus*, **139**, 137–157.

BINDER, A. B. 1982. Post-Imbrium global tectonism: evidence for an initially totally molten Moon. *Moon and Planets*, **26**, 117–133.

BINDER, A. B. 1986. The initial thermal state of the Moon. *In*: HARTMANN, W. K., PHILLIPS, R. J. & TAYLOR, G. J. (eds) *Origin of the Moon*. Lunar and Planetary Institute, Houston, TX, 425–433.

BINDER, A. B. & GUNGA, H. 1985. Young thrust-fault scarps in the highlands: evidence for an initially totally molten Moon. *Icarus*, **63**, 421–441.

BINDSCHADLER, D. L. & HEAD, J. W. 1991. Tessera terrain, Venus: characterization and models for origin and evolution. *Journal of Geophysical Research*, **96**, 5889–5907.

BINDSCHADLER, D. L. & PARMENTIER, E. M. 1990. Mantle flow tectonics: the influence of a ductile lower crust and implications for the formation of topographic uplands on Venus. *Journal of Geophysical Research*, **95**, 21 329–21 344.

BINDSCHADLER, D. L., SCHUBERT, G. & KAULA, W. M. 1992. Cold spots and hot spots: global tectonics and mantle dynamics of Venus. *Journal of Geophysical Research*, **97**, 13 495–13 532.

BJONNES, E. E., HANSEN, V. L., JAMES, B. & SWENSON, J. B. 2012. Equilibrium resurfacing of Venus: results from new Monte Carlo modeling and implications for Venus surface histories. *Icarus*, **217**, 451–461.

BLAIR, D. M., FREED, A. M. *ET AL.* 2013. The origin of graben and ridges in Rachmaninoff, Raditladi, and Mozart basins, Mercury. *Journal of Geophysical Research*, **118**, 47–58.

BLASIUS, K. R., CUTTS, J. A., GUEST, J. E. & MASURSKY, H. 1977. Geology of the Valles Marineris: first analysis of imaging from the Viking 1 Orbiter primary mission. *Journal of Geophysical Research*, **82**, 4067–4091.

BLEACHER, J. E., GLAZE, L. S. *ET AL.* 2009. Spatial alignment analyses for a field of small volcanic vents south of Pavonis Mons and implications for the Tharsis province, Mars. *Journal of Volcanology and Geothermal Research*, **185**, 96–102.

BLEWETT, D. T., HAWKE, B. R., LUCEY, P. G., TAYLOR, G. J., JAUMANN, R. & SPUDIS, P. D. 1995. Remote sensing and geologic studies of the Schiller-Schickard region of the Moon. *Journal of Geophysical Research*, **100**, 16 959–16 978.

BORG, L. E., CONNELLY, J. N., BOYET, M. & CARLSON, R. W. 2011. Chronological evidence that the Moon is either young or did not have a global magma ocean. *Nature*, **477**, 70–72.

BORGIA, A. 1994. Dynamic basis of volcanic spreading. *Journal of Geophysical Research*, **99**, 17 791–17 804.

BORGIA, A., DELANEY, P. T. & DENLINGER, R. P. 2000. Spreading volcanoes. *Annual Review of Earth and Planetary Science*, **28**, 539–570.

BRADEN, S. E., ROBINSON, M. S., TRAN, T., GENGL, H., LAWRENCE, S. J. & HAWKE, B. R. 2010. Morphology of Gruithuisen and Hortensius domes: Mare v. nonmare volcanism. Abstract 2677 presented at the 41st Lunar & Planetary Science Conference, March 1–5, 2010, The Woodlands, Texas.

BRADLEY, B. A., SAKIMOTO, S. E. H., FREY, H. & ZIMBELMAN, J. R. 2002. Medusae Fossae Formation: new perspectives from Mars Global Surveyor. *Journal of Geophysical Research*, **107**, 5058, http://dx.doi.org/10.1029/2001JE001537

BROWN, C. D. & GRIMM, R. E. 1995. Tectonics of Artemis-Chasma – A Venusian plate boundary. *Icarus*, **117**, 219–249.

BROŽ, P. & HAUBER, E. 2012. An unique volcanic field in Tharsis, Mars: pyroclastic cones as evidence for explosive eruptions. *Icarus*, **218**, 88–99, http://dx.doi.org/10.1016/j.icarus.2011.11.030

BROŽ, P. & HAUBER, E. 2013. Hydrovolcanic tuff rings and cones as indicators for phreatomagmatic explosive eruptions on Mars. *Journal of Geophysical Research*, **118**, 1656–1675.

BRUNO, B. C., LUCEY, P. G. & HAWKE, B. R. 1991. High-resolution UV–visible spectroscopy of lunar red spots. Proceedings of Lunar and Planetary Science, **21**, 405–415.

BUCZKOWSKI, D. L. & WYRICK, D. Y. 2014. Tectonism and magmatism identified on asteroids. *In*: PLATZ, T., MASSIRONI, M., BYRNE, P. K. & HIESINGER, H. (eds) *Volcanism and Tectonism Across the Inner Solar System*. Geological Society, London, Special Publications, **401**. First published online July 28, 2014, http://dx.doi.org/10.1144/SP401.18

BURR, D. M., SOARE, R. J., WAN BUN TSEUNG, J.-M. & EMERY, J. P. 2005. Young (late Amazonian), near surface, ground ice features near the equator, Athabasca Valles, Mars. *Icarus*, **178**, 56–73.

BUSSEY, D. B. J., GUEST, J. E. & SØRENSEN, S.-A. 1997. On the role of thermal conductivity on thermal erosion by lava. *Journal of Geophysical Research*, **102**, 10 905–10 908.

BYRNE, P. K., VAN WYK DE VRIES, B., MURRAY, J. B. & TROLL, V. R. 2009. The geometry of volcano flank terraces on Mars. *Earth and Planetary Science Letters*, **281**, 1–13.

BYRNE, P. K., VAN WYK DE VRIES, B., MURRAY, J. B. & TROLL, V. R. 2012. A volcanotectonic survey of Ascraeus Mons, Mars. *Journal of Geophysical Research*, **117**, E01004.

BYRNE, P. K., HOLOHAN, E. P., KERVYN, M., VAN WYK DE VRIES, B., TROLL, V. R. & MURRAY, J. B. 2013a. A sagging-spreading continuum of large volcano structure. *Geology*, **41**, 339–342.

BYRNE, P. K., KLIMCZAK, C. *ET AL.* 2013b. The origin of Mercury's northern volcanic plains. *Geological Society of America, Abstracts with Programs*, **45**, 851.

BYRNE, P. K., KLIMCZAK, C. *ET AL.* 2013c. An assemblage of lava flow features on Mercury. *Journal of Geophysical Research*, **118**, 1303–1322.

BYRNE, P. K., KLIMCZAK, C. *ET AL.* 2013d. Tectonic complexity within volcanically infilled craters and basins on Mercury. Abstract 1261 presented at the 44th Lunar & Planetary Science Conference, March 18–22, 2013, The Woodlands, Texas.

BYRNE, P. K., HOLOHAN, E. P., KERVYN, M., VAN WYK DE VRIES, B. & TROLL, V. R. 2014a. Analogue modelling of volcano flank terrace formation on Mars. *In*: PLATZ, T., MASSIRONI, M., BYRNE, P. K. & HIESINGER, H. (eds) *Volcanism and Tectonism Across the Inner Solar System*. Geological Society, London, Special Publications, **401**. First published online May 8, 2014, http://dx.doi.org/10.1144/SP401.14

BYRNE, P. K., KLIMCZAK, C., ŞENGÖR, A. M. C., SOLOMON, S. C., WATTERS, T. R. & HAUCK, S. A. 2014b. Mercury's global contraction greatly exceeds earlier measurements. *Nature Geoscience*, **7**, 301–307.

BYRNE, P. K., KLIMCZAK, C., SOLOMON, S. C., MAZARICO, E., NEUMANN, G. A. & ZUBER, M. T. 2014c. Deep-seated contractional tectonics in Mare Crisium, the Moon. Abstract 2,396 presented at the 45th Lunar & Planetary Science Conference, March 17–21, 2014, The Woodlands, Texas.

CAMPBELL, B. A. 1999. Surface formation rates and impact crater densities on Venus. *Journal of Geophysical Research*, **104,** 21 952–21 955.

CAMPBELL, D. B., HEAD, J. W., HARMON, J. K. & HINE, A. A. 1984. Venus volcanism and rift formation in Beta Regio. *Science*, **226**, 167–170.

CARLI, C., SERVENTI, G. & SGAVETTI, M. 2014. VNIR spectral characteristics of terrestrial igneous effusive rocks: mineralogical composition and the influence of texture. *In*: PLATZ, T., MASSIRONI, M., BYRNE, P. K. & HIESINGER, H. (eds) *Volcanism and Tectonism Across the Inner Solar System*. Geological Society, London, Special Publications, **401**. First published online June 17, 2014, http://dx.doi.org/10.1144/SP401.19

CARR, M. H. 1966. *Geologic Map of the Mare Serenitatis Region of the Moon*. United States Geological Survey, Miscellaneous Geologic Investigations Series Map, **I-489 (LAC-42)**.

CARR, M. H., GREELEY, R., BLASIUS, K. R., GUEST, J. E. & MURRAY, J. B. 1977. Some martian volcanic features as viewed from the Viking Orbiters. *Journal of Geophysical Research*, **82**, 3985–4015.

CHABOT, N. L., HOPPA, G. V. & STROM, R. G. 2000. Analysis of lunar lineaments: far side and polar mapping. *Icarus*, **147**, 301–308.

CHARETTE, M. P., MCCORD, T. B., PIETERS, C. & ADAMS, J. B. 1974. Application of remote spectral reflectance measurements to Lunar geology classification and determination of titanium content of lunar soils. *Journal of Geophysical Research*, **74**, 1605–1613, http://dx.doi.org/10.1029/JB079i011p01605

CHETTY, T. R. K., VENKATRAYUDU, M. & VENKATASIVAPPA, V. 2010. Structural architecture and a new tectonic perspective of Ovda Regio, Venus. *Planetary and Space Science*, **58**, 1286–1297.

CHEVREL, S. D., PINET, P. C. & HEAD, J. W. 1999. Gruithuisen domes region: a candidate for an extended non-mare volcanism unit on the Moon. *Journal of Geophysical Research*, **104**, 16 515–16 529.

CHRISTENSEN, P. R., BANDFIELD, J. L. *ET AL.* 2003. Morphology and composition of the surface of Mars: mars Odyssey THEMIS results. *Science*, **300**, 2056–2061.

CHRISTIANSEN, E. H. 1989. Lahars in the Elysium region of Mars. *Geology*, **17**, 203–206.

CLARK, J. D., HURTADO, J. H., HIESINGER, H. & VAN DER BOGERT, C. H. 2014. Investigation of lobate scarps: Implications for the tectonic and thermal evolution of the Moon. Abstract 2048 presented at the 45th Lunar & Planetary Science Conference, March 17–21, 2014, The Woodlands, Texas.

COHEN, B. A., SWINDLE, T. D. & KRING, D. A. 2000. Support for the Lunar Cataclysm Hypothesis from lunar meteorite impact melt ages. *Science*, **290**, 1754–1756, http://dx.doi.org/10.1126/science.290.5497.1754

COMER, R. P., SOLOMON, S. C. & HEAD, J. W. 1985. Mars: thickness of the lithosphere from the tectonic response to volcanic loads. *Reviews of Geophysics*, **23**, 61–92.

COOMBS, C. R., HAWKE, B. R. & GADDIS, L. R. 1987. Explosive volcanism on the Moon. Abstract 197–198 presented at the 18th Lunar & Planetary Science Conference, Houston, Texas.

COOMBS, C. R., HAWKE, B. R., PETERSON, C. A. & ZISK, S. H. 1990. Regional pyroclastic deposits in the north-central portion of the lunar nearside. Abstract 228–229 presented at the 21st Lunar & Planetary Science Conference, Houston, Texas.

CRUMPLER, L. S. & AUBELE, J. 2000. Volcanismon Venus. *In*: SIGURDSON, H., HOUGHTON, B., RYMER, H., STIX, J. & MCNUTT, S. (eds) *Encyclopedia of Volcanoes*. Academic Press, New York, 727–770.

CRUMPLER, L. S., HEAD, J. W. & AUBELE, J. C. 1996. Calderas on Mars: characteristics, structure, and associated flank deformation. *In*: MCGUIRE, W. J., JONES, A. P. & NEUBERG, J. (eds) *Volcano Instabilities on the Earth and Other Planets*. Geological Society, London, Special Publications, **110**, 307–348.

CRUMPLER, L. S., AUBELE, J. C. & ZIMBELMAN, J. R. 2007. Volcanic features of New Mexico analogous to volcanic features on Mars. *In*: CHAPMAN, M. G. (ed.) *The Geology of Mars*. Cambridge University Press, Cambridge, 95–125.

DAVIS, P. A. & TANAKA, K. L. 1993. Small volcanoes in Tempe Terra, Mars: their detailed morphometry and inferred geologic significance. Abstract 379–380 presented at the Lunar & Planetary Science Conference, Houston, Texas.

DELANO, J. W. 1986. Pristine lunar glasses: criteria, data, and implications. *Journal of Geophysical Research*, **91**, 201–213.

DELAUGHTER, J. E. & JURDY, D. M. 1999. Corona classification by evolutionary stage. *Icarus*, **139**, 81–92.

DENEVI, B. W., ROBINSON, M. S. *ET AL.* 2009. The evolution of Mercury's crust: a global perspective from MESSENGER. *Science*, **324**, 613–618.

DENEVI, B. W., ERNST, C. M. *ET AL.* 2013. The distribution and origin of smooth plains on Mercury. *Journal of Geophysical Research*, **118**, 891–907.

DI ACHILLE, G., POPA, C., MASSIRONI, M., MAZZOTTA EPIFANI, E., ZUSI, M., CREMONESE, G. & PALUMBO, P. 2012. Mercury's radius change estimates revisited using MESSENGER data. *Icarus*, **221**, 456–460.

DOGLIONI, C., CARMINATI, E., CUFFARO, M. & SCROCCA, D. 2007. Subduction kinematics and dynamic constraints. *Earth Science Reviews*, **83**, 125–175.

DOGLIONI, C., CARMINATI, E., CRESPI, M., CUFFARO, M., PENATI, M. & RIGUZZI, F. 2014. Tectonically asymmetric Earth: from net rotation to polarized westward drift of the lithosphere. *Geoscience Frontiers*, first published online 18 February 2014, http://dx.doi.org/10.1016/j.gsf.2014.02.001

DOHM, J. M. & TANAKA, K. L. 1999. Geology of the Thaumasia region, Mars: plateau development, valley origin, and magmatic evolution. *Planetary and Space Science*, **47**, 411–431.

DOHM, J. M., WILLIAMS, J.-P. *ET AL.* 2009. New evidence for a magmatic influence on the origin of Valles Marineris, Mars. *Journal of Volcanology and Geothermal Research*, **185**, 12–27.

DOMBARD, A. J. & GILLIS, J. J. 2001. Testing the viability of topographic relaxation as a mechanism for the formation of lunar floor-fractured craters. *Journal of Geophysical Research*, **106**, 27 901–27 910.

DOMBARD, A. J. & HAUCK, S. A., II. 2008. Despinning plus global contraction and the orientation of lobate scarps on Mercury: predictions for MESSENGER. *Icarus*, **198**, 274–276.

DOWTY, E., PRINZ, M. & KEIL, K. 1974*a*. Ferroan anorthosite – a widespread and distinctive lunar rock type. *Earth and Planetary Science Letters*, **24**, 15–25, http://dx.doi.org/10.1016/0012-821X (74)90003-X

DOWTY, E., PRINZ, M. & KEIL, K. 1974*b*. 'Very High Alumina Basalt'. A mixture and not a magma type. *Science*, **183**, 1214–1215, http://dx.doi.org/10.1126/science.183.4130.1214

DZURISIN, D. 1978. The tectonic and volcanic history of Mercury as inferred from studies of scarps, ridges, troughs, and other lineaments. *Journal of Geophysical Research*, **83**, 4883–4906.

EDGETT, K. & BLUMBERG, D. 1994. Star and linear dunes on Mars. *Icarus*, **112**, 448–464, http://dx.doi.org/10.1006/icar.1994.1197

EDGETT, K. & CHRISTENSEN, P. R. 1991. The particle size of Martian aeolian dunes. *Journal of Geophysical Research*, **96**, 22 762–22 776.

EDGETT, K. & CHRISTENSEN, P. R. 1994. Mars aeolian sand: regional variations among dark-hued crater floor features. *Journal of Geophysical Research*, **99**, 1997–2018.

EDGETT, K. S. 1990. Possible cinder cones near the summit of Pavonis Mons, Mars. Abstract 311–312 presented at the 21st Lunar & Planetary Science Conference, Houston, Texas.

EDGETT, K. S. & MALIN, M. C. 2002. Martian sedimentary rock stratigraphy: outcrops and interbedded craters of northwest Sinus Meridiani and southwest Arabia Terra. *Geophysical Research Letters*, **29**, 2179, http://dx.doi.org/10.1029/2002gl016515

ELKINS-TANTON, L. T. & HAGER, B. H. 2005. Giant meteoroid impacts can cause volcanism. *Earth and Planetary Science Letters*, **239**, 219–232.

EL MAARRY, M. R., DOHM, J. M. *ET AL.* 2012. Searching for evidence of hydrothermal activity at Apollinaris Mons, Mars. *Icarus*, **217**, 297–314, http://dx.doi.org/10.1016/j.icarus.2011.10.022

ERNST, R. E., HEAD, J. W., PARFITT, E., GROSFILS, E. & WILSON, L. 1995. Giant radiating dyke swarms on Earth and Venus. *Earth Science Reviews*, **39**, 1–58.

ERNST, R. E., GROSFILS, E. B. & MÈGE, D. 2001. Giant dyke swarms: earth, Venus, and Mars. *Annual Review of Earth and Planetary Sciences*, **29**, 489–534.

FAGENTS, S. A. & GREELEY, R. 2001. Factors influencing lava-substrate heat transfer and implications for thermomechanical erosion. *Bulletin of Volcanology*, **62**, 519–532.

FARRAND, W. H., GADDIS, L. R. & KESZTHELYI, L. 2005. Pitted cones and domes on Mars: observations in Acidalia Planitia and Cydonia Mensae using MOC, THEMIS, and TES data. *Journal of Geophysical Research*, **110**, E05005, http://dx.doi.org/10.1029/2004JE002297

FASSETT, C. I., KADISH, S. J., HEAD, J. W., SOLOMON, S. C. & STROM, R. G. 2011. The global population of large craters on Mercury and comparison with the Moon. *Geophysical Research Letters*, **38**, L10202, http://dx.doi.org/10.1029/2011GL047294

FERNÀNDEZ, C., ANGUITA, F., RUIZ, J., ROMEO, I., MARTÍN-HERRERO, À., RODRÍGUE, A. & PIMENTEL, C. 2010. Structural evolution of Lavinia Planitia, Venus: implications for the tectonics of the lowland plains. *Icarus*, **206**, 210–228.

FERRARI, S., MASSIRONI, M., MARCHI, S., BYRNE, P. K., KLIMCZAK, C., MARTELLATO, E. & CREMONESE, G. 2014. Age relationships of the Rembrandt basin and Enterprise Rupes, Mercury. *In*: PLATZ, T., MASSIRONI, M., BYRNE, P. K. & HIESINGER, H. (eds) *Volcanism and Tectonism Across the Inner Solar System*. Geological Society, London, Special Publications, **401**. First published online July 29, 2014, http://dx.doi.org/10.1144/SP401.20

FISCHER, K. M., FORD, H. A., ABT, D. L. & RYCHERT, C. A. 2010. The lithosphere–asthenosphere boundary. *Annual Reviews of Earth and Planetary Sciences*, **38**, 551–575.

FRANKEL, C. 2005. *Worlds on Fire: Volcanoes on the Earth, the Moon, Mars, Venus and Io*. Cambridge University Press, Cambridge.

FREED, A. M., MELOSH, H. J. & SOLOMON, S. C. 2001. Tectonics of mascon loading: resolution of the strike-slip faulting paradox. *Journal of Geophysical Research*, **106**, 20 603–20 620.

FREED, A. M., SOLOMON, S. C., WATTERS, T. R., PHILLIPS, R. J. & ZUBER, M. T. 2009. Could Pantheon Fossae be the result of the Apollodorus crater-forming impact within the Caloris basin, Mercury? *Earth and Planetary Science Letters*, **285**, 320–327.

FREED, A. M., BLAIR, D. M. ET AL. 2012. On the origin of graben and ridges within and near volcanically buried craters and basins in Mercury's northern plains. *Journal of Geophysical Research*, **117**, E00L06, http://dx.doi.org/10.1029/2012JE004119

FREY, H. 2008. Ages of very large impact basins on Mars: implications for the late heavy bombardment in the inner solar system. *Geophysical Research Letters*, **35**, L13203.

FREY, H. & JAROSEWICH, M. 1982. Subkilometer martian volcanoes–Properties and possible terrestrial analogs. *Journal of Geophysical Research*, **87**, 9867–9879.

FREY, H., EDGAR, L. & LILLIS, R. 2007. Very large visible and buried impact basins on Mars: implications for internal and crustal evolution and the late heavy bombardment in the inner solar system. Abstract 3070 presented at the 7th International Conference on Mars, July 9–13, 2007, California Institute of Technology (Caltech).

GADDIS, L. R., PIETERS, C. M. & HAWKE, B. R. 1985. Remote sensing of lunar pyroclastic mantling deposits. *Icarus*, **61**, 461–489.

GADDIS, L., ROSANOVA, C., HARE, T., HAWKE, B. R., COOMBS, C. & ROBINSON, M. S. 1998. Small lunar pyroclastic deposits: a new global perspective. Abstract 1807–1808 presented at the 29th Lunar & Planetary Science Conference, March 16–20, 1998, The Woodlands, Texas.

GALGANA, G. A., GROSFILS, E. B. & McGOVERN, P. J. 2013. Radial dike formation on Venus: insights from models of uplift, flexure and magmatism. *Icarus*, **225**, 538–547.

GALLUZZI, V., DI ACHILLE, G., FERRANTI, L., POPA, C. & PALUMBO, P. 2014. Faulted craters as indicators for thrust motions on Mercury. *In*: PLATZ, T., MASSIRONI, M., BYRNE, P. K. & HIESINGER, H. (eds) *Volcanism and Tectonism Across the Inner Solar System*. Geological Society, London, Special Publications, **401**. First published online June 12, 2014, http://dx.doi.org/10.1144/SP401.17

GARRY, W. B., ZIMBELMAN, J. R. & GREGG, T. K. P. 2007. Morphology and emplacement of a long channeled lava flow near Ascraeus Mons volcano, Mars. *Journal of Geophysical Research*, **112**, E08007.

GERYA, T. V. 2014. Plume-induced crustal convection: 3D thermomechanical model and implications for the origin of novae and coronae on Venus. *Earth and Planetary Science Letters*, **391**, 183–192.

GHAIL, R. C. & WILSON, L. 2013. A pyroclastic flow deposit on Venus. *In*: PLATZ, T., MASSIRONI, M., BYRNE, P. K. & HIESINGER, H. (eds) *Volcanism and Tectonism Across the Inner Solar System*. Geological Society, London, Special Publications, **401**. First published online November 19, 2013, http://dx.doi.org/10.1144/SP401.1

GHATAN, G. J. & HEAD, J. W., III. 2002. Candidate subglacial volcanoes in the south polar region of Mars: morphology, morphometry, and eruption conditions. *Journal of Geophysical Research*, **107**, 5048, http://dx.doi.org/10.1029/2001JE001519

GHENT, R. & HANSEN, V. 1999. Structural and kinematic analysis of eastern Ovda Regio, Venus: implications for crustal plateau formation. *Icarus*, **139**, 116–136.

GIACOMINI, L., MASSIRONI, M., MARCHI, S., FASSETT, C. I., DI ACHILLE, G. & CREMONESE, G. 2014. Age dating of an extensive thrust system on Mercury: implications for the planet's thermal evolution. *In*: PLATZ, T., MASSIRONI, M., BYRNE, P. K. & HIESINGER, H. (eds) *Volcanism and Tectonism Across the Inner Solar System*. Geological Society, London, Special Publications, **401**. First published online August 13, 2014, http://dx.doi.org/10.1144/SP401.21

GIGUERE, T. A., TAYLOR, G. J., HAWKE, B. R. & LUCEY, P. G. 2000. The titanium contents of lunar mare basalts. *Meteoritics and Planetary Science*, **35**, 193–200, http://dx.doi.org/10.1111/j.1945-5100.2000.tb01985.x

GIGUERE, T. A., HAWKE, B. R. ET AL. 2003. Remote sensing studies of the Lomonosov-Fleming region of the Moon. *Journal of Geophysical Research*, **108**, 5118, http://dx.doi.org/10.1029/2003JE002069

GILLIS-DAVIS, J. J., BLEWETT, D. T. ET AL. 2009. Pit-floor craters on Mercury: evidence of near-surface igneous activity. *Earth and Planetary Science Letters*, **285**, 243–250.

GILMORE, M. S. & HEAD, J. W. 2000. Sequential deformation of plains at the margin of Alpha Regio, Venus: implications for tessera formation. *Meteoritics and Planetary Science*, **35**, 667–687.

GILMORE, M. S., COLLINS, G. C., IVANOV, M. A., MARINANGELI, L. & HEAD, J. W. 1998. Style and sequence of extensional structures in tessera terrain, Venus. *Journal of Geophysical Research*, **103**, 16 813–16 840.

GLAZE, L. S., STOFAN, E. R., SMREKAR, S. E. & BALOGA, S. M. 2002. Insights into corona formation through statistical analyses. *Journal of Geophysical Research*, **107**, 5135.

GLOTCH, T. D., HAGERTY, J. J. ET AL. 2011. The Mairan domes: silicic volcanic constructs on the Moon. *Geophysical Research Letters*, **38**, L21204, http://dx.doi.org/10.1029/2011GL049548

GOLOMBEK, M. 1999. Introduction to the special section: mars Pathfinder. *Journal of Geophysical Research*, **104**, 8521–8522, http://dx.doi.org/10.1029/1998JE900032

GOLOMBEK, M. P., PLESCIA, J. B. & FRANKLIN, B. J. 1991. Faulting and folding in the formation of planetary wrinkle ridges. *Proceedings of Lunar and Planetary Science*, **21**, 679–693.

GOLOMBEK, M. P., ANDERSON, F. S. & ZUBER, M. T. 2000. Martian wrinkle ridge topography: evidence for subsurface faults from MOLA. Abstract 1294 presented at the 31st Lunar & Planetary Science Conference, March 13–17, 2000, The Woodlands, Texas.

GOLOMBEK, M. P., ANDERSON, F. S. & ZUBER, M. T. 2001. Martian wrinkle ridge topography: evidence for subsurface faults from MOLA. *Journal of Geophysical Research*, **106**, 23 811–23 822.

GREELEY, R. 1971. Lava tubes and channels in the lunar Marius Hills. *The Moon*, **3**, 289–314.

GREELEY, R. (ed.) 1974. *Geologic Guide to the Island of Hawaii: A Field Guide for Comparative Planetary Geology*. NASA CR-152416. NASA, Washington, DC.

GREELEY, R. 1982. The Snake River Plains, Idaho: representative of a new category of volcanism. *Journal of Geophysical Research*, **87**, 2705–2712.

GREELEY, R. & GUEST, J. E. 1987. *Geologic Map of the Eastern Equatorial Region of Mars, Scale*

1:15,000,000. United States Geological Survey, Miscellaneous Geologic Investigations Series Map, **I-1802-B.**

GREELEY, R. & SCHNEID, B. D. 1991. Magma generation on Mars – amounts, rates, and comparisons with Earth, Moon, and Venus. *Science,* **254,** 996–998.

GREELEY, R. & SPUDIS, P. D. 1981. Volcanism on Mars. *Reviews of Geophysics and Space Physics,* **19,** 13–41.

GREELEY, R., KADEL, S. D. *ET AL.* 1993. Galileo imaging observations of lunar maria and related deposits. *Journal of Geophysical Research,* **98,** 17 183–17 206.

GRIMM, R. E. & PHILLIPS, R. J. 1991. Gravity anomalies, compensation mechanisms, and the geodynamics of western Ishtar Terra, Venus. *Journal of Geophysical Research,* **96,** 8305–8324.

GRINSPOON, D. H. 1993. Implications of the high D/H ratio for the sources of water in Venus atmosphere. *Nature,* **363,** 428–431.

GROSFILS, E. B. & HEAD, J. W. 1994. The global distribution of giant radiating dike swarms on Venus: implications for the global stress state. *Geophysical Research Letters,* **21,** 701–704.

GROSFILS, E. B., AUBELE, J., CRUMPLER, L., GREGG, T. & SAKIMOTO, S. 1999. Volcanism on Venus and Earth's seafloor. *In:* GREGG, T. & ZIMBELMAN, J. (eds) *Environmental Effects on Volcanic Eruptions: From Deep Ocean to Deep Space.* Plenum, New York, 113–142.

GROSFILS, E. B., MCGOVERN, P. J., GREGG, P. M., GALGANA, G. A., HURWITZ, D. M., LONG, S. M. & CHESTLER, S. R. 2013. Elastic models of magma reservoir mechanics: a key tool for investigating planetary volcanism. *In:* PLATZ, T., MASSIRONI, M., BYRNE, P. K. & HIESINGER, H. (eds) *Volcanism and Tectonism Across the Inner Solar System.* Geological Society, London, Special Publications, **401.** First published online December 11, 2013, http://dx.doi.org/10.1144/SP401.2

GUALLINI, L., PAUSELLI, C., BROZZETTI, F. & MARINANGELI, L. 2014. Physical modelling of large-scale deformational systems in the South Polar Layered Deposits (Promethei Lingula, Mars): new geological constraints and climatic implications. *In:* PLATZ, T., MASSIRONI, M., BYRNE, P. K. & HIESINGER, H. (eds) *Volcanism and Tectonism Across the Inner Solar System.* Geological Society, London, Special Publications, **401.** First published online April 25, 2014, http://dx.doi.org/10.1144/SP401.13

GUEST, J. E. & MURRAY, J. B. 1976. Volcanic features of the nearside equatorial lunar maria. *Journal of the Geological Society, London,* **132,** 251–258.

GUEST, J. E. & STOFAN, E. R. 1999. A new view of the stratigraphic history of Venus. *Icarus,* **139,** 56–66.

GUNG, Y., PANNING, M. & ROMANOWICZ, B. 2003. Global anisotropy and the thickness of continents. *Nature,* **422,** 707–711.

GUSTAFSON, J. O., BELL, J. F., GADDIS, L. R., HAWKE, B. R. & GIGUERE, T. A. 2012. Characterization of previously unidentified lunar pyroclastic deposits using Lunar Reconnaissance Orbiter Camera data. *Journal of Geophysical Research,* **117,** E00H25, http://dx.doi.org/10.1029/2011JE003893

HAGERTY, J. J., LAWRENCE, D. J., HAWKE, B. R., VANIMAN, D. T., ELPHIC, R. C. & FELDMAN, W. C. 2006. Refined thorium abundances for lunar red

spots: implications for evolved, nonmare volcanism on the Moon. *Journal of Geophysical Research,* **111,** E06002, http://dx.doi.org/10.1029/2005JE002592

HAMILTON, C. W., FAGENTS, S. A. & THORDARSON, T. 2011. Lava-ground ice interactions in Elysium Planitia, Mars: geomorphological and geospatial analysis of the Tartarus Colles cone groups. *Journal of Geophysical Research,* **116,** E03004.

HAMILTON, V. & STOFAN, E. R. 1996. The geomorphology and evolution of Hecate Chasma, Venus. *Icarus,* **121,** 171–194.

HANSEN, V. L. 2006. Geological constraints on crustal plateau surface histories, Venus: the lava pond and bolide impact hypotheses. *Journal of Geophysical Research,* **111,** E11010.

HANSEN, V. L. & LÓPEZ, I. 2010. Venus records a rich early history. *Geology,* **38,** 311–314.

HANSEN, V. L. & OLIVE, A. 2010. Artemis, Venus: the largest tectonomagmatic feature in the solar system? *Geology,* **38,** 467–470.

HANSEN, V. L. & PHILLIPS, R. J. 1993. Tectonics and volcanism of eastern Aphrodite Terra, Venus – No subduction, no spreading. *Science,* **260,** 526–530.

HANSEN, V. L. & WILLIS, J. J. 1996. Structural analysis of a sampling of tesserae: implications for Venus geodynamics. *Icarus,* **123,** 296–312.

HANSEN, V. L. & WILLIS, J. J. 1998. Ribbon terrain formation, southwestern Fortuna Tessera, Venus: implications for lithosphere evolution. *Icarus,* **132,** 321–343.

HANSEN, V. L. & YOUNG, D. A. 2007. Venus's evolution: A synthesis. *In:* CLOOS, M., CARLSON, W. D., GILBERT, M. C., LIOU, J. G. & SORENSEN, S. S. (eds) *Convergent Margin Terranes and Associated Regions: A Tribute to W.G. Ernst.* Geological Society of America Special Papers, **419,** 255–273.

HANSEN, V. L., BANKS, B. K. & GHENT, R. R. 1999. Tessera terrain and crustal plateaus, Venus. *Geology,* **27,** 1071–1074.

HANSEN, V. L., PHILLIPS, R. J., WILLIS, J. J. & GHENT, R. R. 2000. Structures in tessera terrain, Venus: issues and answers. *Journal of Geophysical Research,* **105,** 4135–4152.

HARRIS, L. B. & BÉDARD, J. H. 2014a. Chapter 9: Crustal evolution and deformation in a non-plate tectonic Archaean Earth: comparisons with Venus. *In:* DILEK, Y. & FURNES, H. (eds) *Evolution of Archean Crust and Early Life.* Modern Approaches in Solid Earth Sciences, **7.** Springer, Berlin, 215–288.

HARRIS, L. B. & BÉDARD, J. H. 2014b. Interactions between continent-like 'drift', rifting and mantle flow on Venus: gravity interpretations and Earth analogues. *In:* PLATZ, T., MASSIRONI, M., BYRNE, P. K. & HIESINGER, H. (eds) *Volcanism and Tectonism Across the Inner Solar System.* Geological Society, London, Special Publications, **401.** First published online May 19, 2014, http://dx.doi.org/10.1144/SP401.9

HARTMANN, W. K. & WOOD, C. A. 1971. Moon: origin and evolution of multi-ring basins. *The Moon,* **3,** 3–78.

HASHIMOTO, G. L., ROOS-SEROTE, M., SUGITA, S., GILMORE, M. S., KAMP, L. W., CARLSON, R. W. & BAINES, K. H. 2008. Felsic highland crust on Venus suggested by Galileo near infrared mapping spectrometer data. *Journal of Geophysical Research,* **113,** E00B24, http://dx.doi.org/10.1029/2008JE003134

HAUBER, E., BLEACHER, J., GWINNER, K., WILLIAMS, D. & GREELEY, R. 2009. The topography and morphology of low shields and associated landforms of plains volcanism in the Tharsis region of Mars. *Journal of Volcanology and Geothermal Research*, **185**, 69–95.

HAUBER, E., BROŽ, P., JAGERT, F., JODŁOWSKI, P. & PLATZ, T. 2011. Very recent and wide-spread basaltic volcanism on Mars. *Geophysical Research Letters*, **38**, L10201.

HAWKE, B. R. & BELL, J. F. 1981. Remote sensing studies of lunar dark-halo craters. *Bulletin of the American Astronomical Society*, **13**, 712.

HAWKE, B. R. & SPUDIS, P. D. 1980. Geochemical anomalies on the eastern limb and farside of the moon. *Proceedings of the Conference on the Lunar Highlands Crust*, 14–16 November 1979, Houston, Texas. Pergamon Press, Oxford, 467–481.

HAWKE, B. R., MACLASKEY, D., MCCORD, T. B., ADAMS, J. B., HEAD, J. W., PIETERS, C. M. & ZISK, S. 1979. Multispectral mapping of lunar pyroclastic deposits. *Bulletin of the American Astronomical Society*, **12**, 582.

HAWKE, B. R., SPUDIS, P. D. & CLARK, P. E. 1985. The origin of selected lunar geochemical anomalies: implications for early volcanism and the formation of light plains. *Earth, Moon, and Planets*, **32**, 257–273.

HAWKE, B. R., COOMBS, C. R., GADDIS, L. R., LUCEY, P. G. & OWENSBY, P. D. 1989. Remote sensing and geologic studies of localized dark mantle deposits on the Moon. *Proceedings of Lunar and Planetary Science*, **19**, 255–268.

HAWKE, B. R., LAWRENCE, D. J., BLEWETT, D. T., LUCEY, P. G., SMITH, G. A., SPUDIS, P. D. & TAYLOR, G. J. 2003. Hansteen Alpha: a volcanic construct in the lunar highlands. *Journal of Geophysical Research*, **108**, 5069, http://dx.doi.org/10.1029/2002JE002013

HAWKE, B. R., GILLIS, J. J. ET AL. 2005. Remote sensing and geologic studies of the Balmer-Kapteyn region of the Moon. *Journal of Geophysical Research*, **110**, E06004, http://dx.doi.org/10.1029/2004JE002383

HAWKE, B. R., GIGUERRE, T. A. ET AL. 2011. Hansteen Apha: a silicic volcanic construct on the Moon. Abstract 1652 presented at the 42nd Lunar & Planetary Science Conference, March 7–11, 2011, The Woodlands, Texas.

HEAD, J. W. 1976. Lunar volcanism in space and time. *Reviews of Geophysics and Space Physics*, **14**, 265–300.

HEAD, J. W. & CRUMPLER, L. S. 1987. Evidence for divergent plate boundary characteristics and crustal spreading on Venus. *Science*, **238**, 1380–1385.

HEAD, J. W. & GIFFORD, A. 1980. Lunar mare domes: classification and modes of origin. *Earth, Moon, and Planets*, **22**, 235–258.

HEAD, J. W. & WILSON, L. 1979. Alphonsus-type dark-halo craters: morphology, morphometry and eruption conditions. *Proceedings of Lunar and Planetary Science*, **10**, 2861–2897.

HEAD, J. W. & WILSON, L. 1980. The formation of eroded depressions around the sources of lunar sinuous rilles: observations. *Proceedings of Lunar and Planetary Science*, **11**, 426–428.

HEAD, J. W. & WILSON, L. 1986. Volcanic processes and landforms on Venus: theory, predictions, and observations. *Journal of Geophysical Research*, **91**, 9407–9446.

HEAD, J. W. & WILSON, L. 1992. Lunar mare volcanism. Stratigraphy, eruption conditions, and the evolution of secondary crusts. *Geochimica et Cosmochimica Acta*, **56**, 2155–2175.

HEAD, J. W., HESS, P. C. & MCCORD, T. B. 1978. Geologic characteristics of lunar highland volcanic (Gruithuisen and Marina region) and possible eruption conditions. *Proceedings of Lunar and Planetary Science*, **9**, 488–490.

HEAD, J. W., CRUMPLER, L. S., AUBELE, J. C., GUEST, J. E. & SAUNDERS, R. S. 1992. Venus volcanism – classification of volcanic features and structures, associations, and global distribution from Magellan data. *Journal of Geophysical Research*, **97**, 13 153–13 197.

HEAD, J. W., MURCHIE, S. ET AL. 1993. Lunar impact basins: new data for the western limb and far side (Orientale and South Pole-Aitken basins) from the first Galileo flyby. *Journal of Geophysical Research*, **98**, 17 149–17 182.

HEAD, J. W., WILSON, L., ROBINSON, M., HIESINGER, H., WEITZ, C. & YINGST, A. 2000. Moon and Mercury: volcanism in Early Planetary History. *In*: ZIMBELMAN, J. R. & GREGG, T. K. P. (eds) *Environmental Effects on Volcanic Eruptions: From Deep Oceans to deep Space*. Kluwer Academic, New York.

HEAD, J. W., WILSON, L. & WEITZ, C. M. 2002. Dark ring in southwestern Orientale basin: origin as a single pyroclastic eruption. *Journal of Geophysical Research*, **107**, 1-1–1-17, http://dx.doi.org/10.1029/2000JE001438

HEAD, J. W., MURCHIE, S. L. ET AL. 2008. Volcanism on Mercury: evidence from the first MESSENGER flyby. *Science*, **321**, 69–72.

HEAD, J. W., MURCHIE, S. L. ET AL. 2009. Volcanism on Mercury: evidence from the first MESSENGER flyby for extrusive and explosive activity and the volcanic origin of plains. *Earth and Planetary Science Letters*, **285**, 227–242.

HEAD, J. W., CHAPMAN, C. R. ET AL. 2011. Flood volcanism in the northern high latitudes of Mercury revealed by MESSENGER. *Science*, **333**, 1853–1856.

HEATHER, D. J., DUNKIN, S. K. & WILSON, L. 2003. Volcanism on the Marius Hills plateau: observational analyses using Clementine multispectral data. *Journal of Geophysical Research*, **108**, 5017, http://dx.doi.org/10.1029/2002JE001938

HEIKEN, G., MCKAY, D. S. & BROWN, R. W. 1974. Lunar deposits of possible pyroclastic origin. *Geochimica et Cosmochimica Acta*, **38**, 1703–1718.

HELBERT, J., MULLER, N., KOSTAMA, P., MARINANGELI, L., PICCIONI, G. & DROSSART, P. 2008. Surface brightness variations seen by VIRTIS on Venus Express and implications for the evolution of the Lada Terra region, Venus. *Geophysical Research Letters*, **35**, 1–5.

HERRICK, R. R. & PHILLIPS, R. J. 1992. Geological correlations with the interior density structure of Venus. *Journal of Geophysical Research*, **97**, 16 017–16 034.

HIESINGER, H. & HEAD, J. W., III . 2004. The Syrtis Major volcanic province, Mars: synthesis from Mars Global Surveyor data. *Journal of Geophysical Research*, **109**, E01004.

HIESINGER, H. & HEAD, J. W., III . 2006. New views of lunar geoscience: an introduction and overview. *Reviews in Mineralogy and Geochemistry*, **60**, 1–81.

HIESINGER, H. & JAUMANN, R. 2014. Chapter 23: The Moon. *In*: SPOHN, T., BREUER, D. & JOHNSON, T. (eds) *Encyclopedia of the Solar System*, 3rd edn. Elsevier, Amsterdam, 493–538.

HIESINGER, H., JAUMANN, R., NEUKUM, G. & HEAD, J. W. 2000. Ages of mare basalts on the lunar nearside. *Journal of Geophysical Research*, **105**, 29 239–29 276, http://dx.doi.org/10.1029/2000JE001244

HIESINGER, H., HEAD, J. W., III, WOLF, U. & NEUKUM, G. 2001. New age determinations of lunar mare basalts in Mare Cognitum, Mare Nubium, Oceanus Procellarum, and other nearside mare. Abstract 1815 presented at the 32nd Lunar & Planetary Science Conference, Houston, Texas.

HIESINGER, H., HEAD, J. W., WOLF, U., JAUMANN, R. & NEUKUM, G. 2003. Ages and stratigraphy of mare basalts in Oceanus Procellarum, Mare Nubium, Mare Cognitum, and Mare Insularum. *Journal of Geophysical Research*, **108**, 5065, http://dx.doi.org/10.1029/2002JE001985

HIESINGER, H., HEAD, J. W., WOLF, U., JAUMANN, R. & NEUKUM, G. 2011. Ages and stratigraphy of lunar mare basalts: A synthesis. *In*: AMBROSE, W. A. & WILLIAMS, D. A. (eds) *Recent Advances and Current Research Issues in Lunar Stratigraphy*. Geological Society of America Special Papers, **477**, 1–51.

HIESINGER, H., VAN DER BOGERT, C. H. *ET AL.* 2012. How old are young lunar craters? *Journal of Geophysical Research*, **117**, E00H10, http://dx.doi.org/10.1029/2011JE003935

HODGES, C. A. 1973. *Mare Ridges and Lava Lakes*. Apollo 17 Preliminary Science Report, NASA SP-330.

HODGES, C. A. & MOORE, H. J. 1994. *Atlas of Volcanic Features on Mars*. United States Geological Survey, Professional Papers, **1534**.

HÖRZ, F., GRIEVE, R., HEIKEN, G., SPUDIS, P. & BINDER, A. 1991. Lunar surface processes. *In*: HEIKEN, G., VANIMAN, D. & FRENCH, B. (eds) *Lunar Sourcebook – A User Guide to the Moon*. Cambridge University Press, Cambridge, 61–120.

HUANG, J., YANG, A. & ZHONG, S. 2013. Constraints of the topography, gravity and volcanism on Venusian mantle dynamics and generation of plate tectonics. *Earth and Planetary Science Letters*, **362**, 207–214.

HULME, G. 1973. Turbulent lava flows and the formation of lunar sinuous rilles. *Modern Geology*, **4**, 107–117.

HURWITZ, D. M., HEAD, J. W., WILSON, L. & HIESINGER, H. 2012. Origin of lunar sinuous rilles: modeling effects of gravity, surface slope, and lava composition on erosion rates during the formation of Rima Prinz. *Journal of Geophysical Research*, **117**, E00H14, http://dx.doi.org/10.1029/2011JE004000

HURWITZ, D. M., HEAD, J. W. & HIESINGER, H. 2013*a*. Lunar sinuous rilles: distribution, characteristics, and implications for their origin. *Planetary and Space Science*, **79**, 1–38, http://dx.doi.org/10.1016/j.pss.2012.10.019

HURWITZ, D. M., HEAD, J. W. *ET AL.* 2013*b*. Investigating the origin of candidate lava channels on Mercury with MESSENGER data: theory and observations. *Journal of Geophysical Research*, **118**, 471–486, http://dx.doi.org/10.1029/2012JE004103

HYNEK, B. M., PHILLIPS, R. J. & ARVIDSON, R. E. 2003. Explosive volcanism in the Tharsis region: global evidence in the Martian geologic record. *Journal of Geophysical Research*, **108**, 5111, http://dx.doi.org/10.1029/2003JE002062

ISACKS, B., OLIVER, J. & SYKES, L. R. 1968. Seismology and the new global tectonics. *Journal of Geophysical Research*, **73**, 5855–5899.

IVANOV, B. A. & MELOSH, H. J. 2003. Impacts do not initiate volcanic eruptions: eruptions close to the crater. *Geology*, **31**, 869–872.

IVANOV, M. A. & HEAD, J. W. 1996. Tessera terrain on Venus: a survey of the global distribution, characteristics and relation to surrounding units from Magellan data. *Journal of Geophysical Research*, **101**, 14 861–14 908.

IVANOV, M. A. & HEAD, J. W., III. 2001. *Geologic map of the Lavinia Planitia Quadrangle (V-55), Venus*. United States Geological Survey, Miscellaneous Geologic Investigations Series, **I-2684**.

IVANOV, M. A. & HEAD, J. W. 2008. Formation and evolution of Lakshmi Planum, Venus: assessment of models using observations from geological mapping. *Planetary and Space Science*, **56**, 1949–1966.

IVANOV, M. A. & HEAD, J. W. 2011. Global geological map of Venus. *Planetary Space Science*, **59**, 1559–1600.

IVANOV, M. A. & HEAD, J. W. 2013. The history of volcanism on Venus. *Planetary Space Science*, **84**, 66–92.

JAUMANN, R., HIESINGER, H. *ET AL.* 2012. Geology, geochemistry, and geophysics of the Moon: status of current understanding. *Planetary and Space Science*, **74**, 15–41, http://dx.doi.org/10.1016/j.pss.2012.08.019

JOHNSON, C. L. & SANDWELL, D. T. 1992. Joints in Venusian lava flows. *Journal of Geophysical Research*, **97**, 13 601–13 610.

JOHNSON, J. R., LARSON, S. M. & SINGER, R. B. 1991. A reevaluation of spectral ratios for lunar mare TiO_2 mapping. *Geophysical Research Letters*, **18**, 2153–2156, http://dx.doi.org/10.1029/91GL02094

JOLLIFF, B. L., GILLIS, J. J., HASKIN, L. A., KOROTEV, R. L. & WIECZOREK, M. W. 2000. Major lunar crustal terranes: surface expressions and crust–mantle origins. *Journal of Geophysical Research*, **105**, 4197–4216.

JOLLIFF, B. L., WISEMAN, S. A. *ET AL.* 2011. Non-mare silicic volcanism on the lunar farside at Compton-Belkovich. *Nature Geoscience*, **4**, 566–571, http://dx.doi.org/10.1038/NGEO1212

JOZWIAK, L. M., HEAD, J. W., ZUBER, M. T., SMITH, D. E. & NEUMANN, G. A. 2012. Lunar floor-fractured craters: classification, distribution, origin and implications for magmatism and shallow crustal structure. *Journal of Geophysical Research*, **117**, E11005.

JOZWIAK, L. M., HEAD, J. W. *ET AL.* 2014. Lunar floor-fractured craters: intrusion emplacement and associated gravity anomalies. Abstract 1464 presented at the 45th Lunar & Planetary Science Conference, March 17–21, 2014, The Woodlands, Texas.

JULL, M. G. & ARKANI-HAMED, J. 1995. The implications of basalt in the formation and evolution of mountains on Venus. *Physics of Earth and Planetary Interiors*, **89**, 163–175.

KADDISH, S. J., FASSETT, C. I., HEAD, J. W., SMITH, D. E., ZUBER, M. T., NEUMANN, G. A. & MAZARICO, E. 2011. A global catalog of large lunar craters (≥ 20 km) from the Lunar Orbiter Laser Altimeter. Abstract 1006 presented at the 42nd Lunar & Planetary Science Conference, March 7–11, 2011, The Woodlands, Texas.

KARGEL, J. S., KIRK, R. L., FEGLEY, B., JR. & TREIMAN, A. H. 1994. Carbonate-sulfate volcanism on Venus? *Icarus*, **112**, 219–252.

KAULA, W. M. 1990. Venus: a contrast in evolution to Earth. *Science*, **247**, 1191–1196.

KEDDIE, S. & HEAD, J. 1994. Sapas Mons, Venus: evolution of a large shield volcano. *Earth, Moon, and Planets*, **65**, 129–190.

KERBER, L., HEAD, J. W., SOLOMON, S. C., MURCHIE, S. L., BLEWETT, D. T. & WILSON, L. 2009. Explosive volcanic eruptions on Mercury: eruption conditions, magma volatile content, and implications for interior volatile abundances. *Earth and Planetary Science Letters*, **285**, 263–271.

KERBER, L., HEAD, J. W., MADELEINE, J.-B., FORGET, F. & WILSON, L. 2011a. The dispersal of pyroclasts from Apollinaris Patera, Mars: implications for the origin of the Medusae Fossae Formation. *Icarus*, **216**, 212–220, http://dx.doi.org/10.1016/j.icarus.2011.07.035

KERBER, L., HEAD, J. W. ET AL. 2011b. The global distribution of pyroclastic deposits on Mercury: the view from MESSENGER flybys 1–3. *Planetary and Space Science*, **59**, 1895–1909.

KESZTHELYI, L., THORDARSON, T., MCEWEN, A., HAACK, H., GUILBAUD, M. N., SELF, S. & ROSSI, M. J. 2004. Icelandic analogs to Martian flood lavas. *Geochemistry Geophysics Geosystems*, **5**, Q11014.

KESZTHELYI, L., JAEGER, W., MCEWEN, A., TORNABENE, L., BEYER, R. A., DUNDAS, C. & MILAZZO, M. 2008. High Resolution Imaging Science Experiment (HiRISE) images of volcanic terrains from the first 6 months of the Mars Reconnaissance Orbiter Primary Science Phase. *Journal of Geophysical Research*, **113**, E04005, http://dx.doi.org/10.1029/2007JE002968

KESZTHELYI, L. P., JAEGER, W. L., DUNDAS, C. M., MARTÍNEZ-ALONSO, S., MCEWEN, A. S. & MILAZZO, M. P. 2010. Hydrovolcanic features on Mars: preliminary observations from the first Mars year of HiRISE imaging. *Icarus*, **205**, 211–229.

KIDDER, J. G. & PHILLIPS, R. J. 1996. Convection-driven subsolidus crustal thickening on Venus. *Journal of Geophysical Research*, **101**, 23 181–23 194.

KIEFER, W. S. & HAGER, B. H. 1991. A mantle plume model for the equatorial highlands of Venus. *Journal of Geophysical Research*, **96**, 20 947–20 966.

KIEFER, W. S. & MURRAY, B. C. 1987. The formation of Mercury's smooth plains. *Icarus*, **72**, 477–491.

KLIMCZAK, C. 2013. Igneous dikes on the Moon: evidence from Lunar Orbiter Laser Altimeter topography. Abstract 1391 presented at the 44th Lunar & Planetary Science Conference, March 18–22, 2013, The Woodlands, Texas.

KLIMCZAK, C., WATTERS, T. R. ET AL. 2012. Deformation associated with ghost craters and basins in volcanic smooth plains on Mercury: strain analysis and implications for plains evolution. *Journal of Geophysical Research*, **117**, E00L03, http://dx.doi.org/10.1029/2012JE004L03

KLIMCZAK, C., ERNST, C. M. ET AL. 2013. Insights into the subsurface structure of the Caloris basin, Mercury, from assessments of mechanical layering and changes in long-wavelength topography. *Journal of Geophysical Research*, **118**, 2030–2044, http://dx.doi.org/10.1002/jgre.20157

KNAPMEYER, M., OBERST, J., HAUBER, E., WÄHLISCH, M., DEUCHLER, C. & WAGNER, R. 2006. Working models for spatial distribution and level of Mars' seismicity. *Journal of Geophysical Research*, **111**, E11006.

KOENIG, E. & AYDIN, A. 1998. Evidence for large-scale strike-slip faulting on Venus. *Geology*, **26**, 551–554.

KOHLSTEDT, D. L. & MACKWELL, S. J. 2010. Strength and deformation of planetary lithospheres. *In*: WATTERS, T. R. & SCHULTZ, R. A. (eds) *Planetary Tectonics*. Cambridge University Press, Cambridge, 397–456.

KOMATSU, G., GULICK, V. C. & BAKER, V. R. 2001. Valley networks on Venus. *Geomorphology*, **37**, 225–240.

KOROTEV, R. L. 1996. On the relationship between the Apollo 16 ancient regolith breccias and feldspathic fragmental breccias, and the composition of the prebasin crust in the central highlands of the Moon. *Meteoritics & Planetary Science*, **31**, 403–412.

KOROTEV, R. L. 1997. Some things we can infer about the Moon from the composition of the Apollo 16 regolith. *Meteoritics & Planetary Science*, **32**, 447–478.

KOROTEV, R. L. 2000. A retrospective look at KREEP. *Meteoritics & Planetary Science*, **35**, A91.

KOROTEV, R. L., JOLIFF, B. L., ZEIGLER, R. A., GILLIS, J. J. & HASKIN, L. A. 2003. Feldspathic lunar meteorites and their implications for compositional remote sensing of the lunar surface and the composition of the lunar crust. *Geochimica et Cosmochimica Acta*, **67**, 4895–4923, http://dx.doi.org/10.1016/j.gca.2003.08.001

KRASSILNIKOV, A. S. & HEAD, J. W. 2003. Novae on Venus: geology, classification, and evolution. *Journal of Geophysical Research*, **108**, 5108, http://dx.doi.org/10.1029/2002JE001983

KRONBERG, P., HAUBER, E. ET AL. 2007. Acheron Fossae, Mars: tetonic rifting, volcanism, and implications for lithospheric thickness. *Journal of Geophysical Research*, **112**, E04005.

KUCINSKAS, A. B., TURCOTTE, D. L. & ARKANI-HAMED, J. 1996. Isostatic compensation of Ishtar Terra, Venus. *Journal of Geophysical Research*, **101**, 4725–4736.

KUMAR, P. S. 2005. An alternative kinematic interpretation of Thetis Boundary Shear Zone, Venus: evidence for strike-slip ductile duplexes. *Journal of Geophysical Research*, **110**, E07001, http://dx.doi.org/10.1029/2004JE002387

LANG, N. P. & HANSEN, V. L. 2006. Venusian channel formation as a subsurface process. *Journal of Geophysical Research*, **111**, E04001, http://dx.doi.org/10.1029/2005JE002629

LANG, N. P. & LÓPEZ, I. 2013. The magmatic evolution of three Venusian coronae. *In*: PLATZ, T., MASSIRONI, M., BYRNE, P. K. & HIESINGER, H. (eds) *Volcanism and Tectonism Across the Inner Solar System*. Geological Society, London, Special Publications, **401**. First published online December 16, 2013, http://dx.doi.org/10.1144/SP401.3

LANGSETH, M. G., KEIHM, S. J. & PETERS, K. 1976. Revised lunar heat-flow values. *Proceedings of Lunar and Planetary Science*, **3**, 3143–3171.

LANZ, J. K., WAGNER, R., WOLF, U., KRÖCHERT, J. & NEUKUM, G. 2010. Rift zone volcanism and associated cinder cone field in Utopia Planitia, Mars. *Journal of Geophysical Research*, **115**, E12019, http://dx.doi.org/10.1029/2010JE003578

LARSEN, K. W., ARVIDSON, R. E., JOLLIFF, B. L. & CLARK, B. C. 2000. Correspondence and least squares analyses of soil and rock compositions for the Viking Lander 1 and pathfinder landing sites. *Journal of Geophysical Research*, **105**, 29 207–29 221.

LAWRENCE, D. J., ELPHIC, R. C., FELDMAN, W. C., PRETTYMAN, T. H., GASNAULT, O. & MAURICE, S. 2003. Small-area thorium features on the lunar surface. *Journal of Geophysical Research*, **108**, 5102, http://dx.doi.org/10.1029/2003JE002050

LAWRENCE, D. J., HAWKE, B. R., HAGERTY, J. J., ELPHIC, R. C., FELDMAN, W. C., PRETTYMAN, T. H. & VANIMAN, D. T. 2005. Evidence for a high-Th, evolved lithology on the Moon at Hansteen Alpha. *Geophysical Research Letters*, **32**, L07201, http://dx.doi.org/10.1029/2004GL022022

LAWRENCE, S. J., STOPAR, J. D. *ET AL.* 2010. LROC observations of the Marius Hills. Abstract 1906 presented at the 41st Lunar & Planetary Science Conference, March 1–5, 2010, The Woodlands, Texas.

LEHMANN, T. R., PLATZ, T. & MICHAEL, G. G. 2012. Ages of lava flows in the Hesperia volcanic province, Mars. Abstract 2526 presented at the 43rd Lunar & Planetary Science Conference, March 19–23, 2012, The Woodlands, Texas.

LENARDIC, W., KAULA, W. M. & BINDSCHADLER, D. L. 1995. Some effects of a dry crustal flow law on numerical simulations of coupled crustal deformation and mantle convection on Venus. *Journal of Geophysical Research*, **100**, 16 949–16 957.

LEONE, G. 2014. A network of lava tubes as the origin of Labyrinthus Noctic and Valles Marineris on Mars. *Journal of Volcanology and Geothermal Research*, **277**, 1–8.

LEWIS, J. S. 2004. *Physics and Chemistry of the Solar System*. 2nd edn. Elsevier, Amsterdam.

LINGENFELTER, R. E., PEALE, S. J. & SCHUBERT, G. 1968. Lunar rivers. *Science*, **161**, 266–269.

LONGHI, J. 1980. Lunar crust, Achondrites. *Geotimes*, **25**, 19–20.

LONGHI, J. 2003. A new view of lunar ferroan anorthosites: postmagma ocean petrogenesis. *Journal of Geophysical Research*, **108**, 5083, http://dx.doi.org/10.1029/2002JE001941

LONGHI, J. & ASHWAL, L. D. 1985. Two-stage models for lunar and terrestrial anorthosites: petrogenesis without a magma ocean. *Journal of Geophysical Research*, **90**, C571–C584.

LOPES, F. C., CASELLI, A. T., MACHADO, A. & BARATA, M. T. 2014. The development of the Deception Island volcano caldera under control of the Bransfield Basin sinistral strike-slip tectonic regime (NW Antarctica). *In*: PLATZ, T., MASSIRONI, M., BYRNE, P. K. & HIESINGER, H. (eds) *Volcanism and Tectonism Across the Inner Solar System*. Geological Society, London, Special Publications, **401**. First published online February 6, 2014, http://dx.doi.org/10.1144/SP401.6

LOPES, R., GUEST, J. E., HILLER, K. & NEUKUM, G. 1982. Further evidence for a mass movement origin of the Olympus Mons aureole. *Journal of Geophysical Research*, **87**, 9917–9928.

LUCCHITTA, B. K. 1976. Analysis of scarps on the rim of mare Serenitatis. Abstract 507–508 presented at the 7th Lunar & Planetary Science Conference, Houston, TX.

LUCCHITTA, B. K., MCEWEN, A. S., CLOW, G. D., GEISSLER, P. E., SINGER, R. B., SCHULTZ, R. A. & SQUYRES, S. W. 1992. The canyon system on Mars. *In*: KIEFFER, H. H., JAKOSKY, B. M., SNYDER, C. W. & MATTHEWS, M. S. (eds) *Mars*. University of Arizona Press, Tucson, AZ, 453–492.

LUCCHITTA, B. K., ISBELL, N. K. & HOWINGTON-KRAUS, A. 1994. Topography of Valles Marineris: implications for erosional and structural history. *Journal of Geophysical Research*, **99**, 3783–3798.

LUCEY, P., KOROTEV, R. L. *ET AL.* 2006. Understanding the lunar surface and space–Moon interactions. *Reviews in Mineralogy & Geochemistry*, **60**, 83–219.

LUCEY, P. G., BLEWETT, D. T. & JOLLIFF, B. L. 2000. Lunar iron and titanium abundance algorithms based on final processing of Clementine ultraviolet-visible images. *Journal of Geophysical Research*, **105**, 20 297–20 306, http://dx.doi.org/10.1029/1999JE001117

MAGEE, K. P. & HEAD, J. W. 2001. Large flow fields on Venus: implications for plumes, rift associations, and resurfacing. *In*: ERNST, R. E. & BUCHAN, K. L. (eds) *Mantle Plumes: Their Identification Through Time*. Geological Society of America Special Papers, **352**, 81–101.

MALIN, M. 1974. Lunar red spots: possible pre-mare materials. *Earth and Planetary Science Letters*, **21**, 331–341.

MALIN, M. C. & EDGETT, K. S. 2000. Sedimentary rocks of early Mars. *Science*, **290**, 1927–1937.

MANCINELLI, P., MINELLI, F., MONDINI, A., PAUSELLI, C. & FEDERICO, C. 2014. A downscaling approach for geological characterization of the Raditladi basin of Mercury. *In*: PLATZ, T., MASSIRONI, M., BYRNE, P. K. & HIESINGER, H. (eds) *Volcanism and Tectonism Across the Inner Solar System*. Geological Society, London, Special Publications, **401**. First published online March 11, 2014, http://dx.doi.org/10.1144/SP401.10

MANDT, K. E., DE SILVA, S. L., ZIMBELMAN, J. R. & CROWN, D. A. 2008. Origin of the Medusae Fossae Formation, Mars: insights from a synoptic approach. *Journal of Geophysical Research*, **113**, E12011.

MANFREDI, L., PLATZ, T., CLARKE, A. B. & WILLIAMS, D. A. 2012. Volcanic history of the Tempe Volcanic Province. Paper presented at the American Geophysical Union, Fall Meeting 2012, Abstract V33B-2873.

MANGOLD, N., ALLEMAND, P. & THOMAS, P. G. 1998. Wrinkle ridges of Mars: structural analysis and evidence for shallow deformation controlled by ice-rich décollements. *Planetary and Space Science*, **46**, 345–356.

MARCHI, S., MASSIRONI, M., CREMONESE, G., MARTELLATO, E., GIACOMINI, L. & PROCKTER, L. M. 2011. The effects of the target material properties and layering on the crater chronology: the case of Raditladi and Rachmaninoff basins on Mercury. *Planetary and Space Science*, **59**, 1968–1980.

MARCHI, S., CHAPMAN, C. R., FASSETT, C. I., HEAD, J. W., BOTTKE, W. F. & STROM, R. G. 2013. Global resurfacing of Mercury 4.0–4.1 billion years ago by heavy bombardment and volcanism. *Nature*, **499**, 59–61.

MARINANGELI, L. & GILMORE, M. S. 2000. Geologic evolution of the Akna Montes-Atropos Tessera region, Venus. *Journal of Geophysical Research*, **105**, 12 053–12 075.

MARINOVA, M. M., AHARONSON, O. & ASPHAUG, E. 2008. Mega-impact formation of the Mars hemispheric dichotomy. *Nature*, **453**, 1216–1219.

MASSIRONI, M., DI ACHILLE, G. *ET AL.* 2014. Lateral ramps and strike-slip kinematics on Mercury. *In*: PLATZ, T., MASSIRONI, M., BYRNE, P. K. & HIESINGER, H. (eds) *Volcanism and Tectonism Across the Inner Solar System*. Geological Society, London, Special Publications, **401**. First published online June 5, 2014, http://dx.doi.org/10.1144/SP401.16

MAXWELL, T. A. 1978. Origin of multi-ring basin ridge systems: an upper limit to elastic deformation based on a finite-element model. *Proceedings of Lunar and Planetary Science*, **9**, 3541–3559.

MCCAULEY, J. F., CARR, M. H. *ET AL.* 1972. Preliminary Mariner-9 report on the geology of Mars. *Icarus*, **17**, 289–327.

MCGILL, G. E. 1993. Wrinkle ridges, stress domains, and kinematics of Venusian plains. *Geophysical Research Letters*, **20**, 2407–2410.

MCGILL, G. E. 2004. Tectonic and stratigraphic implications of the relative ages of Venusian plains and wrinkle ridges. *Icarus*, **172**, 603–612.

MCGILL, G. E., STOFAN, E. R. & SMREKAR, S. E. 2010. Venus tectonics. *In*: WATTERS, T. R. & SCHULTZ, R. A. (eds) *Planetary Tectonics*. Cambridge University Press, Cambridge, 81–120.

MCGOVERN, P. J. & MORGAN, J. K. 2009. Volcanic spreading and lateral variations in the structure of Olympus Mons, Mars. *Geology*, **37**, 139–142.

MCGOVERN, P. J. & SOLOMON, S. C. 1993. State of stress, faulting and eruption characteristics of large volcanoes on Mars. *Journal of Geophysical Research*, **98**, 23 553–23 579.

MCGOVERN, P. J., SMITH, J. R., MORGAN, J. K. & BULMER, M. H. 2004. The Olympus Mons aureole deposits: new evidence for a flank-failure origin. *Journal of Geophysical Research*, **109**, E08008, http://dx.doi.org/10.1029/2004JE002258

MCGOVERN, P. J., GROSFILS, E. B., GALGANA, G. A., MORGAN, J. K., RUMPF, M. E., SMITH, J. R. & ZIMBELMAN, J. R. 2014. Lithospheric flexure and volcano basal boundary conditions: keys to the structural evolution of large volcanic edifices on the terrestrial planets. *In*: PLATZ, T., MASSIRONI, M., BYRNE, P. K. & HIESINGER, H. (eds) *Volcanism and Tectonism Across the Inner Solar System*. Geological Society, London, Special Publications, **401**. First published online March 17, 2014, http://dx.doi.org/10.1144/SP401.7

MCKENZIE, D. & NIMMO, F. 1999. The generation of Martian floods by the melting of ground ice above dikes. *Nature*, **397**, 231–233.

MCKENZIE, D. P. 1967. Some remarks on heat flow and gravity anomalies. *Journal of Geophysical Research*, **72**, 6261–6273.

MCKINNON, W. B., ZAHNLE, K. J., IVANOV, B. A. & MELOSH, H. J. 1997. Cratering on Venus: models and observations. *In*: GOUGHER, S. W., HUNTEN, D. M. & PHILLIPS, R. J. (eds) *Venus II: Geology, Geophysics, Atmosphere, and Solar Wind Environment*. University of Arizona Press, Tucson, AZ, 969–1014.

MCSWEEN, H. Y. 1994. What we have learned about Mars from SNC meteorites. *Meteoritics*, **29**, 757–779.

MCSWEEN, H. Y. 2002. The rocks of Mars, from far and near. *Meteoritics and Planetary Science*, **37**, 7–25.

MCSWEEN, H. Y., ARVIDSON, R. E. *ET AL.* 2004. Basaltic rocks analyzed by the Spirit rover in Gusev Crater. *Science*, **305**, 842–845.

MÈGE, D. & MASSON, P. 1996. A plume tectonics model for the Tharsis province, Mars. *Planetary and Space Sciences*, **44**, 1499–1546.

MELENDREZ, D. E., JOHNSON, J. R., LARSON, S. M. & SINGER, R. B. 1994. Remote sensing of potential lunar resources. 2: high spatial resolution mapping of spectral reflectance ratios and implications for nearside mare TiO_2 content. *Journal of Geophysical Research*, **99**, 5601–5619, http://dx.doi.org/10.1029/93JE03430

MELOSH, H. J. 1978. The tectonics of mascon loading. *Proceedings of Lunar and Planetary Science*, **9**, 3513–3525.

MELOSH, H. J. & DZURISIN, D. 1978. Mercurian global tectonics: a consequence of tidal despinning? *Icarus*, **35**, 227–236.

MELOSH, H. J. & MCKINNON, W. B. 1988. The tectonics of Mercury. *In*: VILAS, F., CHAPMAN, C. R. & MATTHEWS, M. S. (eds) *Mercury*. University of Arizona Press, Tucson, AZ, 374–400.

MEYZEN, C. M., MASSIRONI, M., POZZOBON, R. & DAL ZILIO, L. 2014. Are terrestrial plumes from motionless plates analogues to Martian plumes feeding the giant shield volcanoes? *In*: PLATZ, T., MASSIRONI, M., BYRNE, P. K. & HIESINGER, H. (eds) *Volcanism and Tectonism Across the Inner Solar System*. Geological Society, London, Special Publications, **401**. First published online February 21, 2014, http://dx.doi.org/10.1144/SP401.8

MICHALSKI, J. R. & BLEACHER, J. E. 2013. Supervolcanoes within an ancient volcanic province in Arabia Terra, Mars. *Nature*, **502**, 47–52, http://dx.doi.org/10.1038/nature12482

MONTÉSI, L. G. & ZUBER, M. T. 2003. Clues to the lithospheric structure of Mars from wrinkle ridge sets and localization instability. *Journal of Geophysical Research*, **108**, 5048, http://dx.doi.org/10.1029/2002JE001974

MONTGOMERY, D. R. & GILLESPIE, A. 2005. Formation of Martian outflow channels by catastrophic dewatering of evaporite deposits. *Geology*, **33**, 625–628.

MORESI, L. N. & SOLOMATOV, V. S. 1998. Mantle convection with a brittle lithosphere: thoughts on the global tectonic style of the Earth and Venus. *Geophysical Journal*, **133**, 669–682.

MORGAN, J. K. & MCGOVERN, P. J. 2005. Discrete element simulations of gravitational volcanic deformation: 1. Deformation structures and geometries. *Journal of Geophysical Research*, **110**, B05402.

MORGAN, J. W. 1968. Rises trenches, great faults and crustal blocks 1968. *Journal of Geophysical Research*, **73**, 1959–1982.

MORGAN, P. & PHILLIPS, R. J. 1983. Hot spot heat transfer: its application to Venus and implications to Venus and Earth. *Journal of Geophysical Research*, **88**, 8305–8317.

MOROTA, T., HARUYAMA, J. *ET AL.* 2011. Timing and characteristics of the latest mare eruption on the Moon. *Earth and Planetary Science Letters*, **302**, 255–266.

MOROTA, T., ISHIHARA, Y. *ET AL.* 2014. Lunar mare volcanism: lateral heterogeneities in volcanic activity and relationship with crustal structure. *In*: PLATZ, T., MASSIRONI, M., BYRNE, P. K. & HIESINGER, H. (eds) *Volcanism and Tectonism Across the Inner Solar System*. Geological Society, London, Special Publications, **401**. First published online March 20, 2014, http://dx.doi.org/10.1144/SP401.11

MOUGINIS-MARK, P. J. & CHRISTENSEN, P. R. 2005. New observations of volcanic features on Mars from THEMIS instrument. *Journal of Geophysical Research*, **110**, E08007, http://dx.doi.org/10.1029/2005JE002421

MOUGINIS-MARK, P. J. & ROWLAND, S. K. 2001. The geomorphology of planetary calderas. *Geomorphology*, **37**, 201–223.

MUEHLBERGER, W. R. 1974. Structural history of southeastern Mare Serenitatis and adjacent highlands. *Proceedings of Lunar and Planetary Science*, **5**, 101–110.

MUELLER, K. & GOLOMBEK, M. 2004. Compressional structures on Mars. *Annual Review of Earth and Planetary Science*, **32**, 435–464.

MUELLER, N., HELBERT, J., HASHIMOTO, G. L., TSANG, C. C. C., ERARD, S., PICCIONI, G. & DROSSART, P. 2008. Venus surface thermal emission at 1 μm in VIRTIS imaging observations: evidence for variation of crust and mantle differentiation conditions. *Journal of Geophysical Research*, **113**, E00B17.

MURCHIE, S. L., WATTERS, T. R. *ET AL.* 2008. Geology of the Caloris Basin, Mercury: a View from MESSENGER. *Science*, **321**, 73–76.

MURRAY, B. C., STROM, R. G., TRASK, N. J. & GAULT, D. E. 1975. Surface history of Mercury: implications for terrestrial planets. *Journal of Geophysical Research*, **80**, 2508–2514.

MUSTARD, J. F. & HEAD, J. W. 1996. Buried stratigraphic relationships along the southwestern shores of Oceanus Procellarum: implications for early lunar volcanism. *Journal of Geophysical Research*, **101**, 18 913–18 926.

NAHM, A. L. & SCHULTZ, R. A. 2013. Rupes Recta and the geological history of the Mare Nubium region of the Moon: insights from forward mechanical modelling of the 'Straight Wall'. *In*: PLATZ, T., MASSIRONI, M., BYRNE, P. K. & HIESINGER, H. (eds) *Volcanism and Tectonism Across the Inner Solar System*. Geological Society, London, Special Publications, **401**. First published online December 16, 2013, http://dx.doi.org/10.1144/SP401.4

NAKAMURA, Y., LAMMLEIN, D., LATHAM, G., EWING, M., DORMAN, J., PRESS, F. & TOKSÖZ, M. N. 1973. New seismic data on the state of the deep lunar interior. *Science*, **181**, 49–51.

NEAL, C. R. & TAYLOR, L. A. 1992. *Using Apollo 17 High-Ti Mare Basalts as Windows to the Lunar Mantle. Workshop on Geology of the Apollo 17 Landing Site*. Lunar Science Institute, Houston, TX, 40–44.

NEUKUM, G. & HILLER, K. 1981. Martian ages. *Journal of Geophysical Research*, **86**, 3097–3121.

NEUKUM, G. & IVANOV, B. A. 1994. Crater size distributions and impact probabilities on Earth from lunar, terrestrial-planet, and asteroid cratering data. *In*: GEHRELS, T. (ed.) *Hazard Due to Comets and Asteroids*. University of Arizona Press, Tucson, AZ, 359–416.

NEUKUM, G., KÖNIG, B. & ARKANI-HAMED, J. 1975*a*. A study of lunar impact crater size distributions. *The Moon*, **12**, 201–229.

NEUKUM, G., KÖNIG, B., FECHTIG, H. & STORZER, D. 1975*b*. Cratering in the Earth–Moon system. Consequences for age determination by crater counting. *Proceedings of Lunar and Planetary Science*, **6**, 2597–2620.

NEUKUM, G., IVANOV, B. A. & HARTMANN, W. K. 2001. Cratering records in the inner solar system in relation to the lunar reference system. *Space Science Reviews*, **96**, 55–86.

NEUKUM, G., JAUMANN, R. *ET AL.* 2004. Recent and episodic volcanic and glacial activity on Mars revealed by the High Resolution Stereo Camera. *Nature*, **432**, 971–979.

NEUKUM, G., BASILEVSKY, A. T. *ET AL.* 2010. The geologic evolution ofMars: episodicity of resurfacing events and ages from cratering analysis of image data and correlation with radiometric ages of Martian meteorites. *Earth and Planetary Science Letters*, **294**, 204–220.

NIKOLAEVA, O., IVANOV, M. & BOROZDIN, V. 1992. Evidence on the crustal dichotomy of Venus. *In*: *Venus Geology, Geochemistry, and Geophysics – Research Results from the USSR*. University of Arizona Press, Tucson, AZ.

NIMMO, F. & MCKENZIE, D. 1998. Volcanism and tectonics on Venus. *Annual Review of Earth and Planetary Sciences*, **26**, 23–51.

NITTLER, L. R., STARR, R. D. *ET AL.* 2011. The major-element composition of Mercury's surface from MESSENGER X-Ray Spectrometer. *Science*, **333**, 1847–1850.

OBERBECK, V. R., QUAIDE, W. L., ARVIDSON, R. E. & AGGARWAL, H. R. 1977. Comparative studies of lunar, Martian, and Mercurian craters and plains. *Journal of Geophysical Research*, **82**, 1681–1698.

OHTAKE, M., MATSUNAGA, T. *ET AL.* 2009. The global distribution of pure anorthosite on the Moon. *Nature*, **461**, 236–240, http://dx.doi.org/10.1038/nature08317

ÖPIK, E. 1960. The lunar surface as an impact counter. *Monthly Notices of the Royal Astronomical Society*, **120**, 404–411.

O'ROURKE, J. G. & KORENAGA, J. 2012. Terrestrial planet evolution in the stagnant-lid regime: size effects and the formation of self-destabilizing crust. *Icarus*, **221**, 1043–1060.

PAGE, D. P. 2007. Recent low-latitude freeze–thaw on Mars. *Icarus*, **189**, 83–117.

PAPIKE, J. J. 1998. Comparative planetary mineralogy: chemistry of melt-derived pyroxene, feldspar, and olivine. Abstract 1008 presented at the 27th Lunar & Planetary Science Conference, Houston, TX.

PAPIKE, J. J. & VANIMAN, D. T. 1978. The lunar mare basalt suite. *Geophysical Research Letters*, **5**, 433–436.

PAPIKE, J. J., HODGES, F. N., BENCE, A. E., CAMERON, M. & RHODES, J. M. 1976. Mare basalts – Crystal chemistry, mineralogy, and petrology. *Reviews of Geophysics and Space Physics*, **14**, 475–540.

PARMENTIER, E. M. & HESS, P. C. 1992. Chemical differentiation of a convecting planetary interior: consequences for a one-plate planet. *Geophysical Research Letters*, **19**, 2015–2018.

PASCKERT, J. H., HIESINGER, H. & REISS, D. 2012. Rheologies and ages of lava flows on Elysium Mons, Mars. *Icarus*, **219**, 443–457.

PASCKERT, J. H., HIESINGER, H. & VAN DER BOGERT, C. H. 2014. Lunar mare basalts in- and outside of the South Pole-Aitken basin. Abstract 1968 presented at the 45th Lunar & Planetary Science Conference, March 17–21, 2014, The Woodlands, Texas.

PAVRI, B., HEAD, J. W., KLOSE, K. B. & WILSON, L. 1992. Steep-sided domes on Venus: characteristics, geologic settings, and eruption conditions from Magellan data. *Journal of Geophysical Research*, **97**, 13 445–13 478.

PEALE, S., SCHUBERT, G. & LINGENFELTER, R. E. 1968. Distribution of sinuous rilles and water on the Moon. *Nature*, **220**, 1222–1225.

PHILLIPS, R. J. & HANSEN, V. L. 1998. Geological evolution of Venus: rises, plains, plumes, and plateaus. *Science*, **279**, 1492–1497.

PHILLIPS, R. J. & MALIN, M. C. 1983. The interior of Venus and tectonic implications. *In*: HUNTEN, D. M., COLIN, L., DONAHUE, T. M. & MOROZ, V. I. (eds) *Venus*. University of Arizona Press, Tucson, AZ, 159–214.

PHILLIPS, R. J., GRIMM, R. E. & MALIN, M. C. 1991. Hot-spot evolution and the global tectonics of Venus. *Science*, **252**, 651–658.

PHILLIPS, R. J., RAUBERTAS, R. F., ARVIDSON, R. E., SARKAR, I. C., HERRICK, R. R., IZENBERG, N. & GRIMM, R. E. 1992. Impact craters and Venus resurfacing history. *Journal of Geophysical Research*, **97**, 15 923–15 948.

PHILLIPS, R. J., ZUBER, M. T. ET AL. 2001. Ancient geodynamics and global-scale hydrology on Mars. *Science*, **291**, 2587–2591.

PIETERS, C. M. 1978. Mare basalt types on the front side of the moon – a summary of spectral reflectance data. *Proceedings of Lunar and Planetary Science*, **9**, 2825–2849.

PIETERS, C. M., HEAD, J. W., WHITFORD-STARK, J. L., ADAMS, J. B., McCORD, T. B. & ZISK, S. H. 1980. Late high-titanium basalts of the western maria – Geology of the Flamsteed region of Oceanus Procellarum. *Journal of Geophysical Research*, **85**, 3913–3938, http://dx.doi.org/10.1029/JB085iB07 p03913

PLATZ, T. & MICHAEL, G. G. 2011. Eruption history of the Elysium Volcanic Province, Mars. *Earth and Planetary Science Letters*, **312**, 140–151.

PLATZ, T., KNEISSL, T., HAUBER, E., LE DEIT, L., MICHAEL, G. G. & NEUKUM, G. 2010. Total volume estimates of volcanic material and outgassing in the Elysium Volcanic Region. Abstract 2476 presented at the 41st Lunar & Planetary Science Conference, March 1–5, 2010, The Woodlands, Texas.

PLATZ, T., MÜNN, S., WALTER, T. R., McGUIRE, P. C., DUMKE, A. & NEUKUM, G. 2011. Vertical and lateral collapse of Tharsis Tholus, Mars. *Earth and Planetary Science Letters*, **305**, 445–455, http://dx.doi.org/10.1016/j.epsl.2011.03.012

PLATZ, T., JODLOWSKI, P., FAWDON, P., MICHAEL, G. G. & TANAKA, K. L. 2014. Amazonian volcanic activity at the Syrtis Major volcanic province, Mars. Abstract 2524 presented at the 45th Lunar & Planetary Science Conference, March 17–21, 2014, The Woodlands, Texas.

PLESCIA, J. B. 1993. Wrinkle ridges of Arcadia Planitia, Mars. *Journal of Geophysical Research*, **98**, 15 049–15 059.

PLESCIA, J. B. 1994. Geology of the small Tharsis volcanoes: Jovis Tholus, Ulysses Patera, Biblis Patera, Mars. *Icarus*, **111**, 246–269.

PLESCIA, J. 2003. Cerberus fossae, Elysium, Mars: a source for lava and water. *Icarus*, **164**, 79–95.

PLESCIA, J. B. 2004. Morphometric properties of Martian volcanoes. *Journal of Geophysical Research*, **109**, E03003.

PLESCIA, J. B. & GOLOMBEK, M. P. 1986. Origin of planetary wrinkle ridges based on the study of terrestrial analogs. *Geological Society of America Bulletin*, **97**, 1289–1299.

PLESCIA, J. B. & SAUNDERS, R. S. 1982. Tectonic history of the Tharsis Region, Mars. *Journal of Geophysical Research*, **87**, 9775–9791.

POULET, F., GOMEZ, C. ET AL. 2007. Martian surface mineralogy from Observatoire pour la Minéralogie, l'Eau, les Glaces et l'Activité on board the Mars Express spacecraft (OMEGA/MEX): global mineral maps. *Journal of Geophysical Research*, **112**, E08S02, http://dx.doi.org/10.1029/2006JE002840

POZZOBON, R., MAZZARINI, F., MASSIRONI, M. & MARINANGELI, L. 2014. Self-similar clustering distribution of structural features on Ascraeus Mons (Mars): implications for magma chamber depth. *In*: PLATZ, T., MASSIRONI, M., BYRNE, P. K. & HIESINGER, H. (eds) *Volcanism and Tectonism Across the Inner Solar System*. Geological Society, London, Special Publications, **401**. First published online March 25, 2014, http://dx.doi.org/10.1144/SP401.12

PRICE, M. & SUPPE, J. 1995. Constraints on the resurfacing history of Venus from the hypsometry and distribution of volcanism, tectonism and impact craters. *Earth, Moon, and Planets*, **71**, 99–145.

PRITCHARD, M. E. & STEVENSON, D. J. 2000. The thermochemical history of the Moon: constraints and major questions. Abstract 1878 presented at the 31st Lunar & Planetary Science Conference, March 13–17, 2000, The Woodlands, Texas.

PROCKTER, L. M., ERNST, C. M. ET AL. 2010. Evidence for young volcanism on Mercury from the third MESSENGER flyby. *Science*, **329**, 668–671.

RAITALA, J. 1994. Main fault tectonics of Meshkenet Tessera on Venus. *Earth, Moon and Planets*, **65**, 55–70.

RANALLI, G. 1995. *Rheology of the Earth*. Chapman & Hall, London.

RANALLI, G. 1997. Rheology of the lithosphere in space and time. *In*: BURG, J.-P. & FORD, M. (eds) *Orogeny Through Time*. Geological Society, London, Special Publications, **121**, 19–37.

REESE, C. C., SOLOMATOV, V. S. & MORESI, L. N. 1999. Non-Newtonian stagnant lid convection and magmatic resurfacing on Venus. *Icarus*, **80**, 67–80.

RICHARDSON, J. A., BLEACHER, J. E. & GLAZE, L. S. 2013. The volcanic history of Syria Planum, Mars.

Journal of Volcanology and Geothermal Research,
252, 1–13.

ROBBINS, S. J., DI ACHILLE, G. & HYNEK, B. M. 2011. The
volcanic history of Mars: high-resolution crater-based
studies of the calderas of 20 volcanoes. *Icarus,* **211**,
1179–1203.

ROBERTS, J. H. & ZHONG, S. 2006. Degree-1 convection in
the Martian mantle and the origin of the hemispheric
dichotomy. *Journal of Geophysical Research,* **111**,
E06013.

ROBINSON, M. S. & LUCEY, P. G. 1997. Recalibrated
Mariner 10 color mosaics: implications for Mercurian
volcanism. *Science,* **275**, 197–200.

ROBINSON, M. S., MOUGINIS-MARK, P. J., ZIMBELMAN, J.
R., WU, S. S. C., ABLIN, K. K. & HOWINGTON-KRAUS,
A. E. 1993. Chronology, eruption duration, and atmos-
pheric contribution of the Martian volcano Apollinaris
Patera. *Icarus,* **104**, 301–323.

RODRIGUEZ, J. A. P., KARGEL, J. *ET AL.* 2006. Head-ward
growth of chasmata by volatile outbursts, collapse, and
drainage: evidence from Ganges Chaos, Mars. *Geophysical Research Letters,* **33**, L18203.

ROMEO, I. & TURCOTTE, D. L. 2008. Pulsating continents
on Venus: an explanation for crustal plateaus and
tessera terrains. *Earth and Planetary Science Letters,*
276, 85–97.

ROMEO, I., CAPOTE, R. & ANGUITA, F. 2005. Tectonic and
kinematic study of a strike–slip zone along the
southern margin of Central Ovda Regio, Venus: geodynamical implications for crustal plateaux formation
and evolution. *Icarus,* **175**, 320–334.

ROSENBERG, E. & MCGILL, G. E. 2001. *Geologic Map of
the Pandrosos Dorsa Quadrangle (V-5), Venus.*
United States Geological Survey, Miscellaneous Geologic Investigations Series, **I-2721**.

ROTHERY, D. A., THOMAS, R. J. & KERBER, L. 2014. Prolonged eruptive history of a compound volcano on
Mercury: volcanic and tectonic implications. *Earth
and Planetary Science Letters,* **385**, 59–67.

RYAN, M. R. 1987. Elasticity and contractancy of Hawaiian olivine tholeiite and its role in the stability and
structural evolution of subcaldera magma reservoirs
and rift systems. *In:* DECKER, I. R. W., WRIGHT,
T. L. & STAUFFER, P. H. (eds) *Volcanism in Hawaii.*
United States Geological Survey, Professional
Papers, **1350**, 1395–1447.

SANDWELL, D. T., JOHNSON, C. L., BILOTTI, F. & SUPPE, J.
1997. Driving forces for limited tectonics on Venus.
Icarus, **129**, 232–244.

SAUTTER, V., FABRE, C. *ET AL.* 2014. Igneous mineralogy
at Bradbury Rise: the first ChemCam campaign at
Gale crater. *Journal of Geophysical Research,* **119**,
30–46.

SCHABER, G. 1982. Limited extension and volcanism along
zones of lithospheric weakness. *Geophysical Research
Letters,* **9**, 499–502.

SCHABER, G. G. 1973. Lava flows in Mare Imbrium:
geologic evaluation from Apollo orbital photography.
Proceedings of Lunar and Planetary Science, **4**,
73–92.

SCHABER, G. G., BOYCE, J. M. & MOORE, H. J. 1976. The
scarcity of mappable flow lobes on the lunar maria:
unique morphology of the Imbrium flows. *Proceedings
of Lunar and Planetary Science,* **7**, 2783–2800.

SCHABER, G. G., STROM, R. G. *ET AL.* 1992. Geology and
distribution of impact craters on Venus: what are
they telling us? *Journal of Geophysical Research,* **97**,
13 257–13 301.

SCHUBERT, G., LINGENFELTER, R. E. & PEALE, S. J. 1970.
The morphology, distribution and origin of lunar
sinuous rilles. *Reviews of Geophysics and Space
Physics,* **8**, 199–224.

SCHUBERT, G., ROSS, M. N., STEVENSON, D. J. & SPOHN,
T. 1988. Mercury's thermal history and the generation
of its magnetic field. *In:* VILAS, F., CHAPMAN, C. R. &
MATTHEWS, M. S. (eds) *Mercury.* University of
Arizona Press, Tucson, AZ, 429–460.

SCHUBERT, G., SOLOMATOV, V. S., TACKLEY, P. J. &
TURCOTTE, D. L. 1997. Mantle convection and the
thermal evolution of Venus. *In:* BROUGHER, S. W.,
HUNTEN, D. M. & PHILLIPS, R. J. (eds) *Venus II:
Geology, Geophysics, Atmosphere, and Solar Wind
Environment.* University of Arizona Press, Tucson,
AZ, 1245–1287.

SCHULTZ, P. H. 1972. *A preliminary morphologic study
of the lunar surface.* PhD thesis, University of Texas
at Austin, Austin, TX.

SCHULTZ, P. H. 1976. Floor-fractured lunar craters. *The
Moon,* **15**, 241–273.

SCHULTZ, P. H. & GAULT, D. E. 1975. Seismic effects
from major basin formations on the Moon and
Mercury. *The Moon,* **12**, 159–177.

SCHULTZ, P. H. & SPUDIS, P. D. 1979. Evidence for
ancient mare volcanism. *Proceedings of Lunar and
Planetary Science,* **10**, 2899–2918.

SCHULTZ, P. H. & SPUDIS, P. D. 1983. Beginning and end
of lunar mare volcanism. *Nature,* **302**, 233–236.

SCHULTZ, R. A. 1998. Multiple-process origin of Valles
Ma-rineris basins and troughs, Mars. *Planetary and
Space Sciences,* **46**, 827–834.

SCHULTZ, R. A. 2000. Localization of bedding plane slip
and backthrust faults above blind thrust faults: keys
to wrinkle ridge structure. *Journal of Geophysical
Research,* **105**, 12 035–12 052.

SCHULTZ, R. A. & LIN, J. 2001. Three-dimensional normal
faulting models of the Valles Marineris, Mars, and geodynamic implications. *Journal of Geophysical
Research,* **106**, 16 549–16 566.

SCHULTZ, R. A. & TANAKA, K. L. 1994. Lithospheric-scale
buckling and thrust structures on Mars: the Coprates
rise and south Tharsis ridge belt. *Journal of Geophysical Research,* **99**, 8371–8385.

SCHULTZ, R. A. & ZUBER, M. T. 1994. Observations,
models, and mechanisms of failure of surface rocks
surrounding planetary surface loads. *Journal of Geophysical Research,* **99**, 14691–14702, http://dx.doi.
org/10.1029/94JE01140

SCHULTZ, R. A., OKUBO, C. H., GOUDY, C. L. & WILKINS,
S. J. 2004. Igneous dikes on Mars revealed by Mars
Orbiter Laser Altimeter topography. *Geology,* **32**,
889–892.

SCHULTZ, R. A., MOORE, J. M., GROSFILS, E. B., TANAKA,
K. L. & MÈGE, D. 2007. The Canyonlands model for
planetary grabens: revised physical basis and implications. *In:* CHAPMAN, M. G. (ed.) *The Geology of Mars.*
Cambridge University Press, Cambridge, 371–399.

SCOPPOLA, B., BOCCALETTI, D., BEVIS, M., CARMINATI, E.
& DOGLIONI, C. 2006. The westward drift of the

lithosphere: a rotational drag? *Geological Society of America Bulletin*, **118**, 199–209.

SCOTT, D. H. & TANAKA, K. L. 1982. Ignimbrites of Amazonis Planitia region of Mars. *Journal of Geophysical Research*, **87**, 1179–1190.

SENSKE, D. A., SCHABER, G. G. & STOFAN, E. R. 1992. Regional topographic rises on Venus: geology of western Eistla Regio and comparisons to Beta Regio and Atla Regio. *Journal of Geophysical Research*, **97**, 13 395–13 420.

SHARP, R. P. 1973. Mars: troughed terrain. *Journal of Geophysical Research*, **78**, 4063–4072.

SHEARER, C. K. & NEWSOM, H. E. 2000. W-Hf abundances and the early origin and evolution of the Earth–Moon system. *Geochimica et Cosmochimica Acta*, **64**, 3599–3613.

SHEARER, C. K., HESS, P. C. ET AL. 2006. Thermal and magmatic evolution of the Moon. *Reviews in Mineralogy & Geochemistry*, **60**, 365–518.

SHELLNUTT, J. G. 2013. Petrological modeling of basaltic rocks from Venus: a case for the presence of silicic rocks. *Journal of Geophysical Research*, **118**, 1350–1364.

SHKURATOV, Y. G., KAYDASH, V. G. & OPANASENKO, N. V. 1999. Iron and titanium abundance and maturity degree distribution on the lunar nearside. *Icarus*, **137**, 222–234, http://dx.doi.org/10.1006/icar.1999.6046

SHOEMAKER, E. M. & HACKMAN, R. 1962. Stratigraphic basis for a lunar time scale. *In*: *The Moon. Proceedings of Symposium No. 14 of the International Astronomical Union held at Pulkovo Observatory, Leningrad, December 1960*. Academic Press, San Diego, CA, 289–300.

SILVESTRO, S., FENTON, L. K., VAZ, D. A., BRIDGES, N. T. & ORI, G. G. 2010. Ripple migration and dune activity on Mars: evidence for dynamic wind processes. *Geophysical Research Letters*, **37**, L20203.

SIMONS, M., SOLOMON, S. C. & HAGER, B. H. 1997. Localization of gravity and topography: constraints on the tectonics and mantle dynamics of Venus. *Geophysical Journal International*, **131**, 24–44.

SKINNER, J. A. & TANAKA, K. L. 2007. Evidence for and implications of sedimentary diapirism and mud volcanism in the southern Utopia highland–lowland boundary plain, Mars. *Icarus*, **186**, 41–59, http://dx.doi.org/10.1016/j.icarus.2006.08.013

SKOK, J. R., MUSTARD, J. F., EHLMANN, B. L., MILLIKEN, R. E. & MURCHIE, S. L. 2010. Silica deposits in the Nili Patera caldera on the Syrtis Major volcanic complex on Mars. *Nature Geoscience*, **3**, 838–841.

SLEEP, N. H. 1994. Martian plate tectonics. *Journal of Geophysical Research*, **99**, 5639–5655.

SMITH, J. V., ANDERSON, A. T. ET AL. 1970. Petrologic history of the Moon inferred from petrography, mineralogy, and petrogenesis of Apollo 11 rocks. *In*: LEVINSON, A. A. (ed.) *Proceedings of the Apollo 11 Lunar Science Conference, Volume* 1. Pergamon Press, New York, 897–925.

SMREKAR, S. E. 1994. Evidence for active hotspots on Venus from analysis of Magellan gravity data. *Icarus*, **112**, 2–26.

SMREKAR, S. E. & PARMENTIER, E. M. 1996. Interactions of mantle plumes with thermal and chemical boundary

layers: application to hotspots on Venus. *Journal of Geophysical Research*, **101**, 5397–5410.

SMREKAR, S. E. & PHILLIPS, R. J. 1991. Venusian highlands – geoid to topography ratios and their implications. *Earth and Planetary Science Letters*, **107**, 582–597.

SMREKAR, S. E. & SOTIN, C. 2012. Constraints on mantle plumes on Venus: implications for volatile history. *Icarus*, **217**, 510–523.

SMREKAR, S. E. & STOFAN, E. R. 1999. Origin of corona-dominated topographic rises on Venus. *Icarus*, **139**, 100–115.

SMREKAR, S. E., STOFAN, E. R. & KIEFER, W. S. 1997. Large volcanic rises on Venus. *In*: BROUGHER, S. W., HUNTEN, D. M. & PHILLIPS, R. J. (eds) *Venus II: Geology, Geophysics, Atmosphere, and Solar Wind Environment*. University of Arizona Press, Tucson, AZ, 845–878.

SMREKAR, S. E., MOREELS, P. & FRANKLIN, J. B. 2002. Characterization and formation of polygonal fractures on Venus. *Journal of Geophysical Research*, **107**, 5098.

SMREKAR, S. E., ELKINS–TANTON, L. ET AL. 2007. Tectonic and thermal evolution of Venus and the role of volatiles: implications for understanding the terrestrial planets. *In*: ESPOSITO, L. W., STOFAN, E. R. & CRAVENS, T. E. (eds) *Exploring Venus as a Terrestrial Planet*. American Geophysical Union, Geophysical Monograph Series, **176**, 45–71.

SODERBLOM, L. A., BOYCE, J. M. & ARNOLD, J. R. 1977. Regional variations in the lunar maria – Age, remanent magnetism, and chemistry. *Proceedings of Lunar and Planetary Science*, **8**, 1191–1199.

SOLOMATOV, V. S. & MORESI, L. N. 2000. Scaling of time-dependent stagnant lid convection: application to small-scale convection on Earth and other terrestrial planets. *Journal of Geophysical Research*, **105**, 21 795–21 817.

SOLOMON, S. C. 1977. The relationship between crustal tectonics and internal evolution in the Moon and Mercury. *Physics of the Earth and Planetary Interiors*, **15**, 135–145.

SOLOMON, S. C. 1978. The nature of isostasy on the Moon: how big of a Pratt-fall for Airy methods. *Proceedings of Lunar and Planetary Science*, **9**, 3499–3511.

SOLOMON, S. C. 1993. The geophysics of Venus. *Physics Today*, **46**, 48–55.

SOLOMON, S. C. & CHAIKEN, J. 1976. Thermal expansion and thermal stress in the moon and terrestrial planets – Clues to early thermal history. *Proceedings of Lunar and Planetary Science*, **7**, 3229–3243.

SOLOMON, S. C. & HEAD, J. W. 1979. Vertical movement in mare basins: relation to mare emplacement, basin tectonics, and lunar thermal history. *Journal of Geophysical Research*, **84**, 1667–1682.

SOLOMON, S. C. & HEAD, J. W. 1980. Lunar mascon basins: lava filling, tectonics, and evolution of the lithosphere. *Reviews of Geophysics and Space Physics*, **18**, 107–141.

SOLOMON, S. C. & HEAD, J. W. 1982. Mechanisms for lithospheric heat transport on Venus: implications for tectonic style and volcanism. *Journal of Geophysical Research*, **87**, 9236–9246.

SOLOMON, S. C. & LONGHI, J. 1977. Magma oceanography. I-Thermal evolution. *Proceedings of Lunar and Planetary Science*, **8**, 583–599.

SOLOMON, S. C., HEAD, J. W. *ET AL.* 1991. Venus tectonics: initial analysis from Magellan. *Science*, **252**, 297–312.

SOLOMON, S. C., SMREKAR, S. E. *ET AL.* 1992. Venus tectonics: an overview of Magellan observations. *Journal of Geophysical Research*, **97**, 13 199–13 255.

SPENCER, J. R. & FANALE, F. P. 1990. New models for the origin of Valles Marineris closed depressions. *Journal of Geophysical Research*, **95**, 14 301–14 313.

SPOHN, T., KONRAD, W., BREUER, D. & ZIETHE, R. 2001. The longevity of lunar volcanism. implications of thermal evolution calculations with 2D and 3D mantle convection models. *Icarus*, **149**, 54–65.

SPUDIS, P. D., MCGOVERN, P. J. & KIEFER, W. S. 2013. Large shield volcanoes on the Moon. *Journal of Geophysical Research*, **118**, 1–19, http://dx.doi.org/10.1002/jgre.20059

SQUYRES, S. W., AHARONSON, O. *ET AL.* 2007. Pyroclastic activity at Home Plate in Gusev crater, Mars. *Science*, **316**, 738–742.

SQUYRES, S. W., JANES, D. M., BINDSCHADLER, L., SHARPTON, V. L. & STOFAN, R. 1992. The morphology and evolution of coronae on Venus. *Journal of Geophysical Research*, **97**, 13 611–13 634.

SQUYRES, S. W., ARVIDSON, R. E. *ET AL.* 2006. Rocks of Columbia Hills. *Journal of Geophysical Research*, **111**, E02S11.

STEWART, E. M. & HEAD, J. W. 2001. Ancient Martian volcanoes in the Aeolis region: new evidence from MOLA data. *Journal of Geophysical Research*, **106**, 17 505–17 513.

STOCKSTILL-CAHILL, K. R., MCCOY, T. J., NITTLER, L. R., WEIDER, S. Z. & HAUCK, S. A., II. 2012. Magnesium-rich crustal compositions on Mercury: implications for magmatism from petrologic modeling. *Journal of Geophysical Research*, **117**, E00L15, http://dx.doi.org/ 10.1029/2012JE004140

STOFAN, E. R., BINDSCHADLER, D. L., HEAD, J. W. & PARMENTIER, E. M. 1991. Corona structures on Venus: models of origin. *Journal of Geophysical Research*, **96**, 20 933–20 946.

STOFAN, E. R., SHARPTON, V. L., SCHUBERT, G., BAER, G., BINDSCHADLER, D. L., JANES, D. M. & SQUYRES, S. W. 1992. Global distribution and characteristics of coronae and related features on Venus: implications for origin and relation to mantle processes. *Journal of Geophysical Research*, **97**, 13 347–13 378.

STOFAN, E. R., SMREKAR, S. E., BINDSCHADLER, D. L. & SENSKE, D. A. 1995. Large topographic rises on Venus: implications for mantle upwellings. *Journal of Geophysical Research*, **100**, 23 317–23 327.

STOFAN, E. R., BINDSCHADLER, D. L., HAMILTON, V. E., JANES, D. M. & SMREKAR, S. E. 1997. Coronae on Venus: morphology and origin. *In*: BROUGHER, S. W., HUNTEN, D. M. & PHILLIPS, R. J. (eds) *Venus II: Geology, Geophysics, Atmosphere, and Solar Wind Environment*. University of Arizona Press, Tucson, AZ, 931–965.

STOFAN, E. R., ANDERSON, W., CROWN, A. & PLAUT, J. 2000. Emplacement and composition of steep-sided domes on Venus. *Journal of Geophysical Research*, **105**, 26 757–26 771.

STOFAN, E. R., TAPPER, S. V., GUEST, J. E., GRINROD, P. & SMREAKAR, S. E. 2001. Preliminary analysis of an expanded corona database for Venus. *Geophysical Research Letters*, **28**, 4267–4270.

STROM, G., SCHABER, G. & DAWSON, D. D. 1994. The global resurfacing of Venus. *Journal of Geophysical Research*, **99**, 10 899–10 926.

STROM, R. G. 1964. Analysis of lunar lineaments, I: tectonic maps of the Moon. *University of Arizona Lunar and Planetary Laboratory Communications*, **2**, 205–216.

STROM, R. G. & NEUKUM, G. 1988. The cratering record on Mercury and the origin of impacting objects. *In*: VILAS, F., CHAPMAN, C. R. & MATTHEWS, M. S. (eds) *Mercury*. University of Arizona Press, Tucson, AZ, 336–373.

STROM, R. G., TRASK, N. J. & GUEST, J. E. 1975. Tectonism and volcanism on Mercury. *Journal of Geophysical Research*, **80**, 2478–2507.

STUDD, D., ERNST, R. E. & SAMSON, C. 2011. Radiating graben–fissure systems in the Ulfrun Regio area, Venus. *Icarus*, **215**, 279–291.

SULLIVAN, K. & HEAD, J. W. 1984. Geology of the Venus lowlands: Guinevere and Sedna Planitia. *Proceedings of the 15th Lunar and Planetary Science Conference*, 12–16 March 1984, Houston, Texas, 836–837.

SURKOV, Y. A. 1983. Studies of Venus rocks by Veneras 8, 9 and 10. *In*: HUNTEN, D. M., COLIN, L., DONAHUE, T. M. & MOROZ, V. I. (eds) *Venus*. University of Arizona Press, Tucson, 154–158.

TANAKA, K. L. & GOLOMBEK, M. P. 1989. Martian tension fractures and the formation of grabens and collapse features at Valles Marineris. *Proceedings of Lunar and Planetary Science*, **19**, 383–396.

TANAKA, K. L., SENSKE, D. A., PRICE, M. & KIRK, R. L. 1997. Physiography, geomorphic/geologic mapping, and stratigraphy of Venus. *In*: BROUGHER, S. W., HUNTEN, D. M. & PHILLIPS, R. J. (eds) *Venus II: Geology, Geophysics, Atmosphere, and Solar Wind Environment*. University of Arizona Press, Tucson, AZ, 667–694.

TANAKA, K. L., SKINNER, J. A., JR & HARE, T. M. 2005. *Geologic Map of the Northern Plains of Mars*. United States Geological Survey, Miscellaneous Geologic Investigations Series Map, **SIM-2888**.

TANAKA, K. L., SKINNER, J. A., JR. *ET AL.* 2014. *Geologic Map of Mars, Scale 1:20,000,000*. United States Geological Survey, Miscellaneous Geologic Investigations Series Map, **SIM 3292**, http://dx.doi.org/10.3133/sim3292

TAYLOR, G. J., WARREN, P., RYDER, G., DELANO, J., PIETERS, C. & LOFGREN, G. 1991. Lunar rocks. *In*: HEIKEN, G., VANIMAN, D. & FRENCH, B. (eds) *The Lunar Source Book: A User's Guide to the Moon*. Cambridge University Press, New York, 183–284.

TAYLOR, L. A. 1982. Lunar and terrestrial crusts – A contrast in origin and evolution. *Physics of the Earth and Planetary Interiors*, **29**, 233–241.

TAYLOR, S. R., NORMAN, M. D. & ESAT, T. 1993. The lunar highland crust: The origin of the Mg suite. *Meteoritics*, **28**, 448.

TERADA, K., ANAND, M., SOKOL, A., BISCHOFF, A. & SANO, Y. 2007. Cryptomare magmatism 4.35 Gyr ago recorded in lunar meteorite Kalahari 009. *Nature*, **450**, 849–852, http://dx.doi.org/10.1038/nature 06356

THOMAS, P. 1984. Martian intracrater spotches: occurrence, morphology, and colors. *Icarus*, **57**, 205–227.

THOMAS, P. J., SQUYRES, S. W. & CARR, M. H. 1990. Flank tectonics of Martian volcanoes. *Journal of Geophysical Research*, **95**, 14 345–14 355.

THORARINSSON, S. 1953. The crater groups in Iceland. *Bulletin of Volcanology*, **14**, 3–44.

THOREY, C. & MICHAUT, C. 2014. A model for the dynamics of crater-centered intrusion: application to lunar floor-fractured craters. *Journal of Geophysical Research*, **119**, 286–312.

TIRSCH, D., JAUMANN, R., PACIFICI, A. & POULET, F. 2011. Dark aeolian sediments in Martian craters: composition and sources. *Journal of Geophysical Research*, **116**, E03002.

TRAN, T., ROBINSON, M. S. ET AL. 2011. Morphometry of lunar volcanic domes from LROC. Abstract 2228 presented at the 42nd Lunar & Planetary Science Conference, March 7–11, 2011, The Woodlands, Texas.

TRASK, N. J. & GUEST, J. E. 1975. Preliminary geologic terrain map of Mercury. *Journal of Geophysical Research*, **80**, 2461–2477.

TUCKWELL, G. W. & GHAIL, R. C. 2003. A 400-km-scale strike-slip zone near the boundary of Thetis Regio, Venus. *Earth and Planetary Science Letters*, **211**, 45–45.

TURCOTTE, D. L. 1993. An episodic hypothesis for Venusian tectonics. *Journal of Geophysical Research*, **98**, 17 061–17 068.

TURCOTTE, D. L., MOREIN, G., ROBERTS, D. & MALAMUD, B. D. 1999. Catastrophic resurfacing and episodic subduction on Venus. *Icarus*, **139**, 49–54.

VAN DER BOGERT, C. H., HIESINGER, H., BANKS, M. E., WATTERS, T. R. & ROBINSON, M. S. 2012. Derivation of absolute model ages for lunar lobate scarps. Abstract 1847 presented at the 43rd Lunar & Planetary Science Conference, March 19–23, 2012, The Woodlands, Texas.

VAUCHER, J., BARATOUX, D., MANGOLD, N., PINET, P., KURITA, K. & GREGOIRE, M. 2009. The volcanic history of central Elysium Planitia: implications for Martian magmatism. *Icarus*, **204**, 418–442.

WAGNER, R. J., HEAD, J. W., WOLF, U. & NEUKUM, G. 1996. Age relations of geologic units in the Gruithuisen region of the moon based on crater size-frequency measurements. *27th Lunar & Planetary Science Conference*, 18–22 March 1996, Houston, Texas, 1367–1368, http://www.lpi.usra.edu/meetings/lpsc1996/pdf/1684.pdf

WAGNER, R., HEAD, J. W., III, WOLF, U. & NEUKUM, G. 2002. Stratigraphic sequence and ages of volcanic units in the Gruithuisen region of the Moon. *Journal of Geophysical Research*, **107**, 14-1–14-15.

WAGNER, R., HEAD, J. W., III, WOLF, U. & NEUKUM, G. 2010. Lunar red spots: stratigraphic sequence and ages of domes and plains in the Hansteen and Helmet regions on the lunar nearside. *Journal of Geophysical Research*, **115**, E06015, http://dx.doi.org/10.1029/2009JE003359

WALKER, G. 1983. Ignimbrite types and ignimbrite problems. *Journal of Volcanology and Geothermal Research*, **17**, 65–88, http://dx.doi.org/10.1016/0377-0273(83)90062-8

WARNER, J. L., SIMONDS, C. H. & PHINNEY, W. C. 1976. Genetic distinction between anorthosites and Mg-rich plutonic rocks: new data from 76255. *In*: Abstracts of the Lunar and Planetary Science Conference, Volume 7. Lunar and Planetary Institute, Houston, TX, 915–917.

WARREN, P. H. 1990. Lunar anorthosites and the magma-ocean plagioclase-flotation hypotheses: importance of FeO enrichment in the parent magma. *American Mineralogist*, **75**, 46–58.

WARREN, P. H. 1993. A concise compilation of petrologic information on possibly pristine nonmare moon rocks. *American Mineralogist*, **78**, 360–376.

WARREN, P. H. & WASSON, J. T. 1979. The origin of KREEP. *Reviews of Geophysics and Space Physics*, **17**, 73–88.

WATTERS, T. R. 1991. Origin of periodically spaced wrinkle ridges on the Tharsis Plateau of Mars. *Journal of Geophysical Research*, **96**, 15 599–15 616.

WATTERS, T. R. & JOHNSON, C. L. 2010. Lunar tectonics. *In*: WATTERS, T. R. & SCHULTZ, R. A. (eds) *Planetary Tectonics*. Cambridge University Press, New York, 121–182.

WATTERS, T. R., ROBINSON, M. S. & COOK, A. C. 1998. Topography of lobate scarps on Mercury: new constraints on the planet's contraction. *Geology*, **26**, 991–994.

WATTERS, T. R., NIMMO, F. & ROBINSON, M. S. 2005. Extensional troughs in the Caloris basin of Mercury: evidence of lateral crustal flow. *Geology*, **33**, 669–672.

WATTERS, T. R., SOLOMON, S. C. ET AL. 2009. The tectonics of Mercury: the view after MESSENGER's first flyby. *Earth and Planetary Science Letters*, **285**, 283–296.

WATTERS, T. R., ROBINSON, M. S. ET AL. 2010. Evidence of recent thrust faulting on the Moon revealed by the Lunar Reconnaissance Orbiter Camera. *Science*, **329**, 936–940.

WATTERS, T. R., SOLOMON, S. C. ET AL. 2012. Extension and contraction within volcanically buried impact craters and basins on Mercury. *Geology*, **40**, 1123–1126.

WEIDER, S. Z., NITTLER, L. R. ET AL. 2012. Compositional heterogeneity on Mercury's surface revealed by the MESSENGER X-Ray Spectrometer. *Journal of Geophysical Research*, **117**, E00L05, http://dx.doi.org/10.1029/2012JE004153

WEITZ, C. & HEAD, J. W. 1999. Spectral properties of the Marius Hills volcanic complex and implications for the formation of lunar domes and cones. *Journal of Geophysical Research*, **104**, 18 933–18 956.

WEITZ, C. M., HEAD, J. W. & PIETERS, C. M. 1998. Lunar regional dark mantle deposits: geologic, multispectral, and modeling studies. *Journal of Geophysical Research*, **103**, 22 725–22 760.

WERNER, S. C. 2009. The global Martian volcanic evolutionary history. *Icarus*, **201**, 44–68.

WERNER, S. C., VAN GASSELT, S. & NEUKUM, G. 2003. Continual geological activity in Athabasca Valles, Mars. *Journal of Geophysical Research*, **108**, E128081, http://dx.doi.org/10.1029/2002JE002020

WICHMAN, R. W. & SCHULTZ, P. H. 1995. Floor-fractured impact craters on Venus: implications for igneous crater modification and local mechanism. *Journal of Geophysical Research*, **100**, 3233–3244.

WICHMAN, R. W. & SCHULTZ, P. H. 1996. Crater-centered laccoliths on the Moon: modeling intrusion depth and magmatic pressure at the crater Taruntius. *Icarus*, **122**, 193–199.

WIECZOREK, M. A., JOLLIFF, B. L. *ET AL.* 2006. The constitution and structure of the lunar interior. *Reviews in Mineralogy and Geochemistry*, **60**, 221–364, http://dx.doi.org/10.2138/rmg.2006.60.3, 2006

WILHELMS, D. E. 1976. Mercurian volcanism questioned. *Icarus*, **28**, 551–558.

WILHELMS, D. E. 1987. *The Geologic History of the Moon*. United States Geologic Survey, Professional Papers, **1348**.

WILHELMS, D. E. 1979. *Relative Ages of Lunar Basins*. Reports of the Planetary Geology Program.

WILHELMS, D. E. & MCCAULEY, J. F. 1971. *Geologic Map of the Near Side of the Moon*. United States Geological Survey, Miscellaneous Geologic Investigations Series Map, **I-703**.

WILHELMS, D. E. & SQUYRES, S. W. 1984. The martian hemispheric dichotomy may be due to a giant impact. *Nature*, **309**, 138–140.

WILLIAMS, D. A., FAGENTS, S. A. & GREELEY, R. 2000. A reevaluation of the emplacement and erosional potential of turbulent, low-viscosity lavas on the Moon. *Journal of Geophysical Research*, **105**, 20 189–20 206.

WILLIAMS, D. A., GREELEY, R. *ET AL.* 2007. Hadriaca Patera: insights into its volcanic history from Mars Express High Resolution Stereo Camera. *Journal of Geophysical Research*, **112**, E10004, http://dx.doi.org/10.1029/2007JE002924

WILLIAMS, D. A., GREELEY, R., WERNER, S. C., MICHAEL, G., CROWN, D. A., NEUKUM, G. & RAITALA, J. 2008. Tyrrhena Patera: geologic history derived from Mars Express High Resolution Stereo Camera. *Journal of Geophysical Research*, **113**, E11005, http://dx.doi.org/10.1029/2008JE003104

WILLIAMS, D. A., GREELEY, R. *ET AL.* 2009. The Circum-Hellas volcanic province: overview. *Planetary and Space Science*, **57**, 895–916, http://dx.doi.org/10.1016/j.pss.2008.08.010

WILLIAMS, D. A., GREELEY, R., MANFREDI, L., RAITALA, J. & NEUKUM, G. 2010. The Circum-Hellas volcanic province, Mars: assessment of wrinkle-ridged plains. *Earth and Planetary Science Letters*, **294**, 492–505.

WILLIAMS, N. R., WATTERS, T. R., PRITCHARD, M. E., BANKS, M. E. & BELL, J. F. 2013. Fault dislocation modeled structure of lobate scarps from Lunar Reconnaissance Orbiter Camera digital terrain models. *Journal of Geophysical Research*, **118**, 224–233, http://dx.doi.org/10.1002/jgre.20051

WILLIAMS-JONES, G., WILLIAMS-JONES, A. E. & STIX, J. 1998. The nature and origin of Venusian canali. *Journal of Geophysical Research*, **103**, 8545–8555.

WILSON, L. & HEAD, J. W. 1981. Ascent and eruption of basaltic magma on the earth and moon. *Journal of Geophysical Research*, **86**, 2971–3001.

WILSON, L. & HEAD, J. W. 1983. A comparison of volcanic eruption processes on Earth, Moon, Mars, Io, and Venus. *Nature*, **302**, 663–669.

WILSON, L. & HEAD, J. W. 1990. Factors controlling the structures of magma chambers in basaltic volcanoes.

Abstract 1343–1344 presented at the 21st Lunar & Planetary Science Conference, Houston, Texas.

WILSON, L. & HEAD, J. W. 2002. Tharsis-radial graben systems as the surface manifestations of plume-related dike intrusion complexes: models and implications. *Journal of Geophysical Research*, **107**, 5019.

WILSON, L. & HEAD, J. W. 2003. Lunar Gruithuisen and Mairan domes: rheology and mode of emplacement. *Journal of Geophysical Research*, **108**, 5012, http://dx.doi.org/10.1029/2002JE001909

WISE, D. U., GOLOMBEK, M. P. & MCGILL, G. E. 1979. Tectonic evolution of Mars. *Journal of Geophysical Research*, **84**, 7934–7939.

WOOD, C. A. & HEAD, J. W. 1975. Geologically setting and provenance of spectrally distinct pre-mare material of possible volcanic origin. Paper presented at the Conference on Origins of Mare Basalts and their Implications for Lunar Evolution, November 17–19, 1975, Houston, Texas.

WOOD, J. A., DICKEY, J. S., MARVIN, U. B. & POWELL, B. N. 1970. Lunar anorthosites and a geophysical model of the Moon. *In*: LEVINSON, A. A. (ed.) *Proceedings of the Apollo 11 Lunar Science Conference, Volume 1*. Pergamon Press, New York, 965–988.

WYATT, M. B. & MCSWEEN, H. Y. 2002. Spectral evidence for weathered basalt as an alternative to andesite in the northern lowlands of Mars. *Nature*, **417**, 263–266.

WYRICK, D. Y., MORRIS, A. P., TODT, M. K. & WATSON-MORRIS, M. J. 2014. Physical analogue modelling of Martian dyke-induced deformation. *In*: PLATZ, T., MASSIRONI, M., BYRNE, P. K. & HIESINGER, H. (eds) *Volcanism and Tectonism Across the Inner Solar System*. Geological Society, London, Special Publications, **401**. First published online June 23, 2014, http://dx.doi.org/10.1144/SP401.15

XIAO, L., HUANG, J., CHRISTENSEN, P. R., GREELEY, R., WILLIAMS, D. A., ZHAO, J. & HE, Q. 2012. Ancient volcanism and its implication for thermal evolution on Mars. *Earth and Planetary Science Letters*, **323–324**, 9–18.

YIN, A. 2012. Structural analysis of the Valles Marineris fault zone: possible evidence for large-scale strike-slip faulting on Mars. *Lithosphere*, **4**, 286–330.

ZHONG, S. & ZUBER, M. T. 2001. Degree-1 mantle convection and the crustal dichotomy on Mars. *Earth and Planetary Science Letters*, **189**, 75–84.

ZIETHE, R., SEIFERLEIN, K. & HIESINGER, H. 2009. Duration and extent of lunar volcanism: comparison of 3D convection models to mare basalt ages. *Planetary and Space Science*, **57**, 784–796.

ZIMBELMAN, J. R. 1985. Estimates of rheologic properties for flows on the Martian volcano Ascraeus Mons. *Journal of Geophysical Research*, **90**, 157–162.

ZUBER, M. T. 1995. Wrinkle ridges, reverse faulting, and the depth penetration of lithospheric strain in Lunae Planum, Mars. *Icarus*, **114**, 80–92.

ZUBER, M. T. & MOUGINIS-MARK, P. J. 1992. Caldera subsidence and magma chamber depth of the Olympus Mons volcano, Mars. *Journal of Geophysical Research*, **97**, 18 295–18 307.

A downscaling approach for geological characterization of the Raditladi basin of Mercury

PAOLO MANCINELLI[1]*, FRANCESCO MINELLI[1], ALESSANDRO MONDINI[2], CRISTINA PAUSELLI[1] & COSTANZO FEDERICO[1]

[1]*Sezione di Geologia Strutturale e Geofisica, Dipartimento di Scienze della Terra, Università degli Studi di Perugia, Piazza Università 1, 06100 Perugia, Italy*

[2]*CNR IRPI Perugia, Strada della Madonna Alta 126, 06128 Perugia, Italy*

*Corresponding author (e-mail: pamancinelli@gmail.com)

Abstract: In this work, we combined multi-scale geological maps of Mercury to produce a new global map where geological units are classified based on albedo, crater density and morphological relationships with other units. To create this map, we used the 250 m/pixel mosaic of images acquired by the narrow- and wide-angle cameras onboard the MErcury Surface, Space ENvironment, GEochemistry, and Ranging (MESSENGER) spacecraft during its orbital phase. The geological mapping is supported by digital terrain model data and surface mineralogical variation from the global mosaic of MESSENGER Mercury Atmospheric and Surface Composition Spectrometer observations. This map comprises the global-scale intercrater plains, smooth plains and Odin-type units as reported in previous studies, as well as units we term bright intercrater plains, Caloris rough ejecta and dark material deposits. We mapped a portion of the Raditladi quadrangle (19–35°N, 106–133°E) at a regional scale at a resolution of 166 m/pixel. We characterized the geological context of the area and evaluated the stratigraphic relationships between the units. To obtain a representative geological section, we analysed and corrected available topographical data. The geological cross-section derived from our regional mapping suggests that volcanic emplacement of Raditladi's inner plains followed the topography of the basin after the deposition of impact-related units (i.e. melts, breccias and rim collapse) and was driven by low-viscosity flows. Hollows that appear on Raditladi's peak ring were possibly formed from low-reflectance intercrater plains materials exposed through the peak ring unit.

Supplementary material: Cleaner, larger version of the global-scale geological map and a local-scale map for comparison are available at http://www.geolsoc.org.uk/SUP18741.

Images gathered by the Mercury Dual Imaging System (MDIS) on board the MErcury Surface, Space ENvironment, GEochemistry, and Ranging (MESSENGER) spacecraft of Mercury's surface allow observation of a complex surface, and constraints from the spectrometer data permit evaluation of the planet's surficial composition. The observed morphologies, mineralogy and structures together suggest an active geological history. The quality and volume of data obtained by MESSENGER's instruments allow us to approach the geological analysis with sufficient resolutions at different spatial scales. Creation of a detailed geological map is one of the goals of planetary exploration, and intermediate steps (e.g. partial or regional maps) are relevant to both the final map and to comprehension of the studied areas by providing constraints for other investigations.

Since the Mariner 10 spacecraft returned images of Mercury during its flybys, several authors have mapped local regions for many purposes using different datasets. In fact, the first global-scale geological map of Mercury, developed from Mariner 10

and MESSENGER flyby data, was merged from smaller quadrangles (Frigeri *et al.* 2009). More recently, by exploiting the higher resolution of MESSENGER orbital data, local mapping has been used for a wide range of applications; for example, as a constraint for dating impact-related volcanism (Marchi *et al.* 2011) or to evaluate the nature of the units that fill some peak-ring impact basins (Blair *et al.* 2013). The resolution and coverage of the data gathered by the MESSENGER spacecraft allow the definition of boundaries and shape of landforms with unprecedented detail, and thus it is possible to integrate different datasets to achieve a more complete and robust geological model. For geological mapping, the derived products from the MESSENGER cameras provide higher-resolution images that facilitate the detection of unit boundaries.

Although the nature of certain deposits has been clearly defined – for example, northern plains volcanic flows (Head *et al.* 2011) – uncertainties still exist with respect to the origin of older deposits and the processes that produced such deposits as a

From: PLATZ, T., MASSIRONI, M., BYRNE, P. K. & HIESINGER, H. (eds) 2015. *Volcanism and Tectonism Across the Inner Solar System*. Geological Society, London, Special Publications, **401**, 57–75.
First published online March 11, 2014, http://dx.doi.org/10.1144/SP401.10

whole. However, the uncertainties that appear in a global-scale approach are less relevant when a limited region is observed. In fact, it is easier to fit temporal limits for the deposition of observed units at a local scale, both relative to other units and in absolute terms (e.g. Martellato *et al.* 2010; Marchi *et al.* 2011). This is because small areas are relatively easier to investigate owing to the limited number of processes that have acted on the region relative to a larger area, and to the wider range of data available that allow these processes to be more easily defined in detail. For example, topography and high-resolution images offer the possibility of evaluating depositional relationships between deposits in detail and thus hypothesize a local stratigraphic order.

In maps derived from Mariner 10 data, geological units were defined primarily by their crater density, colour properties and morphological relationships with the surrounding deposits, both at the global and local scale. Such maps covered approximately 45% of the surface with nine quadrangles (Schaber & McCauley 1980; De Hon *et al.* 1981; Guest & Greeley 1983; McGill & King 1983; Grolier & Boyce 1984; Spudis & Prosser 1984; Trask & Dzurisin 1984; King & Scott 1990; Strom *et al.* 1990). After MESSENGER flybys this coverage increased significantly, with 90% of the planet observed and mapped (Denevi *et al.* 2009). Global coverage was achieved during the MESSENGER orbital phase, and high-resolution images allowed better definition of the extent and characterization of two primary units on Mercury. The smooth plains (SP) is the youngest unit observable at a global scale (Denevi *et al.* 2013; Marchi *et al.* 2013). These features cover approximately 27% of Mercury's surface, and were possibly deposited within 200 myr between 3.7 and 3.9 Ga (Denevi *et al.* 2013). SP have also been shown to have an iron-poor yet basalt-like composition, suggesting that they are primarily volcanic in origin (Denevi *et al.* 2013).

The other unit that covers a substantial amount of the planet's surface is generally termed intercrater plains: this unit is primarily composed of plains that are more heavily cratered and lower in albedo than the smooth plains, and are thus likely to be older. This type of terrain usually displays undulating to hummocky surfaces, and has been interpreted as at least partly volcanic in origin based on Mariner 10 and MESSENGER data (Malin 1976; Strom 1977; Spudis & Guest 1988; Strom *et al.* 2011; Stockstill-Cahill *et al.* 2012; Marchi *et al.* 2013).

Other units that are clearly observable at the global scale are primarily related to the impact that formed the Caloris basin, particularly the Odin and the Caloris rough ejecta units. Including all of the above units, approximately 65% of the

smooth plains as mapped by Denevi *et al.* (2013) are interpreted to be volcanic; recent work (Denevi *et al.* 2013) suggested that differences between these deposits are likely to be representative of a different timing of deposition rather than changes in the crustal or subcrustal activity.

On a regional scale, the region incorporating the Raditladi basin offers an ideal area for detailed mapping due to its geological variety. In fact, in this region, we observe all units that are manifest at a global scale, indicating that the surface geology of the Raditladi region is representative of large-scale processes of resurfacing on Mercury across a significant time range. None the less, the Raditladi impact itself significantly altered the local geology by producing complex structures and covering a large number of the pre-impact surface units with its ejecta.

MESSENGER first imaged the Raditladi basin during its initial flyby in January 2008 (Solomon *et al.* 2008). The impact feature has a diameter of 258 km, and is centred at 27°N, 119°E. The impact also produced a peak ring structure approximately 110 km in diameter. The deposits filling the inner basin are marked by concentric circumferential graben that have a distinct topographical signature (Marchi *et al.* 2011; Blair *et al.* 2013).

The units filling the Raditladi basin (i.e. inner plains and annular plains) are dated to 1.1 ± 0.1 Ga ago, and the ejecta blanket has an age of 1.3 ± 0.1 Ga (Martellato *et al.* 2010; Marchi *et al.* 2011), indicating that the basin created by the impact was filled by later deposits within about 200 myr.

Data and methods

To produce a detailed geological cross-section for the Raditladi region, we started by extending the global geological map of Denevi *et al.* (2013) to regions not covered by smooth plains or Odin-type terrain. We then produced a detailed map of the Raditladi region (19–35°N, 106–133°E). Finally, we integrated the surface geology with corrected topographical data derived from the digital terrain model (DTM) of Preusker *et al.* (2011) to produce a geological cross-section of the entire region. This cross-section allows the thickness of the units, and the stratigraphic relationships between overlying deposits, impact-related deposits and pre-impact terrains, to be inferred. We also examined how impacts have affected the local geology.

Global mapping

The first release of a geological map with global coverage came from Denevi *et al.* (2009), in which they mapped reflectance variability using data from Mariner 10 and the first two flybys of Mercury by the MESSENGER spacecraft. In our map, however, we enlarged the coverage to the areas observed after MESSENGER's third flyby and orbital phase. For the distribution of smooth plains and Odin terrains we used data from Denevi *et al.* (2013).

The basemap consisted of a global mosaic released by the MESSENGER team (a link to the Planetary Data System web page hosting this dataset is provided in the References) in which orbital images from the narrow- and wide-angle cameras (NAC and WAC, respectively) were chosen to reduce the influence of observing conditions on the final mosaic, and to highlight morphological variation with average emission and solar incidence angles of 11° and 69°, respectively (Denevi *et al.* 2013). The resulting mosaic has a resolution of 250 m/pixel.

The mapping was performed in a geographical information system (GIS) that allowed integration of the Mariner 10 albedo mosaic and MES-SENGER data. This integration permitted a proper co-registration of different datasets which led to a full database in which geological information is constrained by: (1) local-scale high-resolution images in our region of interest that allowed us to fit the mapping to complex geological features on the surface; and (2) a DTM (Preusker *et al.* 2011) that enabled us to infer the surface composition using topographical relief. This approach was first carried out on a global scale; the geological map obtained covers nearly 100% of the planet (Fig. 1a) and was produced at a scale of 1:1.25 M. In this map, smooth plains and the Odin terrain are reported as were mapped by Denevi *et al.* (2013), all other terrains are mapped by us at 250 m/pixel.

Figure 2 shows examples of the global-scale geological units as mapped in this work. Distinguished by different colours, the units essentially mark albedo variations, the morphology of the deposits and the crater density.

The youngest terrain is the SP unit as mapped by Denevi *et al.* (2013). This unit covers 27% of the surface, and at least 65% of its total exposure is most probably volcanic in origin (Denevi *et al.* 2013). The major SP deposits are found in the northern volcanic plains (Head *et al.* 2011), in the Firdousi plains SE of Rachmaninoff crater and in the Rembrandt, Tolstoj, Beethoven and Caloris impact basins. The SP also covers wide plains surrounding Caloris, but interpretation of the origin for these deposits is still challenging: although their spatial distribution suggests that they are related to the Caloris impact, crater counting shows that they are more recent than the Caloris inner plains (Fassett *et al.* 2009; Denevi *et al.* 2013). The SP

are also mapped in smaller basins often associated with impact events. Owing to their morphology and low crater density, all SP units are believed to be younger than the Late Heavy Bombardment and to have been deposited over a relatively brief time step (*c.* 200 myr) (Denevi *et al.* 2013). Recent studies (Head *et al.* 2011; Stockstill-Cahill *et al.* 2012) have suggested that the SP of the northern volcanic plains were built up to a final thickness of 1 km or more by many thin layers representative of multiple effusive events.

The Odin unit represents the other deposits mapped as smooth plains by Denevi *et al.* (2013) (Figs 1a & 2a). Based on crater density comparison with deposits inside the Caloris basin, this unit was interpreted as being composed of volcanic deposits that partially covered the Caloris ejecta (Fassett *et al.* 2009), but no definitive colour evidence of volcanism was found either by Denevi *et al.* (2009) or in the present study. This unit is distinguished from SP because it encompasses smooth deposits that embayed kilometre-scale blocks of ejecta. The Odin terrains have a gradational transition to surrounding deposits and, among plain units, these deposits show the lowest crater density (Fassett *et al.* 2009; Buczkowski & Seelos 2012; Buczkowski *et al.* 2013; Denevi *et al.* 2013).

The Caloris rough ejecta (CRE) is the only unit that we mapped which is clearly related to a single impact; specifically, it covers the ejecta deposits on the Caloris rim and the undulated regions east of Caloris (Figs 1a & 2a). Based on Mariner 10 data, parts of these regions were previously mapped as the Caloris Montes Formation, the Nervo Formation and the Van Eyck Formation (Schaber & McCauley 1980; Guest & Greeley 1983; Fassett *et al.* 2009); however, in this work, all of these formations have been included in the single CRE unit and mapped at 250 m/pixel. Although the spatial distribution and morphology of the CRE suggest that it was deposited during the Caloris impact event, there is still no unique explanation for the processes that altered the CRE after its emplacement.

The dark material (DM) unit is composed of a few patches of notably low-reflectance material mapped only near or within impact craters. In Table 1 all the regions where we mapped DM units are listed. These deposits display the lowest albedo compared with other deposits at the global scale, show clear boundaries with surrounding units and exhibit variable crater densities. A subset of these deposits is collocated with fresh impacts (e.g. Grainger crater), whereas others are correlated with older impact events (e.g. Tolstoj basin) (Fig. 3, Table 1).

The bright intercrater plains (BIP) unit is possibly the oldest unit found at the global scale, as its

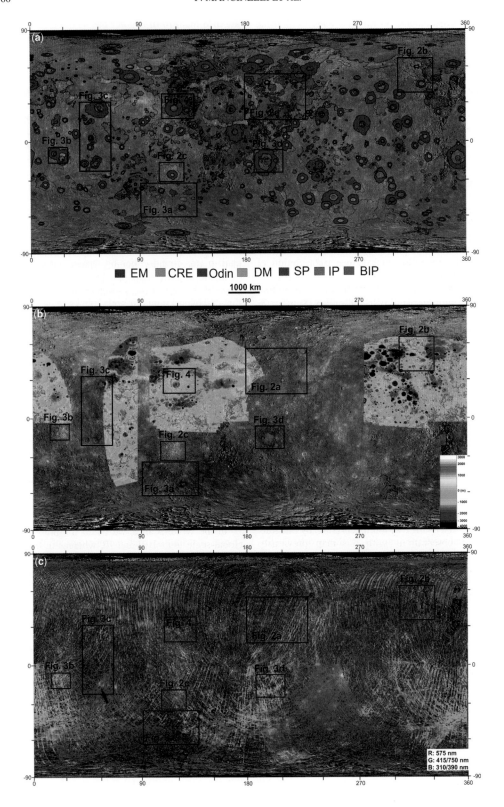

EM CRE Odin DM SP IP BIP

1000 km

crater density and albedo are the highest of all these units. The BIP shows clear morphological boundaries with respect to all other units. These units often have more topographical relief than the surrounding deposits, which may have served to partially preserve the BIP from later volcanic resurfacing events (Fig. 2b). Locally, the BIP is exposed by impacts, regardless of the freshness (i.e. age) of the impact. Examples include the ejecta of the Amaral crater (26.5°S, 118°E) and material on the rims of two unnamed craters nearby (Fig. 2c).

The intercrater plains (IP) is the largest unit in our mapping. It covers wide regions with undulating to hummocky morphologies, and displays a high crater density suggesting that it is an older unit than most others on the planet. In many of the nine quadrangles mapped after Mariner 10, the IP was interpreted as possibly volcanic in origin (De Hon *et al.* 1981; Grolier & Boyce 1984; Spudis & Prosser 1984; Trask & Dzurisin 1984; King & Scott 1990; Strom *et al.* 1990) or as including material with probable volcanic origin (Schaber & McCauley 1980; McGill & King 1983). In this unit, we included all formations with low- to very low-reflectance materials ('LRM' in Denevi *et al.* 2009) and formations with intermediate crater density; that is, 'intermediate plains' as defined by Trask & Dzurisin (1984) that are not mapped as smooth plains by Denevi *et al.* (2013). This unit shows clear boundaries with the SP unit but no evident morphological transition with the bright intercrater plains unit. Both the SP and IP appear to be formed by layered sequences of possibly low-viscosity material, and thus are found in relatively thin and broad deposits (Stockstill-Cahill *et al.* 2012) (Fig. 2a).

To characterize the local geological evolution of the surface, we also mapped the material ballistically ejected by the major impacts (EM) by integrating our observations with the map previously produced by Denevi *et al.* (2013). Figure 1 shows EM deposits larger than 50 km; ejecta deposits were mapped by considering the morphology of the deposit and the presence of radial secondary cratering respect to the main impact.

Raditladi mapping

The map in Figure 1 is representative only of a global scale in which the geology is influenced by large-scale processes. In fact, at the local scale, local processes produce many features that are less readily observable from a global perspective, such as smaller impacts, local volcanism or local tectonics.

In the region surrounding the Raditladi crater (27°N, 119°E), we found both units related to global-scale processes (i.e. global-scale units) and those related to local resurfacing events (e.g. impact craters and volcanism) that produced smaller (local-scale) deposits. Figure 4 shows the geological map produced for this region (19–35°N, 106–133°E). The basemap we used for the geological mapping is a mosaic of two Map Projected Basemap Reduced Data Record (RDRBDR) images (from the NASA/Planetary Data System Geosciences Node 2012); each product is corrected to a solar incidence angle of 30° and an emission angle of 0°, with a resolution of 166 m/pixel. This resolution allows for the mapping of geological features in detail, and the consideration of both the deposits and structures.

A total of 60% of the Raditladi regional study area is covered by smooth plains (SP), and is thus thought to be volcanic in origin (Head *et al.* 2011; Denevi *et al.* 2013) and younger than the Late Heavy Bombardment (Strom *et al.* 1975, 2008; Trask & Guest 1975; Spudis & Guest 1988; Denevi *et al.* 2009; Head *et al.* 2011; Marchi *et al.* 2013). The other unit mapped by Denevi *et al.* (2013) that is also found in our regional mapping is the Odin-type plains, which we found in a number of spots in the eastern and northern regions of the study area.

In the western area, we noted the presence of global-scale units BIP and IP, whereas in the NW regions we observed deposits not clearly identifiable as BIP or IP that were thus mapped as bright crater plains (BCP). This unit is not classifiable as either BIP or as IP because of its roughness, crater density and high albedo (Fig. 4f). There is some

---

Fig. 1. (**a**) Global coverage geological map, showing ejecta material (EM), Caloris rough ejecta (CRE), Odin-type units (Odin) as mapped in Denevi *et al.* (2013), dark material (DM), smooth plains (SP) as mapped in Denevi *et al.* (2013), intercrater plains (IP) and bright intercrater plains (BIP). (**b**) Digital terrain model (DTM) after Preusker *et al.* (2011). (**c**) Global mosaic of MASCS observations in the visible and near-infrared spectrum, with R: 575 nm, G: 415 nm/750 nm and B: 310 nm/390 nm (NASA/JHUAPL/CIW 2012). The black arrow indicates the Nabokov crater (14°S, 56°E). Best-fit areas between the geological map and MASCS mosaic are found mainly where MASCS orbits are denser, for example areas east and SE of Tolstoj (16°S, 200°E), areas south of Picasso (3°N, 49°E) and the area of the Rembrandt basin (32°S, 87°E). All images are shown in equidistant cylindrical projections centred at 0°N, 180°E; the basemap is a monochrome mosaic of Mercury after the MESSENGER orbital phase (at 250 m/pixel) (NASA/JHUAPL/CIW 2013). The black boxes delimit the areas of the subsequent figures and are labelled accordingly.

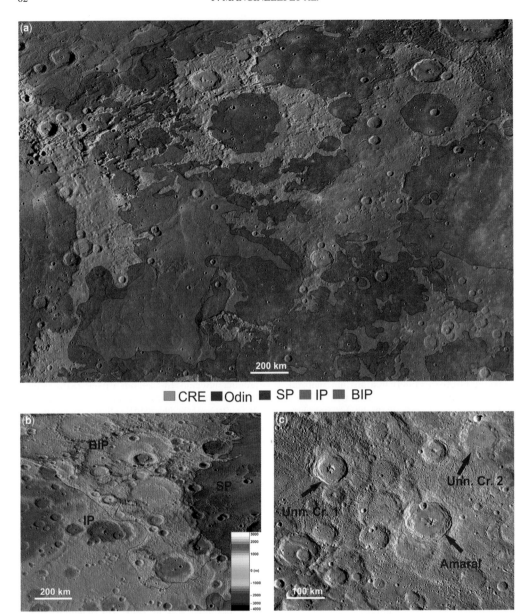

Fig. 2. (a) Sample of the global-scale geological map for an area east of Caloris (the ejecta material is not shown to highlight morphologies of the main units). (b) Where possible, the geological mapping is constrained with topography; an example is given for the bright intercrater plains (BIP), surrounding intercrater plains (IP) and smooth plains (SP). (c) Monochrome image for the Amaral crater and surrounding terrains (26.5°S, 118°E). See Figure 1 for detailed positions.

evidence that this unit is representative of old deposits (i.e. older than the BIP) and is also present globally, encompassing deposits that cannot unequivocally be interpreted as bright intercrater plains or intercrater plains (Whitten *et al.* 2012). Other formations belonging to global-scale units

are found in the eastern and southern regions where we mapped the CRE terrain.

The ejecta of major craters (EM) covers a substantial amount of the SP that constitutes the pre-impact bedrock. In particular, we mapped the ejecta of the Raditladi basin and an unnamed

Table 1. *List of the regions where dark material terrains (DM) were found*

Dark material deposits
External ring of the Rachmaninoff basin (28°N, 57°E)
Ejecta of the Tolstoj crater (16°S, 200°E)
Southeast of the Rembrandt basin (32°S, 87°E)
Ejecta and inside of the Neruda crater (52°S, 125°E)
Rim and ejecta of the Sher-Gil crater (45°S, 134°E)
Ejecta east of the Picasso crater (3°N, 49°E)
Western external ring of the Derain crater (8°S, 19°E)
External ring and ejecta of the Nabokov crater (14°S, 56°E)
Grainger crater (44°S, 105°E)

See the text for a description of DM, and Figure 3 for images of these regions.

crater with a diameter of 120 km located to the east of the area that we refer to as A Crater (Fig. 4a). We found that the EM of A Crater is superposed by the Raditladi basin's ejecta; these secondary craters are radially distributed with respect to the basin, and are found both on the EM and terrace material (TMA) of A Crater (dark blue lines, Fig. 5b).

The topographical depressions found within the EM of Raditladi are filled with smooth deposits that also show a lower albedo with respect to the surrounding ejected material. These deposits, that we named flat land (FL), are notably low in superposed craters and are primarily distributed on the western region of the ejecta (Figs 4 & 5). Given their distribution, smooth morphology, low reflectance and low crater density, we interpret the FL deposits as younger units outside of the impact basin but that are somehow related to the Raditladi impact. Although these units are confined in small basins on the EM of Raditladi, it is currently difficult to unequivocally address how these deposits were formed and the composition of the infilling material.

We have mapped the Raditladi peak ring and the central peak of A Crater as peak structures (PR). Related to Raditladi's PR are hollows (Blewett *et al.* 2011, 2012) that partially cover the SE portion of the peak ring (Fig. 5d). Although the formation of these features is still debated, the geological contexts in which hollows have been found are often related to impacts and always involve low-reflectance material (Blewett *et al.* 2011, 2012).

For minor impacts (craters with diameters <120 km), we mapped the peak structures as superposed peak deformation (SPD) (Fig. 4g). We notice that the SPD are found only on craters with diameters between 20 and 120 km, whereas craters with diameters less than 20 km have a simple geometry.

This observation is in agreement with previous investigations (e.g. Melosh 1989; Baker *et al.* 2011). The inner plains of craters smaller than 120 km in diameter have been mapped as inner plains superposed (IPS), and these deposits encompass all terrains detected inside the small impact craters. Owing to the limited extent of these deposits, it is difficult to distinguish structures or morphological features at our mapping scale.

Inside the major basins (Raditladi and A Crater), we distinguish several units based on the topography, albedo, morphology and crater density of the terrains. On the collapsed portions of the rim, we mapped the hummocky deposits (HD). This unit is taken to be composed of hummocky and undulated materials formed by gravitational collapse from the scarps of the rim, which has undergone chaotic mixing of the deformed and shocked bedrock (Fig. 5). Locally, the cusps produced by the collapse of the rim are not covered by HD deposits, leaving an exposed terraced surface that we mapped as terrace material (TMR for Raditladi and TMA for A Crater).

Inside the Raditladi basin, we mapped the annular plains (AP) and dark annular plains (DAP). These units are interpreted to have been deposited relatively soon after the impact and so are the product of the collapse of the shocked and deformed bedrock from the scarps of the crater. AP and DAP were distinguished from HD due to their smooth morphology and more widespread distribution with respect to HD units (Fig. 5c). We differentiated the AP from DAP by considering both the albedo of deposits and the roughness of the surface, but no evident morphological transition (i.e. topographic step) is found between these two units. Thus, the change of albedo is interpreted to be probably representative of a compositional variation of the materials emplaced during the collapse. These deposits are locally undulated and are found only outside of the peak ring of the basin. We found no evidence of AP or DAP in either A Crater or in the inner basin of Raditladi; thus, we mapped a stratigraphic contact between the annular plains and inner plains only in the outer basin of Raditladi.

The inner plains are among the youngest units found in our mapping, based on both their low crater density and their stratigraphic position. The IPR (the inner plains of the Raditladi basin) cover the central basin and a large portion of the area outside of the peak ring. The IPR deposits show clear boundaries with respect to other units (Fig. 5c) and have been revealed by a small impact that exposed the entire thickness of these inner deposits near the centre of the inner basin. Despite their smooth morphology, these deposits are not topographically flat but possess elevations that deepen from the outer ring (minimum depth of

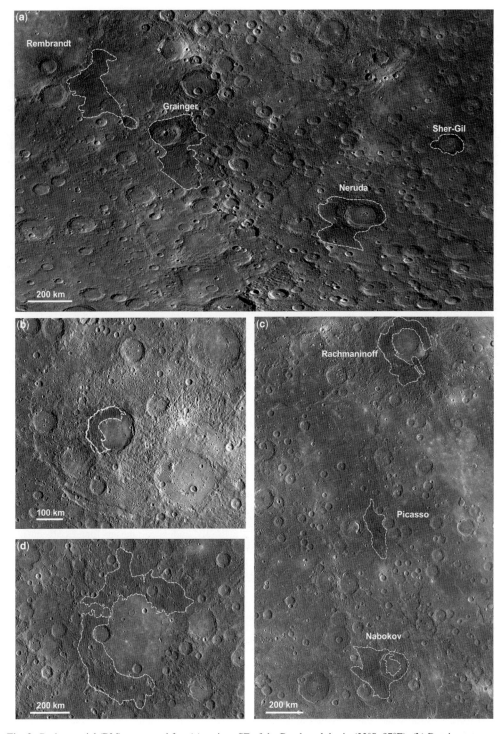

Fig. 3. Dark material (DM) as mapped for: (**a**) regions SE of the Rembrandt basin (32°S, 87°E); (**b**) Derain crater (8°S, 19°E); (**c**) Rachmaninoff (28°N, 57°E), Picasso (3°N, 49°E) and Nabokov (14°S, 56°E) craters; and (**d**) Tolstoj crater (16°S, 200°E). See Figure 1 for detailed positions with respect to the geological map, topography and MASCS mosaic.

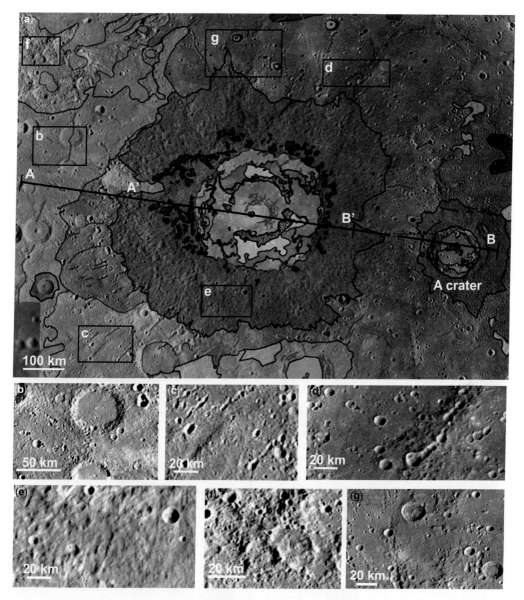

Fig. 4. (a) Geological map for the Raditladi region and trace of the geological cross-section in Figure 8. Also shown are insets of global units mapped in this region: (**b**) bright intercrater plains; (**c**) intercrater plains; (**d**) smooth plains; (**e**) ejecta material; (**f**) bright cratered plains; and (**g**) inner plains superposed (IPS) and superposed peak deformation (SPD). Geological mapping is produced from a mosaic of Map Projected Basemap Reduced Data Record (RDRBDR) files (NASA/Planetary Data System Geosciences Node 2012). These images are corrected to a solar incidence angle of 30° and an emission angle of 0°, with a resolution of 166 m/pixel. See Figure 5 for the colour key to deposits and structures.

c. − 1500 m) towards the inner basin (maximum depth of *c.* − 3000 m).

The inner plains found in A Crater (IPA) (Fig. 5b) have morphological and albedo properties similar to those found in Raditladi, but the latter unit is more highly cratered. We interpret this higher crater density as further support for the hypothesis that A Crater formed prior to the Raditladi basin and that some of the craters found on IPA are possibly caused by secondary cratering from

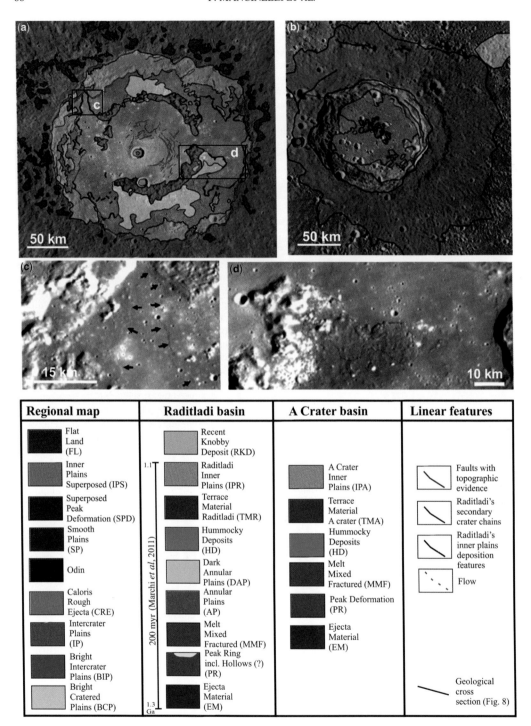

Fig. 5. Details of the regional geological map in Figure 4 for (**a**) the Raditladi crater and (**b**) A Crater. (**c**) The boundaries of the inner plains of Raditladi (IPR) suggest an original low viscosity for these deposits. (**d**) The recent knobby deposits (RKD) are centred in the image. Colour and unit definitions for all units are reported in the lower schematic for the regional map in Figure 4a and for Raditladi basin and A Crater in (a) and (b). The basemap and images are the same as those used in Figure 4.

the Raditladi impact, similar to those found on the EM.

The undulated deposits, that we name the recent knobby deposits (RKD), show almost no superposed craters and are thus interpreted as the youngest unit identified by our mapping. These units are observed only in the eastern portion of the outer Raditladi basin and overlie, and therefore probably post-date, the IPR (Fig. 5d). Despite their small spatial extent, these deposits are clearly observable and have distinct boundaries with the IPR. Topographical data also indicate that the RKD are elevated with respect to the surrounding units, with a maximum height of − 1300 m in contrast to a mean topography of − 1800 m for the surrounding IPR and DAP.

In the regional-scale map (Figs 4 & 5), we also mapped linear features that are observable both from the topography and from MESSENGER images. In particular, we mapped the flow features observed only on the northern deposits, and these features are possibly related to the regional-scale deposition of smooth plains.

We also mapped faults (including the crater rim scarps) that were clearly visible on the images and/ or the DTM. In particular, we mapped Raditladi's circumferential graben that, deforming the IPR unit, are interpreted as the youngest faults of the area (or, at least, as representative of the youngest tectonic activity).

Radially positioned with respect to the Raditladi crater, we identified small chains of secondary craters that we interpret as secondary impacts ballistically emplaced as a result of the Raditladi impact event. These structures are also found on the ejecta material of A Crater (Fig. 5b), further indicating that the Raditladi impact post-dates the A Crater event.

Topographical analysis

The geological map shown in Figure 1 is, in part, representative of global-scale compositional changes and thus of geological features sufficiently large to be detected at the global scale. The Raditladi regional map, however, is imaged at higher resolution (166 m/pixel) and is entirely covered by the M1 portion of the DTM published by Preusker et al. (2011); this work allows the use of topography to evaluate the proper stratigraphic relationships between units during and after mapping. We therefore built a geological cross-section using a corrected topographical profile that was extracted from the M1 portion of the DTM of Preusker et al. (2011).

To evaluate the vertical precision of the DTM, we carried out a preliminary frequency–distribution analysis of its elevation values and found unexpected problems related to the presence of high-frequency signals on the bulk signal. Figure 6a shows the histogram obtained by binning the original data with values of 5 m; the 'solid black' effect stems from having bars at many values between 0 and higher frequencies. This effect disappeared when we sized the bin at 10 m (Fig. 6b) or multiples of 10 m (e.g. 60 m; Fig. 6c) but resurfaced when we chose a bin that is not a multiple of 10 m (e.g. 27.5 m; Fig. 6d). This effect can be caused by undersampling or by aliasing problems (Weibel & Heller 1990) when simplistic gridding techniques are applied; the resulting DTMs could exhibit many erratic features, such as terraces, ghost lines and tiger stripes (Reuter et al. 2009). The effect is confirmed by the Fourier analysis carried out using ENVI 5.0 (Exelis Visual Information Solutions, Boulder, CO, USA). Results are shown in Figure 6e in which pseudo-white elliptic lobes appearing along the axes are due to systematic noise in the DTM.

We overcame this problem by applying simple moving-window (kernel) spatial low-pass filters (LPF) and Fast Fourier filters (FFT). Among these, we chose a 5 × 5 kernel by adding back 20% of the original value to preserve the spatial context. The mean and standard deviation of the distribution do not change (even if the fit to a normal curve is not exact), and elevation minima and maxima change by approximately 0.6 and 1.2%, respectively (Table 2). The filter retains the overall shape of the distribution but removes the high-frequency signal (Fig. 7a, b). Even though LPFs are commonly used to pre-process DTMs, they are mathematical operators that change the original values, and thus potentially drive our interpretations (see the section on 'Discussion of the geological mapping of Raditladi and the geological cross-section' later in this paper). The geological cross-section is obtained by the integration of the surface geology (as determined by mapping) with the topographical profile extracted by the filtered DTM. This profile keeps the general shape shown by the profile extracted from the original DTM (Fig. 7c), thus guaranteeing a sensible geomorphological interpretation but also presenting a more realistic shape in detail (the box in Fig. 7c).

Results and discussion

Discussion for global geology

Figure 1c shows a comparison between the mosaic of observations of the Mercury Atmospheric and Surface Composition Spectrometer (MASCS) and our geological map at the global scale. The image highlights the broad relationship between the compositional differentiation of the surface material and the geological units that we mapped. The

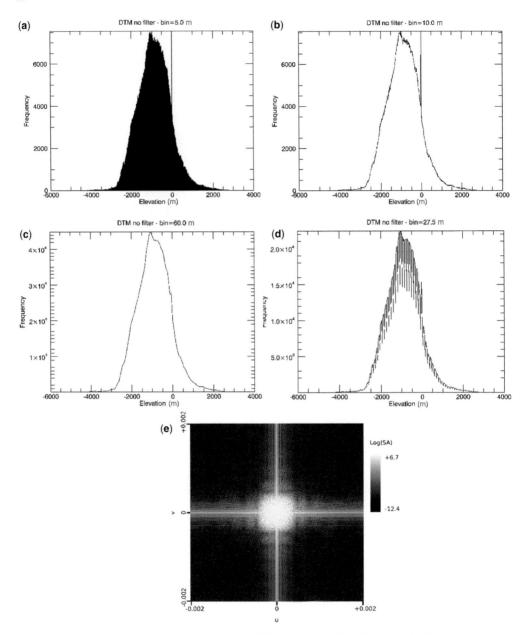

Fig. 6. Histograms of the original DTM estimated with different bins: (**a**) with a bin of 5 m, the histogram shows a signal that repeatedly cycles up to a given value and down to zero, leading to a 'filled in' appearance; (**b**) with original data and a bin of 10 m, the histogram is 'clean' except for a spike at 0 m elevation; (**c**) with a bin multiple of 10 m, the signal is still clean; (**d**) the disturbance reappears with a bin different to 10 m; (**e**) representation of the logarithm of the amplitude of the complex power spectrum of the bi-dimensional Discrete Fourier Transform (DFT) of an area of 1200 × 1200 pixels centred on the Raditladi crater in the DTM. u and v represent, respectively, the dimensionless horizontal and vertical spatial frequencies in the Fourier domain. To centre the (0, 0) values of the frequencies FFT, the image is translated by 0.002 both in the horizontal and vertical spatial frequencies.

monochrome and MASCS mosaics are compiled using MESSENGER orbital data, and are selected to minimize the influence of observing conditions and thus highlight morphological and mineralogical variations, respectively (NASA/JHUAPL/CIW 2012, 2013; Denevi *et al.* 2013).

Table 2. *Minimum, maximum, mean values and standard deviation for the M1 portion of the DTM (Preusker et al. 2011) before (up) and after (down) filtering*

	Minimum (m)	Maximum (m)	Mean (m)	SD (m)
DTM, unfiltered	−4260	3180	−866.6	820.3
DTM, filtered	−4233	3141	−866.6	816.7

The geological units are defined on the basis of surface albedo, boundaries/morphology, topography and crater density of the deposits, but no depth or structural constraints are applied during the mapping or are obtained from the final map. Despite these limitations, we observe in places a good fit between our mapped deposits and the variation in mineralogical composition derived from the MASCS when the two maps are compared (Fig. 1c), in particular for regions where MASCS data are denser; for example, areas east and SE of Tolstoj (16°S, 200°E), the area south of Picasso (3°N, 49°E) and the area where the Rembrandt basin is located (32°S, 87°E). In fact, in the regions where MASCS data overlap, orbits used to compile the mosaic in Figure 1c were chosen to give priority to the data with best illumination and photometric parameters in order to highlight mineralogical variations (NASA/JHUAPL/CIW 2012). This agreement means that the albedo variations (as

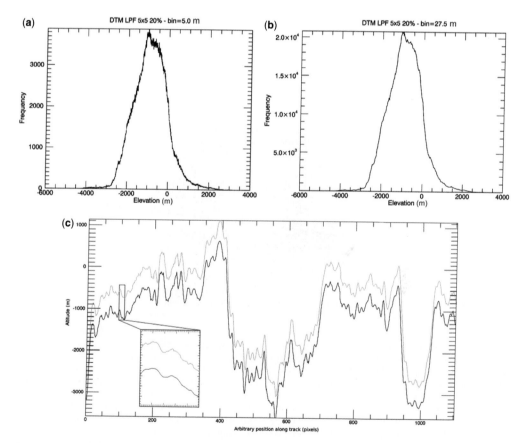

Fig. 7. Filtered DTM: (**a**) 5 m bin histogram of the filtered DTM; and (**b**) 27.5 m bin histogram of the same DTM (both showing no disturbance); (**c**) the grey colour shows the profile extracted from the filtered DTM, and black indicates the original unfiltered DTM, the last is expressly shifted 500 m downwards; in the box, a close-up showing the local effects of the filter.

mapped after being constrained to the morphological features) are representative of a mineralogical change, which may correspond to different units deposited during different phases of resurfacing of the planet Mercury. This inference is supported by data gathered by MESSENGER's X-ray spectrometer (XRS), which confirmed that the reflectance changes correspond at least in part to compositional variations (Nittler *et al.* 2011; Weider *et al.* 2012). The observation that nearly 27% of the surface is covered by late-stage volcanic floods (Denevi *et al.* 2009, 2013; Head *et al.* 2011) confirms that a large portion of Mercury has been resurfaced and is thus a secondary crust. That mineralogical variations are found by MASCS across the whole planet (Fig. 1c) leads us to hypothesize that the majority of the terrains mapped at the global scale is secondary in origin and thus possibly produced by volcanic resurfacing events. This model is supported by work in which a volcanic origin was proposed for the intercrater plains (Malin 1976; Strom 1977; Spudis & Guest 1988; Strom *et al.* 2011; Stockstill-Cahill *et al.* 2012; Marchi *et al.* 2013).

The MASCS mosaic also reinforces the observation that local and relatively small events (e.g. small or medium impacts) can substantially alter the spectral signature of the surface material over a wide region. In fact, from the analysis of the MASCS mosaic, the dark material (DM) deposits appear compositionally distinct with respect to the others (e.g. the Nabokov crater located at 14°S, 55°E) (Figs 1c & 3c). Certain DM deposits that we mapped were previously mapped as the Centre of Low Reflectance Material (LRM_C) by Denevi *et al.* (2009), who proposed that the LRM_C is crustal or mantle material excavated by impacts. This hypothesis is not in agreement (at least for a mantle source) with the more recent crustal thickness map for the northern hemisphere (Smith *et al.* 2012), where no relationship is found between the LRM_C deposits and regions with a thinned crust. Given the morphologies and distribution of the DM deposits, we suggest that these units are volcanic products related to impact events that locally produced the anomalously low-reflectance materials. Another important aspect is that the DM deposits show particularly low crater densities when found near fresh impacts, but such densities increase for those deposits near older impacts, suggesting that impacts directly influenced the emplacement of DM units. Considering that recent works (Blewett *et al.* 2012) found hollows on the majority of the deposits that we mapped as DM (i.e. in the Rachmaninoff, Picasso, Nabokov, Sher-Gil and Grainger basins/craters), and that low-reflectance material is possibly related to the formation of hollows (Xiao *et al.* 2013), we

believe that further investigation is required to unequivocally understand how the DM units were formed and how they relate to other processes, such as impact cratering and hollows formation.

Discussion of the geological mapping of Raditladi and the geological cross-section

The situation prior to the Raditladi impact was likely to have been dominated by SP and IP deposits, which are found in lateral continuity with the BIP and BCP terrains in the western and NW regions of the mapped area. The impact significantly altered the local geology both by depositing ejecta across the surrounding region and by driving impact-related processes such as melt production, peak deformation and rim collapse within the basin. The volcanic processes acted via a later and final resurfacing to produce the inner plains of Raditladi, with probable low-viscosity flows covering a large portion of the outer ring basin and the entire inner basin. Crater counting shows that the IPR were deposited a relatively long time after the impact (i.e. *c.* 200 myr) (Marchi *et al.* 2011). The impact that produced A Crater predates the formation of Raditladi and produced similar deposits (TM, HD and IPA). The albedo and morphological properties of the IPA are similar to the inner plains found in Raditladi and so, despite the lack of clear observation of flow structures or boundaries, volcanic emplacement cannot be excluded.

The geological cross-section considers the morphological features of the impact basins (the Raditladi and A Crater), and evaluates the relationship between the assumed pre-impact morphologies of the deposits (i.e. SP, IP and BIP) and the actual post-impact scenario. The Raditladi crater shows a typical peak-ring morphology (Melosh 1989; Baker *et al.* 2011), with a central basin deeper than the outer parts of the basin and crater walls that partially collapsed after the impact, producing terraces where hummocky deposits are found.

A notable observation pertains to the irregular distribution of ejecta from the Raditladi basin. Although the ejecta deposits are mapped along all azimuths radially outwards from the crater, the topographical relief is much greater in the western and NW deposits than elsewhere. This anomaly may have been caused by an oblique impact that irregularly distributed the ejected mass, or may be due to a pre-impact topographical relief striking SW–NE.

Superposed on the ejecta deposited by the impact that produced A Crater, we find secondary craters probably sourced from the Raditladi impact, indicating that the Raditladi basin was produced after A Crater. Despite it being older than Raditladi,

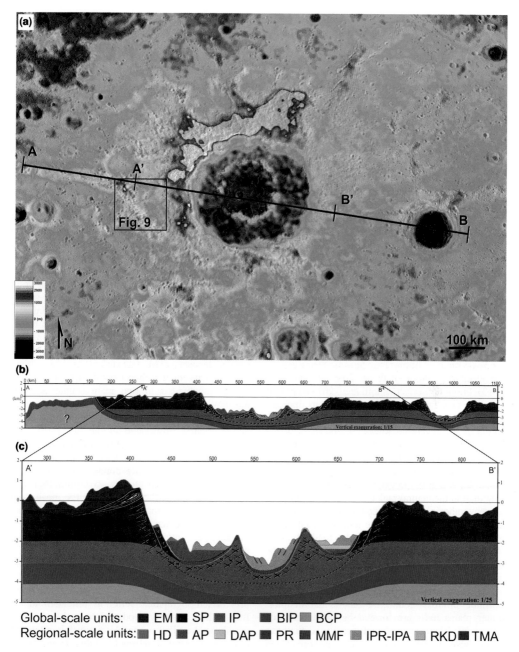

Global-scale units: ■ EM ■ SP ■ IP ■ BIP ■ BCP
Regional-scale units: ■ HD ■ AP ■ DAP ■ PR ■ MMF ■ IPR-IPA ■ RKD ■ TMA

Fig. 8. (**a**) Topography for the area mapped in Figure 4 using the corrected DTM (Preusker *et al.* 2011). The black square indicates an anomalous topographical uplift unrelated to the ejected material. This uplift represents major topographical relief that we suggest is unrelated to impacts and which we interpret as a blind thrust (see Fig. 9 and the text for a further description). (**b**) Geological cross-section for the Raditladi region, with a vertical exaggeration of 15. (**c**) Close-up of the geological cross-section for the Raditladi basin, with a vertical exaggeration of 25. For full unit names and descriptions, see the text and Figure 5.

this crater is well preserved and probably less filled, possibly indicating a lesser intensity of local resurfacing activity after the impact. This observation is reinforced by the homogeneity of the deposits mapped inside its basin (i.e. nearly the entire basin is covered in inner plains material) and by the relatively low thickness of IPA, which does not cover entirely the HD deposits and is possibly excavated by small impacts (Fig. 5b).

The first unit deposited during and immediately after the impact is impact melt, combined with mixed deposits and fractured bedrock; we named this unit 'melt mixed fractured' (MMF). This deposit is buried after the crater formation by breccias collapsed from the walls and the peak ring, which form the AP and DAP units. Considering their spatial distribution and crater density, the DAP are interpreted as being annular plains with distinct reflectance properties.

Units mapped inside the basin of Raditladi are found to overlay in irregular succession; in fact, the eastern sector of the basin is more stratigraphically diversified than the central and western areas. In particular, in the eastern part of the outer basin, we find AP, DAP, IPR and overlying RKD units that are less cratered than the inner plains (Fig. 5d), and thus are interpreted as more recent formations with respect to other units.

Owing to the high resolution of the Map Projected Reduced Data Record (RDRBDR) images (166 m/pixel), we identified morphological features that confirm that Raditladi's IP were deposited by low-viscosity flows (Fig. 5c), as suggested by other authors (Prockter *et al.* 2009; Blair *et al.* 2013). Previous works (Minelli *et al.* 2013) reported a change in the thickness of these deposits, indicating an increase from the annulus outside of the peak ring (an estimated thickness of 700 m) towards the centre of the basin (an estimated thickness of 1200 m). Stratigraphic and topographical data support this hypothesis, as the IP are found at a higher elevation in the annulus outside of the peak ring than in the central basin. The changes in thicknesses are probably caused by some combination of different depths in the pre-flooded basin floor, greater thermal contraction of the thickest flows (Blair *et al.* 2013) and different thicknesses of the underlying units (e.g. underlying hummocky material outside of the peak ring) (Fig. 8).

Given the presence of IP both to the south and north of our cross-section, we assume that the intercrater plains units are located below the smooth plains. We also assumed that the IP and SP have comparable thicknesses (i.e. 1 km or more: Head *et al.* 2011; Stockstill-Cahill *et al.* 2012), an assumption partially supported by the observation that these deposits were produced with similar emplacement mechanisms (Stockstill-Cahill *et al.* 2012). We recognize that further work is necessary to estimate the total thickness of IP deposit more precisely, however.

The geological map (Figs 4 & 5) and cross-section (Fig. 8) show how the SP and IP are in lateral contact with the BIP deposits mapped in the western area. Although interpretation of this context is clearly related to the lack of data at depth and is thus equivocal, we suggest that IP and SP were deposited above the BIP. The pre-flooding topography still remains uncertain and many hypotheses can be made: (1) IP and SP have filled an older topographical depression possibly produced by a regional tilt towards the east to NE; (2) IP and SP filled a regional-scale basin whose origins and lateral extent are unknown; or (3) BIP deformation was driven by the load of the IP and SP deposits, and thus followed their deposition.

Tectonic structures are primarily found in the inner portion of the Raditladi basin where morphological evidence of circumferential graben is clearly detected by both images and the DTM (Figs 5a, 7c & 8). These graben are found only in the northern and eastern quadrants of the inner basin, and they mark IPR with several sets of concentric structures and a mean vertical displacement of a few tens of metres. A recent study (Blair *et al.* 2013) found that the formation of these graben is related to thermal contraction of the inner plains.

In the western portion of the geological cross-section, we found topographical relief whose pattern and morphology does not appear to indicate ejected deposits or structures related to the Raditladi impact. This relief (highest point centred at 12°N, 112°E) is divided in two main segments striking NW–SE, is morphologically distinct from all other topographical reliefs observable in Figure 8a and is clearly observable in the DTM (Fig. 8), although MESSENGER images of this area do not highlight particular reflectance features or tectonic deformation of the surface deposits (Fig. 9). Using the geological cross-section, we note that this structure is elevated with respect to the mean local topography by approximately 400 m and that, except for the ejecta and the rim of Raditladi, it is the highest feature across the entire cross-section (i.e. the only topographical feature above zero). Considering its spatial distribution, the topographical uplift it produced and the morphological context found east and west of the uplift, we interpret this topographical relief as unrelated to the impact and possibly produced by a blind thrust acting on the SP deposits. Although this interpretation is uncertain, analysis of craters found above the uplift may help test our interpretation and the inference that their floors are partially or completely offset (Fig. 9) by the evolution of the thrust after their formation.

We found BIP and BCP deposits in the western portion of the geological map and, under the assumption that the BCP are older than all other units, we placed the BCP at the bottom of our cross-section (Fig. 8). Owing to the lack of information on the intracrustal materials and structures, however,

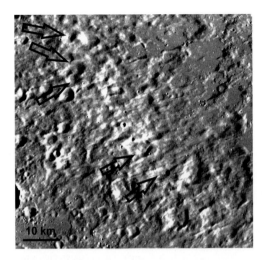

Fig. 9. Monochrome image of the topographical uplift located in Figure 8. Black arrows indicate the crater basins that we interpret to have been deformed by tectonic uplift produced by an underlying blind thrust. The image is taken from the mosaic of images used in Figure 4.

our model cannot be accurately extended below the BIP deposits.

Conclusions and suggestions for further work

Integrating the smooth plains on Mercury, as mapped by Denevi *et al.* (2013), we created a geological map covering nearly the entire planet surface at a resolution of 250 m/pixel. The major units we identified are the intercrater plains (IP) and bright intercrater plains (BIP). Other units found include the Caloris rough ejecta (CRE: Caloris ejecta deposits not covered by later resurfacing) and dark material (DM), which encompass the localized deposits that we interpret as either volcanic or related to impacts.

Given the distribution of the SP and IP and their likely volcanic origin (Malin 1976; Strom 1977; Spudis & Guest 1988; Denevi *et al.* 2013), the volcanic contribution to global resurfacing of the planet Mercury was certainly substantial and probably distributed over a very long time interval. Although we are confident in our global geological map, which is supported by the mosaic of MASCS observations (Fig. 1c), we believe that further analysis is required to achieve a more detailed mapping of units at the local scale and to determine thicknesses and compositions of major units, particularly the IP.

To produce a detailed geological cross-section for the Raditladi region, we focused our detailed

mapping on the area (19–35°N, 106–133°E) where global mapping suggests a particularly varied surface geology and where sufficient data are available (e.g. DTM and crater dating of the ejecta and inner plains), which allowed us to constrain the geological model. As expected, the local map we obtained highlights a diverse collection of surface geology (Figs 4 & 5), including both local-scale deposits and global-scale units ranging from the older bright cratered plains (BCP) to the younger recent knobby deposits (RKD).

The BCP deposits are representative of old terrains that were not buried by later resurfacing, but further detailed investigation will help to resolve their composition and assess whether these deposits are solely localized in this region or are present in other areas of Mercury. The RKD unit cannot currently be uniquely attributed to a definite resurfacing process; in fact, we feel that given their morphology and position, these units can be interpreted either as high-viscosity volcanic deposits or as deposits formed by the gravitational collapse of a portion of the Raditladi peak ring. Future investigation of the RKD unit should also involve crater counting and age determination to quantitatively evaluate its age relative to the inner plains (IPR), annular plains (AP) and dark annular plains (DAP) units found within Raditladi.

We find that the volcanic contribution to local-scale resurfacing is considerable. The IPR appear to be volcanic in origin and to have been deposited in low-viscosity flows (Fig. 5). The geological cross-section we produced (Fig. 8) shows that IPR deposition was influenced by the topography of the basin after the emplacement of the AP and the DAP units, leading to an asymmetric thickness of the IPR in the inner basin with respect to the outer ring of the basin. Later thermal contraction of the IPR volcanic deposits produced the circumferential graben that are observed in the inner basin and exaggerated pre-existing elevation differences (Blair *et al.* 2013).

Hollows found on the Raditladi's peak ring (Blewett *et al.* 2011, 2012) contribute to the complexity and diversification of the surface geology, but may also indicate an increase in the volcanic contribution to resurfacing. Our geological cross-section (Fig. 8) shows that the peak-ring (PR) material of Raditladi originated from IP materials that were uplifted and deformed by the impact, which in turn implies that hollows found on the peak structures possibly originated from IP deposits, and that their formation may be related to the alteration of these deposits induced by the impact. This observation is in agreement with recent studies (Blewett *et al.* 2011, 2012) in which hollows were found to form only on low-reflectance-material host rock (like the IP).

Superposed peak deformation (SPD) regions are found only in craters with diameters between 20 and 120 km, whereas basins with diameters less than 20 km have simple geometries and no SPD units. SPD are mapped only on the basis of morphology, and thus we cannot unequivocally interpret the SPD as representative of impacts that were large enough to have excavated the surficial target unit (i.e. SP) to find the underlying IP terrain (e.g. Melosh 1989), and/or as representative of the simple-to-complex crater transitions (Baker *et al.* 2011). Further imaging of these units at high resolution will help to solve this open question (e.g. by comparing spectral properties of SPD with those of surrounding material).

Although local-scale surface geology is complex, global mapping still offers challenging work best carried out through local- and regional-scale mapping efforts. An example is given by the bright cratered plains (BCP) that we mapped NW of Raditladi (Fig. 4); these deposits possibly represent the oldest units observed in the Raditladi region, and it may be that similar deposits will be found in other regions when investigated under an appropriately scaled mapping project.

We thank the MESSENGER team for making available images and derived products. We thank B. Denevi for making available data from her recent work about smooth plains. We gratefully thank P. K. Byrne, D. M. Blair and an anonymous reviewer for their constructive comments. This work was funded by the ASI-INAF BepiColombo Agreement No. I/022/10/0.

References

BAKER, D. M. H., HEAD, J. W. *ET AL.* 2011. The transition from complex crater to peak-ring basin on Mercury: new observations from MESSENGER flyby data and constraints on basin formation models. *In:* SOLOMON, S. C., MCNUTT, R. L. & PROCKTER, L. M. (eds) *Mercury after the MESSENGER Flybys. Planetary and Space Science*, **59**, 1932–1948.

BLAIR, D. M., FREED, A. M. *ET AL.* 2013. The origin of graben and ridges in Rachmaninoff, Raditladi, and Mozart basins, Mercury. *Journal of Geophysical Research: Planets*, **118**, 1–12.

BLEWETT, D. T., CHABOT, N. L. *ET AL.* 2011. Hollows on Mercury: MESSENGER evidence for geologically recent volatile-related activity. *Science*, **333**, 1856–1859.

BLEWETT, D. T., VAUGHAN, W. M. *ET AL.* 2012. Mercury's hollows: constraints on formation and composition from analysis of geological setting and spectral reflectance reflectance. *Journal of Geophysical Research: Planets*, **118**, 1013–1032, http://dx.doi.org/10.1029/2012JE004174

BUCZKOWSKI, D. L. & SEELOS, K. D. 2012. A map of the Intra-ejecta dark plains of the Caloris basin, Mercury. *In: 43rd Lunar and Planetary Science Conference,*

March 19–23, 2012, The Woodlands, Texas. Lunar and Planetary Institute, Houston, TX, abstract 1844.

BUCZKOWSKI, D. L., EDRICH, S., ACKISS, S. & SEELOS, K. D. 2013. Geomorphic mapping of the Caloris basin, Mercury. *In: Large Meteorite Impacts and Planetary Evolution V, Proceedings of the Conference,* 5–8 August 2013, Sudbury, Canada. Lunar and Planetary Institute, Houston, TX, abstract 3089.

DE HON, R. A., SCOTT, D. H. & UNDERWOOD, J. R. JR. 1981. *Geologic Map of the Kuiper (H-6) Quadrangle of Mercury.* United States Geological Survey, Geologic Investigations Series, **Map I-1233**.

DENEVI, B. W., ROBINSON, M. S. *ET AL.* 2009. The evolution of Mercury's crust: a global perspective from MESSENGER. *Science*, **324**, 613–618.

DENEVI, B. W., ERNST, C. M. *ET AL.* 2013. The distribution and origin of smooth plains on mercury. *Journal of Geophysical Research: Planets*, **118**, 891–907, http://dx.doi.org/10.1002/jgre.20075

FASSETT, C. I., HEAD, J. W. *ET AL.* 2009. Caloris impact basin: exterior geomorphology, stratigraphy, morphometry, radial sculpture, and smooth plains deposits. *Earth and Planetary Science Letters*, **285**, 297–308.

FRIGERI, A., FEDERICO, C., PAUSELLI, C. & CORADINI, A. 2009. Fostering digital geologic maps: the digital geologic map of Mercury from the USGS Atlas of Mercury, Geologic Series. *In: 40th Lunar and Planetary Science Conference*, March 23–27, 2009, The Woodlands, Texas. Lunar and Planetary Institute, Houston, TX, abstract 2417.

GROLIER, M. J. & BOYCE, J. M. 1984. *Geologic Map of the Borealis Region (H-1) of Mercury.* United States Geological Survey, Miscellaneous Investigations Series, **Map I-1660**.

GUEST, J. E. & GREELEY, R. 1983. *Geologic Map of the Shakespeare (H-3) Quadrangle of Mercury.* United States Geological Survey, Miscellaneous Investigations Series, **Map I-1408**.

HEAD, J. W., CHAPMAN, C. R. *ET AL.* 2011. Flood volcanism in the northern high latitudes of Mercury revealed by MESSENGER. *Science*, **333**, 1853.

KING, J. S. & SCOTT, D. H. 1990. *Geologic Map of the Beethoven (H-7) Quadrangle of Mercury.* United States Geological Survey, Miscellaneous Investigations Series, **Map I-2048**.

MALIN, M. C. 1976. Observations of intercrater plains on Mercury. *Geophysical Research Letters*, **3**, 581–584.

MARCHI, S., MASSIRONI, M., CREMONESE, G., MARTELLATO, E., GIACOMINI, L. & PROCKTER, L. 2011. The effects of the target material properties and layering on the crater chronology: the case of Raditladi and Rachmaninoff basins on Mercury. *In:* SOLOMON, S. C., MCNUTT, R. L. & PROCKTER, L. M. (eds) *Mercury after the MESSENGER Flybys. Planetary and Space Science*, **59**, 1968–1980.

MARCHI, S., CHAPMAN, C. R., FASSETT, C. I., HEAD, J. W., BOTTKE, W. F. & STROM, R. G. 2013. Global resurfacing of Mercury 4.0–4.1 billion years ago by heavy bombardment and volcanism. *Nature*, **499**, 59–61, http://dx.doi.org/10.1038/nature12280

MARTELLATO, E., MASSIRONI, M., CREMONESE, G., MARCHI, S., FERRARI, S. & PROCKTER, L. M. 2010.

Age determination of Raditladi and Rembrandt basins and related geological units. *In: 41st Lunar and Planetary Science Conference*, March 1–5, 2010, The Woodlands, Texas. Lunar and Planetary Institute, Houston, TX, abstract 2148.

McGILL, G. E. & KING, E. A. 1983. *Geologic Map of the Victoria (H-2) Quadrangle of Mercury*. United States Geological Survey, Miscellaneous Investigations Series, **Map I-1409**.

MELOSH, H. J. 1989. *Impact Cratering – A Geologic Process*. Oxford Monographs on Geology and Geophysics, **11**. Oxford University Press, New York.

MINELLI, F. *ET AL.* 2013. New mapping of Raditladi basin and detailed analysis of its inner plains. *In: EGU2013: 10th EGU General Assembly 2013*, 7–12 April, 2013, Vienna, Austria Assembly, Volume 15. Copernicus, Göttingen, 9943.

NITTLER, L. R., STARR, R. D. *ET AL.* 2011. The major-element composition of Mercury's surface from MESSENGER X-Ray spectrometry. *Science*, **333**, 1847–1850.

NASA/JHUAPL/CIW 2012. *PIA16668*, taken from http://photojournal.jpl.nasa.gov/catalog/PIA16668. NASA/Johns Hopkins University Applied Physics Laboratory/Carnegie Institute of Washington.

NASA/JHUAPL/CIW 2013. *Global Monochrome Mosaic*, taken from http://messenger.jhuapl.edu/the_mission/mosaics.html. NASA/Johns Hopkins University Applied Physics Laboratory/Carnegie Institute of Washington.

NASA/PLANETARY DATA SYSTEM GEOSCIENCES NODE 2012. Map Projected Basemap Reduced Data Record (RDRBDR) mdis_bdr_256ppd_h04sw0 and mdis_bdr_256ppd_h09ne0.

PREUSKER, F., OBERST, J. *ET AL.* 2011. Stereo topographic models of Mercury after three MESSENGER flybys. *In:* SOLOMON, S. C., McNUTT, R. L. & PROCKTER, L. M. (eds) *Mercury after the MESSENGER Flybys. Planetary and Space Science*, **59**, 1910–1917.

PROCKTER, L. M., WATTERS, T. R. *ET AL.* 2009. The curious case of Raditladi basin. *In: 40th Lunar and Planetary Science Conference*, March 23–27, 2009, The Woodlands, Texas. Lunar and Planetary Institute, Houston, TX, abstract 1758.

REUTER, H. I., HENGL, T., GESSLER, P. & SOILLE, P. 2009. Preparation of DEMs for Geomorphometric analysis. *In:* TOMISLAV, H. & HANNES, I. R. (eds) *Developments in Soil Science*. Elsevier, Amsterdam, 87–120.

SCHABER, G. G. & McCAULEY, J. F. 1980. *Geologic Map of the Tolstoj (H-8) Quadrangle of Mercury*. United States Geological Survey, Miscellaneous Investigations Series, **Map I-1199**.

SMITH, D. E., ZUBER, M. T. *ET AL.* 2012. Gravity field and internal structure of Mercury from MESSENGER. *Science*, **336**, 214–217.

SOLOMON, S. C., McNUTT, R. L. JR. *ET AL.* 2008. Return to Mercury a global perspective on MESSENGER's first Mercury flyby. *Science*, **321**, 59–62.

SPUDIS, P. D. & GUEST, J. E. 1988. Stratigraphy and geologic history of Mercury. *In:* VILAS, F.,

CHAPMAN, C. R. & MATTHEWS, M. S. (eds) *Mercury*. University of Arizona Press, Tucson, AZ, 118–164.

SPUDIS, P. D. & PROSSER, J. G. 1984. *Geologic map of the Michelangelo (H-12) quadrangle of Mercury*. United States Geological Survey, Miscellaneous Investigations Series, **Map I-1659**.

STOCKSTILL-CAHILL, K. R., McCOY, T. J., NITTLER, L. R., WEIDER, S. Z. & HAUCK, S. A., II . 2012. Magnesium-rich crustal compositions on Mercury: Implications for magmatism from petrologic modelling. *Journal of Geophysical Research: Planets*, **117**, E00L15, http://dx.doi.org/10.1029/2012JE004140

STROM, R. G. 1977. Origin and relative age of lunar and mercurian intercrater plains. *Physics of the Earth and Planetary Interiors*, **15**, 156–172.

STROM, R. G., MALIN, M. C. & LEAKE, M. A. 1990. *Geologic map of the Bach (H-15) quadrangle of Mercury*. United States Geological Survey, Miscellaneous Investigations Series, **Map I-2015**.

STROM, R. G., TRASK, N. J. & GUEST, J. E. 1975. Tectonism and volcanism on Mercury. *Journal of Geophysical Research*, **80**, 2478–2507, http://dx.doi.org/10.1029/JB080i017p02478

STROM, R. G., CHAPMAN, C. R., MERLINE, W. J., SOLOMON, S. C. & HEAD, J. W. 2008. Mercury cratering record viewed from MESSENGER's first flyby. *Science*, **321**, 79–81.

STROM, R. G., BANKS, M. E. *ET AL.* 2011. Mercury crater statistics from MESSENGER flybys: implications for stratigraphy and resurfacing history. *In:* SOLOMON, S. C., McNUTT, R. L. & PROCKTER, L. M. (eds) *Mercury after the MESSENGER Flybys. Planetary and Space Science* **59**, 1960–1967.

TRASK, N. J. & DZURISIN, D. 1984. *Geologic map of the Discovery (H-11) quadrangle of Mercury*. United States Geological Survey, Miscellaneous Investigations Series, **Map I-1658**.

TRASK, N. J. & GUEST, J. E. 1975. Preliminary geologic terrain map of Mercury. *Journal of Geophysical Research*, **80**, 2461–2477, http://dx.doi.org/10.1029/JB080i017p02461

WEIBEL, R. & HELLER, M. 1990. Digital terrain modelling. *In:* MAGUIRE, D. J., GOODCHILD, M. F. & RHIND, D. W. (eds) *Geographical Information Systems*, Volume 1. Longman, Harlow, Chapter 19.

WEIDER, S. Z., NITTLER, L. R. *ET AL.* 2012. Chemical heterogeneity on Mercury's surface revealed by the MESSENGER X-Ray Spectrometer. *Journal of Geophysical Research*, **117**, E00L05, http://dx.doi.org/10.1029/2012JE004153

WHITTEN, J. L., HEAD, J. W. *ET AL.* 2012. Intercrater plains on Mercury: topographic assessment with MESSENGER data. *In: 43rd Lunar and Planetary Science Conference*, March 19–23, 2012, The Woodlands, Texas. Lunar and Planetary Institute, Houston, TX, abstract 1479.

XIAO, Z., STROM, R. G. *ET AL.* 2013. Dark spots on Mercury: a distinctive low-reflectance material and its relation to hollows. *Journal of Geophysical Research*, **118**, 1752–1765, http://dx.doi.org/10.1002/jgre.20115

The magmatic evolution of three Venusian coronae

NICHOLAS P. LANG[1]* & IVÁN LÓPEZ[2]

[1]Department of Geology, Mercyhurst University, 501 East 38th Street, Erie, PA 16546, USA

[2]Department of Biology and Geology, Universidad Rey Juan Carlos, Madrid, Spain

*Corresponding author (e-mail: nlang@mercyhurst.edu)

Abstract: The volcanic and tectonic histories of Venusian coronae appear to be intricately linked. We explore that link through the construction of geological maps of three coronae using Magellan synthetic aperture radar (SAR) imagery and altimetry radar data. Each examined corona – Aramaiti, Bhumidevi and Zemire Coronae – is characterized by an annulus of concentric fractures, lava flows, tholi and clusters of small shields. Radial fractures occur at one corona (Bhumidevi Corona), whereas the other two coronae (Aramaiti and Zemire Coronae) occur within linear fracture belts. Based on observed timing relationships, we propose that the evolution of these three coronae is similar to large silicic caldera formation on Earth in that the formation of concentric fractures facilitated a mass evacuation of a magma reservoir residing underneath each corona resulting in subsequent corona collapse. Our model emphasizes the late-stage evolution of these coronae and predicts that all the erupted products could be basaltic. Such predictions are testable through continued geological mapping of coronae, as well as through *in situ* spectral analyses by Venusian landers.

Coronae are quasi-circular, tectonomagmatic features that characterize parts of the Venusian surface. Ranging in diameter from about 50 to >500 km, coronae are distinguished by an annulus of concentric fractures and/or ridges with possible radial fractures that extend from the corona's interior (Squyres *et al.* 1992; Stofan *et al.* 1992). Based on the timing of these structures relative to one another, coronae are largely interpreted to have formed from the interaction of either a mantle plume or diapir with the base of the lithosphere (Janes *et al.* 1992; Squyres *et al.* 1992; Stofan *et al.* 1992; Dombard *et al.* 2007). To elaborate, the general conceptual model of the structural and morphological evolution for coronae is (Squyres *et al.* 1992; Stofan *et al.* 1992): (1) the impingement of a mantle diapir (Hansen 2003) or plume and its associated partial melt (Dombard *et al.* 2007) on the base of the lithosphere causing domical uplift of the planet's surface and the initiation of radial fractures; (2) radial spreading and flattening of the diapir, which transforms the domical uplift to a plateau; and (3) removal of thermal support of the topography causing gravitational relaxation of the plateau to form a moat, rim and/or interior depression or trough resulting in the formation of the annulus of concentric fractures. The timescale to complete this three-step evolutionary sequence is predicted to be of the order of 100–250 Ma (Stofan *et al.* 1997).

In all three evolutionary steps, volcanism is noted to occur in variable amounts (Squyres *et al.* 1992; Stofan *et al.* 1992, 1997; Roberts & Head 1993). Corona-related volcanism manifests itself in a variety of ways including extensive large-scale flows (e.g. Roberts & Head 1993), clusters of small volcanic constructs or shields (e.g. Crumpler *et al.* 1997), and localized intermediate-sized volcanic edifices including steep-sided domes or tholi (e.g. Pavri *et al.* 1992). In an examination of 326 coronae, Roberts & Head (1993) noted that extensive flows occur at 41% of coronae, with a bulk of those flows (47%) having been emplaced early in the structural evolution of the corona (i.e. stage 1); *c.* 45% of the flows were emplaced throughout coronae evolution and 9% of the flows occurred late in coronae evolution (i.e. after stage 3) (Roberts & Head 1993). These large-scale flows typically occur as digitate aprons that either symmetrically or asymmetrically surround a corona and extend for 150–300 km from either the annulus (if the flow field occurred late in the corona's evolution) or an unidentifiable source within the corona (if the flow field was emplaced earlier in the corona's evolution) (Roberts & Head 1993). Over time, corona-related volcanism becomes concentrated towards the corona's interior and decreases in scale (Roberts & Head 1993). Shields associated with coronae typically occur in clusters (termed 'companion shield fields': Crumpler *et al.* 1997) and probably represent the eruption of small batches of magma tapped from the same overall source as that for the corona (Crumpler *et al.* 1997). The timing of shields in relation to corona evolution is unclear (e.g. López & Lang 2008) but, following terrestrial analogies, corona-associated

From: PLATZ, T., MASSIRONI, M., BYRNE, P. K. & HIESINGER, H. (eds) 2015. *Volcanism and Tectonism Across the Inner Solar System*. Geological Society, London, Special Publications, **401**, 77–95.
First published online December 16, 2013, http://dx.doi.org/10.1144/SP401.3

shields may reflect the late-stage waning of magma supply rates to the corona (Crumpler *et al.* 1997). Tholi have also been noted to occur with some coronae where they typically occur on the annulus (Pavri *et al.* 1992), suggesting that they reflect some of the latest stages of corona-related volcanism. Tholi have also been predicted to form from either differentiation of basaltic magma to produce silicic magmas (Pavri *et al.* 1992; Ivanov & Head 1999; Petford 2000; see also McKenzie *et al.* 1992; Fink *et al.* 1993), the melting of basaltic crust to produce more silica-rich magmas (Ivanov & Head 1999) or the eruption of a high-viscosity basaltic foam due to volatile enhancement in the upper part of the magma reservoir (Pavri *et al.* 1992; see also Stofan *et al.* 2000). Clearly, volcanism has played a fundamental role in corona development where the style and volume of volcanism may be largely dictated by magma supply rates related to the corona (e.g. Crumpler *et al.* 1997). Pressure release melting and extension in the initial stages of corona development could, seemingly, produce high magma supply rates resulting in the early extensive flow fields (e.g. Roberts & Head 1993), and subsequent cooling of the diapir may trigger lower magma supply rates creating localized episodes of volcanism in the form of shield fields (e.g. Crumpler *et al.* 1997) and tholi (e.g. Pavri *et al.* 1992). Hence, there exists a seemingly

intimate relationship between the abundance and styles of corona-related volcanism and its evolutionary stage.

Motivated by further understanding of the broad connection between the evolutionary stage of coronae and the styles of volcanism represented at them, we examined the structural and volcanic evolution of three Type 1 (Stofan *et al.* 2001) Venusian coronae. Specifically, using Magellan synthetic aperture radar (SAR) imagery (*c.* 75 m/pixel), we have geologically mapped the extent and distribution of volcanic products at: Aramaiti (AC: 26°S, 81°E), Bhumidevi (BC: 17°S, 343°E) and Zemire Coronae (ZC: 32°N, 312°E) (Fig. 1). The tectonic and volcanic characteristics of the three coronae suggest they have undergone the complete structural evolution sequence proposed for coronae (Squyres *et al.* 1992; Stofan *et al.* 1992) and are consistent with their having formed through the interaction of an upwelled mantle (either diapir or plume and associated partial melt) with the base of the Venusian lithosphere (Squyres *et al.* 1992; Stofan *et al.* 1992; Hansen 2003; Dombard *et al.* 2007). Hence, these coronae are seemingly good candidates with which to explore connections that may exist between the structural and volcanic evolution at coronae. For each corona, we have placed the timing of erupted products into the corona's structural evolution. Our results

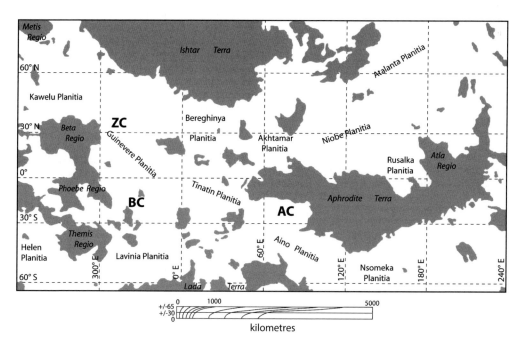

Fig. 1. Sketch map of Venus in Mercator Projection showing the locations of the three examined coronae: (1) Aramaiti Corona (AC: 26°S, 82°E); (2) Bhumidevi Corona (BC: 17°S, 344°E); and (3) Zemire Corona (ZC: 31°N, 314°E). Grey regions represent highlands. After Guest *et al.* (1992).

emphasize the late-stage volcanic and structural evolution of these three coronae and have implications for the petrology of at least some of their erupted deposits.

Data and methodology

Geological mapping and corona analysis conducted in this research involved the use of NASA's Magellan S-band (12.6 cm wavelength) SAR and altimetry data (e.g. Ford *et al.* 1993). Data used for the study include: (1) Full-Resolution Basic Image Data Record (F-BIDR: 75–100 m/pixel, both right and left-illuminated radar images); and (2) Magellan altimetry (*c.* 8 km along-track × 20 km across-track with *c.* 30 m average vertical accuracy, which improves to *c.* 10 m in smooth areas: Ford *et al.* 1993). All images were obtained through the USGS Map-a-Planet website (http://www.mapaplanet.org), and were viewed in both normal and inverted modes to highlight details of primary and secondary structures; structures are typically more apparent in inverted images. For inverted SAR images, low backscatter areas appear bright and high backscatter areas appear dark. In addition, apparent illumination is reversed; hence, left-illuminated inverted images appear right-illuminated. All SAR images presented herein are left-illuminated and inverted. We also constructed a topographical profile for each corona, where the elevation data were extracted from radar altimetry data using NIH-Image macros. Constructing topographical profiles proved useful in elucidating geological–topographical relationships critical for unravelling geological histories for each region. For consistency, the orientation of each profile is from the NW to the SE across each corona; this orientation was randomly chosen due to the inherent symmetry of the examined coronae. With each study area, we mapped an area large enough to place the corona and associated structures and deposits into their appropriate spatial and temporal context. Geological mapping follows the guidelines and cautions of Wilhelms (1990), Tanaka (1994), Hansen (2000) and Skinner & Tanaka (2003). SAR interpretation follows Ford *et al.* (1993).

Study area geology

Aramaiti Corona

Aramaiti is a 350 km-diameter concentric corona (Stofan *et al.* 1997) centred near 26°S, 82°E (Fig. 2a, b) and is located in Aino Planitia approximately 300 km NW of the 175 km-diameter Ohogetsu Corona. Both coronae occur within a composite of low backscatter plains materials (probably flow materials?) that comprise much of Aino Planitia (Stofan & Guest 2003). NW-trending outcrops of regional basement materials (i.e. tesserae terrain and fractured and ridged materials: Bindschadler *et al.* 1992; Basilevsky & Head 1998; Stofan & Guest 2003) occur on the west and east sides of Aramaiti. Aramaiti has previously been examined by Squyres *et al.* (1992), Basilevsky & Head (1998) and Stofan & Guest (2003), and our descriptions of Aramaiti presented below are consistent with their work.

Topography. Figure 2c is a topographical profile with approximately 80× vertical exaggeration highlighting Aramaiti's topography from the NW of the corona to the SE. From the NW, Aramaiti's topography gently rises a couple of hundred metres above the MPR (Mean Planetary Radius) to a relatively narrow rim where it drops about 100 m to create an approximately 30 km-long topographical bench that dips to the NW. From the bench, the topography drops about 500 m into an approximately 50 km-wide depression, or trough, that is concentric around the corona; elevations within the trough are not consistent and appear to be upwards of about 100 m deeper on the northern and southern parts of the trough compared to the eastern and western sides (Basilevsky & Head 1998). A broad approximately 100 km-wide topographical dome rises about 800 m from this trough. From the trough SE of the dome, topography rises around 1 km in <100 km to reach a relatively narrow topographical rim; this rim drops back down into the regional plains. Squyres *et al.* (1992) noted that this topography is essentially inverted from many other coronae and proposed that Aramaiti is topographically more similar to a caldera than a corona; specifically, Aramaiti may have undergone simple downdropping and accompanying isostatic adjustment similar to large terrestrial calderas (Squyres *et al.* 1992). Based on its topography, we agree (and argue below) that Aramaiti has undergone caldera-like processes but, based on its occurrence within a fracture belt (which probably represents the surface expression of dyke emplacement due to mantle upwelling (e.g. Ernst *et al.* 2003)), we contend that Aramaiti Corona must have undergone processes at least very similar to the three-stage evolutionary progression outlined for coronae in general.

Tectonic structures. Tectonically, Aramaiti is characterized by a 100 km-wide annulus of concentric fractures spatially associated with a NE-trending belt of fractures (Fig. 2a, b). Fractures within the annulus appear to be graben with steep (nearly vertical) walls and flat floors approximately 1 km

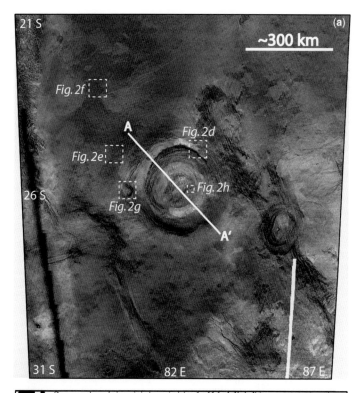

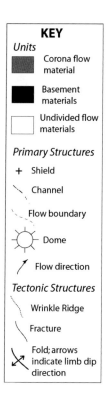

Fig. 2. *Continued.*

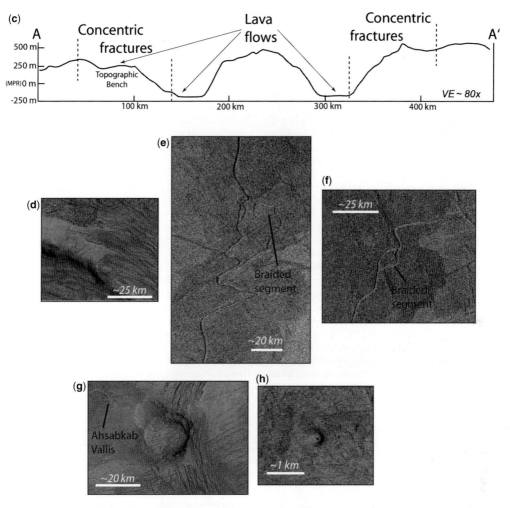

Fig. 2. (*Continued*) Imagery characterizing Aramaiti Corona. (**a**) Inverted SAR image of Aramaiti Corona. Dashed white boxes show the locations of (d)–(h); the white line with A–A' is the location of the topographical profile shown in (c). (**b**) Geological map of Aramaiti Corona. (**c**) Topographical profile of Aramaiti Corona showing the locations of corona-related fractures and lava flows in relation to the corona's topography. (**d**)–(**h**) Enlarged inverted SAR images highlighting specific parts of Aramaiti Corona and surrounding deposits: (d) close-up view of a trough on the north side of Aramaiti showing concentric fractures locally buried by flow material (the lighter-coloured patch in the image); (e) view of a segment of Ahsabkab Vallis highlighting a braided section of the channel; (f) view of part of Ahsabkab Vallis highlighting another braided section of the channel; (g) view of three steep-sided domes (Narina Tholi) that have erupted from Aramaiti's western annulus; these three domes are superposed on one another and post-date an approximately 300 km-long north-trending lava flow Oilule Fluctus and its associated Ahsabkab Vallis that erupted from the same location; (h) small shield volcano located on the inside of Aramaiti's annulus; this shield is similar to the numerous small volcanoes that occur across the map area.

wide (Fig. 2d); these fractures have spacings of around 1–2 km and some appear to be continuous around the corona, whereas others are abruptly truncated. The concentric fractures are spatially associated with the corona rim where they occur along the entire rim crest and slope down into the topographical trough (Fig. 2c); the fractures become sparser in their occurrence on the floor of the trough, but re-occur along the edge of the topographical dome. Fractures within the NE-trending belt have geometries similar to the concentric fractures and we interpret them to also represent graben. The fracture belt is approximately 1000 km wide, with fracture spacings of ≥20 km along the belt edges; fracture spacings decrease to ≤5 km in the middle of the belt in the vicinity of Aramaiti.

Fractures within the belt predominantly cut the concentric fractures, although locally some concentric fractures appear to cut belt fractures. Fractures within the belt also cut Oilule Fluctus and Narina Tholi – materials erupted from Aramaiti (Fig. 2a, b). In addition, fractures also occur in a radial pattern on top of the topographical dome, but are abruptly truncated at the dome's base (Fig. 2a, b).

The map area has also been deformed by regional suites of extensional and contractional structures that appear to have occurred independently from the evolution of Aramaiti. Specifically, NW-trending fractures occur in the western part of the map area and are seemingly associated with activity at Kunapipi Mons (see Stofan & Guest 2003), which is located about 800 km SE of Aramaiti. However, the ridges are orthogonal to the NE-trending fracture belt, suggesting that a kinematic relationship may exist between these two structural suites. Wrinkle ridges approximately 20–100 km long, with wavelengths of ≥ 30 km, deform all plains materials including those comprising Aramaiti's trough. Wrinkle ridges within Aramaiti's trough have lengths of approximately 10–20 km, with wavelengths of ≤ 5–10 km. NE-trending wrinkle ridges occur in the NE corner of the map area and parallel the NE-trending fracture belt. Longer wavelength contraction has also occurred within the map area in the form of NW-trending warps; the warping has a wavelength of around 600 km and broadly parallels the NW-trending wrinkle ridges. Exposed basement materials within the map area also have NW-trending outcrop patterns and may represent the crests of warps.

Volcanic features. Numerous distinct volcanic features are also present at Aramaiti. These include extensive (Oilule Fluctus: Fig. 2b) and trough-filling lava flows, a lava channel (Ahsabkab Vallis: Fig. 2d–f), tholi (Narina Tholi; Fig. 2g), and abundant shields (Fig. 2h).

Oilule Fluctus – the only extensive lava flow seemingly associated with Aramaiti – is an approximately 400 km-long lava flow that has travelled north from a vent in the annulus on Aramaiti's western side (Fig. 2a, b). The flow varies in width from around 50 to >100 km, is cut by NE-trending fractures and contains a prominent channel (Ahsabkab Vallis: Fig. 2b, d, e) that presumably fed the active front of the flow. Oilule Fluctus is denoted by relatively high backscatter (in comparison to surrounding plains materials) (Fig. 2f) indicating that the flow surface is rough at the 12.6 cm scale (e.g. Ford et al. 1993). Radar studies of lava flow surfaces (e.g. Byrnes 2002) suggest that rougher surfaces (or surfaces that have a roughness equal to or greater than the incoming radar

signal: Ford et al. 1993) are typically associated with higher backscatter return and are consistent with them representing 'a'a-textured lava flows, whereas topographically smooth surfaces are associated with smaller amounts of backscatter return and are consistent with them representing pahoehoe-textured lava flows. Based on this analogy, we interpret Oilule Fluctus as an 'a'a flow. If our interpretation is correct, then Oilule Fluctus probably erupted at a high effusion rate, which, in turn, may explain the great length of this flow (Pinkerton & Wilson 1994).

Ahsabkab Vallis is a canali-type channel (Baker et al. 1992) within Oilule Fluctus. It appears to be a constructional channel (Gregg & Greeley 1993) formed during Oilule Fluctus emplacement and extends almost the entire length of the lava flow (Fig. 2b, d–f); the channel first becomes apparent approximately 15 km NE of Narina Tholus (Fig. 2g). It is located predominantly within the centre of Oilule Fluctus, but bends to the NE and extends for about 150 km along the eastern edge of the flow before becoming difficult to trace near Oilule Fluctus' terminus; it is not clear whether Ahsabkab Vallis ends at this point or if it continues further into the surrounding plains materials below the resolution of Magellan imagery or has been subsequently buried. The channel maintains a mostly constant width of about 5 km along its length, but breaks into braids with narrower channel widths at two locations (Fig. 2a, d–f). Sinuosity varies along the length of the channel and ranges from approximately 1.6 near its origin to around 1.0 nearer where the channel becomes difficult to trace. These values are seemingly consistent with a high effusion rate for Oilule Fluctus (Komatsu & Baker 1994) and are also seemingly consistent with Oilule Fluctus being composed of a low-viscosity material (e.g. Baker et al. 1992; Komatsu & Baker 1994; see also Komatsu et al. 1992; Gregg & Greeley 1993).

More localized lava flows occur within Aramaiti's topographical depression, or trough, and are denoted on our map by flow boundary markers (Fig. 2b); we were not able to trace flow boundaries consistently around the topographical trough and were, thus, not able to map out distinct flow units within the trough. The source of these materials is unclear but, based on their location within the trough, may have also originated from the annulus fractures or from other vents located on the floor of the topographical trough (see Squyres et al. 1992).

Superposed on Oilule Fluctus at its source vent are three overlapping tholi named Narina Tholi (Fig. 2g). Based on their location, these tholi probably erupted from the same vent as Oilule Fluctus. Individual domes of Narina Tholi have diameters of around 20 km and have radar back-scatter characteristics similar to the surrounding plains (Stofan & Guest 2003). The tholi are mostly void of structures typically observed on the surfaces of other Venusian tholi (e.g. Pavri *et al.* 1992), although some pits <1 km in diameter and NE- to NNE-trending fractures do occur on the uppermost dome; these fractures have a similar trend to the NE-trending fracture belt and have spacings of <5 km. However, it is not clear whether these fractures formed with the fracture belt or if they represent the formation of a solid crust followed by further growth and cracking due to the intrusion of new magma (e.g. Fink *et al.* 1993; Stofan *et al.* 1997, 2000; Stofan & Guest 2003).

Shields (Fig. 2h) occur in broad clusters around Aaramaiti and are spatially associated with the fracture belt, although some do occur on the topographical dome in the corona's centre. There appears to be a nearly complete absence of shields on the corona's annulus (Fig. 2a, b).

Geological history. Based on cross-cutting and superposition relationships, the following geological history can be developed for the Aramaiti Corona area. Basement materials represent the earliest emplaced units in the map area; deformation of these materials and regional-scale warping followed their emplacement, but most probably predated the emplacement of flow materials associated with Aino Planitia (e.g. Stofan & Guest 2003; see also Ivanov & Head 2011). During the emplacement of Aino Planitia materials was the formation of Aramaiti and Ohegetsu Coronae; given that Ohegetsu's annulus cross-cuts many lineaments in the NE-trending fracture belt and that fractures within this belt cut many of Aramaiti's concentric fractures, Aramaiti may be the older corona. Formation of Aramaiti's annulus appears to be the earliest preserved geological event at the corona and was probably synchronous with the formation of the topographical trough, where the bench-like terrace apparent in the profile may reflect the down-dropping of a coherent fault block into the trough. Oilule Fluctus erupted from the annulus and resulted in the formation of Ahsabkab Vallis. Given that Narina Tholi occur on top of Oilule Fluctus, they probably post-date the flow; if they predated the flow, it seems likely that the eruption of the longer flows would have undermined the tholi resulting in their collapse. The abrupt truncation of concentric fractures within the topographical trough suggests that flows within the trough post-date annulus formation as well. A topographical profile (Fig. 2c) indicates that parts of Oilule Fluctus' current configuration is orientated uphill meaning that it must predate formation

of Aramaiti's topographical depressions; trough-filling flow material must, therefore, post-date both Oilule Fluctus and occurrence of the topographical depressions. The timing of dome formation in Aramaiti's centre in relation to annulus formation is ambiguous, but may have formed after annulus formation (see Squyres *et al.* 1992). Timing of shield formation is also unclear and seemingly could have occurred at any time during, if not throughout, Aramaiti's evolution. Structures within the NE-trending fracture belt cut across Oilule Fluctus, Narina Tholi and the annulus of concentric fractures (Fig. 2a, b, e), meaning that the NE-trending belt must post-date all three features. Wrinkle ridges occurred either synchronously with or after formation of the NE-trending fracture belt. To elaborate, the NE-trending wrinkle ridges parallel, and are spatially associated with, the NE-trending fracture belt, suggesting that the ridges may have originally been fractures that were filled with lava and subsequently experienced contraction, which resulted in ridge formation (e.g. DeShon *et al.* 2000). Further, if a kinematic relationship exists between the NE-trending facture belt and the NW-trending wrinkle ridges, then these two structural suites are probably synchronous or near synchronous in age. If they are not kinematically related, then, given that the ridges deform parts of Oilule Fluctus, the ridges must post-date the latest magmatic stages of Aramaiti.

Bhumidevi Corona

Bhumidevi is an approximately 200 km-diameter concentric corona (Stofan *et al.* 1997) centred near 17°S, 344°E. It is located in Kanykey Planitia between Vasilisa and Alpha Regiones along a NW-trending fracture zone, along which several other coronae and associated materials have occurred (Fig. 3a, b). In fact, there are three other coronae in this map area – Iyatik Corona is located about 250 east of Bhumidevi, and Takus Mana and Qetesh Coronae are both approximately 150 km south of Bhumidevi. Bhumidevi has been previously examined by Squyres *et al.* (1992), Bender *et al.* (2000) and Grindrod & Guest (2006), and our descriptions of Bhumidevi below are generally consistent with their work; we note differences between our analysis and previous workers on Bhumidevi in our description below.

Topography. Figure 3c is a NW–SE-trending topographical profile of Bhumidevi Corona with an approximately 100× vertical exaggeration. Topographically, Bhumidevi is characterized by a concentric, continuous moat that has a consistent depth of about 250 m below the surrounding plains and the MPR. The moat surrounds a continuous,

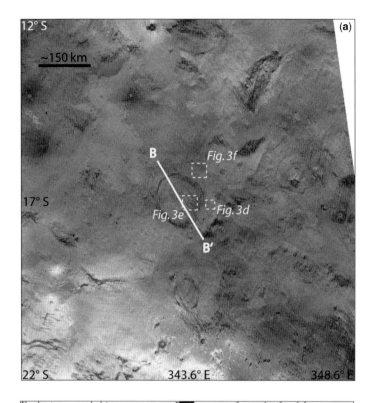

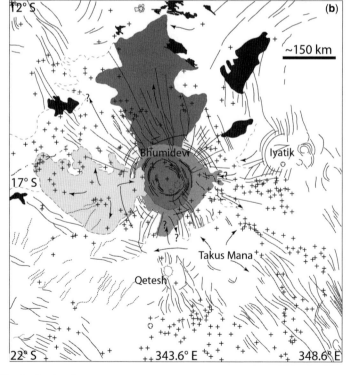

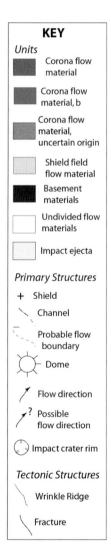

KEY

Units

- Corona flow material
- Corona flow material, b
- Corona flow material, uncertain origin
- Shield field flow material
- Basement materials
- Undivided flow materials
- Impact ejecta

Primary Structures

- **+** Shield
- Channel
- Probable flow boundary
- Dome
- Flow direction
- **?** Possible flow direction
- Impact crater rim

Tectonic Structures

- Wrinkle Ridge
- Fracture

Fig. 3. *Continued.*

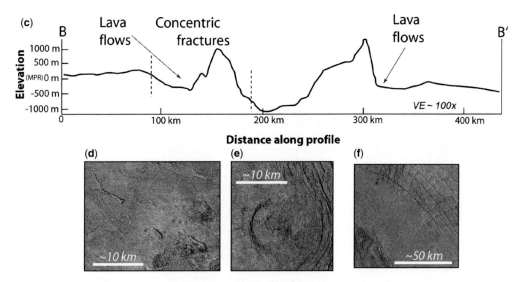

Fig. 3. (*Continued*) Imagery characterizing Bhumidevi Corona. (**a**) Inverted SAR image of Bhumidevi Corona. Dashed white boxes show the locations of (d)–(f); the white line with B–B′ is the location of the topographical profile shown in (c). (**b**) Geological map of Bhumidevi Corona. (**c**) Topographical profile of Bhumidevi Corona showing the locations of corona-related fractures and lava flows in relation to the corona's topography. (**d**)–(**f**) Enlarged inverted SAR images highlighting specific parts of Bhumidevi Corona: (d) view of a steep-sided dome superposed on concentric fractures of Bhumidevi's annulus; (e) view of a trough on the NE side of Bhumidevi showing radial and concentric fractures locally buried by flow material (the lighter-coloured patch in the image); (f) view of flow material that has buried concentric fractures on the eastern side of Bhumidevi; some flow material appears to have erupted from collapsed pits that trend across the image.

asymmetrical narrow rim that ranges in width from around 30 to 75 km and rises approximately 1 km above MPR. This rim surrounds an asymmetrical bowl-shaped depression that resides about 1 km below MPR.

Tectonic structures. Tectonically, the Bhumidevi map area hosts both extensional and contractional structures (Fig. 3b). The corona itself is denoted by a suite of both concentric and radial fractures. Both sets of fractures appear similar to those observed at Aramaiti Corona and, consequently, we interpret these concentric and radial fractures as also probably representing graben. The concentric fractures comprise an approximately 75 km-wide annulus around Bhumidevi that stretches from the outer edge of the moat across the topographical rim to the outer edge of the bowl-shaped depression located in the centre of the corona (Fig. 3c). Concentric fractures on the edge of the bowl-shaped depression are mostly continuous around the bowl and are spaced approximately 1–10 km apart. However, concentric fractures located outside of the topographical rim in the vicinity of the moat are discontinuous and are observed only around the northern two-thirds of the corona where they are spaced from about 1 to *c.* 25 km apart; there is an abrupt truncation of the concentric

fractures at the SE corner of the corona where they appear to be buried by flow materials sourced from various locations. Radial fractures trend from Bhumidevi's annulus in all directions for distances of up to around 300 km. The most prominent and extensive radial fractures trend from the northern part of the corona, whereas radial fractures around the rest of the corona are shorter and appear to be covered by locally sourced flow materials. Some radial fractures extend into the corona's moat, but are completely absent from the bowl-shaped depression in the centre.

Concentric and radial fractures are also observed with other coronae in the map area and some of the radial fractures observed at Bhumidevi could be genetically related to the other coronae. Other tectonic structures observed in the map area include graben associated with the NW-trending fracture belt and a suite of NE-trending wrinkle ridges; the wrinkle ridges are nearly orthogonal to the NW-trending fracture belt suggesting a kinematic relationship between these two suites of tectonic structures. The fracture belt varies between about 150 and 300 km wide, with fracture spacings between approximately 1 and 10 km; the belt is more prominent in the SE corner of the map area in comparison to the NW corner. Wrinkle ridges vary between approximately 10 and 100 km long, with wavelengths of between about 5 and 25 km, confined predominantly to the SW corner of the map area; there is a noticeable absence of wrinkle ridges within Bhumidevi's interior. Fractures also occur in the NE corner of the map area with the same trend as the wrinkle ridges.

Volcanic features. Evidence for volcanic activity at Bhumidevi is similar to that at Aramaiti Corona, and includes long and more localized lava flows, tholi and numerous shields. We have identified only one distinct long lava flow at Bhumidevi – an approximately 325 km-long north-trending flow that appears to have been sourced from a vent associated with the northern part of the corona's annulus. This flow is around 200 km wide and embays basement materials north of Bhumidevi, and is seemingly cut by Bhumidevi-associated north-trending radial fractures. Similar to Oilule Fluctus, this unnamed flow is denoted by relatively high backscatter (in comparison to surrounding plains materials) and we interpret it to likely represent an 'a'a flow that also probably reflects a high effusion rate (e.g. Pinkerton & Wilson 1994). Grindrod & Guest (2006) mapped two other extensive flows extending from Bhumidevi, including one immediately SW and adjacent to the north-trending long flow, and one that extends south from the corona. We note the possibility of these flows in our mapping (as denoted by a flow arrow with a question mark: Fig. 3b), but could not identify any definitive and continuous flow boundaries to these possible flows, possibly due to homogenization of the surface here (e.g. Arvidson *et al.* 1992; López & Hansen 2008). It is possible that these flows comprise the unit 'Corona flow material, uncertain origin' (Fig. 3b), although this unit could also be sourced from Bhumidevi's concentric fractures filling the trough surrounding the bowl-shaped depression. Localized flow materials also occur in Bhumidevi's bowl-shaped depression (Fig. 3b), and may reflect eruptions from fractures within the bowl or possibly even from the few identified shields. Another additional longer flow may extend SE from Bhumidevi from a chain of elongated vents (denoted as a series of channel segments on Fig. 3b and highlighted in Fig. 3d); again, there is an absence of a definitive and continuous flow boundary here. Hence, it is possible that additional extensive lava flows may extend from Bhumidevi Corona.

Two individual tholi also occur at Bhumidevi and are located on the SE edge of the concentric fractures inside of the bowl-shaped depression (Fig. 3d). These tholi are approximately 15–20 km in diameter and have radar backscatter characteristics similar to the surrounding materials. They have relatively steep sides with an apparent concave top. There is an apparent absence of fracturing and pitting with these tholi, although one of the tholi may contain a possible second smaller dome around 5 km in diameter on its top. As these tholi reside on top of the concentric fractures, it seems likely they were sourced from the fractures.

Numerous shields are also spatially associated with Bhumidevi. Minor numbers of shields occur within the bowl-shaped depression and upon the concentric fractures. However, a majority of the shields occur on both the radial fractures and the fractures in the NW-trending fracture belt. A distinct shield field has formed immediately west of Bhumidevi, with material from the shields having flowed down into the moat.

Bhumidevi has also been embayed by flow material from adjacent coronae. Specifically, lava flows from Iyatik, Takus Mana and Qetesh Coronae have all covered parts of Bhumidevi; Bender *et al.* (2000) and Grindrod & Guest (2006) mapped flows from these coronae has having traversed and partially filled Bhumidevi's moat. Although it appears that flow material has partially filled the moat, we have not been able to trace the flow materials back to exact sources. As a result, we have mapped flow material in Bhumidevi's moat as flow materials with an unknown source – it is likely that the flow material represents material from multiple sources including the three surrounding coronae and, possibly, Bhumidevi itself. None the less, flows from these three surrounding coronae appear to have locally covered flow materials from Bhumidevi, radial fractures south and east of Bhumidevi, and the concentric fractures along the south edge of Bhumidevi.

Geological history. Based on cross-cutting and superposition relationships, the following geological history can be determined for the Bhumidevi Corona area. The earliest emplaced geological materials in the map area are the local basement materials, which have been subsequently buried and embayed by volcanic flow materials erupted from the several coronae located within the map area. Bhumidevi was the first corona to form in the map area, and its formation may have predated formation of the NW-trending fracture belt and possibly even the radial fractures extending from the corona. In fact, the earliest preserved activity at Bhumidevi is the formation of the concentric fractures. The single, long, north-trending flow postdates annulus formation, and appears to have erupted from the concentric fractures and must, therefore, post-date concentric fracture formation. Based on the observation that the long flow currently exhibits an uphill direction of travel from the concentric fractures, the long lava flow must also predate initiation of Bhumidevi's topographical trough. Assuming the concentric fractures facilitated trough formation, the concentric fractures, the long, north-trending flow and topographical trough probably all occurred within a short period in relation to one another. Radial fractures post-date eruption of the north-trending flow and probably predate emplacement of the corona flow materials of uncertain origin as well as the unit 'Corona flow material b', although local reactivation of the radial fractures appears to have occurred. Concentric fractures cut, and thus post-date, the unit 'Corona flow material b', suggesting that the bowl-shaped depression probably predates concentric fracture formation. Tholi post-date concentric fracture formation and, due to the overall absence of apparent deformation, must post-date both the formation of the bowl-shaped depression in the centre of Bhumidevi and the unit 'Corona flow material b'. The timing of shield activity is ambiguous, but some timing associations can be deduced from superposition relationships. Specifically, shields located on Bhumidevi's annulus and in the bowl-shaped depression do not appear to be deformed and, therefore, probably post-dated concentric fracture formation and the formation of the current corona topography. Shields spatially associated with the radial fractures probably formed concurrent with or after formation of those fractures; the presence of numerous shields on top of the long, north-trending flow indicates that the shields may largely post-date much of the large-scale volcanic activity here. This interpretation is consistent with the presence of the shield field west of Bhumidevi that has associated flow material that has spilled over into Bhumidevi's moat. This interpretation is further supported by the observation that the shield-sourced flow material appears to stop at the topographical rim surrounding the bowl-shaped depression. Formation of, and associated volcanic activity at, the three surrounding coronae post-dated much of the tectonic and volcanic activity at Bhumidevi. The NE-trending wrinkle ridges are found only in the SW corner of the map area and are not noted to have deformed flow units associated with any of the coronae here, although they do deform some flow material erupted from shields west of Bhumidevi. Given this timing relationship and their orthogonal orientation to the fracture belt, these ridges may have largely formed synchronously with the NW-trending fracture belt and probably post-date formation of Bhumidevi.

Zemire Corona

Zemire Corona is a 200 km-diameter double-ring corona (Stofan *et al.* 1992) located in central Guinevere Planitia near 31°N, 313°E (Fig. 4). It is spatially related to Breksta Linea, a NW–

SE-trending deformation belt that hosts numerous outcrops of basement materials such as tesserae terrain and other tectovolcanic structures (which are located outside of this map area); basement outcrops in this map area have an overall NE-trending pattern (Fig. 4a, b). The area surrounding the corona is characterized by the presence of numerous volcanic features (e.g. paterae and shield fields) that formed in relation with the deformation belt.

Topography. Figure 4c is a NW–SE-trending topographical profile of Zemire Corona with approximately 37× vertical exaggeration. Topographically, Zemire is characterized by a concentric rim that varies in width from about 25 to 75 km, and in elevation from 500 m above MPR to approximately equal with MPR. The outside of the rim is surrounded by a concentric trough that varies in width from around 10 to 30 km. An asymmetric bowl-shaped depression that reaches a depth of approximately 1 km below MPR characterizes the inside of the topographical rim.

Tectonic structures. Similar to the other two field areas, the Zemire map area hosts both extensional and contractional tectonic structures (Fig. 4b). The annulus of Zemire Corona is characterized by a double ring of concentric fractures that appear similar to those observed at Aramaiti and Bhumidevi Coronae and, thus, we also interpret these concentric fractures to represent graben; individual fractures have spacings of approximately 1–10 km. The concentric fractures are absent, however, on the SE corner of the corona where they appear to be covered by extensive flow material (Fig. 4b). Both rings of concentric fractures are around 25 km wide, and are separated by the topographical depression described above and visible in the topographical profile of Zemire Corona (Fig. 4c); where they are present, the inner ring of concentric fractures coincides with the corona's topographical rim (Fig. 4c). Radial fractures also occur at Zemire Corona, but appear to radiate from a point immediately north of the corona's annulus that hosts several steep-sided domes; the fractures cut through the extensive flow materials erupted from Zemire and extend for distances of more than 400 km (Ernst *et al.* 2003). Many of the radial fractures appear to be absent from Zemire's topographical trough and, in some cases, appear to merge with concentric fractures comprising the outer annulus ring.

Zemire Corona occurs within Breksta Linea, which is marked by NW-trending fractures that make a bend to the south at the location of Zemire (Fig. 4b). Fractures within this belt have spacings of around 10–25 km, and are spatially associated

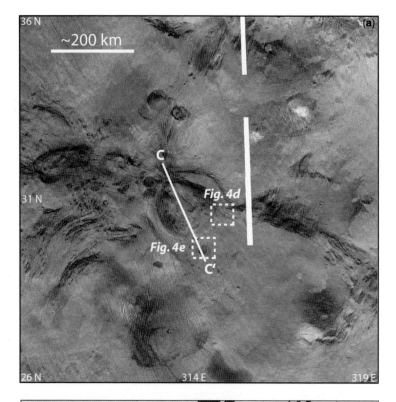

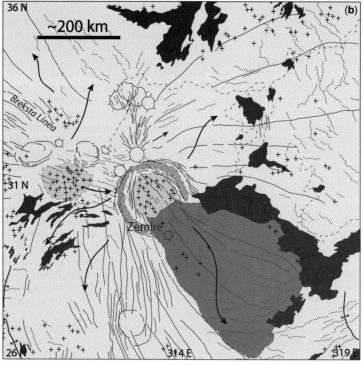

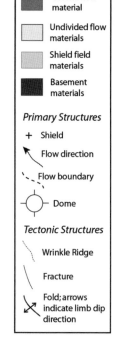

KEY

Units

Trough-filling flow material

Corona flow material

Undivided flow materials

Shield field materials

Basement materials

Primary Structures

+   Shield

Flow direction

Flow boundary

Dome

Tectonic Structures

Wrinkle Ridge

Fracture

Fold; arrows indicate limb dip direction

Fig. 4. *Continued.*

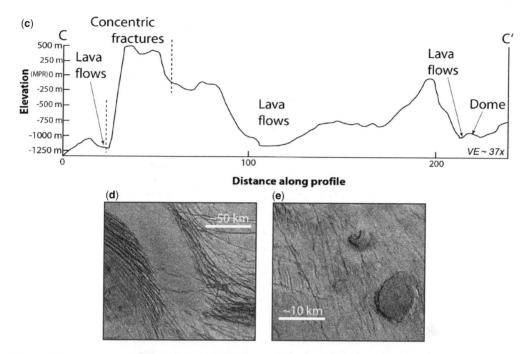

Fig. 4. (*Continued*) Imagery characterizing Zemire Corona. (**a**) Inverted SAR image of Zemire Corona. Dashed white boxes show the locations of (d)–(e); the white line with C–C′ is the location of the topographical profile shown in (c). (**b**) Geological map of Zemire Corona. (**c**) Topographical profile of Zemire Corona showing the locations fractures and lava flows in relation to the corona's topography. (**d**)–(**e**) Enlarged inverted SAR images highlighting specific parts of Zemire Corona: (d) close-up view of a trough on the east side of Zemire Corona showing concentric fractures locally buried by flow material (the lighter-coloured patch in the image); (e) view of steep-sided domes superposed on locally buried concentric and radial fractures on the south side of Zemire; one dome has undergone partial collapse.

with smaller occurrences of concentric fractures that occur to the west and south of Zemire. These concentric fractures appear to mark the locations of smaller, unnamed coronae and caldera-like structures that occur within the fracture belt of Breksta Linea.

Two suites of wrinkle ridges also occur within the map area where they cut across all flow materials associated with Zemire; the suites are orthogonal, with NW and NE trends. The ridges have lengths of approximately 10–200 km and wavelengths of around 10–25 km. Although the ridges appear to deform Zemire-related flow materials, there is a noticeable absence of wrinkle ridges from within much of Zemire itself; the only exception to this is the local occurrence of some NW-trending wrinkle ridges in the NW corner of the corona's topographical moat. The NW-trending ridges parallel, and are spatially associated with, the NW-trending fractures of Breksta Linea.

Volcanic features. Evidence for volcanism at Zemire Corona includes extensive lava flows, shields and several tholi. Corona-related lava flows

have travelled for distances of approximately 300–400 km, which is consistent with them being composed of low-viscosity material. The flows have a lobate morphology and are denoted by relatively higher backscatter (in comparison to the surrounding plains materials); flows are also denoted by the truncation of concentric fractures along the annulus at Zemire's SE corner. The extent of corona-related flows may be greater than what is indicated on the map, but homogenization of the Venusian surface (e.g. Arvidson *et al.* 1992; López & Hansen 2008) may be masking flow boundaries. The specific location of the vents sourcing these flows is unclear, but appears to be probably located either within or inside Zemire's fracture annulus.

Nine tholi also occur in the map area. The tholi range in diameter from around 15 to 50 km and occur in three distinct parts of the map area: (1) three tholi are superposed on the outer fracture ring of Zemire's annulus; (2) four tholi are tightly clustered together approximately 100 km north of Zemire (Anqet Farra); and (3) two smaller tholi occur approximately 50 and 150 km, respectively, NW of Zemire. The three tholi superposed on

Zemire's fracture annulus are circular in map view and do not show evidence for scalloping (e.g. Guest *et al.* 1992) or erosion, although a tholus on the northernmost extent of Zemire's annulus is associated with a series of fractures that radiate from it; fractures with spacings of around 5–10 km trending north from this tholus cut the four tightly clustered tholi north of Zemire (Fig. 4a, b). In addition, the tholus that occurs on the southernmost extent of the fracture annulus is cut by NW-trending fractures with spacings of about 1 km that parallel the regional NW fracture trend (Fig. 4e); each of the three tholi also contain an approximately 1 km-wide central pit in the central portion of their flat-tops. The four tightly clustered tholi north of Zemire show evidence of undergoing scalloping and erosion, and seem to be embayed by the extensive flow material erupted from Zemire. The tops of three of these four tholi appear to be convex-shaped, whereas the top of the fourth tholus is concave; all four tholi contain an approximately 1 km-wide central pit in the centre of their tops. Of the last two tholi, one has experienced extensive scalloping and has undertaken an oval shape, whereas the other is the smallest tholus in the map area and appears to host an approximately 3 km-wide central pit at its summit. Both of these tholi are superposed on the NW-trending fractures comprising Breksta Linea.

Shields occur across the map area, but are most noticeable as two clusters – one to the west of Zemire and one in the centre of the corona. Although some diffuse shield clusters occur away from Breksta Linea, most shield clusters are spatially associated with the fracture belt where they have formed directly on top of specific fractures. A clustering of shields also occurs in the topographical basin in the interior of Zemire; no shields appear to have occurred on both the fracture annulus or in the concentric moat. Although some shields appear to have experienced erosion (more than erosion is tectonic modification, collapse processes) (Fig. 4e), most shields seem to post-date fracture formation and do appear to have been embayed or buried by flow materials.

Geological history. Based on the mapping of tectonic and volcanic features, the following geological history can be outlined for the Zemire Corona area. The earliest emplaced geological materials in the map area are the local basement materials (tessera), which have been subsequently buried and embayed by flow materials comprising Guinevere Planitia, including the formation of Zemire Corona. The earliest preserved structures at Zemire are the concentric fractures comprising its double ring annulus, which presumably would have formed during corona collapse and formation of the topographical bowl in the corona's middle.

Any earlier formed structures such as radial fractures (e.g. Squyres *et al.* 1992; Stofan *et al.* 1992) would, presumably, have been buried by Zemire-related flow materials. Most Zemire-sourced flows bury the concentric fractures, indicating their occurrence after concentric fracture formation. Flows are also broadly warped by the Zemire's topographical rim where it has buried the concentric fractures, suggesting that dome collapse at the collapse of the corona here post-dated the emplacement of corona-related flows. Flow material located within Zemire's topographical trough most probably also post-dated formation of both the concentric fractures and the trough. Also post-dating concentric fracture formation was the emplacement of the three tholi located on Zemire's annulus; the radial fractures extending from the northernmost of these three tholi was probably synchronous with its formation. The relative timing of these three tholi with corona-flow material, however, is unclear. It is possible that the tholi predated these flows and were subsequently embayed, or they may post-date longer lava flow emplacement and are superposed on the flow. Also post-dating Zemire formation are the radial fractures located immediately north of the corona; at their northernmost extent, these radial fractures cut the four overlapping tholi meaning that the tholi must predate fracture formation. Structures comprising Breksta Linea cut flow materials seemingly sourced from Zemire, indicating that Breksta Linea was probably active after corona formation and collapse. However, Zemire is located within the fracture belt (specifically, at a bend in the belt), suggesting that, perhaps, the initial occurrence of Breksta Linea may have been synchronous with the formation of Zemire. Alternatively, Zemire may have served as a regional stress perturbation that caused a bend in the fracture belt. Formation of the three smaller, unnamed coronae within Breksta Linea are also likely to have occurred during active phase within the fracture belt. Wrinkle ridges in the NW corner of Zemire's topographical moat are spatially associated with parallel Breksta Linea fractures, suggesting that these NW-trending ridges may have been fractures that were filled with lava and subsequently experienced contraction resulting in ridge formation (e.g. DeShon *et al.* 2000). This implies that these wrinkle ridges probably post-date Breksta Linea formation, as well as the eruption of flow materials from Zemire.

Discussion

Coronae analysis

The three coronae described here share similar characteristics and histories. Specifically, each corona: (1) occurs in a region that contains tessera terrain as the local basement material; (2) hosts an annulus of concentric fractures that predate and sourced large-volume eruptions that either extended for hundreds of km from the coronae or filled-in topographical troughs; the concentric fractures also step-down into either a central depression or concentric trough, suggesting that the fractures facilitated at least some of the deflated topography; eruption of the extensive flows was post-dated by the final formation of the topographical depressions (both the troughs and central depressions); (3) hosts tholi on its annulus that also post-dates concentric fracture formation; (4) hosts only minor numbers of shields, which also post-dated concentric fracture formation; and (5) are spatially associated with either a fracture belt or suite of radial fractures that probably represent the surface expression of dyke emplacement (e.g. Ernst et al. 2003) and that were active after concentric fracture formation.

These five characteristics largely emphasize the late stages of each corona's evolution and fit within the general coronae evolutionary models described earlier (Squyres et al. 1992; Stofan et al. 1992, 1997). What is noteworthy, though, is the abundance of volcanism that occurred later in each corona's history; that is, volcanism that occurred shortly prior to and concurrent with (extensive volcanism), and subsequent to (trough-filling flows and tholi), corona collapse. The occurrence of extensive lava flows prior to the initiation of corona collapse is consistent with the generation and eruption of magma due to large-volume decompression partial melting in the mantle (Roberts & Head 1993). However, if the mantle source for corona formation dissipated and corona collapse had initiated (i.e. formation of the annulus of concentric fractures), then the source of the late-stage flow materials would most likely not be the mantle upwelling, diapir or plume, but a magma reservoir within the lithosphere. Because much of the late-stage erupted products appear to be stemming from the concentric fractures, the fractures must extend down to the magma reservoir and have facilitated the upwards migration, eruption and seemingly rapid emplacement (at least with Aramaiti and Bhumidevi Coronae) of large volumes of magma (i.e. extensive and localized lava flows) that temporally transitioned into tholi emplacement. Although this temporal transition from extensive lava flows to more localized tholi emplacement is seemingly readily apparent at Aramaiti and Bhumidevi Coronae, we do admit that such timing relations are more ambiguous for Zemire Corona owing to the lack of unambiguous contact relationships between the tholi and extensive corona flows. The dearth of shields on coronae annuli

implies that the concentric fractures facilitated predominantly large-volume eruptions, but may have also fed few local smaller-scale eruptions. Based on compositional measurements of the Venusian surface (Surkov et al. 1984), together with the morphology of larger Venusian volcanic edifices (Head et al. 1992), we argue that the extensive flow deposits extending from each corona are most likely to be basaltic in composition. Such a composition would also be consistent with the long nature of the lava flows (Zimbelman 1998) and means that, when compared with the tholi, the extensive lava flows probably represent a lower-viscosity fluid. If true, then the temporal transition from the extensive lava flows to the tholi reflects a change in the viscosity of the materials erupting from the coronae annuli. Further, because the extensive flows and tholi are erupting from similar vents, then the source for both is most likely to be the same. This implies that the magma in the source temporally transitioned to become more viscous. These changes could reflect either: (1) a compositional evolution of the magma where it became more felsic; or (2) that a more crystal- or bubble-rich material (Pavri et al. 1992) was erupted. The possibility of felsic products at coronae has been demonstrated by Pavri et al. (1992) and Petford (2000), and makes geological sense if a magma reservoir within the lithosphere underwent fractional crystallization over a long enough time period or melted surrounding basement materials. If true, this would suggest that magma reservoirs at each corona were compositionally stratified in a manner similar to some large magmatic centres on Earth (e.g. Hildreth 1981; Druitt & Bacon 1988), where the more felsic (compositionally evolved and, consequently, less dense and more viscous) magma resides at the top of the reservoir and the more mafic (compositionally primitive and, consequently, more dense and less viscous) compositions reside lower in the reservoir. Although tapping from the more primitive and stratigraphically lower parts of the reservoir may occur before tapping of the stratigraphically upper parts of the reservoir, such a process is interpreted as reflecting a migration of vent openings around the magma system (Gardner et al. 1991). Typically, the first erupted products from a compositionally stratified reservoir are the stratigrapically highest and most evolved magmas followed by the stratigraphically lower and compositionally primitive parts of the reservoir (Hildreth 1981). The observation that at least a large percentage of the late-stage flows at the three examined coronae do extend from concentric fractures makes it plausible that more primitive parts of the magma reservoir could be tapped before the compositionally evolved parts. However, the presence of the more viscous magma (i.e.

the tholi) erupting from the same vent as the less viscous magma (i.e. the extensive and trough-filling flows) indicates that the vent is tapping the same part of the magma reservoir. This, in turn, suggests that, with these three coronae, we are observing the stratigraphic top of the reservoir having erupted before the stratigraphically lower parts; consequently, the stratigraphy of the erupted products observed at these coronae reflects an inversion of the magma reservoir. If true, the apparently low-viscous character of the lava flows implies a mafic–ultramafic composition (e.g. Head *et al.* 1992), although compositions such as carbonatite cannot be excluded (e.g. Komatsu *et al.* 1992; Gregg & Greeley 1993); the later erupted products may, therefore, also represent mafic–ultramafic compositions. Consequently, the tholi do not have to represent highly evolved felsic compositions nor do they also need to represent the melting of basaltic crust (e.g. Ivanov & Head 1999) owing to the fact that the heat needed to melt the crust probably went away with the dissipation of mantle upwelling. Ultimately, the timing of the erupted products with corona collapse suggests an intimate connection between the two processes, and any model that explains the volcanic evolution of these three coronae must take this into account.

Evolutionary model

According to our observations and the general models for the formation of Venusian coronae (e.g. Janes *et al.* 1992; Squyres *et al.* 1992; Stofan *et al.* 1992), we propose the following scenario to explain the late-stage structural and volcanic evolution of these three Venusian coronae (Fig. 5):

- Mantle upwelling and/or occurrence of a mantle diapir or plume results in the generation of a zone of partial melt (Dombard *et al.* 2007) in the upper mantle beneath the lithosphere that consists of tracts of tessera terrain. Impingement upon the base of the lithosphere will result in lithospheric bowing (Stofan *et al.* 1992; Squyres *et al.* 1992) and may be associated with dyke emplacement within the lithosphere, resulting in the formation of radial fractures and/or a fracture belt (e.g. Ernst *et al.* 2003); some of this magma may erupt as fissure eruptions and/or create small shields.
- Magma will continue to be injected into the lithosphere but, owing to ascension rates and possible volatile contents, some of the ascending magma will stall, resulting in the formation of a shallow magma reservoir (e.g. Head & Wilson 1992). Creation of neutral buoyancy zones may be enhanced by the presence of widespread occurrence of tessera terrain in

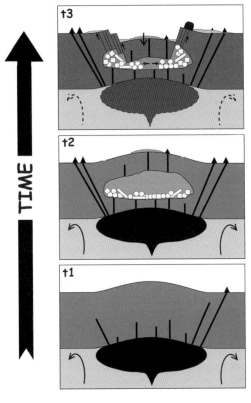

Fig. 5. Schematic diagram outlining our proposed volcanic evolution of the three coronae discussed in this paper. In t1, upwelling in the Venusian mantle causes the impingement of a mantle diapir (or plume) on the base of the lithosphere causing the lithosphere to bulge upwards and initiate radial fracturing; magmatic dykes (thick black lines) will extend from the diapir into the lithosphere; some dykes will propagate to the surface where they may erupt as small shields (black triangles). In t2, magmatic dykes continue to propagate to the surface where they erupt and small shields continue to form; some magma stalls in the lithosphere creating a magma chamber that may undergo fractional crystallization (white rectangles and circles); some magma from this chamber may erupt at the surface (light grey layer at the surface). In t3, loss of support of the lithosphere, probably due to loss of thermal input from the diapir, results in the collapse of the corona and the creation of concentric fractures (corona annulus); lithosphere may drop along the concentric fractures where blocks of it fall into the magma chamber, resulting in the squeezing of the magma up the concentric fractures and onto the surface, where it may flow locally into topographical troughs or extend for long distances from the corona; continued collapse of lithospheric blocks may ultimately squeeze up crystals from the base of the magma chamber, where they erupt at the surface from the concentric fractures as viscous, crystal-rich mushes taking on the form of a steep-sided dome or tholus (black bulb). Alternatively, tholi could represent 'basaltic foam'.

each map area (e.g. Head & Wilson 1992), especially if tesserae were composed of a more felsic composition (e.g. Hashimoto *et al.* 2008; Mueller *et al.* 2008) where the tessera terrain created a density boundary that impeded the rising magma. Fractional crystallization should occur within this shallow reservoir, potentially resulting in a compositional stratification within the reservoir. It is possible that melting of the surrounding, potentially more felsic (Hashimoto *et al.* 2008; Mueller *et al.* 2008), tessera terrain could drive compositional change within the reservoir as well.

• Eventually, the lithosphere located above the shallow magma reservoir will become unstable and will begin to collapse. Initial collapse of the lithosphere will result in the formation of concentric fractures that may extend down to the magma reservoir. These fractures could then facilitate the dropping of the lithosphere into that reservoir, squeezing the magma up into the fractures where it erupts on the surface and, ultimately, drains the magma chamber. Forcing magma from a reservoir in such a fashion could plausibly result in high effusion rates as is postulated to have occurred at Aramaiti and Bhumidevi Coronae. Continual draining of the chamber could induce sagging of the already dropped lithosphere. If the magma chamber was, indeed, stratified with a crystal mush at the bottom, then enough sinking of the roof could potentially drive that mush to erupt as well. Given a presumably higher viscosity of a crystal-rich flow, it would probably not flow far (e.g. Saar *et al.* 2001) and would, instead, result in a more localized eruption; such a scenario could result in the formation of the tholi. Alternatively, the tholi could represent highly vesiculated magma (e.g. Head & Wilson 1992; Pavri *et al.* 1992) where formation of a shallow reservoir favoured by the presence of an underlying tessera basement near the depth of nucleation (Head & Wilson 1992) created a magma 'enriched' in bubbles, resulting in 'basaltic foam'.

This qualitative model is similar to that of large silicic caldera formation on Earth where initial collapse and concentric fracture formation would be similar to piston and/or downsag-style caldera formation (e.g. Cole *et al.* 2005). However, we do not propose that coronae are associated with violently explosive eruptions; the apparent absence of ash deposits around each of the coronae, together with Venus' high surface pressure of 95.6 bars (at MPR) (Ford & Pettengill 1992), seemingly argues against the likelihood of catastrophically explosive eruptions occurring at these three Venusian

coronae. Our model also predicts that steep-sided domes may not necessarily reflect the existence of a highly evolved magma (e.g. highly siliceous: Pavri *et al.* 1992; Petford 2000) or even the melting of basaltic crust (e.g. Ivanov & Head 1999), although these possibilities of tholi formation cannot be ruled out, especially if a tholus did not form in association with a corona (Ivanov & Head 1999, 2004) and/or melting of more felsic crustal material occurred. Instead, the presence of steep-sided domes that occur at coronae could be explained as mafic crystal mushes or even highly vesiculated mafic magma (e.g. Head & Wilson 1992; Pavri *et al.* 1992). Such scenarios are more in line with both compositional measurements (Surkov *et al.* 1984) and the morphology of larger volcanic constructs (Head *et al.* 1992) on the Venusian surface.

Ultimately, however, our model addresses only three coronae and may not apply to all coronae; direct tests of this model are possible, however, through detailed geological mapping of additional coronae and spectral analyses conducted *in situ* by Venusian landers.

Summary and conclusions

Our analysis presented here provides constraints on the late-stage volcanic evolution of three Venusian coronae. Geological mapping of the Aramaiti, Bhumidevi and Zemire Coronae suggests an intimate relationship between the late-stage erupted products and the collapse of each corona. Loss of thermal support of the corona due to the dissipation of mantle upwelling and/or a mantle diapir or plume may result in the initiation of corona collapse and concentric fracture formation, which may, in turn, drive eruption of extensive lava flows. Continual corona collapse would then restrict lava flow emplacement into troughs and depressions within the corona's interior and eventually drive the eruption of a crystal-rich mush from the bottom of the chamber, resulting in the emplacement of tholi along a corona's annulus. Such a model is purely qualitative, but likens the last stages of these coronae's activity to a quiescent form of terrestrial silicic caldera formation and predicts that all erupted products at these coronae could potentially be mafic, although tholi representing melts derived from a potentially felsic basement (i.e. tessera terrain) cannot be fully excluded. We stress that this model currently applies only to the three coronae examined here, but further testing of this model may be conducted through continued geological mapping of other coronae, as well as through *in situ* analyses of coronae-related deposits by Venusian landers.

Many thanks to F. Mazzarini and M.A. Ivanov, whose thoughtful and thorough reviews enhanced and improved the quality of this manuscript. Discussions with C. Fedo, H.Y. McSween and T. Usui were helpful in developing the initial ideas for this paper. We finally want to thank M. Massironi, P. K. Byrne, H. Hiesinger and T. Platz for their work as editors on this volume.

References

ARVIDSON, R. E., GREELY, R. ET AL. 1992. Surface modification of Venus as inferred from Magellan observations of plains. *Journal of Geophysical Research*, **97**, 13 303–13 318.

BAKER, V. R., KOMATSU, G., PARKER, T. J., GULICK, V. C., KARGEL, J. S. & LEWIS, J. S. 1992. Channels and valleys on Venus: preliminary analysis of Magellan data. *Journal of Geophysical Research*, **97**, 13 421–13 444.

BASILEVSKY, A. T. & HEAD, J. W. 1998. Onset time and duration of corona activity on Venus: stratigraphy and history from photogeologic study of stereo images. *Earth, Moon, Planets*, **76**, 67–115.

BENDER, K. C., SENSKE, D. A. & GREELEY, R. 2000. *Geologic Map of the Carson Quadrangle (V-43), Venus (1:5 Million Scale)*. United States Geological Survey, Geologic Investigations Series, **Map 2620**.

BINDSCHADLER, D. L., DECHARON, A., BERATAN, K. K., SMREKAR, S. E. & HEAD, J. W. 1992. Magellan observations of Alpha Regio: implications for formation of complex ridged terrains on Venus. *Journal of Geophysical Research*, **97**, 13 563–13 577.

BYRNES, J. M. 2002. *Lava flow field emplacement studies of Mauna Ulu (Kilauea Volcano, Hawai'I, USA) and Venus, using field and remote sensing analyses*. PhD thesis, University of Pittsburgh, PA.

COLE, J. W., MILNER, D. M. & SPINKS, K. D. 2005. Calderas and caldera structures: a review. *Earth-Science Reviews*, **69**, 1–26.

CRUMPLER, L. S., AUBELE, J. C., SENSKE, D. A., KEDDIE, S. T., MAGEE, K. P. & HEAD, J. W. 1997. Volcanoes and centers of volcanism on Venus. *In*: BOUGHER, S. W., HUNTEN, D. M. & PHILLIPS, R. J. (eds) *Venus II*. University of Arizona Press, Tucson, AZ, 697–756.

DESHON, H. R., YOUNG, D. A. & HANSEN, V. L. 2000. Geologic evolution of southern Rusalka Planitia, Venus. *Journal of Geophysical Research*, **105**, 6983–6995.

DRUITT, T. H. & BACON, C. R. 1988. Compositional zonation and cumulus processes in the Mount Mazama magma chamber, Crater Lake, Oregon. *Transactions of the Royal Society of Edinburgh: Earth Sciences*, **79**, 289–297.

DOMBARD, A. J., JOHNSON, C. L., RICHARDS, M. A. & SOLOMON, S. C. 2007. A magmatic loading model for coronae on Venus. *Journal of Geophysical Research*, **112**, E04006, http://dx.doi.org/10.1029/2006JE002731

ERNST, R. E., DESNOYERS, D. W., HEAD, J. W. & GROSFILS, E. B. 2003. Graben-fissure systems in Guinevere Planitia and Beta Regio (264°–312°E, 24°–60°N), Venus, and implications for regional stratigraphy and mantle plumes. *Icarus*, **164**, 282–316.

FINK, J. H., BRIDGES, N. T. & GRIMM, R. E. 1993. Shapes of Venusian 'pancake' domes imply episodic emplacement and silicic composition. *Geophysical Research Letters*, **20**, 261–264.

FORD, P. G. & PETTENGILL, G. H. 1992. Venus topography and kilometer-scale slopes. *Journal of Geophysical Research*, **97**, 13,103–13,114.

FORD, J. P., PLAUT, J. J., WEITZ, C. M., FARR, T. G., SENSKE, D. A., STOFAN, E. R., MICHAELS, G. & PARKER, T. J. 1993. *Guide to Magellan Image Interpretation*. JPL Publications, **1**. National Aeronautics and Space Administration, Jet Propulsion Laboratory, Pasadena, CA, 93–24.

GARDNER, J. E., SIGURDSSON, H. & CAREY, S. N. 1991. Eruption dynamics and magma withdrawl during the plinian phase of the Bishop Tuff eruption. *Journal of Geophysical Research*, **96**, 8097–8111.

GREGG, T. K. P. & GREELEY, R. 1993. Formation of Venusian canali: considerations of lava types and their thermal behaviors. *Journal of Geophysical Research*, **98**, 10 873–10 882.

GRINDROD, P. M. & GUEST, J. E. 2006. 1:1.5,000,000 geological map of the Aglaonice Region on Venus. *Journal of Maps*, **2006**, 103–117.

GUEST, J. E., BULMER, M. H., AUBELE, J., BERATAN, K., GREELEY, R., HEAD, J. W., MICHAELS, G., WEITZ, C. & WILES, C. 1992. Small volcanic edifices and volcanism in the plains of Venus. *Journal of Geophysical Research*, **97**, 15 949–15 966.

HANSEN, V. L. 2000. Geologic mapping of tectonic planets. *Earth and Planetary Science Letters*, **176**, 527–542.

HANSEN, V. L. 2003. Venus diapirs: thermal or compositional? *Geological Society of America Bulletin*, **115**, 1040–1052.

HASHIMOTO, G. L., ROOS-SEROTE, M., SUGITA, S., GILMORE, M. S., KAMP, L. W., CARLSON, R. W. & BAINES, K. H. 2008. Felsic highland crust on Venus suggested by Galileo Near-Infrared Mapping Spectrometer data. *Journal of Geophysical Research*, **113**, E00B24, http://dx.doi.org/10.1029/2008JE003134

HEAD, J. W. & WILSON, L. 1992. Magma reservoirs and neutral buoyancy zones on Venus: implications for the formation and evolution of volcanic landforms. *Journal of Geophysical Research*, **97**, 3877–3903.

HEAD, J. W., CRUMPLER, L. S., AUBELE, J. C., GUEST, J. E. & SAUNDERS, R. S. 1992. Venus volcanism: classification of volcanic features and structures, associations, and global distribution from Magellan data. *Journal of Geophysical Research*, **97**, 13 153–13 197.

HILDRETH, W. 1981. Gradients in silicic magma chambers: implications for lithospheric magmatism. *Journal of Geophysical Research*, **86**, 10 153–10 192.

IVANOV, M. A. & HEAD, J. W. 1999. Stratigraphic and geographic distribution of steep-sided domes on Venus. Preliminary results from geological mapping and implications for their origin. *Journal of Geophysical Research*, **104**, 18 907–18 924.

IVANOV, M. A. & HEAD, J. W. 2004. Stratigraphy of small shield volcanoes on Venus: criteria for determining stratigraphic relationships and assessment of relative age and temporal abundance. *Journal of Geophysical Research*, **109**, E10001, http://dx.doi.org/10.1029/2004JE002252

IVANOV, M. A. & HEAD, J. W. 2011. Global geologic map of Venus. *Planetary and Space Science*, **59**, 1559–1600.

JANES, D. M., SQUYRES, S. W., BINDSCHADLER, D. L., BAER, G., SCHUBERT, G., SHARPTON, V. L. & STOFAN, E. R. 1992. Geophysical models for the formation and evolution of coronae on Venus. *Journal of Geophysical Research*, **97**, 16 055–16 067.

KOMATSU, G. & BAKER, V. R. 1994. Meander properties of Venusian channels. *Geology*, **22**, 67–70.

KOMATSU, G., KARGEL, J. & BAKER, V. R. 1992. Canalitype channels on Venus: some genetic constraints. *Geophysical Research Letters*, **19**, 1415–1418.

LÓPEZ, I. & HANSEN, V. L. 2008. *Geologic Map of the Helen Planitia Quadrangle (V-52), Venus (1:5 Million Scale)*. United States Geological Survey, Geologic Investigations Series, **Map 3026**.

LÓPEZ, I. & LANG, N. P. 2008. Distribution of small shield volcanoes at Venusian coronae. *Geological Society of America Abstracts with Programs*, **40**, 114.

MCKENZIE, D. P., FORD, P. G., LIU, F. & PETTENGILL, G. H. 1992. Pancake-like domes on Venus. *Journal of Geophysical Research*, **97**, 15,967–15,976.

MUELLER, N., HELBERT, J., HASHIMOTO, G. L., TSANG, C. C. C., ERARD, S., PICCIONI, G. & DROSSART, P. 2008. Venus surface thermal emission at 1 μm in VIRTIS imaging observations: evidence for variation of crust and mantle differentiation conditions. *Journal of Geophysical Research*, **113**, E00B17, http://dx.doi.org/10.1029/2008JE003118

PAVRI, B., HEAD, J. W., KLOSE, K. B. & WILSON, L. 1992. Steep-sided domes on Venus: characteristics, geologic setting, and eruption conditions from Magellan data. *Journal of Geophysical Research*, **97**, 13 445–13 478.

PETFORD, N. 2000. Dyke widths and ascent rates of silicic magmas on Venus. *In*: BARBARIN, B., STEPHENS, W. E., BONIN, B., BOUCHEZ, J.-L., CLARKE, D. B., CUNEY, M. & MARTIN, H. (eds) *The Fourth Hutton Symposium on the Origin of Granites and Related Rocks*. Geological Society of America Special Papers, **350**, 87–95, http://dx.doi.org/10.1130/0-8137-2350-7.87

PINKERTON, H. & WILSON, L. 1994. Factors controlling the lengths of channel-fed lava flows. *Bulletin of Volcanology*, **56**, 108–120.

ROBERTS, K. M. & HEAD, J. W. 1993. Large-scale volcanism associated with coronae on Venus: implications for formation and evolution. *Geophysical Research Letters*, **12**, 1111–1114.

SAAR, M. O., MANGA, M., CASHMAN, K. & FREMOUW, S. 2001. Numerical models of the onset of yield strength in crystal-melt suspensions. *Earth and Planetary Science Letters*, **187**, 367–379.

SKINNER, J. A. & TANAKA, K. L. 2003. How should planetary map units be defined? *In*: MACKWELL, S. & STANSBERY, E. (eds) *Lunar and Planetary Institute Science Conference Abstracts Volume 34*. Lunar and Planetary Institute (LPI), Houston, TX, Abstract 2100.

SQUYRES, S. W., JANES, D. M., BAER, G., BINDSCHADLER, D. L., SCHUBERT, G., SHARPTON, V. L. & STOFAN, E. R. 1992. The morphology and evolution of coronae on Venus. *Journal of Geophysical Research*, **97**, 13 611–13 634.

STOFAN, E. R. & GUEST, J. E. 2003. *Geologic Map of the Aino Planitia Quadrangle (V-46), Venus (1:5 Million Scale)*. United States Geological Survey, Geologic Investigations Series, **Map. 2779**.

STOFAN, E. R., SHARPTON, V. L., SCHUBERT, G., BAER, G., BINDSCHADLER, D. L., JANES, D. M. & SQUYRES, S. W. 1992. Global distribution and characteristics of coronae and related features on Venus: implications for origin and relation to mantle processes. *Journal of Geophysical Research*, **97**, 13,347–13,378.

STOFAN, E. R., HAMILTON, V. E., JANES, D. M. & SMREKAR, S. E. 1997. Coronae on Venus: morphology and origin. *In*: BOUGER, S. W., HUNTEN, D. M. & PHILLIPS, R. J. (eds) *Venus II*. University of Arizona Press, Tucson, 931–968.

STOFAN, E. R., ANDERSON, S. W., CROWN, D. A. & PLAUT, J. J. 2000. Emplacement and composition of steep-sided domes on Venus. *Journal of Geophysical Research*, **105**, 26 757–26 771.

STOFAN, E. R., TAPPER, S. W., GUEST, J. E., GRINROD, P. & SMREKAR, S. E. 2001. Preliminary analysis of an expanded corona database for Venus. *Geophysical Research Letters*, **28**, 4267–4270.

SURKOV, Y. A., BARSUKOV, V. L., MOSKALYEVA, L. P., KHARYUKORVA, V. P. & KERMURDZHIAN, A. L. 1984. New data on the composition, structure, and properties of Venus rocks obtained by Venera 13 and Venera 14. *Journal of Geophysical Research*, **89**, (Suppl. S02), B393–B402, http://dx.doi.org/10.1029/JB089iS02p0B393

TANAKA, K. L. 1994. *The Venus Geologic Mapper's Handbook*, 2nd edn. United States Geological Survey, Open-File Report, 94–438.

WILHELMS, D. E. 1990. Geologic mapping. *In*: GREELEY, R. & BATSON, R. M. (eds) *Planetary Mapping*. Cambridge University Press, New York, 208–260.

ZIMBELMAN, J. R. 1998. Emplacement of long lava flows on planetary surfaces. *Journal of Geophysical Research*, **103**, 27 503–27 516.

A pyroclastic flow deposit on Venus

RICHARD C. GHAIL[1]* & LIONEL WILSON[2]

[1]*Department of Civil and Environmental Engineering, Imperial College London, London SW7 2AZ, UK*

[2]*Lancaster Environment Centre, Lancaster University, Lancaster LA1 4YQ, UK*

*Corresponding author (e-mail: r.ghail@imperial.ac.uk)

Abstract: Explosive volcanism on Venus is severely inhibited by its high atmospheric pressure and lack of water. This paper shows that a deposit located near 16°S, 145°E, here referred to as Scathach Fluctus, displays a number of morphological characteristics consistent with a pyroclastic flow deposit. These characteristics, particularly the lack of channelization and evidence for momentum- rather than cooling-limited flow length, contrast with fissure-fed lava flow deposits. The total erupted volume is estimated to have been between 225 and 875 km^3 but this may have been emplaced in more than one event. Interaction between Scathach Fluctus and a small volcanic cone constrains the flow velocity to 48 m s^{-1}, and plausible volatile concentrations to at least 1.8 wt% H$_2$O, 4.3 wt% CO$_2$ or 6.1 wt% SO$_2$, the latter two values implying that magma was sourced directly from the mantle. The deposit has radar characteristics, particularly an exponential backscatter function, that are similar to those of nearly half the planetary surface, implying that pyroclastic deposits may be much more common on Venus than has been recognized to date, and suggesting both a relatively volatile-rich mantle and a volcanic source for atmospheric SO$_2$.

Volcanic processes dominate the surface of Venus, with features including plains-forming flood lavas, large shield volcanoes, calderas, clusters of small volcanoes, and sinuous lava channels (Head *et al.* 1992). There is a notable absence of pyroclastic activity. This must, in part, be a result of the inhibiting effect of high atmospheric pressure (Head & Wilson 1986), but it is also consistent with a lack of volatiles in volcanic magmas and, therefore, in the interior of Venus. Many terrestrial volcanoes lie within a compositional spectrum between primary, relatively volatile-poor effusive basalts and evolved, relatively volatile-rich, often explosive, andesites and other silicic magmas. The major volatile component is usually water, which may be sourced directly from fertile mantle in basalts (Green 1973) or enriched in evolved magmas through dehydration of subducted crust at convergent plate margins. Magellan imagery and Venera lander data (Surkov & Barsukov 1985) indicate that the majority of Venus' volcanics are basaltic in composition, although there is some evidence for compositions richer in silica (Fink *et al.* 1993; Bridges 1997; Shellnutt 2013), sulphate (Kargel *et al.* 1994) or carbonate (Williams-Jones *et al.* 1998; Komatsu *et al.* 2001), the latter of which implies silica-poor evolved magmas (Hess & Head 1990) that erupt effusive low-viscosity flows. The planet's interior is extremely dry with respect to water, with, perhaps, only 50 ppm water in

basaltic magmas (Grinspoon 1993). The limited and equivocal observational evidence for explosive eruptions on Venus (Guest *et al.* 1992; Campbell 1994; Grosfils *et al.* 2011) is consistent with this inference of a dry interior. The abundance of other major volatiles, CO$_2$ and SO$_2$, is unknown but assumed to be Earth-like (Bullock & Grinspoon 2001). This paper presents evidence for a volatile-rich pyroclastic flow deposit located near Diana Chasma at 16°S, 145°E. We name this deposit Scathach Fluctus, after a Celtic destroyer goddess meaning 'the shadowy one'.

Identification

Scathach Fluctus is identified as a semi-circular doughnut-shaped deposit located on the boundary between plains north of Diana Chasma and fractured terrain west of Ceres Corona (Fig. 1). The perspective view in Figure 1a shows that the deposit is relatively flat and not associated with a volcanic rise. The false colour images in Figures 1b–d & 2 use data from Magellan Cycle 1 (left-looking), Cycle 2 (right-looking) and passive emissivity combined to enhance the impression of relief in the grey-scale image, overlain with colour-coded derived asperity height (roughness) at the scale-length of the Magellan radar wavelength (126 mm). Asperity is defined as the surface roughness, *h*, on

From: PLATZ, T., MASSIRONI, M., BYRNE, P. K. & HIESINGER, H. (eds) 2015. *Volcanism and Tectonism Across the Inner Solar System.* Geological Society, London, Special Publications, **401**, 97–106.
First published online November 19, 2013, http://dx.doi.org/10.1144/SP401.1

(a) Localities

False colour Magellan SAR image of eastern Aphrodite Terra, in the equatorial region of Venus. The localities referred to in the text, and illustrated in detail below, are indicated along with the names of the principal geographical features in the region.

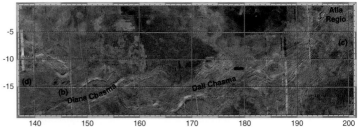

(b) 16s145 Scathach Fluctus

(c) 7s199 Anemone

(i) Central vent with smooth or flow-free margin

(ii) Discrete flows with lobate margins at a uniform source distance

(i) Smooth or flow-free interior lacks an obvious source

(ii) Straight flows with abrupt straight margins and termini. Flows lack a channel morphorphology

(iii) Flow deposits on rough terrain are difficult to identify and lack clear morphological characteristics

N

Mercator Projection

0   25   50 km

Asperity Height

0   20   40   60 mm

(b), (c) and (d) Common Scale and Key

(d) 15s140 Lobate Flows

(i) Lobate margins except where fault-bounded

(ii) Irregular interior gaps

(iii) Curvilinear channels with smooth interiors and rougher lanes and margins the scale of the radar wavelength, and is calculated using the Small Perturbation Model (Chen & Fung 1988) for Cycle 1 data at the appropriate incidence angle. Since Cycle 1 incidence angle varies with latitude, the asperity values derived are not strictly comparable between different images. In practice, however, the differences are small at low latitudes and, for data in the latitude of the images shown, asperity can range from less than 10 mm (very smooth) to about 40 mm (very rough). The surface may, of course, be rougher than this but not on the length-scale of the radar signal (126 mm); that is, the surface may be modulated by larger-amplitude undulations but they will be on a longer length-scale. Thus, asperity is not sensitive to larger-scale features, such as the approximately 100 mm undulations across pahoehoe-type flow lobes, but is sensitive to the roughness of individual lobe surfaces.

The deposit itself (Fig. 1b) is superficially similar to, and intermediate in size between, an anemone (Head et al. 1992), an example of which is shown in Figure 1c, and the lobate flows shown in Figure 1d, both of which appear to originate at fissures. At first sight Scathach Fluctus also appears to originate at a fissure (the yellow/red linear feature orientated NE–SW across the centre of Fig. 1b, immediately to the right of the Fig. 1b.i arrowhead) but this feature cuts across part of the deposit and a number of other features, and may therefore be a more recent tectonic structure. All that can be stated with confidence is that the source lies somewhere in the region of the fractured, but otherwise smooth, surfaces in the vicinity of 15.8°S, 145.3°E. Although most apparent on the plains west of this source region, the deposit is also discernable across fractured terrain to the south, but with a clear right-angled gap in the SW, which, together with other smaller gaps and irregularities, may indicate multiple emplacement events.

Figure 2 shows two sections of the western deposit, where it lies on the plains, printed at the highest resolution of the Magellan data (which at this latitude is about 110 m in both range and azimuth), as well as part of the anemone featured in Figure 1. The smoothest areas of the images have an asperity of less than 10 mm; the roughest parts of the flow surface have an asperity of about 30 mm and, in places, up to 40 mm.

The smoothest areas to the east of the deposit (Fig. 2a.i), nearer the source region, are similar in roughness to the plains and may simply be a smoother part of the deposit, buried by fine-grained sediment (which need only be a few centimetres thick), or may be an area altogether free from flow material. The anemone likewise has a smooth region between the source fissure (Fig. 2b.i) and the first appearance of rough surfaces away from the vent (Fig. 2b.ii). However, at the anemone, some slight roughness variations can be traced directly from the fissure to the area of channel-like flows (Fig. 2b.iii). This morphological arrangement is that expected for a fissure-fed lava eruption. Initially, lava spreads from the fissure as a near-uniform sheet flow; as lava gets further from the fissure more of it cools and its effective yield strength increases. Topographic undulations cause thickness variations in the flow, so that the increasing yield strength is not the same everywhere, being greatest where the flow is thinnest and it is here that incipient stationary levees start to form. These have a finite thickness initially just equal to the current thickness of the sheet flow but, with further cooling, the yield strengths increase and the levee-like structures get thicker. This process forces the hot lava flowing between any two levees to get deeper, to keep up with the thickening levees, and so to flow faster; but this is prevented by the constant volume flux per unit length along strike of the fissure. The only option is for the flowing lava to form narrower streams and so the levee-like zones split, each half of each one forming the levee on one side of each of a series of now discrete, thicker, faster, narrower channelized flow units growing alongside one another. As expected at a cooling-controlled terminus of a lava flow (Borgia et al. 1983; Bruno & Taylor 1995; Blake & Bruno 2000), the anemone flow units terminate in lobate margins (Fig. 2b.iv) that have an asperity similar to that of the channel levees along the flow margins. The distances from the inferred vent region to the points where flows first diverge is 4.5–5.5 km, and the distance from the vent to the flow termini is 19–23 km. Using data on cooling-limited flows on Mt Etna (Pinkerton & Wilson 1994) modified by later observations on Hawaiian flows (Soule et al. 2004), these distances are consistent with volume fluxes feeding individual flow lobes within a factor of three of 300 m^3 s^{-1}.

Fig. 1. Three volcanic deposits are shown at the same scale to illustrate the similarities and differences between them. (**a**) The deposits are all located in the equatorial region of Venus close to eastern Aphrodite Terra. Scathach Fluctus (**b**) is intermediate in size between an anemone (**c**) and a lobate flow deposit (**d**). All three are smoother nearer their presumed source and have rough termini. The anemone has a clearly defined central vent but there is no obvious source for either Scathach Fluctus or the lobate flows. Although very different in scale, both the anemone and lobate flows have curvilinear channel-like morphologies that are absent at Scathach Fluctus.

(a) 16s145 Scathach Fluctus

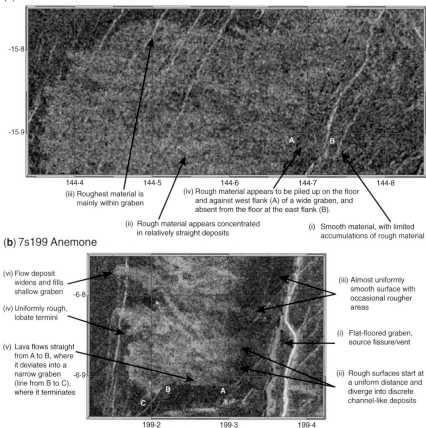

(iii) Roughest material is
mainly within graben

(iv) Rough material appears to be piled up on the floor
and against west flank (A) of a wide graben, and
absent from the floor at the east flank (B).

(ii) Rough material appears concentrated
in relatively straight deposits

(i) Smooth material, with limited
accumulations of rough material

(b) 7s199 Anemone

(vi) Flow deposit
widens and fills
shallow graben

(iv) Uniformly rough,
lobate termini

(v) Lava flows straight
from A to B, where
it deviates into a
narrow graben
(line from B to C),
where it terminates

(iii) Almost uniformly
smooth surface with
occasional rougher
areas

(i) Flat-floored graben,
source fissure/vent

(ii) Rough surfaces start at
a uniform distance and
diverge into discrete
channel-like deposits

(c) 16s145 Scathach Fluctus interaction with a volcano

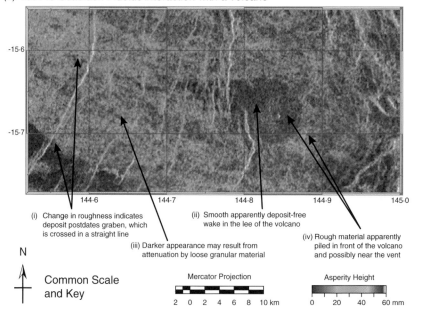

(i) Change in roughness indicates
deposit postdates graben, which
is crossed in a straight line

(ii) Smooth apparently deposit-free
wake in the lee of the volcano

(iii) Darker appearance may result from
attenuation by loose granular material

(iv) Rough material apparently
piled in front of the volcano
and possibly near the vent

N

Common Scale
and Key

Mercator Projection

2   0   2   4   6   8   10 km

Asperity Height

0   20   40   60 mm

Within Scathach Fluctus, there is some evidence for rougher material towards the east in the smooth area but there is no evidence for channelization in the rougher part of the deposit, and the morphology of this rougher part is very different from the channelized area of the anemone. Rather than being channelized, the main part of the deposit appears to undulate in thickness from north to south (Fig. 2a.ii), with these undulations forming more-or-less linear deposits from the east to the western margin. The deposit appears to thin towards the western margin but, other than a regional slope, there is no measurable change in elevation.

The two deposits contrast particularly in their interaction with local graben. The flows from the anemone deviate into a narrow graben (Fig. 2b.v) as they cross it. The flows also enter a graben towards their terminus (Fig. 2b.vi); the flow widens and just fills the graben but does not flow along it, again indicative of a cooling-limited deposit.

By contrast, the margins of Scathach Fluctus do not deviate in flow direction across the graben (Fig. 2c.i shows the clearest example but the same can be seen in Fig. 2a), indicating a high-momentum flow. The graben often remain partially visible through the flows, which may indicate that the graben are more recent, but, in at least some instances, the flows interact with graben. The deposit is often rougher within the graben (Fig. 2a.iii & c.i), having an asperity up to 50% greater than average. These features indicate that the flows often only partially fill any graben that are crossed and, perhaps, do so preferentially with coarser material. The implications are that the flows may be stratified (Fisher & Heiken 1982; Fujii & Nakada 1999) with a coarser base, and that as the flows cross a graben the lower, coarser portion falls under gravity and deposits material on to the graben floor on a parabolic trajectory, while the upper, finer part continues across the graben without any noticeable change, indicative for pyroclastic flows where portions of the ground-hugging avalanche is hindered from travelling further. At obstacles, such as fault scarps, the basal avalanche

can be deflected, while the overriding dilute gas-fine particle cloud does not change its path. These features are most clearly visible in the very large graben at the eastern margin of the deposit (Fig. 2a.iv), in which flow material is visible piled up against the western wall of the graben. At their margins, the flows often terminate within graben, which they apparently flow along, indicating that in these places no part of the flow has sufficient momentum to cross the graben.

These characteristics are interpreted as representing a low-density, high-velocity pyroclastic flow deposit (Wright et al. 1980). It appears that the flow first slowed sufficiently to deposit material as it crossed the easternmost wide graben, after which material was continuously deposited until momentum and forward velocity was lost at the terminus of the deposit. The largest fragments were most probably deposited first and, hence, the thickest and roughest part of the flow is towards the east, and it has relatively thin and smooth margins. The long, near-linear undulations represent random differences in flow particle concentration with distance from the source, stretched out east–west by the velocity of the flow.

Characterization

Scathach Fluctus is fortuitously located in an area of Venus covered by all three Magellan imaging cycles (left, right and stereo), as well as the full range of ancillary data derived from the altimeter. These have been used to enhance the image data with shaded relief and asperity information but, in addition, these data can be used to provide parallax measurements, and some constraints on the physical nature of the deposit and the surrounding terrain.

The patchy appearance of the deposit and the presence of the partly buried graben imply a thin deposit. Parallax measurements can only determine the height of the surface of the deposit and not its thickness (for a full description on how parallax measurements are obtained, see Leberl et al. 1992) but, where the flow crosses and completely

Fig. 2. The highest-resolution Magellan imagery reveals the differences in interaction of the Scathach Fluctus (**a, c**) and anemone (**b**) flows with pre-existing topographical features. In (a), Scathach Fluctus flows are smooth to the east of a wide graben (with flanks marked A and B). Within the graben, the flow appears to be absent from the eastern flank (B) and piled up against the western flank (A), indicative of a parabolic trajectory across the graben and implying a flow with considerable forward velocity and momentum. Similarly, the flow is piled up on the eastern flank of a small volcano (c) and bifurcates, leaving a flow-free tail to the west of the volcano. There is some evidence for a thin veneer of pyroclastic flow material on the upper part of the volcano near the vent (c.iv). This may be material deposited from the upper part of the pyroclastic cloud or material that partially travelled up the edifice leaving a thin veneer at the top, whereas elsewhere on the flanks it has been eroded, not preserved or is not detectable. The relatively straight flow lines and lack of channel morphology in these flows contrasts with the non-linear channel-like flows of the anemone, which also appear to be more strongly influenced by topographical features, notably the narrow graben visible as a pale line from B to C.

buries a graben (e.g. in the lower part of Fig. 2a), an estimate of its local thickness can be obtained from estimates of the depth of the graben adjacent to the flow. Although the majority of the graben are too narrow for reliable parallax measurements, their depths may be estimated from the width of their flanks, using the radar incidence angle to obtain a minimum depth estimate and probable slope angles to obtain a maximum estimate (McGill & Campbell 2006), assuming axially symmetrical slopes. These measurements indicate that the graben are commonly between 100 and 200 m deep, but may be up to twice this depth locally. Therefore, the thickest parts of the deposit in the graben may be up to 200 m thick. Away from the graben the flow will be thinner; a reasonable estimate may be 50 m. Taking this estimate and only the area of the rough part of the flows on the plains (4500 km^2) provides a minimum estimate of the erupted volume: 225 km^3, about twice that of the 1815 Tambora eruption (Sigurdsson & Carey 1989; Self et al. 2004). As noted earlier, however, this Venus deposit may not have been produced by a single event. A further 3500 km^2 can be traced across the fractured terrain in the south. In the west, the deposit has a distinct emissivity (dotted outline in Fig. 3a) relative to surrounding features; although it is less distinctive in the east, the assumption of radial symmetry provides a maximum area estimate of 17 500 km^2. Assuming 50 m as an upper estimate for its average thickness yields a volume of 875 km^3, approaching the volume of the Taupo, Yellowstone and Toba caldera-forming events (Mason et al. 2004), but note again that Scathach Fluctus may represent more than one event.

An estimate of the flow velocity may be obtained from analysis of the interaction between the flow and a small volcanic cone (Fig. 2c). The deposit appears to be piled up on the eastern flank of the cone and to have left a hydrodynamic (tear-drop shaped) wake to the west of it. Assuming the vent of this 7 km diameter volcanic cone is at its centre, the height reached by the deposit on the eastern flank can be determined by parallax. Accounting for the 0.9° regional slope, the flow deposits cover the lower 130 m of the eastern flank of the 540 m-high cone. Assuming that this equates to the 'just stop' obstacle height, $H = v^2/2g$ (where H is height, v is velocity and g is acceleration due to gravity), obtained by equating the kinetic energy of the approaching flow to the potential energy required to rise onto the obstacle, then the flow velocity was 48 m s^{-1} at this location, approximately 40 km from the centre of the emissivity feature. This is low in comparison to velocities inferred from similar-sized terrestrial events; for example, a velocity of 250 m s^{-1} at a distance of approximately 13 km from the vent was calculated for the approximately AD 186 Taupo ignimbrite (Wilson 1985). The low speed is directly attributable to the high atmospheric pressure on Venus, which necessitates a higher volatile concentration to generate an explosive eruption. Via the link between magma volatile content and ejecta velocity in explosive eruptions (Wilson 1980), the velocity tightly constrains the concentration of the three plausible volatiles, H_2O, CO_2 and SO_2 to 1.8, 4.3 and 6.1 wt%, respectively, calculated using the same relationships as those in Head & Wilson (1986). These values have been calculated independently; that is, not for a mixture of these gases. These concentrations are high in comparison to terrestrial magma values and the latter two are more than double the saturation level (Leone & Wilson 2001) in the crust, implying that explosive activity from a long-lived magma chamber in the crust must be primarily water-rich. Since crustal or subducted water is unlikely under Venusian conditions, its source is most probably primordial water in the mantle, concentrated through fractionation. Although crustal sources of carbonate or sulphate are possible (Kargel et al. 1994), which might explosively interact with ascending magma, CO_2- or SO_2-rich magmas may imply a direct mantle source (Green & Gueguen 1974) and a kimberlite-type explosive eruption event (Wilson & Head 2007).

Significance

Is Scathach Fluctus unique or are pyroclastic flow deposits common on Venus but unrecognized? The distinctive emissivity of the deposit is noticeable only because the adjacent plains have an unusually low emissivity of 0.751 ± 0.005, significantly lower than is normal for Venus. Together with a Fresnel reflectivity of 0.224 ± 0.007 and a back-scatter coefficient of −13.9 ± 2.9 dB, the unusual properties of these plains are consistent with high dielectric bedrock mantled by fine-grained granular sediment. Adopting the same procedure and assumptions as Campbell et al. (1992), we calculate a sediment depth of 0.46 m and a rock substrate dielectric constant of 11.5, higher than expected for a normal basalt (c. 7.5) and most probably a result of a high-dielectric or ferroelectric phase. Approximately 2 wt% TiO_2 in the form of the high-dielectric phase perovskite, $CaTiO_3$, is sufficient to account for the inferred dielectric constant. This is only slightly higher than the 1.6 wt% measured by Venera 13 (Surkov & Barsukov 1985) and the 1.19–1.77 wt% range of mid-ocean ridge basalts (Basaltic Volcanism Study Project 1981). Similar attenuations and inferred sediment depths are obtained from the central area of Scathach Fluctus.

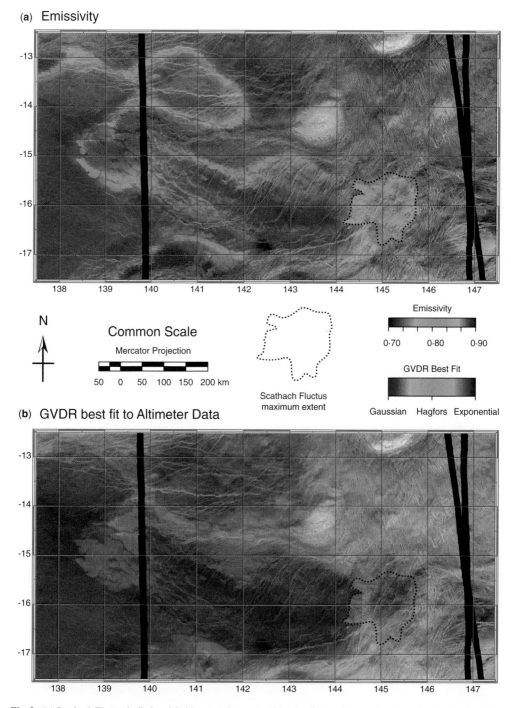

Fig. 3. (a) Scathach Fluctus is distinguishable as an almost circular green feature intermediate in emissivity between the high emissivity (*c*. 0.9) of the densely fractured terrain and the low emissivity (*c*. 0.7) of the plains. (b) The Global Vector Data Record (GVDR) shows the scattering law that most closely fits the altimeter backscatter data; the plains mainly have a Gaussian distribution indicative of a smooth surface with discrete blocks, while the highlands are closer to a Hagfors distribution indicative of a rough, blocky surface. Parts of Scathach Fluctus are closer to an exponential distribution, indicative of a flat or gently undulating surface.

While the granular sediment may plausibly have originated from the nearby impact crater Halle, it is tempting to suggest an air fall origin for the material from the turbulent upper part of the pyroclastic flow (e.g. as illustrated in Fisher & Heiken 1982; Fujii & Nakada 1999).

The emissivity of the visible (rougher) parts of Scathach Fluctus is 0.847 ± 0.002, similar to that of the average Venusian surface, and corresponds to a density close to 2000 kg m^{-3} based on relationships in Campbell (1994) and Rust et al. (1999), which is towards the upper end of the range of pyroclastic flow deposit densities (1240–2360 kg m^{-3}: Lepetit et al. 2009). It implies that the flow must be welded with little or no remaining porosity, giving it the radar characteristics of a solid surface.

The surface scattering function (Fig. 3b), recorded in the Magellan Global Vector Data Record (GVDR), may be used to characterize the metre-scale nature of the surface (see Tyler et al. 1992 for further details on the processing and data products). The plains adjacent to Scathach Fluctus are characterized best by a Gaussian scattering function, implying an undulating surface, whereas the fractured terrain is characterized very well by all three (Gaussian, Hagfors and Exponential) scattering functions, indicating a spectrum of surface types. The main area of Scathach Fluctus is characterized best by an exponential function, implying very flat surfaces, which may be tilted at the metre scale, as might be expected for a pyroclastic deposit. However, this observation is by no means definitive. More than half the surface of Venus is characterized by an exponential scattering function (Tyler et al. 1992), mainly in the plains. Thus, it is possible that a large fraction of the Venusian surface may have originated as pyroclastic flow deposits; however, identifying such deposits convincingly may prove extremely difficult, particularly where flow boundaries are no longer visible. Observations should focus on the identification of high-momentum features, and be supported by analyses of emissivity and reflectivity data. Even then, it may be impossible to properly distinguish between lava flows and pyroclastic flow deposits that are weathered or otherwise altered. Unwelded air fall deposits were tentatively identified on the southern flank of Tepev Mons, located at 15°S, 45°E, on the basis of their radar attenuation (Campbell 1994) but Scathach Fluctus is the first welded flow deposit to be identified. If there are significant numbers of other unidentified pyroclastic deposits then the interior of Venus may be relatively volatile-rich, with significant implications for interior chemistry (Falloon & Green 1990) and atmospheric evolution (Marcq et al. 2011, 2013).

Conclusions

A large pyroclastic flow deposit has confidently been identified on Venus on the basis of its flow morphology. Its size indicates a large, or series of large, explosive events on a scale similar to those of the largest such terrestrial events in historical times, and implies a volatile content much greater than expected. The radar properties of the flow indicate that it is a welded, solid material that may be difficult to distinguish from other materials on the Venusian surface. However, its exponential backscatter function implies that welded pyroclastic flow deposits may be much more common than previously realized, covering up to half the planetary surface, and indicative of a volatile-rich mantle and a volcanic source for atmospheric SO$_2$. Unfortunately, because the plains usually lack clear flow boundaries and structures, the features diagnostic of a high momentum flow – linear undulating deposits that lack channel morphology, the crossing of narrow graben without deviation, the climbing of obstacles and evidence for parabolic flow out from steep drops – may not be identifiable. Scathach Fluctus may, indeed, prove unique on Venus.

The authors would like to thank M. Ivanov and T. Mather, and the editor T. Platz, for their helpful and instructive reviews. All image data were obtained from Map-a-Planet through the PDS Imaging Node, and the GVDR data were retrieved from the PDS Geosciences Node. The image maps were produced in MAPublisher, supplied for academic use by Avenza Systems Inc.

References

BASALTIC VOLCANISM STUDY PROJECT 1981. *Basaltic Volcanism on the Terrestrial Planets*. Pergamon Press, Oxford.

BLAKE, S. & BRUNO, B. C. 2000. Modelling the emplacement of compound lava flows. *Earth and Planetary Science Letters*, **184**, 181–197.

BORGIA, A., LINNEMAN, S., SPENCER, D., MORALES, L. D. & ANDRE, J. B. 1983. Dynamics of lava flow fronts, Arenal Volcano, Costa Rica. *Journal of Volcanology and Geothermal Research*, **19**, 303–329.

BRIDGES, N. T. 1997. Ambient effects on basalt and rhyolite lavas under Venusian, subaerial, and subaqueous conditions. *Journal of Geophysical Research: Planets*, **102**, 9243–9255.

BRUNO, B. C. & TAYLOR, G. J. 1995. Morphologic identification of Venusian lavas. *Geophysical Research Letters*, **22**, 1897–1900.

BULLOCK, M. A. & GRINSPOON, D. H. 2001. The recent evolution of Climate on Venus. *Icarus*, **150**, 19–37.

CAMPBELL, B. A. 1994. Merging Magellan Emissivity and SAR Data for Analysis of Venus surface dielectric properties. *Icarus*, **112**, 187–203.

CAMPBELL, D. B., STACY, N. J. S. *ET AL.* 1992. Magellan observations of extended impact crater related features on the surface of Venus. *Journal of Geophysical Research: Planets*, **97**, 16 249–16 277.

CHEN, M. F. & FUNG, A. K. 1988. A numerical study of the regions of validity of the Kirchhoff and small-perturbation rough surface scattering models. *Radio Science*, **23**, 163–170.

FALLOON, T. J. & GREEN, D. H. 1990. Solidus of carbonated fertile peridotite under fluid-saturated conditions. *Geology*, **18**, 195–199.

FINK, J. H., BRIDGES, N. T. & GRIMM, R. E. 1993. Shapes of Venusian 'pancake' domes imply episodic emplacement and silicic composition. *Geophysical Research Letters*, **20**, 261–264.

FISHER, R. V. & HEIKEN, G. 1982. Mt. Pelée, martinique: may 8 and 20, 1902, pyroclastic flows and surges. *Journal of Volcanology and Geothermal Research*, **13**, 339–371.

FUJII, T. & NAKADA, S. 1999. The 15 September 1991 pyroclastic flows at Unzen Volcano (Japan): a flow model for associated ash-cloud surges. *Journal of Volcanology and Geothermal Research*, **89**, 159–172.

GREEN, D. H. 1973. Experimental melting studies on a model upper mantle composition at high pressure under water-saturated and water-undersaturated conditions. *Earth and Planetary Science Letters*, **19**, 37–53.

GREEN, H. W. & GUEGUEN, Y. 1974. Origin of kimberlite pipes by diapiric upwelling in upper mantle. *Nature*, **249**, 617–620.

GRINSPOON, D. H. 1993. Implications of the high D/H ratio for the sources of water in Venus' atmosphere. *Nature*, **363**, 428–431.

GROSFILS, E. B., LONG, S. M. *ET AL.* 2011. Geologic Map of the Ganiki Planitia Quadrangle (V–14), Venus. United States Geological Survey Scientific Investigations Map **3121**.

GUEST, J. E., BULMER, M. H. *ET AL.* 1992. Small volcanic edifices and volcanism in the plains of Venus. *Journal of Geophysical Research: Planets*, **97**, 15,949–15,966.

HEAD, J. W. & WILSON, L. 1986. Volcanic processes and landforms on Venus: theory, predictions, and observations. *Journal of Geophysical Research: Solid Earth*, **91**, 9407–9446.

HEAD, J. W., CRUMPLER, L. S., AUBELE, J. C., GUEST, J. E. & SAUNDERS, R. S. 1992. Venus volcanism: Classification of volcanic features and structures, associations, and global distribution from Magellan data. *Journal of Geophysical Research: Planets*, **97**, 13,153–13,197.

HESS, P. C. & HEAD, J. W. 1990. Derivation of primary magmas and melting of crustal materials on Venus: Some preliminary petrogenetic considerations. *Earth, Moon, and Planets*, **50–51**, 57–80.

KARGEL, J. S., KIRK, R. L., FEGLEY, B. JR. & TREIMAN, A. H. 1994. Carbonate-sulfate volcanism on Venus? *Icarus*, **112**, 219–252.

KOMATSU, G., GULICK, V. C. & BAKER, V. R. 2001. Valley networks on Venus. *Geomorphology*, **37**, 225–240.

LEBERL, F. W., THOMAS, J. K. & MAURICE, K. E. 1992. Initial results from the Magellan stereo experiment. *Journal of Geophysical Research: Planets*, **97**, 13 675–13 689.

LEONE, G. & WILSON, L. 2001. Density structure of Io and the migration of magma through its lithosphere. *Journal of Geophysical Research: Planets*, **106**, 32 983–32 995.

LEPETIT, P., VIERECK-GOETTE, L., SCHUMACHER, R., MUES-SCHUMACHER, U. & ABRATIS, M. 2009. Parameters controlling the density of welded ignimbrites – a case study on the Incesu Ignimbrite, Cappadocia, Central Anatolia. *Chemie der Erde – Geochemistry*, **69**, 341–357.

MARCQ, E., BELYAEV, D., MONTMESSIN, F., FEDOROVA, A., BERTAUX, J.-L., VANDAELE, A. C. & NEEFS, E. 2011. An investigation of the SO2 content of the venusian mesosphere using SPICAV-UV in nadir mode. *Icarus*, **211**, 58–69.

MARCQ, E., BERTAUX, J.-L., MONTMESSIN, F. & BELYAEV, D. 2013. Variations of sulphur dioxide at the cloud top of Venus's dynamic atmosphere. *Nature Geoscience*, **6**, 25–28.

MASON, B. G., PYLE, D. M. & OPPENHEIMER, C. 2004. The size and frequency of the largest explosive eruptions on Earth. *Bulletin of Volcanology*, **66**, 735–748.

MCGILL, G. E. & CAMPBELL, B. A. 2006. Radar properties as clues to relative ages of ridge belts and plains on Venus. *Journal of Geophysical Research*, **111**, E12006, http://dx.doi.org/10.1029/2006JE002705

PINKERTON, H. & WILSON, L. 1994. Factors controlling the lengths of channel-fed lava flows. *Bulletin of Volcanology*, **56**, 108–120.

RUST, A. C., RUSSELL, J. K. & KNIGHT, R. J. 1999. Dielectric constant as a predictor of porosity in dry volcanic rocks. *Journal of Volcanology and Geothermal Research*, **91**, 79–96.

SELF, S., GERTISSER, R., THORDARSON, T., RAMPINO, M. R. & WOLFF, J. A. 2004. Magma volume, volatile emissions, and stratospheric aerosols from the 1815 eruption of Tambora. *Geophysical Research Letters*, **31**, L20608, http://dx.doi.org/10.1029/2004GL020925

SHELLNUTT, J. G. 2013. Petrological modeling of basaltic rocks from Venus: a case for the presence of silicic rocks. *Journal of Geophysical Research: Planets*, **118**, 1–15.

SIGURDSSON, H. & CAREY, S. 1989. Plinian and co-ignimbrite tephra fall from the 1815 eruption of Tambora volcano. *Bulletin of Volcanology*, **51**, 243–270.

SOULE, S. A., CASHMAN, K. V. & KAUAHIKAUA, J. P. 2004. Examining flow emplacement through the surface morphology of three rapidly emplaced, solidified lava flows, Kilauea Volcano, Hawai'i. *Bulletin of Volcanology*, **66**, 1–14.

SURKOV, Y. A. & BARSUKOV, V. L. 1985. Composition, structure and properties of Venus rocks. *Advances in Space Research*, **5**, 17–29.

TYLER, G. L., SIMPSON, R. A., MAURER, M. J. & HOLMANN, E. 1992. Scattering properties of the Venusian surface: preliminary results from Magellan. *Journal of Geophysical Research: Planets*, **97**, 13 115–13 139.

WILLIAMS-JONES, G., WILLIAMS-JONES, A. E. & STIX, J. 1998. The nature and origin of Venusian canali.

Journal of Geophysical Research: Planets, **103**, 8545–8555.

WILSON, C. J. N. 1985. The Taupo eruption, New Zealand. 2. The Taupo ignimbrite. *Philosophical Transactions of the Royal Society Series A*, **314**, 229–310.

WILSON, L. 1980. Relationships between pressure, volatile content and ejecta velocity in three types of volcanic explosion. *Journal of Volcanology and Geothermal Research*, **8**, 297–313.

WILSON, L. & HEAD, J. W., III 2007. An integrated model of kimberlite ascent and eruption. *Nature*, **447**, 53–57.

WRIGHT, J. V., SMITH, A. L. & SELF, S. 1980. A working terminology of pyroc lstic deposits. *Journal of Volcanology and Geothermal Research*, **8**, 315–336.

Are terrestrial plumes from motionless plates analogues to Martian plumes feeding the giant shield volcanoes?

CHRISTINE M. MEYZEN[1]*, MATTEO MASSIRONI[1,2],
RICCARDO POZZOBON[1,3] & LUCA DAL ZILIO[1]

[1]*Dipartimento di Geoscienze, Università degli Studi di Padova,
via G. Gradenigo, 6, 35131 Padova, Italy*

[2]*INAF, Osservatorio Astronomico di Padova, Vicolo dell'Osservatorio 3, 35122 Padova, Italy*

[3]*IRSPS-DISPUTer, Universita' G. d'Annunzio, Via Vestini 31, 66013 Chieti, Italy*

*Corresponding author (e-mail: christine.meyzen@unipd.it)

Abstract: On Earth, most tectonic plates are regenerated and recycled through convection. However, the Nubian and Antarctic plates could be considered as poorly mobile surfaces of various thicknesses that are acting as conductive lids on top of Earth's deeper convective system. Here, volcanoes do not show any linear age progression, at least not for the last 30 myr, but constitute the sites of persistent, focused, long-term magmatic activity rather than a chain of volcanoes, as observed in fast-moving plate plume environments. The melt products vertically accrete into huge accumulations. The residual depleted roots left behind by melting processes cannot be dragged away from the melting loci underlying the volcanoes, which may contribute to producing an unusually shallow depth of oceanic swells. The persistence of a stationary thick depleted lid slows down the efficiency of melting processes at shallow depths. Numerous characteristics of these volcanoes located on motionless plates may be shared by those of the giant volcanoes of the Tharsis province, as Mars is a one-plate planet. The aim of this chapter is to undertake a first inventory of these common features, in order to improve our knowledge of the construction processes of Martian volcanoes.

The near-'one-plate planet' evolutionary history of Mars has led to the formation of its long-lasting giant shield volcanoes, which dominate the topography of its western hemisphere. Unlike Earth, Mars would have been a transient convecting planet, where plate tectonics would have possibly acted only during the first few hundreds of millions of years of its history (e.g. Sleep 1994; Bouvier *et al.* 2009). Most Martian magmatic activity is probably very old (Noachian) and has been preserved in the near-absence of crustal recycling, in contrast to Earth (Phillips *et al.* 2001; Taylor *et al.* 2006; Werner 2009; Carr & Head 2010). Recent volcanic resurgence occurred in the main Martian volcanic provinces, Tharsis (Fig. 1) and Elysium, during the Amazonian Period (Werner 2009).

Although the large igneous provinces of Mars bear some geomorphological similarities with terrestrial oceanic plateaus, they distinguish themself by their larger scales. Martian volcano heights commonly reach three times those of Hawaiian volcanoes (Plescia 2004; Carr 2006), while they are generally many hundreds of kilometres in breadth. Their accretion, which is vertical due to the absence of lateral plate movement and crustal recycling, loads the lithosphere, and causes lithospheric flexure, deformation and edifice flank compressional

failure (e.g. McGovern & Solomon 1993; Phillips *et al.* 2001; McGovern *et al.* 2002; McGovern & Morgan 2009; Byrne *et al.* 2013). Underlying these huge volcanoes, the lithosphere might reach a thickness of up to 150 km to provide isostatic support (Zuber *et al.* 2000; McGovern *et al.* 2002). Thickening of the lithosphere (17–25 km Ga^{-1}) occurs over time as a result of decreasing mantle potential temperature (30–40 K Ga^{-1}) (Baratoux *et al.* 2011). This thickening of the lithospheric lid increases the final depth of melting and thereby reduces the mean degree of melting. Such a process leads to less efficient melt extraction and subsequent mixing, favouring mantle–melt interaction and high-pressure melt fractionation. Such lithospheric processes are commonly observed in near-stationary plate plume settings (velocity <25 mm a^{-1}).

We note, however, that while Hawaiian islands are located on a fast-moving plate (i.e. the Pacific plate, 65 mm a^{-1}: Argus *et al.* 2011), Hawaiian hotspot intraplate magmatic activity has commonly been considered as some of the best analogues to that observed on Mars (e.g. Zuber & Mougins-Mark 1992; McGovern & Solomon 1993; Plescia 2004; Carr 2006; Bleacher *et al.* 2007; Byrne *et al.* 2013). However, intraplate oceanic volcanism of

From: PLATZ, T., MASSIRONI, M., BYRNE, P. K. & HIESINGER, H. (eds) 2015. *Volcanism and Tectonism Across the Inner Solar System.* Geological Society, London, Special Publications, **401**, 107–126.
First published online February 21, 2014, http://dx.doi.org/10.1144/SP401.8

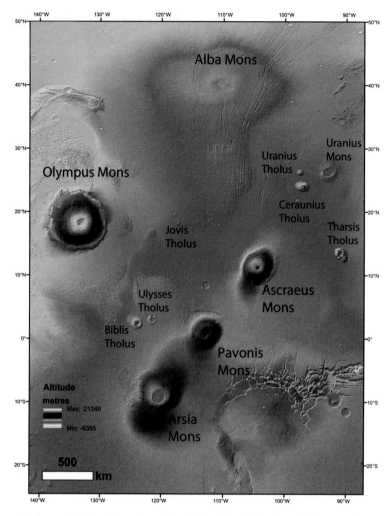

Fig. 1. Topographical map of the Tharsis province produced from Mars Orbiter Laser Altimeter (MOLA) data, with the labels showing the location of major volcanic edifices.

slow-moving plates such as the Crozet or Cape Verde islands might represent a better analogue. As on Mars, volcanoes at slow-moving plates do not show any linear age progression, but constitute the sites of persistent, long-term magmatic activity. Because the lithosphere lid is near-stagnant in these areas, the melting mantle region concentrates its products in a single area rather than them being spread out, as observed in fast-moving plate plume environments such as the Hawaiian–Emperor seamount chain on the Pacific seafloor. The melt products vertically accrete into huge accumulations, and oceanic swell heights are unusually shallow for their ages (Fig. 2). Processes of thermal reheating and mechanical weakening of the lithosphere might also be enhanced. The loading of the

lithosphere by the volcanic accumulations might ultimately cause edifice collapse and/or flexural deformation. These processes are probably similar to those observed on Mars.

The goal of this paper is to describe the essential characteristics of intra-oceanic plumes on slow-moving plates on Earth and to compare them to large shield volcanoes in the Tharsis region of Mars. Specific similarities will be described, whereas some peculiar features, either compositional or morphological and common among these terrestrial volcanoes but not yet detected on Mars, will be emphasized as main objectives for future research. Speculations on deep processes that could have controlled the construction and development of large shield volcanoes on Mars are also

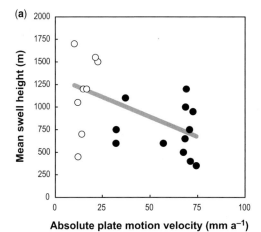

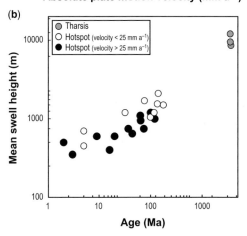

Fig. 2. (**a**) Mean swell height (m) as a function of absolute plate motion velocity (mm a^{-1}) for 21 oceanic islands. The velocities were calculated with the NNR-MORVEL56 model of Argus *et al.* (2011). Filled dots denote locations with plate velocities of >25 mm a^{-1}, while open dots denote those of <25 mm a^{-1}. The mean swell heights are from Monnereau & Cazenave (1990). (**b**) Mean swell height (m) as a function of the age of oceanic lithosphere (Ma). Mean swell heights and ages are from Monnereau & Cazenave (1990). The mean swell height for Tharsis was calculated from Mars Orbiter Laser Altimeter (MOLA) data, as described in Figure 9.

made on the basis of what is known and soundly hypothesized for plumes located on slow-moving oceanic plates.

The Tharsis magmatic province

The Tharsis volcanic province is one of the most noticeable global-scale physiographical feature on

Mars, covering about 20% of its surface area in the western hemisphere (Fig. 1). Its areal extent (>6.5 × 10^6 km^2) is far greater than that of the largest terrestrial igneous provinces. This magmatic province is located near the boundary of the Martian crustal dichotomy between the northern lowland and the southern uplands. It extends from the northern lowland plains southwards to Solis Planum, and from Arcadia Planitia eastwards to Lunae Planum. Its topography describes a bulge-like structure, above which rest massive shield volcanoes (e.g. >500 km wide and up to 25 km high: Fig. 1). Both effusive lava flows and pyroclastic activity have built up these edifices (e.g. Wilson & Head 1994).

These large shield volcanoes, such as Olympus Mons, may have been emplaced on very thick elastic lithosphere (T_e of more than 100 km) (Zuber *et al.* 2000; McGovern *et al.* 2002; Grott *et al.* 2013), meaning that the Tharsis region overlies some of the thickest lithosphere of Mars. But lithospheric thicknesses of less than 50 km also support a few volcanoes in the Tharsis region (Grott & Breuer 2009; Grott *et al.* 2013). Their crustal thickness is highly variable at a regional scale (Zuber *et al.* 2000). The old Alba Patera volcano (3.02 Ga: Robbins *et al.* 2011) is underlain by thick crust, whereas the region beneath the younger shields such as Olympus and Tharsis Montes (Arsia, Pavonis and Ascraeus) would have experienced apparent crustal thinning (Zuber *et al.* 2000). The major volcanic loading exerted by Tharsis volcanoes, with concomitant effects on deformation of the Martian lithosphere that is mirrored by a broad free-air anomaly, would help define the present-day long-wavelength gravity field of Mars (Phillips *et al.* 2001; Golle *et al.* 2012).

The gravity, topography and tectonic attributes of this region have been ascribed to one or more of the following processes: (1) a large-scale mantle convective upwelling (e.g. Schubert *et al.* 1990; Zhong & Zuber 2001; Roberts & Zhong 2006; Keller & Tackley 2009), inducing dynamic topography; (2) regional uplift due to lateral migration and intrusion of material thermally eroded from the base of the crust of the lowlands (Wise *et al.* 1979) or due to massive intrusion (i.e. 85% of the total magmatic products; Phillips *et al.* 1990) in the crust and upper mantle; (3) flexural loading from volcanic construction (Solomon & Head 1982; Phillips *et al.* 2001), possibly compensated by a depleted root (Finnerty *et al.* 1988); (4) preferential concentration of volcanism along early impact-basin ring structures (Schultz *et al.* 1982); and (5) edge-driven convection (King & Redmond 2005).

Among these processes, the first appears to be the main building mechanism for the Tharsis province through time. The formation of large-scale mantle plumes may be driven by viscosity layering in the mid-mantle (Zhong & Zuber 2001; Roberts & Zhong 2006), an endothermic phase change near the core–mantle boundary (Zhong 2009; Šrámek & Zhong 2012) or an early impact-induced thermal anomaly (Golabek *et al.* 2011). Large-scale upwelling driven by degree-1 mantle convection (i.e. whereby one hemisphere is dominated by an upwelling, while the other encompasses a downwelling), which is induced by layered mantle viscosity, is predicted to be short lived with a timescale development ranging from 1 to several hundred million years (myr) (Zhong & Zuber 2001; Roberts & Zhong 2006). These large-scale upwellings would tend to migrate towards regions of lower lithospheric thickness, which can account for the location of the Tharsis province at the dichotomy boundary (Zhong 2009; Šrámek & Zhong 2012). Convection driven by endothermic phase transition near the core–mantle boundary may also promote the formation and focusing of long-lived plumes at Tharsis, as well as late-stage volcanism (Harder 2000; Grott *et al.* 2013). In this case, the development of a degree-1 pattern would require at least 5 billion years (Wenzel *et al.* 2004; Roberts & Zhong 2006), but recent three-dimensional (3D) convection models with temperature-dependent rheology instead require a shorter formation timescale (Buske 2006). Another potential mechanism for stabilizing long-lived plumes at Tharsis is provided by a thick, low-density depleted mantle layer with non-linear conductivity (Schott *et al.* 2001), where minima in thermal conductivity values promote the reoccurrence of thermal instabilities. However, such a model fails to predict the generation of the Tharsis province at the dichotomy boundary (Grott *et al.* 2013). Excess mantle heat arising from an early giant impact, which would have first created the southern highlands crust, might also have triggered the formation of a transient superplume underneath the impacted hemisphere, inducing massive volcanism at Tharsis (Golabek *et al.* 2011). After more than 100 myr, the impact-induced plume is predicted to slowly regress, while new upwellings would form throughout the mantle (Golabek *et al.* 2011).

Most magmatic activity in the Tharsis region is Noachian (>3.7 Ga) in age, but the large shield volcanoes continued to grow up to the Amazonian (<3 Ga) to a lesser degree (Hartmann & Neukum 2001; Werner 2009). In particular, the average eruption rate would have dropped from approximately 0.04 km^3 a^{-1} in the Noachian and Hesperian to approximately 0.01 km^3 a^{-1} in the Amazonian (Greeley & Schneid 1991). Hence, Tharsis volcanoes were probably built up by eruptions widely spaced over a large timespan (Fig. 3). The clustering

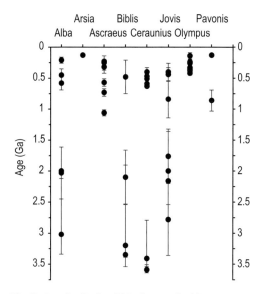

Fig. 3. Age distribution (Ga) of exposed calderas across the Tharsis province. Data are from Robbins *et al.* (2011).

of ages defined by lava flows and caldera floors (Neukum *et al.* 2004; Werner 2009; Robbins *et al.* 2011) confirms that magma supply was highly episodic (Fig. 3). The large shield volcanoes were characterized by very active periods with high effusion rates (as indicated by the large volume of many lava flows), separated by long periods of quiescence (e.g. Hiesinger *et al.* 2007; Giacomini *et al.* 2009; Hauber *et al.* 2011). During the Amazonian, a focusing of volcanism has been recorded between 1.6 and 1 Ga (Werner 2009), whereas the most recently erupted lava flows and the last caldera collapses are, indeed, very young (200–100 Ma) (Neukum *et al.* 2004; Werner 2009; Robbins *et al.* 2011; Pozzobon *et al.* 2014).

Shallow magma plumbing architectures fed by dykes probably constructed the major volcanic edifices of Tharsis (Wilson & Head 1994). Early estimates of magma chamber depths in Tharsis shields range between 9 and 16 km below the summit (Zuber & Mouginis-Mark 1992; Wilson & Head 1998). However, a recent study of Ascreaus Mons suggests a two-level architecture of the underground magma plumbing system, with some fractionation of magmas occurring within the uppermost mantle (Pozzobon *et al.* 2014). In addition, some of the nested calderas on the Tharsis Montes imply the existence of several shallow and large magma reservoirs, which are thermally viable for only a limited period of time (Wilson *et al.* 2001). The morphology of some late-stage lavas erupted near Jovis Tholus also suggests short-duration crustal storage (Wilson *et al.* 2009). Other evidence for multiple stages of magma ascent and withdrawal are provided by the complex geometry of summit calderas at Olympus and Ascraeus Montes (Mouginis-Mark & Robinson 1992; Crumpler *et al.* 1996; Scott & Wilson 2000; Byrne *et al.* 2012).

Average surface heat flux values for individual Tharsis volcanoes derived from the effective elastic thickness since the time of loading range from 14 to 35 mW m^{-2} (McGovern *et al.* 2002). When considering the superficial distribution of heat-producing elements, other estimates of heat fluxes encompass this range and show a similar geographical pattern, with gradually increasing values from 22 to 24 mW m^{-2} in the north to more than 26 mW m^{-2} in the south, along the Tharsis Montes (Grott & Breuer 2010). It is interesting to note that Olympus Mons exhibits the smallest heat flux of the volcanoes from the Tharsis province (McGovern *et al.* 2002; Grott *et al.* 2013). However, when hot mantle upwelling underneath these regions is taken into consideration, these local heat flow values are predicted to be at least twice as great (Grott & Breuer 2010; Grott *et al.* 2013).

At the planetary scale, recent comprehensive geochemical mapping of the Tharsis region by gamma ray spectrometer (GRS: Boynton *et al.* 2007; Newsom *et al.* 2007; El Maarry *et al.* 2009; Taylor *et al.* 2010; Baratoux *et al.* 2011) has shown that this region distinguishes itself in exhibiting some of the most depleted patterns in K, Th, Fe, Si and Ca among the geochemical composition spectrum exhibited by the six provinces of the Martian crust (Taylor *et al.* 2010). Although Si and Fe concentrations are depleted relative to those of other regions on Mars, they are lower and higher, respectively, than those of any terrestrial primary magma compositions (El Maarry *et al.* 2009). GRS chemical imaging reveals two distinct provinces (Boynton *et al.* 2007; El Maarry *et al.* 2009, Gasnault *et al.* 2010; Taylor *et al.* 2010; Baratoux *et al.* 2011), the boundary of which would lie to the north or south of Pavonis Mons. The Arsia, Pavonis and Olympus Montes have a lower Si, Th and Fe content than has Ascraeus (Fig. 4). Alba Mons is also distinct in that it exhibits the lowest Th, and some of the highest Fe, content of the Tharsis volcanoes (Fig. 4). The modelling of SiO$_2$–FeO systematics of these edifices shows that Ascraeus lavas can be produced by a slightly lower melting pressure (*c.* 1.6 GPa) and a higher degree of melting (*c.* 9%) than other Tharsis volcanoes (1.7–1.9 GPa, 5–8%). This implies the presence of a slightly higher mantle potential temperature underneath Ascraeus Mons (1384 °C) relative to other individual edifices from the Tharsis province (1340–1382 °C: Baratoux *et al.* 2011).

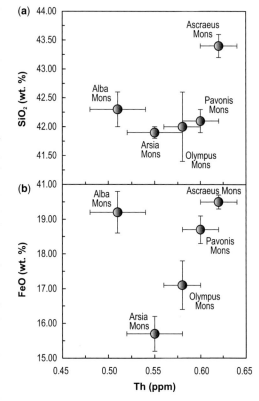

Fig. 4. Abundances of silica, thorium and iron oxide for five volcanoes from the Tharsis province: (**a**) SiO$_2$ v. Th; (**b**) FeO v. Th. Data are from Baratoux *et al.* (2011).

Terrestrial volcanoes at near-stagnant plates

On Earth, oceanic magmatism on a near-stagnant plate (less than 20 mm a^{-1}: Argus *et al.* 2011) is geographically restricted to the Antarctic plate, which has moved at an extremely slow velocity for the past 30 myr (15.2 mm a^{-1}: Argus *et al.* 2011) (Fig. 5). However, with the exception of the Kerguelen archipelago, very few data are available for most volcanoes emplaced on this plate (i.e. Bouvet, Marion and Crozet). We thus extend our study to volcanoes from the Nubian plate, which has an absolute motion of less than 25 mm a^{-1} according to Argus *et al.* 2011 (Fig. 5). This rate might be much lower, as another investigation suggests that the Nubian plate's absolute motion to the NE abruptly slowed at 30 Ma from 22 to about 10 mm a^{-1} (Silver *et al.* 1998). In addition, our interest includes volcanoes from archipelagos, for which geophysical and geochemical characteristics have been fully investigated (e.g. Canary, Madeira and Cape Verde). The emphasis will be placed

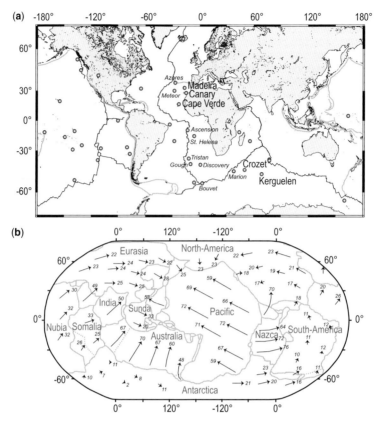

Fig. 5. (a) World map showing the distribution of prominent hotspots (modified from Lin 1998). (b) Plate map boundaries and their absolute horizontal velocities (mm a⁻¹, black arrows) in the NNR-MORVEL56 frame. Adapted and simplified from Argus *et al.* (2011).

mainly on those volcanoes atop an old and thick oceanic lithosphere (>50 Ma).

A near-stationary-lid regime, such as that of the Antarctic plate, should probably be much less efficient in transferring heat out of the planetary interior than a mobile-lid regime. The low buoyancy fluxes of some plumes located on slow-moving plates will further reinforce this effect. Plumes with less than 0.5 Mg s⁻¹ (i.e. Crozet: Sleep 1990), if rising from the core–mantle boundary, should have cooled so much by heat diffusion that they would not melt beneath an old lithosphere (Albers & Christensen 1996). Weak plumes with a buoyancy flux of about 0.5 Mg s⁻¹ have a 150 K or lower temperature anomaly in the upper mantle due to thermal diffusion (Steinberger & Antretter 2006). Therefore, such plumes may be sourced from shallower depths than the core–mantle boundary.

The majority of hotspots on slow-moving plates (i.e. less than 25 mm a⁻¹) exhibit topographical swell heights greater than those observed at faster plates for a given oceanic lithospheric age (Fig. 2).

(Monnereau & Cazenave 1990). Sitting atop these swells, their islands in each archipelago often define a horseshoe-like shape (Cape Verde, Crozet and Canary for the last 30 myr). As the melting region remains geographically fixed with time, the melts are focused and seem to be pooled, forming huge accumulations. Consequently, the Azores, Cape Verde and Crozet bathymetric swells possess some of the greatest amplitudes among hotspot-influenced regions (Monnereau & Cazenave 1990) (Fig. 2). The scatter of gravity and bathymetric data for the Northern Kerguelen Plateau also provides evidence for a swell's signature (Sandwell & MacKenzie 1989; Wallace 2002). Other hotspots, such as those from the Canary and Madeira islands, appear to be associated with less substantial swell anomalies (Filmer & McNutt 1989; Monnereau & Cazenave 1990; Marks & Sandwell 1991; Canales & Dañobeitia 1998; Grevemeyer 1999; Ito & van Keken 2007) (Fig. 2) but with prominent geoid anomalies (Jung & Rabinowitz 1986; Cazenave *et al.* 1988). However, in the Canary Islands, as shown by seismic data, a thick sedimentary cover masks the buried moat (Watts *et al.* 1997; Canales & Dañobeitia 1998).

The building and evolution modes of bathymetric swells have essentially been modelled in Hawaii, and have been ascribed to one or more of the following processes: (1) thermal rejuvenation (e.g. Detrick & Crough 1978), whereby the oceanic lithosphere is heated as it passes over the hotspot and is therefore thinned, producing isostatic uplift of the swell; (2) dynamic pressure (e.g. Olson 1990; Sleep 1990), whereby the upwards buoyancy flux through the asthenosphere from the swell centre supports it; and (3) compositional and thermal buoyancy in a higher-viscosity depleted swell root (Morgan *et al.* 1995; Yamamoto *et al.* 2007). Upon melting, the dense minerals garnet and clinopyroxene in the residue are exhausted, leading to a decrease in Fe relative to Mg, which forms an intrinsically less dense depleted root and produces the swell uplift.

Morgan *et al.* (1995) ascribed the unusually shallow oceanic swells such as those of Cape Verde and Crozet (Fig. 2) to the persistence of a stationary, thick and depleted root, which has not been dragged away from the hotspot in the virtual absence of plate motion since 30 Ma. Indeed, evidence of the presence of low-velocity, low-density upper-mantle material at depths ranging from 40 to 81 km at Cape Verde, Canary, Kerguelen and Crozet islands are provided by deep seismic refraction and gravity data (Marks & Sandwell 1991; Recq & Charvis 1987; Charvis *et al.* 1995; Canales & Dañobeitia 1998; Lodge & Helffrich 2006). These anomalies were interpreted by Morgan *et al.* (1995) to reflect the spreading, thinning and slight melting of a thick sublithospheric swell root.

However, the mechanisms that govern the generation and evolution of swells under a slow-moving plate have been investigated in detail in only a few localities. In the Cape Verde Islands, Lodge & Helffrich (2006), using seismic data to a depth of 120 km, highlighted the existence of low-density material with a non-radial flow distribution, spreading from separate melting loci under the swell. In the case of a dynamic pressure model, a flow field of homogeneous-viscosity material would emanate from the buoyancy flux centre with a radial anisotropic fabric. However, some investigators (Pim *et al.* 2008; Wilson *et al.* 2010) argue against the depleted root assumption, as no lateral variations in P-wave velocities are recorded in the upper 2 km of the mantle. Using measured heat flow, and geoid, seismic and bathymetric data, Wilson *et al.* (2010) have recently favoured dynamic upwelling (50–70%) as the main mechanism to sustain the swell, with minor contributions from thickened oceanic crust and partial thermal rejuvenation of the lithospheric mantle. We note that all studies of the Cape Verde Islands concluded that lithosphere reheating processes alone cannot account for the observed uplift (Courtney & White 1986; Ali *et al.* 2003; Lodge & Helffrich 2006; Pim *et al.* 2008). Several investigators at other locations (e.g. Liu & Chase 1989; Ribe & Christensen 1994) have also pointed out the inefficiency of thermomechanical erosion processes in sustaining a bathymetric swell. Even if endogenous growth of volcanoes is accompanied by intrusive accumulation of neutrally buoyant magma at the base of the crust, underplating also appears to play a minor role at best (although may be absent entirely) in elevating the seafloor (Grevemeyer 1999; Pim *et al.* 2008). In turn, seismic modelling does not show any evidence for underplating at the base of the crust (Pim *et al.* 2008). At the Crozet Islands, measured heat flow, geoid and bathymetric data suggest that the swell is dynamically supported, with a minor contribution coming from crustal thickening (Courtney & Recq 1986; Recq *et al.* 1998). Although the swell is masked by some sedimentary cover at the Canary Islands, Ye *et al.* (1999) identified underplating processes under Gran Canaria at depths of 17–26 km, with a thickness ranging from 8–10 km, but it is unclear whether thermal lithospheric rejuvenation played a role in its development (Canales & Dañobeitia 1998; Grevemeyer 1999). The weak swell observed at Madeira has been ascribed to reheating of the lower lithosphere (Sandwell & MacKenzie 1989; Grevemeyer 1999). At least for the archipelagos, for which the swells are well pronounced, dynamic pressure is favoured over thermal rejuvenation as the main mechanism to account for their formation. However, we caution that the role of a buoyant depleted swell root has been poorly investigated, although this mechanism may be far from negligible in such motionless environments.

For volcanoes emplaced on motionless plates, widely spaced heat-flow determinations have only been published for two oceanic swells: Crozet (96 mW m^{-2}: Courtney & Recq 1986) and Cape Verde (63 mW m^{-2}: Courtney & White 1986). These values are the only measurements of worldwide oceanic swells that are higher than the predicted values of reference models for the thermal evolution of oceanic lithosphere (Harris & McNutt 2007). If loss of heat by advection occurs, these values may underestimate the actual mantle flux (Harris & McNutt 2007). In addition, at low-degree melting environments, where a thick lithosphere lid is present, heat flow anomalies are less affected by the buffering effect of melting (Courtney & White 1986). When a rock melts, the latent heat of fusion consumes energy, inducing a temperature drop and causing a decrease in heat flow at the surface.

Wide-angle refraction seismic data for the Northern Kerguelen Plateau, Cape Verde, Crozet and the Canary Islands suggest the presence of a thickened crust for islands located on slow-moving plates ranging in thickness from 11–19 km (Banda *et al.* 1981; Recq & Charvis 1987; Recq *et al.* 1990, 1998; Charvis *et al.* 1995; Lodge & Helffrich 2006). At Tenerife, seismic and gravity data suggest that up to 1.5×10^5 km^3 of magmatic material has been added to the surface of the flexed oceanic crust that, assuming an age of 6–16 Ma for the shield-building stage on Tenerife, implies a magma generation rate of about $6 \times 10^{-3} - 2 \times 10^{-2}$ km^3 a^{-1} (Watts *et al.* 1997). Using K–Ar ages, the average eruption rate for the entire Canary archipelago has been estimated at 6.8×10^{-3} km^3 a^{-1} by Hoernle & Schmincke (1993) (Fig. 6). Based on bathymetric, gravity and radiometric data, the mean magmatic rate at Cape Verde has been estimated at about $2 \times 10^{-2} - 2.6 \times 10^{-2}$ km^3 a^{-1} (Fig. 6) (Morgan *et al.* 1995; Holm *et al.* 2008). Madeira is characterized by some of the lowest eruption rates, ranging from 2×10^{-6} to 1.5×10^{-4} km^3 a^{-1}, with an average rate of 9.5×10^{-5} km^3 a^{-1} for the subaerial part of the shield stage at the Madeira and Desertas islands (Fig. 6) (Geldmacher *et al.* 2000). The rates determined for the Canary and Madeira islands are substantially lower than those calculated for the Hawaiian volcanoes (Fig. 6) (Bargar & Jackson 1974).

Although their eruption rates are among the lowest observed on Earth, the magmatic activity at least at the Canaries and Madeira islands distinguishes itself by its longevity, with volcanoes often remaining active for tens of millions of years (Fig. 7) (Carracedo 1999; Geldmacher *et al.* 2000, 2001). However, this activity is sporadic, with chronological gaps of up to 12 myr at Selvagen Islands (the Canary archipelago), periods of quiescence that are much longer than those observed in Hawaii (Fig. 7) (Geldmacher *et al.* 2001). Large time gaps also constitute a noteable feature of the Cape Verde Islands, where a period of 5 myr without activity has been documented at Maio (Fig. 7) (Holm *et al.* 2008).

Owing to high rates of subsidence, the individual islands of the Hawaiian archipelago have a relatively short lifespan. Most of the Hawaiian volcanoes have subsided by between 2 and 4 km since their emergence (Moore & Campbell 1987). The bulk of this subsidence occurred rapidly, probably within 1 myr of the end of the shield-building phase (Moore & Campbell 1987). In contrast, islands on motionless plates can be characterized by a near-absence of post-emergence subsidence, as observed at the Kerguelen, Cape Verde and Canary islands (Filmer & McNutt 1989; Carracedo 1999; Wallace 2002). Those islands of the Canary Islands older than 20 Ma are still emergent (e.g. Fuerteventura, at 25 Ma: Geldmacher *et al.* 2001), while islands in the Hawaiian–Emperor volcanic chain submerge by subsidence after about 7 myr (Fig. 7). All of these islands rest on a lithosphere older than that of Hawaii. As proposed by Funck & Schmincke (1998) for the Canary Islands, this stability may be due to a combination of minor recent volcanic loading and to a greater flexural rigidity of the old lithosphere beneath these islands. In addition, the compositional and thermal buoyancy

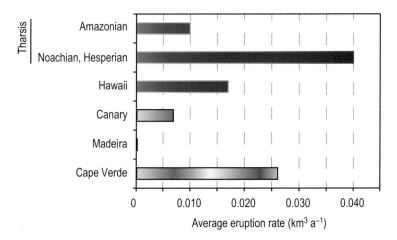

Fig. 6. Eruption rates (km^3 a^{-1}) from different oceanic islands and from the Tharsis province. Eruption rates for Hawaii, Canary, Madeira and Cape Verde are, respectively, from Bargar & Jackson (1974), Hoernle & Schmincke (1993), Geldmacher *et al.* (2000) and Holm *et al.* (2008). Those for the Tharsis province are from Greeley & Schneid (1991).

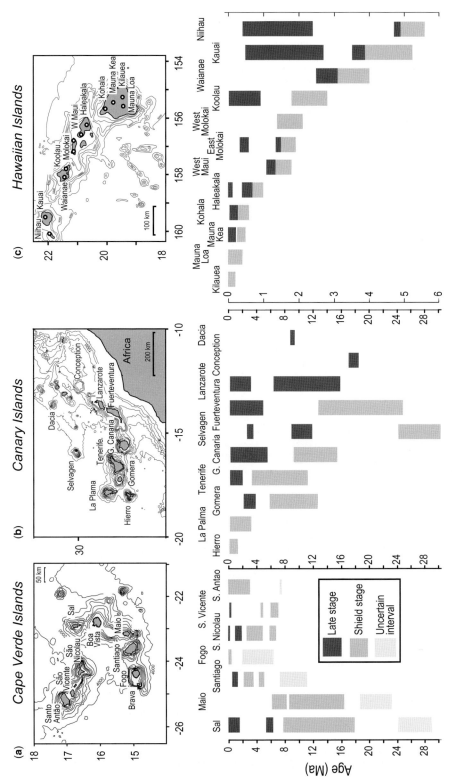

Fig. 7. Archipelago maps and age distribution (Ma) of magmatic rocks across the (**a**) Cape Verde, (**b**) Canary and (**c**) Hawaiian islands. Age distribution for the Cape Verde, Canary and Hawaiian islands modified and simplified from Holm *et al.* (2008), Geldmacher *et al.* (2001) and Clague & Dalrymple (1988), respectively. Archipelago maps for the Cape Verde, Canary Islands and Hawaiian islands are modified from Barker *et al.* (2012), Geldmacher *et al.* (2001) and Carracedo (1999), respectively.

in a higher-viscosity depleted swell root (Morgan *et al.* 1995; Yamamoto *et al.* 2007) may aid in maintaining a low subsidence rate. The emergent phase of these islands will end because of catastrophic mass-wasting processes and subsequent erosion (Carracedo 1999). Indeed, their long-term activity may result in the formation of high-relief volcanoes with steep slopes, which may evolve with time towards unstable configurations (Carracedo 1999). When unloading the lithosphere, such mass-wasting events will also strongly alter the lava chemistry and the plumbing system geometry (Manconi *et al.* 2009).

Intraplate volcanoes, such as those of the Kerguelen, Madeira and Canary islands (e.g. La Palma, El Hierro and Fogo), which are located on slow-moving plates, have a similar underground magma plumbing system geometry, with major fractionation of magmas occurring within the uppermost mantle (Fig. 8) (Amelung & Day 2002; Damasceno *et al.* 2002; Schwarz *et al.* 2004; Klugel *et al.* 2005; Scoates *et al.* 2006; Longpré *et al.* 2008; Stroncik *et al.* 2009). Magma storage and fractionation in the crust play only a minor role.

Their low magma supply rates govern the geometry and the short-term longevity of the plumbing systems by controlling the thermomechanical properties of the lithosphere (Stroncik *et al.* 2009). All of these intraplate volcanoes are characterized by multistage magma ascent, with most of the fractionation processes occurring in the uppermost mantle and only short-term stagnation at shallower levels. Hence, at the Madeira and Canary islands, the plumbing systems are manifest as a plexus of small, ephemeral, partly interconnected magma chambers, sills and dykes extending from uppermost to crustal-level depths (at 430–1100 MPa: Klugel *et al.* 2005; Klugel & Klein 2006; Stroncik *et al.* 2009) (Fig. 8a).

The Cape Verde and Canary archipelagos have primary magmas, which are characterized by a wide spectrum of alkali magmatism, with a predominance of basanite–tephrite and alkali basaltic magmas, and insignificant amounts (if any) of tholeiitic melts (Kogarko & Asavin 2007). On Madeira Island, limited tholeiitic magmatism also occurs, and primary melts are mainly alkali basalts (Kogarko & Asavin 2007). The same seems to be

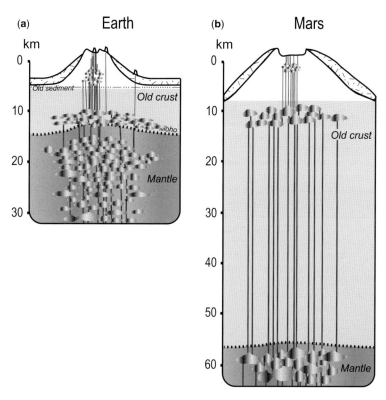

Fig. 8. Plumbing system model beneath (**a**) a terrestrial volcano from a slow-moving plate and (**b**) a Martian volcano. The magma storage systems are represented as plexuses of interconnected magma pockets and sills/dykes in the uppermost mantle and crust.

true for the Crozet (e.g. Gunn *et al.* 1970) and Kerguelen islands (e.g. Scoates *et al.* 2006). Most slow-moving plate hotspots seem to predominantly produce alkali lavas, which may be related to the presence of a thicker lithosphere in these environments. The lithosphere might be thicker as a result of its age and/or the heat loss experienced by the upwelling diapir due to intrinsically low buoyancy. Alkalic basalts may, thus, be produced by lower degrees of melting at higher pressures relative to tholeiitic basalts (Chen & Frey 1983), as the final depth of melting under a thick lithosphere will be deeper. The differentiation of tholeiitic magmas at high pressure may also be another mechanism for producing alkalic basalts under specific conditions (Albarède *et al.* 1997). Alternatively, partial melting of silica-deficient garnet pyroxenite could also be responsible for producing strongly nepheline-normative compositions (Hirschmann *et al.* 2003).

Discussion

On Earth, most of the tectonic plates are continuously regenerated and recycled through convection. The Antarctic and Nubian plates could be considered as slow surfaces of various thicknesses that are essentially not participating in the convection process but, instead, act as conductive lids on top of the much deeper convective system. This situation resembles that observed on Mars, which is characterized by a stagnant-lid regime that has been present for at least the past 4 gyr (Moresi & Solomatov 1998; Lenardic *et al.* 2004). The presence of an immobile layer, which can be modelled as a response to large viscosity contrasts in a temperature-dependent viscosity mantle, makes heat transfer out of the planetary interior dramatically less efficient than in a mobile plate tectonic regime (Reese *et al.* 1998; Hauck & Phillips 2002). Its hampering effect on mantle overturn would mainly depend on the buoyancy ratio of the system (i.e. a proxy of the relative importance of chemical to thermal buoyancy), whose consistency over the planetary evolution of Mars is doubtful. During the primordial evolution of Mars, high buoyancy values (>0.8) would have led to a thickening of the stagnant lid in response to decreasing conductive heat transfer, drastically reducing mixing efficiency compared to an isoviscous case (Tosi *et al.* 2013). However, such a stage would not last owing to the increase in compositional instability at the bottom-heated boundary layer causing a new rise in temperature, and hence activating convection until the mantle overturn stage (Elkins-Tanton *et al.* 2003; Tosi *et al.* 2013). Recent Martian mantle evolution is expected to be characterized by a lower buoyancy ratio (<0.5). When considering

such buoyancy values and Rayleigh number expected for Mars (7.26×10^6: Folkner *et al.* 1997; McGovern *et al.* 2002), mixing times are longer for a stagnant-lid regime in respect to an isoviscous case (Tosi *et al.* 2013). A recent investigation of mixing times in thermochemical models of mantle convection, using the finite-element thermal convection code Ellipsis, confirmed that the recycled mixing time of a compositional anomaly, also with a strong contrast in viscosity (>50) compared with the ambient mantle, is over an order of magnitude less efficient in a stagnant-lid mode rather than in a mobile-lid mode (Debaille *et al.* 2013). On Earth, the near-immobility of the Nubian and Antarctic plates might, thus, account for the less chemically homogenized nature of South Atlantic and Indian upper-mantle reservoirs. In turn, the isotopic variability of ocean island basalts in these two provinces is much higher than that of their Pacific counterpart (Meyzen *et al.* 2007). In addition, without significant plate motion, mantle flow is primarily laminar (Bercovici *et al.* 2000). Such flow in stabilizing convective cells reduces their lateral mixing (Schmalzl *et al.* 1996). On Mars, several chemical provinces have been identified by GRS measurements at the surface (Boynton *et al.* 2007; Taylor *et al.* 2010; Baratoux *et al.* 2011), suggesting poor lateral mixing between convection cells. Such a distribution could be a direct consequence of the stagnant-lid regime. The existence of two isotopically distinct mantle provinces (South Atlantic and Indian) over the Antarctic and Nubian plates begs the question of whether this distribution also arises from a more laminar than toroidal mantle flow pattern in this area.

Owing to the stagnant-lid regime on Mars, residual depleted mantle that is left behind by the melting processes cannot be swept away from the melting locus underlying the giant Martian volcanoes. As suggested by 2D thermodynamical modelling, the residual mantle buoyancy should oppose its advection away from the top of the plume (Manglik & Christensen 1997). This situation is similar to what has been deduced from gravity and bathymetric studies for slow-moving plate hotspots by several investigators (Morgan *et al.* 1995; Lodge & Helffrich 2006). The bulge observed at Tharsis could be generated due to the persistence of a stationary thick depleted layer, as suggested by Finnerty *et al.* (1988). The coupled association of both a residual layer and a thick lithosphere acts as a resistant barrier to melting processes. The final depth of melting over time is forced to higher depths and, hence, the melt production rate is reduced. Another concomitant effect of this depleted keel could be a lateral flow deflection, which would shift future melting sites (Manglik & Christensen 1997). Such a mechanism could account for the widespread distribution of volcanic centres within the Tharsis region. Finally, the formation of a depleted root at the bottom of the lithosphere might inhibit thermal erosion of the plate over time (Manglik & Christensen 1997).

Alternatively, the generation of the Tharsis bulge may be related to edge-driven convection processes. Both the Canary and Cape Verde islands lie on the same continuous basement ridge paralleling the African craton line (Patriat & Labails 2006), the formation of which has been ascribed to edge-driven convection processes (King & Ritsema 2000). Such a form of convection develops as a response to large lateral gradients in lithospheric thickness and thermal structure (King & Redmond 2005). Interestingly, the Tharsis province exhibits a NE–SW-trending rise (from 50°S to 20°N) supporting the Tharsis Montes (Arsia, Pavonis and Ascraeus Montes), which is roughly parallel to the dichotomy boundary (Zuber *et al.* 2000). Its association with a broad, high aeroid anomaly (Zuber *et al.* 2000) probably mirrors the locus of upwelling of material hotter than average mantle (King & Redmond 2005). In addition, the sharpness of the dichotomy boundary suggests a short length-scale variation in crustal thickness, as observed at the West African cratonic boundary. Hence, if the early formation of the southern hemisphere domain on Mars led to the underlying development of a cratonic keel-like structure, the difference in lithospheric thickness between the northern and southern hemispheres may have led to the nucleation of small-scale, edge-driven convective instabilities, resulting in the formation of the Tharsis bulge (King & Redmond 2005).

When compared with the mean heights of Earth's hotspots as a function of lithospheric age, the Tharsis Montes (Ascraeus, Pavonis and Arsia) lie in the extension of the trend defined by motionless plate hotspots (less than 25 mm a^{-1}: Figs 2 & 9). To first order, one could infer that in the absence of plate movement, the melting region remains geographically fixed with time, and melts accumulate to form large swell heights (Figs 2 & 9). However, the crustal thickness is variable beneath the Tharsis Montes. It is more likely that these large swells, as well as those observed for motionless plate hotspots, are supported both by the presence of a thick, stationary, residual-depleted layer/keel, produced by melting processes, and by large-scale mantle upwellings. Therefore, understanding the mechanisms that generate oceanic swells at motionless plates is pivotal to future progress in our knowledge of plume-fed building processes at Tharsis.

Owing to the low gravity on Mars, fluid convection flows (and consequent convective heat transfer) and crystal settling processes will be slower than on

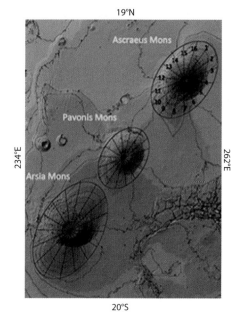

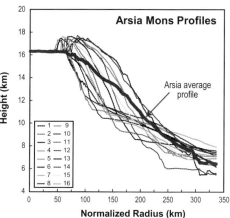

Fig. 9. (**a**) Profile locations across Pavonis, Ascraeus and Arsia Montes. (**b**) Topographical profiles for Arsia Mons. Sixteen profiles with the axis of symmetry centred on the main caldera were extracted for each volcano. In order to eliminate the ellipticity of the edifices, the profiles were normalized to the average radius. An average height profile was, thus, extracted using Origin Pro software. This procedure was used to obtain the most representative topographical profile for each volcano.

Earth (Wilson & Head 1994). Consequently, upwellings will have an intrinsically low buoyancy flux, and their ascent rates will be slow. Such features are shared by some slow-moving plate plumes, such as the Crozet hotspot, because of their low buoyancy fluxes. Uranium disequilibrium studies indicate that these weaker hotspots will also be characterized by lower excess temperatures than stronger hotspots such as Hawaii (Bourdon *et al.* 2006), as they cool more during upwelling (Albers & Christensen 1996). Their upwelling ascension rates will even be more reduced if water is present. Melting induces a dehydration of the mantle upwelling, which increases its viscosity by a factor of 7 and drastically reduces its velocity (Bourdon *et al.* 2006). Such inferences may also partially hold for the Tharsis volcanoes, for which the estimated mantle potential temperatures (i.e. the temperature that mantle peridotite would have if it ascended adiabatically to the surface without melting; 1335–1380 °C: Baratoux *et al.* 2011) fall in the range of those calculated for average ambient mantle and weak plumes (Herzberg & Gazel 2009). This factor could then be superimposed on Martian mantle cooling over time (Baratoux *et al.* 2011).

A major difference between terrestrial and Martian edifices concerns the greater depth interval between the depths of melting and crystallization for volcanoes on Mars relative to those on Earth. On Mars, melting terminates at much greater depths owing to the presence of a thick lithosphere (110–160 km: Baratoux *et al.* 2011). Two processes may contribute to the thickening of the Martian lithosphere: (1) the lithosphere may thicken over time due to the formation of a residual depleted layer in the absence of plate movement; and (2) the low buoyancy of upwelling leads to enhanced heat loss and, thus, to a termination of melting at deeper levels and the formation of a thick lithosphere. Such a situation would be similar to that of terrestrial plumes rising below an old and thick oceanic lithosphere (e.g. at the Canary or Crozet archipelago). As a consequence, quantifying the lithospheric thickness and its influence on melting processes in these environments will help to understand the shape of the mantle melting regime on Mars.

Again, owing to the lower gravity on Mars, magma chambers are expected to be much larger and deeper than their counterparts on Earth in order to avoid cooling and solidification, and to supply magma to the surface (Wilson & Head 1994). Intrusives are expected to be common, but have yet to be discovered. The lower buoyancy forces and the deeper levels of neutral buoyancy would seem to favour the development of magma diapirs and dykes at depth. Loss of heat by diffusion of the rising diapir will also favour the retention and crystallization of melt in the uppermost mantle. Such a mechanism has been inferred for Ascraeus Mons by Pozzobon *et al.* (2014) through fractal analysis of alignments of pit craters. Interestingly, the plumbing system of Ascraeus appears to be similar to that observed for motionless plate volcanoes, where most of the fractionation occurs in the uppermost mantle. A second level of fractionation proceeds in the lower crust and minor crystallization develops at shallow depths within the volcano, just underneath the summit (Fig. 8).

Another common characteristic between volcanoes on slow-moving plates and those on Tharsis is their great volcanic longevity, inherited from their stationary position relative to the assumed melting source. Hence, some Canary (e.g. Gomera: Geldmacher *et al.* 2001) and Cape Verde volcanoes (e.g. Sal: Holm *et al.* 2008) are active for more than 10 myr (Fig. 7), while the longest period of prolonged though episodic activity for volcanoes of Tharsis is up to 1 gyr (Robbins *et al.* 2011; Grott *et al.* 2013) (Fig. 3). The volcanic lifetimes of Martian shields are much longer than those of the prototypical Hawaiian Islands (Fig. 7). The early history of Mars was marked by intense volcanism at least during the late Noachian, with the activity gradually weakening over time. This overall decline was punctuated by episodic periods of high volcanic intensity (Wilson *et al.* 2001; Neukum *et al.* 2004). In particular, Tharsis volcanoes might have been built episodically with active phases lasting less than 1 myr alternating with 100 myr quiet phases (Wilson *et al.* 2001). The occurrence of long periods of quiescence, which separate magmatic stages, also seems to be a defining characteristic of the evolutionary histories of volcanoes from motionless plates. Indeed, most volcanoes at Cape Verde (e.g. Holm *et al.* 2008) and the Canary Islands (e.g. Geldmacher *et al.* 2001) have histories of sporadic activity with chronological gaps of up to 12 myr (e.g. Selvagen Island: Geldmacher *et al.* 2001), much longer than those observed at Hawaii (Fig. 7). Understanding the evolution of volcanoes at motionless plates on Earth will, thus, help to decipher the mechanisms governing the effusion history at Tharsis.

Effusion rates primarily reflect spatial and temporal variations in melting extents, which can be induced by changes in mantle potential temperature, thickness of the overlying lithosphere and/or mantle composition. The effusive rate of approximately $0.04 \text{ km}^3 \text{ a}^{-1}$ (Greeley & Schneid 1991) determined for early activity (Noachian–Hesperian) in Martian volcanic provinces is higher than that calculated during the Amazonian (c. $0.01 \text{ km}^3 \text{ a}^{-1}$: Fig. 6). When the eruption rate for this latter period is compared to those of present-day plumes on Earth, it is found to be drastically lower than that of Hawaiian Islands and closer to some of the values observed for slow-moving plate hotspots (e.g. Canary, Madeira: Fig. 6).

Some volcanoes at near-stationary plates tend to be rugged with steep slopes, such as those observed on the Crozet and Canary islands. Owing to their steeper slopes, these high-relief volcanoes may develop with time towards unstable configurations (e.g. Carracedo 1999). Catastrophic flank failures and collapses may thus occur more frequently than in the Hawaiian Islands, with their concomitant effects on lava chemistry and plumbing system geometry (Manconi *et al.* 2009). Conversely, large Martian volcanoes were primarily affected by pronounced sagging due to load-induced lithospheric flexure (McGovern & Solomon 1993; Byrne *et al.* 2009, 2013). This led to a general compressional environment on the flanks of the volcanoes, preventing mass-wasting events and generating typical imbricate lobate terraces bounded by thrust faults (Thomas *et al.* 1990; Byrne *et al.* 2009). Actually, prominent gravitational collapses have been documented only at the foot of Olympus Mons, which might have spread over a basal décollement causing oversteepening of lower slopes; meanwhile, upper flanks continued to be shortened due to edifice sagging (Byrne *et al.* 2013). However, such characteristic net compression on Martian volcanoes is not likely in the later stages of their development (Amazonian age) as a result of the lithosphere thickening over time. This is well proven for Ascraeus Mons, where compression transitioned to extension (not necessarily related to gravitational collapse) through time (Byrne *et al.* 2012).

Average surface heat fluxes for Tharsis volcanoes ($22-26$ mW m^{-2}: Grott *et al.* 2013) are much lower than those observed on Earth for slow-moving plate hotspots ($63-96$ mW m^{-2}: Harris & McNutt 2007). However, if hot mantle wells up under the Tharsis province, the size and strength of such a plume, as predicted from the modelling of the elastic thickness and maximum average heat flux of Tharsis Montes ($8-24$ mW m^{-2}), would yield central peak heat flux values ranging from 40 up to 120 mW m^{-2} (Grott & Breuer 2010). These calculated values clearly encompass the range of those measured at terrestrial hotspots located on motionless plates. The Discovery-class mission *InSight* (Banerdt *et al.* 2012), scheduled to launch in 2016 to perform *in situ* measurements of heat flow at the Martian surface, may allow the validity of this assumption to be tested (Spohn *et al.* 2012).

Although *in situ* measurements do not exist for the Tharsis province, widespread GRS and thermal emission spectrometer (TES) data and soil analyses (e.g. Gusev Crater, Meridiani Planum: McSween *et al.* 2009) suggest that volcanism on Mars is predominantly of tholeiitic nature, and results from extensive partial melting. Such a composition constitutes a primary difference to intraplate volcanism at slow-moving plates, which mostly consists of erupted Ne-normative alkali basalts (Kogarko & Asavin 2007). However, recent analyses performed by the Mars Science Laboratory at Gale Crater may challenge this view, as a mugearite-like rock was identified (Stolper *et al.* 2013). This rock may have fractionated either from a primary alkaline or a transitional magma, which has been produced by the melting of a mantle source distinct from those of other known Martian basalts (Stolper *et al.* 2013). Future exploration may document other

Table 1. *Similarities and differences between volcanoes from slow- and fast-motion plates and the volcanic provinces on Mars*

Features	Slow-motion plate hotspots	Mars volcanic provinces (Tharsis)	Fast-motion plate hotspots (Hawaii)
Great height of volcanic bulge/swell	Yes	Yes	No
Sagging processes due to self-loading and lithospheric flexure	No?	Yes (only at the early stage)	Yes
Age-progressive chains of volcanoes	No (<30 Ma)	No	Yes
Long-lasting volcanic activity punctuated by large temporal gaps	Yes	Yes	No
Alkali volcanism	Yes	Possible	Yes (except for the shield stage)
Two-level architecture of the plumbing system during the shield stage with most fractionation in the uppermost mantle	Yes	Yes	No
Lithosphere thickening due to:			
– The formation of a stationary residual mantle lid	Yes	Yes	No
– Low buoyancy of upwelling			
Slow convection mixing	Yes	Yes	No
High heat fluxes	Yes	Yes	No

alkaline suites on Mars, which could further reinforce the appropriateness of viewing volcanoes on motionless plates as analogues to Martian volcanoes.

Conclusion

On Earth, many geodynamic features of young volcanism (less than 30 myr old) in motionless intraplate oceanic island environments resemble those of Mars, where plate tectonics plays no meaningful role in shaping planetary geodynamics. Mars can effectively be considered as a one-plate planet, where geodynamics processes occur within a stagnant-lid regime. Several features of volcanism both on Earth and on Mars can be related to the (near-) absence of plate tectonic motion (Table 1).

The dividing of the mantle into several distinct chemical provinces, as proposed for Mars and over the Antarctic and Nubian plates on Earth, reflects a more efficient intracell rather than cross-cell convection mixing, allowing preservation of large-scale mantle chemical heterogeneities. Such a pattern reflects a mantle flow regime dominated by laminar flow, as expected in the (near-) absence of plate tectonic motion.

In these geodynamic environments, poor residual mantle lateral-flowing traction from the melting site will lead to the formation of a near-stationary depleted layer, which will thicken with time. The low buoyancy of upwelling in enhancing conductive heat loss during mantle rising will also lead to the formation of a thicker lithosphere relative to that of overlying higher-buoyancy plumes, such as Hawaii. These two thickening effects add to that due to planetary cooling (Baratoux et al. 2011). Both planetary cooling and depleted lid formation will lead to a cessation of melting, progressively forced towards greater depths over time, while the average extent of melting will be reduced. In both environments, low melt supply will lead to the development of a two-level fractionation architecture of the plumbing system, with most of the fractionation occurring in the uppermost mantle.

The widespread distribution of the Tharsis volcanoes, as well as the horseshoe-like shape of some volcanic islands on slow-moving plates (e.g. Cape Verde, Crozet), could be inherited from lateral-flow deflection induced by the presence of a residual mantle keel, shifting future melting loci around the keel. The pronounced topographical swells/bulges observed in this environment may also be supported both by large-scale mantle upwelling and by their residual mantle roots.

Another similarity between volcanoes on slow-moving plates and those in Tharsis is their great longevity, which is due to their stationary position relative to the melting source. Both sets of volcanoes also have a history of protracted activity punctuated by long periods of quiescence. The accumulation of refractory material at the rim of their melting zones might play a role in the fluctuations of their long-term volcanic activity.

Until now, volcanism on Mars was defined as being predominantly of tholeiitic composition. However, recent in situ analyses at Gale Crater have identified a mugearite-like rock. This raises the possibility that both intraplate volcanism on Earth at slow-moving plates and that on Mars would, instead, be of an alkali nature.

Our knowledge of terrestrial volcanoes from motionless plates can thus help us to better understand the nature and significance of large-scale melting and differentiation processes of volcanoes on Mars. However, the extent of this knowledge is still insufficiently detailed in many respects, and requires further investigation. New data from the Martian planetary record will help to provide new perspectives on the processes and evolution of volcanoes on near-stationary plates. On Mars, the large range of scales of upwelling, and the influence of crustal and lithospheric thicknesses in space and time, are thus relevant to studies of hotspots on motionless plates.

Two anonymous reviewers are gratefully acknowledged for their thorough and constructive reviews that helped to improve the manuscript. We also thank Thomas Platz and Paul K. Byrne for their editorial comments. This project was supported by an ATENEO grant CPDA114007/11.

References

ALBARÈDE, F., LUAIS, B. ET AL. 1997. The geochemical regime of Piton de la Fournaise volcano (Réunion) during the last 530 000 years. Journal of Petrology, **38**, 171–201.

ALBERS, M. & CHRISTENSEN, U. R. 1996. The excess temperature of plumes rising from the core-mantle boundary. Geophysical Research Letters, **23**, 3567–3570.

ALI, M. Y., WATTS, A. B. & HILL, I. 2003. A seismic reflection profile study of lithospheric flexure in the vicinity of the Cape Verde islands. Journal of Geophysical Research, **108**, 2239, http://dx.doi.org/10.1029/2002JB002155

AMELUNG, F. & DAY, S. 2002. Insar observations of the 1995 Fogo, Cape Verde, eruption: implications for the effects of collapse events upon island volcanoes. Geophysical Research Letters, **29**, 1606, http://dx.doi.org/10.1029/2001GL013760

ARGUS, D. F., GORDON, R. G. & DEMETS, C. 2011. Geologically current motion of 56 plates relative to the no-net-rotation reference frame. Geochemistry, Geophysics, Geosystems, **12**, Q11001, http://dx.doi.org/10.1029/2011GC003751

BANDA, E., DAFIOBEITIA, J. J., SURIFIACH, E. & ANSORGE, J. 1981. Features of crustal structure under the Canary islands. *Earth and Planetary Science Letters*, **55**, 11–24.

BANERDT, W. B., SMREKAR, S. E. *ET AL.* 2012. Insight: an integrated exploration of the interior of Mars. Paper 2838, presented at the 43rd Lunar and Planetary Science Conference, 19–23 March 2012, The Woodlands, TX.

BARATOUX, D., TOPLIS, M. J., MONNEREAU, M. & GASNAULT, O. 2011. Thermal history of Mars inferred from orbital geochemistry of volcanic provinces. *Nature*, **472**, 338–341.

BARGAR, K. E. & JACKSON, E. D. 1974. Calculated volumes of individual shield volcanoes along the Hawaiian–Emperor Chain. *Journal of Research of the United States Geological Survey*, **2**, 545–550.

BARKER, A. K., TROLL, V. R., ELLAM, R. M., HANSTEEN, T. H., HARRIS, C., STILLMAN, C. J. & ANDERSSON, A. 2012. Magmatic evolution of the Cadamosto seamount, Cape Verde: beyond the spatial extent of EM1. *Contributions to Mineralogy and Petrology*, **163**, 949–965.

BERCOVICI, D., RICARD, Y. & RICHARDS, M. A. 2000. The relation between mantle dynamics and plate tectonics: a primer in the history and dynamics of global plate motions. *In*: RICHARDS, M. A., GORDON, R. G. & VAN DER HILST, R. D. (eds). *The History and Dynamics of Global Plate Motions*. American Geophysical Union, Geophysical Monograph, **121**, 5–46.

BLEACHER, J. E., GREELEY, R., WILLIAMS, D. A., WERNER, S. C., HAUBER, E. & NEUKUM, G. 2007. Olympus Mons, Mars: inferred changes in late Amazonian aged effusive activity from lava flow mapping of Mars express high resolution stereo camera data. *Journal of Geophysical Research*, **112**, E04003, http://dx.doi.org/10.1029/2006JE002826

BOURDON, B., RIBE, N. M., STRACKE, A., SAAL, A. E. & TURNER, S. P. 2006. Insights into the dynamics of mantle plumes from uranium-series geochemistry. *Nature*, **444**, 713–7.

BOUVIER, A., BLICHERT-TOFT, J. & ALBARÈDE, F. 2009. Martian meteorite chronology and the evolution of the interior of Mars. *Earth and Planetary Science Letters*, **280**, 285–295.

BOYNTON, W. V., TAYLOR, G. J. *ET AL.* 2007. Concentration of H, Si, Cl, K, Fe, and Th in the low- and mid-latitude regions of Mars. *Journal of Geophysical Research*, **112**, E12, http://dx.doi.org/10.1029/2007JE002887

BUSKE, M. 2006. *Three-dimensional thermal evolution models for the interior of Mars and Mercury*. PhD thesis, Georg-August-University, Göttingen.

BYRNE, P. K., VAN WYK DE VRIES, B., MURRAY, J. B. & TROLL, V. R. 2009. The geometry of volcano flank terraces on Mars. *Earth and Planetary Science Letters*, **281**, 1–13.

BYRNE, P. K., VAN WYK DE VRIES, B., MURRAY, J. B. & TROLL, V. R. 2012. A volcanotectonic survey of Ascraeus Mons, Mars. *Journal of Geophysical Research*, **117**, E01004, http://dx.doi.org/10.1029/2011JE003825

BYRNE, P. K., HOLOHAN, E. P., KERVYN, M., VAN WYK DE VRIES, B., TROLL, V. R. & MURRAY, J. B. 2013. A sagging-spreading continuum of large volcano structure. *Geology*, **41**, 339–342, http://dx.doi.org/10.1130/G33990.1

CANALES, J. P. & DAÑOBEITIA, J. J. 1998. The Canary Islands swell: a coherence analysis of bathymetry and gravity. *Geophysical Journal International*, **132**, 479–488.

CARR, M. H. 2006. *The Surface of Mars*. Cambridge University Press, Cambridge.

CARR, M. H. & HEAD, J. W., III . 2010. Geologic history of Mars. *Earth and Planetary Science Letters*, **294**, 185–203.

CARRACEDO, J. C. 1999. Growth, structure, instability and collapse of Canarian volcanoes and comparisons with Hawaiian volcanoes. *Journal of Volcanology and Geothermal Research*, **94**, 1–19.

CAZENAVE, A., DOMINH, K., RABINOWICZ, M. & CEULENEER, G. 1988. Geoid and depth anomalies over ocean swells and troughs: evidence of an increasing trend of the geoid to depth ratio with age of plate. *Journal of Geophysical Research*, **93**, 8064–8077.

CHARVIS, P., RECQ, M., OPERTO, S. & BREFORT, D. 1995. Deep structure of the northern Kerguelen plateau and hotspot-related activity. *Geophysical Journal International*, **122**, 899–924.

CHEN, C.-Y. & FREY, F. A. 1983. Origin of Hawaiian tholeiite and alkali basalt. *Nature*, **302**, 785–789.

CLAGUE, D. A. & DALRYMPLE, G. B. 1988. Age and petrology of alkalic postshield and rejuvenated-stage lava from Kauai, Hawaii. *Contributions to Mineralogy and Petrology*, **99**, 202–218.

COURTNEY, R. & RECQ, M. 1986. Anomalous heat flow near the Crozet plateau and mantle convection. *Earth and Planetary Science Letters*, **79**, 373–384.

COURTNEY, R. C. & WHITE, R. S. 1986. Anomalous heat flow and geoid across the Cape Verde rise: evidence for dynamic support from a thermal plume in the mantle. *Geophysical Journal of the Royal Astronomical Society*, **87**, 815–867.

CRUMPLER, L. S., HEAD, J. W. & AUBELE, J. C. 1996. Calderas on Mars: characteristics, structure, and associated flank deformation. *In*: MCGUIRE, W. J., JONES, A. P. & NEUBERG, J. (eds) *Volcano Instability on the Earth and Other Planets*. Geological Society, London, Special Publications, **110**, 307–348.

DAMASCENO, D., SCOATES, J. S., WEISS, D., FREY, F. A. & GIRET, A. 2002. Mineral chemistry of mildly alkalic basalts from the 25 ma Mont Crozier section, Kerguelen archipelago: constraints on phenocryst crystallization environments. *Journal of Petrology*, **43**, 1389–1413.

DEBAILLE, V., O'NEILL, C., BRANDON, A. D., HAENECOUR, P., YIN, Q.-Z., MATTIELLI, N. & TREIMAN, A. H. 2013. Stagnant-lid tectonics in early Earth revealed by [142]Nd variations in late Archean rocks. *Earth and Planetary Science Letters*, **373**, 83–92.

DETRICK, R. S. & CROUGH, S. T. 1978. Island subsidence, hot spots, and lithospheric thinning. *Journal of Geophysical Research*, **83**, 1236–1244.

ELKINS-TANTON, L., PARMENTIER, E. & HESS, P. 2003. Magma ocean fractional crystallization and cumulate overturn in terrestrial planets: implications for Mars. *Meteoritics and Planetary Science*, **38**, 1753–1771.

EL MAARRY, M. R., GASNAULT, O. ET AL. 2009. Gamma-ray constraints on the chemical composition of the Martian surface in the Tharsis region: a signature of partial melting of the mantle? In: BLEACHER, J. E. & DOHM, J. M. (eds) Tectonic and Volcanic History of the Tharsis Province, Mars. Journal of Volcanology and Geothermal Research, 185, 116–122.

FILMER, P. E. & McNUTT, M. K. 1989. Geoid anomalies over the Canary Islands group. Marine Geophysical Researches, 11, 77–87.

FINNERTY, A. A., PHILLIPS, R. J. & BANERDT, W. B. 1988. Igneous processes and closed system evolution of the Tharsis region of Mars. Journal of Geophysical Research, 93, 10,225–10,235.

FOLKNER, W. M., YODER, C. F., YUAN, D. N., STANDISH, E. M. & PRESTON, R. A. 1997. Interior structure and seasonal mass redistribution of Mars from radio tracking of Mars Pathfinder. Science, 278, 1749–1752.

FUNCK, T. & SCHMINCKE, H.-U. 1998. Growth and destruction of Gran Canaria deduced from seismic reflection and bathymetric data. Journal of Geophysical Research, 103, 15,393–15,407.

GASNAULT, O., TAYLOR, G. J. ET AL. 2010. Quantitative geochemical mapping of Martian elemental provinces. Icarus, 207, 226–247.

GELDMACHER, J., VAN DEN BOGAARD, P., HOERNLE, K. & SCHMINCKE, H.-U. 2000. The ^{40}Ar/^{39}Ar age dating of the Madeira archipelago and hotspot track (Eastern North Atlantic). Geochemistry, Geophysics, Geosystems, 1, 1008, http://dx.doi.org/10.1029/1999GC000018

GELDMACHER, J., HOERNLE, K., VAN DEN BOGAARD, P., ZANKL, G. & GARBE-SCHÖNBERG, D. 2001. Earlier history of the ≤70-ma-old Canary hotspot based on the temporal and geochemical evolution of the Selvagen archipelago and neighboring seamounts in the Eastern North Atlantic. Journal of Volcanology and Geothermal Research, 111, 55–87.

GIACOMINI, L., MASSIRONI, M., MARTELLATO, E., PASQUARE, G., FRIGERI, A. & CREMONESE, G. 2009. Inflated flowson Daedalia Planum (Mars)? Clues from a comparative analysis with the Payen volcanic complex (Argentina). Planetary and Space Science, 57, 556–570.

GOLABEK, G. J., KELLER, T., GERYA, T. V., ZHU, G., TACKLEY, P. J. & CONNOLLY, J. A. D. 2011. Origin of the Martian dichotomy and Tharsis from a giant impact causing massive magmatism. Icarus, 215, 346–357.

GOLLE, O., DUMOULIN, C., CHOBLET, G. & CADEK, O. 2012. Topography and geoid induced by a convecting mantle beneath an elastic lithosphere. Geophysical Journal International, 189, 55–72.

GREELEY, R. & SCHNEID, B. D. 1991. Magma generation on Mars – amounts, rates, and comparisons with Earth, Moon, and Venus. Science, 254, 996–998.

GREVEMEYER, I. 1999. Isostatic geoid anomalies over midplate swells in the central north Atlantic. Geodynamics, 28, 41–50.

GROTT, M. & BREUER, D. 2009. Implications of large elastic thicknesses for the composition and current thermal state of Mars. Icarus, 201, 540–548.

GROTT, M. & BREUER, D. 2010. On the spatial variability of the Martian elastic lithosphere thickness: evidence

for mantle plumes? Journal of Geophysical Research, 115, 1–16.

GROTT, M., BARATOUX, D. ET AL. 2013. Long-term evolution of the Martian crust-mantle system. Space Science Reviews, 174, 49–111.

GUNN, B. M., COY-YLL, R., WATKINS, N. D., ABRANSON, C. E. & NOUGIER, J. 1970. Geochemistry of an oceanite-ankaramite-basalt suite from East Island, Crozet archipelago. Contributions to Mineralogy and Petrology, 28, 319–339.

HARDER, H. 2000. Mantle convection and the dynamic geoid of Mars. Geophysical Research Letters, 27, 301–304.

HARRIS, R. N. & McNUTT, M. K. 2007. Heat flow on hot spot swells: evidence for fluid flow. Journal of Geophysical Research, 112, B3, http://dx.doi.org/10.1029/2006JB004299

HARTMANN, W. K. & NEUKUM, G. 2001. Cratering chronology and the evolution of Mars. Space Science Reviews, 96, 165–194.

HAUBER, E., BROŽ, P., JAGERT, F., JODLOWSKI, P. & PLATZ, T. 2011. Very recent and wide-spread basaltic volcanism on Mars. Geophysical Research Letters, 38, 10, http://dx.doi.org/10.1029/2011GL047310

HAUCK, S. A., II & PHILLIPS, R. J. 2002. Thermal and crustal evolution of Mars. Journal of Geophysical Research, 107, 5052, http://dx.doi.org/10.1029/2001JE001801

HERZBERG, C. & GAZEL, E. 2009. Petrological evidence for secular cooling in mantle plumes. Nature, 458, 619–622.

HIESINGER, H., HEAD, J. W., III & NEUKUM, G. 2007. Young lava flows on the eastern flank of Ascraeus Mons: rheological properties derived from high resolution stereo camera (HRSC) images and Mars orbiter laser altimeter (MOLA) data. Journal of Geophysical Research, 112, E5, http://dx.doi.org/10.1029/2006JE002717

HIRSCHMANN, M. M., KOGISO, T., BAKER, M. B. & STOLPER, E. M. 2003. Alkalic magmas generated by partial melting of garnet pyroxenite. Geology, 31, 481–484.

HOERNLE, K. & SCHMINCKE, H.-U. 1993. The petrology of the tholeiites through melilite nephelinites on Gran Canaria, Canary Islands: crystal fractionation, accumulation, and depth of melting. Journal of Petrology, 34, 573–578.

HOLM, P. M., GRANDVUINET, T., FRIIS, J., WILSON, J. R., BARKER, A. K. & PLESNER, S. 2008. An ^{40}Ar–^{39}Ar study of the Cape Verde hot spot: temporal evolution in a semistationary plate environment. Journal of Geophysical Research, 113, B08201, http://dx.doi.org/10.1029/2007JB005339

ITO, G. & VAN KEKEN, P. E. 2007. Hotspots and melting anomalies. In: SCHUBERT, G. (ed.) Mantle dynamics. Treatise on Geophysics, 7, Elsevier, 371–435.

JUNG, W.-Y. & RABINOWITZ, P. D. 1986. Residual geoid anomalies of the North Atlantic Ocean and their tectonic implications. Journal of Geophysical Research, 91, 10,383–10,396.

KELLER, T. & TACKLEY, P. J. 2009. Towards self-consistent modeling of the Martian dichotomy: the influence of one-ridge convection on crustal thickness distribution. Icarus, 202, 429–443.

KING, S. D. & REDMOND, H. L. 2005. The crustal dichotomy and edge driven convection: a mechanism for Tharsis rise volcanism? Abstract 1960, presented at the 36th Annual Lunar and Planetary Science Conference, 14–18 March 2005, League City, TX.

KING, S. D. & RITSEMA, J. 2000. African hot spot volcanism: small-scale convection in the upper mantle beneath cratons. *Science*, **290**, 1137–1140.

KLUGEL, A. & KLEIN, F. 2006. Complex magma storage and ascent at embryonic submarine volcanoes from Madeira archipelago. *Geology*, **34**, 337–340.

KLUGEL, A., HANSTEEN, T. H. & GALIPP, K. 2005. Magma storage and underplating beneath Cumbre Vieja volcano, la Palma (Canary islands). *Earth and Planetary Science Letters*, **236**, 211–226.

KOGARKO, L. N. & ASAVIN, A. M. 2007. Regional features of primary alkaline magmas of the Atlantic Ocean. *Geochemistry International*, **45**, 841–856.

LENARDIC, A., NIMMO, F. & MORESI, L. 2004. Growth of the hemispheric dichotomy and the cessation of plate tectonics on Mars. *Journal of Geophysical Research*, **109**, E02, http://dx.doi.org/10.1029/2003JE002172

LIN, J. 1998. Hitting the hotspots. *Oceanus*, **41**, 34–37.

LIU, M. & CHASE, C. G. 1989. Evolution of midplate hotspot swells: numerical solutions. *Journal of Geophysical Research*, **94**, 5571–5584.

LODGE, A. & HELFFRICH, G. 2006. Depleted swell root beneath the Cape Verde islands. *Geology*, **34**, 449–452.

LONGPRÉ, M.-A., TROLL, V. R. & HANSTEEN, T. H. 2008. Upper mantle magma storage and transport under a Canarian shield-volcano, Teno, Tenerife (Spain). *Journal of Geophysical Research*, **113**, B08203, http://dx.doi.org/10.1029/2007JB005422

MANCONI, A., LONGPRÉ, M.-A., WALTER, T. R., TROLL, V. R. & HANSTEEN, T. H. 2009. The effects of flank collapses on volcano plumbing systems. *Geology*, **37**, 1099–1102.

MANGLIK, A. & CHRISTENSEN, U. R. 1997. Effect of mantle depletion buoyancy on plume flow and melting beneath a stationary plate. *Journal of Geophysical Research*, **102**, 5019–5028.

MARKS, K. M. & SANDWELL, D. T. 1991. Analysis of geoid height versus topography for oceanic plateaus and swells using nonbiased linear regression. *Journal of Geophysical Research*, **96**, 8045–8055.

MCGOVERN, P. J. & MORGAN, J. K. 2009. Volcanic spreading and lateral variations in the structure of Olympus Mons, Mars. *Geology*, **37**, 139–142.

MCGOVERN, P. J. & SOLOMON, S. C. 1993. State of stress, faulting, and eruption characteristics of large volcanoes on Mars. *Journal of Geophysical Research*, **98**, 23,553–23,579.

MCGOVERN, P. J., SOLOMON, S. C. *ET AL.* 2002. Localized gravity/topography admittance and correlation spectra on Mars: implications for regional and global evolution. *Journal of Geophysical Research*, **107**, E12, http://dx.doi.org/10.1029/2002JE001854

MCSWEEN, H. Y. J., TAYLOR, G. J. & WYATT, M. B. 2009. Elemental composition of the Martian crust. *Science*, **324**, 736–739.

MEYZEN, C. M., LUDDEN, J., HUMLER, E., MÉVEL, C. & ALBARÈDE, F. 2007. Isotopic portrayal of the upper mantle flow field. *Nature*, **447**, 1069–1074.

MONNEREAU, M. & CAZENAVE, A. 1990. Depth and geoid anomalies over oceanic hotspot swells: a global survey. *Journal of Geophysical Research*, **95**, 15,429–15,438.

MOORE, J. G. & CAMPBELL, J. F. 1987. Age of tilted reefs, Hawaii. *Journal of Geophysical Research*, **92**, 2641–2646.

MORESI, L. & SOLOMATOV, V. S. 1998. Mantle convection with a brittle lithosphere: thoughts on the global tectonic style of the Earth and Venus. *Geophysical Journal International*, **133**, 669–682.

MORGAN, J. P., MORGAN, W. J. & PRICE, E. 1995. Hotspot melting generates both hotspot volcanism and a hotspot swell? *Journal of Geophysical Research*, **100**, 8045–8062.

MOUGINIS-MARK, P. J. & ROBINSON, M. S. 1992. Evolution of the Olympus Mons Caldera, Mars. *Bulletin of Volcanology*, **54**, 347–360.

NEUKUM, G., JAUMANN, R. *ET AL.* 2004. Recent and episodic volcanic and glacial activity on Mars revealed by the high resolution stereo camera. *Nature*, **432**, 971–979.

NEWSOM, H. E., CRUMPLER, L. S. *ET AL.* 2007. Geochemistry of Martian soil and bedrock in mantled and less mantled terrains with gamma ray data from Mars odyssey. *Journal of Geophysical Research*, **112**, E3, http://dx.doi.org/10.1029/2006JE002680

OLSON, P. 1990. Hot spots, swells, and mantle plumes. *In*: RYAN, M. P. (ed.) *Magma Transport and Storage*. Wiley, New York, 33–51.

PATRIAT, M. & LABAILS, C. 2006. Linking the Canary and Cape-Verde hot-spots, northwest Africa. *Marine Geophysical Researches*, **27**, 201–215.

PHILLIPS, R. J., SLEEP, N. H. & BANERDT, W. B. 1990. Permanent uplift in magmatic systems with application to the Tharsis region of Mars. *Journal of Geophysical Research*, **95**, 5089–5100.

PHILLIPS, R. J., ZUBER, M. T. *ET AL.* 2001. Ancient geodynamics and global-scale hydrology on Mars. *Science*, **291**, 2587–2591.

PIM, J., PEIRCE, C., WATTS, A. B., GREVEMEYER, I. & KRABBENHOEFT, A. 2008. Crustal structure and origin of the Cape Verde rise. *Earth and Planetary Science Letters*, **272**, 422–428.

POZZOBON, R., MAZZARINI, F., MASSIRONI, M. & MARINANGELI, L. 2014. Self-similar clustering distribution of structural features on Ascraeus Mons (Mars): implications for magma chamber depth. *In*: PLATZ, T., MASSIRONI, M., BYRNE, P. K. & HIESINGER, H. (eds) *Volcanism and Tectonism Across the Inner Solar System*. Geological Society, London, Special Publications, **401**. First published online March 25, 2014, http://dx.doi.org/10.1144/SP401.12

PLESCIA, J. B. 2004. Morphometric properties of Martian volcanoes. *Journal of Geophysical Research*, **109**, E3, http://dx.doi.org/10.1029/2002JE002031

RECQ, M. & CHARVIS, P. 1987. La ride asismique de Kerguelen-Heard – anomalie du géoïde et compensation isostatique. *Marine Geology*, **76**, 301–311.

RECQ, M., LE ROY, I., CHARVIS, P., GOSLIN, J. & BREFORT, D. 1990. Structure profonde du Mont Ross d'après la réfraction sismique (les Kerguelen, Océan Indien Austral). *Canadian Journal of Earth Sciences*, **31**, 1806–1821.

RECQ, M., GOSLIN, J., CHARVIS, P. & OPERTO, S. 1998. Small-scale crustal variability within an intraplate structure: the Crozet bank (Southern Indian Ocean). *Geophysical Journal International*, **134**, 145–156.

REESE, C. C., SOLOMATOV, V. S. & MORESI, L.-N. 1998. Heat transport efficiency for stagnant lid convection with dislocation viscosity: application to Mars and Venus. *Journal of Geophysical Research*, **103**, 13 643–13 657.

RIBE, N. M. & CHRISTENSEN, U. R. 1994. 3-dimensional modeling of plume lithosphere interaction. *Journal of Geophysical Research*, **99**, 669–682.

ROBBINS, S. J., DI ACHILLE, G. & HYNEK, B. M. 2011. The volcanic history of Mars: high-resolution crater-based studies of the calderas of 20 volcanoes. *Icarus*, **211**, 1179–1203.

ROBERTS, J. H. & ZHONG, S. J. 2006. Degree-1 convection in the Martian mantle and the origin of the hemispheric dichotomy. *Journal of Geophysical Research*, **111**, E06013, http://dx.doi.org/10.1029/2005JE002668

SANDWELL, D. T. & MACKENZIE, K. R. 1989. Geoid height versus topography for oceanic plateaus and swells. *Journal of Geophysical Research*, **94**, 7403–7418.

SCHMALZL, J., HOUSEMAN, G. A. & HANSEN, U. 1996. Mixing in vigorous time-dependent three-dimensional convection and application to earth's mantle. *Journal of Geophysical Research*, **101**, 21,847–21,858.

SCHOTT, B., VAN DEN BERG, A. P. & YUEN, D. A. 2001. Focussed time-dependent Martian volcanism from chemical differentiation coupled with variable thermal conductivity. *Geophysical Research Letters*, **28**, 4271–4274.

SCHUBERT, G., SOLOMON, S. C., TURCOTTE, D. L., DRAKE, M. J. & SLEEP, N. H. 1990. Origin and thermal evolution of Mars. *In*: KIEFFER, H., JAKOSKY, B., SNYDER, C. & MATTHEWS, M. (eds) *Mars*. University of Arizona Press, Tucson, AZ, 147–183.

SCHULTZ, P. H., SCHULTZ, R. A. & ROGERS, J. 1982. The structure and evolution of ancient impact basins on Mars. *Journal of Geophysical Research*, **87**, 9803–9820.

SCHWARZ, S., KLUGEL, A. & WOHLGEMUTH-UEBERWASSER, C. 2004. Melt extraction pathways and stagnation depths beneath the Madeira and Desertas rift zones (NE Atlantic) inferred from barometric studies. *Contributions to Mineralogy and Petrology*, **147**, 228–240.

SCOATES, J. S., CASCIO, M. L., WEIS, D. & LINDSLEY, D. H. 2006. Experimental constraints on the origin and evolution of mildly alkalic basalts from the Kerguelen archipelago, southeast Indian Ocean. *Contributions to Mineralogy and Petrology*, **151**, 582–599.

SCOTT, E. D. & WILSON, L. 2000. Cyclical summit collapse events at Ascraeus Mons, Mars. *Journal of the Geological Society, London*, **157**, 1101–1106.

SILVER, P. G., RUSSO, R. M. & LITHGOW-BERTELLONI, C. 1998. Coupling of South American and African plate motion and plate deformation. *Science*, **279**, 60–63.

SLEEP, N. H. 1990. Hotspots and mantle plumes: some phenomenology. *Journal of Geophysical Research*, **95**, 6715–6736.

SLEEP, N. H. 1994. Martian plate tectonics. *Journal of Geophysical Research*, **99**, 5639–5655.

SOLOMON, S. C. & HEAD, J. W. 1982. Evolution of the Tharsis province of Mars – the importance of heterogeneous lithospheric thickness and volcanic construction. *Journal of Geophysical Research*, **87**, 9755–9774.

SPOHN, T., GROTT, M. ET AL. 2012. INSIGHT: measuring the Martian heat flow using the heat flow and physical properties package (HP3). Paper 1445, presented at the 43rd Lunar and Planetary Science Conference, 19–23 March 2012, The Woodlands, TX.

ŠRÁMEK, O. & ZHONG, S. 2012. Martian crustal dichotomy and Tharsis formation by partial melting coupled to early plume migration. *Journal of Geophysical Research – Planets*, **117**, E01005, http://dx.doi.org/10.1029/2011JE003867

STEINBERGER, B. & ANTRETTER, M. 2006. Conduit diameter and buoyant rising speed of mantle plumes: implications for the motion of hot spots and shape of plume conduits. *Geochemistry, Geophysics, Geosystems*, **7**, Q11018, http://dx.doi.org/10.1029/2006GC001409

STOLPER, E. M., BAKER, M. B. ET AL. 2013. The petrochemistry of Jake-M: a Martian mugearite. *Science*, **341**, http://dx.doi.org/10.1126/science.1239463

STRONCIK, N. A., KLUGEL, A. & HANSTEEN, T. H. 2009. The magmatic plumbing system beneath El Hierro (Canary Islands): constraints from phenocrysts and naturally quenched basaltic glasses in submarine rocks. *Contributions to Mineralogy and Petrology*, **157**, 593–607.

TAYLOR, G. J., BOYNTON, W. ET AL. 2006. Bulk composition and early differentiation of Mars. *Journal of Geophysical Research*, **111**, E3, http://dx.doi.org/10.1029/2005JE002645

THOMAS, P. J., SQUYRES, S. W. & CARR, M. H. 1990. Flank tectonics of Martian volcanoes. *Journal of Geophysical Research*, **95**, 14,345–14,355.

TOSI, N., PLESA, A.-C. & BREUER, D. 2013. Overturn and evolution of a crystallized magma ocean: a numerical parameter study for Mars. *Journal of Geophysical Research*, **118**, 1–17, http://dx.doi.org/10.1002/jgre.20109

WALLACE, P. J. 2002. Volatiles in submarine basaltic glasses from the Northern Kerguelen Plateau (ODP site 1140): implications for source region compositions, magmatic processes and plateau subsidence. *Journal of Petrology*, **43**, 1311–1326.

WATTS, A. B., PIERCE, C., COLLIER, J., DALWOOD, R., CANALES, J. P. & HENSTOCK, T. J. 1997. A seismic study of Tenerife, Canary Islands: implications for volcano growth, lithospheric flexure and magmatic underplating. *Earth and Planetary Science Letters*, **146**, 431–447.

WENZEL, M. J., MANGA, M. & JELLINEK, A. M. 2004. Tharsis as a consequence of Mars' dichotomy and layered mantle. *Geophysical Research Letters*, **31**, 4, http://dx.doi.org/10.1029/2003GL019306

WERNER, S. C. 2009. The global Martian volcanic evolutionary history. *Icarus*, **201**, 44–68.

WILSON, L. & HEAD, J. W. 1994. Mars-review and analysis of volcanic-eruption theory and relationships to observed landforms. *Reviews of Geophysics*, **32**, 221–263.

WILSON, L. & HEAD, J. W. 1998. Evolution of magma reservoirs within shield volcanoes on Mars. Abstract 1128, presented at the 29th Annual Lunar and Planetary Science Conference, 16–20 March 1998, Houston, TX.

WILSON, L., SCOTT, E. D. & HEAD, J. W. 2001. Evidence for episodicity in the magma supply to the large Tharsis volcanoes. *Journal of Geophysical Research*, **106**, 1423–1433.

WILSON, L., MOUGINIS-MARK, P. J., TYSON, S., MACKOWN, J. & GARBEIL, H. 2009. Fissure eruptions in Tharsis, Mars: implications for eruption conditions and magma sources. *Journal of Volcanology and Geothermal Research*, **185**, 28–46.

WILSON, D. J., PEIRCE, C., WATTS, A. B., GREVEMEYER, I. & KRABBENHOEFT, A. 2010. Uplift at lithospheric swells-I: seismic and gravity constraints on the crust and uppermost mantle structure of the Cape Verde mid-plate swell. *Geophysical Journal International*, **182**, 531–550.

WISE, D. U., GOLOMBEK, M. P. & McGILL, G. E. 1979. Tectonic evolution of Mars. *Journal of Geophysical Research*, **84**, 7934–7939.

YAMAMOTO, M., MORGAN, J. P. & MORGAN, W. J. 2007. Global plume-fed asthenosphere flow-1: motivation and model development. *In*: FOULGER, G. R. & JURDY, D. M. (eds) *Plates, Plumes and Planetary Processes*. Geological Society of America Special Papers, **430**, 165–188.

YE, S., CANALES, J. P., RHIM, R., DANOBEITIA, J. J. & GALLART, J. 1999. A crustal transect through the northern and northeastern part of the volcanic edifice of Gran Canaria, Canary Islands. *Journal of Geodynamics*, **28**, 3–26.

ZHONG, S. 2009. Migration of Tharsis volcanism on Mars caused by differential rotation of the lithosphere. *Nature Geoscience*, **2**, 19–23.

ZHONG, S. J. & ZUBER, M. T. 2001. Degree-1 mantle convection and the crustal dichotomy on Mars. *Earth and Planetary Science Letters*, **189**, 75–84.

ZUBER, M. T. & MOUGINIS-MARK, P. J. 1992. Caldera subsidence and magma chamber depth of the Olympus Mons volcano. *Journal of Geophysical Research Letters*, **97**, 18 295–18 307.

ZUBER, M. T., SOLOMON, S. C. *ET AL.* 2000. Internal structure and early thermal evolution of Mars from Mars Global Surveyor topography and gravity. *Science*, **287**, 1788–1793.

Lunar mare volcanism: lateral heterogeneities in volcanic activity and relationship with crustal structure

TOMOKATSU MOROTA[1]*, YOSHIAKI ISHIHARA[2], SHO SASAKI[3],
SANDER GOOSSENS[4], KOJI MATSUMOTO[5], HIROTOMO NODA[5], HIROSHI ARAKI[6],
HIDEO HANADA[5], SEIICHI TAZAWA[5], FUYUHIKO KIKUCHI[5], TOSHIAKI ISHIKAWA[5],
SEIITSU TSURUTA[5], SHUNICHI KAMATA[7], HISASHI OTAKE[2],
JUNICHI HARUYAMA[2] & MAKIKO OHTAKE[2]

[1]*Graduate School of Environmental Studies, Nagoya University, Furo-cho,
Chikusa-ku, Nagoya, Aichi 464-8601, Japan*

[2]*Japan Aerospace Exploration Agency, 3-1-1 Yoshinodai, Chuo-ku,
Sagamihara, Kanagawa 252-5210, Japan*

[3]*Graduate School of Science, Osaka University, 1-1 Machikaneyama-cho,
Toyonaka-shi, Osaka 560-0043, Japan*

[4]*CRESST/Planetary Geodynamics Laboratory, NASA/GSFC,
8800 Greenbelt Road, Greenbelt, MD 20771, USA*

[5]*RISE Project, National Astronomical Observatory of Japan,
2-12 Hoshigaoka-cho, Mizusawa-ku, Oshu, Iwate 023-0861, Japan*

[6]*RISE Project, National Astronomical Observatory of Japan, 2-21-1
Osawa, Mitaka, Tokyo 181-8588, Japan*

[7]*Graduate School of Science, Hokkaido University, Kita-10 Nishi-8,
Kita-ku, Sapporo, Hokkaido 060-0810, Japan*

*Corresponding author (e-mail: morota@eps.nagoya-u.ac.jp)

Abstract: Lunar mare basalts are spatially unevenly distributed, and their abundances differ between the nearside and farside of the Moon. Although mare asymmetry has been attributed to thickness variations in the low-density anorthositic crust, the eruptive mechanism of lunar magma remains unknown. In this study, we investigate the relationship between mare distribution and crustal thickness using geological and geophysical data obtained by the SELENE (Kaguya) and the Gravity Recovery and Interior Laboratory spacecraft, and quantitatively re-evaluate the influence of the anorthositic crust on magma eruption. We identify a lateral heterogeneity in the upper limit of crustal thickness that allows magma extrusion to the surface. In the Procellarum KREEP Terrane, where the surface abundances of heat-producing elements are extremely high, magmas can erupt in regions of crustal thickness below about 30 km. In contrast, magma eruptions are limited to regions of crustal thickness below about 20 km in other nearside regions, around 10 km in the South Pole–Aitken Basin and approximately 5 km in the farside Felspathic Highland Terrane. Such heterogeneity may result from lateral variations in magma production in the lunar mantle and/or crustal density.

Lunar mare basalts, which are the most common volcanic features on the Moon, occur preferentially in topographical lows on the nearside, and are rare on the farside (Head & Wilson 1992). The total volume of mare basalts on the nearside has been calculated at about 10^7 km^3 (Head 1975), using thicknesses of extrusive rocks estimated from rim heights of partially flooded pre-mare craters (DeHon 1974).

In contrast, based on an assumption of a mean fill depth of 300 m, the total volume of farside mare basalts was estimated at approximately 10^5 km^3, around 100 times less than that of the nearside (Wasson & Warren 1980).

Gravity and altimetry data obtained by orbiting satellites reveal a 2 km offset between the lunar centres of mass and figure along the Earth–Moon

From: PLATZ, T., MASSIRONI, M., BYRNE, P. K. & HIESINGER, H. (eds) 2015. *Volcanism and Tectonism Across the Inner Solar System*. Geological Society, London, Special Publications, **401**, 127–138.
First published online March 20, 2014, http://dx.doi.org/10.1144/SP401.11
© The Geological Society of London 2015. Publishing disclaimer: www.geolsoc.org.uk/pub_ethics

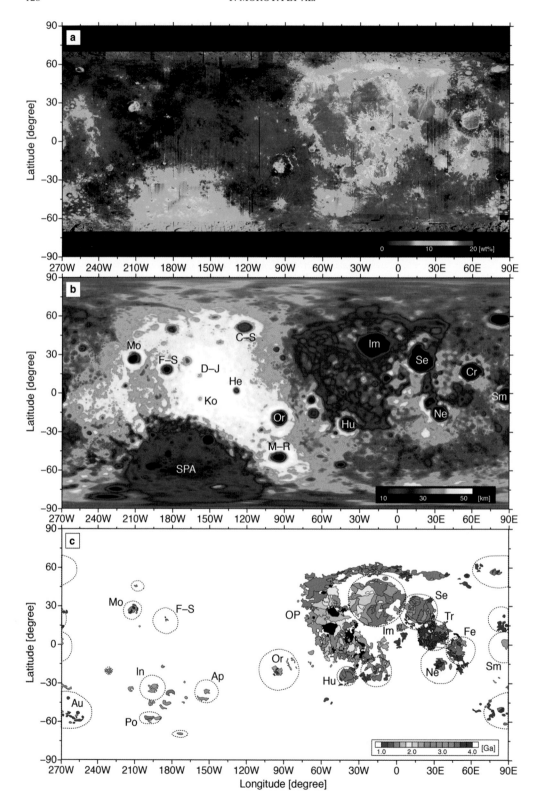

axis (e.g. Kaula *et al.* 1972), requiring that the mean density of the nearside hemisphere is greater than that of the farside. This density dichotomy has been attributed to differences in crustal thickness between two hemispheres (e.g. Neumann *et al.* 1996). On average, the farside anorthositic crust is estimated to be about 10 km thicker than that of the nearside (e.g. Neumann *et al.* 1996; Ishihara *et al.* 2009; Wieczorek *et al.* 2013).

This crustal thickness asymmetry was thought to be responsible for the hemispherical asymmetry of mare distribution. Under this assumption, magmas could be preferentially extruded to the surface in the thin parts of the low-density anorthositic crust by lithostatic pressure at the mare source (e.g. Solomon 1975) or by overpressure in the magma chamber (Head & Wilson 1992) because most lunar magmas were denser than the anorthositic crust. However, this interpretation fails to explain the lack of substantial flooding in the South Pole–Aitken (SPA) Basin (e.g. Lucey *et al.* 1998*b*; Yingst & Head 1997; Pieters *et al.* 2001; Wieczorek *et al.* 2001), corresponding to the deepest region (-8 km with respect to a sphere whose radius is 1738 km and whose origin is set to the centre of mass) on the Moon (e.g. Araki *et al.* 2009). In addition, the farside Moscoviense Basin, whose predicted crustal thickness is close to zero (Ishihara *et al.* 2009; Wieczorek *et al.* 2013), contains 3–10 times less mare basalt (in terms of volume) than the same-sized nearside Humorum Basin (Morota *et al.* 2009). Thus, mare hemispherical asymmetry cannot be explained by differences in crustal thickness alone and, therefore, the quantity of magma produced in the lunar mantle must be laterally heterogeneous (Wieczorek *et al.* 2001). According to numerical simulations of lunar thermal evolution, such lateral heterogeneity in magma production may be the result of the concentration of heat-producing elements on the nearside (Wieczorek & Phillips 2000; Laneuville *et al.* 2013) or a degree-1 mantle flow attributable to overturn of the last-stage cumulates of lunar magma ocean (Zhong *et al.* 2000).

The ratio of intrusive to extrusive activity on the Moon is unknown, but it may be even higher than on Earth because of the contrast between the low-density anorthositic crust and the large-density mare basalts, which are generally rich in FeO and TiO_2. Based on a simple hydrostatic balance model, the absence of abundant farside eruptions implies an upper limit of 37% by volume for the amount of dyke material present in the crust, corresponding to an intrusion to extrusion ratio of about 50:1 (Head & Wilson 1992). If magma production in the nearside mantle was higher than in the farside mantle, crustal intrusions might also be produced in much greater abundance in the nearside than in the farside.

This study aims to quantify the influence of low-density anorthositic crust on magma eruption. We investigate the correlation between mare distribution and crustal thickness using a lunar crustal thickness model (Wieczorek *et al.* 2013), updated by the Gravity Recovery and Interior Laboratory (GRAIL) gravity model and the Lunar Reconnaissance Orbiter (LRO) topography model (Smith *et al.* 2010; Zuber *et al.* 2013). On the basis of the findings, we discuss the extent and pattern of the lateral spatial heterogeneity of magma eruptions, and their relationship with the surface abundances of radioactive elements. We also re-evaluate the effects of resolidification of melt pools generated by large impacts, such as the formation of the SPA Basin, as previous studies have discussed in detail (Wieczorek *et al.* 2001; Shearer *et al.* 2006).

Further, remote-sensing studies of eruption ages of mare basalts have suggested that mare volcanism continued until 1.5–2 Ga, with the total duration estimated to be more than about 2 Gyr since magma eruption began at least as early as 4.0 Ga (Boyce 1976; Hiesinger *et al.* 2003; Morota *et al.* 2011*a*; Cho *et al.* 2012). We therefore also investigate the relationship between the timing of magma eruption and crustal thickness in order to examine whether the conditions that induce magma eruption have varied over time.

Definition of mare area

Lunar mare basalts are distinguishable from non-mare rocks by their high FeO content. The FeO abundance of lunar basaltic samples is >15 wt%, whereas non-mare rocks such as ferroan anorthosites, troctolites and norites contain <10 wt% FeO on average. Lucey *et al.* (1998*a*) developed techniques for estimating FeO and TiO_2 abundances of surface material from multiband image data obtained by orbiting satellites. Adopting Lucey's techniques, Otake *et al.* (2012) mapped the FeO

Fig. 1. (**a**) Global map of FeO content (Otake *et al.* 2012). (**b**) Lunar crustal thickness model (from Wieczorek *et al.* 2013). (**c**) Model ages of mare basalts (Hiesinger *et al.* 2000, 2003, 2006, 2010; Haruyama *et al.* 2009; Morota *et al.* 2009, 2011*a, b*; Cho *et al.* 2012). Ap, Apollo; Au, Australe; C–S, Coulomb–Sarton; Cr, Crisium; D–J, Dirichlet–Jackson; Fe, Fecunditatis; F–S, Freundlich–Sharonov; Hu, Humorum; Im, Imbrium; In, Ingenii; Ko, Kolorev; M–R, Mendel–Rydberg; Mo, Moscoviense; Ne, Nectaris; OP, Oceanus Procellarum; Or, Orientale; Po, Poincaré; Se, Serenitatis; Sm, Smythii; SPA, South Pole–Aitken; Tr, Tranquillitatis.

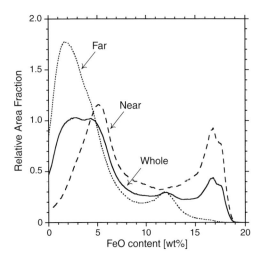

Fig. 2. Area fraction as a function of FeO content.

and TiO$_2$ abundances estimated from Selenological and Engineering Explorer (SELENE) multiband imager data. The global map of FeO abundance is shown in Figure 1a. In this study, the boundary between mare and highland regions is defined by using the FeO abundance of surface material. Our analysis excludes high-latitude regions ($>\pm 70°$) because the photometric correction is inaccurate in these regions.

Figure 2 shows relative area fraction as a function of FeO abundance on the lunar surface. The nearside distribution peaks at approximately 5 and 17 wt%, corresponding to the mean FeO abundances of highland and mare soils, respectively. The FeO abundance of the farside shows a similar bimodal distribution; however, the peak at approximately 12 wt% FeO abundance is contributed mainly by non-mare mafic materials in the floor of the SPA Basin, such as the noritic melt sheet/breccia (Pieters *et al.* 2001), rather than by mare materials. Because the FeO abundances of Apollo soils returned from the mare landing sites exceed about 12 wt%, we adopt approximately 12 wt% FeO as the mare–highland boundary.

Based on this definition for the mare–highland boundary, the fractions of areas covered by maria are estimated as 16.8% for the whole Moon, 31.7% for the nearside and 2.2% for the farside. Consequently, mare basalts cover an area that is about 15 times larger on the nearside than on the farside.

Crustal thickness model

Lunar crustal thicknesses can be estimated from gravity and topography data, assuming that surface topography, surface basalt flows and relief of the

lunar Moho sufficiently explain the lunar gravity field (Neumann *et al.* 1996; Wieczorek & Phillips 1998; Hikida & Wieczorek 2007; Ishihara *et al.* 2009; Wieczorek *et al.* 2013). The lunar gravitational field has been investigated from the Doppler shift in radio tracking signals caused by accumulated accelerations of orbiting spacecraft. Early gravitational field models calculated from Lunar Orbiter (LO), Clementine and Lunar Prospector (LP) data revealed major mass concentrations (mascons) with basins on the lunar nearside, as well as the offset between the lunar centres of mass and figure along the Earth–Moon axis.

The first reliable gravity field of the lunar farside was determined by SELENE using a relay satellite between the main orbiter and the Earth, and a very-long-baseline interferometry (VLBI) subsatellite for differential tracking of the relay satellite (Namiki *et al.* 2009; Matsumoto *et al.* 2010; Goossens *et al.* 2011). On the basis of the SELENE gravity field and topography models, Ishihara *et al.* (2009) computed the crustal thickness map assuming densities for the crust, mantle and mare basalts of 2800, 3360 and 3200 kg m^{-3}, respectively. In the SELENE model, lunar crustal thickness ranges from nearly zero beneath the Moscoviense basin to around 110 km on the southern rim of the Dirichlet–Jackson Basin, with an overall mean thickness of about 53 km.

More recently, high-resolution gravity data were obtained by ranging between the two co-orbiting GRAIL spacecraft (Andrews-Hanna *et al.* 2013; Wieczorek *et al.* 2013; Zuber *et al.* 2013). These data show that the average bulk density of the lunar highland crust is 2550 kg m^{-3}, which is considerably lower than that used in previous studies for constructing crustal thickness models, and that lateral variations in crustal density exist with amplitudes of ± 250 kg m^{-3} (Wieczorek *et al.* 2013). With these constraints on the lateral variations in crustal density, Wieczorek *et al.* (2013) compiled global crustal thickness models from GRAIL gravity and LRO topography data (Fig. 1b). The average crustal thickness was estimated to be 34–43 km. The spatial resolution of the new GRAIL crustal thickness models is 60% better than those of previous models. Here, the GRAIL model is used to evaluate the effect of the crustal structure on magma eruption. It should be noted that because the crustal thickness model takes no account of the surficial layer of mare basalt, the total crustal thicknesses are biased locally, but by no more than a few kilometres (Wieczorek *et al.* 2013).

Figure 3 describes the cumulative fraction of lunar areas as a function of crustal thickness. This figure shows that if crustal thickness were the only factor that determines the global distribution of mare volcanism, and if the upper limit of crustal

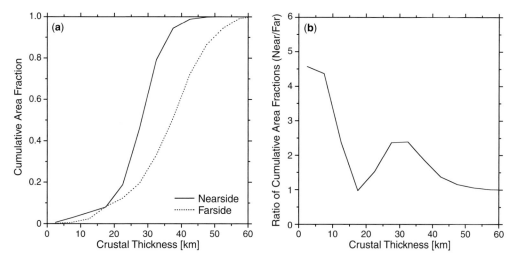

Fig. 3. (a) Cumulative fraction of lunar area as a function of crustal thickness. (b) Ratio of cumulative areas of the nearside and the farside as a function of crustal thickness. The bin size of crustal thickness is 5 km. Data are from Wieczorek *et al.* (2013).

thickness that allows magma extrusion to the near-side surface were the same as that of the farside, mare basalts could cover at most an area 4.5 times larger on the nearside than on the farside. However, this figure is substantially smaller than the observed heterogeneity, approximately some 15 times the areal difference, indicating again that mare hemispherical asymmetry cannot be explained by differences in crustal thickness alone.

Model ages of mare basalts

Using image data from orbital satellites, a considerable number of maria have been dated by various techniques such as crater degradation measurements, crater size–frequency distributions and stratigraphic relationships (Boyce 1976; Wilhelms 1987; Hiesinger *et al.* 2000, 2003, 2006, 2010; Haruyama *et al.* 2009; Morota *et al.* 2009, 2011*a*, *b*; Whitten *et al.* 2011). Hiesinger *et al.* (2000, 2003, 2006, 2010) performed crater counting in numerous nearside maria and in some farside maria using LO images. According to their results, the greatest volume of mare basalts was emplaced in the Late Imbrian Epoch at 3.2–3.8 Ga, consistent with the radiometric ages of basalt samples returned by the Apollo and Lunar missions (e.g. Taylor *et al.* 1983; Nyquist & Shih 1992). The model ages by Hiesinger *et al.* (2003) have also suggested that some mare basalts in Oceanus Procellarum and Mare Imbrium are substantially younger than the returned samples. The youngest mare basalts are located around the Aristarchus Plateau and have

estimated surface ages of approximately 1.2 Ga, expanding the duration of mare volcanism to more than about 3.0 Gyr (Hiesinger *et al.* 2003).

The Terrain Camera (TC) aboard SELENE acquired stereo data over the entire lunar surface with an average resolution of 10 m/pixel (Haruyama *et al.* 2008). From the TC high-resolution images, we performed crater size–frequency distribution measurements on farside mare deposits in the SPA basin and in the Feldspathic Highland Terrane, as well as on young lava flows in the Procellarum KREEP Terrane (Haruyama *et al.* 2009; Morota *et al.* 2009, 2011*a*, *b*; Cho *et al.* 2012). Figure 1c shows a map of the model ages of lava flows, which are compiled from various sources (Hiesinger *et al.* 2000, 2003, 2006, 2010, Haruyama *et al.* 2009; Morota *et al.* 2009, 2011*a*, *b*; Cho *et al.* 2012). Using this map, we investigate the relationship between eruption age and crustal thickness.

Relationship between mare distribution and crustal thickness

Figure 4 shows the fraction of mare and highland regions as a function of lunar crustal thickness. This result shows that magma eruption occurred preferentially in regions of relatively thin crust. On the nearside, mare basalt covers 50% areas of crustal thickness <25 km. In contrast, farside maria cover 50% areas of crustal thickness <10 km. These results affirm once more that the nearside–farside mare asymmetry cannot be solely explained by differences in crustal thickness, although crustal

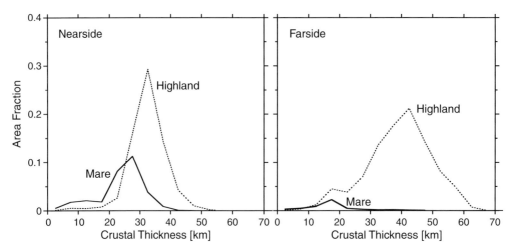

Fig. 4. Area fractions as a function of crustal thickness. The solid and dashed lines indicate the mare and highland areas, respectively. The bin size of crustal thickness is 5 km.

structure may be one of the factors that determined whether basaltic magmas erupted at the surface within mare regions. An additional possible explanation for the mare hemispherical asymmetry is that the farside mantle produced less magma than the nearside mantle (Wieczorek & Phillips 2000; Zhong *et al.* 2000; Wieczorek *et al.* 2001; Morota *et al.* 2009; Laneuville *et al.* 2013).

The nearside–farside hemispherical dichotomy in magma production presented here is consistent with high abundances of radioactive elements in the crust of the Oceanus Procellarum and Mare Imbrium regions (e.g. Lawrence *et al.* 1998; Yamashita *et al.* 2010), and with the Moho temperatures inferred from the compensation states of impact basins. The SELENE gravity field model (Namiki *et al.* 2009; Matsumoto *et al.* 2010) revealed distinct differences between the gravity anomalies of the nearside mascons and the farside basins for the first time. The nearside mascons exhibit a positive gravity plateau surrounded by a weakly negative gravity anomaly. In contrast, large basins on the farside are characterized by concentric rings of positive and negative anomalies: a large central gravity high due to dynamic uplift of the mantle during and after transient cavity growth (e.g. Croft 1981) occurs at the basin centre, a negative anomaly ring along the floor and a positive anomaly that corresponds to the topographical rim of the basin. By simulating the long-term viscoelastic evolution of basin structures and comparing those results with SELENE-derived gravity data for beneath the basins, Kamata *et al.* (2013) determined that the surface temperature gradient on the farside ($<c.$ 25 K km^{-1}) was lower than that on the nearside ($<c.$ 35 K km^{-1}), indicating a very cold farside interior.

The lateral heterogeneity in magma production might also produce much greater abundances of crustal intrusions on the nearside relative to the farside. The accumulation of such intrusions should increase the crustal density and increase the likelihood of later magmas being able to erupt at the surface (Head & Wilson 1992). The formation of crustal intrusions preferentially at the lunar nearside might further promote the lateral hemispherical asymmetry of mare distribution.

To investigate the spatial pattern of volcanic activity, we divided our analysed areas on the nearside and farside into two regions (regions N1 and N2 for the nearside, and regions F1 and F2 for the farside: see Fig. 5). Region N1 encompasses Oceanus Procellarum and Mare Imbrium, the so-called 'Procellarum KREEP Terrane (PKT)' (Jolliff *et al.* 2000). Gamma-ray data obtained by LP and SELENE revealed heavily concentrated radioactive elements on the surface of this region (Lawrence *et al.* 1998; Kobayashi *et al.* 2010; Yamashita *et al.* 2010). Region F1 encompasses the SPA Basin, which, with a diameter of 2500 km, is the largest confirmed impact structure in the solar system and is the oldest impact basin on the Moon (Wilhelms 1987; Hiesinger *et al.* 2012). Region F2 mainly covers the feldspathic highlands on the northern farside, where the lunar crust is thickest (*c.* 100 km), in addition to several mare basins and craters such as Moscoviense, Freundlich–Sharonov, Campbell and Orientale.

Figure 6 shows the fraction of area covered by maria within each bin of crustal thickness. Clearly, magma eruption predominates in regions of thin crust. We also observe from Figure 6 a clear lateral heterogeneity of magma eruption in the

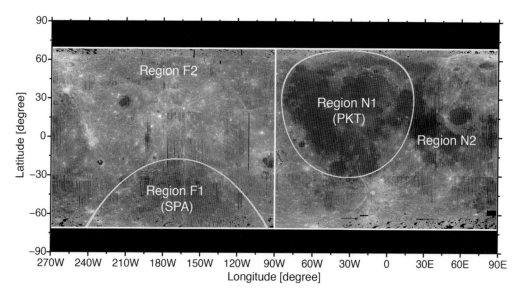

Fig. 5. Regions analysed in this study superposed on a SELENE Multiband Imager 750 nm reflectance map. PKT, Procellarum KREEP Terrane; SPA, South Pole–Aitken.

upper limit of crustal thickness. Where only half of an area is covered by mare basalts, the crustal thickness is around 20 km in Region N2, approximately 10 km in Region F1 and about 5 km in Region F2. However, where mare basalts cover most area, the crustal thickness is about 30 km in Region N1. In this region, the existence of sinuous rilles that originate from within topographically high volcanic centres (e.g. the Aristarchus Plateau) indicates that some mare basalts in Oceanus Procellarum and

Mare Imbrium erupted at high-standing elevations, flowed downhill and then pooled at lower elevations (e.g. Wieczorek *et al.* 2001). This suggests that magma eruption in Region N1 was not limited by a crustal thickness <30 km. In the following sections, we discuss possible causes of this observed spatial variation in volcanic activity.

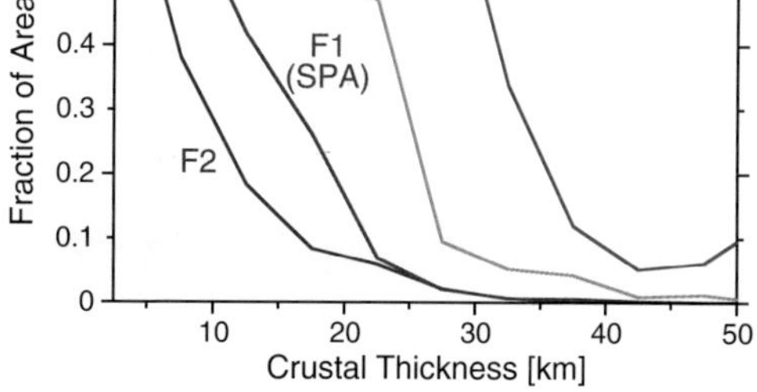

Active magmatism in the Procellarum KREEP Terrane

The large amount of magma erupted in the Procellarum KREEP Terrane is closely correlated with high crustal abundances of radioactive elements in this region, as observed with gamma-ray data from LP and SELENE. Using a simple thermal conduction model, Wieczorek & Phillips (2000) demonstrated that heat-producing elements concentrated in the lower crust of the Procellarum KREEP Terrane could directly cause partial melting in the underlying mantle immediately after crystallization of the magma ocean. This melting zone would have persisted beneath the central region of the Procellarum KREEP Terrane throughout much of lunar history. This idea is consistent with the longevity of mare volcanism around the terrane's centre, revealed by crater-counting studies (e.g. Hiesinger *et al.* 2003; Morota *et al.* 2011*a*). More recently, using a three-dimensional (3D) spherical thermochemical convection code, Laneuville *et al.* (2013) modelled the thermal evolution of the Moon to assess whether the concentration of heat

Fig. 6. The fraction of area covered by maria within each bin of crustal thickness.

sources in the Procellarum KREEP Terrane has affected global lunar evolution. These authors found that the enhanced heat-producing elements in this region could increase the magma production rate to about 10 times that of the farside.

As Wieczorek *et al.* (2001) discussed, the remaining question is why a large volume of magma ascended to the surface in this region, whose anorthositic crust is around 30 km thick. The majority of mare basalts in the central region of the Procellarum KREEP Terrane are rich in FeO and TiO_2. The liquidus density of mare basalts increases with FeO and TiO_2 content. From the relationship between the liquidus densities and the FeO and TiO_2 abundances in the lunar basaltic magmas (Wieczorek *et al.* 2001), liquidus densities of mare basalts in the central region of the Procellarum KREEP Terrane are estimated to be about 3000 kg m^{-3} (Fig. 7), substantially higher than the bulk crustal density (*c.* 2550 kg m^{-3}) revealed by GRAIL data (Wieczorek *et al.* 2013). If magma buoyancy is the only factor determining whether mare basalts erupt at the surface, these facts require the basalts in this region to have been sufficiently superheated to extrude to the surface (Wieczorek *et al.* 2001) and/or for the crust in the Procellarum KREEP Terrane to be denser than the basaltic magma erupted in this region. According to their thermal conduction model, Wieczorek & Phillips (2000) found that mantle temperatures could have reached approximately 2000 K, much higher than the liquidus temperature of lunar basaltic magma, under the Procellarum KREEP Terrane

as a consequence of heat-source concentration in this region. However, this mantle temperature may be an overestimate because the static thermal conduction assumption with low thermal conductivity produces temperatures well in excess of the solidus. Another explanation for eruptions of dense magma in the Procellarum KREEP Terrane is that gases released from magmas helped to drive basaltic eruptions (Wilson *et al.* 1980; Wilson & Head 1981; Wieczorek *et al.* 2001).

Heterogeneity of mare volcanism on the farside

The differences in extrusive volumes between regions F1 and F2 may be attributable to a denser crust beneath the SPA Basin compared to the feldspathic highland area (Huang & Wieczorek 2012; Wieczorek *et al.* 2013). On the bases of crater-scaling laws and numerical simulations of large impacts (e.g. Lucey *et al.* 1998*b*; Potter *et al.* 2012), the SPA-forming impact should have excavated most of the anorthositic crust, and melted the remnant crust and the upper mantle. As a consequence of impact-melt solidification, the SPA Basin was probably coated with a secondary crust of material of higher density and more mafic composition than the pristine anorthositic crust. According to the magma buoyancy hypothesis (Wieczorek *et al.* 2001), such dense crust may promote extrusion of magma to the surface by buoyant force alone. This interpretation is supported by the fact

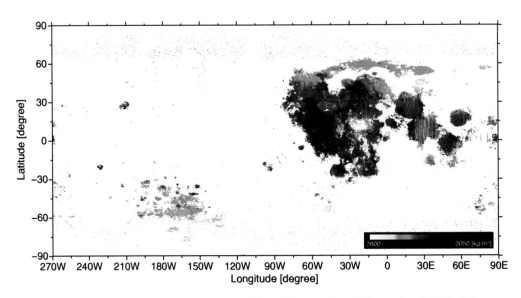

Fig. 7. Liquidus density of mare basalts calculated from FeO and TiO_2 abundances (Wieczorek *et al.* 2001). Only mare areas with >12 wt% FeO are shown.

that no extensive exposure of purest-anorthosite occurs in the SPA Basin (Ohtake *et al.* 2009; Yamamoto *et al.* 2012). In addition, the central peaks of fresh craters in the SPA Basin are commonly composed of an ultramafic assemblage dominated by magnesium-rich orthopyroxene (Nakamura *et al.* 2009). If this interpretation is correct, the crustal density in the central region of the SPA Basin should exceed 2920 kg m^{-3}, the maximum liquidus density of mare basalts in this area (Fig. 7).

Alternatively or additionally, the differences between regions F1 and F2 might arise from lateral heterogeneities in magma production between the interior and exterior of the SPA Basin. The heat sources responsible for any heterogeneity in magma production are probably radioactive elements and/or the SPA-forming impact. SELENE gamma-ray data revealed higher surface abundances of U (0.3–0.9 ppm) and Th (2–3 ppm) on the SPA floor compared to those in the surrounding feldspathic highland regions (*c.* 0.3 ppm U and <1 ppm Th) (Yamashita *et al.* 2010). Furthermore, on the basis of numerical simulations for the long-term viscoelastic evolution of basin structures, Kamata *et al.* (2013) estimated the lower bound on the surface temperature gradient on the farside, which strongly suggested that the deep portion of the thick farside highland crust is highly depleted in radioactive elements (Th ≤0.5 ppm).

Using global spectral data obtained by the SELENE Spectral Profiler, Yamamoto *et al.* (2010) found prominent dunitic olivine exposures around the rims of the Moscoviense Basin. They suggested that the Moscoviense-forming impact excavated some portion of the mantle. However, gamma-ray data (e.g. Lawrence *et al.* 1998; Kobayashi *et al.* 2010; Yamashita *et al.* 2010) do not show a concentration of KREEP elements around the Moscoviense Basin. These observations suggest that KREEP elements are depleted not only at the surface but also at the crust–mantle boundary beneath the Moscoviense region, consistent with the viscoelastic fluid simulation of basin structures (Kamata *et al.* 2013).

The effect of the SPA-forming impact on the thermal evolution of the mantle has yet to be elucidated. A recent version of the impact-induced magmatism model, related to the mantle convection triggered by impact-induced thermal perturbations (Ghods & Arkani-Hamed 2007), requires that the SPA-forming impact caused substantial melting in the mantle. This model also requires a long duration of mare volcanism for the SPA Basin. However, this is incompatible with crater-counting studies of the farside basalts, which indicate no difference in the timing and the duration of mare volcanism within and outside the SPA Basin (Haruyama *et al.* 2009; Morota *et al.* 2011*b*).

Relationship between eruption age and crustal thickness

Lunar thermal evolution models (Solomon 1975; Spohn *et al.* 2001; Ziethe *et al.* 2009) predict that the partial-melting zone in the upper mantle froze from above because of the thickening of the lithosphere, implying that the magma-source region deepened over time. We therefore also investigate the relationship between the timing of magma eruption

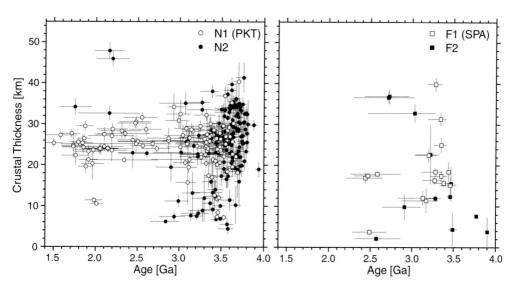

Fig. 8. Model ages of mare basalts in relation to mean crustal thickness in each mare unit.

and crustal thickness to examine whether the conditions that induce magma eruption have varied over time.

Figure 8 shows the relationship between mare basalt age and mean crustal thickness in each mare unit. These results do not show a systematic correlation between the ages of mare basalts and the crustal thicknesses in the nearside (regions N1 and N2). Although the crustal thickness of the area covered by mare basalts in the SPA Basin appears to have decreased with time, whereas crustal thickness in the northern farside appears to have increased with time, these trends are not statistically significant. The probabilities that these trends are generated by statistical fluctuation are estimated to be >0.1 by correlation tests. These results imply that magma-source depth is not the predominant controlling factor of the upper limit of crustal thickness for magma surface extrusion.

Conclusions

Using the geological and geophysical datasets updated by SELENE and GRAIL, we have re-evaluated the correlation between crustal structure and mare distribution. Consistent with previous studies, our results show that magma eruption predominates in regions of relatively thin crust. We also identify a lateral heterogeneity in the upper limit of crustal thickness that allows the surface extrusion of magma. In the Procellarum KREEP Terrane, magmas can erupt where the crust is thinner than about 30 km. Conversely, magma eruptions are limited to areas of crustal thickness below approximately 20 km in other nearside regions, around 10 km in the SPA Basin and about 5 km in the farside highland region. The observed spatial pattern of this lateral heterogeneity is consistent with the Moho temperature distribution inferred from the compensation states of impact basins (Kamata *et al.* 2013). Such lateral heterogeneity can be explained by spatial variations of magma production in the lunar mantle and/or in crustal density. To properly interpret the causes of lateral heterogeneity in maria-forming volcanic activity, we must retrieve mare basalts from unsampled sites, such as from the farside maria and from young maria in the central region of the Procellarum KREEP Terrane, and investigate their petrology. We also require a more detailed understanding of the lateral and vertical density structure of the lunar crust, and of the volumes of intrusive rocks it contains.

We are grateful to two anonymous reviewers for thoughtful and constructive reviews of this manuscript. We also thank T. Platz and P. Byrne for their helpful comments. The GRAIL crustal thickness model used in this study is available on the GRAIL Crustal Thickness Archive (http://www.ipgp.fr/~wieczor/GRAILCrustalThickness Archive/GRAILCrustalThicknessArchive.html). This work was supported by the Japan Society for the Promotion of Science under Grant-in-Aid for Young Scientists (B) (25870314: T. Morota).

References

ANDREWS-HANNA, J. C., ASMAR, S. W. *ET AL.* 2013. Ancient ligneous intrusions and early expansion of the Moon revealed by GRAIL gravity gradiometry. *Science*, **339**, 675–678.

ARAKI, H., TAZAWA, S. *ET AL.* 2009. Lunar global shape and polar topography derived from Kaguya-LALT laser altimetry. *Science*, **323**, 897–900.

BOYCE, J. M. 1976. Ages of flow units in the Lunar nearside Maria based on Lunar Orbiter IV photographs. *In*: *Proceedings of the Seventh Lunar Science Conference, Houston, Texas, March 15–19, 1976.* Lunar and Planetary Institute, Houston, TX, 2717–2728.

CHO, Y., MOROTA, T., HARUYAMA, J., YASUI, M., HIRATA, N. & SUGITA, S. 2012. Young mare volcanism in the Orientale region contemporary with the Procellarum KREEP Terrane (PKT) volcanism peak period ~2 billion years ago. *Geophysical Research Letters*, **39**, L11203, http://dx.doi.org/10.1029/2012GL05 1838

CROFT, S. K. 1981. The modification stage of basin formation – conditions of ring formation. *In*: SCHULTZ, P. H. & MERRILL, R. B. (eds) *Multi-Ring Basins: Formation and Evolution*. Pergamon Press, New York, 227–257.

DEHON, R. A. 1974. Thickness of mare material in the Transquillitatis and Nectaris basins. *In*: *Proceedings of the Fifth Lunar Science Conference*, Houston, Texas, March 18–22, 1974. Lunar and Planetary Institute, Houston, TX, 53–59.

GHODS, A. & ARKANI-HAMED, J. 2007. Impact-induced convection as the main mechanism for formation of lunar mare basalts. *Journal of Geophysical Research*, **112**, E03005, http://dx.doi.org/10.1029/2006JE0 02709

GOOSSENS, S., MATSUMOTO, K. *ET AL.* 2011. Lunar gravity field determination using SELENE same-beam differential VLBI tracking data. *Journal of Geodesy*, **85**, 205–228.

HARUYAMA, J., MATSUNAGA, T. *ET AL.* 2008. Global lunar-surface mapping experiment using the Lunar Imager/ Spectrometer on SELENE. *Earth, Planets, Space*, **60**, 243–256.

HARUYAMA, J., OHTAKE, M. *ET AL.* 2009. Long-lived volcanism on the lunar farside revealed by SELENE Terrain Camera. *Science*, **323**, 905–908.

HEAD, J. W. 1975. Lunar mare deposits: Areas, volumes, sequence, and implication for melting in source areas. *In*: *Conference on Origins of Mare Basalts and their Implications for Lunar Evolution*. Lunar Science Institute, Houston, TX, 66–69.

HEAD, J. W. & WILSON, L. 1992. Lunar mare volcanism: stratigraphy, eruption, conditions, and the evolution of secondary crusts. *Geochimica et Cosmochimica Acta*, **56**, 2155–2174.

HIESINGER, H., JAUMANN, R., NEUKUM, G. & HEAD, J. W., III. 2000. Age of mare basalts on the lunar nearside. *Journal of Geophysical Research*, **105**, 29 239–29 275.

HIESINGER, H., HEAD, J. W., III, WOLF, U., JAUMANN, R. & NEUKUM, G. 2003. Ages and stratigraphy of mare basalts in Oceanus Procellarum, Mare Nubium, Mare Cognitum, and Mare Insularum. *Journal of Geophysical Research*, **108** (E7), 5065–5091, http://dx.doi.org/10.1029/2002JE001985

HIESINGER, H., HEAD, J. W., III, WOLF, U., JAUMANN, R. & NEUKUM, G. 2006. New ages for basalts in Mare Fecunditatis based on crater size–frequency measurements. *In*: 37th Annual Lunar and Planetary Science Conference, March 13–17, 2006, League City, Texas. Lunar Science Institute, Houston, TX, abstract 1151.

HIESINGER, H., HEAD, J. W., III, WOLF, U., JAUMANN, R. & NEUKUM, G. 2010. Ages and stratigraphy of lunar mare basalts in Mare Frigoris and other nearside maria based on crater size-frequency distribution measurements. *Journal of Geophysical Research*, **115**, E03003, http://dx.doi.org/10.1029/2009JE003380

HIESINGER, H., VAN DER BOGERT, C. H., PASCKERT, J. H., SCHMEDEMANN, N., ROBINSON, M. S., JOLLIFF, B. & PETRO, N. 2012. New crater size–frequency distribution measurements of the South Pole–Aitken basin. *In*: 43rd Lunar and Planetary Science Conference, March 19–23, 2012, The Woodlands, Texas. Lunar and Planetary Institute, Houston, TX, abstract 2863.

HIKIDA, H. & WIECZOREK, M. A. 2007. Crustal thickness of the Moon: new constraints from gravity inversions using polyhedral shape models. *Icarus*, **192**, 150–166.

HUANG, Q. & WIECZOREK, M. A. 2012. Density and porosity of the lunar crust from gravity and topography. *Journal of Geophysical Research*, **117**, E05003, http://dx.doi.org/10.1029/2012JE004062

ISHIHARA, Y., GOOSSENS, S. ET AL. 2009. Crustal thickness of the Moon: implications for farside basin structures. *Geophysical Research Letters*, **36**, L19202, http://dx.doi.org/10.1029/2009GL039708

JOLLIFF, B. L., GILLIS, J. J., HASKIN, L. A., KOROTEV, R. L. & WIECZOREK, M. A. 2000. Major lunar crustal terranes: surface expressions and crust–mantle origins. *Journal of Geophysical Research*, **105**, 4197–4216.

KAMATA, S., SUGITA, S. ET AL. 2013. Viscoelastic deformation of lunar impact basins: implications for heterogeneity in the deep crustal paleo-thermal state and radioactive element concentration. *Journal of Geophysical Research*, **118**, 398–415, http://dx.doi.org/10.1002/jgre.20056

KAULA, W. M., SCHUBERT, G., LINGENFELTER, R. E., SJOGREN, W. L. & WOLLENHAUPT, W. R. 1972. Analysis and interpretation of lunar laser altimetry. *In*: KING, E. A., JR, HEYMANN, D. & CRISWELL, D. R. (eds) *Proceedings of the Third Lunar Science Conference. Geochimica et Cosmochimica Acta*, Supplement **3**. MIT Press, Cambridge, MA, 2189–2204.

KOBAYASHI, S., HASEBE, N. ET AL. 2010. Determining the absolute abundances of natural radioactive elements on the lunar surface by the Kaguya Gamma-ray Spectrometer. *Space Science Review*, **154**, 193–218, http://dx.doi.org/10.1007/s11214-010-9650-2

LANEUVILLE, M., WIECZOREK, M. A., BREUER, D. & TOSI, N. 2013. Asymmetric thermal evolution of the Moon. *Journal of Geophysical Research*, **118**, 1435–1452, http://dx.doi.org/10.1002/jgre.20103

LAWRENCE, D., FELDMAN, W. C., BARRACLOUGH, B. L., BINDER, A. B., ELPHIC, R. C., MAURICE, S. & THOMSEN, D. R. 1998. Global elemental maps of the Moon: the Lunar Prospector gamma-ray spectrometer. *Science*, **281**, 1484–1489.

LUCEY, P. G., BLEWETT, D. T. & HAWKE, B. R. 1998a. Mapping the FeO and TiO$_2$ content of the lunar surface with multispectral imagery. *Journal of Geophysical Research*, **103**, 3679–3699.

LUCEY, P. G., TAYLOR, G. J., HAWKE, B. R. & SPUDIS, P. D. 1998b. FeO and TiO$_2$ concentrations in the South Pole-Aitken basin: implications for mantle composition and basin formation. *Journal of Geophysical Research*, **103**, 3701–3708.

MATSUMOTO, K., GOOSSENS, S. ET AL. 2010. An improved lunar gravity field model from SELENE and historical tracking data: revealing the farside gravity features. *Journal of Geophysical Research*, **115**, E06007, http://dx.doi.org/10.1029/2009JE003499

MOROTA, T., HARUYAMA, J. ET AL. 2009. Mare volcanism in the lunar farside Moscoviense region: implication for lateral variation in magma production of the Moon. *Geophysical Research Letters*, **36**, L21202, http://dx.doi.org/10.1029/2009GL040472

MOROTA, T., HARUYAMA, J. ET AL. 2011a. Timing and characteristics of the latest mare eruption on the Moon. *Earth and Planetary Science Letters*, **302**, 255–266.

MOROTA, T., HARUYAMA, J. ET AL. 2011b. Timing and duration of mare volcanism in the central region of the northern farside of the Moon. *Earth, Planets and Space*, **63**, 5–13.

NAKAMURA, R., MATSUNAGA, T. ET AL. 2009. Ultramafic impact melt sheet beneath the South Pole-Aitken basin on the Moon. *Geophysical Research Letters*, **36**, L22202, http://dx.doi.org/10.1029/2009GL040765

NAMIKI, N., IWATA, T. ET AL. 2009. Farside gravity field of the Moon from four-way Doppler measurements of SELENE (Kaguya). *Science*, **323**, 900–905.

NEUMANN, G. A., ZUBER, M. T., SMITH, D. E. & LEMOINE, F. G. 1996. The lunar crust: global structure and signature of major basins. *Journal of Geophysical Research*, **101**, 16 841–16 863.

NYQUIST, L. E. & SHIH, C.-Y. 1992. The isotopic record of lunar volcanism. *Geochimica et Cosmochimica Acta*, **56**, 2213–2234.

OHTAKE, M., MATSUNAGA, T. ET AL. 2009. The global distribution of pure anorthosite on the Moon. *Nature*, **461**, 236–241.

OTAKE, H., OHTAKE, M. & HIRATA, N. 2012. Lunar iron and titanium abundance algorithms based on SELENE (Kaguya) Multiband Imager data. *In*: 43rd Lunar and Planetary Science Conference, March 19–23, 2012, The Woodlands, Texas. Lunar Science Institute, Houston, TX, abstract 1905.

PIETERS, C. M., HEAD, J. W., III, GADDIS, L., JOLLIFF, B. & DUKE, M. 2001. Rock types of South Pole-Aitken basin and extent of basaltic volcanism. *Journal of Geophysical Research*, **106**, 28 001–28 022.

POTTER, R. W. K., COLLINS, G. S., KIEFER, W. S., McGO-
VERN, P. J. & KRING, D. A. 2012. Constraining the size
of the South Pole-Aitken basin impact. *Icarus*, **220**,
730–743.

SHEARER, C. K., HESS, P. C. *ET AL.* 2006. Thermal
and magmatic evolution of the Moon. *In*: JOLLIFF,
B. L., WIECZOREK, M. A., SHEARER, C. K. & NEAL,
C. R. (eds) *Review of Mineralogy and Geochemistry* **60**. Mineralogical Society of America, Virginia,
365–518.

SMITH, D. E., ZUBER, M. T. *ET AL.* 2010. Initial observations from the Lunar Orbiter Laser Altimeter
(LOLA). *Geophysical Research Letters*, **37**, L18204,
http://dx.doi.org/10.1029/2010GL043751

SOLOMON, S. C. 1975. Mare volcanism and crustal structure. *In*: MERRILL, R. B. (eds) *Proceedings of the
Sixth Lunar Science Conference. Geochimica et Cosmochimica Acta*, Suppl. **6**, 1021–1042.

SPOHN, T., KONARD, W., BREUER, D. & ZIETHE, R. 2001.
The longevity of lunar volcanism: implicarions of
thermal evolution calculations with 2D and 3D
mantle convection models. *Icarus*, **149**, 54–65.

TAYLOR, L. A., SHERVAIS, J. W. *ET AL.* 1983. Pre-4.2
AEmare basalt volcanismin the lunar highlands.
Earth and Planetary Science Letters, **66**, 33–47.

WASSON, J. T. & WARREN, P. H. 1980. Contribution of
the Mantle to the lunar asymmetry. *Icarus*, **44**,
752–771.

WHITTEN, J., HEAD, J. W. *ET AL.* 2011. Lunar mare deposits associated with the Orientale impact basin: new
insights into mineralogy, history, mode of emplacement, and relation to Orientale Basin evolution
from Moon Mineralogy Mapper (M3) data from Chan-
drayaan-1. *Journal of Geophysical Research*, **116**,
E00G09, http://dx.doi.org/10.1029/2010JE003736

WIECZOREK, M. A. & PHILLIPS, R. J. 1998. Potential
anomalies on a sphere: applications to the thickness
of the lunar crust. *Journal of Geophysical Research*,
103, 1715–1724.

WIECZOREK, M. A. & PHILLIPS, R. J. 2000. The "Procellarum KREEP Terrane": implications for mare volcanism and lunar evolution. *Journal of Geophysical
Research*, **105**, 20 417–20 430.

WIECZOREK, M. A., ZUBER, M. T. & PHILLIPS, R. J. 2001.
The role of magma buoyancy on the eruption of lunar

basalts. *Earth and Planetary Science Letters*, **185**,
71–83.

WIECZOREK, M. A., NEUMANN, G. A. *ET AL.* 2013. The
crust of the Moon as seen by GRAIL. *Science*, **339**,
671–675.

WILHELMS, D. E. 1987. *The Geologic History of the Moon*.
United States Geological Survey, Professional Papers,
1348.

WILSON, L. & HEAD, J. W. 1981. Ascent and eruption of
basaltic magma on the Earth and Moon. *Journal of
Geophysical Research*, **86**, 2971–3001.

WILSON, L., SPARKS, R. S. J. & WALKER, G. P. L. 1980.
Explosive volcanic eruptions – IV. The control of
magma properties and conduit geometry on eruption
column behavior. *Geophysical Journal International*,
63, 117–140.

YAMAMOTO, S., NAKAMURA, R. *ET AL.* 2010. Possible
mantle origin of olivine around lunar impact basins
detected by SELENE. *Nature Geoscience*, **3**,
533–536, http://dx.doi.org/10.1038/ngeo897

YAMAMOTO, S., NAKAMURA, R. *ET AL.* 2012. Massive
layer of pure anorthosite on the Moon. *Geophysical
Research Letters*, **39**, L13201, http://dx.doi.org/10.
1029/2012GL052098

YAMASHITA, N., HASEBE, N. *ET AL.* 2010. Uranium on the
Moon: global distribution and U/Th ratio. *Geophysical
Research Letters*, **37**, L10201, http://dx.doi.org/10.
1029/2010GL043061

YINGST, R. A. & HEAD, J. W. 1997. Volumes of lunar lava
ponds in South Pole–Aitken and Orientale basins:
implications for eruption conditions, transport mechanisms and magma source regions. *Journal of Geophysical Research*, **102**, 10 909–10 931, http://dx.
doi.org/10.1029/97JE00717

ZHONG, S., PARMENTIER, E. M. & ZUBER, M. T. 2000. A
dynamic origin for the global asymmetry of lunar
mare basalts. *Earth and Planetary Science Letters*,
177, 131–140.

ZIETHE, R., SEIFERLIN, K. & HIESINGER, H. 2009. Duration and extent of lunar volcanism: comparison of
3D convection models to mare basalt ages. *Planetary
Space Science*, **57**, 784–796.

ZUBER, M. T., SMITH, D. E. *ET AL.* 2013. Gravity field of
the Moon from the Gravity Recovery and Interior Laboratory (GRAIL) mission. *Science*, **339**, 668–671.

VNIR spectral characteristics of terrestrial igneous effusive rocks: mineralogical composition and the influence of texture

C. CARLI[1]*, G. SERVENTI[2] & M. SGAVETTI[2,3]

[1]*Istituto di Astrofisica e Planetologia Spaziali-INAF Roma, Via fosso del cavaliere 100, 00133, Rome, Italy*

[2]*Dipartimento di Fisica e Scienze della Terra, Macedonio Meloni, Università degli studi di Parma, Parma, Italy*

[3]*International Research School of Planetary Sciences, Chieti-Pescara, Italy*

*Corresponding author (e-mail: cristian.carli@iaps.inaf.it)

Abstract: Visible and Near-Infrared (VNIR) reflectance spectroscopy is an important technique with which to map mineralogy and mineralogical variations across planetary surfaces using remotely sensed data. Absorption bands in this spectral range are due to electronic or molecular processes directly related to mineral families or specific compositions. Effusive igneous rocks are widely recognized materials distributed on the surfaces of terrestrial planets, and are formed by primary minerals that can be discriminated by electronic absorptions (e.g. crystal field absorption). In this paper, we review the current knowledge of effusive rock compositions obtained by crystal field absorption in VNIR reflectance spectroscopy, and consider how different petrographical characteristics influence the mineralogical interpretation of such rock compositions. We show that: (1) the dominant mineralogy can be clearly recognized for crystalline material, especially with relatively large crystal dimension groundmass or high porphyritic index; (2) both grain and crystal size are important factors that influence the spectra of effusive rocks where groundmass is generally characterized by microscopic crystals; and (3) glassy dark components in the groundmass reduce or hide the crystal field absorption of mafic minerals or plagioclase otherwise expected to be present.

The inner solar system hosts numerous differentiated terrestrial bodies that have been shaped by widespread and often sustained volcanic activity, with a high degree of compositional variation. From Mercury to the main asteroid belt (e.g. 4Vesta), morphological and spectroscopic data provide evidence for the presence of such activity, with extrusive volcanism identified as a major process for crustal formation, consistent with the volcanic nature of the most abundantly distributed rocks. Volcanic rocks are characterized by an aphanitic texture: that is, a very fine-grained groundmass in which most of the individual crystals cannot be distinguished with the naked eye, and which is presumed to have formed by relatively fast cooling (Le Maitre *et al.* 2002). The classification of volcanic rocks is defined either by modal (QAPF diagram, indicating quartz, alkali feldspar, plagioclase and feldspathoid: Streckeisen 1978) or, more commonly, by chemical (the TAS (total alkali silica) diagram: Le Maitre *et al.* 2002) analysis. Mineral names, textural terms or other terms can also be used to distinguish further the rock type (Le Maitre *et al.* 2002). The abundance of glassy phases is also important, as is the presence of phenocrysts, which

are the first crystals that form in the lava at depth, and xenocrysts, which are ripped from the crust through which lava rises.

The geological development of a planet is strongly influenced by its thermal history, which is driven by heat production, transport and loss through time. Clear phenomenological descriptions of planetary thermal histories were proposed by Sleep (2000), with those descriptions a function of primordial accretion heat loss and the radiogenic heat production budget, but also dependent on initial planetary composition and physical parameters such as lithospheric thickness (relative to planet size) and rheology. Different modes of convection related to distinct lithosphere types, including magma ocean, plate tectonics and stagnant lid, can either occur sequentially or recur cyclically throughout a planet's history, and some can be observed on different terrestrial planets today. Different forms of volcanism, as a function of different lithosphere types, are possible and are strongly related to magma composition.

Unlike other terrestrial planets, most of Earth's evolution has been, and still is, dominated by plate tectonics (e.g. Middlemost 1997). This process

From: PLATZ, T., MASSIRONI, M., BYRNE, P. K. & HIESINGER, H. (eds) 2015. *Volcanism and Tectonism Across the Inner Solar System.* Geological Society, London, Special Publications, **401**, 139–158.
First published online June 17, 2014, http://dx.doi.org/10.1144/SP401.19

produces a broad variety of magma compositions that depend on the different conditions in which magma can differentiate, accumulate and interact with country rocks. The result is the wide compositional variation of volcanic rocks observed on Earth, from ultrabasic to acidic, and from calc-alkaline to alkaline.

However, volcanism on other bodies in the solar system is generally regarded as less diverse than that of Earth because of thermal histories that do not necessarily involve plate tectonics. For example, volcanic products on Mars are associated with basic magma compositions characterized by very long lava flows, up to around 2000 km in length (e.g. Keszthelyi *et al.* 2000). Features like lava tubes or processes such as lava flow inflation, consistent with basaltic flows, are indicative of specific emplacement mechanisms and are comparable with those recognized in terrestrial analogues (e.g. Keszthelyi 1995; Sakimoto *et al.* 1997; Sakimoto & Zuber 1998; Peitersen & Crown 1999; Giacomini *et al.* 2009).

Basic–ultrabasic compositions have been inferred for some surface units – for example Mercury's northern volcanic plains (Head *et al.* 2011; Weider *et al.* 2012) or intercrater plains – at least a portion of which is probably volcanic (Nittler *et al.* 2011; Denevi *et al.* 2013). The volcanism of these regions consists of plains with abundant flow features and buried impact craters, forming complex, kilometres-thick sets of lavas, probably emplaced in several phases (Head *et al.* 2011; Byrne *et al.* 2013). Thermal erosion (by lava, which melts, assimilates and carries away the ground rock) has been suggested as being responsible for the morphology associated with some of this volcanism, indicative of a very effusive, flood-lava-style emplacement typical of turbulent komatiitic or high-temperature mafic lavas (Groves *et al.* 1986; Head *et al.* 2011; Byrne *et al.* 2013).

The volcanic history of the Moon, like Mars and Mercury a one-plate planetary body, must be associated with melting of mantle rocks without contamination by recycled crust. The lunar surface shows a wide range of compositions generally compatible with basalt (Hiesinger & Head 2006), consistent with early Apollo observations, and the Apollo and Luna samples, which determined a basaltic nature for the maria. Lunar basaltic flows are several tens of metres thick and extend for hundreds to thousands of kilometres, with lobate fronts 10–60 m high (Schaber *et al.* 1976; Gifford & El Baz 1978). Other surficial volcanic features that may reflect a mafic origin include sinuous rilles, lava terraces, cinder cones and pyroclastic deposits (see Hiesinger & Head 2006 and references therein). Recently, Spudis *et al.* (2013) interpreted large topographical features in the lunar maria as

shield volcanoes comparable to basaltic shield volcanoes on other terrestrial planets. A few domes have dimensions and structures that could be representative of more silicic lavas, of intrusion of shallow laccoliths or of large rock blocks mantled by younger lavas (e.g. Heater *et al.* 2003; Lawrence *et al.* 2005).

Compositional data from rover missions (e.g. on Mars) and laboratory sample analyses (lunar return samples or meteorite samples) provide detailed information on mineralogy, mineral chemistry and bulk-rock chemical composition, thus widening the range of recognized compositions for planetary volcanic systems. However, the analysis and mapping of surface compositions of extraterrestrial bodies in the solar system is principally based on remote-sensing surveys. Presently, the large amount of data acquired by hyperspectral sensors in the visible and near-infrared (VNIR, e.g. OMEGA spectrometer, MarsExpress; CRISM spectrometer, Mars Reconnaissance Orbiter; M^3 spectrometer, Chandrayann; VIR spectrometer, Dawn) and in the thermal infrared (TIR, e.g. TES spectrometer, Mars Global Surveyor) permit us to map these bodies at high spatial resolutions from orbit.

Most planetary bodies in the solar system have extremely tenuous or absent atmospheres and are therefore subject to space weathering at various intensities. Some effects of space weathering on the reflectance spectra of rock and regolith component minerals were experimentally analysed (e.g. Hiroi & Sasaki 2001; Sasaki *et al.* 2001; Brunetto & Strazzulla 2005; Strazzulla *et al.* 2005). For example, space weathering acts as a darkening and reddening agent of silicates, critically influencing the identification of diagnostic absorption bands in spectra of space-weathered material (e.g. Pieters *et al.* 1993; Moroz *et al.* 1996; Yamada *et al.* 1999; Pieters *et al.* 2000; Hapke 2001). Nevertheless, the variety of rock compositions in planetary crusts and regoliths raises a broader question about the intrinsic complexity of the reflectance spectra of planetary surfaces, which form the basis for investigating the spectral effects of space weathering.

In this paper, we review the compositional information that can be recognized in the VNIR spectral range for Earth's effusive basic–ultrabasic volcanic rocks. In particular, we examine to what extent absorptions indicative of the mineralogical rock-forming phases can be identified, while also considering other aspects (e.g. petrography) that can affect rock spectra in light of new data. Furthermore, we discuss both the possibilities and limits in the spectral interpretation of rocks, and present questions that are still open regarding the analysis of effusive rocks of mafic compositions. The purpose of this work is to contribute to the understanding of the spectroscopic complexity of planetary surface compositions, and to provide a framework that can aid in the interpretation of spectra of space-weathered minerals and rocks.

The composition of effusive volcanic rocks on terrestrial bodies in the inner solar system

Initially, basalts were thought to be the primary rock type present in Martian volcanic complexes like the Tharsis province that, among other volcanic landforms, features long lava flows. Analysis of data returned by the Thermal Emission Spectrometer (TES) instrument suggested a possible compositional dichotomy between Mars' southern and northern terrains, and confirmed that the southern areas are likely to correspond to a basaltic crust, whereas the northern areas contain more evolved volcanic rocks, probably of andesitic compositions (Bandfield et al. 2000; Hamilton et al. 2001). The analysis of rocks at the Mars Pathfinder landing site also provided evidence for the presence of andesitic rocks in the Martian crust (Reider et al. 1997; McSween et al. 1999; Waenke et al. 2001; Foley et al. 2003).

Further, analysis by the Spirit rover (MER-A) of basaltic rocks in Gusev crater identified them as picritic-basalts, with evidence of olivine (ol), pyroxene (px), plagioclase (pl) and accessory oxides (McSween et al. 2004). Data from excavated rocks provided evidence of uniform compositions, similar to ol-phyric shergotites (McSween et al. 2006a). In addition, Spirit encountered unaltered rocks in the Columbia Hills that have been identified as members of the Wishstone, Irvine and Backstay classes (McSween et al. 2006b). These rocks are enriched in alkalis and characterized by major mineral phases including px (both low and high in Ca), sodic pl, ferroan ol and Fe–Ti–Cr oxides, as well as minor phases such as apatite, and so have been classified as tephrite, alkaline basalts and hawaiites. Some of the samples of these rocks contain glassy material (McSween et al. 2006b).

Recently, McSween et al. (2009) reviewed the mineral composition of Martian volcanic crust by analysing data gathered by the Spirit and Opportunity rovers, and by the gamma-ray spectrometer (GRS) on the Mars Odyssey orbiter, and concluded that the composition of the Martian crust is probably a mix of fresh and altered basaltic materials. These authors suggested that TES data could have overestimated the SiO_2 content due to superficial weathering, as no areas dominated by siliceous rocks were apparent in the GRS silica distribution map (Boynton et al. 2007). Mars has also been fully mapped in the VNIR at different spatial and spectral resolutions using data gathered by the OMEGA and CRISM spectrometers, and the maps of several mineral phases, including px and ol, have been published (e.g. Poulet et al. 2007, 2009). However, although these minerals are common on the Martian surface and constitute an important spectroscopic component of effusive rocks, there are no px or ol compositional maps available for several equatorial volcanic regions on Mars (e.g. Olympus Mons, Daedalia Planum and Elysium Mons). The presence of, at times, substantial amounts of surface dust (Ruff & Christensen 2002) and Fe^{3+} minerals (Christensen et al. 2000; Poulet et al. 2007) in these regions have both been used to explain this anomaly. Recently, in their discussion of the composition of the Daedalia Planum lava flows, Giacomini et al. (2012) gave evidence of px absorption bands, which suggests a variation in px composition among the different constituent lava flows. In addition, the authors demonstrated how combined spectral characteristics and morphology can be used to improve the geological mapping of the region.

The lunar crust has also been studied in detail thanks to the Apollo and Luna missions, which brought back volcanic samples containing several different rock types. Lunar rocks are generally classified into four groups. One of these groups includes a pristine volcanic basaltic rock containing both effusive and pyroclastic material (Hiesinger & Head 2006). Mare basalts are characterized by effusive textures, a high abundance of ol and px (particularly clinopyroxene (cpx)), and a relatively low content of pl as major mineral phases (Hiesinger & Head 2006). These rocks are generally enriched in FeO and TiO_2, depleted in Al_2O_3 and have high CaO/Al_2O_3 ratios with respect to other lunar compositions (Taylor et al. 1991 and references therein). Lunar volcanism is also characterized by pyroclastic products, which are composed of glasses with colours that vary as a function of composition and the presence of skeletal crystals (see Lucey et al. 2006). This volcanism is similar to terrestrial fire fountains, formed by gasses (probably CO_2 plus other minor components) contained in the rising magmas that are explosively released as they approach the surface (Nicholis & Rutherford 2005). Other volcanic products have also been identified: KREEP (potassium, rare earth elements, phosphorous) basalts enriched in incompatible elements (Warren & Wasson 1979), high potassium basalts and high-alumina basalts. KREEPs are present as small rock fragments or clasts in breccias dominated by px and pl (Shearer et al. 2006 and references therein). Although the origin of this material is still a matter of debate, it is thought to be the product of the melting of a hybrid lunar mantle (Shearer et al. 2006 and references therein).

From a geochemical point of view, mare basalts can be broadly subdivided into three groups based on TiO_2 variation (Neal & Taylor 1992): high Ti; low Ti; and very low Ti. Several papers that investigated the petrographical and mineralogical variation of the different lunar mare samples have described large variations in textures and mineralogy even within the same rock type family.

Recently, M3 hyperspectral data collected from orbit were used to map the Moon's surface, and revealed the presence of a variety of materials. In particular, mafic minerals, pl, glass and spinels were detected using crystal field (C.F.) absorption. Different maria were mapped on the basis of their different spectral signatures, which are characterized by px bands with subtle minimum shifts, and band intensity and symmetry variations. Mare Serenitas was mapped as 13 spectral units following an Integrated Band Depth analysis (Kaur *et al.* 2013). Kaur *et al.* (2013) found that mafic mineralogy varies from low- to intermediate-Ca px and, although these authors suggested that px could vary from a sub-calcic to calcic augite composition, no substantial spatial variation in composition was observed (Kaur *et al.* 2013). The band area ratio varies, indicating possible spectral variation due to the presence of ol or px with variable 1.2 μm bands.

The asteroids form a large family of distinct bodies, some of which are thought to be differentiated, which may in some cases have led to the formation of a volcanic crust. In particular, 4Vesta, recently investigated by the DAWN mission (Russell & Raymond 2011), has a heterogeneous surface. Visual and Infrared Spectrometer (VIR) data of its surface confirmed earlier Earth-based observations that linked the composition of this body with HED (howardite, eucrite and diogenite) meteorites (De Sanctis *et al.* 2012). HED meteorites are characterized by mafic achondrites with textures similar to igneous rocks. Diogenitic and cumulate eucritic samples are similar to intrusive samples. Basaltic eucrites are equivalent to effusive terrestrial rocks, whereas howardites represent brecciated samples (Mittelfehldt *et al.* 1998 and references therein). The basaltic eucrites generally show a pigeonite–pl mineralogy with textures that vary from subophitic to ophitic. Pigeonite frequently shows subsolidus exsolution of augite and an iron-rich composition. Some eucrite samples also show a relatively high ol content. Plagioclase is calcic in composition, ranging from bytownite to anorthite (Mittelfehldt *et al.* 1998 and references therein). Despite this variation in HED mineralogy, the first analyses by the DAWN mission revealed only clear px absorptions with a small minimum shift. In addition, Ammannito *et al.* (2013) recently published evidence for the presence of ol in some of Vesta's regions.

Mercury is the innermost and, until recently, the least-studied planet of the inner solar system. The importance of volcanic products on its surface was debated until remote-sensing data from the MESSENGER mission revealed the presence of extensive volcanism across its the surface (Solomon *et al.* 2008; Head *et al.* 2009; Watters *et al.* 2009). Moreover, the morphological characteristics in some regions suggested the presence of flood magmatism (Head *et al.* 2011) with possible ultramafic compositions (Nittler *et al.* 2011; Weider *et al.* 2012). However, the iron content of the rocks is still ambiguous, as reflectance and X-ray data have shown very a low FeO content (Klima *et al.* 2013; Weider *et al.* 2013), while neutron spectrometer data show substantial neutron absorption (Lawrence *et al.* 2010; Riner *et al.* 2011). Unfortunately, absorption features have not yet been identified in reflectance spectra of Mercury's surface. Therefore, the only means available with which to discriminate different lava flows and deposits are variations in surface albedo and spectral slope, which makes the task of distinguishing surface units very difficult.

On Earth, remotely sensed hyperspectral data require extensive calibration due to interference from the atmosphere, vegetation and surface water. Nevertheless, volcanic regions that have little or no vegetation have been mapped (e.g. Hawaii, Etna). Remote-sensing data, in general, show low reflectances and low spectral contrasts for mafic compositions, although different lava flows can be discriminated thanks to variations in the spectral signature of rocks with age (Abrams *et al.* 1991) and with different surface textures (Sgavetti *et al.* 2003). In addition, Sgavetti *et al.* (2003) showed that oxidation products can also be used to discriminate the distal and proximal portions of lava flows.

Spectral analyses of planetary surfaces, gathered *in situ* by rovers or in the laboratory from returned samples, have shown a wider compositional range than those predicted by remote-sensing data alone. Although indicative of a smaller range of differentiation than that of the Earth, this variation suggests a more complex volcanic history for planetary surfaces than previously thought. Focusing on VNIR spectral variations that can be linked to differences in lava compositions, textures or weathering effects permits the classification and mapping of different volcanic regions on planetary bodies.

VNIR spectroscopy of terrestrial igneous effusive rocks

The compositions of volcanic products on the different inner bodies of our solar system have been investigated by three discrete methods: (1) remote-sensing spectroscopy; (2) *in situ* rover measurements; and (3) laboratory analysis of meteorites and lunar samples. Particular attention has been paid in such analyses to the implications of volcanism to our understanding of the evolution of planetary crusts (e.g. Mittlefehldt *et al.* 1998 and references therein; McSween *et al.* 2006a, b; Shearer *et al.* 2006 and references therein; Poulet *et al.* 2009; Kaur *et al.* 2013). The VNIR spectra of basic and ultrabasic rocks were first investigated by Hunt *et al.* (1974) using rock powder samples of different grain sizes. The spectra of intrusive rocks show well-defined features characteristic of gabbro, anorthosite and peridotite, especially for smaller grain sizes (0–74 μm), whereas effusive rocks are generally featureless with low albedo. For example, the spectral characteristics of basalts include the presence of opaque minerals and oxidation states (Hunt *et al.* 1974). In the VNIR spectra of acidic (Hunt *et al.* 1973a) and intermediate (Hunt *et al.* 1973b) igneous rocks, only a few features generally characteristic of mafic mineralogies or alteration, varying systematically across the different rock types, can be distinguished. Volcanic rocks with ultrabasic–basic compositions show a SiO_2 variation of approximately 30–52%, which can be correlated with Al_2O_3 variation, low alkali content (c. <5%) and a variable amount of MgO–FeO. This chemical variation reflects a mineralogy that ranges from high levels of mafic minerals (px, ol) to high levels of pl, from high magnesium to high iron, and from low to high Ca.

In this section, we summarize what has been learned about mineralogy from VNIR reflectance spectroscopy of volcanic rocks, with particular attention being paid to C.F. absorption bands of single phases, composite absorption bands due to complex mineralogies and our ability to distinguish between different absorption processes. Moreover, we discuss how textures can influence the identification of mineral phases from reflectance spectra. Further, in the following sections, 'particles' refer to synthetic mixtures of minerals prepared by weighing of individual mineral end members, where each particle represents a monomineralic sample. For rock powder samples obtained by grinding, the term 'grain' is used to refer to individual fragments, which can consist of one or more mineral ('crystal') or amorphous ('glass') phases. A grain can be mono- or multi-crystalline for wholly crystalline rocks, or mono- or multi-phase for rocks not completely crystalline. Particle or grain sizes are used to determine the granulometric range of mixtures and powder samples, respectively, whereas crystal size is used to define the size of the crystals in individual grains or in a rock slab.

Mineralogical information

In VNIR spectra, C.F. absorption bands resulting from the presence of transitional elements in well-defined coordination sites within the crystal lattice of major rock-forming minerals are recognizable (Burns 1993). In addition, other useful absorption bands indicative of compositional variation due, for example, to charge transfer (intervalence charge transfer, IVCT, between cations or between cations and ligands: see e.g. Clark 1999) can be observed in this wavelength range, as can some vibrational overtones in the OH^-, CO_3^{2-} and $S_3O_4^{2-}$ functional groups.

Volcanic rocks are generally composed of mafic silicates like px and ol, as well as some sialic minerals like feldspars, specifically calcic pl. These minerals can be characterized by C.F. absorptions due to FeO, principally, as well as TiO_2. Other transitional elements are not generally abundant enough to produce detectable absorption bands. Opaque minerals, such as Fe–Ti oxide (e.g. magnetite, ilmenite) or spinels (Cr, Mg, Al, Fe oxide), as well as Fe^{3+} or Fe^0, in the form of hematite or iron phase particles, can be present in volcanic rocks as a result of secondary processes acting in different fO_2 conditions on a planetary surface. Even if opaque minerals are generally considered neutral phases, due to their spectral characteristics and their relatively low abundances in effusive rocks, they can substantially affect albedo and spectral slope, and so mask absorption features (e.g. Cloutis *et al.* 1990a; Cloutis & Gaffey 1991b). Fe^0 is a strong darkening agent and, when present in very small, nano-sized particles, introduces a reddening component (e.g. Hapke 2001; Lucey & Riner 2011). Glassy phases, which are also present in both Martian and lunar samples, as well as in some meteorites, produce a darkening (general reduction of reflectance) and/or a reddening (positive spectral slope) effect in the VNIR spectra with superimposed C.F. absorption, which is thus a marker for the presence of effusive lava or pyroclastics materials (Adams & McCord 1971; Adams *et al.* 1974; Bell *et al.* 1976; McSween & Treiman 1998 and references therein; Minitti *et al.* 2002; McSween *et al.* 2004; Tompkins & Pieters 2010). This component, which is used to define textural characteristics of volcanic rocks, can also influence mineralogical information obtained from VNIR.

Regolith that is composed of volcanic material can contain large multi-crystal or multi-phase grains that are aggregates of phenocryst fragments and submicroscopic (tens of microns) groundmass. Thus, their spectra appear similar to those acquired on cut-surfaces of rocks (i.e. slab) and are similarly affected by texture (Carli & Sgavetti 2011). In multi-crystal and multi-phase grain spectra, the absorption features are difficult to resolve, and the continua are less predictable and poorly understood owing to the optical properties of different minerals that are aggregated. All of these phases interact with incident light, and the absorptions are the product of intimate or intraparticle mixtures. In these types of mixtures, the resulting spectra are a non-linear combination (e.g. Hapke 1993; Clark 1999). Moreover, the texture of volcanic rocks can also be characterized by iso-orientation of minerals, and the different relationships between phenocrysts, xenocrysts and the groundmass can affect the spectral signature. As a result, the spectral contrast is reduced and the known absorption structures are modified.

Detectable C.F. absorptions in ultrabasic–basic volcanic rocks

C.F. absorptions in rock-forming minerals. Rock-forming minerals (e.g. px, ol and pl) can be easily identified and used to differentiate volcanic products in the infrared portion of VNIR spectra. In addition, the spectral characteristics of these minerals and their mixtures are well known, and were reviewed in a number of papers (e.g. Crown & Pieters 1987; Cloutis & Gaffey 1991*a*, *b*; Sunshine & Pieters 1991; Burns 1993; Hiroi & Pieters 1994; Klima *et al.* 2007, 2008, 2011; Serventi *et al.* 2013*b*). All of these phases have characteristic C.F. absorptions related to the presence of Fe^{2+} in the crystal lattice (Burns 1993). In particular, mafic minerals have lower albedos and stronger absorptions than pl, which often has less than

1 wt% FeO. All of these minerals contribute to the albedo and absorption structures of rocks in the VNIR (see examples in Fig. 1).

Pyroxene is characterized by high spectral variability, which is an expression of great variability in the crystal structure, and which is, in turn, related to compositional variation. Pyroxenes can be divided into orthopyroxenes, characterized by an orthorhombic symmetry (Pbca to $P2_1/c$ structure, from high Mg to high Fe), and clinopyroxene, characterized by monoclinic symmetry (from $P2_1/c$ to $C2/c$ structure with increasing Ca) (Deer *et al.* 1992). In spectroscopy, pxs are characterized on the basis of M1 and M2 site occupancy of Fe^{2+}. Fe^{2+} prefers the M2 site in opx and in low-Ca cpx (Burns 1993), whereas it prefers the M1 site in cpx with intermediate and high Ca (and Fe) content. This behaviour means that pxs can be subdivided into two types with clearly distinct spectral signatures (Cloutis & Gaffey 1991*a*): type B, which includes phases with low–intermediate Ca content (low/intermediate Ca px) characterized by two well-defined absorptions at 1 and 2 µm; and type A, which includes high-Ca compositions (high-Ca px) characterized by a complex band at 1 µm and an absent or weak absorption at 2 µm (Cloutis & Gaffey 1991*a*; Schade *et al.* 2004). The Ca abundance is expressed by the wollastonite content (Wo%). Low-Ca px (Wo < 11%) shows the two C.F. absorption bands at approximately 0.92 and 1.90 µm, characteristic of Fe^{2+} in the M2 site. The centre of this absorption shifts from short to long wavelengths with increasing total FeO (Klima *et al.* 2007) or ferrosilite (fs%) content (Cloutis & Gaffey 1991*a*). In addition, the

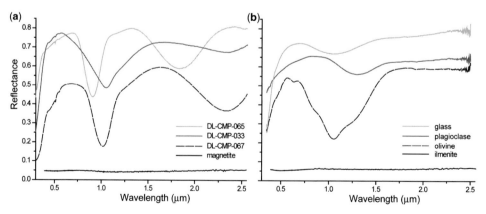

Fig. 1. Examples of mineral spectra. (**a**) pxs (opx DL-CMP-065, En_{90}; cpx DL-CMP-067, Wo_{39} En_{52} and DL-CMP-033, Wo_{49} En_{42}; particle size <40 µm) from Klima–RELAB library and magnetite from the USGS mineral library. (**b**) Anorthositic glass (FeO 1.8 wt%, g.s. particle size 20–50 µm; Carli *et al.* 2013), plagioclase (An80, FeO 0.5 wt%, 36–63 µm; Serventi *et al.* 2013*a*), olivine (Fo88, 36–63 µm; Serventi *et al.* 2013*a*) and ilmenite from the USGS library. In this box, the reflectance is shifted for clarity as follows: glass +0.1, pl +0.5 and ol −0.5.

iron content is responsible for a general decrease of albedo and the appearance of a band at about 1.2 μm (Klima *et al.* 2007).

Recently, Klima *et al.* (2011) discussed in detail the spectral variability of synthetic cpx with different Ca and Mg–Fe content. High-Ca pxs show a complex, wide and asymmetric absorption at 1 μm, and an absorption structure at approximately 1.2 μm (Klima *et al.* 2011). These authors stated that a minimum position around 1 μm could be seen for at least 2% of FeO in M2 (Klima *et al.* 2011). The band at 2 μm becomes weaker for very high wollastonite content (*c.* 50%) and is almost featureless for very high FeO content. For all cpx, increasing Fe^{2+} produces a shift in the positions of the absorption bands to longer wavelengths. An exception is high-Ca px, for which the 1 μm band position varies in a linear fashion with Wo% amount, independently of fs% (Klima *et al.* 2011).

Pyroxene C.F. absorptions were initially identified in transmittance spectra and modelled using a Gaussian distribution for each of the three different crystal orientations (Burns *et al.* 1972). Two dominant Gaussian peaks at approximately 1 and 2 μm, corresponding to Fe^{2+} absorption in the M2 site, were used for all compositions with the exception of hedembergitic compositions, which are characterized by two Gaussian peaks related to absorptions at about 0.98 and 1.2 μm, and the absence of the 2 μm absorption (Burns 1993 and references therein). In reflectance spectra, absorptions of mafic minerals have also been modelled by Gaussian distributions (e.g. MGM, Modified Gaussian Model: Sunshine *et al.* 1990; Sunshine & Pieters 1991). Two modified Gaussian peaks at 1 and 2 μm, associated with Fe^{2+} in the M2 site, plus a third Gaussian peak in the 1.2 μm region associated with absorption due to Fe^{2+} in the M1 site, are generally used (Sunshine *et al.* 1990; Klima *et al.* 2007). Klima *et al.* (2011) pointed out that for high-Ca px (e.g. augitic or diopsidic px), the 1 μm band can produce up to three different Gaussian peaks, two associated with FeO in the M1 site and one with FeO in the M2 site.

Olivine shows a wide composite absorption band in the 0.7–1.5 μm spectral range (Burns 1993) due to the presence of iron in both the M1 and M2 octahedral sites. Four different absorption positions have been identified for the different crystal orientations. For each orientation, the positions of the reflectance minima shift to longer wavelengths in a linear manner with increasing ferrous iron content (Cloutis *et al.* 1986; Burns 1993). Using an MGM-based deconvolution, Sunshine & Pieters (1998) produced a composite absorption band with three different Gaussian peaks, and reproduced the band centre shift observed by Burns (1993).

Plagioclase (pl) shows a clear absorption at around 1.25 μm, attributed to iron substituting for Ca^{2+} in the crystal lattice, that is detectable even for very low iron content (Cheek *et al.* 2011; Serventi *et al.* 2013b). Since pl spectra show higher albedo than those of mafic minerals, pl was generally considered a spectrally neutral phase. Moroz & Arnold (1999) discussed the effects of pl as a neutral component when mixed with an absorbing component. However, the almost 0.5 wt% FeO content of pl produces an intense absorption that can easily be recognized (Serventi *et al.* 2013b). Basaltic pl generally contains even higher levels of FeO (*c.* 1.0–1.5 wt%), which can substantially contribute to the reflectance spectra of effusive rocks.

Opaque minerals, particularly Fe–Ti oxide, are also present in ultrabasic and basic volcanic rocks. Oxides are generally a darkening agent in VNIR spectra, where they appear as a broad C.F. absorption band at approximately 1 μm due to FeO, and a weak absorption at about 0.6 μm due to Ti^{3+} (Burns 1993). Glass, another important component of effusive rocks, is generally characterized by an absorption structure at around 1.1 μm that is related to the presence of FeO (Bell *et al.* 1976; Dyarn & Burns 1981; Cloutis *et al.* 1990b; Burns 1993). In addition, glasses can also be darkening or reddening agents, as seen in the lunar glassy component (Gills-Davis *et al.* 2007, 2008; Tompkins & Pieters 2010) and in terrestrial volcanic rocks (Carli & Sgavetti 2011). The spectral variability of the glassy component is linked to composition and oxygen fugacity, although its effects on the identification of absorption structures of minerals in volcanic rocks have not been sufficiently discussed in the literature. In this paper we do not discuss hydrated mineral phases present as alteration or secondary phases seen in terrestrial rocks.

C.F. absorptions in mineral mixtures and volcanic rocks. In VNIR, volcanic rock-forming minerals absorb in a narrow spectral range in the 1 μm region (composite band) indicative of iron in M2 or M1 octahedral sites of mafic phases or in pl. There have been several studies focusing on the characterization of the spectral characteristics of mixtures between some of these compositions, by analysing mixtures of separate mineral phases of one grain size. Both the analysis of absorption band spectral parameters (reflectance, continuum slope and absorption spectral contrast) and band deconvolution by MGM (Sunshine *et al.* 1990) were used to establish possible relationships between spectral characteristics and compositions of the mineral mixtures. Distinct mixtures of two pxs (Cloutis & Gaffey 1991b), opx and ol (Cloutis *et al.* 1986; Moroz *et al.* 2000), pl and px (Crown &

Pieters 1987; Pompilio *et al.* 2007), and px and oxide (Cloutis *et al.* 1990a; Pompilio *et al.* 2007) were investigated, as well as a limited number of mixtures of three or more components (see Cloutis & Gaffey 1991b and references therein). These papers reported semi-quantitative trends relating spectral properties to mineral compositions. It was possible to deduce these relationships because the parameters of the spectral end members are known. Trends or variations were identified for px and ol mixtures, whereas mixtures with pl were considered in only a few cases (Crown & Pieters 1987; Pompilio *et al.* 2007; Serventi *et al.* 2013a, b).

The influence of pl in mixtures with mafic minerals was investigated in detail by Serventi *et al.* (2013a, b). These authors showed the unexpected spectroscopic effects of pl chemistry superimposed over the expected effects due to mafic minerals. The greatest spectral parameter variations are seen in mixtures containing pl with high volumetric FeO concentrations, which can increase with both pl abundance and/or FeO content in pl. Moreover, these spectral variations can even be correlated to the particle size of the material.

MGM-based deconvolution of composite bands has allowed for a quantitative evaluation of the different mineral compositions in mixtures. Opx and cpx absorptions have been clearly resolved from 1.0 and 2.0 μm composite bands of these mineral mixtures by modelling the components

due to Fe^{2+} in the M2 (at 1 and 2 μm) and M1 (at 1.2 μm) sites (Sunshine & Pieters 1993). In addition, a clear relationship was identified both between Gaussian peak positions and mineral compositions, and between band depths (intensity) and mineral abundances (Sunshine & Pieters 1993). Some papers have recently implemented a systematic approach using MGM for more complicated materials (e.g. Clenet *et al.* 2011), even if no papers have discussed in detailed the application of MGM to more complex composite bands due to the presence of high-Ca px, ol and pl, or more than two mineral phases.

Synthetic mixtures of minerals generally show spectral characteristics that are comparable with those of intrusive rock powders, in which crystals range in size from hundreds of microns to sizes visible to the naked eye (Fig. 2). In contrast, powder spectra of volcanic rocks show a lower reflectance and reduced spectral contrast than intrusive rock spectra for analogue grain-size ranges and similar bulk-rock compositions. In the example in Figure 3, a noritic sample and a basalt with similar SiO_2 content (see Table 1) show different bidirectional reflectance and spectral contrast but similar spectral signatures, which is in agreement with the similar amounts of pxs. The minima shifts are in agreement with the lower Ca content of noritic px.

The differences in reflectance and band intensity can be interpreted as a consequence of the increasing particle size of minerals (e.g. Craig

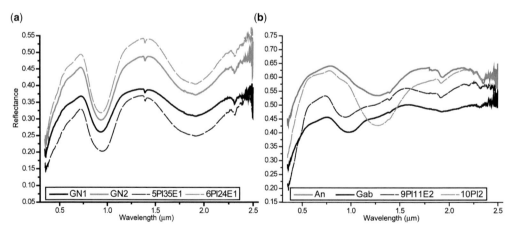

Fig. 2. An example of reflectance spectra from mineral mixtures, with particle size 250–125 μm (Serventi *et al.* 2013b), and rock powders, with grain size <250 μm (Carli 2009). Left box: GN1 55% pl (An80–FeO 0.36 wt%); 10% cpx ($En_{44}Wo_{46}$); 35% opx (En_{75}); GN2 61% pl (An81–FeO 0.37 wt%); 17% cpx ($En_{46}Wo_{45}$); 22% opx (En_{78}). These samples are spectrally similar (e.g. position and intensity) to mineral mixtures 5Pl35E1 and 6Pl24E1, respectively (50% Pl3, 50% E1 and 60% Pl2, 40% E1, see Serventi *et al.* 2013b). Right box: An (94% pl (An79–FeO 0.45 wt%), 4% cpx ($En_{42}Wo_{45}$)); Gab (81% pl (An79–FeO 0.36 wt%), 16% cpx ($En_{42}Wo_{45}$); 3% opx (En_{82})). These samples are spectrally similar (e.g. in terms of position and intensity) to mineral mixtures 10Pl2 and 9Pl11E2, respectively (100% Pl2 and 90% Pl1, 10% E2; see Serventi *et al.* 2013b). For clarity, the reflectance of the 6Pl24E1 and An spectra have been shifted, by +2.5% and 5%, respectively.

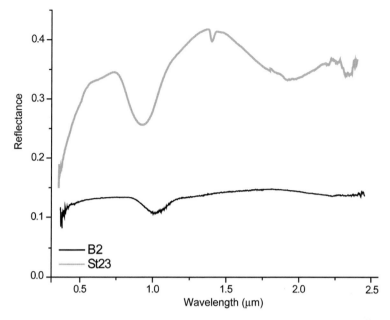

Fig. 3. An example of an intrusive cumulate mafic rock (St23 from Carli 2009) and an effusive basalt (B2; Carli & Sgavetti 2011). Both samples are characterized by px absorptions but the spectral contrast and reflectance are reduced in the effusive basalt despite similar high amount of pxs (low-Ca in St23 and intermediate-Ca in B2).

et al. 2008) and in cut-rock (slab) spectra (Sgavetti et al. 2006; Pompilio et al. 2007) rather than due to different mineralogical associations.

Spectroscopic effects of composition and petrographical parameters. In regolith, the presence of different lithologies in the same image pixel, as well as variations in grain size (from very fine powder to outcrop scale) and compositions, can influence the spectral signature in an unpredictable way. In particular, if the original rock had an

aphanitic texture with a groundmass wholly characterized by crystals a few tens of microns in size, even fine-grained regoliths are spectrally controlled by multi-crystal grains that each reflect the characteristics of the rock texture (Carli & Sgavetti 2011). In such material, spectral behaviour is controlled by the effect of optical coupling (Hapke 1993 and references therein). This phenomenon is caused by the extremely close proximity of the particles, which has considerable implications for both grains and rock slabs. In this subsection we

Table 1. *XRF analysis of major elements of rocks presented in Figures 3–5*

	A2	Sal22	Et02	Et26	B2	St23	Et12	Et13	Et14	Et15
SiO_2	48.79	48.40	46.97	47.99	49.29	50.44	46.90	46.55	46.35	47.09
TiO_2	1.54	1.84	1.80	1.57	2.85	0.19	1.78	1.77	1.79	1.79
Al_2O_3	14.89	16.21	16.77	18.41	13.45	17.29	16.82	17.07	16.39	16.85
Fe_2O_{3tot}	11.95	11.86	11.69	10.30	14.82	11.24	11.47	11.35	11.62	11.51
MnO	0.19	0.16	0.19	0.17	0.22	0.14	0.19	0.19	0.19	0.19
MgO	7.53	7.66	5.35	4.59	5.60	9.75	5.63	5.26	5.82	5.64
CaO	12.75	8.95	10.20	10.18	10.19	8.76	10.40	10.03	10.45	10.47
Na_2O	2.04	3.45	3.53	3.88	2.71	1.10	3.54	3.75	3.52	3.64
K_2O	0.15	0.87	1.95	1.49	0.43	0.38	1.83	1.93	1.81	1.88
P_2O_5	0.13	0.35	0.57	0.54	0.28	0.02	0.54	0.55	0.52	0.54
Total	99.96	99.75	99.02	99.12	99.84	99.31	99.10	98.45	98.46	99.60

These measurements were made using an X-ray fluorescence spectrometer at the XRF laboratory at the Geoscience Department of the University of Padova.

discuss how it can influence a measured spectral signature.

- *Texture effects in wholly crystalline rock samples: differences between rock slab and powders.* The influence of original rock texture on the spectral features of rock powders can be clearly observed if the grain size of the powdered material is larger than the crystal size of the original rock. In this case, each grain is a multi-crystal grain and will spectrally behave like the original rock. For intrusive rock, a rock slab sample larger than the illuminated spot (which must, in turn, be larger than the individual crystals) will have a spectrum affected by the spectroscopic interaction of electromagnetic energy with the rock component minerals (i.e. single crystals) along the optical path of the light. This generally results in a blue slope, and substantially reduced reflectance and spectral contrast, which further decrease at longer wavelengths resulting in the suppression of some diagnostic absorptions (Yon & Pieters 1988; Pompilio *et al.* 2007). In contrast, the same rock powder sample with grain size close to or smaller than the crystal size will behave similarly to an intimate mixture of single mineral particles, producing a spectrum in which the spectral bands of the distinct phases can be recognized.

The change in slope between slab and powder rock sample spectra, although well documented by Harloff & Arnold (2001) and Pompilio *et al.* (2007) for both effusive and intrusive rocks, is still unexplained. Increasing opaque mineral concentrations, variation in iron-bearing silicate abundances and alteration due to weathering of some minerals (e.g. serpentinization) have been proposed as possible causes of the reduction in compositional information to the point of producing featureless spectra for intrusive slab samples (Carli & Sgavetti 2011). Moreover, rocks with very different compositions can converge to very similar spectral features, whereas powders of the same rocks can display clearly distinct spectral characteristics. Pompilio *et al.* (2007) also stated that absorption minima in slab spectra shift to slightly higher wavelengths with respect to those of powder spectra, relating this shift to the strongly negative slope observed in slab spectra. Carli *et al.* (2012) clearly showed that it is possible to distinguish the 1 μm absorption bands due to px in both powder and slab spectra of different rocks belonging to the gabbro−norite series by applying Gaussian models (e.g. MGM). The absorption positions are also comparable with those measured for pure minerals, apart for slight differences in positions between powders and

slabs (Cloutis & Gaffey 1991*a*). In addition, the intensity of the absorption band in slab samples is reduced, whereas attenuation increases with the iron content.

- *Texture effects in wholly crystalline rock samples: effects related to grain size and roughness.* Regolith crystal and grain size can play a crucial role in the interpretation of absorption features in the spectra of effusive rock samples. However, Figure 4 shows that effusive rocks with different crystal sizes, but with holocrystalline texture and similar bulk-rock compositions, can have different spectral characteristics. The spectra in Figure 4 were acquired from slab samples at S.Lab. (Spectroscopy Laboratory, at IAPS−INAF in Rome), using a Fieldspec Pro mounted on a goniometer, with incidence angle $(i) = 30°$, emission angle $(e) = 0°$ and an illuminated spot of approximately 0.5 cm², at standard conditions. The samples were illuminated with a QTH (quartz tungsten halogen) lamp (see Carli & Sgavetti 2011 for more details on sample preparation and laboratory set-up).

The grain-size effect in single mineral phases is well studied (e.g. Craig *et al.* 2008), whereas only a few papers have addressed the effects of grain size in rock powders and roughness variation in slabs. Harloff & Arnold (2001) measured the specular reflectance for both different pxs and for basalts, and related the reflectance, the continuum and the band depth to the powder grain size and to slab roughness in all the samples. These authors also showed that reflectance increases with decreasing grain size and roughness, whereas the continuum and the band depth increase with increasing grain size and roughness up to a maximum before dropping off. Carli & Sgavetti (2011) discussed spectral characteristics measured in bidirectional reflectance with $i = 30°$ and $e = 0°$ for basaltic samples with similar textures and different compositions, as well as for basaltic samples with different textures and similar compositions. For all samples, the slab spectra showed the lowest reflectance, although samples with larger grain sizes (<2.00 mm) displayed spectral characteristics (including reflectance, slope and band intensity) very similar to those of the slab spectra. These results suggest that rock-related optical coupling (Hapke 1993) must be taken into account even for regolith with grain sizes on the order of millimetres.

It is therefore evident that, for effusive rock powders with a grain size that is comparable to crystal size, the spectral signature is controlled by mineralogy and characterized by a reflectance that varies as a function of particle size. However, if the crystal size is smaller than the

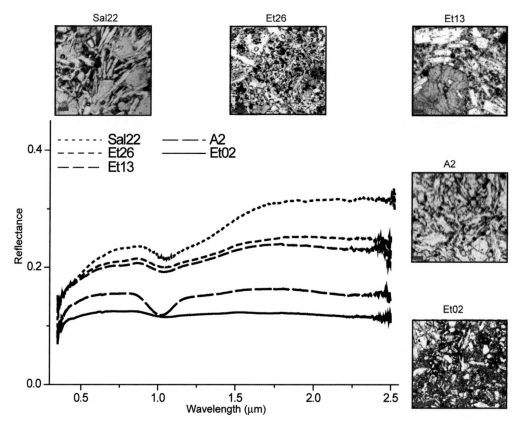

Fig. 4. Volcanic rocks with very similar bulk-rock compositions (see Table 1) but different crystal sizes and/or P.I. and/or relative mineral abundances. In the thin-section images, spectral variation depends not only on composition but also on the different volcanic conditions that generate mineral and texture variations. The A2 and Sal22 samples are from Iceland and Payun Matru (Argentina), respectively (see fig. 1 in Carli & Sgavetti 2011). The Et02 sample is from Etna's 2002 lava flow, Et26 is from Etna's 1665 lava flow and Et13 is from Etna's 1983 lava flow. In all thin section images, the red scale bar denotes 200 μm.

maximum grain size, a large number of mixed grains are expected to be present, resulting in spectral information that is strongly affected by optical coupling.

- *Texture effects in wholly crystalline rock samples: influence of pheno- and xenocryst.* Another important factor to consider when interpreting the spectra of effusive rocks is the relationship between groundmass, phenocrysts and xenocrysts. Phenocrysts are minerals that crystalize in equilibrium with magma, generally in the deep crust. Xenocrysts, however, are minerals that have been dislodged from the walls of the magma conduit during magma ascent. The % abundance of phenocrysts with respect to the groundmass defines the porphyritic index (P.I.) of a lava, an important parameter that indicates the degree of crystallization of the rising magma.

From a spectral point of view, lavas with a porphyritic texture (high P.I.) are expected to have a reflectance and a spectral contrast that are very close to those of intrusive rocks. Unfortunately, no studies addressing this topic in detail have yet been published. Carli & Sgavetti (2011) pointed out that fine-grained rock samples with a small percentage of millimetre-size phenocrysts have spectral signatures dominated by absorptions characteristic of the mineral phases composing the phenocrysts. In contrast, spectra of the same rocks taken from coarse-grained powder samples are dominated by the groundmass. This is probably due to a pre-eminence of multi-crystal grains in coarser-grained samples, which produce a strong signal that greatly subdues or completely masks the phenocryst signature.

- *Texture effects in partly amorphous rock samples: how groundmass, glassy component and crystal size work.* The groundmass composition and crystallinity of volcanic rocks can strongly affect the mineralogical information that we can retrieve from reflectance spectra. The ground-mass is the fine-grained matrix within which larger crystals are embedded. Mineral crystal size can vary from a few microns to tens of microns or larger. Glassy material can also be an important component because it defines the groundmass texture, which can be holocrystal-line, hypocrystalline, vitrophyric or holohyaline (in order of increasing glass content). More-over, this component is frequently characterized by oxide microphases (tachylitic glass), which are unresolvable even under high-magnification optical microscopy. Both the glassy component and the fine-grained groundmass contribute to a general darkening of the effusive samples and a lowering in the intensity of C.F. absorp-tions, at times resulting in featureless spectra, which complicates the spectroscopic analysis of these rocks. However, Carli & Sgavetti (2011) pointed out that two basaltic samples with simi-lar compositions but different groundmass tex-tures have different spectral characteristics in the VNIR.

Here, we discuss in detail some effects of rock texture in four different samples belonging to a Mt Etna lava flow from the 1983 event. The spectra of rock powder samples (<0.250 mm) are shown in

Figure 5. Powders were prepared by grinding and sieving representative rock portions in a 0.250 mm sieve. The spectra were acquired with the same experimental set-up as that described for Figure 4 (see Carli & Sgavetti 2011 for more details on sample preparation and laboratory set-up). The original rock samples were collected at different locations along a vertical section going from the inner to the outer part of a 1.5 m-thick lava flow. The four samples (Et13, Et14, Et12 and Et15 from in to outside) show similar mineral assemblages: pl, cpx, minor ol and oxides. The groundmass varies from holocrystalline to hyalopilitic and vitro-phyric, with a tachylitic glass.

The bulk-rock compositions reported in Table 1 show homogeneous major-element chemistry. Table 2 and Figure 6 give the compositions of the major mineral phases, and show overlapping mineral chemistries of all the samples, with small variations between phenocrysts and microcrystals. Clinopyroxene is a diopsidic augite with very little variation in the Wo_{44-46} and En_{37-40} components, whereas pl has a compositional range that varies from An50 to An80, with average FeO close to 0.7 wt% for all the samples. Olivine compositions are also very similar: Fo70 for all of the samples except Et12, which is slightly richer in Mg (Fo77). SEM images of thin sections were collected to determine the P.I. and to qualitatively describe the

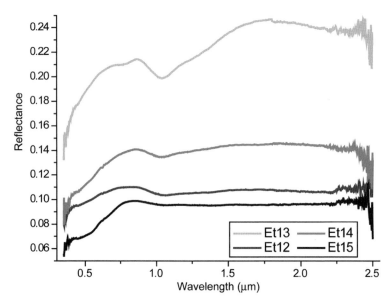

Fig. 5. Spectra of four different samples from a Mount Etna lava flow. The samples have very similar P.I., with similar relative mineral associations, but with different groundmass textures. The intensity of the C.F. composite band, which is due to mafic and pl mineralogy, is strongly reduced from sample Et13 to Et15, increasing the tachylitic glass content in the groundmass of these samples (see also Fig. 6 and Tables 2 & 3). These spectra were acquired with a FieldspecPro® $i = 30°$, $e = 0°$, white Spectralon standard Labsphere®, QTH lamp, at SLAB, IAPS–INAF, Rome.

Table 2. *Average composition for the major mineral phases present as pheno- and microcrysts*

	Clinopyroxene				Olivine					Plagioclase			
	Et12	Et13	Et14	Et15	Et12	Et13	Et14	Et15		Et12	Et13	Et14	Et15
SiO_2	48.21	48.60	46.98	48.34	38.72	37.35	37.60	37.94		51.92	51.57	50.58	50.93
TiO_2	1.60	1.65	2.07	1.63	0.05	0.03	0.04	0.03		0.10	0.09	0.08	0.10
Al_2O_3	4.53	4.70	4.81	5.36	0.03	0.31	0.02	0.02		28.79	29.76	30.58	29.67
Cr_2O_3	0.02	0.02	0.01	0.01	0.00	0.01	0.01	0.02		0.01	0.02	0.01	0.01
FeO_{tot}	8.57	8.66	9.97	8.05	20.42	27.71	26.12	25.33		0.70	0.68	0.62	0.70
Mno	0.19	0.21	0.21	0.23	0.32	0.67	0.60	0.60		0.02	0.01	0.01	0.01
MgO	13.26	13.15	12.94	13.63	40.56	33.78	36.14	37.37		0.08	0.07	0.07	0.09
CaO	21.73	21.46	20.88	21.84	0.26	0.49	0.35	0.35		11.92	12.45	13.29	13.11
Na_2O	0.59	0.63	0.63	0.59	0.01	0.06	0.02	0.02		4.24	4.06	3.67	3.73
K_2O	0.05	0.03	0.03	0.02						0.61	0.37	0.29	0.37
Total	99.40	99.44	99.08	100.15	100.37	100.42	100.91	101.69		98.39	99.09	99.19	98.71
Wo	46.37	46.15	44.76	46.41					Ab	37.73	36.25	32.73	33.21
En	39.34	39.29	38.60	40.23	Fo	77.95	68.14	71.09	72.45	An	58.67	61.56	65.56	64.61
Fs	14.29	14.56	16.64	13.36	Fa	22.05	31.86	28.91	27.55	Or	3.60	2.19	1.71	2.18

The mineral chemistry was determined by electron microprobe analyses with a CAMECA SX50 (EMP) at the microprobe laboratory of CNR–IGG, Padova.

groundmass characteristics over a representative area of each sample. Phenocryst abundances were calculated from compositional maps collected for major elements (Si, Al, Ca, Na, Mg, Fe and Ti). Red–green–blue (RGB) images and single-channel images were generated using the ENVI® software, and phenocrysts with different mineral chemistries were distinguished from the sample's cavity by processing the images with ImageJ®. The areal abundances of each phase were calculated and are reported in Table 3. The P.I. varies from 36 to 41%. Moreover, pl abundances are 21–27%, whereas those of mafic minerals, composed almost entirely of cpx (*c.* 90%), have abundances of

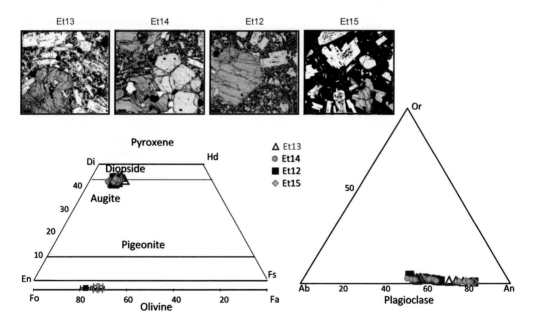

Fig. 6. The px, ol and pl composition of Etna samples (see also Table 3). Thin-section images show the groundmass texture variation from holocrystaline (Et13) to vitrophyric (Et15). As before, the red scale bar in the thin section images denotes 200 μm.

Table 3. *Areal abundances of phenocrysts present on Etna samples and holes*

	Pl (%)	Cpx + ol (%)	Ox (%)	Hole (%)	Σ (PhX%)
PhX _Et13	26.91	11.26	1.12		39.28
PhX _Et14	22.10	12.35	0.90	1.50	35.35
PhX _Et12	21.00	13.46	1.83	5.10	36.28
PhX _Et15	24.47	16.11	0.81	8.07	41.39

Holes indicate sample cavity. The rest is the groundmass, which varies from all microcrystalline to very high glass component from Et13 to Et15. A threshold of 20 pixels was used as a minimum area to be considered.

11–16%. All of the samples show a very similar phenocryst abundance that is consistent with the spectral variation observed, which is mainly controlled by differences in the groundmasses.

In the sample spectra (Fig. 5), despite identical mineralogical and bulk compositions and similar phenocryst distributions, reflectance varies from approximately 22 to 9% at 0.8 μm. In sample Et13, the wide absorption centred at 1.037 μm is consistent with the presence of cpx and ol. Samples Et14 and Et12 also show a minimum in a similar position, but the spectral contrast between the shoulders and the minimum is greatly reduced. Finally, Et15 shows a very weak, almost featureless, band at wavelengths longer than 0.80 μm, with a minimum at 1.05 μm. The 1 μm Band Area was calculated as the integrated area delimited by a linear continuum between the band onset and offset. The Band Area varies from 4% for sample Et13, to 2% for Et14 and Et12, and 1% for Et15, indicating a clear reduction of spectral information due to the glassy component of the groundmass.

Discussion

Several papers have been published that focus on spectral reflectance characteristics of rock-forming minerals (e.g. px, ol and pl) from effusive rocks, with some also discussing the spectral characteristics of these bulk rocks. Several points must still be clarified, however, to allow for a better interpretation of the information contained in remote-sensing data. On Mars, ol or px spectroscopic indexes of regions with volcanic edifices have not been mapped despite well-established evidence for the presence of basalts on the planet. In contrast, the spectra of lunar maria and the surface of Vesta clearly show the presence of px absorptions but not the presence of ol, although ol eucrite and ol lunar basalts are well documented by rock sample and meteorite analyses. However, compositional interpretation has been generally based on spectroscopic criteria that do not take into consideration the specific petrographical characteristics of volcanic rocks. Basic–ultrabasic volcanic rocks

are, in fact, characterized by relatively high levels of mafic minerals with variable amounts of pl, which increase from ultrabasic to basic compositions. However, texture can influence spectra by modifying the spectral slope and reducing or extinguishing band intensities, which facilitates the recognition of some mineral phases over others.

Spectral reflectance studies of Earth's basic–ultrabasic rocks have contributed to a substantial improvement in the understanding of this topic. Spectral knowledge of these rocks to date, from literature and this study, can be summarized as follows:

- For volcanic rocks with holocrystalline texture:
  - the presence of the 1 and 2 μm bands is indicative of the occurrence of px, whereas a wide composite band at 1 μm indicates the presence of ol. The absence of the 2 μm band in a spectrum can be interpreted as the result of the presence of high-Ca px ± ol (e.g. Hunt *et al.* 1974; Harloff & Arnold 2001; Carli & Sgavetti 2011). Similar bulk-rock chemical compositions have spectra with different absorption features (e.g. Figs 4 & 5) because different magma chamber depths, crystal fractionation and cooling histories lead to different mineral abundances and/or chemistries (i.e. the presence of px with different composition and/or ol, as ground mass and/or phenocrysts);
  - the spectra of powders of effusive rocks with groundmass crystal sizes of hundreds of microns are similar to those of intrusive rock powders, and are also similar to those of synthetic mixtures of minerals (i.e. to gabbro or to gabbronorite ± ol) (see the subsection on 'C.F. absorptions in mineral mixtures and volcanic rocks'). If the crystals in the groundmass are tens of microns in size, even the fine-grained powders will contain multi-crystal grains. This results in a reduction in albedo and in spectral contrast, although the predominant mafic mineralogy can still be resolved (e.g. see Et13 in Fig. 5: see also Carli & Sgavetti 2011);

- the P.I. can also play an important role: with increasing P.I., phenocryst absorptions become more apparent, and mafic minerals with abundances of just a few per cent dominate the spectra of intermediate- to fine-grained powders (see the subsection on 'Spectroscopic effects of composition and petrographical parameters': see also Carli & Sgavetti 2011). In contrast, when the powder grain size is considerably larger than the crystal size, optical coupling strongly influences the spectral signature and so the groundmass dominates the spectral characteristics (see 'Spectroscopic effects of composition and petrographic parameters').

- The presence of glass (i.e. tachylite) in a groundmass substantially modifies its petrographical texture and strongly affects the spectral information, as reported by Carli & Sgavetti (2011) and discussed in detail in the subsection on 'Spectroscopic effects of composition and petrographical parameters'. The sample set discussed there comes from the same lava flow, and is thus characterized by very similar P.I., mineralogical associations and identical bulk-rock chemical compositions. Nevertheless, with increasing glass contents in the sample groundmass, from low (e.g. Et14: hyalopilitic groundmass) to high (e.g. Et15: vitrophyric groundmass) glass, the spectral shape is modified from a low-reflectance spectrum with a strongly reduced 1 μm band to a darker, almost featureless spectrum.

As mentioned above, remotely acquired spectral reflectance data are often analysed using specific indexes to identify diagnostic absorptions. However, these indexes can be difficult to apply when several compositional and petrographical factors in outcropping rocks and in regolith interplay, resulting in a reduction of the mineralogical information contained in the VNIR reflectance spectra. In order to characterize the composition of an area, VNIR reflectance spectroscopy should be used in concert with geomorphological studies to identify end-member geomorphic units over which spectral units containing the mineralogical information can be draped, so as to substantially increase the geological significance of the resulting geomorphic–spectral units (e.g. Giacomini et al. 2012). Future missions will carry hyperspectral sensors for in situ rock analyses (e.g. the Ma_MISS (Corradini et al. 2011) and MicrOmega (Pilorget et al. 2012) instruments on the ExoMars 2018 rover). Any spectroscopic data acquired by future rover missions will be strongly affected by the rock petrographical characteristic of the surface rocks, particularly by volcanic rocks. Since these rocks are expected to be the most abundant materials on terrestrial planets, more studies aimed at exploring the effects of texture on mineralogy at different spatial resolutions are needed in order to better interpret spectral data collected both from orbit and by rovers. As a final consideration, improvements in the spectroscopic characterization of volcanic rocks, by taking into consideration different chemical compositions and different mineral assemblages, will help further inform the analysis and identification of effusive lithologies using both direct (e.g. specific absorption bands) and indirect (e.g. typical reflectance, spectral slope and mineral alterations) information.

Implications

The surfaces of inner solar system bodies are extensively covered by volcanic products with variable compositions and petrographical characteristics that can be analysed using high-resolution spectroscopic data. VNIR absorptions permit the identification of some of the spectroscopically dominant minerals in the rocks. This allows us to discriminate px-rich (pigeonitic or augitic) and ol-rich volcanic rock compositions within basic–ultrabasic effusive materials with holocrystalline groundmasses. The relationship between the 1 and 2 μm bands can also be helpful, by signalling the presence or absence of low- to intermediate-Ca px. Reflectance and spectral contrast can be directly related to the P.I. and the crystal size of the groundmass. For samples with a very fine groundmass and with tachylitic glass, the spectral absorptions are reduced to a featureless spectrum, masking the expected absorptions of the mineral phases.

Because of the improved spatial resolution of remote-sensing data, and the use of hyperspectral sensors on rovers to investigate volcanic rocks in greater detail, further work should be focused on finding both direct and indirect evidence of compositional evolution of magmatic processes. To accomplish this, as well as to better explore planetary volcanic regions using VNIR spectroscopy and to discriminate among the effusive products, major effort should be made to:

- understand the 'abundance limits' for mafic mineralogies (by separating e.g. ol-poor from ol-rich rocks, alkaline from subalkaline rocks, and iron-poor from iron-rich magmatism);
- understand the limit (e.g. the chemistry and relative abundance) beyond which pheno- or xenocrystal compositions have a greater effect on the spectral signature than groundmass mineralogy – for example, future work should investigate how the spectral signature varies between massive basalts with similar compositions and

mineralogical assemblages, but different P.I., in order to understand how spectral information is influenced by different crystal distribution in the rock; and

- investigate the relationship between crystal size and grain size in particulate soils (like regoliths) derived from effusive igneous rocks in order to determine the most suitable conditions for obtaining diagnostic spectral information that would be more reliable than information provided by synthetic mixtures of minerals, which are more comparable with intrusive igneous rocks.

These efforts could produce a standard of information for volcanic compositions from which to begin the investigation of weathering effects on the spectra of effusive igneous rocks. We suggest that integrating and comparing geomorphic volcanic units and spectral units of compositional significance is the most reliable approach for geological mapping of volcanic regions on planetary surfaces, as already demonstrated by Giacomini *et al.* (2012).

The authors would like to thank L. Peruzzo for a useful introduction to the use of the SEM at the GeoScience Department of the University of Padua. Financial support was perovided by Agenzia Spaziale Italiana, SIMBIO-SYS project. The authors would also like to thank two anonymous reviewers, and P. K. Byrne and T. Platz, for their helpful suggestions that improved this manuscript.

References

ABRAMS, M., ABBOTT, E. & KAHLE, A. 1991. Combined use of visible, reflected infrared, and thermal infrared images for mapping Hawaiian lava flows. *Journal of Geophysical Research*, **96**, 475–484.

ADAMS, J. B. & McCORD, T. B. 1971. Optical properties of mineral separates, glass, and anorthositic fragments from Apollo mare samples. *In*: *Proceedings of the 2nd Lunar Science Conference, Volume 3*. MIT Press, Cambridge, MA, 2183–2195.

ADAMS, J. B., PIETERS, C. M. & McCORD, T. B. 1974. Orange Glass: evidence for regional deposits of pyroclastic origin on the Moon. *In*: *Proceedings of the 5th Lunar Science Conference, Houston, Texas, March 18–22, 1974, Volume 1*. Pergamon Press, New York, 177–186.

AMMANNITO, E., DE SANCTIS, M. C. *ET AL.* 2013. Olivine in an unexpected location on Vesta's surface. *Nature*, **504**, 122–125, http://dx.doi.org/10.1038/nature12665

BANDFIELD, J. L., HAMILTON, V. E. & CHRISTENSEN, P. R. 2000. A global view of Martian surface compositions from MGS-TES. *Science*, **287**, 1626–1630.

BELL, P. M., MAO, H. K. & WEEKS, R. A. 1976. Optical spectra and electron paramagnetic resonance of lunar and synthetic glasses: a study of the effects of

controlled atmosphere, composition, and temperature. *In*: *Proceedings of the 7th Lunar Science Conference, Houston, Texas, March 15–19, 1976, Volume 3*. Pergamon Press, New York, 2543–2559.

BOYNTON, W. V., TAYLOR, G. J. *ET AL.* 2007. Concentration of H, Si, Cl, K, Fe, and Th in the low- and mid-latitude regions of Mars. *Journal of Geophysical Research*, **112**, E12S99, http://dx.doi.org/10.1029/2007JE002887

BRUNETTO, R. & STRAZZULLA, G. 2005. Elastic collisions in ion irradiation experiments: a mechanism for space weathering of silicates. *Icarus*, **179**, 265–273.

BURNS, R. G. 1993. *Mineralogical Applications of Crystal Field Theory*. Cambridge University Press, Cambridge.

BURNS, R. G., ABU-EDI, R. M. & HUGGINS, F. E. 1972. Crystal field spectra of lunar pyroxenes. *In*: *Proceedings of the 3rd Lunar Science Conference, Volume 1*. MIT Press, Cambridge, MA, 533–543.

BYRNE, P. K., KLIMCZAK, C. *ET AL.* 2013. An Assemblage of lava flow features on Mercury. *Journal of Geophysical Research*, **118**, 1303–1322, http://dx.doi.org/10.1002/jgre.20052

CARLI, C. 2009. *Analisi spettroscopica nel VNIR di rocce ignee: Caratterizzazione composizionale della superficie dei pianeti terrestri*. PhD thesis, University of Parma.

CARLI, C. & SGAVETTI, M. 2011. Spectral characteristics of rocks: effects of composition and texture and implications for the interpretation of planet surface compositions. *Icarus*, **211**, 1034–1048.

CARLI, C., SGAVETTI, M., CAPACCIONI, F. & SERVENTI, G. 2012. Studying Spectral Variability of an Igneous Stratified Complex as a Tool to Maps Lunar Highlands. Abstract 9007, presented at the Second Conference on the Lunar Highland Crust, Bozeman, Montana, July 13–15, 2012.

CARLI, C., ROUSH, T. & CAPACCIONI, F. 2013. Retrieving Optical Constants of Glasses with Variable Iron Abundance. Abstract 1918, presented at the 44th Conference on Lunar Planetary Science, Woodlands, Texas. March 18–22, 2013.

CHEEK, L. C., PIETERS, C. M., PARMAN, S. W., DYAR, M. D., SPEICHER, E. A. & COOPER, R. F. 2011. Spectral Characteristics of Pl with Variable Iron Content: Application to the Remote Sensing of the Lunar Crust. Abstract 1617, presented at the 42nd Conference on Lunar Planetary Science, Woodlands, Texas, March 7–11, 2011.

CHRISTENSEN, P. R., BANDFIELD, J. L. *ET AL.* 2000. Detection of crystalline hematite mineralization on Mars by the Thermal Emission Spectrometer: evidence for near-surface water. *Journal of Geophysical Research*, **105**, 9623–9642.

CLARK, R. N. 1999. Chapter 1: Spectroscopy of rocks and minerals, and principles of spectroscopy. *In*: RENCZ, A. Z. (ed.) *Manual of Remote Sensing, Volume 3, Remote Sensing for the Earth Sciences*. Wiley, New York, 3–58.

CLENET, H., PINET, P., DYDOU, Y., HEURIPEAU, F., ROSEMBERG, C., BARATOUX, D. & CHEVREL, S. 2011. A new systematic approach using the Modified Gaussian Model: insight for the characterization of chemical composition of olivines, pyroxenes and olivine–

pyroxene mixtures. *Icarus*, **213**, 404–422, http://dx. doi.org/10.1016/j.icarus.2011.03.002

CLOUTIS, E. A. & GAFFEY, M. J. 1991a. Pyroxene spectroscopy revisited: spectral-compositional combinations and relationships to geothermometry. *Journal of Geophysical Research*, **96**, 22 809–22 826.

CLOUTIS, E. A. & GAFFEY, M. J. 1991b. Spectral compositional variations in the constituent minerals of mafic and ultramafic assembleages and remote sensing implications. *Earth, Moon, and Planets*, **53**, 11–53.

CLOUTIS, E. A., GAFFEY, M. J., JACKOWSKI, T. L. & REED, K. L. 1986. Calibrations of phase abundance, composition, and particle size distribution of Olivine-Orthopyroxene mixtures from reflectance spectra. *Journal of Geophysical Research*, **91**, 11 641–11 653.

CLOUTIS, E. A., GAFFEY, M. J., SMITH, D. G. W. & LAMBERT, R. S. J. 1990a. Reflectance spectra of mafic silicate-opaque assemblages with applications to meteorite spectra. *Icarus*, **84**, 315–333.

CLOUTIS, E. A., GAFFEY, M. J., SMITH, D. G. W. & LAMBERT, R. S. J. 1990b. Reflectance spectra of glass-bearing mafic silicate mixtures and spectral deconvolution procedures. *Icarus*, **86**, 383–401.

CORRADINI, A., AMMANNITO, E. ET AL. 2011. Ma_Miss Experiment: Miniaturized Imaging Spectrometer for Subsurface Studies. Abstract 1125, presented at the EPSC–DPS 6 Joint Meeting, Nantes, France, 2–7 October 2011.

CRAIG, M. A., CLOUTIS, E. A., REDDY, V., BAILEY, D. T. & GAFFEY, M. J. 2008. The Effects of Grain Size, <10 μm–4.75 mm, on the Reflectance Spectrum of Planetary Analogs from 0.35–2.5 μm. Abstract 1356, presented at the 38th Conference on Lunar Planetary Science, League City, Texas, March 12–16, 2007.

CROWN, D. A. & PIETERS, C. M. 1987. Spectral properties of PL and pyroxene mixtures and the interpretation of lunar soil spectra. *Icarus*, **72**, 492–506.

DE SANCTIS, M. C., AMMANNITO, E. ET AL. 2012. Spectroscopic characterization of mineralogy and its diversity across Vesta. *Science*, **336**, 697–700, http://dx.doi.org/10.1126/science.1219270

DEER, W. A., HOWIE, R. A. & ZUSSMAN, J. 1992. *An Introduction to the Rock-Forming Minerals*. Longman, Harlow.

DENEVI, B. W., ERNST, C. M. ET AL. 2013. The distribution and origin of smooth plains on Mercury. *Journal of Geophysical Research*, **118**, 891–907, http://dx.doi.org/10.1002/jgre.20075

DYAR, M. D. & BURNS, R. G. 1981. Coordination chemistry of iron in glasses contributing to remote sensed spectra of the moon. *In: Proceedings of the 12th Lunar Science Conference, Houston, Texas, March 16–20, 1981*. Pergamon Press, New York, 695–702.

FOLEY, C. N., ECONOMOU, T. & CLAYTON, R. N. 2003. Final chemical results from the Mars Pathfinder alpha proton X-ray spectrometer. *Journal of Geophysical Research*, **108**, 8096.

GIACOMINI, L., MASSIRONI, M., MARTELLATO, E., PASQUARÈ, G., FRIGERI, A. & CREMONESE, G. 2009. Inflated flows on Daedalia Planum (Mars)? Clues from a comparative analysis with the payen volcanic complex (Argentina). *Planetary Space Science*, **57**, 556–570.

GIACOMINI, L., CARLI, C., MASSIRONI, M. & SGAVETTI, M. 2012. Spectral analysis and geological mapping of the Daedalia Planum lava field (Mars) using OMEGA data. *Icarus*, **220**, 679–693, http://dx.doi.org/10.1016/j.icarus.2012.06.010

GIFFORD, A. W. & EL-BAZ, F. 1978. Thickness of mare flow fronts. *In: Proceedings of the 9th Lunar Planetary Science Conference, Houston, Texas, March 13–17, 1978*. Lunar and Planetary Institute, Houston, TX, 382–384.

GILLIS-DAVIS, J. J., LUCEY, P. G., HAMMER, J. E. & WILCOX, B. B. 2007. Syntheses and Reflectance Analyses of Lunar Red Glass Compositions: Information to Improve Understanding of Remotely Sensed Spectral Data. Abstract 1443, presented at the 38th Conference on Lunar Planetary Science, League City, Texas, March 12–16, 2007.

GILLIS-DAVIS, J. J., LUCEY, P. G., HAMMER, J. E. & DENEVI, B. B. 2008. Syntheses and Reflectance Analyses of Lunar Green Glass Compositions: Information to Improve Understanding of Remotely Sensed Spectral Data. Abstract 1535, presented at the 39th Conference on Lunar Planetary Science, League City, Texas, March 10–14, 2008.

GROVES, D. I., KORKIAKOSKI, E. A., MCNAUGHTON, N. J., LESHER, C. M. & COWDEN, A. 1986. Thermal erosion by komatiites at Kambalda, Western Australia and the genesis of nickel ores. *Nature*, **319**, 136–137.

HAMILTON, V. E., WYATT, M. B., MCSWEEN, H. Y., JR. & CHRISTENSEN, P. R. 2001. Analysis of terrestrial and martian volcanic compositions using thermal emission spectroscopy: 2. Application to martian surface spectra from the Mars Global Surveyor Thermal Emission Spectrometer. *Journal of Geophysical Research*, **106**, 14 733–14 746.

HAPKE, B. 1993. *Theory of Reflectance and Emittance Spectroscopy*. Topics in Remote Sensing, **3**. Cambridge University Press, Cambridge.

HAPKE, B. W. 2001. Space weathering from Mercury to the asteroid belt. *Journal of Geophysical Research*, **106**, 10 039–10 073.

HARLOFF, J. & ARNOLD, G. 2001. Near-infrared reflectance spectroscopy of bulk analogue materials for planetary crust. *Planetary Space Science*, **49**, 191–211.

HEAD, J. W., MURCHIE, S. L. ET AL. 2009. Volcanism on Mercury: evidence from the first MESSENGER flyby for extrusive and explosive activity and the volcanic origin of plains. *Earth and Planetary Science Letters*, **285**, 227–242.

HEAD, J. W., CHAPMAN, C. R. ET AL. 2011. Flood volcanism in the northern high latitudes of Mercury revealed by MESSENGER. *Science*, **333**, 1853–1856, http://dx.doi.org/10.1126/science.1211997

HEATER, D. J., DUNKIN, S. K. & WILSON, L. 2003. Volcanism on the Marious Hills plateau: Observational analyses using Clementine multispectral data. *Journal of Geophysical Research*, **108**, 5017, http://dx.doi.org/10.1029/2002JE001938

HIESINGER, H. & HEAD, J. W., III 2006. New views of lunar geoscience: an introduction and overview. *In:* ROSS, J. J. (ed.) *New Views of the Moon. Reviews in Mineralogy and Geochemistry*, **60**, 1–81.

HIROI, T. & PIETERS, C. M. 1994. Estimation of grain size and mixing ratios of fine powder misture of common geologic minerals. *Journal of Geophysical Research*, **99**, 10 867–10 879.

HIROI, T. & SASAKI, S. 2001. Importance of space weathering simulation products in compositional modeling of asteroids: 349 Dembowska and 446 Aeternitas as examples. *Meteoritics & Planetary Science*, **36**, 1587–1596.

HUNT, G. R., SALISBURY, J. W. & LENHOFF, C. J. 1973*a*. Visible and near infrared spectra of minerals and rocks: VII. Acid igneous rocks. *Modern Geology*, **4**, 217–224.

HUNT, G. R., SALISBURY, J. W. & LENHOFF, C. J. 1973*b*. Visible and near infrared spectra of minerals and rocks: VIII. Intermediate igneous rocks. *Modern Geology*, **4**, 237–244.

HUNT, G. R., SALISBURY, J. W. & LENHOFF, C. J. 1974. Visible and near infrared spectra of minerals and rocks: IX. Basic and Ultrabasic igneous rocks. *Modern Geology*, **5**, 15–22.

KAUR, P., BHATTACHARYA, S., CHAUHAN, P., AJAI, & KIRAN KUMAR, A. S. 2013. Mineralogy of Mare Serenitatis on the near side of the Moon based on Chandrayaan-1 Moon mineralogy mapper (M3) observations. *Icarus*, **222**, 137–148.

KESZTHELYI, L. P. 1995. A preliminary thermal budget for lava tubes on the Earth and planets. *Journal of Geophysical Research*, **100**, 20 411–20 420.

KESZTHELYI, L., MCEWEN, A. S. & THORDARSON, T. 2000. Terrestrial analogs and thermal models for Martian flood lavas. *Journal of Geophysical Research*, **105**, 15 027–15 050.

KLIMA, R. L., PIETERS, C. M. & DYAR, M. D. 2007. Spectroscopy of synthetic Mg-Fe pyroxenes I: spinallowed and spin-forbidden crystal field bands in the visible and near-infrared. *Meteoritics & Planetary Science*, **42**, 235–253.

KLIMA, R. L., PIETERS, C. M. & DYAR, M. D. 2008. Characterization of the 1.2 μm M1 pyroxene band: extracting cooling history from near-IR spectra of pyroxenes and pyroxene-dominated rocks. *Meteoritics & Planetary Science*, **43**, 1591–1604.

KLIMA, R. L., DYAR, M. D. & PIETERS, C. M. 2011. Near-infrared spectra of clinopyroxenes: effects of calcium content and crystal structure. *Meteoritics & Planetary Science*, **46**, 379–395.

KLIMA, R. L., IZENBERG, N. R. *ET AL.* 2013. Training the Ferrous Iron Content of Silicate Minerals in Mercury's Crust. Abstract 1602, presented at the 44th Conference on Lunar Planetary Science, Woodlands, Texas, March 18–22, 2013.

LAWRENCE, D. J., HAWKE, B. R., HAGERTY, J. J., ELPHIC, R. C., FELDMAN, W. C., PRETTYMANN, T. H. & VANIMAN, D. T. 2005. Evidence for a high-Th, evolved lithology on the Moon at Hansteen Alpha. *Geophysical Research Letters*, **32**, http://dx.doi.org/10.1029/2004GL022022

LAWRENCE, D. J., FELDMAN, W. C. *ET AL.* 2010. Identification and measurement of neutron-absorbing elements on Mercury's surface. *Icarus*, **209**, 195–209.

LE MAITRE, R. W., STRECKEISEN, A. *ET AL.* (eds). 2002. *Igneous Rocks—A Classification and Glossary of Terms.* Cambridge University Press, Cambridge.

LUCEY, P., KOROTEV, R. L. *ET AL.* 2006. Understanding the lunar surface and space–moon interactions.

In: ROSS, J. J. (ed.) *New Views of the Moon. Reviews in Mineralogy and Geochemistry*, **60**, 83–219.

LUCEY, P. G. & RINER, M. A. 2011. The optical effects of small iron particles that darken but do not redden: evidence of intense space weathering on Mercury. *Icarus*, **212**, 451–462.

MCSWEEN, H. Y., JR. & TREIMAN, A. H. 1998. Martian meteorites. *In:* PAPIKE, J. J. (ed.) *Planetary Materials.* Reviews in Mineralogy, **36**. Mineralogical Society of America, Washington, DC, 6-1–6-53.

MCSWEEN, H. Y., JR., MURCHIE, S. L. *ET AL.* 1999. Chemical, multispectral, and textural constraints on the composition and origin of rocks at the Mars Pathfinder landing site. *Journal of Geophysical Research*, **104**, 8679–8715.

MCSWEEN, H. Y., ARVIDSON, R. E. *ET AL.* 2004. Basaltic Rocks analyzed by the Spirit Rover in Gusev Crater. *Science*, **305**, 842–845, http://dx.doi.org/10.1126/science.3050842

MCSWEEN, H. Y., RUFF, S. W. *ET AL.* 2006*a*. Alkaline volcanic rocks from the Columbia Hills, Gusev Crater, Mars. *Journal of Geophysical Research*, **111**, E09S91, http://dx.doi.org/10.1029/2006JE002698

MCSWEEN, H. Y., WYATT, M. B. *ET AL.* 2006*b*. Characterization and petrologic interpretation of olivine-rich basalts at Gusev Crater, Mars. *Journal of Geophysical Research*, **111**, E02S10, http://dx.doi.org/10.1029/2005JE002477

MCSWEEN, H. Y., TAYLOR, G. J. & WYATT, M. B. 2009. Elemental composition of the Martian crust. *Science*, **324**, 736–739.

MIDDLEMOST, E. A. K. 1997. *Magmas, Rocks and Planetary Development: A Survey of Magma/Igneous Rock Systems.* Longman, Harlow.

MINITTI, M. E., MUSTARD, J. F. & RUTHERFORD, M. J. 2002. Effects of glass content and oxidation on the spectra of SNC-like basalts: applications to Mars remote sensing. *Journal of Geophysical Research*, **107**, 6-1–6-14, http://dx.doi.org/10.1029/2001JE00 1518

MITTLEFEHLDT, D. W., MCCOY, T. J., GOODRICH, C. & KRACHER, A.. 1998. Non-chondritic meteorites from asteroidal bodies. *In:* PAPIKE, J. J. (eds) *Planetary Materials.* Reviews in Mineralogy, **36**. Mineralogical Society of America, Washington, DC, 4-1–4-195.

MOROZ, L. V. & ARNOLD, G. 1999. Influence of neutral components on relative band contrasts in reflectance spectra of intimate mixtures: implications for remote sensing 1. Nonlinear mixing modeling. *Journal of Geophysical Research*, **104**, 14 109–14 122.

MOROZ, L. V., FISENKO, A. V., SEMIONOVA, L. F., PIETERS, C. M. & KOROTAEVA, N. N. 1996. Optical effects of regolith processes on S-Asteroids as simulated by Laser Shots on ordinary chondrite and other mafic materials. *Icarus*, **122**, 366–382.

MOROZ, L., SCHADE, U. & WASCH, R. 2000. Reflectance Spectra of olivine-orthopyroxene-bearing assembleages at decreased temperatures: implications for remote sensing of asteroids. *Icarus*, **147**, 79–93, http://dx.doi.org/10.1006/icar.2000.6430

NEAL, C. R. & TAYLOR, L. A. 1992. Petrogenesis of mare nasalts—A record of lunar volcanism. *Geochimica et Cosmochimica Acta*, **56**, 2177–2211.

NICHOLIS, M. G. & RUTHERFORD, M. J. 2005. Pressure Dependence of Graphite–C–O Phase Equilibria and its Role in Lunar Volcanism. Abstract 1732, presented at the 36th Conference on Lunar Planetary Science, League City, Texas, March 14–18, 2005.

NITTLER, L. R., STARR, R. D. ET AL. 2011. The major-element composition of Mercury's surface from MESSENGER X-ray spectrometry. *Science*, **333**, 1847–1850.

PEITERSEN, M. N. & CROWN, D. A. 1999. Down flow width behaviour of Martian and terrestrial lava flows. *Journal of Geophysical Research*, **104**, 8473–8488.

PIETERS, C. M., FISCHER, E. M., RODE, O. D. & BASU, A. 1993. Optical effects of space weathering on lunar soils and the role of the finest fraction. *Journal of Geophysical Research*, **98**, 20 817–20 824.

PIETERS, C. M., TAYLOR, L. A. ET AL. 2000. Space weathering on airless bodies: resolving a mystery with lunar samples. *Meteoritics & Planetary Science*, **35**, 1101–1107.

PILORGET, C., BIBRING, J.-P. THE MICROMEGA TEAM 2012. The MicrOmega Instrument Onboard ExoMars and Future Missions: An IR Hyperspectral Microscope to Analyze Samples at the Grain Scale and Characterize Early Mars Processes. Abstract 7006, presented at the Third International Conference on Early Mars: Geologic and Hydrologic Evolution, Physical and Chemical Environments, and the Implications for Life, Lake Tahoe, Nevada, May 21–25, 2012.

POMPILIO, L., SGAVETTI, M. & PEDRAZZI, G. 2007. Visible and near-infrared reflectance spectroscopy of pyroxene-bearing rocks: new constraints for understanding planetary surface compositions. *Journal of Geophysical Research*, **112**, E01004, http://dx.doi.org/10.1029/2006JE002737

POULET, F., GOMEZ, C. ET AL. 2007. Martian surface mineralogy from Observatoire pour la Mineralogie, l'Eau, les Glaces et l'Activite on board the Mars Express spacecraft (OMEGA/MEx): global mineral maps. *Journal of Geophysical Research*, **112**, E08S02, http://dx.doi.org/10.1029/2006JE002840

POULET, F., MANGOLD, N. ET AL. 2009. Quantitative compositional analysis of Martian mafic regions using the MEx/OMEGA reflectance data: 2. Petrological implications. *Icarus*, **210**, 84–101.

REIDER, R., ECONOMOU, T. ET AL. 1997. The chemical composition of Martian soil and rocks returned by the mobile alpha proton X-ray spectrometer: preliminary results in X-ray mode. *Science*, **278**, 1771–1774.

RINER, M. A., LUCEY, P. G., MCCUBBIN, F. M. & TAYLOR, G. J. 2011. Constraints on Mercury's surface composition from MESSENGER neutron spectrometer data. *Earth and Planetary Science Letters*, **308**, 107–114, http://dx.doi.org/10.1016/j.epsl.2011.05.042

RUFF, S. W. & CHRISTENSEN, P. R. 2002. Bright and dark regions on Mars: particle size and mineralogical characteristics based on thermal emission spectrometer data. *Journal of Geophysical Research*, **107**, 5127, http://dx.doi.org/10.1029/2001JE001580

RUSSELL, C. T. & RAYMOND, C. A. 2011. The Dawn Mission to Vesta and Ceres. *Space Science Review*, **163**, 3–23, http://dx.doi.org/10.1007/s11214-011-9836-2

SAKIMOTO, S. E. H. & ZUBER, M. T. 1998. Flow and convection cooling in lava tubes. *Journal of Geophysical Research*, **103**, 27 465–27 487.

SAKIMOTO, S. E. H., CRISP, J. & BALOGA, S. M. 1997. Eruption constraints on tube-fed planetary lava flows. *Journal of Geophysical Research*, **102**, 6597–6613.

SASAKI, S., NAKAMURA, K., HAMABE, Y., KURAHASHI, E. & HIROI, T. 2001. Production of iron nanoparticles by laser irradiation in a simulation of lunar-like space weathering. *Nature*, **410**, 555–557.

SCHABER, G. G., BOYCE, J. M. & MOORE, H. J. 1976. The scarcity of mapable flow lobes on the lunar maria: unique morphology of the imbrium flows. *In: Proceedings of the 7th Lunar Science Conference, Houston, Texas, March 15–19, 1976, Volume 3*. Pergamon Press, New York, 2783–2800.

SCHADE, U., WÄSCH, R. & MOROZ, L. 2004. Near-infrared reflectance spectroscopy of Ca-rich clinopyroxenes and prospects for remote spectral characterization of planetary surfaces. *Icarus*, **168**, 80–92.

SERVENTI, G., CARLI, C. & SGAVETTI, M. 2013a. Plagioclase Influence in Mixtures with Very Low Mafic Mineral Content. Abstract 1490, presented at the 44th Conference on Lunar Planetary Science, Woodlands, Texas, March 18–22, 2013.

SERVENTI, G., CARLI, C., SGAVETTI, M., CIARNIELLO, M., CAPACCIONI, F. & PEDRAZZI, G. 2013b. Spectral variability of plagioclase-mafic mixtures 1): effects of chemistry and modal abundance in reflectance spectra of rocks and mineral mixtures. *Icarus*, **226**, 282–298, http://dx.doi.org/10.1016/j.icarus.2013.05.041

SGAVETTI, M., TRAMUTOLI, V., POMPILIO, L. & LONGHI, I. 2003. Field spectroscopy on Mount Etna. *In*: BLANCO, A., DOTTO, E. & OROFINO, V. (eds) *Planetary Science, Proceedings of the Fifth Italian Meeting*, Gallipoli (Lecce), Italy, September 15–19, 2003. Alenia Spazio, Rome, 246–252.

SGAVETTI, M., POMPILIO, L. & MELI, S. 2006. Reflectance spectroscopy (0.3–2.5 um) at various scales for bulk-rock identification. *Geosphere*, **2**, 142–160.

SHEARER, C. K., HESS, P. C. ET AL. 2006. Thermal and magmatic evolution of the Moon. *In*: Ross, J. J. (ed.) *New Views of the Moon Reviews in Mineralogy and Geochemistry*, 365–518.

SLEEP, N. H. 2000. Evolution of the mode of convection within terrestrial planets. *Journal of Geophysical Research*, **105**, 17 563–17 578, http://dx.doi.org/10.1029/2000JE001240

SOLOMON, S. C., MCNUTT, R. L., JR. ET AL. 2008. Return to Mercury: a global perspective on MESSENGER's first Mercury flyby. *Science*, **321**, 59–62.

SPUDIS, P. D., MCGOVERN, P. J. & KIEFER, W. S. 2013. Large shield volcanoes on the Moon. *Journal of Geophysical Research*, **118**, 1063–1081, http://dx.doi.org/10.1002/jgre.20059

STRAZZULLA, G., DOTTO, E., BINZEL, R., BRUNETTO, R., BARUCCI, M. A., BLANCO, A. & OROFINO, V. 2005. Spectral alteration of the Meteorite Epinal (H5) induced by heavy ion irradiation: a simulation of space weathering effects on near-Earth asteroids. *Icarus*, **174**, 31–35.

STRECKEISEN, A. 1978. IUGS Subcommission on the Systematics of Igneous Rocks. Classification and nomenclature of volcanic rocks. Recommendations and suggestions. *Neues Jahrbuch fur Mineralogy*, **143**, 1–14.

SUNSHINE, J. M. & PIETERS, C. M. 1991. Identification of modal abundances in the spectra of natural and laboratory pyroxene mixtures: a key component for remote analysis of lunar basalts. *Abstracts of the Lunar and Planetary Science Conference*, **22**, 1361–1362.

SUNSHINE, J. M. & PIETERS, C. M. 1993. Estimating modal abundances from the spectra of natural and laboratory pyroxene mixtures using the modified Gaussian model. *Journal of Geophysical Research*, **98**, 9075–9087.

SUNSHINE, J. M. & PIETERS, C. M. 1998. Determining the composition of olivine from reflectance spectroscopy. *Journal of Geophysical Research*, **103**, 13 675–13 688.

SUNSHINE, J. M., PIETERS, C. M. & PRATT, S. F. 1990. Deconvolution of mineral absorption bands: an improved approach. *Journal of Geophysical Research*, **95**, 6955–6966.

TAYLOR, G. J., WARREN, P., RYDER, G., DELANO, J., PIETERS, C. & LOFGREN, G. 1991. Lunar rocks. *In*: HEIKEN, G., VANIMAN, D. & FRENCH, B. (eds) *The Lunar Source Book: A User's Guide to the Moon*. Cambridge University Press, Cambridge, 183–284.

TOMPKINS, S. & PIETERS, C. M. 2010. Spectral characteristics of lunar impact melts and inferred mineralogy. *Meteoritics & Planetary Science*, **45**, 1152–1169.

WAENKE, H., BRUCKNER, J., DREIBUS, G., RIEDER, R. & RYABCHIKOV, I. 2001. Chemical composition of rocks and soils at the pathfinder site. *Space Science Review*, **96**, 317–330.

WARREN, P. H. & WASSON, J. T. 1979. The origin of KREEP. *Reviews of Geophysics and Space Physics*, **17**, 73–88.

WATTERS, T. R., MURCHIE, S. L., ROBINSON, M. S., SOLOMON, S. C., DENEVI, B. W., ANDRÉ, S. L. & HEAD, J. W. 2009. Emplacement and tectonic deformation of smooth plains in the Caloris basin, Mercury. *In*: SOLOMON, S. C., PROCKTER, L. M. & BLEWETT, D. T. (eds) *MESSENGER's First Flyby of Mercury. Earth and Planetary Science Letters*, **285**, 309–319.

WEIDER, S. Z., NITTLER, L. R. *ET AL.* 2012. Chemical heterogenity on Mercury's surface revealed by the MESSENGER X-ray spectrometer. *Journal of Geophysical Research*, **117**, E00L05, http://dx.doi.org/10.1029/2012JE004153

WEIDER, S. Z., NITTLER, L. R., STARR, R. D. & SOLOMON, S. C. 2013. Distribution of Iron on the Surface of Mercury from MESSENGER x-ray Spectrometer Measurements. Abstract 2189, presented at the 44th Conference on Lunar Planetary Science, Woodlands, Texas, March 18–22, 2013.

YAMADA, M., SASAKI, S. *ET AL.* 1999. Simulation of space weathering of planet-forming materials: nanosecond pulse laser irradiation and proton implantation on olivine and pyroxene samples. *Earth Planets Space*, **51**, 1255–1265.

YON, S. A. & PIETERS, C. M. 1988. Interactions of light with rough dielectric surfaces: spectral reflectance and polarimetric properties. *In*: RYDER, G. (ed.) *Proceedings of the 19th Lunar Science Conference*. Cambridge University Press, New York, 581–592.

Age relationships of the Rembrandt basin and Enterprise Rupes, Mercury

SABRINA FERRARI[1]*, MATTEO MASSIRONI[1,2], SIMONE MARCHI[3], PAUL
K. BYRNE[4,5], CHRISTIAN KLIMCZAK[4], ELENA MARTELLATO[2,6] &
GABRIELE CREMONESE[2]

[1]*Department of Geosciences, University of Padua, via Gradenigo
6, 35131 Padova, Italy*

[2]*Astronomical Observatory of Padua, INAF, Vic. Osservatorio 5, 35122 Padova, Italy*

[3]*NASA Lunar Science Institute, Southwest Research Institute,
1050 Walnut Street, Boulder, CO 80302, USA*

[4]*Department of Terrestrial Magnetism, Carnegie Institution of Washington,
5241 Broad Branch Road NW, Washington, DC 20015-1305, USA*

[5]*Lunar and Planetary Institute, Universities Space Research Association,
3600 Bay Area Boulevard, Houston, TX 77058, USA*

[6]*Department of Physics and Astronomy, University of Padua,
via Marzolo 8, 35131 Padova, Italy*

*Corresponding author (e-mail: sabrina.ferrari@studenti.unipd.it)

Abstract: The Rembrandt basin is the largest well-preserved impact feature in the southern hemisphere of Mercury. Its smooth volcanic infill hosts wrinkle ridges and graben, and the entire basin is cross-cut by the Enterprise Rupes scarp system. On the basis of the Model Production Function crater chronology, our analysis shows that the formation of the Rembrandt basin occurred at 3.8 ± 0.1 Ga during the Late Heavy Bombardment, consistent with previous studies. We also find that the smooth plains interior to the basin were emplaced between 3.7 and 3.6 ± 0.1 Ga, indicative of a resurfacing event within the Rembrandt basin that is consistent with the presence of partially buried craters. These youngest plains appear temporally unrelated to basin formation, and so we regard their origin as likely to be due to volcanism. We identify the same chronological relationship for the terrain cross-cut by Enterprise Rupes to the west of the basin. Therefore, volcanic activity affected both the basin and its surroundings, but ended prior to the majority of basin- and regional-scale tectonic deformation. If Enterprise Rupes formed prior to the Rembrandt basin, then regional-scale tectonic activity along this structure might have lasted at least 200 myr.

Our current understanding of the early surface of Mercury involves a primitive crust that experienced intense cratering during the Late Heavy Bombardment (LHB) approximately 4.1–3.6 Ga ago (e.g. Tera *et al.* 1974; Strom & Neukum 1988; Marchi *et al.* 2013a). Concurrent crustal shortening due to the global cooling-induced contraction of the planet led to the formation of thrust fault-related landforms (lobate scarps and wrinkle ridges: e.g. Watters *et al.* 2004). Extensive smooth deposits, the majority of which are interpreted to be volcanic (Head *et al.* 2009 and references therein; Prockter *et al.* 2010), that covered some 27% of the planet's surface (Denevi *et al.* 2013) were also emplaced. Activity on some lobate scarps continued beyond the emplacement of the youngest volcanic smooth plains (Strom *et al.* 1975; Melosh & McKinnon 1988; Watters *et al.* 2009c; Head *et al.* 2009).

Notably, some tectonic deformation on Mercury is manifest in spatially coherent patterns. The distinctive arrangement of structures observed within volcanically infilled and buried impact basins and craters on Mercury (Watters *et al.* 2009b, 2012; Head *et al.* 2011; Fassett *et al.* 2012; Klimczak *et al.* 2012) may have resulted from the cooling of low-viscosity lava flows emplaced as rapidly accumulated, thick units against a backdrop of pervasive global contraction (Freed *et al.* 2012; Watters *et al.* 2012). Within the largest basins in particular, the load of volcanic infill could have induced subsidence and, consequently, the formation of radial and concentric wrinkle ridges (Watters *et al.*

From: PLATZ, T., MASSIRONI, M., BYRNE, P. K. & HIESINGER, H. (eds) 2015. *Volcanism and Tectonism Across the Inner Solar System*. Geological Society, London, Special Publications, **401**, 159–172.
First published online July 29, 2014, http://dx.doi.org/10.1144/SP401.20

2009*b*, 2012), whereas vertical motion (Dzurisin 1978; Melosh & Dzurisin 1978; Blair *et al.* 2013) or inward flow of lower crustal material triggered by basin formation (Watters *et al.* 2005; Watters & Nimmo 2010) could have promoted subsequent uplift of basin centres and the formation of radial and concentric graben (Watters *et al.* 2009*c*; Klimczak *et al.* 2010; Byrne *et al.* 2013). Alternatively, such distinctive patterns of extensional features in Mercury's largest impact basins may have formed due to the thermal contraction of rapidly emplaced lava flow on the basin floor (Freed *et al.* 2012; Byrne *et al.* 2013). Nevertheless, patterns of deformation that include basin-radial and -concentric faulting were probably controlled by the shape of the impact structure itself.

Such a systematic pattern is preserved in the 715 km-diameter Rembrandt basin. Imaged for the first time during the second flyby of the MErcury Surface, Space ENvironment, Geochemistry, and Ranging (MESSENGER) spacecraft, most of the basin interior is covered by smooth, high-reflectance plains interpreted to be of volcanic origin (Denevi *et al.* 2009; Watters *et al.* 2009*a*). Similar high-reflectance plains are broadly distributed across the surface of Mercury, and have also been recognized in the surroundings of the Rembrandt basin (Denevi *et al.* 2013).

The smooth-plains infill of the Rembrandt basin hosts sets of basin-radial and -concentric contractional and extensional tectonic structures (Fig. 1a, b). This pattern resembles to first order the arrangement of structures within the Caloris basin, the largest preserved impact structure on Mercury (Murchie *et al.* 2008; Watters *et al.* 2009*b*; Byrne *et al.* 2012), in which individual sets of radial and concentric landforms are most probably due to multiple episodes of deformation (Strom *et al.* 1975; Melosh & McKinnon 1988; Watters *et al.* 2005, 2009*c*; Klimczak *et al.* 2010). Notably, the Rembrandt basin and its smooth plains are crosscut by an 820 km-long scarp system (Watters *et al.* 2009*a*; Byrne *et al.* 2014), which trends approximately east–west outside the basin before bending towards the NE within the basin itself (white arrows in Fig. 1a). The individual faults of this system, collectively named Enterprise Rupes, accommodated crustal shortening that likely resulted from the global contraction of Mercury as its interior cooled (Watters *et al.* 2009*c*), perhaps augmented by other processes such as mantle convection and tidal de-spinning (Massironi *et al.* 2014). It is probable that the current shape of the reverse fault system, and particularly its northwards bend, was influenced by the formation of the Rembrandt basin. Ferrari *et al.* (2012) suggested that the branch of Enterprise Rupes outside the Rembrandt basin predates the basin-forming impact, as part of a

pre-existing scarp system. The basin-forming impact partially overprinted that system, and induced a perturbation of the prevailing stress field that reorientated the strike of the new branch as it grew within the new basin. This scenario accounts for the different orientations of the portions of Enterprise Rupes outside and inside the basin, and predicts differences in vertical displacements between those segments that have been observed (Ferrari *et al.* 2012).

The Rembrandt basin and its surrounding region were probably affected by many of the processes that modified the surface of Mercury (i.e. subsequent impacts in general, global- and basin-scale tectonic deformation, and volcanic resurfacing), and so it is a suitable case-study location for understanding the sequence and duration of such processes. In this study, we aim to establish a chronological sequence of formation for the surface units and structures of the Rembrandt basin and its surroundings. We therefore performed crater counts on different type units in the area in order to estimate the absolute model ages by applying the Model Production Function (MPF) of Marchi *et al.* (2009, 2012, 2013*b*), complemented by identified cross-cutting and superposition relationships, to determine the relative timing of terrain and structure formation within and surrounding the Rembrandt basin.

Geological units and cross-cutting relationships in and around the Rembrandt basin

The geological sketch of the region that includes the Rembrandt basin and Enterprise Rupes in Figure 1b was compiled from data returned by the MESSENGER spacecraft during its flyby and orbital phases. The map consists of internal basin units, the basin's proximal ejecta, superposed impact craters and their ejecta deposits, and prominent tectonic structures. The MESSENGER Mercury Dual Imaging System (MDIS) solar-day-one monochrome base map mosaic (with an average spatial resolution of 250 m/pixel) (Hawkins *et al.* 2007), together with the wide-angle camera (WAC) multispectral mosaic base map (bands 1000, 750, and 430 nm; average spatial resolution 1000 m/pixel) were analysed in the ESRI® ArcMap geographical information system (GIS) environment. The MDIS orbital datasets were complemented with flyby narrow-angle camera (NAC) images taken under illumination conditions that better show landforms with subtle topographical expressions. Single images were processed with the open-access USGS Integrated Software for Imagers and Spectrometers (ISIS) software. Mapped units were classified on the basis of their surface textures and stratigraphic

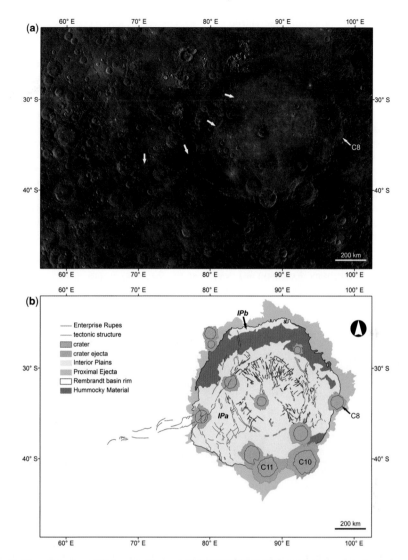

Fig. 1. The Rembrandt basin and Enterprise Rupes. (**a**) MESSENGER MDIS monochrome base map, shown with a Mercator projection centred on Rembrandt (33°S, 88°E); white arrows point to Enterprise Rupes and black arrows indicate radially lineated basin ejecta. (**b**) Geological sketch map showing the variety of units of the Rembrandt basin and its superposed structures.

relationships: the 'Hummocky Material' and the 'Interior Plains' units fill the basin, whereas the 'Proximal Ejecta' unit surrounds the basin rim (Fig. 1b).

The Hummocky Material unit (81 000 km^2 in area) probably consists of a mixture of impact melt and ejecta deposits (breccias) that formed hills, depressions and high-standing knobs, morphological features characteristic of ejecta deposited during the formation of large basins of the Moon (e.g. Cintala & Grieve 1998) and the Caloris basin on Mercury (Fassett *et al.* 2009; Denevi *et al.* 2013).

This texture was initially recognized within the Rembrandt basin by Watters *et al.* (2009*a*). The Hummocky Material unit is mostly buried by the smooth Interior Plains deposits or, along the southern rim, appears strongly reworked by subsequent impact events (e.g. craters C10 and C11 in Fig. 1b) that make a clear demarcation of the unit difficult.

Beyond the basin rim, radially lineated terrain (black arrows in Fig. 1a) has been interpreted as basin ejecta (Watters *et al.* 2009*a*), of which the Proximal Ejecta unit (132 000 km^2 in area) represents the coherent material that surrounds almost the entire rim crest. It was probably emplaced contemporaneously with the Hummocky Material unit.

The smooth Interior Plains deposits (370 000 km^2 in cumulative area) have a higher reflectance than the basin formation-related units, and are likely to be volcanic in origin (Denevi *et al.* 2009; Watters *et al.* 2009*a*). This unit covers most of the basin floor but occurs as two discontinuous exposures (which we label as '*IPa*' and '*IPb*' in Fig. 1b). IPa is the most extensive exposure (357 000 km^2) and covers the central portion of the basin, extending from the western to the eastern rims; IPb occurs as a much smaller patch (12 000 km^2) that embayed the northern basin rim wall.

MESSENGER colour data (Fig. 2) indicate that the IPa sub-unit shows subtle variations in tone. Colour variations on Mercury can be attributed to different surface compositions (Robinson & Lucey 1997; Blewett *et al.* 2007; Robinson *et al.* 2008;

Murchie *et al.* 2008). On the SW basin rim, the IPa sub-unit darkens and becomes texturally indistinguishable from the surface immediately exterior to the basin (the dark cross symbols in Fig. 2 indicate the point where the IPa sub-unit gradually darkens towards the basin perimeter), suggesting that the interior smooth plains and the surrounding terrain may be compositionally similar. This portion of terrain exterior to the basin also displays several patches of higher reflectance smooth plains, mainly associated with large craters and within a prominent trough radial to Rembrandt (white arrows in Fig. 2). The reflectance of those patches is similar to those within the Interior Plains, and both host several partly buried craters (craters marked with black arrows in Fig. 2), which attest to substantial volcanic (Denevi *et al.* 2009, 2013) resurfacing after their formation. A pit-floor crater situated in the 60 km-diameter crater C8 of Figure 1a (Denevi *et al.* 2013) may represent a source vent from which effusive volcanism erupted, at least for the

Fig. 2. MESSENGER MDIS WAC multispectral mosaic (bands 1000, 750 and 430 nm; average spatial resolution is 1000 m/pixel) of the SW boundary of Rembrandt basin. The arcuate line of black crosses indicates the transition of surface reflectance within the expansive Interior Plains sub-unit (IPa, white boundary): while retaining the same texture, the surface becomes darker towards the basin perimeter. High-reflectance plains, such as the Interior Plains and several smaller, outlying patches (white arrows), partially filled craters (black arrows) and reduced ejecta proximal to the basin (white dashed boundary), suggest resurfacing for both the Interior Plains deposits and the units to the west/SW. Enterprise Rupes (black dashed line) deformed most of the units shown here.

basin interior. It is entirely possible, however, that the vents from which these lavas issued may be located both within and outside the basin, but have been obscured by subsequent flows.

Moreover, at the footwall of Enterprise Rupes (black dotted line, Fig. 2), the Proximal Ejecta unit appears either to be texturally smooth or entirely absent (white stars, Fig. 2), which suggests that volcanic resurfacing of the Exterior Plains unit also occurred. However, both the basin-radial and -concentric contractional and extensional tectonic structures, and Enterprise Rupes itself, deform the Rembrandt basin's Interior Plains unit, and so at least the most recent activity of each of these tectonic systems must post-date the emplacement of the Interior Plains unit (Fig. 1b).

The Model Production Function chronology applied to the surface of Mercury

The age estimation of geological units on airless planetary bodies using crater size–frequency distributions (SFDs) is based on the tenet that older surfaces have accumulated more craters than younger surfaces. Ages for the Rembrandt basin units have been calculated by means of the Model Production Function (MPF) chronology for Mercury (Marchi et al. 2009; Massironi et al. 2009), which relies on models of the impactor flux on the planet. This flux is converted into the expected crater SFD per unit time and per unit surface by the crater-scaling law of Holsapple & Housen (2007), which takes into account the properties of the materials involved. Then the cumulative number of 1 km-diameter craters is calibrated using radiometric ages and crater densities observed at the Apollo and Luna landing sites on the Moon, in order to determine the absolute age calibration.

Two key features of the MPF method are: (i) the distinction between Main Belt Asteroids (MBA) and Near Earth Objects (NEO) as impactor size–frequency distributions; and (ii) the use of a crater-scaling law, which allows for the computation of the transient crater diameter, D_t, as a function of impactor size, d, and impact parameters, as well as of target properties. The equation for the crater-scaling law adopted from Holsapple & Housen (2007) is:

$$D_t = kd \left[\frac{gd}{2v_\downarrow^2} \left(\frac{\rho}{\delta}\right)^{\frac{2v}{\mu}} + \left(\frac{Y}{\rho v_\downarrow^2}\right)^{\frac{2+\mu}{2}} \left(\frac{\rho}{\delta}\right)^{\frac{v(2+\mu)}{\mu}} \right]^{-\left(\frac{\mu}{2+\mu}\right)}$$

(1)

where g is the target gravitational acceleration of the planet, $v_\downarrow$ is the perpendicular component of the impactor velocity, δ is the projectile density, ρ and Y are the density and tensile strength of the target, respectively, k and μ depend on the cohesion of the target material, and v depends on the porosity of the target material (Table 1). As discussed by Marchi et al. (2009, 2011), the density, strength and cohesion of Mercury's crust are expected to vary as a function of depth, as has been inferred for these properties on the Moon (Toksöz et al. 1973; Hörz et al. 1991). Hence, different ranges of strength and density of the target material must be considered according to the size of the impactor, and a transition of the crater-scaling law from shallower 'cohesive soil' (i.e. reworked by impact gardening) to deeper 'hard rock' can be used. Impact and target parameters used in this study for both regimes are listed in Table 1.

The transient crater diameter, D_t, is converted to a *final* crater diameter, D, according to the following expressions:

$$D = 1.3D_t \quad \text{if } D_t \leq D_*/1.3 \qquad (2)$$

and

$$D = 1.4D_t^{1.18}/D_*^{0.18} \quad \text{if } D_t > D_*/1.3. \qquad (3)$$

Here, D_* is the observed simple-to-complex transition crater diameter, which for Mercury corresponds to 11 km (Pike 1988). The factors adopted for these estimates are described in detail in Marchi et al. (2011).

Based on these calculations, age assessments are performed assuming a specific rheological layering

Table 1. *Parameters used for the crater-scaling law of Holsapple & Housen (2007)*

	Impactor density (g cm^{-3})	Constant	Scaling exponent	Scaling exponent	Target density (g cm^{-3})	Target tensile strength (MPa)
	δ	k	μ	v	ρ	Y
Cohesive soil (gravity regime)	2.6	1.03	0.41	0.4	2.8	2
Hard rock (strength regime)		0.93	0.55			20

for each mapped unit. For instance, the mixture of impact melt and ejecta formed during a giant impact should be less cohesive than the basement on which it lies (i.e. ejecta over a deeply fractured horizon). Conversely, a basin's volcanic infill probably partly or even entirely strengthens the pre-existing fractured layer at the basin floor, such that a cohesionless stratum might be confined to a thin regolith cover. The depth of transition from one regime to another can be considered to be reflected by an S-shaped kink in crater SFDs, to which an appropriate MPF curve can be fitted (Massironi *et al.* 2009). The depth of transition (H) can be inferred by the crater size corresponding to the centre of the S-shaped kink. Following Marchi *et al.* (2011), we assume that the kink occurs at $D_t \sim 4H$.

However, S-shaped kinks can also arise from a loss of craters due to different resurfacing processes (Platz *et al.* 2013). Neukum & Horn (1975) assumed that observable kinks in crater SFDs determined for lunar surfaces are always markers

of resurfacing, even in those cases where the morphological signature of later lava flows (and thus resurfacing) is not observed. In such cases, kinks are regarded as indications of composite SFDs in which smaller craters reflect the age of younger units and larger craters indicate the age of older units. Neukum & Horn (1975), however, considered that crater SFD S-shaped kinks depend solely on volcanic resurfacing, whereas MPFs attribute the kinks to either rheological layering (i.e. fractured material over hard rock) or to resurfacing. Thus, in order to derive a more accurate age estimate using MPF, it is a fundamental requirement that the crater production function be placed in the geological context of the units being investigated.

MPF applied on the Rembrandt area

We performed crater counts on the geological units described above, in the areas shown in Figure 3, using a portion of the global monochrome

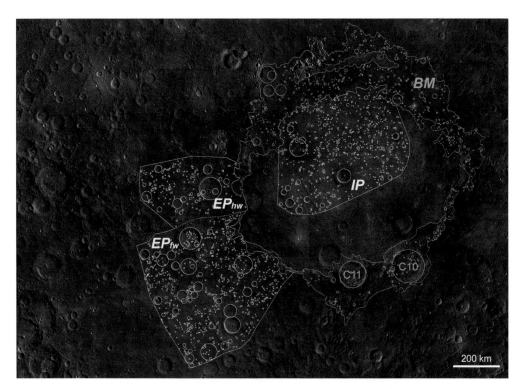

Fig. 3. Units in the Rembrandt basin region for which crater counts were performed, overlain on the MDIS 750 nm mosaic (average spatial resolution: 250 m/pixel). The red dashed line marks the leading edge of the Enterprise Rupes system. The green line bounds the crater-counting area of basin-related materials (i.e. the Hummocky Material and Proximal Ejecta units) labelled as *BM*, and green circles indicate craters counted within this area. White lines bound the crater-counting areas of units affected by resurfacing: the IPa sub-unit counting area within the Rembrandt basin is labelled as *IP*; counting areas located exterior to the basin are labelled EP_{fw} at the footwall of Enterprise Rupes and EP_{hw} at the hanging wall of the scarp; white circles indicate craters counted within this unit.

MESSENGER mosaic (average spatial resolution of 250 m/pixel). We outlined craters in our study areas using the ESRI ArcMap CraterTools extension developed by Kneissl et al. (2011), which accommodates distortion in crater shape due to non-shape-preserving map projections. Since the crater-age determination is based on primary craters (i.e. those formed by impacts with objects in heliocentric orbits; Melosh 1989), we avoided the inclusion of clusters and chains of craters that may be the result of secondary impacts. The effect of the contribution of far-field secondaries, which are normally not distinguishable from primary craters (e.g. McEwen & Bierhaus 2006), has not been considered.

In order to assess the Calorian age of the Rembrandt basin (c. 3.9 Ga) suggested by previous workers (Watters et al. 2009a; Fassett et al. 2012), we first performed a count of the primary craters of the Rembrandt basin-related units. Basin-proximal ejecta are presumed to have formed instantaneously within the Mercurian geological record (Spudis & Guest 1988), implying a production

of fractured material comparable to the production within the basin rim. Thus, we merged the Hummocky Material and Proximal Ejecta into a single allostratigraphic unit, which served to increase the statistical robustness of our count. Craters counted in this unit are shown in Figure 3. The resultant crater SFD shows a distinct S-shaped kink for crater diameters between 17 and 30 km (Fig. 4a), which in this specific case cannot be correlated to any observable volcanic resurfacing.

The age of the basin's formation, as suggested by Watters et al. (2009a), has justified our use of the population of Main Belt Asteroids (MBA) as the prime source of impactors in our MPF analysis as it best accounts for the cratering distributions found for the oldest terrains of Mercury (>3.8 Ga: Strom et al. 2005; Marchi et al. 2009, 2013b). The MPF curve that fits the crater SFD best (Fig. 4a) indicates an age of 3.8 ± 0.1 Ga and, as a function of the crater diameter range enclosed within the S-shaped kink of the crater SFD (i.e. 17–30 km), suggests a thickness of the upper heavily fractured layer of approximately 3.5 km (see Marchi et al.

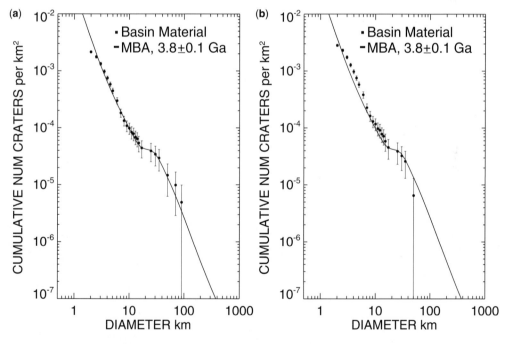

Fig. 4. MPF minimum χ^2 best fit of the cumulative crater-count distributions (applied binning scheme: 18 bins per diameter decade or steps 1.0, 1.1, 1.2, 1.3, 1.4, 1.5, 1.7, 2.0, 2.5, 3.0, 3.5, 4.0, 4.5, 5.0, 6.0, 7.0, 8.0, 9.0, 10.0 in the interval $1 \leq D \leq 10$) determined for the ejecta-related units (Hummocky Material and Proximal Ejecta) of the Rembrandt basin. For each unit we report the age assessment for Main Belt Asteroid populations. Error bars correspond to a variation of the minimum χ^2 of ±50%. (**a**) The best-fit line obtained in the diameter range 5–100 km for the combined Hummocky Material and Proximal Ejecta units when considering fractured layers (i.e. cohesive material) of 3.5 km thickness over hard basement. (**b**) The best-fit line of these basin-related materials obtained after subtracting the two major craters (C10 and C11) in the diameter range 7–60 km, using the same target assumptions as for the best-fit line shown in (a).

2011). This is consistent with depth values of the transition between fractured near-surface rocks and the underlying intact material suggested by Schultz (1993) for conditions applicable to Mercury.

The occurrence of several very large craters in a relatively small area could give rise to a statistical effect that would appear as a deflection of the SFD curve, which, in the context of MPF analysis, may lead to misleading conclusions regarding the target rheology and layering (see Kirchoff *et al.* 2013 for a detailed discussion). In order to exclude this effect, we therefore removed the two major craters present in the counting area of the basin-related material (C10 and C11 in Fig. 3, which are 97 and 79 km in diameter, respectively) from the crater SFD shown in Figure 4a. As shown in Figure 4b, the deflection in the crater SFD still persists, and so we therefore regard it as reflecting a genuine change in stratigraphy, with the Hummocky Material and Proximal Ejecta units probably comprised of mechanically weaker surface units overlying more competent units. Although we observe no evidence in support of resurfacing as the cause of the kink in the SFD, we nevertheless also applied the Neukum Production Function method (NPF: Neukum *et al.* 2001) to the complete basin crater SFD, which returned a broadly similar cratering model age of 3.85 Ga for the age of Rembrandt basin (Fig. 5).

Next, we applied the MPF method to a portion of the central Interior Plains unit (Fig. 3). For the purposes of age calculation, we included the broadest portion of the Interior Plains but we did not consider the southern part of the unit, which hosts chains and clusters of secondary craters associated with the C10 and C11 impacts. The craters counted for this unit are shown in Figure 3. We identified, and included in our counting, several craters larger than 6–8 km in diameter that show evidence of partial burial, which we attribute to the resurfacing of the basin floor. Since this volcanic material is likely to be mechanically more cohesive than the upper fractured layer of the basin floor, we assumed the 'hard rock' scaling law (Holsapple & Housen 2007) for our MPF analysis.

The crater SFD of the portion of the Interior Plains unit we investigated shows a slight S-shaped kink for crater diameters between 8 and 12 km (Fig. 6a). The observational evidence for resurfacing (i.e. partially buried craters in Fig. 2) suggests that there may be two geologically distinct units – one atop the other – present within what we term the IPa sub-unit. Thus, we interpret this S-shaped kink as the product of a composite crater SFD that describes the resurfacing of an older unit by a younger, overlying unit. We therefore fit MPF curves both for craters larger than 12 km in diameter

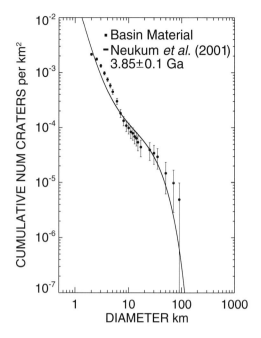

Fig. 5. The best-fit cratering model age of Rembrandt basin materials (i.e. Hummocky Material and Proximal Ejecta units) obtained in the diameter range 7–100 km using the production function and the chronology model of Neukum *et al.* (2001).

and for craters smaller than 8 km in diameter. The best MPF fit for larger craters corresponds to an age of 3.7 ± 0.1 Ga (Fig. 6a), whereas that for smaller craters corresponds to an age of 3.6 ± 0.1 Ga (Fig. 6b). Although the younger age of 3.6 ± 0.1 Ga is most probably overestimated – smaller crater bins are directly influenced by the larger craters of the distribution (Michael & Neukum 2010) – the obtained ages imply no statistical difference between the inferred layers.

This result does not clarify the nature of the inferred older layer, which, according to its MPF age (3.7 ± 0.1 Ga), could be either melt formed during the impact event or lava emplaced soon after basin formation. In any case, the crater population that formed on the older layer after basin formation defines the early production size–frequency to which the MPF is fit in Figure 6a. A subsequent resurfacing event was able to completely cover, and thus remove from the count, those craters whose depths were sufficiently shallow to be buried by this unit, but this phase of volcanism was of insufficient volume to completely bury craters above a given diameter (i.e. such that craters like those shown in Fig. 2 remain visible).

Hiesinger *et al.* (2002) suggested that the crater diameters in which S-shaped kinks exist in SFDs

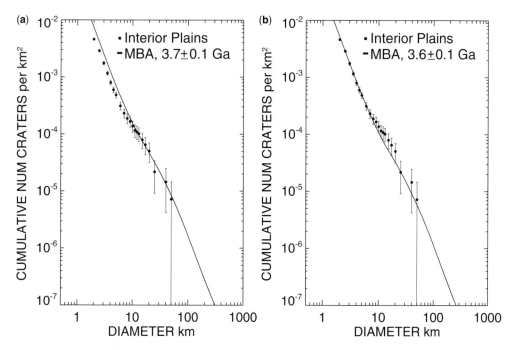

Fig. 6. MPF minimum χ^2 best fit determined for smooth plains interior to the basin. For each unit we report the age assessment for Main Belt Asteroid populations. Error bars correspond to a variation of the minimum χ^2 of $\pm 50\%$. (a) MPF best-fit line for the IPa sub-unit counting area using larger craters (diameter fit range 10–60 km) obtained using hard rock as the target material. (b) MPF best-fit line for the same counting area, using smaller craters (diameter fit range 2–12 km) and, again, assuming a hard-rock target material.

where resurfacing has occurred are related to the thickness of the younger units. A deflection of an S-shaped kink that occurs at larger diameters corresponds to the extinction of smaller craters of the early crater population, whereas a deflection that occurs at smaller diameters corresponds to the post-flooding crater population (Hiesinger *et al.* 2002 and reference therein). Considering the rim height/diameter (h/D) relationship fixed for Mercury by Pike (1988), the crater diameters where these deflections occur can be used to estimate the minimum (h_{min}) and maximum (h_{max}) thicknesses of the younger IPa sub-unit. The h/D relationship (Pike 1988) depends on the interior morphology of the craters themselves, however, yielding two near-isometric trends on the basis of crater size:

$$h = 0.052\, D^{0.930} \quad \text{(for bowl-shaped craters}$$
$$\text{with } 2.4\,\text{km} \le D \le 12\,\text{km)} \quad (4)$$

$$h = 0.150\, D^{0.487} \quad \text{(for immature-complex craters}$$
$$\text{with } 13\,\text{km} \le D \le 43\,\text{km)}. \quad (5)$$

The standard errors of these relationships are ± 0.08 km for equation (4) and ± 0.13 km for

equation (5) (Pike 1988). Using this approach, we obtained a minimum thickness, h_{min}, of 0.36 ± 0.08 km for the younger unit on the basis of first crater SFD deflection at a crater diameter of 8 km (equation 4), and a maximum thickness, h_{max}, of 0.52 ± 0.13 km for the same unit when considering the second crater SFD deflection at crater diameters of 12 km (equation 5). However, the smallest partially buried craters we identified within the IPa sub-unit measure approximately 6 km in diameter. Therefore, given that craters less than 6 km in diameter should be considered to be totally buried, using equation (4) with $D = 6$ km, we derived a minimum thickness of the overlying infill of 0.28 ± 0.08 km, which is broadly consistent with previous results: Ernst *et al.* (2011) estimated a thickness of around 2 km for the high-reflectance plains that overlie the centre of the Rembrandt basin and that we mapped as IPa. This value was inferred on the basis of the maximum depth of post-resurfacing craters that do not appear to have exhumed low-reflectance material, which is assumed to underlie the high-reflectance plains (Ernst *et al.* 2010). We also performed crater counts on the terrains bordering the Rembrandt basin at its SW rim and at the footwall of Enterprise

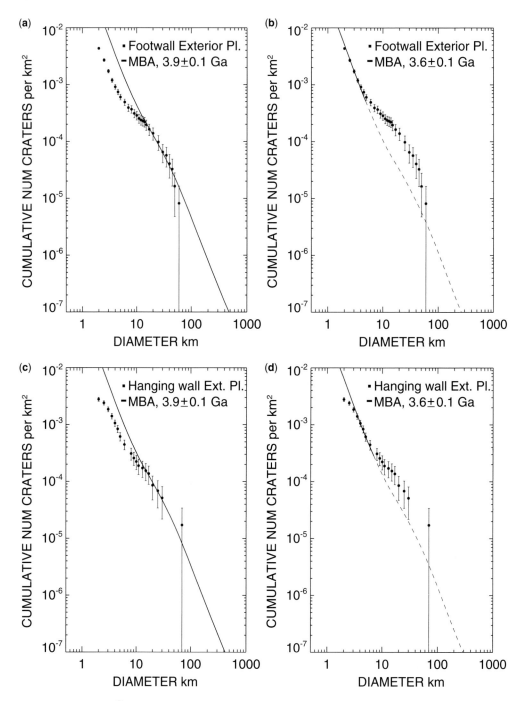

Fig. 7. MPF minimum χ^2 best fit determined for units exterior to the basin. As for previous figures, we report the age assessment for the Main Belt Asteroid populations. Error bars correspond to a variation of the minimum χ^2 of $\pm50\%$. (**a**) The MPF best fit of larger craters of the EP_{fw} counting area (diameter fit range 12–70 km) obtained using hard rock as the target material. (**b**) The MPF best fit of smaller craters of the EP_{fw} counting area (diameter fit range 2–5 km) obtained using hard rock as the target material. (**c**) The MPF best fit of larger craters of the EP_{hw} counting area (diameter fit range 13–71 km) obtained using hard rock as the target material. (**d**) The MPF best fit of smaller craters of the EP_{hw} counting area (diameter fit range 3–6 km) obtained assuming hard rock as the target material.

Rupes ('Footwall' Exterior Plains, EP_{fw}: Fig. 3), which display textural and reflectance similarities to the Interior Plains unit. Craters counted in this area are shown in Figure 3, and probably include the largest secondary craters associated with the formation of the Rembrandt basin.

Since the EP_{fw} unit shows evidence of resurfacing (i.e. partially filled craters, shown with black arrows in Fig. 2), and given that the corresponding crater SFD shows an S-shaped kink for crater diameters between 4 and 14 km (Figs 7a & b), we applied the same analysis (i.e. under the assumption that a composite SFD is due to resurfacing) as for the IPa sub-unit. Owing to the broad kink, only a reduced portion of the SFD can be fitted by isochrones. We found that the best-fit MPF for larger craters returns an age of 3.9 ± 0.1 Ga (Fig. 7a), whereas, for smaller craters, the best-fit MPF yields an age of 3.6 ± 0.1 Ga (Fig. 7b).

As for the EP_{fw} unit, partially buried craters were identified atop the Enterprise Rupes contractional system (black arrows, Fig. 2). This terrain displays high-reflectance smooth surfaces associated with large craters (white arrows, Fig. 2), similar to the terrain at the base of Enterprise Rupes. We therefore applied the same approach as for EP_{fw} to the smooth area atop the Rupes ('Hanging wall' Exterior Plains, EP_{hw}: Fig. 3). This analysis returned an S-shaped crater SFD, kinked between 6 and 15 km, thus giving ages of 3.9 ± 0.1 Ga (Fig. 7c) for the older layer and 3.6 ± 0.1 Ga (Fig. 7d) for the younger layer.

The values for the EP_{hw} unit are identical (within error) to those found for the EP_{fw} unit. Using the crater h/D relationship (Pike 1988), we obtained a value for h_{min} for the overlying EP_{fw} unit of 0.19 ± 0.08 km, on the basis of the first crater SFD deflection at a crater diameter of 4 km (equation 4), and a h_{max} value for the EP_{fw} unit of 0.54 ± 0.13 km, on the basis of the second deflection at a crater diameter of 14 km (equation 5). A similar calculation for the overlying EP_{hw} unit yields a minimum thickness of 0.28 ± 0.08 km for the first crater SFD deflection at a crater diameter 6 km (equation 4), and a maximum thickness of 0.56 ± 0.13 km using the second deflection at a crater diameter of 15 km (equation 5). The results for the Exterior Plains units, especially the h_{max} estimates, are in close agreement with those calculated for the younger volcanic sub-unit of the Rembrandt basin's Interior Plains deposits (i.e. 0.36–0.52 km).

Discussion and conclusions

Using the Model Production Function method, we date the formation of the Rembrandt basin to 3.8 ± 0.1 Ga, in agreement with the age suggested by previous works (Watters *et al.* 2009*a*; Fassett *et al.* 2012). On the basis of MPF results, both of the ages of the units exterior to the Rembrandt basin we investigate (3.9 ± 0.1 and 3.6 ± 0.1 Ga) are within the uncertainty of the basin's age itself (3.8 ± 0.1 Ga). Terrains dated as 3.9 ± 0.1 Ga, therefore, may correspond to distal basin ejecta that partially covered pre-existing heavily cratered terrain (cf. Trask & Guest 1975). Since resurfacing has been recognized within the Rembrandt basin, all layers dated as 3.6 ± 0.1 Ga may be due to a later stage of volcanism. Within and surrounding the Caloris basin, MESSENGER data have revealed smooth plains that appear to be younger than the basin itself (Spudis & Guest 1988; Strom *et al.* 2008; Head *et al.* 2009; Fassett *et al.* 2012; Denevi *et al.* 2013), although definitive evidence that these plains are not of impact origin has yet to be found. The thermochemical convection models of Roberts & Barnouin (2012) suggest that the thermal impulse resulting from large impacts, such as that responsible for the Caloris basin, can alter the underlying mantle dynamics, producing subsequent volcanism far from the impact site. Such a scenario may also apply to the Rembrandt basin as the causative mechanism for the later phase of volcanism inferred for those units exterior to the basin rim.

However, on Mercury as on the Moon, it is likely that late-stage lavas also erupted from vents inside the basin, in particular along the margins of the basin floor, because of the likelihood of intense fracturing there that favours the ascent and eruption of magmas (Melosh 2011). In this case, it is possible that lavas that erupted within the basin may have spread beyond the rim, predating the development of parts of the system comprising Enterprise Rupes. Within the basin, the depletion of underlying magma reservoirs, coupled with the load of emplaced lavas, may have caused the basin floor to subside. Subsidence could have induced a first stage of horizontal shortening of the basin floor, forming contractional structures in the central Interior Plains unit (Watters *et al.* 2009*a*). In addition, the lava flows emplaced within the basin may have rapidly cooled, undergoing thermal contraction that could have produced antithetic normal faults within the basin, as has been found for smaller basins and craters (Freed *et al.* 2012; Blair *et al.* 2013).

While we provide observational evidence for volcanic resurfacing that affected the Rembrandt basin and its surroundings after basin formation, volcanism ended before the cessation of basin- and regional-scale tectonic deformation at 3.6 ± 0.1 Ga. Notably, deformation along other age-dated large scarp systems appears to have ceased at 3.7–3.6 Ga (e.g. Di Achille *et al.* 2012; Giacomini *et al.* 2014), somewhat earlier than that along Enterprise Rupes.

The strain rate of global contraction of Mercury is still unclear. However, should the large Enterprise Rupes system have developed before the formation of the Rembrandt basin (e.g. Ferrari *et al.* 2012) during the LHB, regional-scale tectonism was active in this area from >3.8 ± 0.1 Ga to beyond 3.6 ± 0.1 Ga. Recent analysis of Mercury's radius change using MESSENGER data indicates that Mercury experienced far more contraction than previously recognized (Byrne *et al.* 2014). Dating surface units associated with large tectonic structures, as we have done here, elsewhere on Mercury will provide a more detailed understanding of the global tectonic history of the innermost planet.

This research was supported by the Italian Space Agency (ASI) within the SIMBIOSYS Project (ASI-INAF agreement No. I/022/10/0) and by P. di Ateneo CPDA 112213/11 of the University of Padua. P.K. Byrne and C. Klimczak acknowledge support from the MESSENGER project, which in turn is supported by the NASA Discovery Program under contracts NASW-00002 to the Carnegie Institution of Washington and NAS5–97271 to the Johns Hopkins University Applied Physics Laboratory. We thank Mario D'Amore for improving the layout of plots. Careful reviews by B.W. Denevi, T. Platz and an anonymous reviewer substantially improved this manuscript.

References

BLAIR, D. M., FREED, A. M. *ET AL.* 2013. The origin of graben and ridges in Rachmaninoff, Raditladi, and Mozart basins, Mercury. *Journal of Geophysical Research: Planets*, **118/1**, 47–58, http://dx.doi.org/10.1029/2012JE004198

BLEWETT, D. T., HAWKE, B. R., LUCEY, P. G. & ROBINSON, M. S. 2007. A Mariner 10 color study of mercurian craters. *Journal of Geophysical Research: Planets*, **112**(E2), E02005, http://dx.doi.org/10.1029/2006JE002713

BYRNE, P. K., WATTERS, T. R., MURCHIE, S. L., KLIMCZAK, C., SOLOMON, S. C., PROCKTER, L. M. & FREED, A. M. 2012. A tectonic survey of the Caloris basin, Mercury. *In: 43rd Lunar and Planetary Science Conference, held March 19–23, 2012 at The Woodlands, Texas.* Lunar and Planetary Institute, Houston, TX, Abstract 1722.

BYRNE, P. K., KLIMCZAK, C. *ET AL.* 2013. Tectonic complexity within volcanically infilled craters and basins on Mercury. *In: 44th Lunar and Planetary Science Conference, held March 18–22, 2013 at The Woodlands, Texas.* Lunar and Planetary Institute, Houston, TX, Abstract 1261.

BYRNE, P. K., KLIMCZAK, C., ŞENGOR, A. M. C., SOLOMON, S. C., WATTERS, T. R. & HAUCK, S. A., II 2014. Mercury's global contraction much greater than earlier estimates. *Nature Geoscience*, **7**, 301–307.

CINTALA, M. & GRIEVE, R. 1998. Scaling impact melting and crater dimensions: Implications for the lunar cratering record. *Meteoritics and Planetary Science*, **33**, 889–912.

DENEVI, B. W., ROBINSON, M. S. *ET AL.* 2009. The evolution of Mercury's crust: a global perspective from MESSENGER. *Science*, **324**, 613–618.

DENEVI, B. W., ERNST, C. M. *ET AL.* 2013. The distribution and origin of smooth plains on Mercury. *Journal of Geophysical Research: Planets*, **118**, 891–907, http://dx.doi.org/10.1002/jgre.20075

DI ACHILLE, G., POPA, C., MASSIRONI, M., MAZZOTTA EPIFANI, E., ZUSI, M., CREMONESE, G. & PALUMBO, P. 2012. Mercury's radius change estimates revisited using MESSENGER data. *Icarus*, **221**, 456–460.

DZURISIN, D. 1978. Tectonic and volcanic chronology of Mercury as inferred from studies of scarps, ridges, troughs, and other lineaments. *Journal of Geophysical Research*, **83**, 4883–4906.

ERNST, C. M., MURCHIE, S. L. *ET AL.* 2010. Exposure of spectrally distinct material by impact craters on Mercury: Implications for global stratigraphy. *In:* BLEWETT, D. T., HAUCK, S. A. & KORTH, H. (eds) *Mercury after Two MESSENGER Flybys. Icarus*, **209**, 210–223.

ERNST, C. M., MURCHIE, S. L. *ET AL.* 2011. Thickness of volcanic fill in impact basin on Mercury. *In: Geological Society of America 2011 Annual Meeting: Archean to Anthropocene: The Past is the Key to the Future, 9–12 October 2011, Minneapolis, Minnesota.* Geological Society of America, Boulder, CO, Abstract 142-8.

FASSETT, C. I., HEAD, J. W. *ET AL.* 2009. Caloris impact basin: exterior geomorphology, stratigraphy, morphometry, radial sculpture, and smooth plains deposits. *Earth and Planetary Science Letters*, **285**, 297–308.

FASSETT, C. I., HEAD, J. W. *ET AL.* 2012. Large impact basins on Mercury: global distribution, characteristics, and modification history from MESSENGER orbital data. *Journal of Geophysical Research: Planets*, **117/E12**, E00L08, http://dx.doi.org/10.1029/2012JE004154

FERRARI, S., MASSIRONI, M., KLIMCZAK, C., BYRNE, P. K., CREMONESE, G. & SOLOMON, S. C. 2012. Complex history of the Rembrandt basin and scarp system, Mercury. *In: European Planetary Science Congress 2012, held 23–28 September, 2012, Madrid, Spain.* Copernicus, Göttingen, Abstract EPSC-2012-874.

FREED, A. M., BLAIR, D. M. *ET AL.* 2012. On the origin of graben and ridges within and near volcanically buried craters and basins in Mercury's northern plains. *Journal of Geophysical Research: Planets*, **117/E12**, E00L06, http://dx.doi.org/10.1029/2012JE004119

GIACOMINI, L., MASSIRONI, M., MARCHI, S., FASSETT, C. I., DI ACHILLE, G. & CREMONESE, G. 2014. Age dating of an extensive thrust system on Mercury: implications for the planet's thermal evolution. *In:* PLATZ, T., MASSIRONI, M., BYRNE, P. K. & HIESINGER, H. (eds) *Volcanism and Tectonism Across the Inner Solar System.* Geological Society, London, Special Publications, **401**. First published online August 13, 2014, http://dx.doi.org/10.1144/SP401.21

HAWKINS, S. E., BOLDT, J. D. ET AL. 2007. The Mercury Dual Imaging System on the MESSENGER Spacecraft. *Space Science Review*, **131**, 247–338, http://dx.doi.org/10.1007/s11214-007-9266-3

HEAD, J. W., MURCHIE, S. L. ET AL. 2009. Volcanism on Mercury: evidence from the first MESSENGER flyby for extrusive and explosive activity and the volcanic origin of plains. *Earth and Planetary Science Letters*, **285**, 227–242.

HEAD, J. W., CHAPMAN, C. R. ET AL. 2011. Flood volcanism in the northern high latitudes of Mercury revealed by MESSENGER. *Science*, **333**, 1853–1856.

HIESINGER, H., HEAD, J. W., WOLF, U., JAUMANN, R. & NEUKUM, G. 2002. Lunar mare basalt flow units: thicknesses determined from crater size-frequency distributions. *Geophysical Research Letters*, **29**, 8, http://dx.doi.org/10.1029/2002GL014847

HOLSAPPLE, K. A. & HOUSEN, K. R. 2007. A crater and its ejecta: an interpretation of Deep Impact. *In*: COMBI, M. R. & A'HEARN, M. F. (eds) *Deep Impact Mission to Comet 9P/Tempel 1, Part 1. Icarus*, **187**(1), 345–356.

HÖRZ, F., GRIEVE, R., HEIKEN, G., SPUDIS, P. & BINDER, A. 1991. Lunar surface processes. *In*: HEIKEN, G. H., VANIMAN, D. T. & FRENCH, B. M. (eds) *Lunar Source Book: A User's Guide to the Moon*. Cambridge University Press, Cambridge, 61–120.

KIRCHOFF, M. R., CHAPMAN, C. R., MARCHI, S., CURTIS, K. M., ENKE, B. & BOTTKE, W. F. 2013. Ages of large lunar impact craters and implications for bombardment during the Moon's middle age. *Icarus*, **225**, 325–341.

KLIMCZAK, C., SCHULTZ, R. A. & NAHM, A. L. 2010. Evaluation of the origin hypotheses of Pantheon Fossae, central Caloris basin, Mercury. *Icarus*, **209**, 262–270, http://dx.doi.org/10.1016/j.icarus.2010.04.014

KLIMCZAK, C., WATTERS, T. R. ET AL. 2012. Deformation associated with ghost craters and basins in volcanic smooth plains on Mercury: strain analysis and implications for plains evolution. *Journal of Geophysical Research: Planets*, **117** (E12), E00L03, http://dx.doi.org/10.1029/2012JE004100

KNEISSL, T., VAN GASSELT, S. & NEUKUM, G. 2011. Map-projection-independent crater size-frequency determination in GIS environments – new software tool for Arc-GIS. *Planetary and Space Science*, **59**, 1243–1254.

MARCHI, S., MOTTOLA, S., CREMONESE, G., MASSIRONI, M. & MARTELLATO, E. 2009. A new chronology for the Moon and Mercury. *Astronomical Journal*, **137**, 4936–4948.

MARCHI, S., MASSIRONI, M., CREMONESE, G., MARTELLATO, E., GIACOMINI, L. & PROCKTER, L. 2011. The effects of the target material properties and layering on the crater chronology: the case of Raditladi and Rachmaninoff basins on Mercury. *Planetary and Space Science*, **59**, 1968–1980.

MARCHI, S., BOTTKE, W. F., KRING, D. A. & MORBIDELLI, A. 2012. The onset of the lunar cataclysm as recorded in its ancient crater populations. *Earth and Planetary Science Letters*, **325**, 27–38.

MARCHI, S., BOTTKE, W. F. ET AL. 2013a. High-velocity collisions from the lunar cataclysm recorded in asteroidal meteorites. *Nature Geoscience*, **6**, 303–307.

MARCHI, S., CHAPMAN, R. C., FASSETT, C. I., HEAD, J. W., BOTTKE, W. F. & STROM, R. G. 2013b. Global resurfacing of Mercury 4.0–4.1 billion years ago by heavy bombardment and volcanism. *Nature*, **499**, 59–61.

MASSIRONI, M., CREMONESE, G., MARCHI, S., MARTELLATO, E., MOTTOLA, S. & WAGNER, R. J. 2009. Mercury's geochronology revised by applying Model Production Function to Mariner 10 data: geological implications. *Geophysical Research Letters*, **36**, L21204, http://dx.doi.org/10.1029/2009GL040353

MASSIRONI, M., DI ACHILLE, G. ET AL. 2014. Lateral ramps and strike-slip kinematics on Mercury. *In*: PLATZ, T., MASSIRONI, M., BYRNE, P. K. & HIESINGER, H. (eds) *Volcanism and Tectonism Across the Inner Solar System*. Geological Society, London, Special Publications, **401**. First published online June 5, 2014, http://dx.doi.org/10.1144/SP401.16

MCEWEN, A. S. & BIERHAUS, E. B. 2006. The importance of secondary cratering to age constraints on planetary surfaces. *Annual Reviews of Earth and Planetary Sciences*, **34**, 535–567.

MELOSH, H. J. 1989. *Impact Cratering: A Geological Process*. Oxford Monographs on Geology and Geophysics, **11**. Oxford University Press, New York.

MELOSH, H. J. 2011. Impact cratering. *In*: MELOSH, H. J. (ed.) *Planetary Surface Processes*. Purdue University, West Lafayette, IN, 222–275.

MELOSH, H. J. & DZURISIN, D. 1978. Mercurian global tectonics: a consequence of tidal despinning? *Icarus*, **35**, 227–236.

MELOSH, H. J. & MCKINNON, W. B. 1988. The tectonics of Mercury. *In*: VILAS, F., CHAPMAN, C. R. & MATTHEWS, M. S. (eds) *Mercury*. University of Arizona Press, Tucson, AZ, 374–400.

MICHAEL, G. G. & NEUKUM, G. 2010. Planetary surface dating from crater size-frequency distribution measurements: partial resurfacing events and statistical age uncertainty. *Earth and Planetary Science Letters*, **294**, 223–229.

MURCHIE, S. L., WATTERS, T. R. ET AL. 2008. Geology of the Caloris basin, Mercury: a new view from MESSENGER. *Science*, **321**, 73–76.

NEUKUM, G. & HORN, P. 1975. Effects of lava flows on lunar crater populations. *The Moon*, **15**, 205–222.

NEUKUM, G., OBERST, J., HOFFMANN, H., WAGNER, R. & IVANOV, B. A. 2001. Geologic evolution and cratering history of Mercury. *Planetary and Space Science*, **49**, 1507–1521.

PIKE, R. J. 1988. Geomorphology of impact craters on Mercury. *In*: VILAS, F., CHAPMAN, C. R. & MATTHEWS, M. S. (eds) *Mercury*. University of Arizona Press, Tucson, AZ, 165–273.

PLATZ, T., MICHAEL, G., TANAKA, K. L., SKINNER, J. A. & FORTEZZO, C. M. 2013. Crater-based dating of geological units on Mars: methods and application for the new global geological map. *Icarus*, **225**, 806–825.

PROCKTER, M. L., ERNST, C. M. ET AL. 2010. Evidence for young volcanism on Mercury from the third MESSENGER flyby. *Science*, **329**, 668–671.

ROBERTS, J. H. & BARNOUIN, O. S. 2012. The effect of the Caloris impact on the mantle dynamics and volcanism of Mercury. *Journal of Geophysical Research: Planets*, **117**/**E2**, E02007, http://dx.doi.org/10.1029/2011JE003876

ROBINSON, M. S. & LUCEY, P. G. 1997. Recalibrated Mariner 10 color mosaics: implications for Mercurian volcanism. *Science*, **275**, 197–200.

ROBINSON, M. S., MURCHIE, S. L. *ET AL.* 2008. Reflectance and color variations on Mercury: indicators of regolith processes and compositional heterogeneity. *Science*, **321**, 66–69.

SCHULTZ, R. A. 1993. Brittle strength of basaltic rock masses with applications to Venus. *Journal of Geophysical Research: Planets*, **98/E6**, 10 883–10 895, http://dx.doi.org/10.1029/93JE00691

SPUDIS, P. D. & GUEST, J. E. 1988. Stratigraphy and geologic history of Mercury. *In*: VILAS, F., CHAPMAN, C. R. & MATTHEWS, M. S. (eds) *Mercury*. University of Arizona Press, Tucson, AZ, 118–164.

STROM, R. G. & NEUKUM, G. 1988. The cratering record on Mercury and the origin of impacting objects. *In*: VILAS, F., CHAPMAN, C. R. & MATTHEWS, M. S. (eds) *Mercury*. University of Arizona Press, Tucson, AZ, 336–373.

STROM, R. G., TRASK, N. J. & GUEST, J. E. 1975. Tectonism and volcanism on Mercury. *Journal of Geophysical Research*, **80**, 2478–2507.

STROM, R. G., MALHOTRA, R., ITO, T., YOSHIDA, F. & KRING, D. A. 2005. The origin of planetary impactors in the inner solar system. *Science*, **309**, (5742), 1847–1850.

STROM, R. G., CHAPMAN, C. R., MERLINE, W. J., SOLOMON, S. C. & HEAD, J. W. 2008. Mercury cratering record viewed from MESSENGER's first flyby. *Science*, **321**, 79–81.

TERA, F., PAPANASTASSIOU, D. A. & WASSERBURG, G. J. 1974. Isotopic evidence for a terminal lunar cataclysm. *Earth and Planetary Science Letters*, **22**, 1–21.

TOKSÖZ, M. N., PRESS, F. *ET AL.* 1972. Structure composition and properties of lunar crust. *In*: *Proceedings of the Third Lunar Science Conference, Houston, Texas, January 10–13, 1972, Sponsored by the Lunar Science Institute, Volume 3*. MIT Press, Cambridge, MA, 2527–2544.

TRASK, N. J. & GUEST, J. E. 1975. Preliminary geologic terrain map of Mercury. *Journal of Geophysical Research*, **80**, 2461–2477.

WATTERS, T. R. & NIMMO, F. 2010. The tectonics of Mercury. *In*: WATTERS, T. R. & SCHULTZ, R. A. (eds) *Planetary Tectonics*. Cambridge University Press, Cambridge, 15–79.

WATTERS, T. R., ROBINSON, M. S., BINA, C. R. & SPUDIS, P. D. 2004. Thrust faults and the global contraction of Mercury. *Geophysical Research Letters*, **31**, L04071.

WATTERS, T. R., NIMMO, F. & ROBINSON, M. S. 2005. Extensional troughs in the Caloris Basin of Mercury: evidence of lateral crustal flow. *Geology*, **33**, 669–672.

WATTERS, T. R., HEAD, J. W. *ET AL.* 2009a. Evolution of the Rembrandt Impact Basin on Mercury. *Science*, **324**, 618–621.

WATTERS, T. R., MURCHIE, S. L., ROBINSON, M. S., SOLOMON, S. C., DENEVI, B. W., ANDRÉ, S. L. & HEAD, J. W. 2009b. Emplacement and tectonic deformation of smooth plains in the Caloris basin, Mercury. *Earth and Planetary Science Letters*, **285**, 309–319.

WATTERS, T. R., SOLOMON, S. C., ROBINSON, M. S., HEAD, J. W., ANDRÈ, S. L., HAUCK, S. A. & MURCHIE, S. L. 2009c. The tectonics of Mercury: the view after MESSENGER's first flyby. *Earth and Planetary Science Letters*, **285**, 283–296.

WATTERS, T. R., SOLOMON, S. C. *ET AL.* 2012. Extension and contraction within volcanically buried impact craters and basins on Mercury. *Geology*, **40**, 1123–1126, http://dx.doi.org/10.1130/G33725.1

The development of the Deception Island volcano caldera under control of the Bransfield Basin sinistral strike-slip tectonic regime (NW Antarctica)

F. C. LOPES[1,2]*, A. T. CASELLI[3], A. MACHADO[1] & M. T. BARATA[1]

[1]Centre for Geophysics of the University of Coimbra, Coimbra, Portugal

[2]Department of Earth Sciences, University of Coimbra, 3000-272 Coimbra, Portugal

[3]Grupo de Estudio y Seguimiento de Volcanes Activos, Int. Güiraldes 2160 – Ciudad Universitaria, Pab.2. C1428EHA, Buenos Aires, Argentina

*Corresponding author (e-mail: fcarlos@dct.uc.pt)

Abstract: Deception Island is a small and volcanically active caldera volcano of Quaternary age, located in the marginal basin of Bransfield Strait, NW Antarctica. The distribution and orientation of fracture and fault systems that have affected the Deception volcanic edifice, and the elongated geometry of its volcanic caldera, are consistent with a model of Riedel deformation induced by a regional left-lateral simple shear zone. It is suggested that this caldera was formed above a magma chamber stretched under the influence of the regional transtensional regime with left-lateral simple shear. The collapse may have occurred in at least two phases: first, a small volume event occurred along the compressed flanks of the volcano edifice; and second, a large collapse event affected the stretched flanks of the volcano edifice.

Regional shear zone corridors control magma movement, deform volcanoes and may destabilize their flanks. They can also contribute to the formation of volcanic calderas. Through analogue experiments, Holohan et al. (2008) demonstrated that volcanic calderas that form in regional strike-slip tectonic environments can be structurally controlled by the interaction between structures associated with regional deformation and volcanotectonic subsidence. The final geometry of the caldera may be a result of magma chamber elongation through simple shearing in a pre-collapse phase and its slightly sigmoidal stretching in a direction roughly parallel to the regional distension (orientation of the major axis of the deformation ellipse). Regional pre-collapse faulting, generally resulting from Riedel deformation with orientations tangential to the core of the magma chamber, together with the faults associated with the edges of the magma chamber, would reactivate in order to accommodate subsidence of the caldera bottom. Reverse ring faults, formed at the ends of the shortening axis, would spread towards the lengthening axis. Initially, the collapse takes place, on a small scale, on the flanks that are in compression (along the shortening axis). However, a large collapse can occur on the flanks that are in extension (along the extended axis; Mathieu et al. 2011).

Such a structural behaviour may be applicable to Deception Island, a volcanically active shield volcano of Quaternary age (less than 780 ka; Valencio et al. 1979; Smellie 2002; Baraldo et al. 2003) and of small volume, with a basal diameter of c. 30 km and a height of 1500 m above the surrounding seafloor. It is located in the marginal basin of Bransfield Strait (Bransfield Trough; 62°57′S; 60°37′W), which separates the South Shetland Islands from the Antarctic Peninsula (Smellie 2002; Fig. 1). The emerged top of the island has an elongated horseshoe shape, 8 × 12 km in size, inside which there is a 5–9 km diameter volcanic caldera of unknown age, with a narrow opening to the sea on the SE side named Neptune's Bellows (Fig. 2; Baker et al. 1975). The caldera is mostly flooded by the 'Port Foster Bay' with a perimeter of c. 32 km and maximum depth of c. 160 m (Barclay et al. 2009). Most of the recent volcanic activity has occurred in the close neighbourhood of the rims of this depression. The historic eruptions occurred in 1842, 1912, 1917, 1967, 1969 and 1970 (Orheim 1971; Smellie 2002; Torrecillas et al. 2012), producing considerable volumetric changes and modification to the coastline (Torrecillas et al. 2012). The highest point of the island (542 m) lies on the eastern side of the caldera (cf. Mount Pond; Fig. 2). More than half of the island is covered by glaciers (López-Martínez & Serrano 2002).

The origin of the Deception caldera remains controversial and no geochronological data are

From: PLATZ, T., MASSIRONI, M., BYRNE, P. K. & HIESINGER, H. (eds) 2015. Volcanism and Tectonism Across the Inner Solar System. Geological Society, London, Special Publications, **401**, 173–184.
First published online February 6, 2014, http://dx.doi.org/10.1144/SP401.6

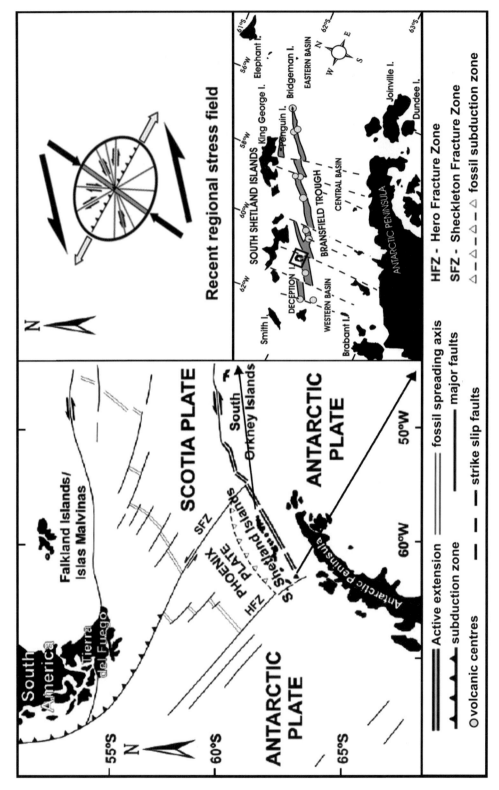

Fig. 1. Regional tectonic framework and main morphotectonic features of the western Scotia Arch and Deception Island regions, and current regional stress field (redrawn from Maestro *et al.* 2007; Torrecillas *et al.* 2012).

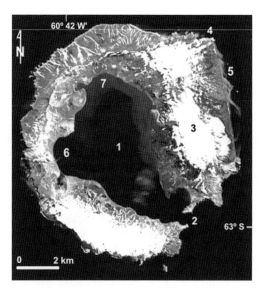

1. Port Foster

2. Neptunes Bellows

3. Mount Pond

4. Macaroni Point

5. Costa Recta

6. Fumarole Bay

7. Telefon Bay

Fig. 2. Geographical setting of Deception Island (background image from USGS 2005).

available. For some authors, the large interior bay is the result of the passive collapse of the volcanic edifice along orthogonal faults (Marti & Baraldo 1990). However, the evolution and collapse of the volcanic edifice could have been influenced by the kinematics of the great NE–SW-trending regional faults (Smellie 1989; Rey et al. 1995; Marti et al. 1996; Smellie 2002). Gravimetric and magnetic anomalies identified in the Deception Island area reveal a NE–SW linear trend in the crustal mass (Ortiz et al. 1992a, b; Muñoz-Martín et al. 2005). Cooper et al. (1998) and Smellie (2002) suggested that the caldera is currently affected by resurgence, although this has been refuted by Barclay et al. (2009). The resurgence hypothesis was supported by increased seismic activity during early 1992 (Ortiz et al. 1992a, b), although there was no eruption. The last period of intense tectonic seismicity was 1998–1999 (Ibáñez et al. 2000).

This work proposes a generalized interpretative model for the development of the Deception Island volcano caldera, by relating its geometry to the elongation of its underlying magma chamber through simple shear controlled by the regional transtensional regime, through comparison to the results of analogue models of caldera collapse in strike-slip tectonic regimes constructed by Holohan et al. (2008) and Mathieu et al. (2011).

Regional tectonic context

Bransfield Strait (62°57′S; 60°37′W) is a narrow, 500 km-long, NE–SW-trending basin (Bransfield Basin or Trough) located SE of the South Shetland Trench and Islands. It separates the South Shetland Islands from the Antarctic Peninsula, at the southwestern edge of the Scotia plate (Fig. 1). This basin ends abruptly towards the SW along the Hero Fracture Zone, whereas the NE limit is more gradual and is defined by the Shackleton Fracture Zone, Southwestern Scotia plate (González-Casado et al. 2000). Morphologically, it is divided into the western, central and eastern Bransfield sub-basins, which are separated by the highs of Deception and Bridgeman islands, respectively (Jeffers & Anderson 1991; Fig. 1).

The opening of Bransfield Trough began during the Late Cenozoic (probably about 4 myr ago) and could be the result of two processes: (a) a back-arc process related to the South Shetland Islands volcanic arc, resulting from the slow oblique subduction or the roll-back of the former Phoenix plate under the Antarctic plate along the northern margin of the South Shetland Block (at South Shetland Trench; Maldonado et al. 1994; Lawver et al. 1995, 1996); or (b) a transtensional basin related to the left-lateral motion between the Antarctic and the Scotia plates along the Shackleton Fracture Zone and the South Scotia Ridge, and between the Antarctic plate and the South Shetland Block along the South Shetland Trench (Grácia et al. 1996, 1997; Klepeis & Lawver 1996; González-Casado et al. 2000; Giner-Robles et al. 2003; Maestro et al. 2007). The orientation of the macrostructures that define the present-day morphology of this basin corresponds to a left-lateral strike-slip zone with simple shear, formed within a stress field whose maximum horizontal compressive stress (σ_1) trends N30°E, a result also confirmed by seismotectonics of focal mechanism solutions (Fig. 1; Grácia *et al.* 1996; Maestro *et al.* 2007; Pérez-López *et al.* 2007). Nevertheless, it is suggested that, during the Late Cenozoic, a first stage occurred, characterized by σ_1 and σ_3 trending NW–SE and NE–SW, respectively, that caused the subduction of the Phoenix oceanic plate under the Antarctic plate (Barker & Lawver 1988). Recent volcanic and seismic activity (Forsyth 1975; Pelayo & Wiens 1989; Fisk 1990; Grácia *et al.* 1996; Robertson Maurice *et al.* 2001), high heat flow (Nagihara & Lawver 1989), a positive magnetic anomaly (Roach 1978) and a large negative gravity anomaly (Garrett 1990), recorded along the axis of Central Bransfield Basin, suggest that the basin is still opening (Saunders & Tarney 1984; Fisk 1990; Lawver *et al.* 1995). The recent crustal movements or spreading rates in Bransfield Trough and South Shetland Islands, based on geodetic studies, are estimated to be 5–20 mm a^{-1} in the NW–SE direction (Dietrich *et al.* 1996). However, the motion of the South Shetland Block has been calculated to amount to 17 mm a^{-1} in a N20°E direction (Dietrich *et al.* 2001), which is consistent with sinistral movement between the Phoenix and Antarctic plates.

Geological setting

Lithology

Considering the volcano-tectonic evolution of Deception Island and only a single episode of caldera collapse, the volcanic deposits that cover the island can be divided into two groups (Smellie 2001, 2002; Fernández-Ibáñez *et al.* 2005; Fig. 3):

(1)   pre-caldera deposits (Port Foster Group) – mainly composed of pillow-lavas (currently below sea-level), explosive hydrovolcanic eruption material and pyroclastic density current deposits;

(2)   post-caldera deposits (Mount Pond Group) – composed of subaerial lavas and agglutinate in scoria cones, tuff cones and scarce tuff rings. All the exposed post-caldera rocks appear to be late Pleistocene and recent, probably <100 kyr in age (Shultz 1970; Smellie 2001).

According to Wit *et al.* (1991) and Smellie (2002), the pre-caldera phase is represented by effusive rocks varying in composition from basalt to andesite. In the post-caldera phase, basaltic andesites predominate and only a few dacites are observed. The most recent date estimated for the volcano collapse event is 100 ka, on the basis of dating of the youngest rock before the formation

of the caldera (Smellie 2001; carried out by Shultz 1970).

Structural framework

The volcanic structure and morphology of Deception Island are controlled, in several areas, by systems of faults and fractures, inferred on the basis of field observations, interpretation of satellite image/digital elevation model (DEM) interpretation and bathymetric data, seismic reflection profile and seismic tomography studies (Smellie 1989; Lawver *et al.* 1996; Grácia *et al.* 1997; Fernández-Ibáñez *et al.* 2005; Pérez-López *et al.* 2007; Ben-Zvi *et al.* 2009; Lopes *et al.* 2012). The satellite lineaments have the same orientations as the mesoscopic fractures measured in the field. One of the most remarkable geological and geomorphological features is the linearity of the easternmost, NNW–SSE-trending coastal landform of the island named the Costa Recta Beach. According to Fernández-Ibáñez *et al.* (2005), it is the geomorphological expression of a retreated scarp produced by a submarine fault orientated NNW–SSE. Another major feature interpreted from satellite image data is the sinistral displacement of the Mount Pond Crest, on the eastern flank of the island, controlled by an ENE–WSW-trending left-lateral strike-slip fault. Also, the morphology of the coast at Macaroni Point may indicate the existence of a main trend of faults parallel to N50–60°E.

Analysis of the primary fracture sets on Deception Island shows that the main structures have the same NE–SW and NW–SE orientations as the macrostructures that define the morphology of the Bransfield Trench (Rey *et al.* 1995; Grácia *et al.* 1996; González-Casado *et al.* 1999; Maestro *et al.* 2007; Pérez-López *et al.* 2007). Gravimetric and magnetic anomalies also reveal a NE–SW linear trend in the crustal structure beneath the island (Ortiz *et al.* 1992a, b; Muñoz-Martín *et al.* 2005), which is compatible with the dominant regional tectonic trend. Other structural orientations were observed: north–south, NNE–SSW, ENE–WSW to east–west, WNW–ESSE and NNW–SSE (Paredes *et al.* 2007). The dominant regional tectonic regime is sinistral transtensional, with σ_1 oriented N30°E. This orientation of σ_1 is compatible with the dominant pattern established from earthquakes and other fault studies in the Scotia Arc and NW Antarctic Peninsula regions.

From the gravity and magnetic models, two types of crust were identified in the area of Deception Island (Muñoz-Martín *et al.* 2005). These were interpreted as continental crust (located north of Deception Island) and more basic crust (south of Deception Island). The transition between these

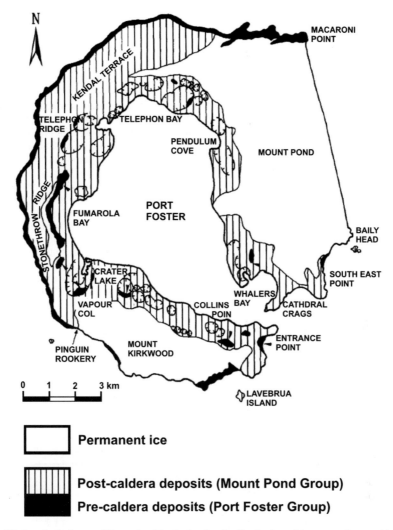

Fig. 3. Simplified geological map of Deception Island, showing the distribution of the pre- and post-caldera volcanic deposits (modified from Marti & Baraldo 1990).

crustal types corresponds to an elevated Bouguer anomaly trending NE–SW.

Methodology

A re-evaluation of the findings published by other authors was made so that those interpretations could be appraised through comparison with the results obtained from analogue modelling of caldera collapse in strike-slip tectonic regimes conducted by Holohan *et al.* (2008) and Mathieu *et al.* (2011). For the development of our proposal it is important to know: (a) the geometry and depth of the Deception magma chamber; (b) the orientation of the major regional faults and local faults and fractures; and (c) the shape of the pre-collapse Deception volcanic edifice.

Geophysical studies indicating the geometry, location and depth of the Deception magma chamber were taken from Ben-Zvi *et al.* (2009). Field data of the primary fault and fracture systems and their Riedel interpretations were taken from Maestro *et al.* (2007). Caldera-related tectonic features were evaluated through the use of satellite imagery and a DEM from Fernández-Ibáñez *et al.* (2005), based upon a 1:25 000 topographic database. The identification of the main structural lineaments was done by visual interpretation, following the interpretation proposed by Maestro *et al.* (2007), Paredes *et al.* (2007), Pérez-López *et al.* (2007) and Lopes *et al.* (2012). Riedel's regional deformation model is based on the geometrical relationship of the interpreted main structural lineaments with the orientation of the Bransfield basin regional macrostructures. The palaeo-shape of Deception Island volcano was considered on the basis of the reconstruction of the palaeo-volcanic edifice prior to the formation of the present caldera made by Torrecillas *et al.* (2013).

Results and discussion

Relevant aspects

The orientation of the macrostructures that define the morphology of Bransfield Strait suggests that this basin is a regional, N60°E-to-N75°E, left-lateral simple shear zone corridor, whose movements are controlled by a stress field with σ_1 oriented N30°E (Fig. 1). This scenario was interpreted by different authors on the basis of several structural analyses of the Bransfield Strait and Deception Island, using fault population analysis, seismotectonics of focal mechanism solutions and tectonic geomorphology (Rey *et al.* 1995; Grácia

et al. 1996; González-Casado *et al.* 1997, 2000; Smellie 2002; Fernández-Ibáñez *et al.* 2005; Maestro *et al.* 2007).

The orientation of the faults and fractures observed in the field (Maestro *et al.* 2007) and the orientation of structural lineaments interpreted from morphostructural analysis of DEM (Maestro *et al.* 2007; Pérez-López *et al.* 2007; Lopes *et al.* 2012) are similar to the orientations of fractures that characterize Riedel's deformation model. This interpretation is based on their geometrical and kinematic relationship with the orientation of the regional macrostructures (Fig. 4).

Riedel deformation is a geometric structural pattern commonly associated with strike-slip shear zones (Riedel 1929; Tchalenko 1970; Harris & Cobbold 1984), and may correspond to en-echelon shear zones within and adjacent to the principal shear zone (Fig. 4b). These short planar structures can be of many types: (a) synthetic strike-slip faults, or Riedel (R) shears; (b) antithetic strike-slip faults, or conjugate Riedel (R′) shears; (c) secondary synthetic faults, or P-shears; (d) extension or tension (T) fractures; and (e) strike-slip faults parallel to the principal displacement zone, or Y-shears. R-shears and R′-shears form at about 15 and 75° to the

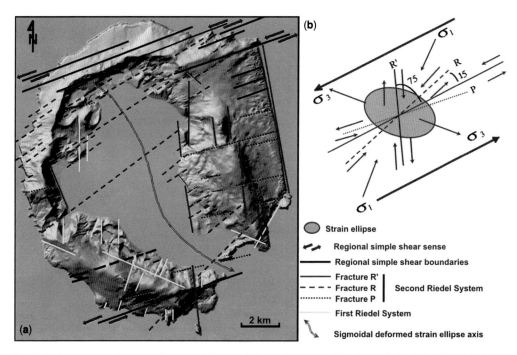

Fig. 4. Main structural alignments that control the morphology of Deception Island, and their relationship with the sinistral simple shear zone dominant in the Bransfield Strait. (**a**) Main structural alignment patterns interpreted from a digital elevation model, following the interpretations proposed by Pérez-López *et al.* (2007), Maestro *et al.* (2007) and Lopes *et al.* (2012). (**b**) Comparison with a theoretical model of Riedel deformation, adapted to the trend and kinematics of the regional simple shear zone. σ_1 denotes maximum compressive stress; σ_3, maximum extensive stress.

main fault zone, respectively. These two fault orientations can be understood as conjugate fault sets at 30° to the short axis of the instantaneous strain ellipse associated with the simple shear strain field. P-shears are roughly symmetrical to the R-shears with respect to the overall shear direction and they form with further displacement of the principal shear zone. T-fractures are often normal faults arranged parallel to the short axis of the strain ellipse and form at about 45° to the main fault zone. Riedel shear zones formed in a given stress regime are rapidly abandoned and rotated by the displacement of the main shear zone, and new Riedel shear zones develop.

According to Maestro et al. (2007), on the basis of the geometrical and kinematic relationship between the location and orientation of the faults and fractures in Deception Island, it is possible to distinguish two evolutionary Riedel stages. These stages can be related to an inferred counterclockwise rotation of island.

Stage 1 (first Riedel system in Fig. 4a) – considering that the current maximum principal horizontal stress direction in Deception Island is approximately N30°E, the fault system of this Riedel first stage is unfavourably oriented with respect to the present-day stress field. According to Maestro et al. (2007), these structures particularly affect the oldest volcanic deposits outcropping in

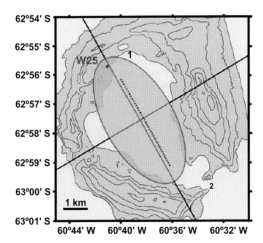

Fig. 5. A map of Deception Island showing its elliptical caldera. 1, Telefon Bay; 2, Neptune's Bellows (modified from Ben-Zvi et al. 2009).

the SE side of the island (e.g. Entrance Point and Cathedral Crags). The synthetic Riedel shears (R) and the antithetic Riedel shears (R′) are oriented north–south and WNW–ESSE, respectively. The secondary P-shears are oriented NNE–SSW.

Stage 2 (second Riedel system in Fig. 4a) – this second Riedel stage is compatible with the

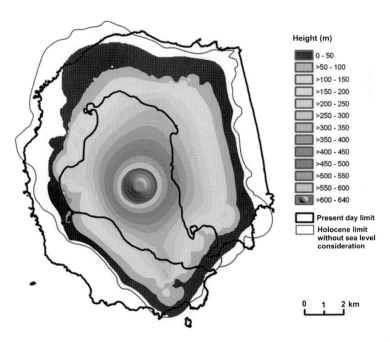

Fig. 6. Palaeo-digital elevation model (DEM) of the Deception volcano edifice before the caldera formation, with superposition of present-day and Holocene caldera limit without considering the differences in sea-level (modified from Torrecillas et al. 2013).

present-day stress field orientation. Its structures, which affect all volcanic deposits, display several orientations. The synthetic Riedel shears (R) are oriented NE–SW (the faults that delimit the deepest area of Port Foster) and the antithetic Riedel shears (R′) are oriented NNW–SSE (Costa Recta Fault). The P-shears are oriented ENE–SWS and show a left-lateral strike-slip movement (the faults that displaces Mount Pond Ridge). The Y-shears present left-lateral strike-slip movement and are visible in the NE part of the island (Macaroni Point), where the coast shows a series of steps toward the west.

Deception Island has an elliptical caldera, with semi-major and semi-minor axes of 4.5 and 2.7 km, respectively. This geometry has been determined from field mapping, satellite imagery, the locations of vents and fissure eruptions and fractal analysis (Pérez-López *et al.* 2007), and confirmed by seismic tomography (Ben-Zvi *et al.* 2009). The semi-major axis, which extends from Telefon Bay (the northwestern-most point) to Neptune's Bellows (the southeastern-most point), is oriented N25–30°W, roughly parallel to the direction of the regional extension (Fig. 5). The seismic tomography also reveals the existence of an

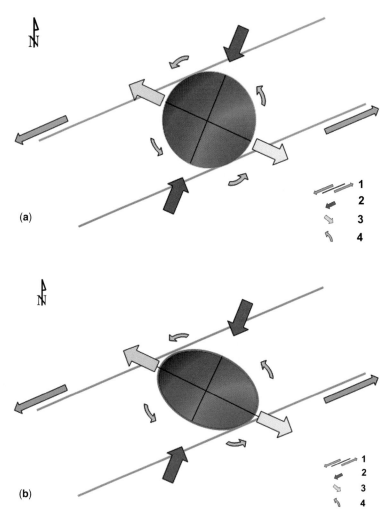

Fig. 7. Simplified sketch of the pre-collapse evolution of Deception magma chamber through the left-lateral simple shear zone motion dominant in Bransfield Strait. (**a**) The initial circular magma chamber. (**b**) As the magma chamber undergoes a counterclockwise rotation, inside the left-lateral shear zone, it acquires an elliptical shape, with stretching roughly parallel to the NW–SE direction of the regional extension, and shortening along the NE–SW direction of the regional maximum compressive stress (σ_1). 1, Regional simple shear; 2, direction of compression; 3, direction of extension; 4, sense of rotation.

elliptical (5 and 3 km across along the NNW–SSE and ENE–WSW directions, respectively) magma chamber that extend downwards from ≤2 km below the seafloor to >4 km depth and that may contain a substantial volume of melt (Ben-Zvi *et al.* 2009). According to Lipman (1997), information on underlying structures, the depth of subsidence or any relation to the source magma reservoir is commonly ambiguous at old and eroded calderas but some consistent relations have emerged between the collapse geometry of little-eroded young calderas and the shape of underlying magma chambers. As Deception Island has a young volcano caldera (estimated age <100 kyr; Smellie 2001), it is possible to infer a relationship between its elliptical geometry and the elliptical shape of its underlying shallow magma chamber.

Torrecillas *et al.* (2013) presented a palaeoreconstruction of the pre-caldera volcano on Deception Island, using a Geodynamic Regression Model, taking into account the combined regional and local geodynamic deformation in both directions, NE–SW (compression) and NW–SE (extension). A pre-collapse 100 kyr time period was selected for the reconstruction, which is the most recent age estimated for the volcano collapse event, based on the dating of the youngest pre-collapse rocks. The resulting model shows an old stratovolcano with an elliptical shape at the stage immediately before its collapse and caldera formation (Fig. 6). The palaeo-elliptical shape was explained by Torrecillas *et al.* (2013) as being the result of the regional geodynamic deformation suffered by the Deception volcano since the time of its formation, 600 ka ago (Valencio *et al.* 1979; Smellie 2002; Baraldo *et al.* 2003), until the selected data of reconstruction.

The glacier that covers the ENE flank of the island displays, along the Costa Recta Cliff (Fig. 2), a set of folds whose vergence appears to reflect the orientation of the caldera's semi-minor axis (Caselli, pers. comm. 2011). Although these folds could simply be a consequence of the movement of the ice layers that compose the glacier, their location in relation to the current stress field suggests that they may also include rock material and that they might have been influenced by regional deformation.

We suggest that the collapse of the Deception caldera results from volcanotectonic deformation above the underlying magma chamber that was stretched under the influence of the regional left-lateral transtensional regime. This inferred structural outcome matches that proposed by Holohan *et al.* (2008) and Mathieu *et al.* (2011) using analogue models of volcanic caldera formation in strike-slip regimes.

The collapse of deception caldera in a left-lateral shear zone

The following description summarizes the hypothetical formation process of the Deception Island caldera.

(1) Formation of Deception Island *c.* 780 ka, by volcanic activity related to the widespread extension and faulting that marked the Bransfield Trough.

(2) Counterclockwise rotation and the consequent elliptical deformation of Deception magma chamber through the left-lateral simple shear zone motion dominant in Bransfield Strait, created by a stress field whose σ_1 trends N30°E (Fig. 7a).

(3) Development of the first stage of Riedel deformation (first Riedel system in Fig. 4a) under the influence of this regional tectonism.

(4) As the Deception magma chamber undergoes a counterclockwise rotation in the regional left-lateral shear zone, it acquires an elliptical shape, with stretching roughly parallel to the NW–SE direction of the regional extension, and shortening along the NE–SW direction of the regional maximum compressive stress (σ_1; Fig. 7b). Perpendicular to the axes of maximum extension and compression, respectively, normal faults and reverse faults/folds are formed.

(5) As the rotation continues, it changes the orientations of the Riedel first-stage deformation

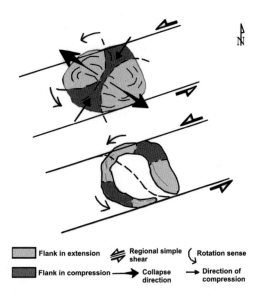

Fig. 8. Sketches summarizing the hypothetical formation process of Deception Island caldera, above an elliptically elongated magma chamber.

structures and leads to the development of a Riedel second-stage system of structures (second Riedel system in Fig. 4a). According to Maestro *et al.* (2007), the angle between the direction of the new and old faults in this scenario is dependent on the cohesive strength values, the coefficient of friction on young and mature faults under the Coulomb failure criterion and the rotation angle of the external boundaries on different intervening blocks.

(6) Owing to the combination of continuous regional stretching and volcano-tectonic fracturing, *c.* 100 ka, the caldera subsides above the elliptical elongated magma chamber (Fig. 8). The tectonic fracturing allows the lateral drainage of underlying magma to lower (often submarine) elevations. Volcanic eruptions empty out of the magma chamber over long periods of time. This results in chamber roof instability and, consequently, the caldera collapses. The caldera collapse may have occurred in at least two phases: initially, a small-scale collapse occurred along the compressed flanks (i.e. along the shortening axis). Subsequently, the main collapse event occurred, affecting the flanks under extension (i.e. along the lengthening axis). Deception caldera can be classified as a drainage caldera (Lockwood & Hazlett 2010), a kind of caldera typical of basaltic oceanic volcanoes.

The chronological relationship between the period of volcanic collapse and the passage of Riedel stage 1 to Riedel stage 2 is unclear. It is not possible at present to know when the second-stage Riedel structures began to develop. According to Maestro *et al.* (2007), the structures belonging to this stage affect all deposits, whereas the structures belonging to the first stage affect only pre-caldera deposits. Thus, the beginning of the second stage may lie either just before or during the volcanic collapse. In any case:

(7) Continued development of the post-collapse regional shear zone. The Riedel faults that cut across the central area may serve as magma ascent conduits. If, as determined by Dietrich *et al.* (2001), the recent left-lateral movement rate along the axis of Bransfield Trough is 17 mm a^{-1}, then the counterclockwise rotation rate of Deception Island is about 2.3 mm a^{-1}.

Conclusions

Following the evidence described above, we propose that the subsidence of Deception Island caldera may have been strongly controlled by regional pre-collapse tectonics. Under the control of the Bransfield Basin sinistral strike-slip tectonic regime with simple shear, the Deception magma chamber experienced a counterclockwise rotation and consequent elliptical deformation, with stretching roughly parallel to the NW–SE direction of the regional extension, and shortening along the NE–SW direction of the regional maximum compressive stress. Owing to the combination of continuous regional stretching and volcano-tectonic fracturing, the caldera subsided above the sigmoidally elongated magma chamber. The collapse may have occurred at least in two phases: first, at a small scale, along the compressed flanks of the volcano edifice, and second, in a large collapse event, which affected the stretched flanks of the volcano edifice. Currently, the temporal relationship between the period of volcanic collapse and the passage of Riedel stage 1 to Riedel stage 2 is unclear. However, the absence of structures of the first stage that affect post-caldera deposits suggests that the beginning of the second stage may lie either just before or during the volcanic collapse.

This research is funded by FCT, the Portuguese Science Foundation, under the contract PEst-OE/CTE/UI0611/2012 – Center for Geophysics. We would like to thank the Dirección Nacional del Antártico and Instituto Antártico Argentino for helping us in the logistic and subsistence arrangements for the field seasons. We are grateful to R. Vegas for reviewing an early version of the work. Our thanks go to P. K. Byrne (Editor) and to two anonymous reviewers for their comments and suggestions at various stages of the manuscript.

References

BAKER, P., MC REATH, I., HARVEY, M., ROOBAL, M. & DAVIES, T. 1975. *The Geology of the South Shetland Islands, V: Volcanic Evolution of Deception Island.* British Antarctic Survey Scientific Report, **78**.

BARALDO, A., RAPALINI, A. E., BOHNEL, H. & MENA, M. 2003. Paleomagnetic study of Deception Island, South Shetland Islands, Antarctica. *Geophysical Journal International,* **153**, 1–11.

BARCLAY, A. H., WILCOCK, W. S. D. & IBÁÑEZ, J. M. 2009. Bathymetric constraints on the tectonic and volcanic evolution of Deception Island Volcano, South Shetland Islands. *Antarctic Science,* **21**, 153–167.

BARKER, P. F. & LAWVER, L. A. 1988. South American– Antarctic plate motion over the past 50 Myr, and the evolution of the South American–Antarctic Ridge. *Geophysical Journal,* **94**, 377–386.

BEN-ZVI, T., WILCOCK, W. S. D., BARCLAY, A. H., ZANDOMENEGHI, D., IBÁÑEZ, J. M. & ALMENDROS, J. 2009. The P-wave velocity structure of Deception Island, Antarctica, from two-dimensional seismic tomography. *Journal of Volcanology and Geothermal Research,* **180**, 67–80.

COOPER, A. P. R., SMELLIE, J. L. & MAYLIN, J. 1998. Evidence for shallowing from bathymetric records of Deception Island. *Antarctic Science*, **10**, 455–461.

DIETRICH, R., DACH, R. ET AL. 1996. The SCAR 95 GPS Campaign: objectives, data analysis and final solution. In: DIETRICH, R. (ed.) *The Geodetic Antarctic Project GAP95 German Contributions to the SCAR 95 Epoch Campaign.* Deutsche Geodätische Kommission bei der Bayerischen Akademie der Wissenschaften, Reithe B, Heft Nr.

DIETRICH, R., DACH, R. ET AL. 2001. ITRF coordinates and plate velocities from repeated GPS campaigns in Antarctica – an analysis based on different individual solutions. *Journal of Geodesy*, **74**, 756–766.

FERNÁNDEZ-IBÁÑEZ, F., PEREZ-LOPEZ, R., MARTÍNEZ-DÍAZ, J. J., PAREDES, C., GINER-ROBLES, J. L., CASELLI, A. T. & IBÁÑEZ, J. M. 2005. Costa Recta beach, Deception Island, West Antarctica: a retreated scarp of a submarine fault? *Antarctic Science*, **17**, 418–426.

FISK, M. R. 1990. Volcanism in the Bransfield Strait, Antarctica. *Journal of South American Earth Sciences*, **3**, 91–101.

FORSYTH, D. W. 1975. Fault plane solutions and tectonics of South Atlantic and Scotia Sea. *Journal of Geophysical Research*, **80**, 1424–1443.

GARRETT, S. W. 1990. Interpretation of reconnaissance gravity and aeromagnetic surveys of the Antarctic Peninsula. *Journal of Geophysical Research*, **95**, 6759–6777.

GINER-ROBLES, J. L., GONZÁLEZ-CASADO, J. M., GUMIEL, P., MARTÍN, S. & GARCÍA, C. 2003. A kinematic model of the Scotia plate (SW Atlantic Ocean). *Journal of South American Earth Sciences*, **16**, 179–191.

GONZÁLEZ-CASADO, J. M., LÓPEZ-MARTÍNEZ, J., DURÁN, J. J. & BERGAMÍN, J. F. 1997. Fracturación y campos de esfuerzos recientes en el entorno del Estrecho de Bransfield, Antártida Occidental. *Boletín da la Real Sociedad Española de Historia Natural (Sección Geológica)*, **93**, 181–188.

GONZÁLEZ-CASADO, J. M., LOPEZ-MARTINEZ, J., GINER, J., DURAN, J. J. & GUMIEL, P. 1999. Analisis de la microfracturacion en la Isla Decepcion, Antartida Occidental. *Geogaceta*, **26**, 27–30.

GONZÁLEZ-CASADO, J. M. & LÓPEZ-MARTÍNEZ, J. 2000. Bransfield Basin, Antarctic Peninsula: not a normal backarc basin. *Geology*, **28**, 1043–1046.

GRÁCIA, E., CANALS, M., FARRÁN, M., PRIETO, M. J., SORRIBAS, J. & GEBRA TEAM. 1996. Morphostructure and evolution of the Central and Eastern Bransfield Basins (NW Antarctic). *Marine Geophysical Research*, **18**, 429–448.

GRÁCIA, E., CANALS, M., FARRÁN, M., SORRIBAS, J. & PALLÀS, R. 1997. The Central and Eastern Bransfield basins (Antarctica) from high-resolution swath-bathymetry data. *Antarctic Science*, **9**, 168–180.

HARRIS, L. B. & COBBOLD, P. R. 1984. Development of conjugate shear bands during simple shearing. *Journal of Structural Geology*, **7**, 37–44.

HOLOHAN, E. P., WYK DE VRIES, B. & TROLL, V. R. 2008. Analogue models of caldera collapse in strike-slip tectonic regime. *Bulletin of Volcanology*, **70**, 773–796.

IBÁÑEZ, J. M., DEL PEZZO, E., ALMENDROS, J., LA ROCCA, M., ALGUACIL, G., ORTIZ, R. & GARCIA, A. 2000.

Seismovolcanic signals at Deception Island volcano, Antarctica: wave field analysis and source modeling. *Journal of Geophysical Research*, **105**, 13905–13931.

JEFFERS, J. D. & ANDERSON, J. B. 1991. Sequence stratigraphy of the Bransfield Basin, Antarctica: implications for tectonic history and hydrocarbon potential. In: ST JOHN, B. (ed.) *Antarctica as an Exploration Frontier: Hydrocarbon Potential, Geology and Hazards.* American Association of Petroleum Geologists, Studies in Geology, **31**, 13–29.

KLEPEIS, R. A. & LAWVER, L. A. 1996. Tectonics of the Antarctic–Scotia plate boundary near Elephant and Clarence Islands, West Antarctica. *Journal of Geophysical Research*, **101**, 20 211–20 231.

LAWVER, L. A., KELLER, R. A., FISK, M. R. & STRELIN, J. A. 1995. Bransfield Basin, Antarctic Peninsula: active extension behind a dead arc. In: TAYLOR, B. (ed.) *Backarc Basins: Tectonic and Magmatism.* Plenum Press, Amsterdam, 315–342.

LAWVER, L. A., SLOAN, B. J. ET AL. 1996. Distributed, active extension in Bransfield Basin, Antarctic Peninsula: evidence from multibeam bathymetry. *GSA Today*, **6–11**, 1–6.

LIPMAN, P. W. 1997. Subsidence of ash-flow calderas: relation to caldera size and magma-chamber geometry. *Bulletin of Volcanology*, **59**, 198–218.

LOCKWOOD, J. P. & HAZLETT, R. W. 2010. *Volcanoes: Global Perspectives.* Wiley-Blackwell, Chichester.

LOPES, F. C., CASELLI, A. T., MACHADO, A. & BARATA, M. T. 2012. A importância do contexto tectónico em desligamento esquerdo na morfo-estrutura da caldeira vulcânica da Ilha de Deception. In: LOPES, F. C., ANDRADE, A. I., HENRIQUES, M. H., QUINTA-FERREIRA, M., BARATA, M. T. & PENA DOS REIS, R. (eds) *Para Conhecer a Terra: Memórias e Notícias de Geociências no Espaço Lusófono.* Imprensa da Universidade de Coimbra, Coimbra.

LÓPEZ-MARTÍNEZ, J. & SERRANO, E. 2002. Geomorphology. In: LÓPEZ-MARTÍNEZ, J., SMELLIE, J. L., THOMSON, J. W. & THOMSON, M. R. A. (eds) *Geology and Geomorphology of Deception Island.* BAS Geomap Series, 6-A and 6-B. British Antarctic Survey, Cambridge, 31–39.

MAESTRO, A., SOMOZA, L., REY, J., MARTÍNEZ-FRIAS, J. & LÓPEZ-MARTÍNEZ, J. 2007. Active tectonics, fault patterns, and stress field of Deception Island: a response to oblique convergence between the Pacific and Antarctic plates. *Journal of South American Earth Sciences*, **23**, 256–268.

MALDONADO, A., LARTER, R. D. & ALDAYA, F. 1994. Forearc tectonic evolution of the South Shetland margin, Antarctic Peninsula. *Tectonics*, **13**, 1345–1370.

MARTI, J. & BARALDO, A. 1990. Pre-caldera pyroclastic deposits of Deception Island (South Shetland Islands). *Antarctic Science*, **2**, 345–352.

MARTI, J., VILA, J. & REY, J. 1996. Deception Island (Bransfield Strait, Antarctica): an example of a volcanic caldera developed by extensional tectonic. In: McGUIRE, W. C., JONES, A. P. & NEUBERG, J. (eds) *Volcano Instabilities on the Earth and Other Planets.* Geological Society, London, Special Publications, **10**, 253–265.

MATHIEU, L., VAN WYK DE VRIES, B., PILATO, M. & TROLL, V. R. 2011. The interaction between volcanoes and strike-slip, transtensional and transpressional fault zones: analogue models and natural examples. *Journal of Structural Geology*, **33**, 898–906.

MUÑOZ-MARTÍN, A., CATALÁN, M., MARTÍN-DÁVILA, J. & CARBÓ, A. 2005. Upper crustal structure of Deception Island area (Bransfield Strait, Antarctica) from gravity and magnetic modeling. *Antarctic Science*, **17**, 213–224.

NAGIHARA, S. & LAWVER, L. A. 1989. Heat flow measurements in the King George Basin, Bransfield Strait. *Antarctic Journal of Science*, **23**, 123–125.

ORHEIM, O. 1971. Volcanic activity on Deception Island. *In*: ADIE, R. J. (ed.) *Antarctic Geology and Geophysics*. Universitetsforlaget, Oslo, 117–120.

ORTIZ, R. R., GARCIA, A. & RISSO, C. 1992*a*. *Seismic and volcanic activity in the South Shetlands Islands and the Antarctic Peninsula environment*. Museo Nacional de Ciencias Naturales, Madrid [unpublished report].

ORTIZ, R., VILA, J. *ET AL*. 1992*b*. Geophysical features of Deception. *In*: YOSHIDA, Y., KAMINUMA, K. & SHIRAISHI, K. (eds) *Recent Progress in Antarctic Earth Sciences*. Terra Scientific Publishing Company, Tokyo, 443–448.

PAREDES, C., DE LA VEGA, R., PÉREZ-LÓPEZ, R., GINER-ROBLES, J. L. & MARTÍNEZ-DÍAZ, J. J. 2007. Descomposición fractal en subdominios morfotectónicos del mapa de lineamientos morfológicos en la isla Deception (Shetland del Sur, Antártida). *Boletín Geológico y Minero*, **118**, 775–787.

PELAYO, A. M. & WIENS, D. A. 1989. Seismotectonics and relative plate motions in the Scotia Sea region. *Journal of Geophysical Research*, **94**, 7293–7320.

PÉREZ-LÓPEZ, R., GINER-ROBLES, J. L., MARTÍNEZ-DÍAZ, J. J., RODRÍGUEZ-PASCUA, M. A., BEJAR, M., PAREDES, C. & GONZÁLEZ-CASADO, J. M. 2007. Active tectonics on Deception Island (West-Antarctica): a new approach by using the fractal anisotropy of lineaments, fault slip measurements and the caldera collapse shape. *In*: COOPER, A. K., RAYMOND, C. R. *ET AL*. (eds) *Antarctica: A Keystone in a Changing World*. Online Proceedings of the 10th ISAES, USGS Open-File Report 2007-1047, Short Research Paper 086.

REY, J., SOMOZA, L. & MARTINEZ-FRIAS, J. 1995. Tectonic volcanic and hydrothermal event sequence on Deception Island (Antarctica). *Geo-Marine Letters*, **15**, 1–8.

RIEDEL, W. 1929. Zur mechanik geologischer brucherscheinungen. Ein beitrag zum problem der 'Fiederspalten'. *Centralblatt fur Mineralogie, Geologie, und Paleontologie, Part B*, 354–368.

ROACH, P. J. 1978. The nature of back-arc extension in Bransfield Strait. *Geophysical Journal of the Royal Astronomical Society*, **53**, 165.

ROBERTSON MAURICE, S. D., WIENS, D. A., SHORE, P. J., DORMAN, L., ADAROS, R. & VERA, E. 2001. Seismicity and tectonics of the South Shetland Islands Region from a combined land-sea seismograph deployment. *Journal of Geophysical Research*, **108**, 4-1–4-12.

SAUNDERS, A. D. & TARNEY, J. 1984. Geochemical characteristics of basaltic volcanism within backarc basins. *In*: KOKELAAR, B. P. & HOWELLS, M. F. (eds) *Marginal Basin Geology: Volcanism and Associated Sedimentary and Tectonic Processes in Modern and Ancient Marginal Basins*. Geological Society, London, Special Publications, **16**, 59–76.

SHULTZ, C. H. 1970. Petrology of the Deception Island volcano, Antarctica. *Antarctic Journal U.S.*, **5**, 97–98.

SMELLIE, J. L. 1989. Deception Island. *In*: DALZIEL, I. W. D. (ed.) *Tectonics of the Scotia Arc, Antarctica. 28th International Geological Congress, Field Trip Guide Book*. American Geophysical Union, Washington, DC, **T180**, 146–152.

SMELLIE, J. L. 2001. Lithostratigraphy and volcanic evolution of Deception Island, South Shetland Islands. *Antarctic Science*, **13**, 188–209.

SMELLIE, J. L. 2002. Geology. *In*: SMELLIE, J. L. & LÓPEZ-MARTÍNEZ, J. (eds) *Geology and Geomorphology of Deception Island*. Sheets 6-A and 6-B, 1:25 000. BAS GEOMAP series. British Antarctic Survey, Cambridge, 11–30.

TCHALENKO, J. S. 1970. Similarities between shear zones of different magnitudes. *Geological Society of America Bulletin*, **81**, 1625–1640.

TORRECILLAS, C., BERROCOSO, M., PÉREZ-LÓPEZ, R. & TORRECILLAS, M. D. 2012. Determination of volumetric variations and coastal changes due to historical volcanic eruptions using historical maps and remote-sensing at Deception Island (West-Antarctica). *Geomorfology*, **136**, 6–14.

TORRECILLAS, C., BERROCOSO, M., FELPETO, A., TORRECILLAS, M. D. & GARCIA, A. 2013. Reconstructing palaeo-volcanic geometries using a Geodynamic Regression Model (GRM): application to Deception Island volcano (South Shetland Islands, Antarctica). *Geomorphology*, **182**, 79–88.

USGS. 2005. *Global Land Survey*. UTM + LANDSAT, Antarctica, United States Geological Survey.

VALENCIO, D. A., MENDIA, J. E. & VILAS, J. F. 1979. Paleomagnetism and K–Ar age of Mesozoic and Cenozoic igneous rocks from Antarctica. *Earth and Planetary Science Letters*, **45**, 61–68.

WIT, H. E., VAN ENST, J. W. A. & LABAN, L. 1991. Deception Island Volcanism (South Shetland Islands, Antarctica): results from thin-section investigations. *Polarforschung*, **59**, 173–178.

Analogue modelling of volcano flank terrace formation on Mars

PAUL K. BYRNE[1,2]*, EOGHAN P. HOLOHAN[3], MATTHIEU KERVYN[4],
BENJAMIN VAN WYK DE VRIES[5] & VALENTIN R. TROLL[6]

[1]*Lunar and Planetary Institute, Universities Space Research Association,
Houston, TX 77058, USA*

[2]*Department of Terrestrial Magnetism, Carnegie Institution of Washington,
Washington, DC 20015, USA*

[3]*German Research Center for Geosciences (GFZ Potsdam), Section 2.1: Physics of
Earthquakes and Volcanoes, Telegrafenberg, Potsdam 14473, Germany*

[4]*Department of Geography, Vrije Universiteit Brussel, B-1050 Brussels, Belgium*

[5]*Laboratoire Magmas et Volcans, Blaise Pascal Université, 63038 Clermont-Ferrand, France*

[6]*Department of Earth Sciences, CEMPEG, Uppsala University, 752 36 Uppsala, Sweden*

Corresponding author (e-mail: byrne@lpi.usra.edu)

Abstract: Of the features that characterize large shield volcanoes on Mars, flank terraces remain
the most enigmatic. Several competing mechanisms have been proposed for these laterally expansive, topographically subtle landforms. Here we test the hypothesis that horizontal contraction of a
volcano in response to the down-flexing of its underlying basement leads to flank terracing. We
performed a series of analogue models consisting of a conical sand–plaster load emplaced on a
basement comprising a layer of brittle sand–plaster atop a reservoir of viscoelastic silicone.
Our experiments consistently produced a suite of structures that included a zone of concentric
extension distal to the conical load, a flexural trough adjacent to the load base and convexities (terraces) on the cone's flanks. The effects of variations in the thickness of the brittle basal layer, as
well as in the volume, slope and planform eccentricity of the cone, were also investigated. For a
given cone geometry, we find that terrace formation is enhanced as the brittle basement thickness
decreases, but that a sufficiently thick brittle layer can enhance the basement's resistance to loading
such that terracing of the cone is reduced or even inhibited altogether. For a given brittle basement
thickness, terracing is reduced with decreasing cone slope and/or volume. Our experimental results
compare well morphologically to observations of terraced edifices on Mars, and so provide a
framework with which to understand the developmental history of large shield volcanoes on the
Red Planet.

The most distinctive of Mars' volcanic attributes
is arguably its population of enormous shield volcanoes – the largest of which, Olympus Mons,
stands some 22 km above its surrounding plains
(Plescia 2004). These shields host a range of recognized volcanic features, including summit caldera
complexes, abundant lava flows, parasitic shields
and pit craters (e.g. Crumpler & Aubele 1978;
Mouginis-Mark et al. 1984; Zimbleman & Edgett
1992; Hodges & Moore 1994; Plescia 2000;
Wyrick et al. 2004; Bleacher et al. 2007; Byrne
et al. 2012). Interestingly, of the 22 largest Martian
shields, a subset also possesses laterally expansive,
topographically subtle undulations on their slopes,
termed 'flank terraces'.

Terraces were first observed on the upper flanks
of Olympus Mons (Carr et al. 1977; Morris 1981),
and were later reported on the flanks of the three
Tharsis Montes, Arsia, Pavonis and Ascraeus
(Thomas et al. 1990). Their low relief and broad
planform render flank terraces difficult to see with
conventional photogeological data, however, and
so a detailed assessment of their morphology and
distribution (both on a given volcano and across
Mars) was not possible until the availability of the
near-global Mars Orbiter Laser Altimeter (MOLA)
digital elevation model (DEM) dataset (Smith
et al. 2001).

Using MOLA-derived slope maps, Byrne et al.
(2009) identified terraces on the flanks of at least
nine Martian shields – Alba, Arsia, Ascraeus, Elysium, Olympus, Pavonis and Uranius Montes, and
Albor and Hecates Tholi – arranged systematically
in a partially or fully axisymmetrical imbricate

From: PLATZ, T., MASSIRONI, M., BYRNE, P. K. & HIESINGER, H. (eds) 2015. *Volcanism and Tectonism
Across the Inner Solar System*. Geological Society, London, Special Publications, **401**, 185–202.
First published online May 8, 2014, http://dx.doi.org/10.1144/SP401.14

'fishscale' pattern about each volcano (Fig. 1). Terraces are convex-outwards, convex-upwards landforms with a near-flat upper surface, the slope of which increases towards the terrace base. Scale-invariant features, terraces are expressed across a range of geometrical sizes: the average terrace length for each of these nine volcanoes spans 12–51 km, and the vertical relief ranges from 0.1 to 1.3 km (Byrne et al. 2009).

Their morphological similarity to the lobate surface traces of thrust faults motivated early workers to interpret flank terraces as shortening structures. In their study of Olympus Mons, Thomas et al. (1990) concluded that compression of the volcano as a result of self-loading led to terrace formation via thrust faulting. McGovern & Solomon (1993) also argued for thrust-fault-related terracing, but concluded that this deformation resulted from compression of a volcano as its weight caused sagging, or down-flexing, of the underlying lithosphere (e.g. Zucca et al. 1982). Alternatively, Crumpler et al. (1996) suggested that inflation of a magma chamber situated high in a volcano could steepen its upper flanks through uplift of material along inwards-dipping reverse faults and so produce terraces. Later, Morgan & McGovern (2005) and McGovern & Morgan (2009) suggested that the terraces on Olympus Mons are due to a combination of thrust faulting and slumping as the edifice spread. Following analysis of flank terrace morphology and distribution across nine volcanoes on Mars, Byrne et al. (2009) found no convincing evidence for self-loading, magma chamber tumescence or volcano spreading as terrace formation mechanisms, but considered volcano-induced lithospheric flexure (i.e. volcano sagging) to be a plausible candidate process.

To further test the hypothesis of an origin for flank terracing through lithospheric flexure requires a modelling approach. Past studies have mainly used numerical models (e.g. McGovern & Solomon 1993), but these have been continuum-based and so did not directly reproduce the discontinuous structures that terraces might represent. Analogue models surmount this difficulty and so allow for the formation of such structures, such that their spatial and temporal development can be characterized. Williams & Zuber (1995) incrementally emplaced a conical load upon a basement composed of an agar–sand–gelatine plate overlying a substrate of corn syrup. This produced a flexural trough and annular fissures around the conical load, but the material used for the load (lead gunshot) was too coarse to reveal the structural effect of flexure upon the load itself. In contrast, Kervyn et al. (2010) and Byrne et al. (2013) emplaced conical sand–plaster loads upon a basement of the same material, overlying a silicone

polymer substrate. This model set-up produced convex-outwards, convex-upwards structures on the conical load's flanks.

The purpose of this chapter is to complement those of Kervyn et al. (2010) and Byrne et al. (2013) by reporting more detailed experimental observations of flank terrace formation in analogue models of a volcano down-flexing its lithospheric basement. In particular, we report new observations of the kinematics of terrace development, and on how model terracing is affected by changes in the geometry of the basement and the cone. We also compare the model structures in detail to those observed on Mars. Our results provide an improved basis for understanding gravitational deformation of large shield volcanoes on Mars in particular, and on terrestrial planets in general.

Experimental methods

In this study, we simulated volcano-induced flexure by emplacing a brittle sand–plaster cone atop a basement consisting of brittle upper and ductile lower substrata (Fig. 2). The cone and basement serve as analogues to a Martian shield and its underlying lithosphere, respectively. The system deformed in response to the load of the cone. Several parameters in the experimental set-up, including the thickness of the brittle upper layer, as well as the volume, slope and plan-view aspect ratio of the cone, were then varied.

Analogue materials

Fine (c. 300 μm mean grain diameter) aeolian quartzose sand served as a granular analogue to rocks of the Martian volcanic edifices and of the planet's upper lithosphere. Dry sand is a suitable analogue for such rocks as it undergoes plastic deformation that, at model normal stresses, is characterized by brittle failure in approximate accordance with a Navier–Coulomb failure criterion (Hubbert 1951; Merle & Borgia 1996; Schellart 2000; Lohrmann et al. 2003). When mixed in a 10:1 ratio with powdered gypsum to increase its cohesion, the material had a bulk density of approximately 1400 kg m^{-3}, an angle of repose of 33° and cohesion of 50–100 Pa (Donnadieu & Merle 1998; Walter & Troll 2001; Cecchi et al. 2005; Delcamp et al. 2008).

A transparent silicone putty (polydimethylsiloxane (PMDS), commercially produced by Dow Corning® as Silastic SGM 36: ten Grotenhuis et al. 2002) simulated the rocks of the lower, viscoelastic lithosphere. At low experimental strain rates, silicone putty is an effective analogue to material with a rheology that is ductile over geological

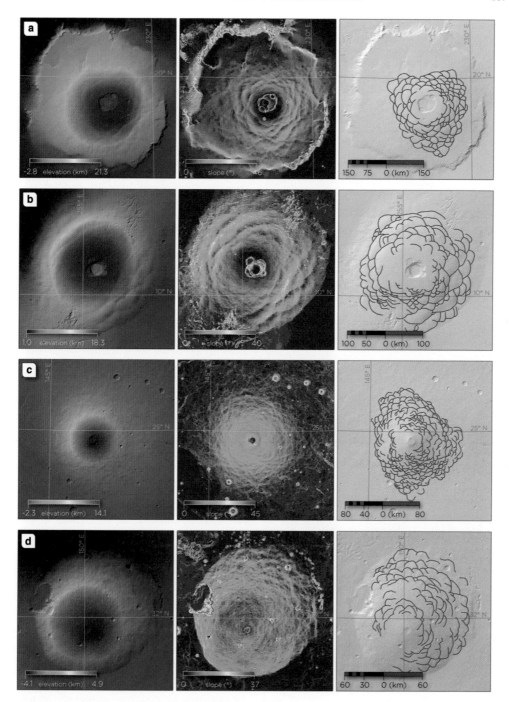

Fig. 1. Four exemplar Martian shield volcanoes that display flank terraces, shown in composite colour-coded elevation and hillshade maps (left column), in slope maps (centre column) and as hillshade maps with terrace outlines in purple (right column). (**a**) Olympus Mons, (**b**) Ascraeus Mons, (**c**) Elysium Mons and (**d**) Hecates Tholus. Elevation, slope and hillshade data are from the 128 pixel-per-degree MOLA dataset (Smith *et al.* 2001). Terrace maps are from Byrne *et al.* (2009). Hillshade maps are shown with illumination from the NW and at a declination of 45°. Each map is shown in an azimuthal equidistant projection, centred at (a) 18.5°N, 226°E, (b) 11.5°N, 255.5°E, (c) 25°N, 147°E and (d) 32°N, 150°E.

timescales (e.g. Weijermars *et al.* 1993; Merle & Borgia 1996; ten Grotenhuis *et al.* 2002; Oehler *et al.* 2005). Mixed with about 1% by weight of quartzose sand, the putty had a bulk density of approximately 1200 kg m^{-3} and a viscosity of

4 × 10^4 Pa s at experimental conditions (Delcamp *et al.* 2008; Byrne *et al.* 2013).

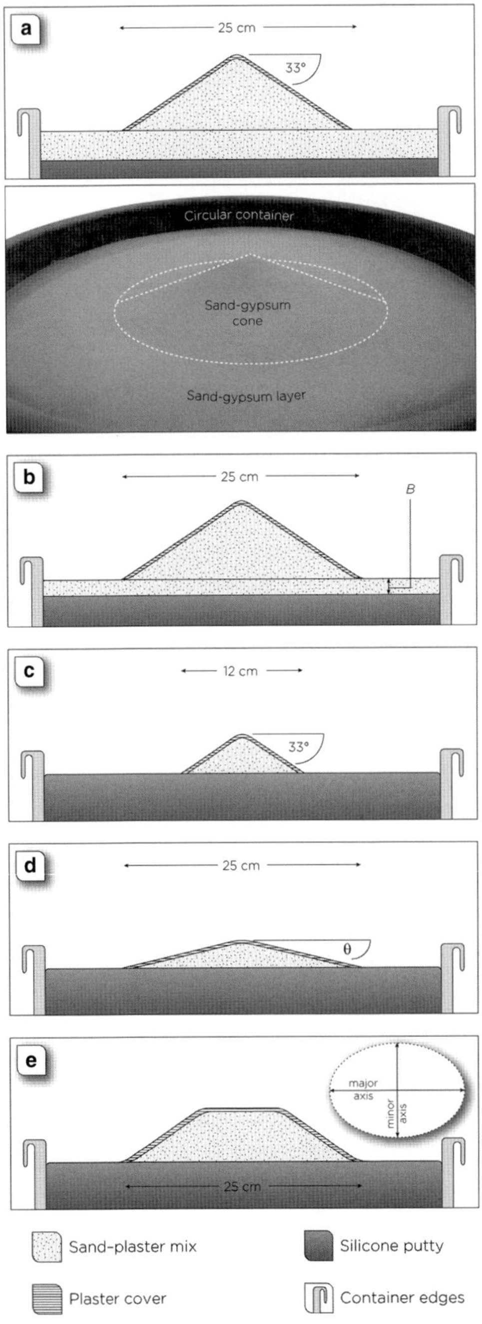

Standard experimental procedure

The experimental apparatus consisted of a 1 m-high, 48 cm-diameter cylindrical container, into which an 80 cm-thick silicone layer was placed first, followed by the addition, via sieving, of a 30 mm-thick sand–plaster substratum (Fig. 2a) (Table 1). Sand–plaster mix of sufficient volume to form a cone 25 cm in diameter (*c.* 1325 cm^3) was then poured directly upon the sand–plaster layer, its angle of repose resulting in flank slopes of around 33° and a height of approximately 8 cm (see the 'Model scaling' subsection below). The cone was smoothed by sieving more sand onto its flanks, before a thin (<1 mm) layer of plaster was sieved over the entire model to preserve any small-scale structures that might form.

The basement responded to the cone's weight almost immediately, sometimes resulting in a flexural depression even before delivery of the cone was complete. Consequently, the time for cone emplacement was restricted to less than 60 s. Cone emplacement was therefore effectively instantaneous, corresponding to only the very fastest effusion rates theorized for Mars (e.g. Isherwood *et al.* 2013) and to numerical models with instantaneous loads (e.g. McGovern & Solomon 1993).

All experiments were incrementally photographed in high spatial resolution from plan view. High-incidence-angle illumination (relative to the surface normal) was used, to highlight any subtle deformation. In most cases, structures had formed within 600 s of cone emplacement, with very little new surface strain apparent after 1200 s. Silicon carbide (SiC) particles, the dark colour of which strongly contrasted with the lighter sand–plaster mix, were added to the cone surfaces to help focus the overhead camera.

Fig. 2. (**a**) Schematic cross-section (top) and photograph (second from top) of the standard experimental set-up. The cross-section shows the sand–plaster cone, plaster cover, brittle upper-basal layer, silicone putty and the edges of the container in which the model was constructed. The photograph shows the inner wall of the container, the surface of the brittle upper layer and the model cone itself (its outline is dashed in white). Below the photograph, schematic cross-sections are shown for experiments where (**b**) brittle-layer thickness (*B*), (**c**) cone volume, (**d**) cone slope (θ) and (**e**) cone eccentricity (with major and minor axes labelled on a schematic of the cone's planform) were varied. The cone in (**e**) is drawn along its major axis; its slopes are at the angle of repose. Sketches show only the upper part of the (typically 80 cm-deep) experimental set-up.

Table 1. *List of model parameters in experiments of (a) standard volcano loading, as well as models in which (b) basal brittle-layer thickness,* B, *(c) cone volume, (d) cone slope,* θ *or (e) cone aspect ratio was varied**

Experiment	Cone diameter (cm)	Cone slope (°)	Cone volume (cm^3)	Cone height (cm)	Brittle-layer thickness (mm)	Number of experiments
(a) Volcano loading	25	33	1325	8.1	**30**	5
	25	33	1325	8.1	**70**	1
	25	33	1325	8.1	**20**	1
(b) Variation in basal sand–plaster layer thickness, *B*	25	33	1325	8.1	**15**	1
	25	33	1325	8.1	**10**	1
	25	33	1325	8.1	**5**	2
	25	33	1325	8.1	**2**	1
	25	33	1325	8.1	**0**	4
(c) Variation in cone volume	12	33	**147**	3.9	0	4
	25	**5**	1325	1.1	0	1
	25	**10**	1325	2.2	0	1
(d) Variation in cone slope, *θ*	25	**20**	1325	4.5	0	2
	25	**25**	1325	6.4	0	1
	25	**30**	1325	7.2	0	1
(e) Variation in cone aspect ratio	**25/17**	33	**1770**	8.1	0	3

*Changes in a model parameter from that described as part of the standard experimental set-up are shown in bold. The total number of experiments performed in this study is 29.

To quantify horizontal displacements and strain fields in the models, particle image velocimetry (PIV) analyses of the experiment photographs were performed by using LaVision DaVis 7.2 digital image correlation software (cf. Walter 2011). The PIV results are displayed as incremental horizontal displacement fields, together with dilational strain derived from the components of the two-dimensional (2D) infinitesimal displacement-gradient tensor (Holohan *et al.* 2013). Here, we generally interpret dilation, a measure of the increase or decrease in horizontal surface area, in terms of overall horizontal extension and contraction, respectively.

Variations in experimental parameters

Key experimental parameters were varied to assess the effects of flexure under a range of geometrical scenarios representative of those observed or inferred for Martian shields (Table 1). First, we chose to model variations in thickness of the brittle upper Martian lithosphere, *B*. For the oceanic lithosphere on Earth, which is thought to be chemically and structurally similar to the Martian lithosphere, the seismogenic (i.e. brittle) thickness is commonly similar to estimates of effective elastic thickness, T_e (cf. Watts & Burov 2003). A range of T_e values for Mars has been calculated by examining crustal deformation related to large surface loads (e.g. Comer *et al.* 1985) or by modelling gravity/topography admittances from gravity

and topographical data (e.g. Zuber *et al.* 2000; McGovern *et al.* 2004; Belleguic *et al.* 2005). Given then the calculated differences in T_e beneath large Martian shields (e.g. table 3 in McGovern *et al.* 2004) and the expected commensurate effect upon flexural response, we performed a series of experiments in which *B* was varied from 70 to 2 mm by uniformly sieving sand–plaster layers of different thicknesses directly upon the silicone prior to cone emplacement (Fig. 2b, Table 1), as well as a number of experiments where *B* = 0 (i.e. the brittle upper layer was absent altogether).

Edifice volume on Mars varies by almost four orders of magnitude (Plescia 2004), and so several experiments were conducted with a poured cone volume (*c.* 147 cm^3) that was approximately 11% of that in the standard set-up. These lower-volume cones had a basal diameter of 12 cm (Fig. 2c), a height of approximately 4 cm and a slope of around 33° (see the subsection on 'Model scaling' below). To account for an order-of-magnitude variation in Martian flank slopes (Plescia 2004), we also varied model cone slopes (*θ*) from 30° to 5° in increments of 5° (Fig. 2d). These cones were constructed by sieving rather than by pouring. Finally, volcanic loads with non-axisymmetrical planforms were simulated. Whilst some Martian shields appear relatively circular in plan view (e.g. Elysium Mons: Plescia 2004), others have a pronounced ellipticity. An example is Ascraeus Mons, the construction of which has been influenced by a SW–NE regional structural trend (Byrne *et al.*

2012) (contrast Fig. 1b, c). We therefore additionally modelled cones with an eccentricity of 1:1.5 (short:long axes in plan view) (Fig. 2e). These latter experimental parameters (edifice volume, slope and planform) were investigated by using a model set-up in which $B = 0$. This experimental configuration was motivated by the observation that the suite of structures formed during flexure was essentially unchanged compared with models that featured higher, more realistic values of B.

Model scaling

Material properties and forces in analogue experiments should geometrically, kinematically and dynamically scale to nature (Hubbert 1951; Ramberg 1981) (Table 2). The scaling ratio for any physical property is defined as:

$$X^* = X_{\text{model}} \times (X_{\text{nature}})^{-1} \quad (1)$$

The length ratio ($l^* = l_{\text{model}} \times (l_{\text{nature}})^{-1}$) is the geometric scaling factor for these experiments and is 10^{-6}, such that 1 cm in the models represented 10 km in nature. A model radius of 12.5 cm therefore corresponded to a volcano 250 km in diameter, whilst a cone radius of 6 cm scaled to a volcano diameter of 120 km, representative, respectively, of a medium-sized (e.g. Uranius Patera) and small (e.g. Tharsis Tholus) shield volcano on Mars (Plescia 2004).

However, this length ratio scaled our approximately 8 and 4 cm-high model cones (Table 1) to natural cones of around 80 and 40 km in height, values that are in excess of the true natural range (Plescia 2004) (even accounting for underestimation due to infilling of flexural troughs: e.g. Zucca *et al.* 1982; Byrne *et al.* 2013). Nevertheless, experiments with model slopes comparable to Martian volcanoes (i.e. 5°) and heights of about 1 cm (corresponding to a volcano 10 km tall) displayed similar surface deformation patterns to cones at the angle of repose (see the subsection on 'Experimental parameter variations' below). This comparable

behaviour, together with similarities to the predictions of terracing in numerical studies that considered a very low-slope edifice geometry (e.g. McGovern & Solomon 1993; Borgia 1994; Van Wyk de Vries & Matela 1998), lead us to conclude that even with unrealistic scaled model heights, the overall structural resemblance between our analogue model results and natural volcanoes with low slopes are meaningful. Further, we note that although cone height in angle-of-repose models is not geometrically similar to nature, models with very low flank slopes (c.g. 5–10°) accurately scale to Martian shields approximately 20–10 km in relief (Plescia 2004).

A model upper-lithosphere brittle-layer thickness of 2–70 mm (Table 1) correlates to natural values of 2–70 km. This variation in B is comparable to the range of gravity/topography admittance-derived T_e estimates for Arsia Mons (McGovern *et al.* 2004; Belleguic *et al.* 2005), and is similar to those values for Ascraeus and Pavonis Montes, and for the Elysium Rise as a whole (McGovern *et al.* 2004). Our range in B values is also broadly in line with those calculated by Beuthe *et al.* (2012) using top- and bottom-loading models of the Martian lithosphere.

The dynamic scaling factor is the scale ratio for stress and cohesion, and is given by the relationship:

$$\sigma^* = l^* \times g^* \times \rho^* \quad (2)$$

where l^* is the length ratio, g^* is the ratio of accelerations due to gravity and ρ^* is the density ratio. At sea level on Earth, g is 9.81 m s^{-2}, whilst equatorial surface gravity on Mars is 3.71 m s^{-2}; g^* is thus 2.64. The density of the sand–plaster mix used in these experiments is approximately 1400 kg m^{-3}, whereas Martian crustal density is assumed to be about 2600 kg m^{-3}, such that ρ^* is 5.38 × 10^{-1}. The stress ratio, from equation (2), is therefore 1.42 × 10^{-6}. From this, a rock cohesion on Mars of 5 × 10^7 Pa should scale to a model cohesion of around 70 Pa, comparable to the low-cohesion sand–plaster mix used.

Table 2. *Scaling parameters, values for models and Mars and scaling ratios used in this study*

Parameter	Symbol	Units	Model	Nature	Ratio
Length	l	m	1×10^{-2}	1×10^4	1×10^{-6}
Gravity	g	m s^{-2}	9.81×10^0	3.71×10^0	2.64×10^0
Stress	σ	Pa	1.37×10^2	9.65×10^7	1.42×10^{-6}
Time	T	s	2.91×10^2	1.04×10^{12}	2.81×10^{-10}
Density	ρ	kg m^{-3}	1.4×10^3	2.6×10^3	5.38×10^{-1}
Cohesion	C	Pa	7.1×10^1	5×10^7	1.42×10^{-6}
Viscosity	μ	Pa s	4×10^4	1×10^{20}	4×10^{-16}

The behaviour of Coulomb materials is theoretically independent of time, but the deformation of a ductile material is rate-dependent. A time ratio, T^*, can be calculated (Donnadieu & Merle 1998) as:

$$T^* = \mu^* \times (\sigma^*)^{-1} \qquad (3)$$

where μ^* is the viscosity ratio and σ^* is the stress ratio given above. The viscosity of the silicone used in the experiments is 4×10^4 Pa, but the viscosity of the Martian ductile lithosphere is unknown. However, assuming it is composed largely of mafic lithologies (e.g. McSween et al. 2003), we consider a value of 10^{20} Pa to be reasonable (Watts & Zhong 2000). The viscosity ratio, μ^*, is thus 4×10^{-16}, giving a time ratio, T^*, of 2.81×10^{-10}.

We note that the ratio between the densities of the granular material and silicone putty is at the upper end, or slightly in excess, of the natural range (Belleguic et al. 2005), which may have enhanced the 'sink potential' (Borgia et al. 2000) of our models. This experimental limitation is probably minor, however, since the main effect of the greater density of the overlying brittle material would only be to increase the rate of sagging (which is itself poorly understood). An increased sagging rate should not fundamentally alter the associated brittle deformation on the volcano and in the surrounding basement.

Volcano-induced flexure model results

Four distinct structural elements characterized all volcano load-induced flexural experiments in this study: flank convexities; a bowl-shaped depression; a load-centric trough; and an annular zone of extension (cf. Kervyn et al. 2010; Byrne et al. 2013). Here we first report spatial and temporal observations for a standard volcano-loading experiment, and then we describe the results of models in which parameters were varied with respect to those of the standard experimental set-up.

Volcano loading

Where an axisymmetrical cone at the angle of repose was placed directly upon a basement consisting of an upper, brittle stratum 30 mm thick and a lower, ductile stratum (Fig. 3a), the load downflexed, and sagged rapidly into, the basement (Fig. 3b). As a consequence, the cone experienced a reduction in basal diameter and a shortening of flank length.

A set of convex-upwards, outwards-verging terraces, arranged concentrically in an imbricate stacking pattern, formed on the cone as flank shortening proceeded (Fig. 3c: 1). The leading edges of these features were manifest as ridge-like bounding scarps. The longest terraces, laterally continuous for several tens of degrees of arc, occurred at midflank elevations. Vertical relief on terraces was typically of the order of 1 mm, although some examples accumulated relief of several mm along their leading edges. More subtle structures were visible with high-incidence-angle illumination, which showed them to be present on all flank sectors (i.e. at all azimuths). Flank convexities were greatest in number at mid- to lower-flank elevations; few, if any, terraces were observed near the summit.

A broad, shallow, bowl-shaped flexural depression developed in the basement as the load sank, and remained visible upon removal of the cone and the brittle layer. The flexural bowl fully enclosed the cone and was itself bounded by a very subtle topographical rise. The bowl's base was nearflat, with some diapirism of the silicone putty producing a knotted texture at its centre. The opposing slopes of the bowl, and the lower flanks of the cone, together defined a shallow flexural trough around the load (Fig. 3c: 2).

Finally, an annular pattern of arcuate and en echelon fractures developed immediately outside the flexural trough to mark a circular system of extension concentric to the load (Fig. 3c: 3). Intersecting fractures formed discrete blocks within this extensional zone, which transitioned in places to arcuate or wholly load-concentric systems of half-graben or graben bounded by steep scarps. As sagging continued, the annulus tended to broaden, its inner edge moving towards the cone.

Analyses of sequential photographs of a representative standard experiment provide a useful insight into the temporal development of these flexural structures. Whilst unequivocal model terraces are visible within about 5 min of loading, it may be that some formed even sooner. The insets in Figure 3b show terraces at the base (black outline) and at mid-flank elevations (white outline) of a model cone after deformation had ceased. The leading edges of these structures are characterized by arcuate traces of shadowed sand grains (as illumination is from the top left). Corresponding insets in Figure 3a show morphologically similar, albeit much more subtle, shadowed arcuate convexities at the same positions on the cone within 60 s of its delivery.

The PIV analysis of the time-lapse images yields further insight into the development of experimental strain. Figure 3d–f shows deformation in the horizontal plane (i.e. orthogonal to the downwards-looking camera) in 120 s increments at 3, 5 and 7 min from the start of flexure. The PIV shows that the cone experiences an initial overall horizontal contraction (shown in blue), within which there are sharp gradients that clearly demarcate the analogue terrace boundaries. In contrast,

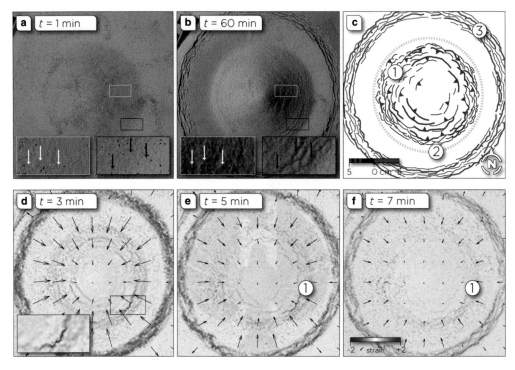

Fig. 3. Photographs of a representative volcano-loading experiment at (**a**) 1 min and (**b**) 60 min after the onset of sagging. The insets show subtle convexities (a) and fully formed terraces (b) on the mid (white outline) and lower (black outline) flanks of the model cone. Illumination in these images is from the NW. (**c**) Structural sketch of the cumulative experiment deformation in (b). (1) Terrace traces are outlined in blue (flags are on the inferred hanging wall), (2) the axis of the flexural trough is marked by a grey dashed line, whilst (3) fissures, normal faults and graben are shown in red (ticks are on the downthrown side of faults). (**d**)–(**f**) PIV images of horizontal displacements (extension denotes red, contraction denotes blue) across the same experiment 3, 5 and 7 min after flexural onset. The inset in (d) shows localized extension behind the leading edges of some model terraces. Note that the longest duration of terrace activity occurs on the lowermost flanks of the cone (e and f: 1).

a broad horizontal extension (shown in red) characterizes the distal concentric fracture system, with sharp gradients in extension closely associated with individual fissures and normal faults. The zone of central horizontal contraction arises from the inwards horizontal motion of the cone–basement system. This inwards motion peaks between the cone and the concentric fracture system, and decreases towards the cone summit. The zone of peripheral horizontal extension results from the divergence in horizontal displacement vectors associated with the formation of a surrounding flexural bulge.

As the experiment progressed, terraces on the cone's upper and mid flanks stopped accumulating strain, whereas those at lower elevations continued to develop. The zone of active deformation therefore appeared to focus with time on the lower flanks. In some models, there was a clear downwards progression of terrace formation, somewhat like the behaviour of an 'in-sequence' propagating

fold-and-thrust belt. Consequently, the terraces towards the cone base stayed active for the greatest length of time. These terraces were often the largest, and commonly showed an area of extension just inside their leading edge (inset in Fig. 3d). The duration and magnitude of this extension appeared to be closely linked to that of the strain accumulation of the associated terraces (Fig. 3e: 1 and Fig. 3f: 1).

Experimental parameter variations

The inclusion of basal sand–plaster layers of thicknesses $B < 30$ mm had a noticeable, if subtle, effect on the expression of flank convexities and extensional zones (Fig. 4). With thinner brittle layers, the spacing between terraces on the cone decreased so that more, although shorter, convexities formed, especially at mid-flank elevations (Fig. 4a: 1; $B = 15$ mm). Smaller terraces continued to populate lower cone elevations (Fig. 4a: 2). Notably, as B

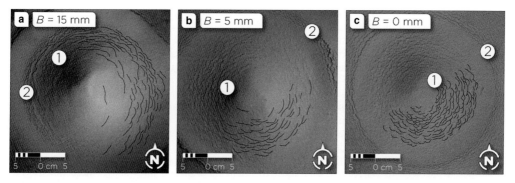

Fig. 4. Model cones 25 cm in diameter and at the angle of repose, but for which brittle-layer thickness, B, was (**a**) 15 mm, (**b**) 5 mm and (**c**) 0 mm (i.e. absent). (a: 1) Few terraces form high on the cone, but more, shorter terraces develop at mid-flank elevations; (a: 2) the lower flanks are substantially terraced. (b: 1) Terraces populate the mid to upper flanks, whilst (b: 2) the width of the annular zone of extension is reduced. (c: 1) Where the brittle layer is not present, terraces are smaller and are distributed across all flank elevations, even close to the summit, and (c: 2) the load-concentric fractures are smaller, but define a wider extensional annulus. In all images, illumination is from the SE or east; black lines are terrace traces on the experiment visible only from other illumination azimuths, and are included here for clarity.

decreased, terraces began to form progressively closer to the summit (Fig. 4b: 1; $B = 5$ mm). This trend continued even when the brittle layer was absent entirely. Where $B = 0$, the greatest difference in how the suite of structures was manifest was the increased number, but decreased size, of flank convexities, which were distributed almost to the summit (Fig 4c: 1; cf. Fig. 3b).

Lower B values also resulted in deeper and narrower flexural depressions, as reflected in a decreased radial distance from the topographical bulge to the load summit. As B tended to zero, the radial width of the annular fracture zone increased (Fig. 4b: 2), although the constituent fractures

individually accumulated less strain (Fig. 4c: 2). Other differences associated with decreasing B values included a tendency for the extensional annulus to occur closer to the cone, reflecting the reduced diameter of the flexural depression, and for the inwards-facing wall of the flexural trough to transition from a state of extension to one of contraction as down-sagging ensued (visible using PIV). Conversely, increasing B inhibited basement flexure, such that, at $B \geq 70$ mm, the related structures (including terraces) were no longer detectable.

Reduced-volume cones at the angle of repose (Fig. 5a) resulted in proportionately reduced

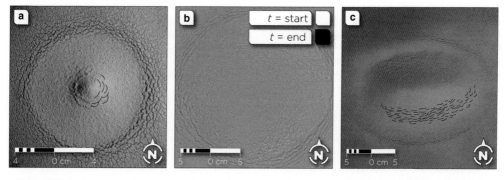

Fig. 5. Models with other geometric parameters varied. (**a**) A low-volume (147 cm^3) cone with slopes at the angle of repose. (**b**) A cone 25 cm in diameter with slopes of 10°, where images of the start and end of the experiment have been overlain, and the colours of the former inverted, to show the relative motion of SiC grains on the cone surface over the course of the model's deformation. The path from white to black particles gives the direction of movement of individual grains as the experiment progressed. (**c**) A cone with a planform eccentricity of 1:1.5 (25 × 17 cm) with slopes at the angle of repose. Illumination in (a) is from the east and in (c) the SE.

expressions of the flexural bowl, the load-concentric trough and the extensional annulus. Notably, the number of individual terraces that formed was consistently lower in these experiments. Otherwise, the characteristic suite of flexural structures developed as for models using the standard experimental set-up.

A greater difference in the expression of flank deformation was apparent in experiments where cone slopes were reduced. Compared with angle-of-repose models with the same diameter, models that featured cones with lower slopes produced shallower depressions in the silicone and showed less overall surface displacement. Flexural bowls, troughs and concentric fracture zones developed as before, but were less well expressed. Terraces were easily visible on 15–30° slopes, but were difficult to detect below this threshold (even with high-incidence-angle illumination). However, photographic sequences of model cone flanks, even to slopes of 5°, show the same gross kinematic behaviour as angle-of-repose cones. Specifically, individual sand and SiC grains inside the boundary of the annular extensional zone moved towards the centre of the load, shortening the radial and concentric distances between each other. This behaviour is shown for a cone with slopes of 15° in Figure 5b, in which photographs of the start and end of the experiment have been overlain, and the colours of the former inverted. The relative motion of SiC grains on the cone's surface is clearly visible, even in the absence of obvious flank terracing. Particles beyond the annular fractures behaved in the opposite manner, moving away from the load and from one another.

In experiments with cones of elliptical planform, the entire suite of flexural structures formed as for the standard set-up, but the distribution of flank terracing differed (Fig. 5c). Rather than forming a generally axisymmetrical pattern, the number and expression (length and vertical relief) of terraces were greatest on the sectors parallel or near-parallel to the long axes of elliptical model cones. PIV data show that the greatest horizontal strains were accommodated on these same sectors.

Discussion of model deformation

Observations of the structural outcomes of each experimental configuration, sequential photography and PIV analyses together provide a framework from which to understand the kinematic development of our volcano-induced flexure models. We first summarize the development of structures in a representative standard volcano-loading experiment, and then review the effects of parameter variations upon model deformation.

Model terrace formation

On the basis of their convex-upwards shape, outwards vergence and formation within a body undergoing net horizontal contraction, we interpret model terraces as shortening structures formed by thrust-fault-related folding (Fig. 6). Under this interpretation, the surface strain on a cone that down-flexes its basement (Fig. 6a) is accommodated by shear fractures that penetrate some distance into the cone and that may be surface-breaking (Fig. 6b: 1). Terraces on the cone's flanks develop as hanging wall anticlines above these thrusts (Fig. 6b: 2). That folding contributes to terrace morphology is supported by the identification (with PIV) of areas in extension inside the leading edge of well-developed terraces in some models. This deformation is consistent with localized outer-arc extension of the terrace hanging wall anticline as the structure continues to accumulate strain, since it is observed on those terraces that are amongst the longest active structures.

The arrangement of terraces in a fishscale pattern (Byrne *et al.* 2009) is probably due to a constrictional strain regime in the central region of the down-sagged volcano–basement system. As shown by PIV (Fig. 3), the horizontal displacements of the sagging cone are directed inwards towards its summit. This displacement field must lead to a combination of both radial and concentric components of shortening in the horizontal plane (see Holohan *et al.* 2013), with a corresponding component of vertical lengthening lying normal to this plane (Fig. 7). Horizontal displacement (heave) along the fishscale arrangement of terrace-bounding thrust faults resolves both the radial and concentric shortening components, whereas the vertical displacement (throw) on the faults and related folds accommodates the associated lengthening along the vertical axis.

The dip angle of the terrace-bounding faults in our models remains an open question, however, because the viscous response of the silicone putty used in our experiments precluded the cross-sectioning of the models. Continued sagging during sectioning of a wetted model would have yielded unrealistic strains. Under a scenario where maximum and intermediate principal stresses lie in the horizontal plane (i.e. σ_3 is vertical), terrace faults probably dip into the cone at angles of approximately 30° (e.g. Fig. 6b: 1). Should these principal stresses be slope-parallel (i.e. σ_3 normal to the cone surface: van Wyk de Vries & Matela 1998), however, and where cone slope is relatively high (>20°), the terrace-bounding faults may be near-horizontal or even inclined down-slope.

Another key question concerns the timing of terrace formation on Mars. Little evidence remains

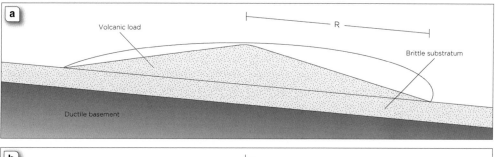

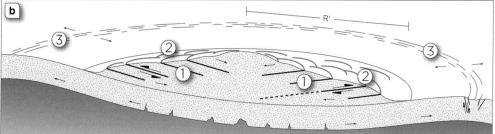

Fig. 6. Schematic block diagrams of a conical volcano atop a brittle substratum, both of which are underlain by a ductile basement, (**a**) before and (**b**) after the onset of sagging. This behaviour results in (1) the development of blind or surface-breaking thrust faults that penetrate some distance into the cone's interior. These thrusts may dip down into the cone (as shown here) or, depending on its slope and the exact orientations of the principal stresses, they may be near-horizontal or even outwards-dipping. (2) Hanging-wall anticlines form above these thrusts and describe a fishscale map pattern on the volcano's flanks. (3) Deformation of the cone is accompanied by the growth of an extensional annulus concentric to the load. In (b), small arrows indicate movement directions; half-arrow pairs indicate the assumed sense of slip on faults. R is the initial cone radius; R' is the final cone radius. Adapted from Byrne *et al.* (2013).

regarding the onset or cessation of flank terracing on Martian volcanoes (Byrne *et al.* 2009). The convexities visible within 1 min of cone emplacement in volcano-loading models (Fig. 3a) may be lobate slumps formed during the construction of the cone (cf. Cecchi *et al.* 2005) that were later (i.e. after *c.* 5 min) reactivated as terrace-bounding faults during sagging. If so, then asperities upon a volcano could influence the subsequent locations of flank-shortening structures. If these faint structures are terraces, however, then remarkably low strains are required for their formation. Terraces may thus rapidly form early in the flexural process, serving to further localize flank shortening amid continued volcano sagging during or after the active lifetime of the edifice.

The time taken for structures to develop in these experiments, about 10 min, is equivalent to approximately 68 000 years using our time scaling ratio, T^*, for ductile material. This is an extremely short time for deformation to occur with respect to the accepted ages of Martian volcanoes (e.g. Plescia & Saunders 1979; Werner 2009; Platz & Michael 2011; Platz *et al.* 2011) and is reliant on the viscosity parameter chosen for the ductile Martian lithosphere. If the choice of viscosity value is correct, then we might infer that flank terraces

form rapidly on volcanoes that were largely complete by the time lithospheric flexure began. However, if our value is too low, then the lithosphere's flexural response may post-date volcanic construction considerably. A viscosity value that is too high implies that flexure, and hence terracing, may have occurred during the main shield-building phase. Since the possible viscosity range for Mars' ductile lithosphere is extremely wide, our temporal scaling factor is not well constrained. Nevertheless, the flank shortening represented by terraces is of the order of only a few per cent (Byrne *et al.* 2009), and so it is likely that neither substantial nor long-lived volcano sagging is required to form the terraces we observe today.

Although terracing may be initially distributed along most of the cone's flanks (and presumably its interior), PIV analysis shows that strain accumulation on the terraces (and, in some cases, the formation of new terraces) becomes increasingly concentrated on the lowermost slopes as sagging progresses. This latter finding is broadly consistent with the results of McGovern & Solomon (1993), who used finite-element numerical models to predict failure within, and proximal to, an edifice down-flexing an elastic plate that overlaid a visco-elastic mantle. In their models of a load emplaced

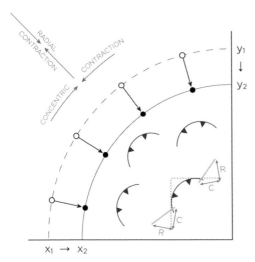

Fig. 7. Schematic sketch of how the fishscale pattern of flank terraces may arise through a combination of horizontal contractions orientated radially and concentrically to the centre of a conical load. Circumferential horizontal shortening, in addition to radial shortening, results as material points on the cone move radially inwards towards the cone summit (note the reduction of the arc length from x_1 to x_2 and from y_1 to y_2). The fishscale arrangement of the terrace-bounding faults enables across-fault contraction (solid light grey lines) to resolve both the radial and concentric components of shortening (dark grey lines, labelled 'R' and 'C', respectively). Modified from Holohan *et al.* (2013).

instantaneously upon a 40 km-thick elastic lithosphere, these authors found a similar down-slope propagation of circumferentially orientated thrust faulting as the underlying lithosphere flexes (see McGovern & Solomon 1993, fig. 8b–d). McGovern & Solomon (1993) inferred that initial faulting is restricted to high flank elevations before shifting down-slope, such that most of the upper edifice surface is in failure. In contrast, we observe more widely distributed terracing at first, with on-going deformation increasingly localized at lower-flank elevations. One reason for this discrepancy could be that model terracing on upper-mid slopes develops so quickly that we are unable to observe its propagation at the temporal resolution of the time-lapse images.

The models of McGovern & Solomon (1993) also predicted normal faulting initially at the base of the cone, which then migrates outwards on the surrounding surface. The outwards movement of extension about the edifice in these authors' models is also similar to the broadening of the load-concentric extensional annulus (Fig. 6b: 3) that we observe in our models as sagging continues. This

behaviour indicates that as the cone down-flexes its basement, the diameter of the flexural bowl is reduced, presumably as it deepens, in a manner consistent with the change in flexural profile associated with numerical models of viscoelastic deformation of Earth's oceanic lithosphere (e.g. Watts & Zhong 2000, fig. 12).

The effects of variations in experimental parameters

The strength and effective elastic thickness of the lithosphere are closely linked to its seismogenic (i.e. brittle) thickness (Watts & Burov 2003). In particular, effective flexural rigidity scales with brittle thickness to the third power (see the supplementary material of Byrne *et al.* 2013), and so even small changes in its brittle thickness may have large effects on the basement's flexural response to loading. This is manifest in our models, where decreasing the basal brittle-layer thickness (B) progressively reduced its capacity to support its superposed load. For a given cone volume (e.g. 1325 cm^3), sand–plaster layers with $B < 30$ mm led to the formation of deeper flexural depressions, as well as narrower annular extensional zones and flexural rises that developed closer to the load, than for models where $B = 30$ mm. Cones atop thinner brittle layers, therefore, experienced commensurately more sagging, and distributed strain into more (albeit smaller) structures, than cones on thicker layers. The change in flexural profile with successively thinner sand–plaster layers increasingly constricted a greater portion of the cone, which probably accounts for why terraces formed at progressively greater flank elevations as B tended to zero. Conversely, and consistent with the findings of Byrne *et al.* (2013) (see below), a sufficiently large value of B with respect to cone height would inhibit flexure, and hence terrace formation, altogether.

The emplacement of a cone with angle-of-repose slopes but with smaller volumes than the standard experimental set-up produced proportionately fewer, smaller terraces, but otherwise terrace formation was little affected. That the number and size of terraces that formed on these models was consistently lower presumably reflects the correspondingly smaller surface areas of the lower-volume cones. This finding suggests that the model structures are scale-invariant, as has been concluded for Martian flank terraces (Byrne *et al.* 2009). Changes to flank slope angle reduced the expression of terraces. Surface contraction was nevertheless observed across all cones during sagging, even when terraces were not visible. The reduced expression or absence of terraces on lower-angle cones may, in part, be an experimental limitation, with the sand–plaster mix partly accommodating the much lower contractional surface strains in these models volumetrically (i.e. via grain flow) only.

Whilst cone height did not scale to natural values in most experiments, comparable deformation behaviour was observed in models with low cone slopes (e.g. 10–20°), as in those with slopes at the angle of repose of the granular material used (e.g. Fig 5b). The shallower depressions in the silicone, and less overall surface displacement, in lower-slope models relative to angle-of-repose cones were probably the result of their lower weight and volume. In addition, the ratios of cone height to brittle basement thickness in our experiments fall within the range of values in which volcano-induced flexure is likely to occur on Earth, according to the dimensionless parameter Π_{Sag} developed by Byrne et al. (2013). This parameter relates a volcano's loading of its basement (as a function of its weight and volume) to the resistance of the basement to flexure (i.e. its flexural rigidity) (see the supplementary material of Byrne et al. 2013 for more detail). The Π_{Sag} values for all experiments in this study (including those of differing volume and cone slope) extend from 9.5×10^{-3} to 1.39×10^{2}, well within the range of about 1×10^{-5}–1.1×10^{3} for Terran volcanoes (Byrne et al. 2013). Although they are not known, similar values for Mars as for Earth are assumed for the purposes of this study.

Cones with eccentric planforms displayed more prominent terracing along flank sectors parallel, and less prominent terracing along sectors perpendicular, to the major axis. Such a pattern of terracing is probably a function of the same subsidence at the cone centre being accommodated over a greater horizontal distance along sectors perpendicular to the major axis than on parallel sectors. Consequently, flanks perpendicular to the cone's long axis undergo lower strains (cf. Holohan et al. 2008), and so less prominent terracing, than flanks lying parallel to the long axis.

Our results therefore indicate that terraces may form across a range of cone geometries, independent of cone volume, flank slope, planform and probably all but the greatest brittle basement thicknesses. On the basis of gross structural similarities between volcanoes with low slopes in nature, numerical studies with low edifice slopes (e.g. McGovern & Solomon 1993; Borgia 1994; van Wyk de Vries & Matela 1998) and the analogue model results shown here, we are confident that the main structural relationships observed in our experiments are geologically realistic representations of real-world, volcano-induced lithospheric flexure. We explore this inference in the next section.

Comparison of model results with natural examples

Flank terraces

The flank structures formed in these flexural experiments share the convex-upwards, outwards-verging morphology and imbricate 'fishscale' plan-view stacking pattern of volcano flank terraces on Mars (compare Figs 1 & 3). Therefore, on the basis of their strong similarity to our model results, we consider Martian volcano flank terraces to be hanging-wall anticlines over blind or surface-breaking thrust faults that formed in a constrictional regime applied to a volcano as it flexed downwards. If this interpretation is correct, then flank terraces are a kinematic volcano equivalent to lobate scarps, landforms occurring widely on Mercury, Mars and the Moon that are thought to form above thrust faults that accommodate crustal shortening (Strom et al. 1975; Nahm & Schultz 2011; Watters et al. 2012).

Moreover, the systematic formation of laboratory terraces on cones of varying slopes and volumes is consistent with observations of Martian volcanoes. Both large (e.g. Olympus Mons) and small (e.g. Hecates Tholus) shields are terraced (Fig. 1a, d), as are volcanoes with different slopes – for example, Elysium Mons and Alba Patera (c. 7° and 1°, respectively) (Plescia 2004). Even in low-slope models where terraces were not observed, these cones still displayed the same general style of surface deformation as cones at the angle of repose.

Further, that fewer, larger terraces form in models that feature intermediate to thick (i.e. $B = 15$–30 mm) brittle basal-layer thicknesses (Figs 3b & 4a) matches observations at volcanoes for which photogeological and gravity/topography admittance studies have returned the highest estimates of T_e (Thurber & Toksöz 1978; Comer et al. 1985; Zuber et al. 2000; McGovern et al. 2004; Belleguic et al. 2005; Beuthe et al. 2012). Whilst estimates for lithosphere thicknesses differ between studies, in all cases the relative differences between high and low T_e values are the same. Specifically, Ascraeus and Olympus Montes are consistently quoted as being supported by the thickest lithosphere of all Martian volcanoes, and these volcanoes display the most prominent and longest terraces (Byrne et al. 2009) (Fig. 1). This observation is consistent with our model results. Although the range of Π_{Sag} values for which lithospheric flexure will occur on Mars remains to be determined, a key test will be to determine whether terraces occur on Mars' smaller shields, and to relate the gravitational load of such volcanoes to the rigidity of their supporting basement.

We note that model terraces lack the snub-nosed roundness characteristic of most observed terraces

(Byrne *et al.* 2009), and feature a generally flattened upper surface instead. However, photogeological data indicate that terraces on Mars have been covered, and their forms tempered, by contemporaneous or subsequent erupted material (Thomas *et al.* 1990). The analogue structures may thus correspond to the terrace form prior to being mantled by erupted material. Moreover, whereas our models generally produced terraces arranged axisymmetrically about a cone, terraces on Martian volcanoes are often less evenly distributed (e.g. Hecates Tholus, Fig. 1d). Aeolian or mass-wasting processes – coupled with differences in edifice geometry, internal architecture, regional slopes and near- and far-field stresses – probably contribute to the observed asymmetry in terrace distributions on Martian shields (Byrne *et al.* 2009, 2013).

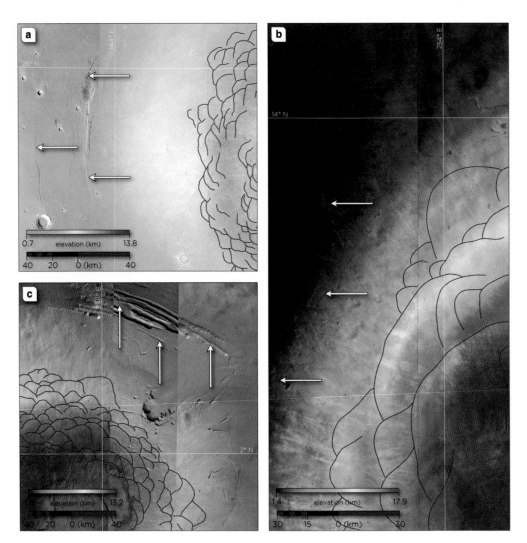

Fig. 8. Examples of volcano-concentric graben (marked by white arrows) centred on or spatially associated with terraced Martian volcanoes (**a**) Elysium, (**b**) Ascraeus and (**c**) Pavonis Montes. The images are portions of High Resolution Stereo Camera (HRSC) images (Neukum *et al.* 2004) from the ESA Mars Express spacecraft: (a) h1295_0000, h1317_0000 and h7424_0000; (b) h1217_0001 and h1206_0001; and (c) h0891_0000, h2175_0000 and h3276_0000 – shown with colour-coded MOLA elevation data and with terrace outlines (purple) from Byrne *et al.* (2009). Each image is shown in an azimuthal equidistant projection, centred at (a) 24.5°N, 145°E, (b) 12°N, 253.7°E and (c) 2°N, 248°E.

Finally, there is precedent for associating extension with the development of shortening structures, as is observed for some terraces (e.g. Fig. 3d). For example, lobate scarps can exhibit extensional faults along the crests of their forelimbs (Mueller & Golombek 2004; Morley 2007; Watters & Johnson 2010). Extension along model terraces may therefore be manifest in nature as faults and/or fissures, which in places might have influenced the locations of later pit crater chains (interpreted to have formed due to dilational faulting: Ferrill et al. 2004; Wyrick et al. 2004) on, for example, the mid and upper flanks of Ascraeus Mons (Byrne et al. 2012).

Flexural trough

One of the distinguishing features of lithospheric loading is the formation of a flexural trough or moat concentric to the load (e.g. McGovern & Solomon 1997). Given their large masses, Martian volcanoes can be expected to display measurable vertical deflections of the lithosphere (Comer et al. 1985; Zuber et al. 1993). Such troughs are not readily visible on photogeological images, but can be identified with geophysical data. On the basis of topographical profiles or gravitational field models, flexural troughs have been identified or inferred for several large crustal loads on Mars, including Olympus Mons (Zuber & Smith 1997), the Tharsis Rise (Solomon & Head 1982) and the planet's entire southern hemisphere (Watters 2003).

The difficulty in resolving these features directly may, in part, be due to infilling by material such as landslide deposits (McGovern & Solomon 1997) and lavas from smaller, nearby shields or from fissures (e.g. Bleacher et al. 2007). Evidence for the latter process in particular is given by the embayed margins of several Tharsis and Elysium volcanoes by younger lavas (e.g. Plescia 2004). A very low amplitude relative to wavelength (Zuber et al. 1993) will also render a flexural trough difficult to detect. In any case, assuming these depressions do form part of the flexural response of the Martian lithosphere to volcano loading, they correspond to the concentric troughs observed in our experiments.

Annular graben

If bending stresses associated with lithospheric flexure due to a major volcanic load are sufficiently high, extensional fractures will form concentric to and outside of the load (Comer et al. 1985; Williams & Zuber 1995). Load-concentric fractures have been observed on Earth (Lambeck & Nakiboglu 1980), associated with mascons on the Moon (Melosh 1978; Solomon & Head 1980) and related to large crustal loads on Mars, including its volcanoes (Comer et al. 1985).

It is likely, therefore, that extensional structures such as the ring fractures about Elysium Mons (Plescia 2004) (Fig. 8a), the arcuate graben that cross-cut the western flank of Ascraeus Mons (Byrne et al. 2012) (Fig. 8b) and the group of prominent troughs at the NE base of Pavonis Mons (Crumpler & Aubele 1978) (Fig. 8c) correspond to load-concentric fractures formed in the laboratory. In the models, extension is spatially associated with the flexural bulge encircling the load, and so it is reasonable to assume that this structural arrangement also applies to the Martian examples.

We note that models of volcano-induced flexure as configured in this study cannot account for the 6–8 km-high basal scarp that encircles Olympus Mons (e.g. Carr et al. 1977). However, Byrne et al. (2013) showed that the addition of a silicone layer beneath the edifice acts to decouple it from its underlying basement, so that although flexure-induced constriction of a cone is accommodated on its mid to upper flanks by distributed terracing, as before, outwards-directed slip (i.e. volcano spreading) of the cone is enabled along the silicone layer (see Byrne et al. 2013, fig. 3b). Flexural slip along such a basal décollement produces a large monoclinal fold around the cone's base, the leading edge of which can oversteepen and collapse. This structural outcome strongly resembles the overall morphology of Olympus Mons (see their Fig. 4b), indicating that the solar system's largest volcano formed through the hybrid action of volcano sagging and spreading.

Summary and concluding remarks

In experiments where a brittle conical load at the angle of repose was emplaced upon a dual-layer (brittle and ductile) basement, the near-surface regions of the load entered a constrictional strain regime as the cone–basement system down-sagged or 'flexed'. The region under the cone developed into a bowl-shaped depression that, in conjunction with the opposite-facing lowermost flank of the cone, formed a load-concentric flexural trough. Beyond the trough an annular zone of extension developed, whereas the cone's mid and lower flanks experienced a net reduction of surface area. Associated horizontal radial and concentric shortening was accommodated by the formation of convex, outwards-verging terrace structures arranged in an axisymmetrical fishscale pattern about the cone.

The expression of flank terraces, in terms of number, size and distribution, was strongly influenced by the thickness of the brittle upper-basement layer (our analogue to the seismogenic lithosphere).

With a sufficiently great brittle-layer thickness, basement flexure and terrace formation was inhibited entirely. However, reducing the thickness of the brittle upper layer, ultimately to zero, promoted greater sagging, with the flexural wavelength of the basement decreasing whilst its amplitude increased. Progressively thinner brittle layers resulted in the formation of smaller terraces at greater flank elevations until, with the upper layer absent, the zone of terracing extended almost to the cone summit. In comparison, variations in cone volume, slope or eccentricity had only minor effects on the development of flank terracing.

Our experimental results bear strong similarities in morphology and spatial distribution to flank terraces on Mars, to systems of arcuate graben attributed to volcano-induced flexure of the Martian lithosphere and to geophysical signatures interpreted to be buried or in-filled flexural troughs surrounding several of the largest shield volcanoes on that planet. These similarities provide compelling support for the conclusion that Martian flank terraces are flexurally induced structures, formed to accommodate surface shortening of a shield volcano as it down-flexes its basement. The horizontal contraction of an edifice during flexure may inhibit the ascent of magma to summit elevations (McGovern 2007; Byrne *et al.* 2012), whilst serving to enhance the overall structural integrity of the volcano (Byrne *et al.* 2013). A full understanding of the structural evolution of large shield volcanoes on Mars therefore requires that they be appraised within the context of lithospheric flexure.

We thank M. Bonini, T. Platz and M. Massironi for helpful reviews that improved this manuscript. P. K. Byrne and V. R. Troll thank the Enterprise Ireland International Collaboration Programme (grant No. IC/2006/37/ED-209) and the TekNat faculty at Uppsala University for support during several stages of this research. E. P. Holohan acknowledges a Marie Curie International Mobility Fellowship co-funded by Marie Curie Actions and the Irish Research Council. M. Kervyn acknowledges the support of a Research Credit from the Fonds voor Wetenschappelijke Onderzoek (FWO) for developing the analogue modelling laboratory at VUB. This research has made use of NASA's Planetary Data System and Astrophysics Data System.

References

BELLEGUIC, V., LOGNONNÉ, P. & WIECZOREK, M. 2005. Constraints on the Martian lithosphere from gravity and topography data. *Journal of Geophysical Research*, **110**, E11005, http://dx.doi.org/10.1029/2005JE002437

BEUTHE, M., LE MAISTRE, S., ROSENBLATT, P., PÄTZOLD, M. & DEHANT, V. 2012. Density and lithospheric thickness of the Tharsis Province from MEX MaRS and MRO gravity data. *Journal of Geophysical Research*, **117**, E04002, http://dx.doi.org/10.1029/2011JE003976

BLEACHER, J., GREELEY, R., WILLIAMS, D., CAVE, S. & NEUKUM, G. 2007. Trends in effusive style at the Tharsis Montes, Mars, and implications for the development of the Tharsis Province. *Journal of Geophysical Research*, **112**, E09005, http://dx.doi.org/10.1029/2006JE002873

BORGIA, A. 1994. Dynamic basis of volcanic spreading. *Journal of Geophysical Research*, **99**, 17 791–17 804.

BORGIA, A., DELANEY, P. & DENLINGER, R. 2000. Spreading volcanoes. *Annual Review of Earth and Planetary Sciences*, **28**, 539–570.

BYRNE, P. K., VAN WYK DE VRIES, B., MURRAY, J. B. & TROLL, V. R. 2009. The geometry of volcano flank terraces on Mars. *Earth and Planetary Science Letters*, **281**, 1–13.

BYRNE, P. K., VAN WYK DE VRIES, B., MURRAY, J. B. & TROLL, V. R. 2012. A volcanotectonic survey of Ascraeus Mons, Mars. *Journal of Geophysical Research*, **117**, E01004, http://dx.doi.org/10.1029/2011JE003825

BYRNE, P. K., HOLOHAN, E. P., KERVYN, M., VAN WYK DE VRIES, B., TROLL, V. R. & MURRAY, J. B. 2013. A sagging-spreading continuum of large volcano structure. *Geology*, **41**, 339–342.

CARR, M. H., GREELEY, R., BLASIUS, K. R., GUEST, J. E. & MURRAY, J. B. 1977. Some Martian volcanic features as viewed from the Viking orbiters. *Journal of Geophysical Research*, **82**, 3985–4015.

CECCHI, E., VAN WYK DE VRIES, B. & LAVEST, J.-M. 2005. Flank spreading and collapse of weak-cored volcanoes. *Bulletin of Volcanology*, **67**, 72–91.

COMER, R. P., SOLOMON, S. C. & HEAD, J. W. 1985. Mars: thickness of the lithosphere from the tectonic response to volcanic loads. *Reviews of Geophysics*, **23**, 61–92.

CRUMPLER, L. S. & AUBELE, J. C. 1978. Structural evolution of Arsia Mons, Pavonis Mons and Ascraeus Mons: Tharsis region of Mars. *Icarus*, **34**, 496–511.

CRUMPLER, L. S., HEAD, J. W. & AUBELE, J. C. 1996. Calderas on Mars: characteristics, structure and associated flank deformation. *In*: MCGUIRE, W. J., JONES, A. P. & NEUBERG, J. (eds) *Volcano Instability on the Earth and Other Planets* Geological Society, London, Special Publications, **110**, 307–348.

DELCAMP, A., VAN WYK DE VRIES, B. & JAMES, M. R. 2008. The influence of edifice slope and substrata on volcano spreading. *Journal of Volcanology and Geothermal Research*, **177**, 925–943.

DONNADIEU, F. & MERLE, O. 1998. Experiments on the indentation process during cryptodome intrusions: new insights into Mount St. Helens deformation. *Geology*, **26**, 79–82.

FERRILL, D. A., WYRICK, D. Y., MORRIS, A. P., SIMS, D. W. & FRANKLIN, N. M. 2004. Dilational fault slip and pit chain formation on Mars. *GSA Today*, **14**, 4–12.

HODGES, C. & MOORE, H. 1994. *Atlas of Volcanic Landforms on Mars* United States Geological Survey, Professional Papers, **1534**.

HOLOHAN, E. P., TROLL, V. R., VAN WYK DE VRIES, B., WALSH, J. J. & WALTER, T. R. 2008. Unzipping Long

Valley: an explanation for vent migration patterns during an elliptical ring fracture explosion. *Geology*, **36**, 323–326.

HOLOHAN, E. P., WALTER, T. R., SCHÖPFER, M. P. J., WALSH, J. J., VAN WYK DE VRIES, B. & TROLL, V. R. 2013. Origins of oblique-slip faulting during caldera subsidence. *Journal of Geophysical Research: Solid Earth*, **118**, 1778–1794.

HUBBERT, M. K. 1951. Mechanical basis for certain familiar geologic structures. *Geological Society of America Bulletin*, **62**, 355–372.

ISHERWOOD, R. J., JOZWIAK, L. M., JANSEN, J. C. & ANDREWS-HANNA, J. C. 2013. The volcanic history of Olympus Mons from paleo-topography and flexural modeling. *Earth and Planetary Science Letters*, **363**, 88–96.

KERVYN, M., BOONE, M. N., VAN WYK DE VRIES, B., LEBAS, E., CNUDDE, V., FONTIJN, K. & JACOBS, P. 2010. 3D imaging of volcano gravitational deformation by computerized X-ray micro-tomography. *Geosphere*, **6**, 482–498.

LAMBECK, K. & NAKIBOGLU, S. M. 1980. Seamount loading and stress in the ocean lithosphere. *Journal of Geophysical Research*, **85**, 6403–6418.

LOHRMANN, J., KUKOWSKI, N., ADAM, J. & ONCKEN, O. 2003. The impact of analogue material properties on the geometry, kinematics, and dynamics of convergent sand wedges. *Journal of Structural Geology*, **25**, 1691–1711.

MCGOVERN, P. J. 2007. Flexural stresses beneath Hawaii: implications for the October 15, 2006, earthquakes and magma ascent. *Geophysical Research Letters*, **34**, L23305, http://dx.doi.org/10.1029/2007GL031305

MCGOVERN, P. J. & MORGAN, J. K. 2009. Volcanic spreading and lateral variations in the structure of Olympus Mons, Mars. *Geology*, **37**, 139–142.

MCGOVERN, P. J. & SOLOMON, S. C. 1993. State of stress, faulting, and eruption characteristics of large volcanoes on Mars. *Journal of Geophysical Research*, **98**, 23 553–23 579.

MCGOVERN, P. J. & SOLOMON, S. C. 1997. Some implications of a basal detachment structural model for Olympus Mons. *In: Proceedings of the 28th Annual Lunar and Planetary Science Conference, March 17–21, 1997, Abstract Volume* LPI Contribution **1090**. Lunar and Planetary Institute, Houston, TX, 913.

MCGOVERN, P. J., SOLOMON, S. C. ET AL. 2004. Correction to: localized gravity/topography admittance and correlation spectra on Mars: Implications for regional and global evolution. *Journal of Geophysical Research*, **109**, E07007, http://dx.doi.org/10.1029/2004JE002286

MCSWEEN, H. Y., JR, GROVE, T. L. & WYATT, M. B. 2003. Constraints on the composition and petrogenesis of the Martian crust. *Journal of Geophysical Research*, **108**, 5137, http://dx.doi.org/10.1029/2003JE002175

MELOSH, H. J. 1978. The tectonics of mascon loading. *In: Proceedings of the 9th Lunar and Planetary Science Conference, Houston, Texas, March 13–17, 1978, Volume 3.* LPI Contribution **1575**. Pergamon Press, New York, 3513–3525.

MERLE, O. & BORGIA, A. 1996. Scaled experiments of volcanic spreading. *Journal of Geophysical Research*, **101**, 13 805–13 817.

MORGAN, J. & MCGOVERN, P. 2005. Discrete element simulations of gravitational volcanic deformation: 1. deformation structures and geometries. *Journal of Geophysical Research*, **110**, B05402, http://dx.doi.org/10.1029/2004JB003252

MORLEY, C. K. 2007. Variations in late Cenozoic-Recent strike-slip and oblique-extensional geometries, within Indochina: the influence of pre-existing fabrics. *Journal of Structural Geology*, **29**, 36–58.

MORRIS, E. 1981. Structure of Olympus Mons and its basal scarp. *In: Abstracts of Papers Presented to the Third International Colloquium on Mars, held in Pasadena, California, August 31–September 2, 1981* LPI Contribution **441**. Lunar and Planetary Institute, Houston, TX, 161.

MOUGINIS-MARK, P., WILSON, L., HEAD, J., BROWN, S., HALL, J. & SULLIVAN, K. 1984. Elysium Planitia, Mars: regional geology, volcanology, and evidence for volcano-ground ice interactions. *Earth, Moon, and Planets*, **30**, 149–173.

MUELLER, K. & GOLOMBEK, M. 2004. Compressional structures on Mars. *Annual Review of Earth and Planetary Sciences*, **32**, 435–464.

NAHM, A. L. & SCHULTZ, R. A. 2011. Magnitude of global contraction on Mars from analysis of surface faults: implications for Martian thermal history. *Icarus*, **211**, 389–400.

NEUKUM, G., JAUMANN, R. ET AL. 2004. *HRSC – The High Resolution Stereo Camera of Mars Express* European Space Agency, Special Publications, **ESA SP-1240**.

OEHLER, J.-F., VAN WYK DE VRIES, B. & LABAZUY, P. 2005. Landslides and spreading of oceanic hotspot and arc shield volcanoes on Low Strength Layers (LSLs): an analogue modeling approach. *Journal of Volcanology and Geothermal Research*, **144**, 169–189.

PLATZ, T. & MICHAEL, G. 2011. Eruption history of the Elysium Volcanic Province, Mars. *Earth and Planetary Science Letters*, **312**, 140–151.

PLATZ, T., MÜNN, S., WALTER, T. R., PROCTER, J. N., MCGUIRE, P. C., DUMKE, A. & NEUKUM, G. 2011. Vertical and lateral collapse of Tharsis Tholus, Mars. *Earth and Planetary Science Letters*, **305**, 445–455.

PLESCIA, J. B. 2000. Geology of the Uranius Group Volcanic Constructs: Uranius Patera, Ceraunius Tholus, and Uranius Tholus. *Icarus*, **143**, 376–396.

PLESCIA, J. B. 2004. Morphometric properties of Martian volcanoes. *Journal of Geophysical Research*, **109**, E03003, http://dx.doi.org/10.1029/2002JE002031

PLESCIA, J. B. & SAUNDERS, R. 1979. The chronology of the Martian volcanoes. *In: Proceedings of the 10th Lunar and Planetary Science Conference, Houston, Texas, March 19–23, 1979, Volume 3.* LPI Contribution **1576**. Pergamon Press, New York, 2841–2859.

RAMBERG, H. 1981. *Gravity, Deformation and the Earth's Crust*, 2nd edn. Academic Press, London.

SCHELLART, W. P. 2000. Shear test results for cohesion and friction coefficients for different granular materials: scaling implications for their usage in analogue modelling. *Tectonophysics*, **324**, 1–16.

SMITH, D. E., ZUBER, M. T. ET AL. 2001. Mars orbiter laser altimeter: experiment summary after the first year of global mapping of Mars. *Journal of Geophysical Research*, **106**, 23,689–23,722.

SOLOMON, S. C. & HEAD, J. W. 1980. Lunar mascon basins: lava filling, tectonics, and evolution of the lithosphere. *Reviews of Geophysics and Space Physics*, **18**, 107–141.

SOLOMON, S. C. & HEAD, J. W. 1982. Evolution of the Tharsis province, Mars: the importance of heterogeneous lithospheric thicknesses on volcanic construction. *Journal of Geophysical Research*, **87**, 9755–9774.

STROM, R. G., TRASK, J. J. & GUEST, J. E. 1975. Tectonism and volcanism on Mercury. *Journal of Geophysical Research*, **80**, 2478–2507.

TEN GROTENHUIS, S. M., PIAZOLO, S., PAKULA, T., PASSCHIER, C. W. & BONS, P. D. 2002. Are polymers suitable rock analogs? *Tectonophysics*, **350**, 35–47.

THOMAS, P. J., SQUYRES, S. W. & CARR, M. H. 1990. Flank tectonics of Martian volcanoes. *Journal of Geophysical Research*, **95**, 14 345–14 355.

THURBER, C. H. & TOKSÖZ, M. N. 1978. Martian lithospheric thickness from elastic flexure theory. *Geophysical Research Letters*, **5**, 977–980.

VAN WYK DE VRIES, B. & MATELA, R. 1998. Styles of volcano-induced deformation: numerical models of substratum flexure, spreading and extrusion. *Journal of Volcanology and Geothermal Research*, **81**, 1–18.

WALTER, T. R. 2011. Structural architecture of the 1980 Mount St. Helens collapse: an analysis of the Rosenquist photo sequence using digital image correlation. *Geology*, **39**, 767–770.

WALTER, T. & TROLL, V. R. 2001. Formation of caldera periphery faults: an experimental study. *Bulletin of Volcanology*, **63**, 191–203.

WATTERS, T. R. 2003. Lithospheric flexure and the origin of the dichotomy boundary on Mars. *Geology*, **31**, 271–274.

WATTERS, T. R. & JOHNSON, C. L. 2010. Lunar tectonics. *In*: WATTERS, T. R. & SCHULTZ, R. A. (eds) *Planetary Tectonics* Cambridge University Press, Cambridge, 121–182.

WATTERS, T. R., ROBINSON, M. S., BANKS, M. E., THANH, T. & DENEVI, B. W. 2012. Recent extensional tectonics on the Moon revealed by the Lunar Reconnaissance Orbiter Camera. *Nature Geoscience*, **5**, 181–185.

WATTS, A. B. & BUROV, E. B. 2003. Lithospheric strength and its relationship to the elastic and seismogenic layer thickness. *Earth and Planetary Science Letters*, **213**, 113–131.

WATTS, A. B. & ZHONG, S. 2000. Observations of flexure and the rheology of oceanic lithosphere. *Geophysical Journal International*, **142**, 855–875.

WEIJERMARS, R., JACKSON, M. P. A. & VENDEVILLE, B. 1993. Rheological and tectonic modeling of salt provinces. *Tectonophysics*, **217**, 143–174.

WERNER, S. C. 2009. The global Martian volcanic evolutionary history. *Icarus*, **201**, 44–68.

WILLIAMS, K. K. & ZUBER, M. T. 1995. An experimental study of incremental surface loading of an elastic plate: application to volcano tectonics. *Geophysical Research Letters*, **22**, 1981–1984.

WYRICK, D., FERRILL, D. A., MORRIS, A. P., COLTON, S. L. & SIMS, D. W. 2004. Distribution, morphology, and origins of Martian pit crater chains. *Journal of Geophysical Research*, **109**, E06005, http://dx.doi.org/10.1029/2004JE002240

ZIMBLEMAN, J. & EDGETT, K. 1992. The Tharsis Montes, Mars: comparison of volcanic and modified landforms. *In*: *Proceedings of the 22nd Lunar and Planetary Science Conference, Houston, Texas, March 18–22, 1991* LPI Contribution **1588**. Lunar and Planetary Institute, Houston, TX, 31–44.

ZUBER, M. T. & SMITH, D. E. 1997. Mars without Tharsis. *Journal of Geophysical Research: Planets*, **10**(E12), 28 673–28 685.

ZUBER, M. T., BILLS, B. G., FREY, H. V., KIEFER, W. S., NEREM, R. S. & ROARK, J. H. 1993. Possible flexural signatures around Olympus and Ascraeus Montes, Mars. *In*: *Proceedings of the 24th Lunar and Planetary Science Conference, March 15–19, 1993, Houston, Texas* LPI Contribution **1590**. Lunar and Planetary Institute, Houston, TX, 1591–1592.

ZUBER, M. T., SOLOMON, S. C. *ET AL.* 2000. Internal structure and early thermal evolution of Mars from Mars Global Surveyor topography and gravity. *Science*, **287**, 1788–1793.

ZUCCA, J., HILL, D. & KOVACH, R. 1982. Crustal structure of Mauna Loa volcano, Hawaii, from seismic refraction and gravity data. *Bulletin of the Seismological Society of America*, **72**, 1535–1550.

Self-similar clustering distribution of structural features on Ascraeus Mons (Mars): implications for magma chamber depth

RICCARDO POZZOBON[1,2]*, FRANCESCO MAZZARINI[3],
MATTEO MASSIRONI[2] & LUCIA MARINANGELI[1]

[1]*IRSPS-DISPUTer, Università degli Studi G. d'Annunzio,
Via dei Vestini 31, I-65127, Chieti, Italy*

[2]*Dipartimento di Geoscienze, Università degli Studi di Padova,
via G. Gradenigo 6, 35131, Padova, Italy*

[3]*Istituto Nazionale di Geofisica e Vulcanologia, Via Della Faggiola 32, 56100, Pisa, Italy*

*Corresponding author (e-mail: r.pozzobon@unich.it)

Abstract: The occurrence and distribution of monogenic eruptive features in volcanic areas testify to the presence of deep-crustal or subcrustal magma reservoirs hydraulically connected to the surface via a fracture network. The spatial distribution of vents can be studied in terms of self-similar (fractal) clustering, described by a fractal exponent D and defined over a range of lengths (l) between a lower and upper cutoff, L_{co} and U_{co}, respectively. The computed U_{co} values for several volcanic fields on Earth match the thickness of the crust between vents and magma reservoirs at depth. This analysis can thus be extended to other volcanic fields and volcanoes on rocky planets in the solar system where features such as vents and dykes occur, and for where complementary geophysical data are currently lacking. We applied this method to the Ascraeus Mons volcano on Mars, which presents hundreds of collapse pits similar to those observed on Earth volcanoes that are most likely related to feeder dykes. Based on structural mapping with High Resolution Stereo Camera data at 12 m/px and Context Camera data at 6 m/px mosaics, more than 2300 collapse pits and dyke traces were analysed, revealing two distinct fractal clustered populations. The obtained U_{co} values reveal the presence and likely depth of both a deep magma reservoir (c. 60 km deep) and a small shallower chamber (c. 11 km deep). This analysis can help to better constrain the depth and time evolution of volcanic processes on Tharsis, and on terrestrial planets' volcanoes in general.

The Tharsis region on Mars, one of the largest volcanic provinces in the solar system, is characterized by the presence of numerous basaltic shield volcanoes. Such volcanoes develop by multiple effusive eruptions above coherent intrusive complexes consisting of dykes and intrusive sheets (e.g. Walker 1992). Their plumbing system can be roughly modelled as deep magma reservoir/s connected to shallow magma chambers and intrusive cone sheet complexes (e.g. Walker 1992; Canon-Tapia & Walker 2004; Canon-Tapia 2008). Understanding the geometry of the plumbing systems of shield volcanoes may thus provide inferences on the structure of the crust and on the depth where transported magma may stall to form large magma chambers.

Magma transport through the crust mainly occurs via dykes and sills (e.g. Lister & Kerr 1991; Petford *et al.* 1993, 2000; Rubin 1995). These fluid-driven fractures or hydro-fractures are generally opening-mode in nature (e.g. Gudmundsson 2002; Gudmundsson & Brenner 2004). The magma ascent rate depends on magma properties (including

viscosity, density, temperature and heat content), country rock properties (including temperature, density, thermal conductivity and permeability) and the stress field. High-density xenolith settling in basalt magmas (Basu 1977; Spera 1980; Petford *et al.* 2000), numerical analysis (Dahm 2000*a*, *b*) and magma cooling rates (Maaloe 1973), all indicate velocities of magma ascent on the Earth of 10^{-2} to 1 m s^{-1}, implying high bulk permeability of the crust. Rock-fracturing processes enhance the bulk permeability of the crust and allow the ascent of magma at rates that are akin to the timescale characterizing magmatic activity (Turcotte 1982; Petford *et al.* 1993, 2000; Rubin 1993; Hutton 1996; Canon-Tapia & Walker 2004). Depending on the relative magnitude of magma buoyancy, the crustal fracture toughness, the presence of mechanical discontinuities in the crust and the magma availability, dykes may stop at some crustal levels, they may construct magmatic chambers or they may directly reach the surface to generate eruptions (e.g. Ida 1999; Dahm 2000*a*; Gudmundsson 2002; Taisne & Tait 2009).

From: PLATZ, T., MASSIRONI, M., BYRNE, P. K. & HIESINGER, H. (eds) 2015. *Volcanism and Tectonism Across the Inner Solar System*. Geological Society, London, Special Publications, **401**, 203–218.
First published online March 25, 2014, http://dx.doi.org/10.1144/SP401.12

As for Earth, magma movement within rocky planets from deep storage levels to the surface broadly occurs via fractures along pressure gradients. The elastic properties of rocks that govern brittle rock deformation (i.e. fracture formation) do not strongly depend on the gravity field (e.g. Grosfils 2007). Moreover, the displacement (D_{max}) to length (L) ratio for faults on Earth and other rocky planets, including Mars and Venus, follows the relationship $D_{max} = cL^n$ (Schultz *et al.* 2008). Differences in gravity do not affect the fractal exponent of the distribution ($n = 1$ for all planets), but do affect the c value that scales with gravity (i.e. the lower the gravity, the lower the D_{max}/L ratio) (Schultz *et al.* 2008). Indeed, gravity ultimately controls the load at depth. Where the elastic properties of lithospheric materials are fixed (i.e. stiffness, tensile strength, fracture toughness, Young's modulus and Poisson's ratio of basalts), the load normal to a dyke is a function of gravity (normal stress = vertical stress $\times$ ($v/1 - v$), where v is Poisson's ratio). Since Mars' gravity is about one-third that of Earth, dykes should propagate to greater lengths on Mars than on Earth. What is common to all rocky planets where volcanism occurs is that fractures form connected networks between magmatic reservoirs at depth and the surface, which magma may exploit. We thus assume that on Earth, Mars and other rocky planets, magma moves through the lithosphere via dykes, and that these dykes are connected to pressurized magma reservoirs at depth.

In general, eruptions of basaltic magmas imply the transfer of melt from deep reservoirs up to intermediate-depth magma chambers in the middle upper crust or directly to the planet's surface. Polygenetic volcanoes repeatedly erupt from the same general vent (summit or crater) and are formed if basaltic magmas stop at intermediate- to shallow-crustal magma chambers, whereas monogenetic volcanoes erupt only once and are constructed when magma directly erupts from feeders. According to Connor & Conway (2000), volcanic fields are dominantly basaltic in composition and are formed by monogenetic vents, each associated with feeder dykes. The occurrence and spatial distribution of monogenetic eruptive structures within volcanic areas on Earth are linked to fracture systems and associated stress fields (Takada 1994*a*). Moreover, morphometric parameters of monogenetic cones, such as cone elongation, breaching direction and cone alignment, as well as pit crater alignment, indicate the direction of fractures that act as magma feeders (Tibaldi 1995). It has been proposed that fractures filled by magma (i.e. dykes) tend to coalesce during their ascent to the surface, thereby controlling the final level of magma emplacement. The actual distribution of volcanic vents at the surface (i.e. the formation of monogenetic and/or polygenetic volcanoes) is mainly controlled by the magma input rate and the crustal strain rate (e.g. Fedotov 1981; Takada 1994*a*, *b*). High strain rates and/or small magma input rates promote the formation of monogenetic volcanoes, whereas low strain rates and high magma input rates mainly generate polygenetic volcanoes (e.g. Takada 1994*a*). It is important to emphasize that basaltic monogenetic vents indicate the presence of deep-crustal or subcrustal magma reservoirs directly connected via a fracture network to the surface, involving a hydraulic connection through the whole crust or a large portion of it. Moreover, the correlation between vent distribution and fracture network properties is such that the spatial distribution of vents may be studied in terms of self-similar (fractal) clustering (Mazzarini & D'Orazio 2003; Mazzarini 2004, 2007; Mazzarini *et al.* 2008), as in the case of fracture networks (Bonnet *et al.* 2001).

Other markers can be used to map the inferred surface distribution of feeder dykes on rocky planets, such as pit craters. Pit craters form by collapse into a subsurface cavity due to, for example, explosive eruptions, hydrothermal activity and dilational faulting (e.g. Ferrill *et al.* 2004), and are also commonly associated with volcanism on Earth (e.g. Okubo & Martel 1998; Carter *et al.* 2007; Wall *et al.* 2010). Pit craters, and pit crater chains and troughs, are common on Mars, and are found both on the flanks of large shield volcanoes and on the floors of large basins (e.g. Ferrill *et al.* 2004). Many eruptions in regions of plains volcanism on Mars are inferred to have started as fissure eruptions characterized by alignment of pit craters (Hauber *et al.* 2009).

Pit craters, monogenic vents and dyke traces have already been mapped for Pavonis Mons, and studied statistically in terms of degree of spatial randomness (Bleacher *et al.* 2009) in an effort to verify a possible structural control and any possible alignments of vent fields. Nearest-neighbour analyses of vents in the Pavonis Mons South Volcanic Field indicated that the spatial distribution of these features is consistent with a random Poisson distribution. The position of an individual vent does not affect the subsequent formation of another vent, and is consistent with the ascent of magma from small, shallow magma reservoirs, rather than a huge plumbing system below Pavonis Mons (Bleacher *et al.* 2009).

However, the spatial relationships between vents and feeder dykes and the pit craters and chains related to them have not been studied in terms of fractal clustering before. We hereby present an analysis of the spatial distribution of pit craters on the Ascreaus Mons shield volcano on Mars. Pit alignments have been interpreted as the surface traces of feeder dykes injected within a crustal fracture network and, by analysing their spatial distribution (self-similar clustering: e.g. Mazzarini *et al.* 2013*a*, *b*), we make inferences for the volcano's deep magma reservoirs and plumbing system.

Tharsis volcanic province and Ascraeus Mons

Ascraeus Mons is located on the Tharsis Rise, the largest positive topographical feature on Mars, a broad elevated region extending over approximately 30×10^6 km^2 that dominates the western hemisphere of the planet. The plateau itself is about 10 km high and 8000 km wide (Phillips *et al.* 2001). The Tharsis region is one of the largest volcanic provinces in the solar system, and is characterized by the presence of almost 20 volcanic edifices. Tharsis is probably a place of regional volcanism and pervasive fracturing resulting from the lithospheric load and voluminous extrusive and intrusive magmatic activity (Phillips *et al.* 2001; Carr 2006). The Tharsis Montes, three large mountains aligned in a SW–NE orientation, have been interpreted as shield volcanoes (Carr 2006), primarily because of their shapes, abundant lava flows and distinct volcanic-shield-like caldera complexes. Wilson & Head (2001) contended that the Tharsis volcanoes were built during volcanic activity lasting several Gyr (Carr 2006), implying average effusion rates of about 10^{-2} km^3 a^{-1} (Meyzen *et al.* 2014). Therefore, it is likely that magma supply from the mantle was episodic, with the growth of the volcanoes being characterized by very active periods with high effusion rates (as indicated by the large volume of many lava flows), separated by long periods of quiescence during which the magma chamber solidified (e.g. Hiesinger *et al.* 2007; Giacomini *et al.* 2009; Hauber *et al.* 2009). The most recent volcanic flows covering the surface of Ascraeus Mons are clearly very young since they show a very low density of impact craters (Neukum *et al.* 2004; Robbins *et al.* 2011; Byrne *et al.* 2012).

The Ascraeus Mons volcano has a main edifice age of approximately 3.6 Ga (Werner 2009) and an estimated volume of 1.1×10^6 km^3 (Carr 2006). Its main edifice has marginal lava aprons extending from the NE and SE flanks. The lava flow apron on the NE flank of Ascraeus Mons begins at an elevation of 6 km, and extends onto the surrounding plains to the north and NE. The apron on the southern side is just slightly smaller than that to the north, consisting of two portions separated at about 8.5 km of elevation. Most of the volcanotectonic structures that characterize Ascraeus Mons occur on these aprons, but narrow crack-like depressions, graben and lines of collapse pits (pit crater chains), approximately circumferential to the shield, are also common on the other flanks. The flanks of Ascraeus Mons have average slopes of 7.4° and show terraces with a broad, convex-upward profile in section, and a systematic 'fish-scale' imbricate stacking pattern in plan view. Byrne *et al.* (2009) suggested that the fish-scale pattern of convex lobate terraces most probably occurred in the presence of lithospheric flexure owing to the increasing lithostatic load during the evolution of large shield volcanoes. Likewise, the presence of circumferential graben and flexural troughs around the major Martian shield volcanoes are consistent with lithospheric flexure (Zuber *et al.* 1993). The graben and pit chains cross-cut and thus appear younger than the lobate flank terraces (thrusts) and concentric graben attributed by Byrne *et al.* (2009) to the effect of lithospheric compensation of the volcanic edifice after the bulk of its formation.

Pit crater chains and faults on Mars are associated with normal faults or tensional fractures (Tanaka & Golombek 1989; Banerdt *et al.* 1992; Ferrill *et al.* 2004). This is taken as an indication that faults can also show pure opening deformation, particularly close to the surface where they can produce void space expressed as a collapse pit (Ferrill *et al.* 2004). Dykes may have injected along these features (Byrne *et al.* 2012).

Structural mapping and interpretation

The mapping of the main structures on Ascraeus Mons was performed using nadir level-3 image mosaics from the High Resolution Stereo Camera (HRSC) experiment onboard the Mars Express spacecraft, using ArcGis 9.3 software. The images used for the Ascraeus Mons mosaic are listed in Table 1. The mosaic covers an area larger than the Ascraeus edifice itself, and was selected in order to also consider the surrounding plains, where structures related to the volcano are likely to be located. The mapped structural and morphological features are concentric pits and pit chains, radial pits and pit chains, concentric and radial graben, normal faults and sinuous rilles. Structures without strong concentric or radial orientations are here termed transitional structures. In this study, all the pit chains and craters were interpreted as the trace of a subsurface feeder dyke injected into fractures. Under this classification scheme, concentric graben related to the collapse and lithospheric flexure of the edifice (Byrne *et al.* 2012) were not considered. To visualize the spatial distribution of, and relationship between, concentric and radial structures, the HRSC mosaic was added to a MOLA (Mars Orbiting Laser Altimeter)-derived hillshade map in a Transverse Mercator projection.

Table 1. *Constituent images of the HRSC mosaic*

HRSC image number	Resolution (m/px)
H0016_0008_ND3	12.3
H0068_0000_ND3	12.4
H1217_0001_ND3	12.1
H2021_0001_ND3	12.5
H2032_0001_ND3	12.2
H2054_0001_ND3	12.2
H3254_0000_ND3	12.5
H3265_0000_ND3	12.5
H3276_0000_ND3	12.5
H5171_0000_ND3	12.5
H5189_0000_ND3	12.5
H5207_0000_ND3	12.5
H5225_0000_ND3	12.5
H5243_0000_ND3	12.5
H5261_0000_ND3	12.5
H5279_0000_ND3	12.5
H5297_0001_ND3	12.5

High-resolution images reveal that most of the pit craters on Ascraeus Mons have conical shapes, with diameters varying from >4 km to <100 m. The smallest pits have apparently flat floors with surface textures similar to the surrounding topographical surface. Most of the largest pits have a conical shape with no evidence of wall erosion or sediment accumulation.

Concentric pit chains and graben are the dominant types of volcanotectonic structure. Pit craters are sometimes organized into single elongated collapse pits (Fig. 1) and long chains (up to 70 km in length) consisting of small pits, in some cases linked by normal faults (Fig. 2a, b). Graben are usually longer than pit chains and, in some cases, contain collapse pits in their inner part. An example of the relationship between graben and pit chains can be observed in Figure 2b. The graben formed first and, after being filled by a lava flow, was reactivated with a dilational component that led to the formation of the pit chains. In some cases, aligned pit craters become wider and deeper (Fig. 2b) and can coalesce, resulting in a unique structure delimited by normal faults.

Model age determination

One lava flow on the eastern flank of the volcano (HRSC images H2054 and H0016; 11°48′N, 102°50′W), displaying the characteristic smooth undulating surface of a pahoehoe flow, is notable because it fills and covers a graben belonging to the concentric system, and is in turn cut by two collapse pits (Fig. 3a). Four other lava flows showing similar cross-cutting relationships were identified on the NE and SW flanks of Ascraeus Mons. Such cross-cutting relationships suggest that the lava flows were emplaced during the formation of the concentric graben and pit chains. Thus, dating these lava flows allows us to define the age of the event responsible for the concentric pit chains and graben.

To improve our crater counts and obtain better statistics for derived crater size–frequency distributions, a Context Camera (CTX) mosaic was produced and overlaid on the HRSC mosaic. CTX images have a higher resolution than HRSC data, approximately 6 m/px (metres per pixel) (and in some areas reaching 5 m/px). With CTX data, smaller, previously unseen craters can be taken into account. The CTX images we used are listed in Table 2. The ArcGis plugin *Cratertools* (Kneissl *et al.* 2011) was used to map and count the craters on the lava flow.

The Interactive Data Language (IDL) tool *Craterstats* (Michael & Neukum 2010) was used to calculate the age of the lava flow. The Neukum Production Function (NPF) we used was proposed by Ivanov (2001) and the chronology curve is by Hartmann & Neukum (2001). All craters with diameter <25 m (i.e. <4 px) were omitted. None the less, some smaller craters tended to show a roll-off in the plots due to their small diameter value. This roll-off was not taken into account in the age fit. The production function was fit to the crater size–frequency distribution of the selected areas, applying a resurfacing correction to take into account crater degradation as a function of the surface age. Figure 3 shows the NPF model plots for the four lava flows, which give ages of between about 130 and 270 Ma. In Table 3 we show the crater-counting data for each lava flow.

Self-similar clustering of dykes

On Earth, a direct genetic and spatial link between subsurface fractures and vent occurrence has previously been observed (e.g. Tibaldi 1995; Connor & Conway 2000; Mazzarini 2004; Mazzarini & Isola 2010). Thus, the scale invariance in vent distribution is thought to reflect the fractal properties of the connected part of a fracture network. The connected fracture network allows basaltic magma to pass from deep reservoirs through the crust to the surface.

A two-point correlation function method was used to measure the fractal dimension of the vent population. For a population of N points (vent centres), the correlation integral $C(l)$ is defined as the correlation sum that accounts for all the points at a distance of less than a given length l (Hentschel & Procaccia 1983; Bonnet *et al.* 2001). Under this approach, the term $C(l)$ is computed as:

$$C(l) = 2N(l)/N(N - 1) \qquad (1)$$

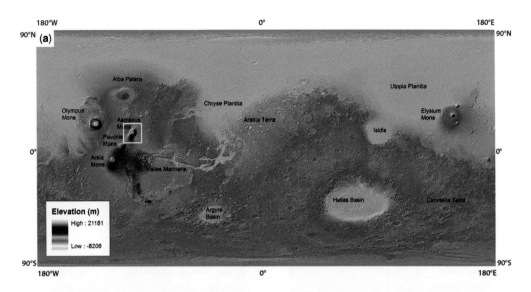

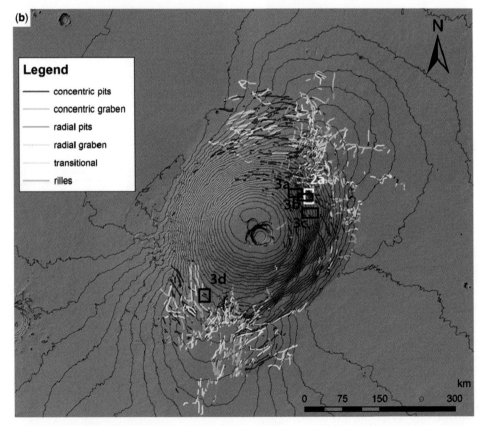

Fig. 1. (a) Global MOLA topography at 128 pixels per degree (ppd). The white square highlights the Ascraeus Mons study area. **(b)** The structural mapping of this study overlaid on a MOLA-derived hillshade map (128 ppd, illuminated from the NW). Contour lines are shown with an interval of 500 m. Most of the structures are located on the two lava aprons, especially the radial structures. The white square highlights the image shown in Figure 2a, while the black squares highlight the images where the lava flows used for crater counts presented in Figure 3a–d are located.

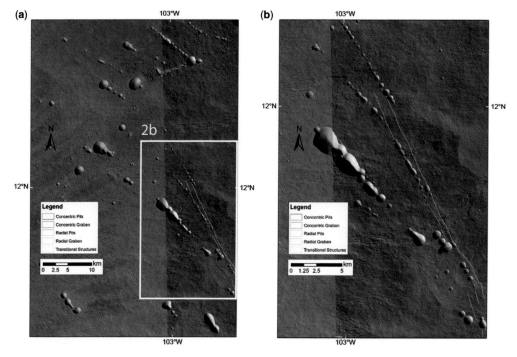

Fig. 2. (**a**) Detail of volcano-concentric structures. Normal faults are inferred to have assumed a dilational behaviour, leading to pit crater formation. The faults and the pit craters' alignment and coalescence along faults are visible more clearly in (**b**).

where $N(l)$ is the number of pairs of points whose distance is less than l. The fractal distribution of $C(l)$ is defined by:

$$C(l) = cl^D \tag{2}$$

where c is a normalization constant and D is the fractal exponent.

If scaling holds, equation (2) is valid, and the slope of the curve in a $\log(C(l))$ v. $\log(l)$ diagram yields the D value. The higher the D value, the less clustered the system. The computed D value is, however, valid for a defined range of distances

Table 2. *Constituent images of the CTX mosaic*

CTX image	Resolution (m/px)
P06_003264_1938_XN_13N104W	5.1
P10_004754_1914_XI_11N104W	5.1
P10_004820_1923_XI_12N104W	5.1
P10_004965_1920_XN_12N103W	5.1
P13_005954_1927_XI_12N105W	5.1
P17_007668_1966_XN_16N105W	5.1
P18_007958_1920_XN_12N103W	5.1

(l) (the size range) over which the scaling of $C(l)$ with l holds. The size range is bound by two values: the lower (L_{co}) and upper (U_{co}) cutoffs, respectively. Within the size range, the linear fit of the curve is well defined and the angular coefficient of the straight line is the fractal exponent. The local slope $(\Delta\log(C(l)/\Delta\log(l)))$ is a point-by-point measure of the slope of the tangent to the curve; where a linear fit holds, the curve is a straight line, and the local slope (the angular coefficient of the regression line) is constant. For each analysis, the size range of samples is in turn defined by a plateau in the local slope v. $\log(l)$ diagram: the wider the range the better the computation of the power-law distribution (Walsh & Watterson 1993).

The choice of the zones where the plateau is well defined and the determination of the cutoffs was carried out following the methods of Mazzarini (2004). The wider length range for which the correlation between $\log(l)$ and local slope is greatest was selected. Heuristically, a size range of at least one order of magnitude and with at least 150 samples is required for extracting robust parameter estimates (Bonnet *et al.* 2001; André-Mayer & Sausse 2007; Clauset *et al.* 2009).

Findings based on this approach suggest that, for basaltic volcanic fields (Mazzarini *et al.* 2013*a, b* and references therein): (i) the distribution of monogenetic vents is linked to the mechanical layering of the crust; (ii) vents tend to cluster according to a power-law distribution; and (iii) the range of lengths over which the power-law distribution is defined is characterized by an upper cutoff that approximates the thickness of the fractured medium (crust) between the source (magma reservoirs) and the sink (surface). This correlation has been studied in volcanic fields located in a range of geodynamic settings: continental rifts such as the East African Rift System (Mazzarini 2004, 2007; Mazzarini & Isola 2010); extensional continental back-arc settings, such as in southernmost Patagonia (Mazzarini & D'Orazio 2003); transtensional settings, such as volcanic fields along the Trans-Mexican Volcanic Belt (Mazzarini et al. 2010); and within compressional continental settings at the rear of the Andes chain in northernmost Patagonia (Mazzarini et al. 2008). The scaling of the U_{co} value with the distance between surface and source for fluids has been observed even for mud volcanoes growing during crustal shortening in Azerbaijan (Bonini & Mazzarini 2010).

Linear scaling between U_{co} (in km) and the source-to-surface distance (T, in km) has been verified for these volcanic fields using independent geophysical methods as well as with seismic tomography. Scaling of U_{co} with depth of the main magma reservoirs has also been proposed for basaltic volcanoes. For example, at Mt Etna volcano (Sicily), the most active volcano in Europe, vents distributed along the volcano's slopes have self-similar clustering with U_{co} at approximately 10 km (Mazzarini & Armienti 2001), and the main intrusive complex beneath the volcano is at an average depth of around 9 km.

Self-similar clustering analyses applied to large datasets (i.e. >200 objects) is statistically robust. Mazzarini & Isola (2010) showed that removing a random sample of 20% of the vents from a large dataset does not affect the estimation of fractal dimension (less than 0.01% of variation) and the error introduced into the estimation of the cutoffs is less than 1–2%. Moreover, Mazzarini et al. (2013a) tested the effect of uncertainties in point locations by adding random errors as high as 5–25 times that of the coarsest image resolution used to locate objects. The lower error randomly added to the vent locations generated fractal exponent and cutoff values identical to those computed for the original dataset. In the case of the highest random error added, the resulting fractal exponent was 3% higher than that computed for the original dataset, and the cutoffs were very similar to those found for the original dataset (with a difference of less than 2%). In our analysis, considering the large number of objects used in the analysis and the

spatial resolution of images used to map them, we conservatively assume an error of 10% in the computation of both the fractal exponent and the distribution of lower and upper cutoffs.

The lengths of fractures and dykes in nature often display a power-law distribution in the form:

$$N(l) \approx l^{-a} \qquad (3)$$

where $N(l)$ is the number of fractures (dykes) whose length is longer than l, and a is the fractal exponent (e.g. Bonnet et al. 2001 and references therein). High values of the a exponent imply a high frequency of short fractures, whereas low values of the a exponent mean that the system is mainly controlled by long fractures. Generally, fracture networks exhibit fractal length distributions with an exponent $a < 3$ (Renshaw 1999) or $a < 4$ (Bonnet et al. 2001). It is worth noting that, according to fractal scaling relationships, longer dykes are also thicker (e.g. Cruden & McCaffrey 2006).

Ascraeus Mons pits self-similar clustering

Under the assumption that pit alignments represent the surface traces of feeder dykes, we applied length distribution and self-similar clustering analyses to pit alignments along the flanks of Ascraeus Mons. About 2000 pit alignments ($N = 2268$) around Ascraeus Mons were mapped (Fig. 4), and their lengths computed. The barycentre of the traces were computed and stored in a geographical information system (GIS) database. Assuming, as before, that the pit alignments' traces correspond to those of dykes, the separation and length distributions of dykes were analysed using the coefficient of variation (CV). CV is defined as the ratio between the standard deviation and the mean values of the analysed distribution (Gillespie et al. 1999 and references therein). CV describes the degree of clustering of the distribution: CV > 1 indicates clustering, CV = 1 indicates a random or Poisson distribution and CV < 1 indicates anti-clustering (a regular distribution). The separation between dykes varies in the 0.2–35.9 km range, with an average of 1.3 km and standard deviation of 1.9 km; the CV is 1.4. The lengths of the mapped dykes vary in the 0.1–13.6 km range, with an average of 0.9 km and standard deviation of 1.2 km; the CV is 1.3. Spatial and size distributions of dykes show CV > 1, which is indicative of clustering.

The analysis of the size distribution of dyke trace lengths, defined by the fractal exponent a in equation (3), clearly shows a cumulative distribution with two distinct populations; the trade-off length of about 1.5 km marks the boundary between both population (Fig. 5). On this basis, we analysed

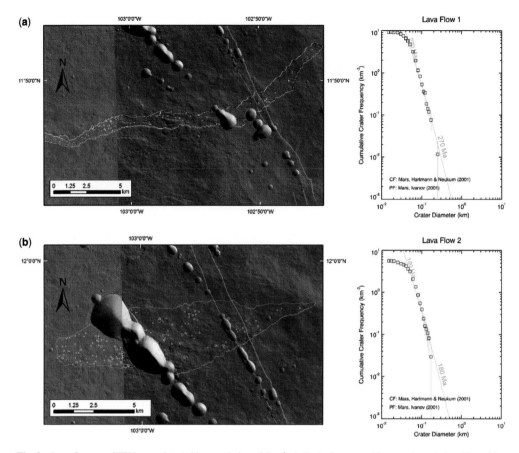

Fig. 3. Lava flows on CTX image data (with a resolution of 6 m/px) displaying mutual intersection relationships with graben and pit chains, together with model age plots (**a**)–(**d**). The yellow circles represent impact craters used to obtain the lava flow model age. The roll-off in the upper part of the plot is due to crater diameter values close to the image resolution.

self-similar clustering of two datasets: dykes with lengths greater than 1.5 km and those with lengths less than 1.5 km (Table 4). Dykes shorter than 1.5 km show self-similar clustering with $D = 1.13$, $L_{co} = 1.5$ km and $U_{co} = 10.5$ km, whereas those dykes longer than 1.5 km have values of $D = 1.51$, $L_{co} = 11$ km and $U_{co} = 60$ km (Fig. 6).

Applying to dykes on Mars that which is observed for feeder conduits on Earth, we propose that the plumbing system of Ascraeus Mons is characterized by two main magmatic sources: a shallow source only about 11 km deep with respect to the Martian datum and linked to the surface via dykes traces shorter than 1.5 km; and a deeper magmatic source located at a depth of approximately 60 km from which dykes traces longer than 1.5 km originate. The importance of a shallow magma reservoir on Tharsis volcanoes was previously discussed by Mouginis-Mark & Robinson (1992) for Olympus

Mons. Moreover, the complex structure of the plumbing system of Ascraeus Mons suggested by our analysis is consistent with the possible migration and evolution of the system as proposed by Scott & Wilson (2000), who analysed the sequence of caldera collapse events on the volcano.

Proposed model for the Ascraeus Mons plumbing system

The depth of the magma reservoirs feeding volcanic fields and vents on central volcanoes has been defined by using the upper cutoff value for the vents' fractal distribution (Mazzarini & Armienti 2001; Mazzarini & Isola 2010; Mazzarini *et al.* 2013*a*, *b*). Notably, spatial clustering of fractures in a given network is described by characteristic power-law exponents, which increase as the fracture

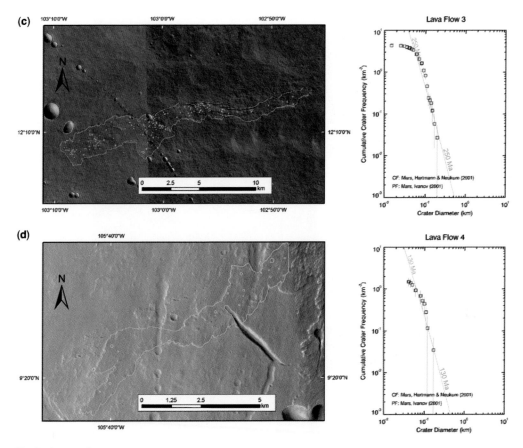

Fig. 3. *Continued.*

network evolves (Barton *et al.* 1995). This model considers that the volume of intact rock decreases with time as new fractures develop because older fracture systems form the boundary for newly forming ones. Consequently, new fractures developing in an already heterogeneous and fractured crust (by eventually reactivating and interconnecting inherited planes of weakness) will show less clustering and higher power-law exponent values than earlier structures. Taking into account the fact that vents develop within a fracture network (as feeder dykes are part of the actual network), we thus propose a relative timing for the formation of the two dyke populations we infer for Ascraeus Mons.

The lower fractal exponent derived from our analysis is $D = 1.13$, with a U_{co} value of 10.5 km, corresponding to a magma source depth of approximately 11 km that fed the first dyke population. The late dyke population (second cluster of pits) has a higher fractal exponent, where $D = 1.5$, and a value of U_{co} of 60 km, giving an average depth for the magma source of 60 km relative to the Martian geoid.

On the basis of the fractal analysis results we thus propose a plumbing system evolution for Ascraeus Mons consisting of four main stages:

- During Stage 1, a fracture network, connected to a large, primary magma source probably present at 60 km below the Martian geoid, developed (Fig. 7a).
- In Stage 2 (Fig. 7b), the formation and coalescence of a shallow system of magma chambers,

Table 3. *Lava flow crater-counting results*

Lava flow	Area (km²)	Craters counted	Coordinates
1	46	431	11°50′N, 102°50′W
2	32.5	140	11°55′N, 102°53′W
3	38.5	217	12°10′N, 103°2′W
4	12.3	18	9°22′N, 105°35′W

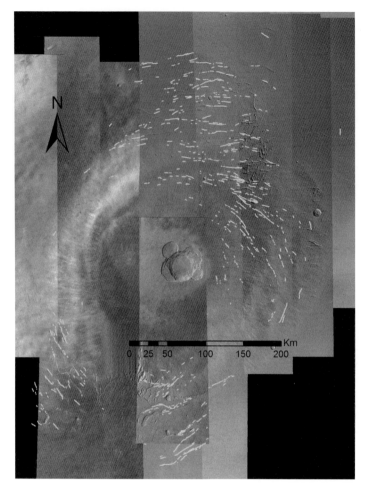

Fig. 4. Regional map of concentrically aligned pit craters used for the fractal clustering analysis, shown on the HRSC (12 m/px) mosaic of Ascraeus Mons.

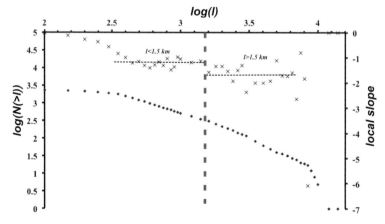

Fig. 5. The cumulative distribution of dyke length, displaying two distinct populations. The local slope is plotted with the '×' symbol. The trade-off length is about 1.5 km, indicating two discrete, self-similar dyke sets.

Table 4. *Parameters of distribution for the population of dykes in Ascraeus Mons*

Dataset	n	a	R^2	D	R^2	L_{co} (km)	U_{co} (km)
$L > 1.5$ km	318	1.81 ± 0.02	0.99	1.51 ± 0.01	0.99	11.0 ± 1.5	60.0 ± 5.1
$L < 1.5$ km	1950	1.22 ± 0.01	0.99	1.13 ± 0.01	0.99	1.5 ± 0.2	10.5 ± 1.0

n, number of samples; a, fractal exponent of length distribution; R^2, goodness of fit; D, fractal exponent of dyke self-similar clustering; L_{co}, lower cutoff; U_{co}, upper cutoff.

between 1.06 Ga and 230 Ma, within Ascraeus undergoes several collapse events (Robbins *et al.* 2011); the location of the constituent magma chambers may have relocated through the period of caldera collapse (Scott & Wilson 2000).

- Stage 3, as indicated by our crater counts (ranging between 270 and 130 Ma), records a new magmatic pulse (probably supplied from the deeper magma source) that intrudes the shallow reservoir and probably triggers feeder dykes from a depth of around 11 km (with a fractal

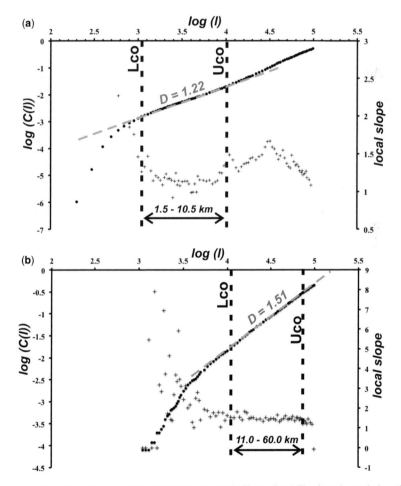

Fig. 6. Self-similar clustering of the two dyke populations shown in Figure 5. (**a**) The fractal correlation of pit craters with a length $L < 1500$ m, correlated with the fractal exponent $D = 1.22$. (**b**) The fractal correlation of the family of pit craters with $L > 1500$ m, and their fractal exponent $D = 1.51$. In both cases the U_{co} value corresponds to the depth of the magmatic source. The population in (a) is related to the approximately 11 km-deep magma source, whereas the population in (b) relates to the approximately 60 km-deep magma source.

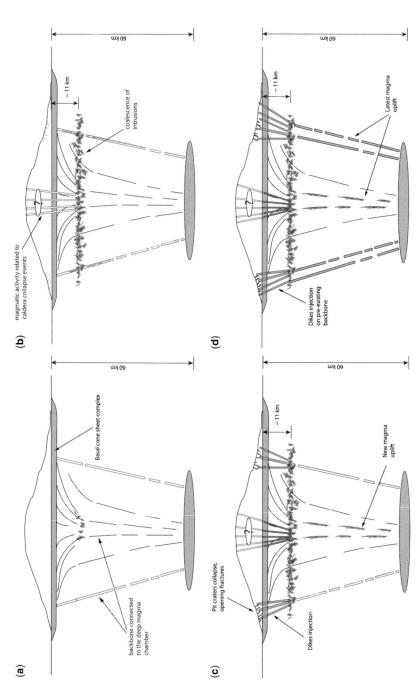

Fig. 7. Conceptual model of the geometry and evolution of the Ascraeus Mons plumbing system, showing feeder dykes that lead to the opening of cracks and pit craters on the volcano's surface (and that possibly filled the caldera) and a probable small magma chamber located inside the edifice. (**a**) The first stage of the magmatic evolution of the Ascraeus plumbing system with shield building (Byrne *et al.* 2012) and associated cone sheet (Canon-Tapia & Walker 2004). The volcanic edifice is entirely formed and linked to the deep magma source. (**b**) The magma pathways allow the vertical migration of intrusions to a depth of around 11 km, which then coalesce to form a shallow magma reservoir. (**c**) A new magmatic pulse (occurring between 270 and 130 Ma) feeds the shallow magma source, from which dykes emanate (with $L < 1.5$ km), reaching the surface and causing the formation of a first set of cracks and pit craters. (**d**) A later magmatic pulse coming directly from the deeper (approximately 60 km) source leads to dykes (with $L > 1.5$ km) that modify the pre-existing fracture network and opens a new pathway to the surface, leading to the formation of a second population of pit craters and pit chains.

exponent $D = 1.13$) that reach the volcano's surface; pit craters and chains are these feeder dykes' surface expressions (Fig. 7c). This evolution is also consistent with the presence of a shallow magma chamber inside the volcanic edifice.

- Stage 4, between 270 and 130 Ma, is the latest large magmatic pulse from the deep reservoir. During this episode, part of the dyke network linked to the approximately 11 km-deep reservoir is reactivated, and the fracture network provides a way for new magma batches sourced from the deep reservoir to ascend. This results in an increase of the fractal exponent, D, from 1.13 to 1.5. The newly formed population of pit craters and pit chains opens on the surface, and the first population is probably modified (e.g. through the coalescence of pits into chains: Fig. 7d).

Conclusion

An analysis of the self-similar clustering of vents has been applied to the distribution of pit chains on Ascraeus Mons. As for volcanic fields and central volcanoes on Earth, such an analysis provides clues as to the depth of the pressurized magmatic reservoirs at the time of dyke emplacement (e.g. Mazzarini & Isola 2010). On the basis of these results, we propose a model of the complex magmatic history of Ascraeus Mons, with periods of relative quiescence and intense activity up to the very recent past due to the build-up of a multi-reservoir plumbing system. In particular, we identify and date, using crater counts and derived size–frequency distributions, lava flows that display cross-cutting relationships with pit chains and related structures that correspond to the four main volcanic phases we describe above. These phases involve the building of the main plumbing structure connected to a deep magma source by a fracture network, followed by the emplacement of a shallower magmatic body beneath the volcano at a depth of approximately 11 km. Next, feeder dykes emanating from the shallow chamber were triggered by a magmatic pulse from the major reservoir between 270 and 130 Ma. The dykes intruded the shallow magma chamber, and formed pit craters and pit chains at the surface. This finding does not preclude magmatic activity along the central conduit and thus the occurrence of very shallow magmatic chambers inside the volcanic edifice. The latest phase involved a magmatic pulse coming directly from the deep source to the surface, which served to reactivate and modify the pre-existing fracture network and associated surface faults, resulting in the formation of a new population of pit chains

and craters. This proposed multi-stage plumbing system evolution is consistent with the geological evolution suggested for Ascraeus Mons by Byrne et al. (2012), who proposed a multi-stage evolution of the volcano including a late-stage phase of surficial extension and pit crater formation.

Finally, the fractal analysis of surface volcanic features we presented may thus provide insights into the deep structure of other volcanoes on Mars for which geophysical, geochemical and petrological observations do not currently exist.

We thank P. K. Byrne, N. Lang and an anonymous reviewer who provided helpful and detailed suggestions to improve the manuscript. This research was supported by ASI research grant No. I/060/10/0.

References

ANDRÉ-MAYER, A.-S. & SAUSSE, J. 2007. Thickness and spatial distribution of veins in a porphyry copper deposit, Rosia Poieni, Romania. Journal of Structural Geology, 29, 1695–1708, http://dx.doi.org/10.1016/j.jsg.2007.06.010

BANERDT, W. B., GOLOMBEK, M. P. & TANAKA, K. L. 1992. Stress and Tectonics on Mars. In: KIEFFER, H., JAKOSKY, B. M., SNYDER, C. W. & MATTHEWS, M. S. (eds) Mars. University of Arizona Press, Tucson, AZ, 249–297.

BARTON, C. A., ZOBACK, M. D. & MOOS, D. 1995. Fluid flow along potentially active faults in crystalline rock. Geology, 23, 683, http://dx.doi.org/10.1130/0091-7613(1995)023<0683:FFAPAF>2.3.CO;2

BASU, A. R. 1977. Textures, microstructures and deformation of ultramafic xenoliths from San Quintin, Baja California. Tectonophysics, 43, 213–246, http://dx.doi.org/10.1016/0040–1951(77)90118-4

BLEACHER, J. E., GLAZE, L. S. ET AL. 2009. Spatial and alignment analyses of a field of small volcanic vents south of Pavonis Mons and implications for the Tharsis province, Mars. Journal of Volcanology and Geothermal Research, 185, 96–102, http://dx.doi.org/10.1016/j.jvolgeores.2009.04.008

BONINI, M. & MAZZARINI, F. 2010. Mud volcanoes as potential indicators of regional stress and pressurized layer depth. Tectonophysics, 494, 32–47, http://dx.doi.org/10.1016/j.tecto.2010.08.006

BONNET, E., BOUR, O., ODLING, N. E., DAVY, P., MAIN, I., COWIE, P. & BERKOWITZ, B. 2001. Scaling of fracture systems in geological media. Reviews of Geophysics, 39, 347–383, http://dx.doi.org/10.1029/1999RG000074

BYRNE, P. K., VAN WYK DE VRIES, B., MURRAY, J. B. & TROLL, V. R. 2009. The geometry of volcano flank terraces on Mars. Earth and Planetary Science Letters, 281, 1–13, http://dx.doi.org/10.1016/j.epsl.2009.01.043

BYRNE, P. K., VAN WYK DE VRIES, B.,, MURRAY, J. B. & TROLL, V. R. 2012. A volcanotectonic survey of Ascraeus Mons, Mars. Journal of Geophysical Research, 117, E01004, http://dx.doi.org/10.1029/2011JE003825

CANON-TAPIA, E. 2008. How deep can be a dike? *Journal of Volcanology and Geothermal Research*, **171**, 215–228, http://dx.doi.org/10.1016/j.jvolgeores.2007.11.021

CANON-TAPIA, E. & WALKER, G. P. L. 2004. Global aspects of volcanism: the perspectives of 'plate tectonics' and 'volcanic systems'. *Earth-Science Reviews*, **66**, 163–182, http://dx.doi.org/10.1016/j.earscirev.2003.11.001

CARR, M. 2006. *The Surface of Mars*. Cambridge University Press, Cambridge.

CARTER, A., VAN WYK DE VRIES, B., KELFOUN, K., BACHÈLERY, P. & BRIOLE, P. 2007. Pits, rifts and slumps: the summit structure of Piton de la Fournaise. *Bulletin of Volcanology*, **69**, 741–756, http://dx.doi.org/10.1007/s00445-006-0103-4

CLAUSET, A., SHALIZI, C. R. & NEWMAN, M. E. J. 2009. Power-law distributions in empirical data. *SIAM Review*, **51**, 661–703, http://dx.doi.org/10.1137/070710111

CONNOR, C. B. & CONWAY, F. M. 2000. Basaltic volcanic fields. *In*: SIGURDSSON, H. (ed.) *Encyclopedia of Volcanoes*. Academic Press, New York, 331–343.

CRUDEN, A. & MCCAFFREY, K. 2006. Dimensional scaling relationships of tabular igneous intrusions and their implications for a size, depth, and compositionally dependent spectrum of emplacement processes in the crust. *Eos, Transactions of the American Geophysical Union*, **87**, Abstract V12B-06.

DAHM, T. 2000*a*. Numerical simulations of the propagation path and the arrest of fluid-filled fractures in the Earth. *Geophysical Journal International*, **141**, 623–638, http://dx.doi.org/10.1046/j.1365-246x.2000.00102.x

DAHM, T. 2000*b*. On the shape and velocity of fluid-filled fractures in the Earth. *Geophysical Journal International*, **142**, 181–192, http://dx.doi.org/10.1046/j.1365-246x.2000.00148.x

FEDOTOV, S. A. 1981. Magma rates in feeding conduits of different volcanic centres. *Journal of Volcanology and Geothermal Research*, **9**, 379–394, http://dx.doi.org/10.1016/0377-0273(81)90045-7

FERRILL, D. A., WYRICK, D. Y., MORRIS, A. P., SIMS, D. W. & FRANKLIN, N. M. 2004. Dilational fault slip and pit chain formation on Mars. *GSA Today*, **14**, 4–12.

GIACOMINI, L., MASSIRONI, M., MARTELLATO, E., PASQUARÈ, G., FRIGERI, A. & CREMONESE, G. 2009. Inflated lava flows on Daedalia Planum (Mars)? Clues from a comparative analysis with the Payen volcanic complex (Argentina). *Planetary and Space Science*, **57**, 556–570.

GILLESPIE, P. A., JOHNSTON, J. D., LORIGA, M. A., MCCAFFREY, K. J. W., WALSH, J. J. & WATTERSON, J. 1999. Influence of layering on vein systematics in line samples. *In*: MCCAFFREY, K. J. W., LONERGAN, L. & WILKINSON, J. J. (eds) *Fractures, Fluid Flow and Mineralization*. Geological Society, London, Special Publications, **155**, 35–56.

GROSFILS, E. B. 2007. Magma reservoir failure on the terrestrial planets: assessing the importance of gravitational loading in simple elastic models. *Journal of Volcanology and Geothermal Research*, **166**, 47–75, http://dx.doi.org/10.1016/j.jvolgeores.2007.06.007

GUDMUNDSSON, A. 2002. Emplacement and arrest of sheets and dykes in central volcanoes. *Journal of Geophysical Research*, **116**, 279–298.

GUDMUNDSSON, A. & BRENNER, S. L. 2004. How mechanical layering affects local stresses, unrests, and eruptions of volcanoes. *Geophysical Research Letters*, **31**, L16606, http://dx.doi.org/10.1029/2004GL020083

HARTMANN, W. K. & NEUKUM, G. 2001. Cratering chronology and the evolution of Mars. *Space Science Review*, **96**, 165–194.

HAUBER, E., BLEACHER, J., GWINNER, K., WILLIAMS, D. & GREELEY, R. 2009. The topography and morphology of low shields and associated landforms of plains volcanism in the Tharsis region of Mars. *In*: BLEACHER, J. E. & DOHM, J. M. (eds) *Tectonic and Volcanic History of the Tharsis Province, Mars. Journal of Volcanology and Geothermal Research*, **185**, 69–95, http://dx.doi.org/10.1016/j.jvolgeores.2009.04.015

HENTSCHEL, H. G. E. & PROCACCIA, I. 1983. The infinite number of generalised dimension of fractals and strange attractors. *Physica*, **8D**, 435–444.

HIESINGER, H., HEAD, J. W. & NEUKUM, G. 2007. Young lava flows on the eastern flank of Ascraeus Mons: rheological properties derived from High Resolution Stereo Camera (HRSC) images and Mars Orbiter Laser Altimeter (MOLA) data. *Journal of Geophysical Research*, **112**, 24.

HUTTON, D. H. W. 1996. The "space problem" in the emplacement of granite. *Episodes*, **19**, 114–119.

IVANOV, B. A. 2001. Mars/Moon cratering rate ratio estimates. *Space Science Review*, **96**, 87–104.

IDA, Y. 1999. Effects of the crustal stress on the growth of dikes: conditions of intrusion and extrusion of magma. *Journal of Geophysical Research*, **104**, 17,897–17,909, http://dx.doi.org/10.1029/1998JB900040

KNEISSL, T., VAN GASSELT, S. & NEUKUM, G. 2011. Map-projection-independent crater size-frequency determination in GIS environments—New software tool for ArcGIS. *In*: PONDRELLI, M., TANAKA, K., ROSSI, A. & FLAMINI, E. (eds) *Geological Mapping of Mars. Planetary and Space Science*, **59**, 1243–1254.

LISTER, J. R. & KERR, R. C. 1991. Fluid-mechanical models of crack propagation and their application to magma transport in dykes. *Journal of Geophysical Research*, **96**, 10,049–10,077, http://dx.doi.org/10.1029/91JB00600

MAALOE, S. 1973. Temperature and pressure relations of ascending primary magmas. *Journal of Geophysical Research*, **78**, 6877–6886, http://dx.doi.org/10.1029/JB078i029p06877

MAZZARINI, F. 2004. Volcanic vent self-similar clustering and crustal thickness in the northern Main Ethiopian Rift. *Geophysical Research Letters*, **31**, L04604, http://dx.doi.org/10.1029/2003GL018574

MAZZARINI, F. 2007. Vent distribution and crustal thickness in stretched continental crust: the case of the Afar Depression (Ethiopia). *Geosphere*, **3**, 152–162, http://dx.doi.org/10.1130/GES00070.1

MAZZARINI, F. & ARMIENTI, P. 2001. Flank Cones at Mount Etna Volcano: Do they have a power-law distribution? *Bulletin of Volcanology*, **62**, 420–430, http://dx.doi.org/10.1007/s004450000109

MAZZARINI, F. & D'ORAZIO, M. 2003. Spatial distribution of cones and satellite-detected lineaments in the Pali Aike Volcanic Field (southernmost Patagonia): insights into the tectonic setting of a Neogene rift system. *Journal of Volcanology and Geothermal Research*, **125**, 291–305, http://dx.doi.org/10.1016/S0377-0273(03)00120-3

MAZZARINI, F. & & ISOLA, I. 2010. Monogenetic vent self-similar clustering in extending continental crust: examples from the East African Rift System. *Geosphere*, **6**, 567–582, http://dx.doi.org/10.1130/GES00569.1

MAZZARINI, F., FORNACIAI, A., BISTACCHI, A. & PASQUARÈ, F. A. 2008. Fissural volcanism, polygenetic volcanic fields, and crustal thickness in the Payen Volcanic Complex on the central Andes foreland (Mendoza, Argentina). *Geochemistry, Geophysics, Geosystems*, **9**, Q09002, http://dx.doi.org/10.1029/2008GC002037

MAZZARINI, F., FERRARI, L. & ISOLA, I. 2010. Self-similar clustering of cinder cones and crust thickness in the Michoacan–Guanajuato and Sierra de Chichinautzin volcanic fields, Trans-Mexican Volcanic Belt. *Tectonophysics*, **486**, 55–64, http://dx.doi.org/10.1016/j.tecto.2010.02.009

MAZZARINI, F., KEIR, D. & ISOLA, I. 2013a. Spatial relationship between earthquakes and volcanic vents in the central-northern Main Ethiopian Rift. *Journal of Geothermal and Volcanological Research*, **262**, 2013, 123–133.

MAZZARINI, F., ROONEY, T. O. & ISOLA, I. 2013b. The intimate relationship between strain and magmatism: a numerical treatment of clustered monogenetic fields in the Main Ethiopian Rift. *Tectonics*, **32**, 49–64, http://dx.doi.org/10.1029/2012TC003146

MEYZEN, C. M., MASSIRONI, M., POZZOBON, R. & DAL ZILIO, L. 2014. Are terrestrial plumes from motionless plates analogues to Martian plumes feeding the giant shield volcanoes? *In*: PLATZ, T., MASSIRONI, M., BYRNE, P. K. & HIESINGER, H. (eds) *Volcanism and Tectonism Across the Inner Solar System*. Geological Society, London, Special Publications, **401**. First published online February 21, 2014, http://dx.doi.org/10.1144/SP401.8

MICHAEL, G. G. & NEUKUM, G. 2010. Planetary surface dating from crater size–frequency distribution measurements: partial resurfacing events and statistical age uncertainty. *Earth and Planetary Science Letters*, **294**, 223–229.

MOUGINIS-MARK, P. J. & ROBINSON, M. S. 1992. Evolution of the Olympus Mons Caldera, Mars. *Bulletin of Volcanology*, **54**, 347–360, http://dx.doi.org/10.1007/BF00312318

NEUKUM, G., JAUMANN, R. ET AL. 2004. Recent and episodic volcanic and glacial activity on Mars revealed by the High Resolution Stereo Camera. *Nature*, **432**, 971–979, http://dx.doi.org/10.1038/nature03231

OKUBO, C. H. & MARTEL, S. J. 1998. Pit crater formation on Kilauea volcano, Hawaii. *Journal of Volcanology and Geothermal Research*, **86**, 1–18.

PETFORD, N., KERR, R. C. & LISTER, J. R. 1993. Dike transport of granitoid magmas. *Geology*, **21**, 845–848, http://dx.doi.org/10.1130/0091-7613(1993)021<0845:DTOGM>2.3.CO;2

PETFORD, N., CRUDEN, A. R., MCCAFFREY, K. J. W. & VIGNERESSE, J. L. 2000. Granite magma formation, transport and emplacement in the Earth's crust. *Nature*, **408**, 669–673, http://dx.doi.org/10.1038/35047000

PHILLIPS, R. J., ZUBER, M. T. ET AL. 2001. Ancient geodynamics and global-scale hydrology on Mars. *Science*, **291**, 2587–2591.

RENSHAW, C. E. 1999. Connectivity of joint networks with power law length distributions. *Water Resources Research*, **35**, 2661–2670, http://dx.doi.org/10.1029/1999WR900170

ROBBINS, S. J., DI ACHILLE, G. & HYNEK, B. M. 2011. The volcanic history of Mars: high-resolution crater-based studies of the calderas of 20 volcanoes. *Icarus*, **211**, 1179–1203.

RUBIN, A. M. 1993. Dikes v. diapirs in viscoelastic rock. *Earth and Planetary Science Letters*, **119**, 641–659, http://dx.doi.org/10.1016/0012-821X(93)90069-L

RUBIN, A. M. 1995. Propagation of magma-filled cracks. *Annual Review of Earth and Planetary Sciences*, **23**, 287–336, http://dx.doi.org/10.1146/annurev.ea.23.050195.001443

SCOTT, E. D. & WILSON, L. 2000. Cyclical summit collapse events at Ascraeus Mons, Mars. *Journal of the Geological Society, London*, **157**, 1101–1106, http://dx.doi.org/10.1144/jgs.157.6.1101

SCHULTZ, R. A., SOLIVA, R., FOSSEN, H., OKUBO, C. H. & REEVES, D. M. 2008. Dependence of displacement-length scaling relations for fractures and deformation bands on the volumetric changes across them. *Journal of Structural Geology*, **30**, 1405–1411.

SPERA, F. J. 1980. Aspects of magma transport. *In*: HARGRAVES, R. B. (ed.) *Physics of Magmatic Processes: Princeton*. Princeton University Press, Princeton, NJ, 265–323.

TAISNE, B. & TAIT, S. 2009. Eruption v. intrusion? Arrest of propagation of constant volume, buoyant, liquid-filled cracks in an elastic, brittle host. *Journal of Geophysical Research*, **114**, B06202, http://dx.doi.org/10.1029/2009JB006297

TAKADA, A. 1994a. The influence of regional stress and magmatic input on styles of monogenetic and polygenetic volcanism. *Journal of Geophysical Research*, **99**, 13,563–13,573, http://dx.doi.org/10.1029/94JB00494

TAKADA, A. 1994b. Development of subvolcanic structure by the interaction of liquid-filled cracks. *Journal of Volcanology and Geothermal Research*, **61**, 207–224, http://dx.doi.org/10.1016/0377-0273(94)90004-3

TANAKA, K. L. & GOLOMBEK, M. P. 1989. Martian tension fractures and the formation of grabens and collapse features at Valles Marineris. *In*: RYDER, G. & SHARPTON, V. L. (eds) *Proceedings of the Nineteenth Lunar and Planetary Science Conference*. Cambridge University Press, Cambridge, 383–396.

TIBALDI, A. 1995. Morphology of pyroclastic cones and Tectonics. *Journal of Geophysical Research*, **100**, 24, 521–24,535, http://dx.doi.org/10.1029/95JB02250

TURCOTTE, D. L. 1982. Magma migration. *Annual Review of Earth and Planetary Sciences*, **10**, 397–408,

http://dx.doi.org/10.1146/annurev.ea.10.050182.
    002145
WALKER, G. P. L. 1992. Coherent intrusion complexes' in
    large basaltic volcanoes—a new structural model.
    Journal of Volcanology and Geothermal Research,
    50, 41–54.
WALL, M., CARTWRIGHT, J., DAVIES, R. & McGRANDLE,
    A. 2010. 3D seismic imaging of a Tertiary Dyke
    Swarm in the Southern North Sea, UK. *Basin Research*,
    22, 181–194, http://dx.doi.org/10.1111/j.1365-2117.
    2009.00416.x
WALSH, J. J. & WATTERSON, J. 1993. Fractal analysis of
    fracture pattern using the standard box-counting tech-
    nique: valid and invalid methodologies. *Journal of*

Structural Geology, **15**, 1509–1512, http://dx.doi.
    org/10.1016/0191–8141(93)90010–8
WERNER, S. C. 2009. The global martian volcanic evol-
    utionary history. *Icarus*, **201**, 44–68, http://dx.doi.
    org/10.1016/j.icarus.2008.12.019
WILSON, L. & HEAD, J. W. 2001. Evidence for episodicity
    in he magma supply to the large Tharsis volcanoes.
    Journal of Geophysical Research, **106**, 1423–1433.
ZUBER, M. T., BILLS, B. G., FREY, H. V., KIEFER, W. S.,
    NEREM, R. S. & ROARK, J. H. 1993. Possible flexural
    signatures around Olympus and Ascraeus Montes,
    Mars. Abstract 1591, presented at the 24th Annual
    Lunar and Planetary Science Conference, March 15–
    19, 1993, Houston, TX.

Lithospheric flexure and volcano basal boundary conditions: keys to the structural evolution of large volcanic edifices on the terrestrial planets

PATRICK J. McGOVERN[1]*, ERIC B. GROSFILS[2], GERALD A. GALGANA[1,3], JULIA K. MORGAN[4], M. ELISE RUMPF[5], JOHN R. SMITH[6] & JAMES R. ZIMBELMAN[7]

[1]*Lunar and Planetary Institute, Universities Space Research Association, Houston, TX 77058, USA*

[2]*Geology Department, Pomona College, Claremont, CA 91711, USA*

[3]*AIR Worldwide, 131 Dartmouth Street, Boston, MA 02116, USA*

[4]*Department of Earth Science, Rice University, Houston, TX 77005, USA*

[5]*Hawaii Institute of Geophysics and Planetology, University of Hawaii at Manoa, Honolulu, HI 96822, USA*

[6]*HURL, University of Hawaii at Manoa, Honolulu, HI 96822, USA*

[7]*MRC 315, Smithsonian Institution, Washington, DC 20013-7012, USA*

*Corresponding author (e-mail: mcgovern@lpi.usra.edu)

Abstract: Large volcanic edifices constitute enormous loads at the surfaces of planets. The lithosphere, the mechanically strong outer layer of a planet, responds to growing edifice loads by flexing. The shape of this lithospheric flexure and the resulting stress state exert critical influences on the structure of the evolving edifices, which in turn feed back into the flexural response. Flexural subsidence of the lithosphere forms topographical moats surrounding volcanoes that are partially to completely filled by landslide debris, volcaniclastic materials and sediments, or by relatively flat aprons of volcanic flows. Flexure creates a characteristic 'dipole' state of stress that influences subsequent magma ascent paths and chamber dynamics in the lithosphere. Compression in the upper lithosphere can inhibit magma ascent and favour the development of oblate magma chambers or sill complexes. This compression can be transferred into the edifice unless a décollement allows the volcano base to slip over the underlying lithosphere; generally, basal décollements are found to operate via high pore-fluid pressure in a clay sediment-based layer. Volcanoes lacking such a layer, regardless of the thickness of the basal sediments, lack basal décollements and, thus, tend to be limited in size by compressive stresses adverse to magma ascent.

Large volcanic edifices are among the most striking geological features on the terrestrial (solid-surface) planets. In this paper, we consider volcanoes that are broad and tall enough to generate a substantial flexural response in the lithosphere. In the solar system, these are generally volcanoes of basaltic composition, perhaps because the characteristic low viscosity of basaltic magmas allows relatively long flows to form, thereby producing broad edifices. On Earth (Fig. 1), these are typically found in hotspot-related chains in ocean basins, such as the Hawaiian, Galápagos and Marquesas chains in the Pacific Ocean, the Canary chain in the Atlantic and La Réunion in the Indian Ocean. On Mars, the tallest and broadest volcanoes are in the Tharsis volcanic province, including Olympus (Fig. 1) and Alba Montes, and the three Ascraeus, Pavonis and Arsia Montes that are collectively known as the Tharsis Montes. The Elysium rise, often considered a 'smaller Tharsis' but actually comparable in size and shape to Alba Mons (McGovern *et al.* 2001), also supports several large shields on its surface. Much shorter (yet broad) edifices like Nili and Meroe Paterae (in the Syrtis Major volcanic province), and Tyrrhenu Mons, occur in the southern highlands, usually on the margins of impact basins such as Isidis and Hellas, respectively. On Venus, there are at least 167 volcanoes with flow unit diameters in excess of 100 km (Crumpler *et al.* 1996; McGovern & Solomon 1998), such as Sapas Mons (Fig. 1). On the Moon, recent high-resolution topography data have revealed the presence of several large basaltic shields, again located on the margins of impact basins (Spudis *et al.* 2013).

From: PLATZ, T., MASSIRONI, M., BYRNE, P. K. & HIESINGER, H. (eds) 2015. *Volcanism and Tectonism Across the Inner Solar System*. Geological Society, London, Special Publications, **401**, 219–237.
First published online March 17, 2014, http://dx.doi.org/10.1144/SP401.7

Lithospheric flexure is the fundamental response of the lithosphere to loads emplaced above, beneath and within it (e.g. Watts 2001). For large volcanoes, typically the surface load (which causes downward flexure) is the most significant, although subsurface loads caused by processes such as buoyant

(a)

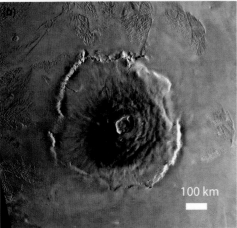

(c)

Fig. 1. Images of solar system volcanoes. (**a**) Hawaiian Islands, Earth; topography and bathymetry from Eakins *et al.* (2003) and Smith & Kelley (2010). (**b**) Olympus Mons, Mars; *Viking Orbiter 1* imaging, from NSSDC Photo Gallery, Mars, http://nssdc.gsfc.nasa.gov/photo_gallery/photogallery-mars.html. (**c**) Sapas Mons, Venus, from Magellan synthetic aperture radar (SAR) imaging.

crustal underplating (e.g. Watts & ten Brink 1989; Wolfe *et al.* 1994) and dense intrusive complexes (e.g. Dieterich 1988; McGovern *et al.* 2001) also are significant in certain settings. Downward flexure of the lithosphere produces topography: annular arches as well as trenches (moats) that contain volcanic materials in various forms, ranging from intact flows to blocky landslide and slump materials, to volcano-derived sediments and volcaniclastic materials. Downward flexure also produces near-surface horizontal compressive stresses that can impede magma ascent (McGovern & Solomon 1993; Galgana *et al.* 2011). These stresses may or may not be transmitted from the lithosphere into the edifice, depending on the nature of the volcano–lithosphere interface. Thus, the basal boundary condition of the edifice plays a critical role in determining edifice size, tectonics and propensity for flank movements/failures. The magnitudes and length scales of these flexural responses depend on the thickness of the elastic lithosphere.

In this paper, we explore the influence of lithospheric flexure and basal boundary condition on the development and structure of large volcanic edifices in the solar system. First, we identify constraints on lithospheric thickness in large volcano settings. Then we describe the effects of lithospheric response in terms of topography and stress, using theory of magma ascent in dykes to connect lithospheric stress patterns with edifice shapes. We then characterize the nature of the basal boundaries of volcanoes, and establish links with size, surface morphology and subsurface structure. In this way, we demonstrate the strong influence of lithospheric flexure and basal boundary conditions on the evolution of planetary volcanoes.

Constraints on lithospheric thickness

On Earth, the oceanic lithosphere setting of large volcanoes dictates the thickness of the mechanical lithosphere via the mechanism of plate cooling with age. For volcanoes emplaced near a spreading ridge, estimates of elastic lithosphere thickness, T_e, are low (see Table 1): for example, T_e is 12 km for the western Galápagos Islands (Feighner & Richards 1994). At locations on colder lithosphere, further from the ridge or on a slow-spreading plate, T_e is thicker, as at Marquesas, Reunion, Hawaii and the Canary Islands (see Table 1). On Mars, lithospheric thickness is largely a function of the time at which the edifice was emplaced (i.e. the age of loading, as inferred from impact crater counts) in the context of a planet cooling with time (McGovern *et al.* 2002, 2004*b*), with many large volcanoes emplaced on relatively thick lithospheres (Table 1). On Venus, lithospheric thickness

Table 1. *Lithosphere and structural parameters for large volcanoes in the solar system. For volcanoes on Earth, the parameters correspond to averages for a volcanic chain*

	Lithosphere age* (Ma)	T_e (km)	Volcano edifice diameter[†] (km)	Volcano height above base (km)
Earth				
Canary	130[1]	35[1] 20[2] 28–36[3]	100[2]	5.4–7.9[4]
Galápagos	<10[5]	12[6]	50–60[7]	3.5[7]
Hawaii	80–90[8] 110[9]	25–35[‡10] 33–44[§10] 30 ± 7[11]	75–210[12]	8.7–9.2[13,14]
Marquesas	50[15]	18[16]	80–120[17]	3.4–4.7[17,18]
La Réunion	55[19]	28 ± 4[19]	220[20]	7.1[20]
Mars				
Olympus Mons	unknown	>70[21] 93 ± 40[22] >70[22]	840 × 640[23]	21.9[23]
Ascraeus Mons	unknown	>20[21] 105 ± 40[22] >50[22]	375 × 870[23]	14.9[23]
Pavonis Mons	unknown	<100[21] >50[22]	380 × 535[23]	14.0[23]
Arsia Mons	unknown	2–80[21] <30[22] <85[22]	461 × 326[23]	11.7[23]
Alba Mons	unknown	38–65[21] 66 ± 20[22] 73 ± 30[22]	1015 × 1150[23]	5.8[23]
Moon				
Marius Shield	unknown	50–75[‖24]	330[25]	2.2[25]
Cauchy Shield	unknown	45 ± 15[¶24]	560[25]	1.8[25]
Venus				
Sif Mons	<800[26]	30–50[27]	142[28]	2.2[28]
Aruru Corona	<800[26]	0–7[29]	260[30]	1.4[30]

*At the time of initiation of the edifice growth
[†]Including submerged parts of the edifice, if any.
[‡]Two-dimensional model.
[§]Three-dimensional spatially variable model (T_m).
[‖]Using an early stage value for the Imbrium Basin.
[¶]Using an early stage value for the Serenitatis Basin.
References: [1]Dañobeitia *et al.* (1994), [2]Watts (1994), [3]Canales & Dañobeitia (1998), [4]Ancochea *et al.* (1994), [5]Hey *et al.* (1977), [6]Feighner & Richards (1994), [7]Nordlie (1973), [8]Clague & Dalrymple (1987), [9]Waggoner (1993), [10]Wessel (1993), [11]Watts & ten Brink (1989), [12]Moore (1987), [13]Peterson & Moore (1987), [14]Fornari & Campbell (1987), [15]Kruse (1988), [16]Filmer *et al.* (1993), [17]Brousse *et al.* (1978), [18]Filmer *et al.* (1994), [19]Bonneville *et al.* (1988), [20]Lénat *et al.* (1989), [21]McGovern *et al.* (2004b), [22]Belleguic *et al.* (2005), [23]Plescia (2004), [24]Solomon & Head (1980), [25]Spudis *et al.* (2013), [26]McKinnon *et al.* (1997), [27]Smrekar (1994), [28]McGovern & Solomon (1997), [29]Hoogenboom *et al.* (2004), [30]McGovern (1996).

may be related to the time elapsed since a proposed global resurfacing event at 500–800 Ma (McKinnon *et al.* 1997). During this event, large outpourings of magma and extensive tectonic deformation were likely to be accompanied by high heat flux from the interior, resulting in low T_e. Subsequent cooling of the interior may have led to substantial thickening of the lithosphere with time (Reese *et al.* 2007). Large volcanoes as a group are younger than the mean surface age of

Venus (Namiki & Solomon 1994), suggesting that some amount of cooling (i.e. lithosphere thickening) was required to allow large edifice construction. The Moon is similar to Mars, with lithosphere thickness related to age in a cooling planet; the shield volcanoes found by Spudis *et al.* (2013) were largely formed during the principal interval of mare volcanism on the Moon 3.9–3.0 Ga, although activity at some edifices may have lasted until as recently as 1 Ga.

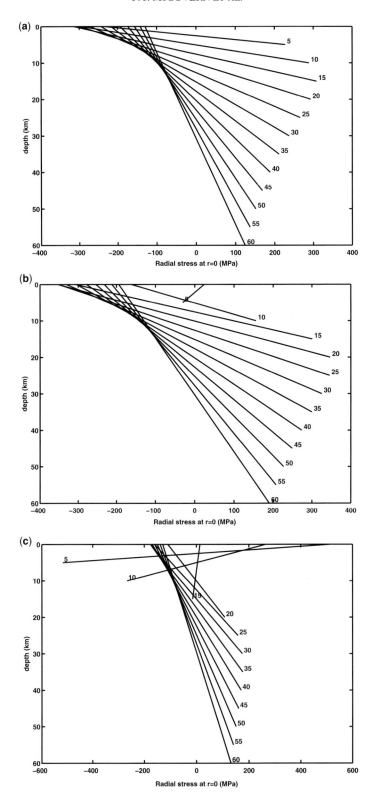

Effects of lithospheric flexure

Flexural containment of volcanic material

Lithospheric flexure creates a topographical depression, or 'moat', surrounding large edifices, which is itself surrounded by a topographical high, or 'arch'. On Earth, flexural moats are filled primarily with volcano-derived landslide materials, turbidites, volcaniclastics and sediments, as observed at the Hawaiian (Moore et al. 1989; Garcia & Hull 1994) and Marquesas (Wolfe et al. 1994) chains. On Mars, the flexural moat at Olympus Mons is filled with a combination of landslide deposits called the Olympus Mons aureole deposits (e.g. Lopes et al. 1982; McGovern et al. 2004a) and lava flows. For the Tharsis Montes, the moats appear to be filled dominantly by lava flows, with small volumes of surficial flanking deposits attributed to glacial activity (Head & Marchant 2003). On Venus, flexural moats around volcanoes are nearly completely filled by predominantly radial lava flows (McGovern & Solomon 1997), with volumes of moat-filling materials 7–10 times that of the topographical expressions of edifices alone. Large shield volcanoes on the Moon (Spudis et al. 2013) are likely to be similar to large volcanoes on Venus and Mars.

Flexural stress state and magma ascent

The flexural response to a volcanic edifice load produces a 'dipole' state of stress in the lithosphere beneath the edifice, with peaks in differential stress (the difference between the greatest and least principal stresses) at the top and bottom of the lithosphere, and a low-stress 'neutral plane' in between (e.g. Watts 2001; Galgana et al. 2011, 2013). The effects of this dipole were manifested physically in the Hawaiian earthquake pair of 15 October 2006 (McGovern 2007), with a deep extensional earthquake followed by a shallow compressional one. To demonstrate such stress states, we calculate lithospheric stresses for large volcanoes on Venus using the thick-plate analytical flexure solution of Comer (1983) in axisymmetrical geometry, with compressive stresses negative. After McGovern et al. (2013), we consider three edifice shapes: a cone; a truncated cone (representing a domical edifice); and an annulus with a triangular cross-section (representing a subset of features on Venus with annular faulting and/or topographical signatures called 'coronae').

Radial normal stresses at the symmetry axis ($r = 0$) for a cone (Fig. 2a) show extension in the lower lithosphere and compression in the upper lithosphere. Flexural stress magnitudes generally increase with decreasing elastic lithosphere thickness, T_e, except that this trend reverses for $T_e \leq 10$ km, a result of the relatively short wavelength of flexural response relative to load size (McGovern et al. 2013). The stress state for a truncated cone (or domical volcano) is similar (Fig. 2b), with the reversal in stress magnitudes occurring for $T_e \leq 15$ km, and there is even a reversal in stress sign for $T_e = 5$ km. For an annular volcano shape (Fig. 2c), the stress magnitude reversal starts at $T_e = 20$ km and the stress sign reversal occurs at $T_e = 15$ km.

The stresses calculated by the means described above can be assessed by two criteria to determine at what locations magma may ascend through the lithosphere in dykes, following the exposition of McGovern et al. (2013). The first criterion simply states the finding of Anderson (1951) – that intrusions tend to form perpendicular to the least compressive principal stress. This 'stress orientation' criterion is:

$$\Delta\sigma_y + \sigma_{local} > 0 \qquad (1)$$

where $\Delta\sigma_y$ is the differential stress, defined as the difference of horizontal normal stress and vertical normal stress ($\sigma_y - \sigma_z$; this is termed the 'tectonic' stress by Rubin 1995), and σ_{local} is a local variation in stress due to factors such as local inhomogeneities and modifications to the lithosphere, regional-scale stress fields and reductions in stress from yielding at the top and bottom of the lithosphere. A second, less well-known ascent criterion is based on pressure balance in vertical dykes, and can be derived from equation (7) of Rubin (1995) (modified to account for our 'tension positive, z positive upwards' sign conventions):

$$d\Delta\sigma_y/dz + d\Delta P/dz - \Delta\rho g > 0 \qquad (2)$$

where P is magma pressure, $\Delta\rho$ is host rock density (ρ_r) minus magma density (ρ_m) and ΔP is local excess magma pressure (McGovern et al. 2013). In practice, when flexure is significant (see below), the term $d\Delta\sigma_y/dz$ – the vertical gradient of tectonic stress – dominates the pressure balance. We collect the local excess magma pressure gradient and buoyancy terms into the term ΔPG, and the stress gradient criterion for ascent becomes $d\Delta\sigma_y/dz + \Delta PG > 0$ (McGovern et al. 2013).

Fig. 2. Radial normal stress (σ_{rr}) beneath the centre of volcanic edifices ($r = 0$) v. depth (z) in the lithosphere for three edifice shapes, all with a maximum height of 1 km and a radius of 200 km. Elastic thickness (T_e) values range from 5 to 60 km in 5 km increments: (**a**) conical edifice; (**b**) truncated cone or domical edifice; (**c**) annular edifice.

For a given vertical lithospheric stress distribution at a given r coordinate, we apply the stress orientation and stress gradient ascent criteria of the previous paragraph to determine whether or not magma can ascend at that location. Given that in elastic lithospheres the stress extrema occur at the top and bottom of the lithosphere, we apply the stress orientation criterion at those locations. For the elastic case, the stress gradient is constant throughout the thickness of the lithosphere. Thus, there are three ascent criteria to be met (McGovern *et al.* 2013). In practice, the 'dipole' nature of flexural stresses (half the lithosphere in compression, half in extension) means that one of the stress orientation criteria will not be satisfied. However, the stress states calculated here do not account for factors that would mitigate conditions nominally adverse to magma ascent, including buoyancy, overpressure, lithospheric uplift, brittle failure, ductile flow and thermomechanical erosion. To account for these mitigating factors, we apply stress and stress gradient 'thresholds' to the stresses calculated by the models described above, in the form of σ_{local} for the stress orientation criterion and ΔPG for the stress gradient criterion. The thresholds allow ascent criteria to be satisfied for modestly adverse values of stress (negative horizontal differential stress) and stress gradient (negative $d\Delta\sigma_y/dz$) (see McGovern *et al.* 2013).

The results of applying ascent criteria to flexural stress states have strong implications for the generation and maintenance of different volcanic edifice shapes. At high elastic thicknesses ($T_e > 50$ km), magma ascent criteria are met over a wide radius range that includes the centre of a conical edifice (Fig. 3). Presuming a magma source concentrated towards the central axis, direct ascent along such pathways would tend to produce an edifice also concentrated towards the centre; that is, a conical shape (McGovern *et al.* 2013). For thinner elastic lithospheres, the region near the centre of the cone experiences high $d\Delta\sigma_y/dz$ (throughout the thickness of the lithosphere) and compression (in the upper lithosphere), thereby shutting off magma ascent near the centre of the edifice (McGovern *et al.* 2013). This regime would tend to shut off summit eruptions, and therefore halt construction of the central part of a cone; instead, flank eruptions would tend to build up the flanks at the expense of the centre, thereby producing a flat-topped edifice or dome. We find that this is consistent with model results for truncated cone edifices at intermediate T_e values (around 25 km, Fig. 4); eruptions from circumferential dykes are predicted on the outer flank of such edifices. Such eruptions would tend to reinforce the domical shape (McGovern *et al.* 2013).

At low T_e (≤ 10 km), conical and truncated-cone edifices return extremely high magnitudes of negative stress gradients and compressive stresses near the top of the lithosphere (McGovern *et al.* 2013); these conditions are inconsistent with the formation of such edifice shapes. However, annular edifices on thin lithospheres (Fig. 5) exhibit two zones of permitted ascent: one on the inside of the ridge and one on the outside. The short-wavelength response of thin lithospheres tends to produce very narrow zones of magma ascent (Fig. 5). This configuration is broadly consistent with long-term maintenance of an annular shape, although radial migration of the eruption centres in response to the growing load is likely and necessary to produce single (rather than paired) annuli. Models of annular edifices emplaced on thick lithospheres ($T_e \geq 50$ km) predict ascent everywhere, requiring a very fortuitous pattern of sublithosphere magma accumulation (essentially, a torus) to generate the observed topography (McGovern *et al.* 2013). On intermediate-thickness lithospheres, annular loads generate a central magma cut-off that is more consistent with a domical rather than an annular shape (McGovern *et al.* 2013). Thus, we argue that annular edifices are the most likely expression of volcanic edifice construction at low T_e (McGovern *et al.* 2013).

Edifice shape on Venus

Drawing on the results above, we argue that lithospheric flexure, through its influence on magma ascent, controls the shape of large volcanic edifices on Venus. Conical volcanoes, such as Sif Mons (Fig. 6c), probably form on relatively thick lithosphere, consistent with T_e values in the 30–50 km range found for the Western Eistla Regio rise by Smrekar (1994). Similar ranges of T_e are found for other volcanic rises that support large volcanoes, like Atla and Bell Regiones (Phillips 1994; Smrekar 1994), although Simons *et al.* (1997) found a T_e as low as 20 km at the latter. Annular-shaped edifices, such as Irnini Mons (Fig. 6c), are predicted to form on relatively thin lithosphere, consistent with the $T_e = 0$–14 km estimate from gravity–topography admittance for a top-loaded (i.e. appropriate for a constructed edifice) model (Hoogenboom *et al.* 2004). A majority of coronae investigated by Hoogenboom *et al.* (2004) with a top-loading model had a lower bound of 0 km for T_e, consistent with very thin lithosphere. Interestingly, almost all coronae with lower bounds on $T_e > 0$ km from top-loading analysis occur in the 'Fracture Belt' geological setting (Hoogenboom *et al.* 2004), suggesting that coronae in such settings may form by a different mechanism to the constructional one proposed by McGovern *et al.* (2013) and espoused here. As argued above, domical edifices such as Tuulikki Mons (Fig. 6c) are likely to form

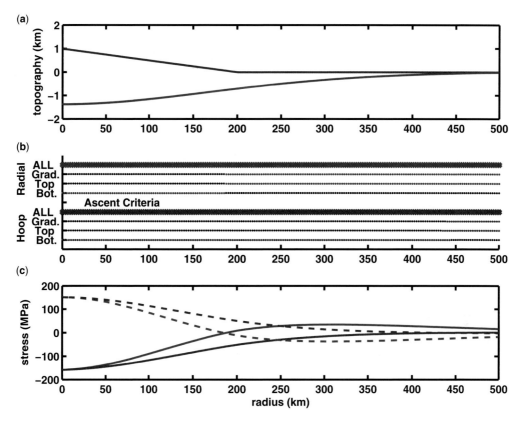

Fig. 3. Flexure solutions v. radius for volcanoes with conical edifice, with a peak height of 1 km and a radius of 200 km, for $T_e = 50$ km. Stress threshold $\sigma_{local} = 160$ MPa, stress gradient threshold $\Delta PG = 10$ MPa km^{-1}. (**a**) Edifice topography (blue) and surface flexure (red). (**b**) Ascent criteria for radial (red) and out-of-plane, or 'hoop', (blue) normal stresses. Individual criteria (stress at the top and bottom of, and stress gradient within, the lithosphere) are shown with thin dot symbols, and the total ascent criteria (the intersection of the individual criteria) are shown with thick lines. For individual criteria, zones where the criteria are totally satisfied are shown in the appropriate colour, and zones where the criteria are not satisfied but stresses are within the thresholds are shown in black. (**c**) Radial (red) and hoop (blue) normal stresses for the top (solid line) and bottom (dashed line) of the elastic lithosphere. Note that, in the absence of significant membrane stress components, the stress gradients in our models have essentially the same shape as the upper lithosphere (solid) stress curves, with magnitudes obtainable by dividing by $T_e/2$.

on intermediate-thickness lithosphere owing to central magma ascent shutoff (Fig. 4). The thickness of the mechanical lithosphere of Venus is limited by the high surface temperature, even for a dry rheology (e.g. Mackwell *et al.* 1998). Thus, in general, Venus cannot support tall edifices like the Tharsis volcanoes (with very limited exceptions such as Maat Mons, the tallest volcanic edifice on Venus at a height of 8 km).

Edifice shape on Mars

By the arguments given above, tall, conical edifices like Olympus Mons and Ascraeus Mons (Fig. 6b) should form on relatively thick lithospheres. This prediction is consistent with estimates of

T_e at these edifices derived from gravity–topography relationships (Table 1), the thickest found at Martian volcanoes. Domical edifices, like Alba Mons (formerly Alba Patera), the Elysium rise and Arsia Mons (Fig. 6b), all formed on lithosphere significantly thinner than that associated with the conical edifices (Table 1). There are no independent estimates of elastic thickness for Acheron Fossae, a half-annulus of elevated terrain partially buried by the moat-filling deposits and aureole of Olympus Mons, but by our reasoning we would predict a low value of T_e (<30 km). We note that the ranges of T_e determined for Mars exceed those for Venus (and Earth), and that the characteristic T_e ranges for conical and domical volcanoes are shifted upwards compared to those for

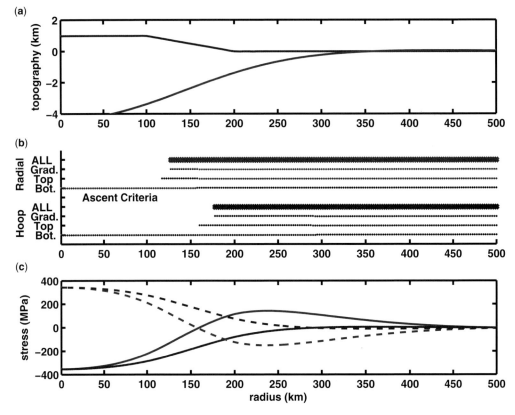

Fig. 4. Flexure solutions v. radius for volcanoes with a truncated cone (or domical) edifice, as in Figure 3, for $T_e = 25$ km.

Venus. This probably occurs because the greater height of Martian volcanoes requires greater lithospheric thickness to minimize conditions adverse to magma ascent (Figs 2–5), although the lower gravitational acceleration of Mars also plays a role in reducing adverse stresses. Thus, only Mars can support giant volcanoes like those in Tharsis.

Edifice shape on Earth

On Earth, estimates of lithospheric thickness for most island chains would, by the arguments presented here, predict conical to domical edifice shapes. However, stress-mitigating factors, such as buoyancy from displaced ocean water, buoyant support from crustal underplating (ten Brink & Brocher 1987; Wolfe *et al.* 1994) and volcanic spreading of edifices on basal décollements (McGovern & Solomon 1993; Borgia 1994), may help volcanoes in affected chains grow larger than they otherwise would. Volcanic spreading also reduces compressive stresses in edifices, facilitating

magma ascent and the formation of extensive elongate and long-lived rift zones, for example, on Hawaiian volcanoes (Fiske & Jackson 1972; Dieterich 1988). The chain volcanoes with the lowest T_e values – the Galápagos Islands – have a distinctive 'overturned soup bowl' topography (McBirney & Williams 1969) that, combined with a deep central depression, is annular in character (see the Fernandina profile in Fig. 6a). This suggests that lithospheric controls on magma ascent are similar to those shaping corona structures on Venus. Superposition of stresses and buttressing from neighbouring volcanoes also play roles in large volcanic edifice evolution (e.g. Fiske & Jackson 1972; Chadwick & Howard 1991).

Effects of inhibition of magma ascent

Compressive stress in the lithosphere and edifice has marked effects on magma ascent. With reference to the criteria of Anderson (1951) for intrusive magma ascent, the lower lithosphere stress state is conducive to magma ascent in vertical dykes; however,

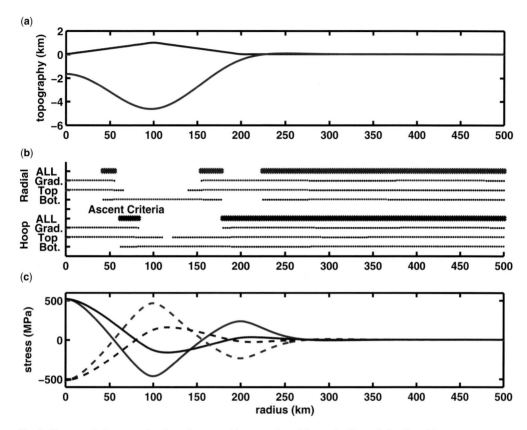

Fig. 5. Flexure solutions v. radius for volcanoes with an annular edifice, as in Figure 3, for $T_e = 5$ km.

in the upper lithosphere, ascending magma will be diverted into subhorizontal sills. Finite-element models of pressurization of magma chambers embedded in flexing lithospheres demonstrate this tendency: chambers in the upper (compressional) lithosphere fail near their mid-sections under a stress state that favours sill formation (Galgana et al. 2011; see also Grosfils et al. 2013). Thus, chambers are likely to assume oblate forms via lateral expansion, and continued magma supply may result in an extensive sill complex. Such complexes may express themselves at the surface by topographical slope breaks and annular zones of faulting (produced both by tectonic stresses and dykes emanating from the distal ends of the sills), as seen at the Alba Mons edifice on Mars, and structures with annular topography and tectonics termed 'coronae' on Venus (McGovern et al. 2001). On Earth, the presence of very shallow sills beneath Galápagos volcano summits (e.g. Amelung et al. 2000) is consistent with predictions for the upper lithosphere stress state from surface loading (Fig. 2).

Effects of basal boundary condition

The basal boundary condition of a volcanic edifice exerts a strong influence on the state of stress, thereby influencing magma ascent and the tectonic state of the edifice. For volcanoes that are welded to the underlying lithosphere, flexural compressive stresses in the upper lithosphere will be transmitted into the edifice (McGovern & Solomon 1993, 1998). However, if the edifice is allowed to spread outwards along a basal décollement, the stress state can be extensional (McGovern & Solomon 1993, 1998). Volcanic spreading is a fundamental process on volcanoes in the solar system (Borgia et al. 1990, 2000; McGovern & Solomon 1993; Borgia 1994; Van Wyk de Vries & Matela 1998; López et al. 2008), and has been documented in the Hawaiian Islands via geological, seismic and geodetic means (Denlinger & Okubo 1995; Morgan et al. 2000), and inferred at Olympus Mons on Mars (Borgia et al. 1990; McGovern & Solomon 1993; McGovern et al. 2004a; McGovern & Morgan 2009).

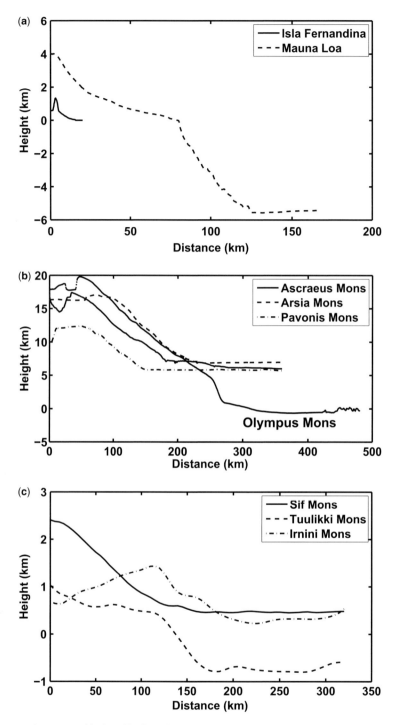

Fig. 6. Representative topographical profiles for volcanoes on Earth, Mars and Venus. (**a**) Earth: topography and bathymetry for Mauna Loa, Hawaii from Eakins *et al.* (2003); topography for Isla Fernandina, Galápagos, from SRTM (Farr & Kobrick 2000; Farr *et al.* 2007); vertical exaggeration is 10:1. (**b**) Mars: Olympus, Ascraeus, Pavonis and Arsia Montes, from MOLA altimetry (Smith *et al.* 2001); vertical exaggeration is 10:1. (**c**) Venus: Sif, Tuulikki and Irnini Montes, from Magellan topography (Ford & Pettengill 1992); vertical exaggeration is 50:1.

The absence or mitigation of flexurally induced compression at Hawaiian edifices by décollement slip accounts, in part, for the prominence of Hawaiian rift zones, which are topographically prominent loci of intrusion and eruption extending deep into the volcanic edifices (Dieterich 1988; Clague & Denlinger 1994). Rift zone intrusions are accommodated by, and drive flank slip along, the décollement. In contrast, volcanoes that lack basal décollements, such as in the Canary, Galápagos and Marquesas chains, lack Hawaiian-style rift zones. While some Canary volcanoes exhibit linear eruptive centres sometimes termed 'rift zones', seismic and geodetic evidence for basal décollement spreading is lacking. Canarian rifts are not related to basal spreading; instead, they have been associated with local uplift (e.g. Carracedo 1994; Carracedo et al. 1999) and sector collapse (Walter & Troll 2003; Walter et al. 2005). Furthermore, flank failures at the Canarian and Marquesan volcanoes are shallowly rooted (e.g. Wolfe et al. 1993; Carracedo et al. 1999; Day et al. 1999), in that they do not extend down to the volcano basal boundary, in contrast to the deep-seated, décollement-enabled large landslides at the Hawaiian chain (Moore et al. 1989; see also Morgan & McGovern 2005).

Lower-flank extensional features and evidence for flank failure at the basal scarp of Olympus Mons indicate volcanic spreading, although mid–upper flank terraces (e.g. Byrne et al. 2009, 2013) show that compression still exists in part of the edifice. This may result from radial variations in basal friction (McGovern & Morgan 2009), perhaps due to removal of fluid from the deepest parts of an overpressurized décollement by loading-induced compaction, or due to contributions from the flexure-induced slope of the basal boundary. Under this scenario, the higher-friction mid-flank part of the décollement would allow partial transmission of compressional flexural stresses into the edifice, while the lower-friction distal décollement would relieve them entirely.

The mechanism of décollement deformation is critical to determining the behaviour of basally detached volcanoes. Scenarios for basal spreading at volcanoes have invoked viscous or ductile flow of substrata beneath the edifice, such as ice at Olympus Mons (Tanaka 1985) and sediments at volcanoes on Earth (Van Wyk de Vries & Borgia 1996); such flow is often held to result from generic sediment 'weakness'. Alternatively, discrete slip surfaces (faults) can be activated in layers with elevated pore-fluid pressures (e.g. Nakamura 1980). Critical taper-wedge models applied to Kilauea's flank indicate that pore pressures about 50% of lithostatic levels would be required to obtain a critical wedge (i.e. a wedge on the verge of failure everywhere: Thurber & Gripp 1988). Iverson (1995) considered the effects of external forces on the stability of a rigid-wedge Hawaiian flank model. He found that forces from magma injection into rift zones (at the head of the wedge) were incapable of driving failure of more than a small part of the wedge. Iverson (1995) demonstrated that a plausible scenario for deep, extensive wedge failure required a thick (c. 200 m) layer of low hydraulic diffusivity (low permeability). Clays have low permeabilities due to small grain sizes and high tortuosity of fluid paths through platy clay particles. Abyssal clays, when compressed, will form rocks (claystones and shales) with lower permeabilities than rocks formed from calcareous (chalks and limestones) and siliceous (cherts) material (Brace 1980).

Constraints on the basal boundary condition at terrestrial volcanoes

On Earth, large volcanoes are emplaced on generally sediment-covered oceanic crust/lithosphere. Core samples returned from boreholes enable us to infer the thickness and composition of these sediment layers. Results from the five terrestrial volcanic chains listed above are compiled in McGovern (1996); we briefly summarize that compilation here (Fig. 7). Boreholes drilled into the oceanic crust west of the Hawaiian Islands (Ocean Drilling Program (ODP) sites 842 and 843: Dziewonski et al. 1992; Shipboard Scientific Party 1992a, b) constrain the thickness of pelagic sediments in this region. This location, beyond the crest of the Hawaiian flexural arch, is far enough from the islands to avoid material deposited by erosion and mass wasting of the volcanoes (cf. Moore et al. 1989), although small amounts of volcanic ash from Hawaiian island eruptions is present (Shipboard Scientific Party 1992a, b). At Site 842, two layers of clay were found, extending to 35.7 m below the seafloor (Shipboard Scientific Party 1992a). Beneath the clays was a layer of chert interspersed with claystone and nanofossil ooze, reaching to 237.7 m below seafloor. At Site 843, a layer of dark clay extended from the seafloor to 121.8 m below, and was underlain by a breccia of reddish brown and red chert to a depth of 228.0 m (Shipboard Scientific Party 1992b). A layer of nanofossil limestone (with chert nodules) and a thin layer of calcareous clay lie on top of the basalt basement, reached at depths of 228.8–242.5 m below the seafloor. Note that these sites were deliberately located above elevated basement topography to ensure that basalt basement would be reached, to enable a seismometer to be installed (Dziewonski et al. 1992). Sites with lower basement topography (valleys) are likely to have substantially greater

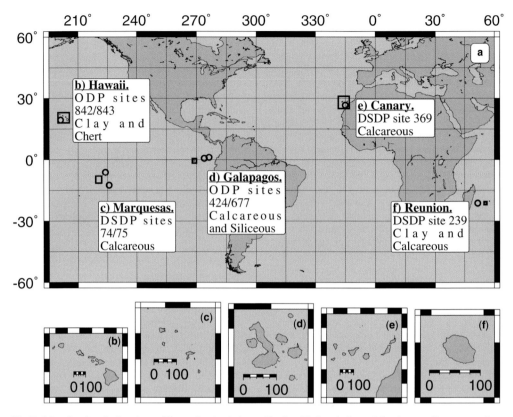

Fig. 7. Map showing the locations of five volcanic chains on Earth, with descriptions of dominant sediment types for each chain. (**a**) Main map. The boxes show the locations of the inset maps. The locations of DSDP and ODP sediment-sampling boreholes are shown as circles. Graticule is shown with 30° increments. (**b**)–(**f**) Inset maps showing the Hawaiian, Marquesas, Galápagos and Canary chains, and La Réunion, respectively. Scale bars have an overall length of 100 km, with increments of 20 km. Graticules are shown with 1° increments.

depths of sediment filling them. Single-channel seismic data (Collins *et al.* 1992) suggest that sediments in the basins near these sites are about 100 m thicker than those above the basement peaks. Thus, the reported measurements are lower bounds on the average thickness of the sediment layer.

The Galápagos Islands fall in a broad near-equatorial belt of high productivity in the Pacific. The biogenic sediment production rate is very high (37–59 mm a^{-1} at ODP Site 677: Alexandro-vich 1989), such that even relatively young (i.e. *c.* 10 Ma) crust may have accumulated hundreds of metres of sediment. At ODP Site 424 (0°36′N, 86°08′W), just south of the Galápagos Spreading Centre, oceanic crust of age 0.6–0.62 Ma is covered with about 30 m of sediment, most of which is siliceous foraminiferous nannofossil ooze (Hoffert *et al.* 1980). At ODP Site 677 (1°12′N, 83°44′W), oceanic crust of age about 5.9 Ma (Shipboard Scientific Party 1988) is covered by

200–300 m of sediment. Three major sedimentary units have been identified: Unit I consists of alternating clayey biogenic calcareous siliceous oozes and clayey biogenic siliceous calcareous oozes; Unit II consists of siliceous nannofossil ooze and siliceous nannofossil chalk; and Unit III consists of cherty limestone and nannofossil chalk (Shipboard Scientific Party 1988).

Near the Marquesas Islands in the eastern Pacific, surface sediments are principally calcareous nannofossil oozes with contributions from high-silica radiolarian ooze and very minor clay constituents; overall sediment thicknesses are 82–100 m (Davies 1985; Shipboard Scientific Party 1971*a*, *b*). Near La Réunion, surface sediments are a complex mix of brown clay, calcareous clays and calcareous ooze (Kolla & Kidd 1982), with the nearest borehole reporting units of silty clay and clay-rich nanno ooze (Unit I) and brown clay and brown silty clay (Unit II), each unit about 160 m in thickness (Shipboard Scientific Party 1972).

Sediment production in the Atlantic Ocean is dominated by organisms that produce calcium carbonate. Thus, surface sediments in most of the Atlantic are calcareous. Deep sediment samples have been obtained from Deep Sea Drilling Program (DSDP) Hole 369, on the edge of the African continental slope just south of the Canary Islands (Shipboard Scientific Party 1977). The sediments fall into three principal lithological groups: a layer of nannofossil marls and siliceous nannofossil marls to depth 346 m below seafloor; a layer of argillaceous nannofossil limestone, argillaceous marls and chalks from 346 to 422 m; and a layer of silty nannofossil marls to 489 m (Shipboard Scientific Party 1977).

The roughness of the basement topography also influences the character of the basal boundary condition of a volcano. High-amplitude topography at the base of the volcano will promote a welded basal boundary condition because basement topographical highs will tend to protrude into the edifice as the volcano grows over them, helping to 'lock' the edifice in place. These highs will also tend to interrupt the continuity of the sediment layers resting on their flanks, inhibiting the formation of broad basal décollements that facilitate movement of large portions of an edifice. Long linear topographical peaks orientated parallel to the spreading axis, called abyssal hills, are the characteristic topographical variations observed near spreading ridges. Abyssal hills form at the ridge axis through a complex interaction of tectonic and volcanic processes (see Goff 1991 and references therein). Abyssal hill roughness and width increase with decreasing spreading rate (Goff 1991; Malinverno 1991; Bird & Pockalny 1994; Macario et al. 1994). Low spreading rates at the Mid-Atlantic Ridge result in high roughness at the base of the Canary Island volcanoes, whereas substantially faster spreading rates in the Pacific and Indian Ocean spreading centres generate topographically smoother bases for Hawaiian, Marquesan and La Réunion volcanoes (see Goff 1991). Relief on crust formed at the Cocos–Nazca (Galápagos) spreading ridge is considerably greater than that for crust formed at Pacific plate spreading centres (Hey et al. 1977), consistent with the intermediate spreading rate at the former. However, volcanic construction of the Galápagos platform (on which the Galápagos volcanoes are emplaced) may reduce the effect of abyssal hill relief on basal boundary condition in that setting.

Basal boundary condition on Mars

Sedimentation has played a large role in the resurfacing of Mars, particularly in the northern lowlands (e.g. Fuller & Head 2002). Abundant evidence for transport by water of material from highlands to lowlands exists in the form of channels, valley networks and chaos terrain (e.g. Carr 1996). Further, the widespread detection of phyllosilicates on Mars (e.g. Bibring et al. 2006; Ehlmann et al. 2011) suggests that clay minerals are a substantial component of this sediment. Clays can form as the result of hydrothermal circulation of water through the Martian crust, perhaps aided by impact heating (e.g. Schwenzer et al. 2012). McGovern & Morgan (2009) proposed that the roughness of the pre-volcano basement at Olympus Mons (due to heavy impact bombardment) was mitigated by subsequent sediment deposition, and that thickening of basal sediments with increasing distance from the centre of the Tharsis rise allowed the NW flank of Olympus Mons to spread more easily than the SE flank.

Basal boundary condition on Venus

The planet Venus lacks evidence for meaningful sedimentation due to the absence of an eroding agent such as liquid water on the surface (with the possible exception of very fluid lava flows that carved the long, sinuous canali). There is very limited aeolian redistribution of impact-generated dust (Arvidson et al. 1992). Thus, volcanoes are likely to be welded to the underlying lithosphere (e.g. McGovern & Solomon 1998) and, therefore, unlikely to exhibit volcanic spreading. Possible exceptions include several intermediate–large volcanoes located on rifts that show evidence for extensional stresses on their flanks (López et al. 2008). However, the deformation producing the stress in these volcanoes is thought to occur as ductile flow in the crust below a brittle layer and, as such, is more similar to the scenarios proposed for ductile 'substratum spreading' (e.g. Van Wyk de Vries & Matela 1998) than to the overpressurized décollement-based movement described here for the Hawaiian Islands and Olympus Mons. In other words, the volcanoes in López et al. (2008) are still welded to a very thin 'lithosphere' but may deform due to sublithospheric flow.

Basal boundary condition on the Moon

At the opposite extreme from Venus, the Moon lacks an atmosphere entirely: evidence for sedimentation other than deposition of impact ejecta is negligible and no viable pore fluid exists at the surface. Therefore, we presume that lunar shield volcanoes are welded to the substrate lithosphere. This is consistent with observations that the lunar shields are constructed by basaltic flows (Spudis et al. 2013), with no evidence of features related to volcanic spreading.

Implications for basal boundary conditions

Given the constraints on sediment thickness and composition, and topographic roughness discussed above, we can now predict which volcanic chains should exhibit evidence of a basally detached boundary condition. Such behaviour is most likely to be found in regions with a thick clay layer and comparatively smooth (low-amplitude) basement topography (Table 2).

In the Canary and Galápagos chains, predicted or observed high-amplitude abyssal hill topography and the lack of a thick clay layer make a welded basal boundary condition likely, and no evidence of basally detached behaviour is observed. Nakamura (1980) proposed that a paucity of sediments at the base of Galápagos volcanoes prevents the basal slip required to accommodate rift-zone formation. However, the sediment productivity in the waters around the Galápagos Islands is very high, such that hundreds of metres of sediments can accumulate even on young (c. 10 Ma) crust (Alexandrovich 1989). In fact, the thickness of the sediment layer in holes drilled near the Galápagos Islands (Shipboard Scientific Party 1988) is similar to that of the much older crust in the vicinity of the Hawaiian Islands (Shipboard Scientific Party 1992a, b), and considerably larger than that at sites near the Marquesas Islands (Shipboard Scientific Party 1983). The Galápagos sediments are dominantly calcareous and siliceous oozes (Shipboard Scientific Party 1988), which form relatively high-permeability rocks when compressed. These rocks are unlikely to sustain the high pore pressures required for the formation of a décollement. Thus, despite the thickness of this inferred basal sediment layer, the Galápagos Islands constitute examples of large welded-base volcanoes. We instead support a

modified version of the hypothesis of Nakamura (1980). Fine-grained abyssal clay sediments are required to form a low-permeability layer that allows build-up of high pore pressures that facilitate décollement formation. By this view, it is the paucity of abyssal clay sediments, not of pelagic sediments in general, that prevents the formation of a décollement beneath the Galápagos volcanoes.

The Marquesas Islands were emplaced on crust formed at a moderately fast spreading ridge; abyssal hill topography should be low in amplitude, not presenting insurmountable obstacles to basal detachment formation. The lack of a thick clay layer in the sediments near the Marquesas, however, makes it unlikely that a décollement will form. This prediction is consistent with the absence of characteristic basally detached features such as well-developed rift zones on the Marquesas volcanoes.

The situation at La Réunion is more complex. The composition of surface sediments is variable (Kolla & Kidd 1982), and so too presumably is the composition of the deeper sediments. Evidence from the closest DSDP site (Shipboard Scientific Party 1972) suggests that abyssal clay layers with thicknesses sufficient to support a décollement layer may be present beneath La Réunion. Given the low-amplitude basal topography predicted by a high spreading rate at the fossil ridge beneath La Réunion (Schlich 1982), basally detached behaviour should be possible. Such behaviour is predicted to have taken place at about 1.8 Ma in the largest of a series of collapse events that have increased in frequency, while decreasing in magnitude (Gillot *et al.* 1994) and collapse basal depth. The decrease in size and depth of such activity may indicate the depletion or exhaustion of a thin or discontinuous clay layer at the base of La Réunion, a layer that

Table 2. *Predictions and observations of basal boundary conditions at large volcanoes or volcanic chains in the solar system*

	Inferred basal topography	Thick clay layer?	Basally detached behaviour predicted?	Basally detached behaviour observed or inferred?
Canary	Rough	No	No	No
Galápagos	Rough	No	No	No
Hawaii	Smooth	Yes	Yes	Yes
Marquesas	Smooth	No	No	No
La Réunion	Smooth	Possibly	Possibly	Limited
Olympus Mons	Rough?	Yes (inferred)	Yes	Yes
Tharsis Montes	Rough?	No	No	No
Venus volcanoes	Rough?	No	No	No*
Lunar volcanoes	Rough?	No	No	No

*Volcanic spreading inferred at several medium–large volcanoes on Venus (López *et al.* 2008) is due to substratum spreading and not a basal detachment between the edifice and lithosphere.

no longer functions as a décollement. However, current flank motion of the eastern part of the Island, including the Enclos-Fouque depression and the Grande Brule slide, has been attributed to a hybrid structure of a hydrothermally altered weak core and a lower-flank décollement (Merle & Lénat 2003).

Among the volcanoes we have discussed in this paper, the volcanoes of the Hawaiian chain are unique in exhibiting well-defined and long-lived structural features indicative of a detached basal boundary condition (and are probably unique among all large, subaereally exposed oceanic volcanoes). Both the topographical and sedimentary characteristics of the basement crust near Hawaii are unambiguously conducive to the formation of a basal décollement, the only chain listed in Table 2 for which this is so. The uncertain and possibly variable nature of basal clay layers at La Réunion seems to be consistent with the evidence for limited detachment activity (see also Merle & Lénat 2003). The absence of clay sediments at the Canary, Galápagos and Marquesas chains is consistent with the lack of evidence at those chains for basally detached behaviour. We therefore conclude that the character of oceanic basement in terms of sediment thickness, composition and roughness is a useful predictor of volcano morphology. We argue that it is not weakness or ductility of basal sediments *per se*, but rather generation of high pore pressures, specifically in clay sediments, that plays the major role in generating décollements on Earth and Mars.

As on Earth, clay sediments appear to play a leading role in facilitating volcanic spreading on Mars. The low hydraulic diffusivity of abyssal clay sediments was invoked by Iverson (1995) to account for the effectiveness of décollement slip at Hawaii, and McGovern *et al.* (2004*a*, *b*) and Morgan & McGovern (2005) suggested a similar scenario beneath Olympus Mons. McGovern & Morgan (2009) proposed that an increase in phyllosilicate sediment thickness with increasing distance from the Tharsis rise (and, thus, decreasing elevation) accounted, in part, for the greater extent from the centre and the extensional nature of the NW flank of Olympus Mons relative to the other flanks. If Noachian sediments were distributed preferentially in lowlands, then volcanoes at low basal elevations elsewhere on Mars are the most likely to have basal detachments. Large volcanoes in the northern lowlands are generally absent (exception: the Elysium rise), but Apollinaris Mons, like Olympus Mons, is located at the highland–lowland transition. Apollinaris Mons is the only other volcano on Mars with a prominent basal escarpment and disrupted terrain surrounding it (Robinson *et al.* 1993), potential analogues to the Olympus Mons basal scarp and aureole, respectively (McGovern & Morgan 2009). Tharsis Tholus exhibits several throughgoing faults indicative of sector collapse (Crumpler *et al.* 1996) attributed to deformation of a ductile (Plescia 2003) or 'weak' (and modelled as viscous: Platz *et al.* 2011) basal layer. Tharsis Tholus is located downslope from Ascraeus Mons, in a saddle between that edifice, Alba Mons and Tempe Terra. This location would be likely to accumulate sediments being transported downslope from Tharsis, and is, therefore, consistent with the phyllosilicate-based, overpressurized basal detachment scenario proposed for Olympus Mons by McGovern & Morgan (2009). Unfortunately, the base of Tharsis Tholus is covered by 0.5–3.5 km of younger regionally derived magmas (Plescia 2003), so potential scarp or aureole structures are obscured.

Conclusions

Lithospheric flexure plays a key role in the structural development of large volcanoes in the solar system. The stress state in the lithosphere influences subsequent magma ascent, leading to a link between volcano shape and lithospheric thickness. Thick lithospheres facilitate the growth of tall, conical volcanoes, whereas thinner lithospheres favour domical to annular shapes. The basal boundary condition of a volcano determines whether or not compressive flexural stresses can be transmitted into the edifice. For planets with water in the near-surface (Earth and Mars), volcanoes emplaced on a layer of sediments with low hydraulic diffusivity will generate high pore pressures in that layer, allowing décollement slip and extensional edifice stress states. Volcanoes lacking a layer with such properties will, in general, generate a welded basal boundary condition, with transmission of compressive stress into the edifice. Volcanoes with a combination of conditions favourable to magma ascent (i.e. thick lithosphere, clay-based basal décollement) can grow to enormous size, consistent with the observation that the Hawaiian Islands on Earth and Olympus Mons on Mars are the largest edifices on their respective planets. Volcanoes with conditions adverse to magma ascent (thin lithosphere, welded bases) will have their growth limited, consistent with the small size of the Galápagos edifices and the relatively small heights of volcanoes on Venus.

We thank K. Harpp and an anonymous reviewer for thorough and helpful reviews. We also thank S.C. Solomon for lengthy discussions on the structure and evolution of large volcanic edifices. We are grateful for ongoing support from NASA's Planetary Geology and Geophysics Program. This is LPI Contribution 1771.

References

ALEXANDROVICH, J. M. 1989. Radiolarian biostratigraphy of ODP leg 111, site 677, eastern equatorial Pacific, late Miocene through Pleistocene. *In*: BECKER, K., SAKAI, H. *ET AL*. (eds) *Proceedings of the ODP, Scientific Results*, **111**. Ocean Drilling Program, College Station, TX, 245–262.

AMELUNG, F., JONSSON, S., ZEBKER, H. & SEGALL, P. 2000. Widespread uplift and 'trapdoor' faulting on Galapagos volcanoes observed with radar interferometry. *Nature*, **407**, 993–996.

ANCOCHEA, E., HERNÁN, F., CENDRERO, A., CANTAGREL, J. M., FÚSTER, J. M., IBARROLA, E. & COELLO, J. 1994. Constructive and destructive episodes in the building of a young oceanic island, La Palma, Canary Islands, and genesis of the Caldera de Taburiente. *Journal of Volcanology and Geothermal Research*, **60**, 243–262.

ANDERSON, E. M. 1951. *The Dynamics of Faulting and Dyke Formation with Applications to Britain*, 2nd edn. Oliver and Boyd, Edinburgh.

ARVIDSON, R. E., GREELEY, R. *ET AL*. 1992. Surface modification of Venus as inferred from Magellan observations of Plains. *Journal of Geophysical Research*, **97**, 13 303–13 318.

BELLEGUIC, V., LOGNONNÉ, P. & WIECZOREK, M. 2005. Constraints on the martian lithosphere from gravity and topography data. *Journal of Geophysical Research*, **110**, E11005, http://dx.doi.org/10.1029/2005JE002437

BIBRING, J. P., LANGEVIN, Y. *ET AL*. 2006. Global mineralogical and aqueous Mars history derived from OMEGA/Mars Express data. *Science*, **312**, 400–404, http://dx.doi.org/10.1126/science.1122659

BIRD, R. T. & POCKALNY, R. A. 1994. Late Cretaceous and Cenozoic seafloor and oceanic basement roughness: spreading rate, crustal age and sediment thickness correlations. *Earth and Planetary Science Letters*, **123**, 239–254.

BONNEVILLE, A., BARRIOT, J. & BAYER, R. 1988. Evidence from geoid data of a hotspot origin for the Southern Mascarene Plateau and Mascarene Islands (Indian Ocean). *Journal of Geophysical Research*, **93**, 4199–4212.

BORGIA, A. 1994. Dynamical basis of volcanic spreading. *Journal of Geophysical Research*, **99**, 17 791–17 804.

BORGIA, A., BURR, J., MONTERO, W., MORALES, L. D. & ALVARADO, G. E. 1990. Fault propagation folds induced by gravitational failure and slumping of the central Costa Rica volcanic range: implications for large terrestrial and Martian volcanic edifices. *Journal of Geophysical Research*, **95**, 14 357–14 382.

BORGIA, A., DELANEY, P. T. & DENLINGER, R. P. 2000. Spreading volcanoes. *Annual Review of Earth and Planetary Science*, **28**, 539–570.

BRACE, W. F. 1980. Permeability of crystaline and argillaceous rocks. *International Journal of Rock Mechanics and Mining Science and Geomechanics Abstracts*, **17**, 241–251.

BROUSSE, R., CHEVALIER, J.-P., DENIZOT, M. & SALVAT, B. 1978. Étude géomorphologie des Isles Marquises. *Cahiers du Pacifique*, **21**, 9–74.

BYRNE, P. K., VAN WYK DE VRIES, B., MURRAY, J. B. & TROLL, V. R. 2009. The geometry of volcano flank terraces on Mars. *Earth and Planetary Science Letters*, **281**, 1–13.

BYRNE, P. K., HOLOHAN, E. P., KERVIN, M., VAN WYK DE VRIES, B., TROLL, V. R. & MURRAY, J. B. 2013. A sagging-spreading continuum of large volcano structure. *Geology*, **41**, 339–342.

CANALES, J. P. & DAÑOBEITIA, J. J. 1998. The Canary Islands swell: a coherence analysis of bathymetry and gravity. *Geophysical Journal International*, **132**, 479–488.

CARR, M. H. 1996. *Water on Mars*. Oxford University Press, New York.

CARRACEDO, J. C. 1994. The Canary Islands: an example of structural control on the growth of large oceanic-island volcanoes. *Journal of Volcanology and Geothermal Research*, **60**, 225–241.

CARRACEDO, J. C., DAY, S. J., GUILLOU, H. & PEREZ-TORRADO, F. J. 1999. Giant Quaternary landslides in the evolution of La Palma and El Hierro, Canary Islands. *Journal of Volcanology and Geothermal Research*, **94**, 169–190.

CHADWICK, W. W. JR. & HOWARD, K. A. 1991. The pattern of circumferential and radial eruptive fissures on the volcanoes of Fernandina and Isabela islands, Galapagos. *Bulletin of Volcanology*, **53**, 259–275.

CLAGUE, D. A. & DALRYMPLE, G. B. 1987. The Hawaiian-Emperor volcanic chain, part 1, Geologic evolution. *In*: DECKER, R. W., WRIGHT, T. L. & STAUFFER, P. H. (eds) *Volcanism in Hawaii*. United States Geological Survey, Professional Papers, **1350**, 5–54.

CLAGUE, D. A. & DENLINGER, R. P. 1994. Role of olivine cumulates in destabilizing the flanks of Hawaiian volcanoes. *Bulletin of Volcanology*, **56**, 425–434.

COLLINS, J. A., DUENNEBIER, F. & SHIPBOARD SCIENTIFIC PARTY 1992. Site survey and underway geophysics. *In*: DZIEWONSKI, A., WILKENS, R. *ET AL*. (eds) *Proceedings of the ODP, Initial Reports*, **136**. Ocean Drilling Program, College Station, TX, 27–34.

COMER, R. P. 1983. Thick plate flexure. *Geophysical Journal of the Royal Astronomical Society*, **72**, 101–113.

CRUMPLER, L. S., HEAD, J. W. & AUBELE, J. C. 1996. Calderas on Mars: characteristics, structure, and associated flank deformation. *In*: McGUIRE, W. J., JONES, A. P. & NEUBERG, J. (eds) *Volcano Instability on the Earth and Other Planets*. Geological Society, London, Special Publications, **110**, 307–348.

DAÑOBEITIA, J. J., CANALES, J. P. & DEHGHANI, G. A. 1994. An estimation of the elastic thickness of the lithosphere in the Canary Archipelago using admittance function. *Geophysical Research Letters*, **21**, 2649–2652.

DAVIES, T. A. 1985. Mesozoic and Cenozoic sedimentation in the Pacific Ocean basin. *In*: NAIRN, A. E. M., STEHLI, F. G. & UYEDA, S. (eds) *The Ocean Basins and Margins*, **7A**. Plenum, New York, 65–88.

DAY, S. J., CARRECEDO, J. C., GUILLOU, H. & GRAVESTOCK, P. 1999. Recent structural evolution of the Cumbre Vieja volcano, La Palma, Canary Islands: Volcanic rift zone reconfiguration as a precursor to volcano flank instability. *Journal of Volcanology and Geothermal Research*, **94**, 135–167.

DENLINGER, R. P. & OKUBO, P. 1995. Structure of the mobile south flank of Kilauea volcano,

Hawaii. *Journal of Geophysical Research*, **100**, 24 499–24 507.

DIETERICH, J. H. 1988. Growth and persistence of Hawaiian volcanic rift zones. *Journal of Geophysical Research*, **93**, 4258–4270.

DZIEWONSKI, A., WILKENS, R. *ET AL*. (eds) 1992. *Proceedings of the ODP, Initial Reports*, **136**. Ocean Drilling Program, College Station, TX.

EAKINS, B. W., ROBINSON, J. E., KANAMATSU, T., NAKA, J., SMITH, J. R., TAKAHASHI, E. & CLAGUE, D. 2003. *Hawaii's Volcanoes Revealed*. United States Geological Survey, Geologic Investigations Series, **I-2809**.

EHLMANN, B. L., MUSTARD, J. F., MURCHIE, S. L., BIBRING, J.-P., MEUNIER, A., FRAEMAN, A. A. & LANGEVIN, Y. 2011. Subsurface water and clay mineral formation during the early history of Mars. *Nature*, **479**, 53–60, http://dx.doi.org/10.1038/nature10582

FARR, T. G. & KOBRICK, M. 2000. Shuttle Radar Topography Mission produces a wealth of data. *Eos, Transactions of the American Geophysical Union*, **81**, 583–585.

FARR, T. G., ROSEN, P. A. *ET AL*. 2007. The shuttle radar topography mission. *Reviews of Geophysics*, **45**, RG2004, http://dx.doi.org/10.1029/2005RG000183

FEIGHNER, M. A. & RICHARDS, M. A. 1994. Lithospheric structure and compensation mechanisms of the Galapagos Archipelago. *Journal of Geophysical Research*, **99**, 6711–6729.

FILMER, P., MCNUTT, M. K. & WOLFE, C. J. 1993. Elastic thickness of the lithosphere in the Marquesas and Society Islands. *Journal of Geophysical Research*, **98**, 19 565–19 577.

FILMER, P., MCNUTT, M. K., WEBB, H. F. & DIXON, D. J. 1994. Volcanism and archipelagic aprons in the Marquesas and Hawaiian Islands. *Marine Geophysical Research*, **16**, 385–406.

FISKE, R. S. & JACKSON, E. D. 1972. Orientation and growth of Hawaiian volcanic rifts: the effect of regional structure and gravitational stresses. *Proceedings of the Royal Society A*, **329**, 299–326.

FORD, P. G. & PETTENGILL, G. H. 1992. Venus topography and kilometer-scale slopes. *Journal of Geophysical Research*, **97**, 13 103–13 114.

FORNARI, D. J. & CAMPBELL, J. F. 1987. Submarine topography around the Hawaiian Islands. *In*: DECKER, R. W., WRIGHT, T. L. & STAUFFER, P. H. (eds) *Volcanism in Hawaii*. United States Geological Survey, Professional Papers, **1350**, 109–124.

FULLER, E. R. & HEAD, J. W. III 2002. Amazonis Planitia: the role of geologically recent volcanism and sedimentation in the formation of the smoothest plains on Mars. *Journal of Geophysical Research*, **107**, (E10), 5081, http://dx.doi.org/10.1029/2002 JE001842

GALGANA, G. A., MCGOVERN, P. J. & GROSFILS, E. B. 2011. Evolution of large Venusian volcanoes: insights from coupled models of lithospheric flexure and magma reservoir pressurization. *Journal of Geophysical Research*, **116**, E03009, http://dx.doi.org/10.1029/2010JE003654

GALGANA, G. A., GROSFILS, E. B. & MCGOVERN, P. J. 2013. Radial dike formation on Venus: insights from models of uplift, flexure and magmatism. *Icarus*, **225**, 538–547.

GARCIA, M. O. & HULL, D. M. 1994. Turbidites from giant Hawaiian landslides: results from Ocean Drilling Program Site 842. *Geology*, **22**, 159–162.

GILLOT, P.-Y., LEFÈVRE, J.-C. & NATIVEL, P.-E. 1994. Model for the structural evolution of the volcanoes of Réunion Island. *Earth and Planetary Science Letters*, **122**, 291–302.

GOFF, J. A. 1991. A global and regional stochastic analysis of near-ridge abyssal hill morphology. *Journal of Geophysical Research*, **96**, 21 713–21 737.

GROSFILS, E. B., MCGOVERN, P. J., GREGG, P. M., GALGANA, G. A., HURWITZ, D. M., LONG, S. M. & CHESTLER, S. R. 2013. Elastic models of magma reservoir mechanics: a key tool for investigating planetary volcanism. *In*: PLATZ, T., MASSIRONI, M., BYRNE, P. K. & HIESINGER, H. (eds) *Volcanism and Tectonism Across the Inner Solar System*. Geological Society, London, Special Publications, **401**. First published online December 11, 2013, http://dx.doi.org/10.1144/SP401.2

HEAD, J. W. & MARCHANT, D. R. 2003. Cold-based mountain glaciers on Mars: western Arsia Mons. *Geology*, **31**, 641–644.

HEY, R., JOHNSON, G. L. & LOWRIE, A. 1977. Recent plate motions in the Galapagos area. *Geological Society of America Bulletin*, **88**, 1385–1403.

HOFFERT, M., PERSON, A., COURTOIS, C., KARPOFF, A. M. & TRAUTH, D. 1980. Sedimentology, mineralogy, and geochemistry of hydrothermal deposits from holes 424, 424A, 424B and 424C (Galapagos Spreading Center). *In*: ROSENDAHL, B. R., HEKINIAN, R. *ET AL*. (eds) *Initial Reports of the Deep Sea Drilling Project*, **54**. United States Government Printing Office, Washington, DC, 339–376.

HOOGENBOOM, T., SMREKAR, S. E., ANDERSON, F. S. & HOUSEMAN, G. 2004. Admittance survey of type I coronae on Venus. *Journal of Geophysical Research*, **109**, E03002, http://dx.doi.org/10.1029/2003JE002171

IVERSON, R. M. 1995. Can magma-injection and groundwater forces cause massive landslides on Hawaiian volcanoes? *Journal of Volcanology and Geothermal Research*, **66**, 295–308.

KOLLA, V. & KIDD, R. B. 1982. Sedimentation and sedimentary processes in the Indian Ocean. *In*: NAIRN, A. E. M. & STEHLI, F. G. (eds) *The Ocean Basins and Margins*, **6**. Plenum, New York, 1–50.

KRUSE, S. 1988. Magnetic lineations on the flanks of the Marquesas Swell: implications for the age of the seafloor. *Geophysical Research Letters*, **15**, 573–576.

LÉNAT, J.-F., VINCENT, P. & BACHÈLERY, P. 1989. The offshore continuation of an active basaltic volcano: Piton de la Fournaise (Reunion Island, Indian Ocean); structural and geomorphological interpretation from Sea Beam mapping. *Journal of Volcanology and Geothermal Research*, **36**, 1–36.

LOPES, R., GUEST, J. E., HILLER, K. & NEUKUM, G. 1982. Further evidence for a mass movement origin of the Olympus Mons aureole. *Journal of Geophysical Research*, **87**, 9917–9928.

LÓPEZ, I., LILLO, J. & HANSEN, V. L. 2008. Regional fracture patterns around volcanoes: possible evidence for volcanic spreading on Venus. *Icarus*, **195**, 523–536.

MACARIO, A., HAXBY, W. F., GOFF, J. A., RYAN, W. B. F., CANDE, S. C. & RAYMOND, C. A. 1994. Flow line variations in abyssal hill morphology for the Pacific-Antarctic Ridge at 65°S. *Journal of Geophysical Research*, **99**, 17 921–17 934.

MACKWELL, S. J., ZIMMERMAN, M. E. & KOHLSTEDT, D. L. 1998. High-temperature deformation of dry diabase with application to tectonics on Venus. *Journal of Geophysical Research*, **103**, 975–984.

MALINVERNO, A. 1991. Inverse square-root dependence of mid-ocean-ridge flank roughness on spreading rate. *Nature*, **352**, 58–60.

McBIRNEY, A. R. & WILLIAMS, H. 1969. *Geology and Petrology of the Galapagos Islands*. Geological Society of America, Memoirs, **118**.

McGOVERN, P. J. 1996. *Studies of large volcanoes on the terrestrial planets: implications for stress state, tectonics, structural evolution, and moat filling*. PhD thesis, Massachusetts Institute of Technology, Cambridge, MA.

McGOVERN, P. J. 2007. Flexural stresses beneath Hawaii: implications for the October 15, 2006, earthquakes and magma ascent. *Geophysical Research Letters*, **34**, L23305, http://dx.doi.org/10.1029/2007GL031305

McGOVERN, P. J. & MORGAN, J. K. 2009. Volcanic spreading and lateral variations in the structure of Olympus Mons, Mars. *Geology*, **37**, 139–142.

McGOVERN, P. J. & SOLOMON, S. C. 1993. State of stress, faulting, and eruption characteristics of large volcanoes on Mars. *Journal of Geophysical Research*, **98**, 23 553–23 579.

McGOVERN, P. & SOLOMON, S. 1997. Filling of flexural moats around large volcanoes on Venus: Implications for volcano structure and global magmatic flux. *Journal of Geophysical Research*, **102**, 16 303–16 318.

McGOVERN, P. J. & SOLOMON, S. C. 1998. Growth of large volcanoes on Venus: mechanical models and implications for structural evolution. *Journal of Geophysical Research*, **103**, 11 071–11 101.

McGOVERN, P. J., SOLOMON, S. C., HEAD, J. W., SMITH, D. E., ZUBER, M. T. & NEUMANN, G. A. 2001. Extension and uplift at Alba Patera, Mars: insights from MOLA observations and loading models. *Journal of Geophysical Research*, **106**, 23 769–23 809.

McGOVERN, P. J., SOLOMON, S. C. ET AL. 2002. Localized gravity/topography admittance and correlation spectra on Mars: implications for regional and global evolution. *Journal of Geophysical Research*, **107**, (E12), 5136, http://dx.doi.org/10.1029/2002JE001854

McGOVERN, P. J., SMITH, J. R., MORGAN, J. K. & BULMER, M. H. 2004a. The Olympus Mons aureole deposits: new evidence for a flank-failure origin. *Journal of Geophysical Research*, **109**, E08008, http://dx.doi.org/10.1029/2004JE002258

McGOVERN, P. J., SOLOMON, S. C. ET AL. 2004b. Correction to 'Localized gravity/topography admittance and correlation spectra on Mars: implications for regional and global evolution'. *Journal of Geophysical Research*, **109**, E07007, http://dx.doi.org/10.1029/2004JE002286

McGOVERN, P. J., RUMPF, M. E. & ZIMBELMAN, J. R. 2013. The influence of lithospheric flexure on magma ascent at large volcanoes on Venus. *Journal of Geophysical Research*, **118**, 2423–2437, http://dx.doi.org/10.1002/2013JE004455

McKINNON, W. B., ZAHNLE, K. J., IVANOV, B. A. & MELOSH, H. J. 1997. Cratering on Venus: models and observations. *In*: BOUGHER, S. W., HUNTEN, D. M. & PHILLIPS, R. J. (eds) *Venus II*. University of Arizona Press, Tucson, AZ.

MERLE, O. & LÉNAT, J.-F. 2003. Hybrid collapse mechanism at Piton de la Fournaise volcano, Reunion Island, Indian Ocean. *Journal of Geophysical Research*, **108**, 2166, http://dx.doi.org/10.1029/2002JB002014

MORGAN, J. K. & McGOVERN, P. J. 2005. Discrete element simulations of gravitational volcanic deformation: 1. Deformation structures and geometries. *Journal of Geophysical Research*, **110**, http://dx.doi.org/10.1029/2004JB003252

MORGAN, J. K., MOORE, G. F., HILLS, D. J. & LESLIE, S. 2000. Overthrusting and sediment accretion along Kilauea's mobile south flank, Hawaii: evidence for volcanic spreading from marine seismic reflection data. *Geology*, **28**, 667–670.

MOORE, J. G. 1987. Subsidence of the Hawaiian Ridge. *In*: DECKER, R. W., WRIGHT, T. L. & STAUFFER, P. H. (eds) *Volcanism in Hawaii*. United States Geological Survey, Professional Papers, **1350**, 85–100.

MOORE, J. G., CLAGUE, D. A., HOLCOMB, R. T., LIPMAN, P. W., NORMARK, W. R. & TORRESAN, M. E. 1989. Prodigious submarine landslides on the Hawaiian Ridge. *Journal of Geophysical Research*, **94**, 17 465–17 484.

NAKAMURA, K. 1980. Why do long rift zones develop in Hawaiian volcanoes—a possible role of thick oceanic sediments. *Bulletin of the Volcanology Society of Japan*, **25**, 255–269.

NAMIKI, N. & SOLOMON, S. C. 1994. Impact crater densities on volcanoes and coronae on Venus: implications for volcanic resurfacing. *Science*, **265**, 929–933.

NORDLIE, B. E. 1973. Morphology and structure of the western Galapagos volcanoes and a model for their origin. *Geological Society of America Bulletin*, **84**, 2391–2956.

PETERSON, D. W. & MOORE, R. B. 1987. Geologic history and evolution of geologic concepts, island of Hawaii. *In*: DECKER, R. W., WRIGHT, T. L. & STAUFFER, P. H. (eds) *Volcanism in Hawaii*. United States Geological Survey, Professional Papers, **1350**, 149–189.

PHILLIPS, R. J. 1994. Estimating lithospheric properties at Atla Regio, Venus. *Icarus*, **112**, 147–170.

PLATZ, T., MÜNN, S., WALTER, T. R., PROCTER, J. N., McGUIRE, P. C., DUMKE, A. & NEUKUM, G. 2011. Vertical and lateral collapse of Tharsis Tholus, Mars. *Earth and Planetary Science Letters*, **305**, 445–455.

PLESCIA, J. B. 2003. Tharsis Tholus: an unusual martian volcano. *Icarus*, **165**, 223–241.

PLESCIA, J. B. 2004. Morphometric properties of Martian volcanoes. *Journal of Geophysical Research*, **109**, E03003, http://dx.doi.org/10.1029/2002JE002031

REESE, C. C., SOLOMATOV, V. S. & ORTH, C. P. 2007. Mechanisms for cessation of magmatic resurfacing on Venus. *Journal of Geophysical Research*, **112**, E04S04, http://dx.doi.org/10.1029/2006JE00 2782

ROBINSON, M. S., MOUGINIS-MARK, P. J., ZIMBELMAN, J. R., WU, S. S. C., ABLIN, K. K. & HOWINGTON-KRAUS, A. E. 1993. Chronology, eruption duration, and atmospheric contribution of the Martian volcano Apollinaris Patera. *Icarus*, **104**, 301–323, http://dx.doi.org/10.1006/icar.1993.1103

RUBIN, A. M. 1995. Propagation of magma-filled cracks. *Annual Review of Earth and Planetary Science*, **23**, 287–336.

SCHLICH, R. 1982. The Indian Ocean: Aseismic ridges, spreading centers, and oceanic basins. *In*: NAIRN, A. E. M. & STEHLI, F. G. (eds) *The Ocean Basins and Margins*, **6**. Plenum, New York, 51–147.

SCHWENZER, S. P., ABRAMOV, O. *ET AL.* 2012. Puncturing Mars: how impact craters interact with the Martian cryosphere forming liquid water and alteration products. *Earth and Planetary Science Letters*, **335–336**, 9–17.

SHIPBOARD SCIENTIFIC PARTY 1971a. Site 74. *In*: TRACEY, J. I., JR, SUTTON, G. H. *ET AL.* (eds) *Initial Reports of the Deep Sea Drilling Project, 8*. United States Government Printing Office, Washington, DC, 621–674.

SHIPBOARD SCIENTIFIC PARTY 1971b. Site 75. *In*: TRACEY, J. I., JR, SUTTON, G. H. *ET AL.* (eds) *Initial Reports of the Deep Sea Drilling Project*, **8**. United States Government Printing Office, Washington, DC, 675–709.

SHIPBOARD SCIENTIFIC PARTY 1972. Site 239. *In*: SIMPSON, E. S. W., SCHLICH, R. *ET AL.* (eds) *Initial Reports of the Deep Sea Drilling Project*, **25**. United States Government Printing Office, Washington, DC, 25–63.

SHIPBOARD SCIENTIFIC PARTY 1977. Site 369. *In*: LANCELOT, Y., SEIBOLD, E. *ET AL.* (eds) *Initial Reports of the Deep Sea Drilling Project*, **41**. United States Government Printing Office, Washington, DC, 327–421.

SHIPBOARD SCIENTIFIC PARTY 1983. Site 505: sediments and ocean crust in an area of low heat flow south of the Costa Rica rift. *In*: CANN, J. R., LANGSETH, M. G. *ET AL.* (eds) *Initial Reports of the Deep Sea Drilling Project*, **69**. United States Government Printing Office, Washington, DC, 175–214.

SHIPBOARD SCIENTIFIC PARTY. 1988. Sites 677 and 678. *In*: BECKER, K., SAKAI, H. *ET AL.* (eds) *Proceedings of the ODP, Initial Reports*, **111**. Ocean Drilling Program, College Station, TX, 253–346.

SHIPBOARD SCIENTIFIC PARTY 1992a. Site 842. *In*: DZIEWONSKI, A., WILKENS, R. *ET AL.* (eds) *Proceedings of the ODP, Initial Reports*, **136**. Ocean Drilling Program, College Station, TX, 37–63.

SHIPBOARD SCIENTIFIC PARTY 1992b. Site 843. *In*: DZIEWONSKI, A., WILKENS, R. *ET AL.* (eds) *Proceedings of the ODP, Initial Reports*, **136**. Ocean Drilling Program, College Station, TX, 65–99.

SIMONS, M., SOLOMON, S. C. & HAGER, B. H. 1997. Localization of gravity and topography: constraints on the tectonics and mantle dynamics of Venus. *Geophysical Journal International*, **131**, 24–44.

SMITH, D. E., ZUBER, M. T. *ET AL.* 2001. Mars Orbiter Laser Altimeter: experiment summary after the first year of global mapping of Mars. *Journal of Geophysical Research*, **106**, 23 689–23 722.

SMITH, J. R. & KELLEY, C. D. 2010. Multibeam synthesis of the northwestern Hawaiian Islands supports diverse research in the Papahānaumokuākea Marine National Monument. Abstract OS13C-1240, presented at the 2010 Fall Meeting of the American Geophysical Union, San Francisco, California, 13–17 December.

SMREKAR, S. E. 1994. Evidence for active hotspots on Venus from analysis of Magellan gravity data. *Icarus*, **112**, 2–26.

SOLOMON, S. C. & HEAD, J. W. 1980. Lunar mascon basins: Lava filling, tectonics, and evolution of the lithosphere. *Reviews of Geophysics and Space Physics*, **18**, 107–141.

SPUDIS, P. D., McGOVERN, P. J. & KIEFER, W. S. 2013. Large shield volcanoes on the Moon. *Journal of Geophysical Research*, **118**, 1063–1081, http://dx.doi.org/10.1002/jgre.20059

TANAKA, K. L. 1985. Ice-lubricated gravity spreading of the Olympus Mons aureole deposits. *Icarus*, **62**, 191–206.

TEN BRINK, U. S. & BROCHER, T. M. 1987. Multichannel seismic evidence for a subcrustal intrusive complex under Oahu and a model for Hawaiian volcanism. *Journal of Geophysical Research*, **92**, 13 687–13 707.

THURBER, C. H. & GRIPP, A. E. 1988. Flexure and seismicity beneath the south flank of Kilauea volcano and tectonic implications. *Journal of Geophysical Research*, **93**, 4271–4278.

VAN WYK DE VRIES, B. & BORGIA, A. 1996. The role of basement in volcano deformation. *In*: McGUIRE, W. J., JONES, A. P. & NEUBERG, J. (eds) *Volcano Instability on the Earth and Other Planets*. Geological Society, London, Special Publications, **110**, 95–110.

VAN WYK DE VRIES, B. & MATELA, R. 1998. Styles of volcano-induced deformation: numerical models of substratum flexure, spreading and extrusion. *Journal of Volcanology and Geothermal Research*, **81**, 1–18.

WAGGONER, D. G. 1993. The age and alteration of central Pacific oceanic crust near Hawaii, site 843. *In*: WILKENS, R. H., FIRTH, J. *ET AL.* (eds) *Proceedings of the Ocean Drilling Program, Scientific Results*, **136**. Ocean Drilling Program, College Station, TX, 119–132.

WALTER, T. R. & TROLL, V. 2003. Experiments on rift zone formation near unstable flanks in volcanic edifices. *Journal of Volcanology and Geothermal Research*, **127**, 107–120.

WALTER, T. R., TROLL, V., CAILLEAU, B., BELOUSOV, A., SCHMINKE, H. U., BOGGARD, P. & AMELUNG, F. 2005. Rift zone reorganization through flank instability on ocean island volcanoes: Tenerife, Canary Islands. *Bulletin of Volcanology*, **67**, 281–291.

WATTS, A. 2001. *Isostasy and Flexure of the Lithosphere*. Cambridge University Press, UK.

WATTS, A. B. 1994. Crustal structure, gravity anomalies and flexure of the lithosphere in the vicinity of the Canary Islands. *Geophysical Journal International*, **119**, 648–666.

WATTS, A. B. & TEN BRINK, U. S. 1989. Crustal structure, flexure, and subsidence history of the Hawaiian Islands. *Journal of Geophysical Research*, **94**, 10 473–10 500.

WESSEL, P. 1993. A reexamination of the flexural deformation beneath the Hawaiian Islands. *Journal of Geophysical Research*, **98**, 12 177–12 190.

WOLFE, C. J., McNUTT, M. K. & DETRICK, R. S. 1994. The Marquesas archipelagic apron: seismic stratigraphy and implications for volcano growth, mass wasting, and crustal underplating. *Journal of Geophysical Research*, **99**, 13 591–13 608.

Elastic models of magma reservoir mechanics: a key tool for investigating planetary volcanism

ERIC B. GROSFILS[1]*, PATRICK J. McGOVERN[2], PATRICIA M. GREGG[3],
GERALD A. GALGANA[2,4], DEBRA M. HURWITZ[1,2],
SYLVAN M. LONG[1,5] & SHELLEY R. CHESTLER[1,6]

[1]*Geology Department, Pomona College, Claremont, CA 91711, USA*

[2]*Lunar and Planetary Institute, USRA, Houston, TX 77058, USA*

[3]*College of Earth, Ocean, and Atmospheric Science, Oregon State University,
Corvallis, OR 97331, USA*

[4]*AIR Worldwide, 131 Dartmouth Street, Boston, MA 02116, USA*

[5]*Leggette, Brashears & Graham Inc., 4 Research Drive, Shelton, CT 06484, USA*

[6]*Department of Earth and Space Sciences, University of Washington, Seattle, WA 98195, USA*

Corresponding author (e-mail: egrosfils@pomona.edu)

Abstract: Understanding how shallow reservoirs store and redirect magma is critical for deciphering the relationship between surface and subsurface volcanic activity on the terrestrial planets. Complementing field, laboratory and remote sensing analyses, elastic models provide key insights into the mechanics of magma reservoir inflation and rupture, and hence into commonly observed volcanic phenomena including edifice growth, circumferential intrusion, radial dyke swarm emplacement and caldera formation. Based on finite element model results, the interplay between volcanic elements – such as magma reservoir geometry, host rock environment (with an emphasis on understanding how host rock pore pressure assumptions affect model predictions), mechanical layering, and edifice loading with and without flexure – dictates the overpressure required for rupture, the location and orientation of initial fracturing and intrusion, and the associated surface uplift. Model results are either insensitive to, or can readily incorporate, material and parameter variations characterizing different planetary environments, and they also compare favourably with predictions derived from rheologically complex, time-dependent formulations for a surprisingly diverse array of volcanic scenarios. These characteristics indicate that elastic models are a powerful and useful tool for exploring many fundamental questions in planetary volcanology.

After decades of planetary exploration it is clear that silicic volcanism has been, and remains, one of the primary geological processes controlling the surface evolution of large rocky bodies in our solar system. From myriad studies and observations of igneous processes on Earth and other bodies, it is known that eruption dynamics are closely linked to the complex physical and chemical evolution magma undergoes during ascent, as well as the environmental conditions it encounters at or near the surface (for recent summaries see Sigurdsson *et al.* 2000; Zimbelman & Gregg 2000; Cashman & Sparks 2013). These factors, in turn, help dictate the morphology of the resulting volcanic landforms.

Magma can move rapidly and directly from where it forms to the surface, but the process of ascent commonly occurs in stages, with molten material pausing at one or more depths along the way. In general, stalled magma will either freeze in place to become part of the lithosphere or, if the flux is sufficient, magma accumulation can form reservoirs (cf. Schöpa & Annen 2013 and references therein). Building on decades of previous analyses (e.g. Sparks *et al.* 1977; Hildreth 1981; Christensen & DePaolo 1993; Bachmann & Bergantz 2004; de Silva *et al.* 2006), a considerable array of field, petrological, geochemical and geophysical evidence gathered in recent years indicates that large bodies of magma form periodically and can become eruptible quite rapidly at shallow depth, often within a few kilometres of the surface (e.g. Burgisser & Bergantz 2011; Druitt *et al.* 2012; Gertisser *et al.* 2012; Gualda *et al.* 2012; Huber *et al.* 2012; Matthews *et al.* 2012). Understanding the mechanics underpinning how shallow reservoirs store and redirect near-surface magma is thus critical for deciphering the links between surface and subsurface volcanic activity.

From: PLATZ, T., MASSIRONI, M., BYRNE, P. K. & HIESINGER, H. (eds) 2015. *Volcanism and Tectonism Across the Inner Solar System.* Geological Society, London, Special Publications, **401**, 239–267.
First published online December 11, 2013, http://dx.doi.org/10.1144/SP401.2

In this contribution, we summarize key results from several recent dry rock numerical modelling studies that re-examine how reservoirs redirect magma through inflation and rupture. Motivated, in part, by the apparent rapidity with which eruptible reservoirs can form, the emphasis here is upon models employing an elastic rheology. Elastic models are easily implemented and, although simple, are none the less capable of providing powerful insight into commonly observed reservoir-derived volcanic features such as central eruptions that feed edifice growth, circumferential intrusions, subsurface systems of radial dykes and equant calderas. These volcanic features exhibit remarkably consistent overarching physical and morphological characteristics irrespective of scale or where in the solar system they are found. Although elastic methods have been used to study the mechanics of internally pressurized magma reservoirs for some time (e.g. Anderson 1936; Mogi 1958; Dieterich & Decker 1975; Muller & Pollard 1977; Blake 1981; Davis 1986; McTigue 1987; Yang *et al.* 1988; Tait *et al.* 1989; Sartoris *et al.* 1990; Zuber & Mouginis-Mark 1992; Parfitt *et al.* 1993; Chadwick & Dieterich 1995; Koenig & Pollard 1998; Bonaccorso & Davis 1999; Fialko *et al.* 2001; Pinel & Jaupart 2003; Trasatti *et al.* 2005; Gudmundsson 2012), systematic re-examination is needed because many previous models adopt an 'overpressure only' formulation that implies specific reservoir-proximal pore pressure conditions that are difficult to create and sustain. Using more conservative dry rock elastic models to investigate common volcanic scenarios establishes a baseline framework critical for advancing our understanding of planetary volcanism and, by creating benchmarks against which more sophisticated models can be calibrated, this framework contributes robustly to the groundwork upon which future quantitative explorations of planetary volcanism can build.

Methods

Finite element models implemented within COMSOL Multiphysics (www.comsol.com), which solves Navier–Cauchy equations for linear elastic stress and displacement in response to applied loads, have been used by the authors to simulate a magma reservoir as an internally pressurized cavity within an axisymmetrical, gravitationally loaded host rock of sufficient lateral and vertical extent that edge effects become negligible (Fig. 1). A general summary of this approach is provided below; full model details, including descriptions of physical parameter ranges explored, can be found in the papers cited.

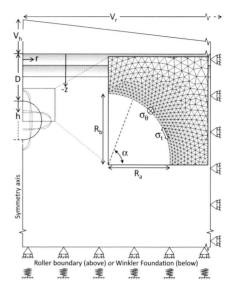

Fig. 1. Illustration of axisymmetrical finite element model geometry using a spherical reservoir example. Inset shows a sample of the mesh near the reservoir wall (at lower resolution than actual, for clarity), the rupture location at angle α, and the two wall-parallel principal stresses σ_θ (out of the plane of the page) and σ_t (within the plane of the page). Part of a conical edifice of axial height V_h and full radius V_r is depicted, as are two mechanical layers (shaded); the edifice and layers are present only in some simulations. The outer and basal edges of the model are subjected to a roller constraint; however, in flexural models, a Winkler foundation is applied to the base instead.

By convention, compressive stresses in the models are negative, and depth z is negative downwards. Mesh elements vary in size, with tens of metres of resolution or less typical near areas of interest (e.g. near the reservoir wall) transitioning into coarser elements in more distal areas. In response to the interplay between the host rock stresses and those applied to the interior of the reservoir wall, COMSOL calculates stresses (σ_r, σ_z, σ_θ, σ_{rz}) throughout the volume in polar cylindrical coordinates, from which the remaining principal stresses normal and tangential to the reservoir wall are derived (Grosfils 2007). Comparing individual applied load components, described below, to standard analytical benchmarks reveals that existing numerical model solutions exhibit maximum percentage errors $\ll 1\%$, which typically translates to absolute displacement differences of the order of 10^{-4}–10^{-5} m and absolute stress differences of the order of 10^{-3} MPa at the reservoir wall (e.g. Grosfils 2007); similar results are reported from calibration tests performed independently (Hobbs 2011).

Host rock boundary conditions and stress state

Gravitational loading in the host rock ($z \leq 0$ m) is implemented via body forces ($\rho_r g$; g is negative) and the application of a lithostatic ($\sigma_z = \sigma_r = \sigma_\theta = -\rho_r g z$) prestress; while alternative stress state end-members exist (e.g. McGarr 1988) and have been assessed to some extent (e.g. Sartoris et al. 1990; McGovern & Solomon 1993; Grosfils 2007), their viability in volcanic terrain remains unclear (Grosfils 2007; Gudmundsson 2012), and discussion is restricted here to lithostatic conditions. The host rock is assumed to be dry; that is, pore pressure is negligible. The left-hand boundary ($r = 0$ m), corresponding to the rotation axis, is free to displace vertically. Similarly, implementation of roller boundary conditions along the outer and lower edges of the model ensures that only vertical and lateral displacements, respectively, occur along these margins. The upper boundary of the model space ($z = 0$ m), representing the horizontal surface of the lithosphere, is free to displace.

An exception to the basic model configuration described above occurs when a conical volcanic edifice (or other topography) is introduced. Instead of applying a downwards-directed stress at the $z = 0$ m surface to simulate the weight of a volcano (e.g. Pinel & Jaupart 2003), each edifice is constructed explicitly to ensure that coupling between the load and response is fully integrated into the model. Body forces continue to be applied throughout the edifice, but the optimal prestress state depends on the application being simulated. End-members range from no prestress (in essence, modelling very rapid construction of the edifice relative to the timescales needed for stress equilibration) to a configuration in which the applied lithostatic prestress incorporates the edifice load as part of the vertical column and, hence, varies as a function of distance from the rotation axis. Within the broader model space, the upper surface, whether defined by the flanks of the edifice or the flat regions surrounding it, remains free of displacement constraints, and boundary conditions at the vertical edges of the model are not altered. The lithosphere can continue to be treated as rigid through sustained use of a roller boundary condition along the lower edge of the domain (e.g. Hurwitz et al. 2009). Alternatively, a flexural response can be introduced, in which case the roller boundary condition along the base is replaced by a combination of two terms: a 'Winkler' buoyant restoring stress (Watts 2001) that acts to resist deflection of the lithosphere into the underlying fluid asthenosphere; and a stress required for equilibrium that is equal and opposite to the vertical component of the edifice-free lithostatic load at

the base of the model (Galgana et al. 2011). These terms can also be augmented with additional basal loading as desired, for instance to simulate plume impingement or underplating processes (Galgana et al. 2013).

Reservoir boundary conditions

To simulate magma pressure, a normal stress P_m is applied to the interior of the reservoir wall; because the intent is to simulate the magma as a simple, potentially eruptible fluid, no shear stress tractions are imposed. This normal stress is defined as:

$$P_m = [\rho_r g D + \Delta P] + \rho_m g h \tag{1}$$

where ρ_r and ρ_m are the host rock and magma densities, g is gravitational acceleration (negative), ΔP is the overpressure (negative) and the remaining terms are defined in Figure 1. The terms in brackets are applied uniformly at all depths, h, within the reservoir, while the final term varies with h and is thus depth dependent. At the crest of the reservoir ($z = D$), where $h = 0$ m, the first term counters the lithostatic load of the rock column; in essence, creating a state of equilibrium with the surrounding host rock (i.e. in the absence of other contributing stress factors the magma chamber is mechanically stable), while the second term introduces an excess uniform pressure, defined here as the overpressure, that disturbs this equilibrium and, hence, will drive reservoir inflation (or deflation). In models that incorporate a volcano at the surface, an additional, edifice geometry-dependent term is included to offset the additional load (Hurwitz et al. 2009; Galgana et al. 2011).

Reservoir failure criteria

In response to inflation, simulated by gradually increasing the magnitude of the overpressure ΔP and thereby P_m, outwards displacement can lead to surface deformation, as well as shear and/or tensile rupture of the reservoir wall; shear failure typically requires less overpressure (Gerbault 2012). Mohr–Coulomb criteria can be used to evaluate when and where initial shear failure will take place. Given an angle of internal friction, ϕ, and cohesion, C, shear failure occurs on a fault plane when:

$$\tau \geq -\sigma_n \tan\phi + C \tag{2}$$

where τ is the shear stress and σ_n is the normal stress (negative). Within the host rock, the expected Mohr–Coulomb fault planes develop at angles of

$$\pm\left(45 - \frac{\phi}{2}\right) \tag{3}$$

relative to the maximum compressive stress direction. If sufficient magma intrudes along a fault at the site of shear failure at the reservoir wall, a possibility proposed to explain observations at some locations (e.g. Gerbault 2012; Xu *et al.* 2013), this could alleviate the overpressure. However, observations of seismicity at actively inflating volcanic systems suggest that shear failure proximal to the reservoir, although common, most often does not lead to significant intrusion (e.g. Moran *et al.* 2011; Parks *et al.* 2012). If the overpressure continues to increase, tensile failure will occur once the depth-dependent, expansion-induced, wall-parallel tension within the host rock at the magma–rock interface equals the sum of the factors resisting failure at a given depth (i.e. the wall-parallel component of the lithostatic stress augmented by the tensile strength, *T*, of the material). Tensile failure is, in some senses, a practical limiting factor when exploring inflation conditions within a reservoir, since release of magma from the reservoir via an intrusion will relieve the overpressure driving outwards displacement (cf. Geyer & Bindeman 2011).

Dry rock model v. 'overpressure only' formulation

Most analytical and numerical simulations of magma reservoir inflation and failure reduce the model system to a cavity inflated solely by overpressure (i.e. $g = 0 \text{ m s}^{-2}$ in equation 1) within an unloaded ($\sigma_z = \sigma_r = \sigma_\theta = 0 \text{ MPa}$) host material. A recent review of magma reservoir mechanics summarizes this 'overpressure only' formulation and, drawing on hydraulic fracturing mathematics and data, offers a three-point critique of the dry rock numerical model approach (Gudmundsson 2012). Before proceeding to a summary of several key results, and discussion of broader implications for understanding reservoir-derived planetary volcanism, it is useful to explore in detail the differences between a dry rock approach and traditional 'overpressure only' model assumptions, and to assess the criticisms levied. In short, different assumptions about host rock conditions are embedded within each modelling approach – this is a critical factor that should be considered carefully before adopting or interpreting the results from either one formulation or the other.

Hydraulic fracture, dry rock. To first order the standard equation for hydraulic failure of a borehole under lithostatic conditions, neglecting thermal stresses and adjusting for the sign convention described above, is given by Zoback (2010, p. 220) as:

$$T = 2\sigma - 2P_p - (P_m - P_p) \qquad (4)$$

where P_p is the pore pressure (negative) in the host rock, T is the tensile strength (positive) and σ denotes the least compressive stress magnitude (negative) in the host rock. In the current inflation application, since lithostatic conditions are applied and far-field tectonic stresses are neglected, at the rock–fluid interface σ can be considered the lithostatic stress magnitude parallel to the borehole/ reservoir wall. If one assumes negligible pore pressure in the host rock ($P_p = 0 \text{ MPa}$), hydraulic failure becomes possible once (cf. section 8.10 of Jaeger & Cook 1979):

$$P_m = 2\sigma - T. \qquad (5)$$

This is the standard failure criterion theory for borehole rupture in dry, unfractured rock, and it indicates that fluid pressure P_m within the borehole must reach approximately twice the lithostatic stress value in the surrounding host rock before rupture of the wall will occur. Complementing this theory, there is both field and laboratory evidence that such overpressures are, indeed, required to rupture rock when negligible pore pressure exists (e.g. Lockner & Byerlee 1977; Jaeger & Cook 1979, p. 226). Recalling that the overpressure is defined here as the uniform pressure in excess of lithostatic (equation 1), another way of presenting this criterion is to state that, for a borehole to rupture:

$$\Delta P = \sigma - T. \qquad (6)$$

This is simply the elastic limit for an internally pressurized cylinder in the absence of other normal loads (p. 60 of Timoshenko & Goodier 1951), that is a dry rock borehole formulation. Proceeding in the same vein, if one examines instead an internally pressurized sphere far from the free surface, then the analogous analytically derived criterion for failure becomes (Timoshenko & Goodier 1951, p. 359; Grosfils 2007):

$$\Delta P = 2(\sigma - T). \qquad (7)$$

The conditions necessary for rupture can be corrected for the presence of the free surface (e.g. McTigue 1987; Gudmundsson 1988; Parfitt *et al.* 1993; Grosfils 2007), but, in essence, the geometric difference between the borehole and the reservoir affects how stress concentrates during inflation: roughly twice as much pressure is required to rupture an inflating sphere.

Hydraulic fracture, elevated pore pressure. As an alternative condition in the host rock, if one assumes very high pore pressure (lithostatic, i.e. $P_p = \sigma$) then, returning to equation (4), it quickly becomes apparent that hydraulic fracture of a borehole will occur once:

$$P_m = \sigma - T, \qquad (8)$$

that is, when

$$\Delta P = -T. \qquad (9)$$

Similarly, if one reassesses the analytical criterion for failure of a spherical reservoir in the presence of lithostatic pore pressure (replacing σ_t with $\sigma - P_p$ in equation 35 of Grosfils 2007), then

$$\Delta P = 2(\sigma - T - P_p) = -2T. \qquad (10)$$

Comparing the high pore pressure results for the borehole and spherical reservoir, it is once again apparent that twice as much overpressure is required to rupture the reservoir. In addition, however, comparison between the equations for failure under dry rock conditions with those that assume lithostatic pore pressure reveals that dry rock conditions require significantly higher fluid overpressure: greater by σ in the case of borehole geometries and greater by 2σ for the spherical reservoir.

Addressing dry rock model critique. Guided by the first-order mathematics describing rupture of inflating boreholes and spherical reservoirs, it is now possible to assess the concerns that have been raised about the dry host rock model approach (Gudmundsson 2012).

Concern 1. The first concern is that dry rock tensile failure models include an 'extra' stress that is somehow superimposed upon (and hence violates) the lithostatic conditions in the host material, even though it has been demonstrated that dry rock models accurately reproduce lithostatic conditions (cf. Grosfils 2007, fig. 2a). Quantitatively, this criticism follows previous lines (e.g. Gudmundsson 1990, 2006) by defining the condition necessary for intrusion initiation from a fluid source, after adjusting signs to match the convention used in the numerical models, as

$$P_m = \sigma - T = \sigma + \Delta P \qquad (11)$$

and hence, for failure, the overpressure required is

$$\Delta P = -T. \qquad (12)$$

The assertion is that this criterion, which denotes failure of a pre-existing, internally pressurized, fluid-filled fracture (Jaeger & Cook 1979, section 17.5), describes both the initiation of hydraulic fractures injected from (vertical) drill holes as well as the overpressure required to fracture the wall of a magma reservoir and commence dyke injection (Gudmundsson 2012).

Comparing this assertion with the equations in the previous subsections, it is apparent that the criterion expressed in equation (12) is correct for a borehole only if the pore pressure in the host rock is assumed to be lithostatic in magnitude. [Note: the criterion is only approximately right (factor of 2 difference), given the same assumption, for a pressurized spherical magma reservoir.] If dry rock is instead employed for the host rock, the overpressure required becomes greater because the effective stress parallel to the borehole wall retains a magnitude of σ when inflation commences. Unlike the case when pore pressure is assumed to be lithostatic, in dry rock models the full wall-parallel components of the lithostatic stress continue to resist inflation-driven expansion and rupture. Contrary to the concern expressed, the difference between dry rock and 'overpressure only' model formulations is thus neither a result of adding an 'extra' wall stress to the former nor a violation of lithostatic conditions in the host rock. Instead, it simply reflects the fact that dry rock models include negligible pore pressure in the host rock, whereas 'overpressure only' formulations assume that the host rock has a lithostatic pore pressure and, thus, exists at the brink of tensile failure (see the subsection on 'Environmental parameters' later) prior to the initiation of magma reservoir inflation.

Concern 2. The second concern is that it is unreasonable to use $T = 0$ MPa (or a value close to it) for the host rock tensile strength. This value is often chosen in part for mathematical convenience (Grosfils 2007), but it also reflects the physical plausibility that any host rock mass will contain distributed fractures and, hence, will have little to no tensile strength (Zoback 2010, p. 121). Concern about this choice for T is an extension of Concern 1, for if equation (12) is used as the criterion for failure then only the tensile strength of the host rock is left to resist rupture. It follows that, if $T = 0$ MPa, nothing whatsoever resists rupture, and the slightest overpressure would cause the wall of a reservoir to fail (Gudmundsson 2012). Of course, this issue is not alleviated significantly if T remains small, of the order of a few MPa as is typical (Schultz 1995), since only slightly greater overpressures could be tolerated. In contrast, however, the failure criterion for dry rock retains the wall-parallel host rock lithostatic stresses (equation 7) because they have not been countered by lithostatic pore pressure. The wall-parallel lithostatic stresses also resist rupture and they are normally of much greater magnitude than T. As such, the decision to assume $T = 0$ MPa for convenience has a negligible effect on the deduced rupture characteristics (overpressure, location and fracture orientation; see Grosfils 2007, fig. 4). Thus, unless pore pressure is assumed to be lithostatic, this second concern is unwarranted. Using a low value for T is physically motivated, internally self-consistent and, furthermore, assuming any plausible value of T (including zero), has been demonstrated to have little to no impact on when/where a magma reservoir in a dry host rock will rupture (cf. Grosfils 2007).

Concern 3. The third concern involves two separate parts that are again linked to and dependent upon the validity of Concern 1. The first part states that dry rock model formulations are faulty because they indicate that overpressures needed to induce rupture of a reservoir or borehole wall can become quite large at depth. This objection is not warranted in and of itself, of course, because, as discussed above, numerous laboratory experiments and borehole field measurements demonstrate that stress magnitudes P_m of approximately 2σ (equation 5) – that is, overpressures ΔP of around σ (equation 6) – are required for borehole hydraulic fracture in dry rock (e.g. Lockner & Byerlee 1977; Jaeger & Cook 1979). While the mathematics are not identical, the extension to magma chamber rupture and dyke injection is clear. The second part, which lies at the heart of the matter, is that the tensile strength of rock appears to be a material property that is effectively independent of confining pressure (i.e. depth). This is indisputable, but it is also irrelevant unless one assumes that pore pressure in the host rock is lithostatic in magnitude, as shown above, since, otherwise, the tensile strength is not the only factor resisting rupture. In the absence of pore pressure, the requirement of high overpressures to rupture the chamber wall, as is observed in borehole and laboratory equivalents, does not imply that the tensile strength must be greater than typical measured values (i.e. $T \leq 10$ MPa: Schultz 1995). The elevated pressure is required to overcome the combination of tensile strength and wall-parallel lithostatic stress (equations 6 & 7). The need for elevated overpressure does not indicate a flaw in the dry rock model formulation. It is, instead, a self-consistent outcome that does not contradict measurements of low tensile strength (as this factor is generally of minimal import among those resisting rupture in the absence of lithostatic pore pressure), and it is consistent with both laboratory and field data.

Summary. It is apparent that the 'overpressure only' and dry rock approaches are rooted in very different assumptions about the host rock conditions near a reservoir as it inflates towards failure. A traditional 'overpressure only' model formulation implies that the pore pressure in the host rock is lithostatic; that is, that the material surrounding the reservoir is in a state of incipient tensile failure at equilibrium. The implications of this assumption are that: (a) overpressure will rupture the reservoir wall once it is close to T in magnitude (with $T < 5$ MPa typical, implying that, at best, only minimal overpressure could ever develop); and (b) the tensile strength of the rock is the only thing

resisting rupture as the reservoir wall expands. In dry rock models, however, no pore pressure is assumed. The implications of this decision are that: (a) the overpressure needed to cause rupture of the reservoir wall can become quite large at depth (although the magnitude required is sensitive to both reservoir aspect ratio and depth, and in many instances is fairly low: see 'Pressure for failure' for the subsections 'Uniform and layered half-space models' and 'Models with edifice loading' in the Results section of this paper); and (b) the tensile strength of the rock is of almost no importance when assessing reservoir rupture. Not only will the choice of pore pressure affect the magnitude of the overpressure required for rupture but, as described in the Results section below, there are other less obvious, yet equally important, volcanic implications – including the location and orientation of failure at the reservoir wall – that affect the geometry of the resulting intrusions and likelihood of surface eruption.

Results

While a great many volcanic scenarios and implications have been re-examined to date using a dry rock model formulation, we focus here on collating and summarizing three central outcomes: the overpressure needed to induce rupture; the location and style of initial failure; and the nature of the surface deformation that occurs in response to reservoir inflation. Each is an area critical to ongoing efforts to understand the nature of volcanic hazards on Earth, and each can provide important insight into the characteristics of volcanic activity on other bodies in the solar system.

Elastic models of magma reservoir pressurization tend to fall into two categories. In the first, the reservoir is treated as a point source (radius ≪depth), and, thus, the pressure and reservoir size are inseparable (e.g. Mogi 1958). This approach remains a powerful tool for modelling aspects of surface displacement, and continues to be widely used to gain insight into areas of active inflation (e.g. Rymer & Williams-Jones 2000; Newman *et al.* 2001, 2006; Pritchard & Simons 2004; Masterlark 2007; Parks *et al.* 2012). However, it also has considerable limitations (McTigue 1987), among which is an inability to model the stresses surrounding the reservoir – a critical factor since assessing the conditions for reservoir rupture in response to inflation is not possible. In the second category, mitigating these limitations, the reservoir is modelled as a finite, internally pressurized body within the lithosphere (e.g. Dieterich & Decker 1975). This is the approach the authors have employed when assessing magma reservoir inflation and rupture as it permits explicit inclusion of free surface and similar effects, variable reservoir sizes and geometries, different host rock stress conditions, and a wide variety of other factors that can be introduced and examined systematically.

As reservoir inflation takes place in response to iterative increases in overpressure, stress magnitudes are monitored around the reservoir perimeter until tensile rupture occurs at the site along the wall where the failure criterion expressed in equation (7) is first met for either σ_t (wall parallel, in the rz-plane) or σ_θ (wall parallel, orthogonal to the rz-plane). Here we report this location using the angle α (Fig. 1), with $\alpha = 90°$, $\alpha = 0°$ and $\alpha = -90°$ denoting the reservoir crest ($h = 0$ m), middle ($h = R_b$) and base ($h = 2R_b$), respectively. In some models, specifically those exploring ring faulting and caldera formation, the location and characteristics of Mohr–Coulomb failure are monitored and assessed in a similar fashion using the criteria defined in the subsection above on 'Reservoir failure criteria'.

Surface uplift and deformation has not been a primary focus of dry rock modelling efforts to date since their emphasis has been on constraining rupture characteristics, but limited assessments of uplift in half-space models have addressed two specific questions. First, how do the uplift predictions of Mogi (1958) compare with those obtained from numerical models when a uniform host rock is considered, and how do these results in turn compare with uplift expected when mechanical layering (i.e. strong and weak materials) is present? Second, do elastic models that invoke inflation to explain a given uplift event predict a stable reservoir, or is the pressure required greater than the rupture limit given the deduced reservoir geometry?

We begin here with an overview of several important insights obtained from dry rock half-space models. This baseline then informs a brief exploration of key differences observed and/or new findings obtained when additional factors such as edifice loading are introduced. Finally, we provide an initial report on the implications for ring faulting and caldera formation.

Uniform and layered half-space models

Pressure for failure. Using a simple half-space approach, with no surface topography but incorporating variable host rock density structures, Grosfils (2007) showed that the total uniform pressure ($\rho_r gD + \Delta P$) needed to induce rupture, when normalized to the lithostatic stress at the centre of a spherical reservoir (radius denoted simply as R when $R_a = R_b$), approaches a value of 3 for scaled reservoir depths R/DtC (where $DtC \equiv D + h$)

approaching 0 (Fig. 2a). The choice to normalize by the lithostatic stress at the reservoir centre is useful when comparing an array of host rock density structures (not all of which vary linearly with depth), magma densities (neutrally buoyant and not), reservoir sizes and reservoir depths. The results show that the total uniform pressure needed for tensile rupture is insensitive to plausible density structures assumed for the host rock, the gravitational acceleration (g) employed and, for a given scaled depth, the reservoir radius.

For many applications, however, it is practical to isolate the overpressure needed for failure, and this is especially valuable for models in which the magma and host rock density are equal and uniform. Under these conditions normal stresses across the reservoir wall, once balanced at the reservoir crest, remain balanced with depth h below the crest. Since overpressure is defined here as the pressure in excess of lithostatic at the reservoir crest, in this type of depiction one typically normalizes the overpressure by the lithostatic stress at the crest. Recasting previously published data in this framework, the normalized overpressure magnitude is, as expected (see the subsection on 'Hydraulic fracture, dry rock'), effectively 2 for spherical reservoirs sufficiently distant from the free surface (Fig. 2b). Only when the reservoir lies at shallow depth ($D < R$), leading to stress modifications around the reservoir in response to considerable free surface displacement, does the overpressure required drop significantly. When compared with other reservoir geometries as well as simulations that include lithostatic pore pressure (Fig. 2c), the overpressure magnitude required to rupture a sphere at depth is the greatest, indicating that this geometry is normally the most stable (i.e. the most resistant to rupture) when responding elastically to overpressure-driven inflation. Although differences occur as the Young's modulus (E) for a uniform host rock is varied, reflecting the fact that the magnitude of surface displacement will increase in response to a given overpressure as Young's modulus decreases, scaled overpressure values varied on the order of 15% or less as E was changed across two orders of magnitude (1–100 GPa).

Many recent half-space studies have also examined the impact of mechanical layering in the host rock, implemented as Young's modulus variations that represent volcanic materials ranging from strong, coherent lava flows to weak ash layers (e.g. Trasatti *et al.* 2003; Gudmundsson 2006; Manconi *et al.* 2007, 2010; Trasatti *et al.* 2008). To examine how layering affects the scaled overpressure needed for rupture, models with a spherical reservoir of fixed radius ($R = 1$ km) are employed with the reservoir placed at different

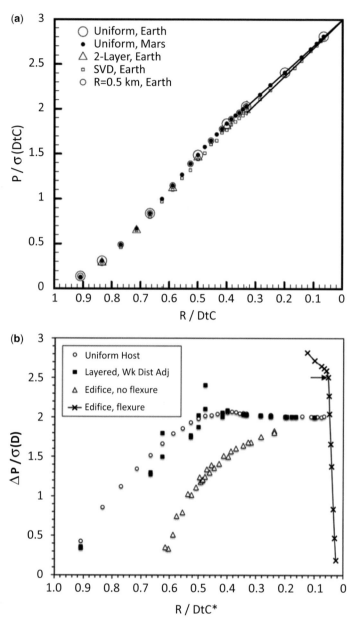

Fig. 2. Magma pressure required to induce tensile failure. (**a**) Uniform pressure P (sum in brackets from equation 1) required, normalized by the lithostatic stress at the depth of the reservoir centre (DtC), for a spherical reservoir as a function of scaled reservoir radius R. Tests of different host-rock density structures (uniform, two layer and smoothly varying (SVD)), gravity values (Earth, Mars) and reservoir radii (1 and 0.5 km) show that the uniform pressure is linearly dependent on the scaled radius but insensitive to the other parameters. From Grosfils (2007). (**b**) Magma overpressure ΔP required to induce rupture of a spherical reservoir, $R = 1$ km. The overpressure is normalized by the lithostatic stress at the top of the reservoir, while the radius is scaled by the distance between the reservoir centre and the nearest overlying weak layer, denoted for clarity by DtC^*. Uniform host rock data ($DtC^* = DtC$) are derived from (**a**). When $DtC > 2R$ then $\Delta P \approx 2\sigma_z(D)$, but at shallower depths the relative ΔP required drops precipitously. In layered models, DtC^* is the distance between the centre of the reservoir and the base of the nearest weak mechanical layer. For example, if the shallowest 500 m-thick near-surface mechanical layer is weak ($E = 1$ GPa) and the other is intermediate (like the host rock) or stronger, then $DtC^* = DtC - 500$ m. Proximity of the reservoir to a mechanically weak layer

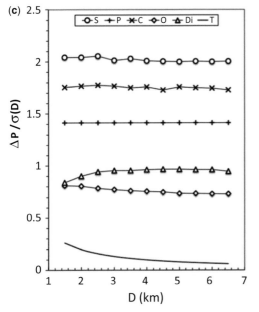

(c)

When the inflating reservoir is centred far below a mechanically weak material – either the free surface or a subsurface layer assigned a low Young's modulus – the scaled overpressure required for rupture hovers around 2 (Fig. 2b). As observed in uniform substrate models, however, the scaled overpressure declines rapidly as the distance between the reservoir and a weak overlying material layer decreases. In essence, reservoir wall displacement due to inflation occurs more readily when host rock deformation readily displaces a weak proximal layer, facilitating rupture (Long & Grosfils 2009). This first-order result indicates that a reservoir at depth will rupture at lower overpressure, when located near a weak subsurface layer, than when no weak subsurface layer is present. It is thus beneficial, when striving to understand the stability of an inflating magma reservoir, to incorporate as much knowledge of subsurface layering as possible into the model, especially since many studies also suggest that weak layers help magma reservoir formation by facilitating dyke–sill transitions (e.g. Gudmundsson 2006; Maccaferri et al. 2011). Further assessment, using reservoirs of different shapes and sizes as well as different subsurface materials, is needed to advance volcanologists' knowledge in this key area.

Fig. 2. (*Continued*) dictates ΔP in a fashion analogous to the uniform host rock case. Models of edifice ($V_h = 800$ m, $V_r = 77.2$ km) loading with a basal roller constraint exhibit a pattern similar to the uniform case for an edifice 800 m tall and 77.2 km in radius; here $DtC^* = DtC + 800$ m. The results demonstrate general resiliency; that is, no strong sensitivity to reservoir radius or depth. Models of edifice ($V_h = 5$ km, $V_r = 200$ km) loading with flexure ($T_e = 40$ km) exhibit very different behaviour: $DtC^* = DtC + 5$ km. Arrow denotes reservoir centred within the flexural neutral plane; at shallower depths, flexural compression makes failure increasingly difficult whereas, at greater depths, flexural extension sharply reduces the scaled overpressure required for rupture. (**c**) The importance of magma reservoir geometry, with all volumes identical to a sphere with $R = 1$ km. Range in D encompasses commonly inferred near-surface reservoir depths, avoiding free surface effects for the reference spherical geometry (S). Prolate (P) and cylindrical (C) reservoirs ($R_b = 1.25$ km) require progressively less overpressure to fail, while oblate (O) and discoidal (Di) reservoirs ($R_a = 1.25$ km) require less than half the overpressure needed to cause rupture of a sphere. Values depicted for non-spherical geometries are illustrative since greater elongation leads to higher stress concentration and lower overpressure. For comparison, (T) illustrates the results when a tensile strength of 5 MPa is the only thing resisting rupture (i.e. if pore pressure is lithostatic) of a spherical reservoir.

Location and orientation of failure. For a spherical reservoir, the location at which initial tensile rupture occurs, and the orientation of the resulting fracture, is sensitive to the scaled reservoir size and the host rock conditions assumed (Grosfils 2007). Since the lithostatic stress components parallel to the reservoir wall are of much less magnitude at the crest of the reservoir than at the base, inflation promotes rupture at the crest unless the reservoir lies very close to the surface, specifically when $DtC < 3R$ (Fig. 3a). When $DtC < 3R$, deformation of the free surface modifies the reservoir wall displacement, with greater displacement of the upper hemisphere shifting the point of failure away from the crest, closely tracking the relocation of the maximum strain deviation (Jaeger & Cook 1979) along the reservoir wall. When rupture takes place at the crest, $\sigma_t = \sigma_\theta$ and vertically ascending radial dykes will form whereas, away from the crest, σ_t reaches a state of tension before σ_θ and circumferential intrusion emplacement will initiate at the reservoir wall. In addition, because numerical models facilitate independent calculation of σ_t and σ_θ (normally only the latter is calculated in analytical treatments), rupture near the mid-point of the reservoir is now known to promote lateral sill emplacement (Grosfils 2007), not radial dyke injection, as is often assumed based on analytical model results (e.g. Parfitt et al. 1993; McLeod & Tait 1999).

depths within host rock ($E = 60$ GPa) overlain by two 500 m-thick layers (Fig. 1), each of which is independently assigned a Young's modulus of either 1 GPa (weak), 60 GPa (control, host rock) or 100 GPa (stiff) magnitude (Long & Grosfils 2009).

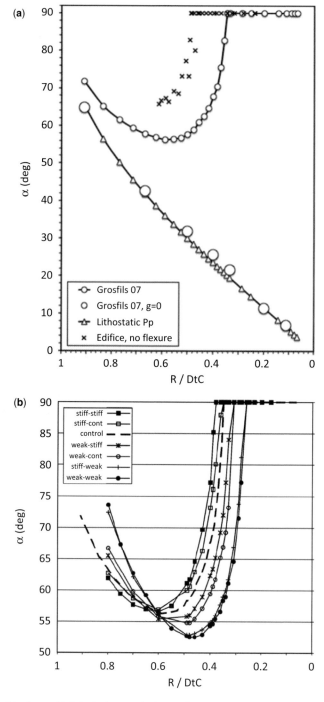

Fig. 3. Initial rupture location α. (**a**) Spherical reservoirs in a uniform host rock; edifice-loaded model matches conditions from Figure 2b. In dry rock models, $DtC > 3R$ promotes rupture at the crest, whereas failure rotates away from the crest for shallower reservoirs. Fracture orientation will initiate vertical dykes from the crest and circumferential intrusions elsewhere. 'Overpressure only' models (McTigue – lithostatic Pp; numerical – Grosfils 07, $g = 0$) that account for the presence of the free surface exhibit a very different pattern of rupture. Modified from the original in Grosfils (2007). (**b**) Effect of mechanical layering implemented by varying Young's modulus.

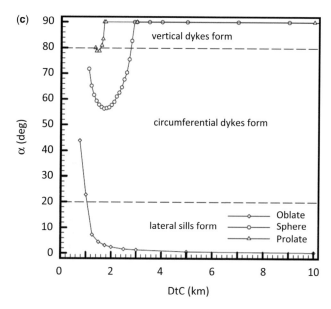

Fig. 3. (*Continued*) The legend lists a shallower layer then a deeper layer, weak is $E = 1$ GPa, control is $E = 60$ GPa (the same as the deeper host rock within which the reservoir lies), and strong is $E = 100$ GPa. 'Weak-stiff' thus denotes a configuration in which the shallowest mechanical later has $E = 1$ GPa, while the layer below this has $E = 100$ GPa; the host rock below has $E = 60$ GPa. For reference, the dashed line depicts uniform model results from (a). When $R/DtC < c.$ 0.6, the presence of strong layers above the reservoir shifts failure closer to the crest than in the uniform case, whereas weaker layers shift failure towards greater depth; the reverse pattern holds for reservoirs at shallower depths. In all models, initial fracture alignment will favour radial dykes at the crest and circumferential intrusions otherwise. From Long & Grosfils (2009). (**c**) Effect of geometry for spherical ($R = 1$ km), oblate ($R_a = 1.25$ km) and prolate ($R_b = 1.25$ km) reservoirs of equal volume. Compared to the spherical reference, stress concentration near the tip of a prolate reservoir pins rupture closer to the crest, whereas the same phenomenon pins rupture of an oblate reservoir to locations near DtC except at very shallow depth. Results shown are intended to be illustrative, as they depend on the degree of reservoir elongation. From Grosfils (2007).

The rupture location pattern observed in dry rock experiments is very different to the results predicted using 'overpressure only' model for mulations (Fig. 3a). Analytical models that take into account stress modifications caused by the free surface (e.g. Jeffrey 1921; McTigue 1987; Gudmundsson 1988; Parfitt *et al.* 1993) predict rupture near the DtC for a deep reservoir, with the rupture location occurring progressively closer to the crest for shallower reservoirs. If, in dry rock models, $g = 0$ m s^{-2}, this is equivalent to using a lithostatic pore pressure to establish an effective stress of 0 (both eliminate gravitational loading effects equivalently when assessing tensile failure), and duplicates the 'overpressure only' results. This demonstrates that dry rock numerical model formulations and those assuming a lithostatic pore pressure are, indeed, equivalent except for this key pore pressure difference, and underscores that the choice of host rock conditions sharply alters where rupture is expected to occur. It is interesting to note that, when lithostatic pore pressure is invoked, even if the top of a 1 km-radius reservoir

lies within 100 m of the surface (i.e. $R/DtC = 0.9$) rupture cannot occur at the crest, and the circumferential intrusions emplaced will dip by approximately 65°. Intrusions initiated at deeper reservoirs will have shallower dips (Fig. 3a). Since vertical dyke ascent from reservoirs is a common element of many volcanic systems, the inability of the 'overpressure only' models to initiate vertical dykes argues against the validity of assuming lithostatic pore pressure in the host rock.

Mechanical layering in the host rock (Fig. 3b) yields the same approximate behaviour observed in the uniform half-space models (Long & Grosfils 2009). For a 1 km-radius spherical reservoir, as DtC decreases, the presence of two 500 m-thick weak (small E) layers at the surface above the reservoir means that deformation of the reservoir wall – and, hence, relocation of the point of failure – occurs more readily than in the control case and, thus, at greater scaled depth. Similarly, the presence of two stiff (high E) layers retards rotation of the point of failure away from the crest, so rupture will remain focused at the crest to shallower scaled depths. Other configurations fall between these end-members. It is important to notice that, for *c.* $0.3 < R/DtC < c.$ 0.45 (if $R = 1$ km, $DtC = c.$ 2–3 km), the values of E selected for the shallow layers can have a significant effect on where rupture is predicted to occur. Thus, while the overarching behaviour resembles the uniform host rock case, incorporating known mechanical variations is of particular value when trying to assess where rupture of a very shallow reservoir, and intrusion, will commence.

As expected, failure location is also highly dependent on reservoir geometry (Fig. 3c). When compared with a spherical reservoir of equivalent volume, prolate reservoirs concentrate stress at their upper and lower tips, and so are less perturbed by the presence of the free surface. As a result, they continue to rupture at the crest when centred at much shallower depths. Similarly, owing to stress concentration at the lateral tips, an oblate reservoir will rupture near the DtC unless it lies very close to the free surface. In this instance, stress modifications caused by the free surface deformation will rotate the point of failure towards the crest. However, even at the shallowest depth depicted in Figure 3c, the failure location has only rotated halfway to the peak.

Surface deformation. Modern measurement of surface displacement using GPS (Global Positioning System), InSAR (Interferometric Synthetic Aperture Radar) and other techniques remains a valuable tool for gaining insight into the inflation (or deflation) of a subsurface magma body (e.g. Owen *et al.* 1995, 2000; Cervelli *et al.* 2001; Bartel *et al.* 2003; Dzurisin 2003; Pritchard & Simons 2004); while discussion here on vertical displacement, simultaneous assessment of radial displacement data provides clearer insight into the magma body than vertical displacements alone (e.g. Newman *et al.* 2001, 2006; Battaglia *et al.* 2003; Sturkell *et al.* 2008). Analytical point source models can duplicate observations readily (e.g. Mogi 1958), leading to estimates of volume change at depth, while analytical models of finite cavity inflation permit more direct insight into overpressure variations and analysis of shallow reservoirs (e.g. Davis 1986; McTigue 1987). For instance, using a spherical reservoir of radius $R = 1$ km and $DtC = 5$ km in a uniform host rock with $E = 60$ GPa, the uplift predicted for an overpressure of 10 MPa using analytical models is just over 12.5 mm (Fig. 4a); as Young's modulus increases, displacement magnitude decreases. Setting $g = 0$ m s^{-2} in dry rock numerical models, equivalent in outcome for these purposes to assuming the pore pressure is lithostatic, the uplift predicted is almost identical to (0.25 mm lower than) the analytical benchmark. To match the analytical uplift requires increasing the overpressure by 0.15 MPa or decreasing the depth by 50 m, indicating an error of 1–2%. If, instead, a dry rock host is used, the overpressure required to produce the same uplift is 112.1 MPa. This is, however, simply the lithostatic pressure at the crest (*c.* 102 MPa) plus the 10 MPa overpressure, once again demonstrating the equivalency of the two sets of mathematics. What is different, however, is the reservoir's proximity to failure. In the 'overpressure only' (lithostatic pore pressure) models, assuming a standard tensile strength value of approximately 5 MPa for the host rock, rupture of the reservoir will either have occurred or be imminent under the model conditions described. This implies that the maximum uplift that can occur prior to rupture is surprisingly small, of the order of 12 mm. In contrast, equivalent models assuming dry rock yield the same sort of radially decaying profile but permit just over 250 mm prior to rupture initiation (Fig. 4a).

Half-space models of uplift (described in 'Pressure for failure' in the subsection 'Uniform and layered half-space models' in the Results section) that incorporate shallow mechanical layers (e.g. Long & Grosfils 2009) demonstrate how surface displacements can vary significantly depending upon the Young's moduli (E) values assigned to them (Fig. 4b), a result that is consistent with the inverse relationship between Young's modulus and displacement. When compared with the control case (both near-surface layers assigned the same E as the host rock), the use instead of stiffer layers tends to reduce uplift above an inflating sphere, while weaker layers have the opposite effect. In addition, when equivalent volume oblate and prolate reservoirs are examined, oblate reservoirs induce more inflation than the spherical case, while prolate reservoirs induce less. A more comprehensive discussion of uplift patterns, the relationship to subsurface strain within a weak layer and other factors is provided by Long & Grosfils (2009), but two first-order points are worth calling attention to here. First, even with the multiplicative factors added by using oblate reservoirs and weak surface layers, 'overpressure only' elastic models are generally incapable of producing commonly observed degrees of uplift without invoking overpressures much greater than those required to induce rupture (e.g. Grosfils 2007; Long & Grosfils 2009). Second, future measurements of surface uplift on Venus or other planets using InSAR-like techniques will be tremendously useful for identifying centres of active magmatism. However, inferring volcanological conditions with confidence requires more detailed knowledge of the subsurface geology than is currently available for many centres of activity on Earth, let alone the other bodies in our solar system.

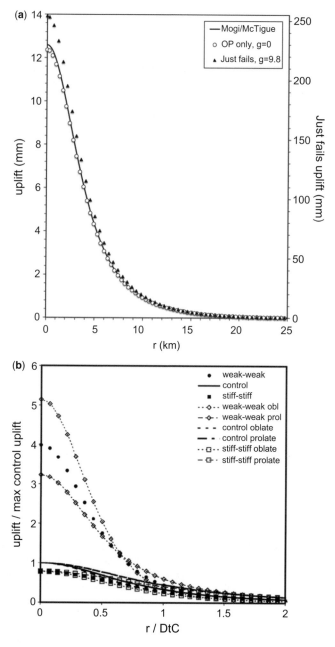

Fig. 4. Surface uplift above an inflating magma reservoir. (**a**) Spherical reservoir; $R = 1$ km, $DtC = 5$ km, $\Delta P = 10$ MPa. The Mogi–McTigue solution is closely matched by 'overpressure only' models. The same uplift profile (not shown) is obtained from a dry rock model, $g = 9.8$ m s^{-2}, when uniform pressure $P = 112.1$ MPa; this is simply the lithostatic load at the crest (102 MPa) plus the same 10 MPa overpressure. In the 'overpressure only' model, approximately 12.5 mm is the maximum uplift the system can sustain before rupture, well below magnitudes commonly observed. However, about 250 mm of uplift can occur prior to rupture in the dry rock model, which more closely resembles measurements from actively inflating volcanic systems. Modified from Grosfils (2007). (**b**) Effects of mechanical layering and reservoir geometry variations within the host rock; the protocol for layer stiffnesses and configuration follows Figure 3b. Weak layers at the surface promote significantly greater uplift, while stiffer layers decrease uplift magnitude. When compared with a sphere, these effects are amplified for oblate reservoirs and depressed when a prolate geometry is employed instead. Original from Long & Grosfils (2009).

Models with edifice loading

Pressure for failure. The overpressure required to rupture a spherical reservoir beneath a conical surface load, a geometry considered a good approximation of a typical edifice (Grosse *et al.* 2009), will depend on the size and shape of the edifice, and the volume (V) relative to that of the reservoir (Pinel & Jaupart 2000; Hurwitz *et al.* 2009). An illustration, using an intermediate-sized Venusian edifice (volcano height $V_h = 0.8$ km, volcano radius $V_r = 77.2$ km), is shown in Figure 2b. For an array of reservoir size–depth combinations, addition of the edifice load decreases the overpressure markedly relative to the half-space model results, but the general pattern is otherwise quite similar. In contrast, when flexural effects are added to the model, the rupture behaviour is quite different (Galgana *et al.* 2011). Flexural stresses created by a surface load create an hourglass pattern in which depths above and below a flexural neutral plane (low differential stress) are subjected to increasing compression or extension, respectively (see also McGovern *et al.* 2013). To illustrate the impact this has on the overpressure needed for reservoir rupture, data from a model of a large Venusian edifice in a region of high elastic thickness ($V_h = 5$ km, $V_r = 200$ km and elastic thickness $T_e = 40$ km) are plotted in Figure 2b; all data are for a spherical reservoir with $R = 1$ km. Even at the neutral plane (c. 14 km depth: Galgana *et al.* 2011), the overpressure needed for rupture is considerably higher than in the half-space models or in models that include an edifice but not flexural adjustments. At shallower depths, as the reservoir moves increasingly towards the surface and into areas of greater flexural compression, the overpressure needed also increases. At greater depths the reverse is true: as the reservoir moves away from the neutral plane and into areas of greater flexural extension, the overpressure drops precipitously, reaching levels comparable to those otherwise achieved only at very shallow scaled depths in half-space models.

Location and orientation of failure. When compared to half-space conditions, the weight of an edifice tends to amplify the expected reservoir rupture patterns (Hurwitz *et al.* 2009). For instance, as shown in Figure 3a, the presence of an edifice pins rupture to the crest of a spherical reservoir unless it is located at very shallow depth, in which case the rupture location is rotated away from the crest by the deformation induced at the free surface. Similar patterns are observed for non-spherical reservoirs. Regardless of reservoir shape, these effects are stronger for steeper stratocone-like edifices (with more mass concentrated near the

rotation axis) than for shield-like volcanoes, and larger reservoirs require larger edifices to have the same impact. As in the half-space models, failure occurs in the σ_t orientation for these edifice-loaded simulations, initiating vertical dykes at the crest, lateral sills near the mid-section, or inclined circumferential intrusions in-between.

When flexure in response to the edifice load is introduced, the rupture patterns are dominated by the resulting hourglass stress state in the lithosphere (Galgana *et al.* 2011). Near the neutral plane, where differential stresses are low, the reservoir behaves as if it were simply located at great depth in a half-space, failing at or near the crest. At greater depths, the least compressive stress is horizontal and, thus, reservoirs will fail at either the crest or the base depending on the balance of effects stemming from their distance below the free surface (which favours failure at the crest) and the depth-dependence of the differential stress increase (which favours failure at the base). Above the neutral plane, the least compressive stress is vertical, and reservoirs will, therefore, fail at or just above their centres to feed lateral sills.

Taking into account both pressure and rupture location data, it is instructive to now examine the differences between equivalent dry rock (Hurwitz *et al.* 2009) and 'overpressure only' (Pinel & Jaupart 2003) models when an edifice load is present (Fig. 5). For this simulation, following Pinel & Jaupart (2003), $R = 4$ km, $DtC = 5$ km, $V_h = 0.4$ km, $V_r = 4$ km and $T = 20$ MPa. Using dry rock models, inflation of the reservoir (thin line defined by the arrows showing displacement, exaggerated 1500× in the figure for clarity) causes initial rupture and circumferential intrusion at $\alpha = 58.9°$ when $\Delta P = 58.5$ MPa (full grey circle). If gravitational effects are eliminated in the host rock to simulate the existence of lithostatic pore pressure, leaving just the overpressure loading the reservoir and the edifice weight applied at the surface, the resulting displacement is the same, demonstrating the equivalence of the two models in this regard. However, because only the tensile strength resists rupture in this case, application of this overpressure magnitude would cause the entire wall to rupture in tension (thick grey line). If the overpressure magnitude is reduced sufficiently, to $\Delta P = 21.2$ MPa, the conditions defining initial failure in an 'overpressure only' formulation are identified, with rupture occurring instead at $\alpha = 90°$ to feed a vertical dyke (half grey circle). This example illustrates that, even though equivalent overpressures will produce equivalent displacements, not only will the overpressure magnitude needed for initial rupture differ substantially in the two model formulations, as discussed in the section on 'Uniform and layered half-space models', but as a result the location of

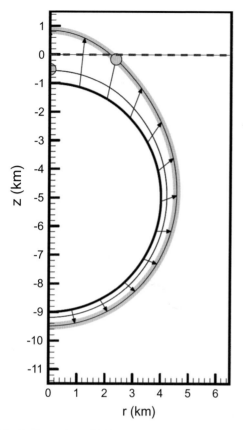

Fig. 5. Comparison of deformation rupture conditions for a spherical reservoir inflated to the point of failure under dry rock and 'overpressure only' host-rock conditions beneath a small conical edifice ($V_h = 400$ m, $V_r = 4$ km). Arrows denote displacement (1500×) resulting from inflation to the point of failure in dry rock, which occurs at $\alpha = 58.9°$ (small grey circle on the outer black line). In 'overpressure only' models, inflated at the same ΔP, identical displacement occurs, but under these conditions failure happens along the entire reservoir margin (thick grey line). Reducing ΔP until the point of first failure occurs, displacement is greatly reduced and $\alpha = 90°$ (half grey circle, on inner black line). In both models, the impact of the small edifice is evident but, when comparing them both, the displacement and the location of failure differ when conditions promoting initial rupture are implemented. From Hurwitz et al. (2009).

failure and style of intrusion can vary significantly as well.

Surface deformation. As noted previously, our emphasis has been on elucidating the conditions needed to initiate rupture of a subsurface reservoir. Surface deformation is challenging to interpret volcanologically, even in simple half-space models

(see 'Surface deformation' in the subsection on 'Uniform and layered half-space models'), and we are not aware that any systematic parameterized study conducted to date has examined the interplay between edifice loading, reservoir inflation and the resulting patterns of surface deformation. This may be sensible given that additional components like mechanical layering (e.g. Manconi et al. 2007, 2010) should also be included if one wishes to deduce subsurface magmatic conditions from observed surface deformation. Instead, researchers have focused their efforts on matching uplift patterns at specific volcanoes where abundant independent constraints on the subsurface host rock geology exist (e.g. Currenti et al. 2010; Ronchin et al. 2013). Such endeavours can provide powerful insight into a specific edifice or situation when the models are formulated carefully. As noted immediately above, the pressures deduced and failure patterns expected will depend quite sensitively on the host rock conditions employed and, since overpressure magnitude is directly linked to both displacement (observed) and rupture (limiting condition), it is critical to verify that the magma reservoir overpressure conditions inferred from displacement modelling do not violate the rupture constraints for the system under study (cf. the discussion of Long Valley inversions: Long & Grosfils 2009).

Ring fault initiation

While the sections above focus on tensile failure of an inflating reservoir, under normal conditions shear failure will occur first (e.g. Gerbault 2012). The extent to which shear failure affects the inflation process by promoting reservoir-fed intrusion remains poorly understood, and more careful future analysis is required to assess this problem. However, it is certain that shear failure and fault slip are necessary elements of ring faulting and caldera formation, and that caldera formation can be accompanied by massive and devastating eruptions.

To better understand how ring faulting initiates above a magma reservoir in response to inflation, basic volcano-tectonic conditions conducive to ring fault formation have been proposed as numerical modelling constraints (cf. Gudmundsson 1998; Folch & Marti 2004). The crux is that ring fault formation is argued to occur when conditions promote the development of (1) steeply inclined normal faults that (2) propagate downwards from regions of maximum tensile stress at the surface and (3) link up with a region of high shear stress at the outer margins of the reservoir. This combination is difficult to create and, therefore, is in conceptual agreement with the observation that slip on caldera-bounding faults occurs only rarely in active systems (Newhall & Dzurisin 1988). The proposed conditions are useful constraints, but there is concern because: (1) in formulating the link between magma reservoir pressure variations and failure in the host rock, models often adopt conditions that imply lithostatic pore pressure, which is not known to be a pervasive effect above shallow, caldera-forming reservoirs; (2) the approach assumes fault behaviour (e.g. geometry and point of origin) that is restrictive given the diverse range of field observations available to constrain fault attitude and, to a limited extent, propagation direction (Geshi *et al.* 2002; Marti *et al.* 2008); and (3) results obtained are different to those from analogue models – which, for instance, predict outwards-dipping ring faults that grow upwards from the reservoir – even when similar circumstances are examined. The latter concern has led many numerical modellers to question the applicability of physical simulations (e.g. Gudmundsson 2007), but it should also undermine confidence in the validity of existing numerical results (Acocella 2008) until agreement between the two approaches can be demonstrated.

The dry rock modelling approach avoids the first two assumptions, letting the model system mechanics dictate the conditions for fault initiation and geometry, and detailed exploration of how evolution of an inflating magmatic system can prime the host rock for ring fault formation is underway. As in the models described in previous sections, stresses parallel to the reservoir wall are monitored for tensile failure. In addition, Mohr–Coulomb failure criteria (cohesion of 10 MPa and angle of internal friction set to 25°) and parameterized Andersonian stress states linked to fault type (Simpson 1997) are used to assess the potential for, and alignment of, shear failure. While this effort remains a work in progress, the basic approach is demonstrated here using a discoidal reservoir subjected to overpressure conditions (Fig. 6). This example duplicates a geometry modelled previously using 'overpressure only' conditions for which it was concluded that 'the maximum tensile and shear stress both occur at a point directly above the centre of the chamber and (conditions) are thus not favourable to ring-fault formation' (Gudmundsson 2007, p. 155).

When overpressurized to the point of initial tensile failure (Fig. 6), assumed here to be the maximum inflation a reservoir will undergo prior to relieving the pressure via intrusion, Mohr–Coulomb failure within the host rock occurs across a broad zone at the surface and along the outer edge of the discoidal reservoir. Above the roof of the reservoir (green shades), stress tensor conditions favour the formation of normal faults according to Andersonian criteria, but examination of optimal concentric Mohr–Coulomb failure plane orientations reveals that Mohr–Coulomb failure permitting ring fault subsidence in this area will require reverse slip. Thus, alignment of the slip planes and the faulting style required by the stress state are in conflict, and ring fault initiation and subsidence during inflation will be very difficult. Beyond the lateral extent of the reservoir (red shades), a similar mismatch between the stress conditions and Mohr–Coulomb failure plane alignments should prevent ring fault initiation and subsidence. Critically, shear rupture will first occur at the reservoir wall within this zone at $h \cong 500$ m but, because conditions do not favour ring fault growth or subsidence at this location, local fault adjustments should dominate the response within host material. This is consistent with earthquake data from active magmatic centres, and we hypothesize that ruptures remaining local in extent will generally modulate neither the reservoir pressure nor the structural evolution significantly. Put another way, the host rock will adjust locally but, unless volumetrically significant intrusion occurs, inflation will continue and the area of Mohr–Coulomb failure will continue to expand until either conditions favour larger, throughgoing faults or tensile rupture occurs.

Between the two extensive regions where ring fault initiation and subsidence appear to be prohibitively difficult there is a narrow, near-vertical zone above the margin of the reservoir where the sense of fault slip and alignment of the Mohr–Coulomb failure planes match. These planes are anchored in an area of Mohr-Coulomb failure, and shear faulting at this site is expected somewhat before tensile failure. The host rock between the nucleation site at the reservoir and the surface has not generally exceeded the Mohr–Coulomb failure limit, raising questions about whether or not a throughgoing ring fault is possible prior to tensile failure, but the conditions are otherwise primed, mechanically, for ring faults to form: the faults can initiate, the stress conditions in the host rock are amenable and alignment of the most likely slip planes is suitable. Stresses associated with dynamic crack growth, or other factors not captured in static continuum models, are likely to play a key role in determining when caldera-like ring faults will be produced, and advancing models that incorporate such factors (cf. Holohan *et al.* 2007; Gerbault 2012) will probably expand the understanding of caldera formation appreciably.

As is true of other ring-faulting assessments, dry rock numerical models help evaluate when conditions are suitable and unsuitable for this type of structure to form. For the example described here, suitable ring-faulting conditions are established readily and without invoking special external

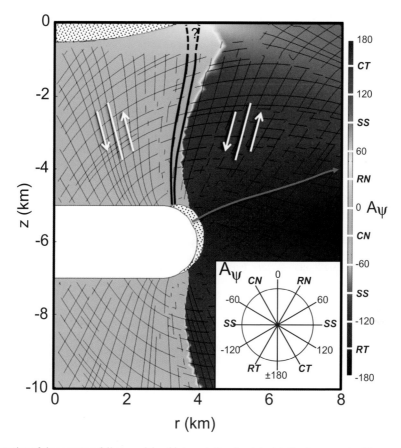

Fig. 6. Illustration of shear rupture failure model and interpretation. Inset depicts the A_ψ parameter (Simpson 1997) for Andersonian faulting; RN, pure radial normal; CN, pure concentric normal; RT, pure radial thrust; CT, pure concentric thrust; SS, pure strike-slip. In this dry rock model with a discoidal reservoir, stress conditions needed to initiate tensile rupture with $T = 5$ MPa are depicted; the path for the sill that would originate just above the mid-depth and relieve overpressure is denoted by a blue arrow – no eruption is expected. Well before tensile rupture occurs, zones of Mohr–Coulomb failure (white stippled pattern) form first at the surface (starting at $r = 0$ km) and then at the reservoir wall (starting along the upper edge at $r = 3.85$ km). Thin black lines show orientations of conjugate Mohr–Coulomb slip planes ($C = 25$ MPa, $\phi = 25°$), plotted only where the planes are concentric; the steep set above the reservoir resembles ring fault geometries observed in the field, while the other does not. Ring fault initiation and caldera subsidence is possible in regions where the A_ψ fault type (colour) and sense of slip needed for ring faulting and caldera subsidence (see white arrows) agree, provided that spatial overlap with a zone of Mohr–Coulomb failure also occurs. A narrow zone at the reservoir margin (thick black lines) meets these conditions and is thus primed for ring fault formation.

continuum conditions such as regional extension or domal uplift, a finding contrary to results reported based on an 'overpressure only' model formulation (Gudmundsson 2007). In addition, while further discussion is not pursued here, a dry host rock model approach accurately duplicates published analogue model results that simulate underpressure conditions (Kennedy *et al.* 2004), providing a new mechanical basis for understanding why initial outwards-dipping ring faults jump laterally to form steep, secondary, inwards-dipping faults (Grosfils 2011).

Discussion

In the Results, we reported several key results from dry rock modelling efforts to date. Below, we draw on these data to summarize why such models are suitable for analysing volcanism in diverse settings before exploring some implications for several types of reservoir-derived volcanic features commonly found on the terrestrial planets. We conclude with a brief discussion of elastic model limitations, and by framing a set of key directions that will benefit from additional future research.

Environmental parameters

Finite element numerical models exploring magma reservoir mechanics are an excellent tool for assessing this aspect of volcanism in a planetary context. Magma and rock characteristics, such as density and strength, which derive from material properties that are invariant (i.e. the tensile strength of a given basalt does not depend on the planet where the basalt is found), are not expected to affect model applicability to different planetary contexts. Similarly, while not discussed at length in the current contribution, neither plausible host rock density structures nor magma density variations play a major role in defining the location or conditions of tensile rupture unless the lithostatic stress is negated (Grosfils 2007), for instance by lithostatic pore pressure. The interplay between density variations will, of course, affect dyke propagation significantly and, with other factors, helps to dictate the depth at which a magma reservoir will form on a given planet (e.g. Wilson & Head 1994).

When inflation drives a reservoir to the point of tensile rupture, the scaled overpressure required is identical whether the gravity of Earth or Mars is employed in the model (Fig. 2), demonstrating that the mechanical response is independent of the planetary gravity applied. While gravity will, in part, dictate the atmospheric pressure applied to the surface, a factor neglected in the model results reported here, as pressure increases (e.g. for Venus) this is equivalent to the reservoir behaving as if it resided at slightly greater depth. However, even for Venus, this is comparable to only a few hundred metres of extra overburden, which does not meaningfully affect the results reported here. Gravity's impact on shear rupture conditions has not yet been assessed systematically, but depth-dependent variations are expected to occur because the depth needed to achieve a given normal stress across a fault (equation 2) will depend on the magnitude of the gravitational acceleration (see the subsection on 'Caldera formation' in the 'Discussion' section).

Critically, it is clear from comparing recent dry rock with traditional 'overpressure only' simulations that the host rock stress state environment assumed becomes a key factor when evaluating the evolution and rupture of a magma system at depth. 'Overpressure only' elastic models imply lithostatic pore pressures in the host rock (see the earlier subsection on 'Hydraulic fracture, elevated pore pressure'). Use of lithostatic pore pressure implies that the effective stress in the host rock ($\sigma - P_p$) is 0 (Hubbert & Rubey 1959; Gerbault 2012); that is, that the entire rock mass must exist in a state close to tensile failure even at equilibrium. This is contrary to what is normally observed even in critically stressed crust (Zoback 2010), although clear exceptions exist (Engelder & Leftwich 1997). Even when ideal conditions occur, however, establishing and maintaining lithostatic pore pressures at shallow, reservoir-hosting depths is believed to be difficult in most geological settings (Brace 1980; Nur & Walder 1990; Swarbrick & Osborne 1998; Simpson 2001; Zoback 2010). Because pore pressures (a) vary widely on Earth and are temporally and spatially variable, (b) are unknown for most active and relict volcanic systems, and (c) are difficult to achieve and sustain at lithostatic values over time, we argue that dry rock models represent an appropriately conservative and logically defensible end-member to use for establishing a baseline on which to build when exploring volcanic system characteristics. This is even more true when considering volcanism in a planetary context, as the dry rock model formulation is not only the most obvious choice for targets such as Venus, where pore fluids that weaken rock are unlikely to exist at shallow depth owing to elevated temperature, but also the best starting point to use for bodies such as Mars where pore pressure of some kind is plausible but unconstrained in magnitude (e.g. Kohlstedt & Mackwell 2010). Key differences between the models, when compared with observations, have several important volcanological implications, summarized below. Observations are generally, although not uniformly, better matched by dry rock model formulations and results.

First, in 'overpressure only' models, rupture of a spherical reservoir wall is resisted solely by the tensile strength of the material (equation 10). With $T \leq 10$ MPa, but normally only 2–3 MPa (Schultz 1995), overpressures of the order of approximately 5 MPa are sufficient to cause rupture, and 20 MPa (c. $2T$) becomes the effective limit of what a reservoir can sustain prior to failure. Even if one ignores pressure increases derived from infusion of new material, known to be a common occurrence and a factor that often triggers eruption, 5 MPa is very low given pressures expected from small amounts of crystallization alone (Tait *et al.* 1989). The implication is that reservoirs of plausible sizes should fail almost immediately either when new magma is injected or shortly after crystallization commences. In contrast, dry rock models require constant scaled overpressures to rupture reservoirs at low R/DtC ratios (Fig. 2), but the magnitude of the overpressure needed decreases significantly at shallower depths or in response to variations in surface loading conditions and/or reservoir geometry. To illustrate, a spherical reservoir in a uniform half-space with $R = 1$ km and $DtC = 3$ km requires $\Delta P = 102$ MPa to rupture (i.e. for $\rho_r = 2600$ kg m^{-3}, the lithostatic load on Earth at $D = 2$ km is 51 MPa), whereas the same reservoir with $DtC = 1.75$ km requires only $\Delta P = 34$ MPa, or $\Delta P = 10$ MPa when located beneath an intermediate-sized edifice (Fig. 2b). A mildly oblate reservoir of equivalent volume in a uniform half-space will fail at small R/DtC when $\Delta P < c.$ 40 MPa (Fig. 2c). These examples demonstrate that overpressures needed to induce failure of reservoirs in dry rock models at commonly inferred reservoir depths can often be quite reasonable. In addition, for a reservoir at a given DtC, increasing R means that the overpressure required to induce failure decreases. If small reservoirs form at depth, they will require fairly high overpressures to rupture given the magnitude of the lithostatic stress resisting failure. Instead, they have the opportunity to expand stably in response to magma influx (e.g. Jellinek & DePaolo 2003). Ultimately, the R/DtC scaling places a depth- and geometry-dependent limit on the reservoir size that can be achieved before rupture occurs.

Second, in 'overpressure only' models, deep-seated spherical reservoirs will fail near their mid-depth, feeding sills (radial dykes are not predicted: Grosfils 2007), while reservoirs at increasingly shallow depths will feed circumferential intrusions with increasingly steep dips (Fig. 3). Deep reservoirs will, therefore, stabilize in place through lateral intrusion and elongation, without feeding magma regularly to shallower depths. Indeed, contrary to what is commonly observed, vertical dyke injection from the crest and direct vertical magma ascent is not strictly possible. Because the effective stress in the host rock is zero, however, differences between the host rock and magma densities can help dictate the failure location (Pinel & Jaupart 2000); if a layer of sufficient thickness accumulates at the top of the reservoir, then rupture nearer the crest can occur, for instance, but for basalts at least sufficient foam accumulation is difficult to achieve under normal magmatic conditions (Parfitt et al. 1993). In contrast, for dry host rock conditions, deep-seated spherical reservoirs will rupture near the crest and feed magma to shallower depths while, at shallow depths, the failure location rotates away from the crest and reservoirs will emplace circumferential intrusions (Fig. 3). Edifice loading will pin the rupture location more firmly to the crest, mechanical layering plays, at most, a limited role and only oblate reservoirs at depth will stabilize in isolation from the surface; that is, without promoting magma ascent upon rupture.

Third, limits placed on the overpressure by rupture conditions also limit the surface deformation that can occur. Uplift magnitude will depend on the stiffness of the host rock and the effects of mechanical layering (cf. Long & Grosfils 2009; Geyer & Gottsmann 2010), but values of a few centimetres are expected for typical reservoir geometries and mechanical layering configurations in 'overpressure only' models (Fig. 4). In essence, any small increase in magma overpressure should lead readily to intrusion. Hence, inflation could drive only minimal surface uplift before the pressure is relieved (see 'Surface deformation' in the subsection on 'Uniform and layered half-space models'), a result that is inconsistent with observations of considerable surface inflation, as well as low intrusion frequencies during inflation events at active magmatic centres world-wide (e.g. Newhall & Dzurisin 1988; Pritchard & Simons 2004; Chaussard & Amelung 2012). Dry rock models, however, permit considerable uplift that, for otherwise identical conditions, is an order of magnitude or so greater than in the 'overpressure only' formulations (Fig. 4). This result is in better agreement with the decimetre- to metre-scale displacements commonly measured in volcanically active areas (Berrino et al. 1984; Pritchard & Simons 2004; Gottsmann et al. 2006; Baker & Amelung 2012; Chaussard & Amelung 2012), and intrusion frequency will be lower than in 'overpressure only' models because rupture is more difficult. Even with these much greater values, however, uplifts such as the 1.5 m observed at Campi Flegrei in 1982–1984 (Berrino et al. 1984) are not easily explained using an elastic model alone (Trasatti et al. 2005; Grosfils 2007), indicating that alternative rheologies or factors must be considered. It is worth noting as well that caution is required when using elastic 'overpressure only' models to infer reservoir conditions from surface inflation events. While excellent fits can be achieved (e.g. Battaglia et al. 2003), the pressures invoked are sometimes well in excess of the limit placed by tensile rupture for the reservoir geometries employed. This indicates that rupture criteria should be used jointly with surface deformation inversions when striving to constrain the depth, geometry and overpressure present within an actively inflating magma reservoir (Long & Grosfils 2009).

Implications for reservoir-related planetary volcanism

Finite element models examining magma inflation and rupture typically assume that the reservoir wall has no unusual stress-concentrating defects, but local inhomogeneities are, of course, likely to alter the nature of the failure process by introducing irregular stress concentrations that are difficult to model (e.g. Letourneur et al. 2008). The key question is, however, do such inhomogeneities play a critical role? Field observations – for instance, persistent emplacement of radial intrusions linked to shallow magma sources – indicate that reservoir inflation and rupture repeatedly produce remarkably similar volcanic landforms across a wide range of scales and geological settings. This implies that observed patterns of magma reservoir volcanism are not generally dictated by local inhomogeneities, and that they, instead, derive from overarching factors that are predictable and linked to the inflation process and response. Idealized analysis of wall-parallel stress magnitudes (cf. Russo *et al.* 1997) using numerical models is thus expected to provide useful insight into the nature of reservoir rupture, and any corresponding magmatic and/or volcanic activity. Here, we briefly explore several implications that dry rock model results have for four common types of reservoir-derived volcanic features.

Summit eruptions and edifice growth. Dry rock model data indicate that many paths will lead to the creation of edifices fed by central eruptions from a near-surface magma reservoir, consistent with abundant observations of such features on the terrestrial planets (e.g. Head & Wilson 1992; Wilson & Head 1994). To illustrate the interplay between magma reservoir rupture and edifice growth, consider a spherical reservoir with $R/DtC = 0.45$. Emplaced beneath a surface with negligible topography, the overpressure required to induce rupture will be roughly twice the lithostatic load at the crest (Fig. 2). Shallow reservoirs will, thus, fail more readily than those located at greater depths and, upon rupture, circumferential intrusions initiated at α of approximately $60°$ will result (Fig. 3), directing magma towards shallower depths. Dykes that breach the surface will begin to emplace lavas and other deposits, perhaps initially in a distributed fashion given the geometry of the intrusions. Whether widespread layers are emplaced – which, in effect, means that the reservoir crest lies at progressively greater depth – or an edifice begins to grow, the surface load will begin to rotate the point of failure towards the reservoir crest. Once vertical dyke injection from the crest is established, magma can be transported with maximum efficiency towards the surface, and growing conical edifices with summit eruptions become the expected volcanic norm unless tectonic stresses or other factors, such as basal slip, modify the edifice stress regime. The edifice size required to produce noticeable rupture point rotation depends on the reservoir radius and the edifice flank slope (i.e. shield- or stratocone-like) but, for reservoirs with R ranging from 0.2 to 4 km, typical edifice radii of the order of 5 km prove sufficient (cf. fig. 4 of Hurwitz *et al.* 2009). As an edifice grows in size, for a time the magma system

stabilizes: failure locked at the reservoir crest will feed material efficiently towards the edifice summit and, simultaneously, the overpressure required for failure decreases (Fig. 2b), which means that rupture and upwards propagation of magma via vertical dykes will occur more readily. As the edifice size increases significantly, the effective depth of the reservoir increases as well, a feedback mechanism that stabilizes the system by elevating the pressure needed for rupture (perhaps allowing the reservoir to increase in size). Eventually, however, if the magma system remains active, the growing load from the volcano can begin to induce significant flexure. The magnitude of the flexural response and the impact on reservoir rupture depends on the elastic thickness of the lithosphere and the reservoir depth due to the hourglass-like pattern of the flexural stresses. Generally speaking, however, an edifice with a radius above several tens of kilometres can induce sufficient flexure to terminate rupture at the crest of a typical kilometre-scale reservoir, aborting direct vertical magma ascent, unless elastic thicknesses are very large, the lithosphere is unusually weak and/or the reservoir depth is very great (Murphy *et al.* 2012). Depending on the specific volcanic configuration, lateral re-direction of magma by flexural stresses can terminate edifice growth and limit edifice size, or it may redirect magma to the surface via flank eruptions that act to modify the edifice morphology (McGovern *et al.* 2013).

Circumferential intrusions. Although observed less often in modern settings, circumferential intrusions are a common element of older eroded volcanic systems (Chadwick *et al.* 2011). It has also been proposed that such features could be responsible for circumferential fracture systems observed on Venus (common) and Mars (rare), but proving this connection has been difficult as circumferential fractures can form via a variety of volcanic and tectonic mechanisms (Ernst *et al.* 2001). Even the clearest evidence of magma involvement – lava flows emerging occasionally from individual structures within sets of circumferential fractures – are problematic, for this could imply either that circumferential dykes ascending to shallow depth generate the fractures (e.g. Mastin & Pollard 1988) or that other dykes encountering a pre-existing tectonic fracture set with a circumferential geometry simply utilize it to reach the surface more easily.

Dry rock model results predict that shallow- to steeply-dipping circumferential intrusions can form in response to inflation of spherical or oblate reservoirs, predominantly at shallow scaled depths (Grosfils 2007; Hurwitz *et al.* 2009); for a spherical reservoir, dips steeper than $55°$ are expected, while oblate reservoirs tend to intrude dykes only at shallower dips (Fig. 3). Edifice loading in the absence of flexure will normally increase the dip expected at a given scaled depth (the exception occurs for more oblate reservoirs, where little change in dip occurs: Hurwitz *et al.* 2009). Older systems exhibiting circumferential intrusions (e.g. Clough *et al.* 1909) may, thus, record the initial or pre-edifice stages of volcanism at a given shallow magmatic centre, only exposed by deep erosion, because resulting edifice growth tends to shut down circumferential intrusion (see the subsection on 'Summit eruptions and edifice growth') while burying older intrusion signatures with younger lavas. Significant downwards flexure will make it exceedingly difficult for circumferential intrusions to feed dykes toward the surface (Galgana *et al.* 2011), but bulk stress states induced by surface loading can promote lateral propagation from the reservoir and then circumferential dyke ascent at greater distances from the symmetry axis, a mechanism that may help explain the growth of domical and annular edifices (McGovern *et al.* 2013).

Lateral radial dyke swarm formation. An unexpected result from dry rock models is that laterally propagating radial dykes, commonly observed on Earth, Venus and Mars, are difficult to explain as a direct product of magma reservoir inflation and rupture. Previous analyses using 'overpressure only' formulations have implied that lateral dyke initiation is straightforward, but the models do not examine the wall-parallel stresses σ_t and σ_θ independently, and generally assume that rupture at the mid-depth leads directly to lateral dyke injection (e.g. Parfitt *et al.* 1993). In fact, duplication of these model conditions reveals that rupture near the mid-section occurs in the σ_t orientation, and will emplace lateral sills (Grosfils 2007). Similarly, dry rock models predict patterns of failure that produce vertical radial dyke ascent from the crest, lateral sill injection or the emplacement of inclined circumferential intrusions. Only in three very specific circumstances have conditions conducive to lateral radial dyke propagation been identified.

First, growth of an edifice at the surface can create a stress trap capable of redirecting magma from vertical ascent to lateral propagation at shallow depth, in essence right along the base of the edifice (Hurwitz *et al.* 2009). The lateral extent of this stress trap is limited by the edifice radius, and will not permit lateral dyke injection to continue beyond this distance unless such propagation is enabled by other factors (e.g. Parfitt & Head 1993; Pinel & Jaupart 2004). Tests striving to match data from Summer Coon (Poland *et al.* 2008) reveal that the model predictions closely simulate measured dyke depth, extent and thickness variations

(Hurwitz *et al.* 2009), indicating that this mechanism can explain typical lateral dyke injection patterns often observed at composite volcanoes. Second, while Hurwitz *et al.* (2009) restrict their analysis to sub-edifice reservoirs, subsequent efforts examining rupture of reservoirs that lie partially within an edifice have shown that volcanologically plausible evolution of a mildly oblate reservoir's aspect ratio can initiate both circumferential and lateral dyke injection within an edifice (Fig. 7). Tests by Chestler & Grosfils (2013) striving to match complex intrusion patterns reported for the Galapagos (e.g. Chadwick *et al.* 2011; Bagnardi *et al.* 2013) do so using a reservoir configuration consistent with existing size and depth constraints but, again, the lateral extent of the radial dykes will be limited. Third, at a much larger scale, lateral dykes can occur when broad flexural upwarping, generated by a zone of melt at the base of the lithosphere that creates conditions conducive to

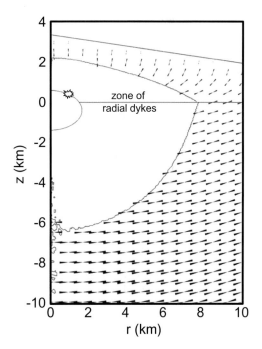

Fig. 7. Model depicting radial dyke initiation within a conical edifice ($V_h = 3.4$ km, $V_r = 23.7$ km; akin to a Galapagos-sized volcano). Reservoir is centred at $z = -0.4$ km, with $R_b = 1$ km and $R_a = 1.7$ km. Short linear symbols denote alignment of the minimum compressive stress, and are perpendicular to the page where missing in the zone surrounding the reservoir. Orientation of the fracture formed when initial failure occurs (at the star, $z = 0.4$ km) will inject a radial dyke laterally into the edifice within a zone where the stress state strongly favours continued radial dyke stability.

surface eruption, is integrated with downwarping introduced as the resulting surface load grows (Galgana *et al.* 2013). For an edifice the size of a typical large Venusian volcano, this combination produces a stress trap at the base of the edifice similar to that identified by Hurwitz *et al.* (2009). It differs, however, in that the basal uplift affects the stress state across a much broader area, creating conditions in which lateral dykes are more likely to propagate across distances akin to those observed in many giant radial systems (Grosfils & Head 1994; Ernst *et al.* 1995).

While the results summarized above are certainly encouraging, all three mechanisms require the presence of an edifice at the surface. In many instances this condition is met, but observations from Venus indicate that, of the 118 giant radiating dyke swarms identified in an initial global survey, roughly half do not exhibit evidence of the necessary elevated (domical or edifice-like) central topography (Grosfils & Head 1994). While constraints upon the topography of similar systems on Earth are limited (e.g. Ernst *et al.* 1995), the Venusian data indicate that roughly half of the giant radial dyke swarms identified cannot be explained using the three mechanisms described above. Further research in this area is clearly required.

Caldera formation. As discussed above, preliminary results demonstrate that the conditions necessary for initiating ring faults develop more readily than previously believed, and that special circumstances such as basal uplift or regional tension are not required. When conditions are primed for throughgoing ring faults to form as a result of magma reservoir inflation, the expected geometry – matching proposed criteria (Gudmundsson 1998; Folch & Marti 2004) – is a steep, inwards-dipping normal fault anchored in a zone of shear failure at depth (Fig. 6). The ring fault formation predicted by dry rock models differs from proposed criteria in that rupture initiates at the reservoir wall, not at the surface; this matches data from caldera formation events monitored in the field (e.g. Geshi *et al.* 2002).

From limited analyses to date there are several implications for the mechanics of caldera formation in a planetary context. First, caldera formation, while almost certainly aided by factors such as basal uplift and regional extension (Gudmundsson 2007) or edifice loading (Pinel & Jaupart 2005), does not require these external drivers. Calderas should occur wherever inflation of a suitable reservoir geometry primes the host rock for ring faulting, and the presence of a caldera does not imply that precursory tectonic activity of a specific kind occurred at the site. Second, given the dependence of Mohr–Coulomb shear failure on normal stress

magnitude and, hence, gravitational loading, an inflating reservoir at identical depths on Earth and Mars is expected to generate ring faults and form calderas more readily on Mars. Third, however, because reservoirs are likely to form at proportionately greater depths on Mars than they do on Earth (e.g. Wilson & Head 1994), which changes the aspect ratio of the roof rock above a reservoir, this should adversely affect the likelihood that a ring fault on Mars will reach the surface (e.g. Roche *et al.* 2000). Thus, unless they form above unusually shallow reservoirs, calderas are expected to be rarer on Mars than on rocky planets with higher gravity (i.e. Earth and Venus). To first order, this is consistent with what is observed (e.g. Radebaugh *et al.* 2001; Sanchez & Shcherbakov 2012), but some caution is warranted as caldera detection capabilities and preservation likelihood are not equivalent on these three bodies, and further work is required to test the validity of the proposed link between reservoir mechanics and caldera observations on the terrestrial planets.

Applicability and limitations of elastic models

Field and geophysical data demonstrate that subsurface magma reservoirs are long-lived mechanically and thermally complex systems (e.g. de Silva *et al.* 2006), and it is reasonable to expect that elastic continuum numerical models can provide only limited insight into their evolution and the volcanic consequences. Complementing physical–chemical model assessments that explore the role played by magma evolution during an eruption (cf. Karlstrom *et al.* 2012; Segall 2013), their primary use when assessing inflation has been to predict the conditions that prime a system for rupture, and to then project the expected outcome based on the assumption that the host rock stress state dictates subsequent rupture evolution and/ or intrusion path. Continuum elastic models are generally unable, however, to track critical factors that probably play a key role in such activity, such as time-dependent fatigue (induced by repeated inflation–deflation cycles, for instance: e.g. Kendrick *et al.* 2013) or the dynamic stress changes associated with intrusion propagation or shear crack displacement (e.g. Meriaux & Lister 2002; Shuler *et al.* 2013). The attraction of elastic numerical models in spite of these concerns is that: (a) they can describe bulk system behaviour adequately, particularly for a process such as inflation, which responds rapidly to changes in the system; (b) they are easy to implement and require minimal time or computing power to run; and (c) the simplicity of their components means that they can be calibrated carefully using analytical

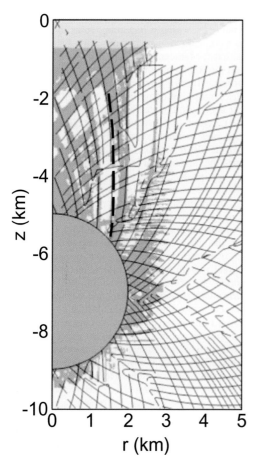

Fig. 8. Mohr–Coulomb failure planes predicted for a spherical reservoir ($R = 2$ km, $DtC = 7$ km, $\Delta P = 130$ MPa) within a dry elastic host rock, superimposed on elasto-plastic strain results reported by Gerbault *et al.* (2012) for the same conditions. Dark blue regions denote areas of plastic shear failure, while the light blue region denotes the area of tensile failure; yellow indicates intact rock. There is excellent agreement between the shear failure orientations predicted by the two approaches. Using the method illustrated in Figure 6, the only plausible location for ring fault development in the dry elastic host rock model results is shown by a heavy dashed line. While offset slightly from the main ring fault-like zone of shear in the elasto-plastic model, the two results agree quite well, and the elastic model yields similar first-order insight into expected ring fault nucleation and geometry.

solutions. Elastic numerical model solutions, in turn, provide an important framework for calibrating advanced models that incorporate factors such as temperature-dependent rheology and plastic strain (e.g. Jellinek & DePaolo 2003; Trasatti *et al.* 2005; Del Negro *et al.* 2009).

Given their obvious limitations and simplicity, how useful are the results from elastic models, and how well do they perform when compared with more advanced solutions? Perhaps surprisingly, the answer is that they do remarkably well for a wide array of volcanologically plausible configurations. For instance, Gregg *et al.* (2012) used a temperature-dependent viscoelastic model formulation to study the overpressure needed to induce throughgoing rupture, and concluded that the solutions did not diverge noticeably from the outcomes of purely elastic dry rock solutions (e.g. Grosfils 2007) until reservoirs reach volumes of the order of 10^2–10^3 km^3 (i.e. comparable to volumes inferred for Taupo caldera, New Zealand: Ellis *et al.* 2007) with small roof aspect ratios ($R_a > 3D$). Similarly, the Mohr–Coulomb failure plane orientations used to assess potential for ring fault development in a dry rock elastic formulation compare well with detailed plane strain elasto-plastic models (Gerbault *et al.* 2012) that track time-dependent shear failure and fracture development above both inflating and deflating magma reservoirs (Fig. 8). In spite of this agreement, elasto-plastic model results represent a significant advance in that they predict which planes are more likely to serve as sites for throughgoing ring fault nucleation and growth. It is also worth noting, however, that analysis of spherical geometries requires computations that individually can take weeks–months to run, a significant limitation considering an elastic model formulation such as ours can predict a similar outcome (Fig. 8) using computations that require only minutes to complete.

Conclusions

Numerical elastic continuum models of magma reservoir inflation can provide insights applicable to a surprisingly diverse array of volcanic conditions. When integrated with field, laboratory and remote-sensing data gathered on Earth and from other planets, such models have already provided important new insight into many common reservoir-related processes such as edifice construction, caldera formation and radial dyke injection. In part, this is because many factors of volcanological importance can be considered within an elastic model framework. Complementing uniform homogeneous simulations, for example, models can easily introduce magma and host rock density variations, or explore the impacts of mechanical layering, edifice loading and flexure. Furthermore, elastic model components and predictions are largely insensitive to, or can readily incorporate and assess, material and parameter variations that characterize different planetary environments. Finally, when compared with models incorporating more advanced behaviour (e.g. Holohan *et al.* 2007; Gerbault *et al.* 2012) that can require weeks or months to run, elastic models are simple to design and test, take minutes or hours to execute, yet often produce results that are directly comparable to those obtained using more complex approaches. As a result, although one must remain cognizant of their limitations, elastic models of magma reservoir inflation and rupture have become a key tool for understanding planetary volcanism.

In spite of recent advances, however, there are three areas where existing model results indicate that additional research will be of particular benefit for addressing outstanding questions and advancing modelling capabilities. The first area pertains to the role of pore pressure variations within the host rock surrounding a magma reservoir. Models that assume dry rock conditions develop a baseline framework for understanding magma reservoir inflation and rupture on this basis. More traditional 'overpressure only' formulations that assume lithostatic pore pressure have been widely used for decades, however, and they make markedly different predictions when assessing the rupture overpressure, location and orientation, as well as the surface uplift magnitude that can occur prior to tensile failure and intrusion. Given these differences, it has become critical to obtain better insight than volcanologists have at present into: (a) the pore pressure conditions that normally occur within host rock surrounding an active magma reservoir; and (b) the range of plausible pore pressures, and what controls their spatial and temporal variability. As noted by Gerbault *et al.* (2012), major variations in pore pressure within the host rock adjacent to a magma reservoir could lead to rupture without any changes to an existing magma overpressure.

The second area we identify for additional research focuses on the need to address specific recognized gaps in existing knowledge, which become apparent when a given type of volcanic feature cannot be reproduced as expected; several key examples are identified in the subsections on 'Uniform and layered half-space models', 'Models with edifice loading' and 'Implications for reservoir-related planetary volcanism'. For example, while we have reported several different ways to initiate laterally propagating radial dykes (see the subsection 'Lateral radial dyke swarm formation'), all require the presence of an edifice or comparable load at the surface. However, many radial dyke swarms clearly formed without this condition in place (Grosfils & Head 1994), which begs the question of what is missing from current model formulations – dry rock and 'overpressure only' alike. Addressing core questions like this example,

although seemingly somewhat narrow in focus, remains exceedingly important. Inability to reproduce common volcanic situations probably indicates that fundamental framing of numerical magma reservoir inflation simulations by the volcanological community may be neglecting some key factor (e.g. magmas are treated as fluids, yet magmas convect and, thus, couple shear stress to the reservoir wall). Returning to the dyke swarm example, identifying how to produce laterally propagating radial dykes across a wider array of conditions may help refine the fundamental framing of magma reservoir inflation models in ways that inform elastic and more complex models alike.

Finally, in spite of the benefits realized to date using elastic models, there are obvious motivations for improving more complex model formulations (e.g. discrete element, elasto-plastic and viscoelastic). Advances in these directions require moving beyond two-dimensional (2D) plane stress–strain models that treat reservoirs as a slice through a horizontal cylinder, however, and beyond axisymmetrical models like those reported on here that permit rapid parameterized analyses of ellipsoidal and similar geometries. Indeed, the benefits of 3D models of individual volcanoes that incorporate spatially variable material properties, irregular topography and regional tectonic stresses are beginning to be exploited (e.g. Bonaccorso *et al.* 2005). Ultimately, once calibrated against carefully framed and understood elastic model results, continued development of advanced 3D model formulations shows exceptional promise as a means of improving geologists' insight into volcanic phenomena on Earth and other rocky bodies in the solar system.

We are grateful for many thought-provoking conversations about magma reservoir mechanics over the years with J. Head, A. Newman, E. Parfitt, M. Parmentier, A. Rubin, S. de Silva and L. Wilson. We also thank two anonymous reviewers for suggestions that enhanced the manuscript. This work was supported in part by NASA grants NAG5-9618, NAG5-10498, NNG05GJ92G and NNX08AL77G, with additional student involvement supported by the Summer Undergraduate Research Program at, and a Mellon Post-Baccalaureate Fellowship from, Pomona College.

References

Acocella, V. 2008. Structural development of calderas: a synthesis from analogue experiments. *In*: Gottsman, J. & Marti, J. (eds) *Caldera Volcanism: Analysis, Modelling and Response. Developments in Volcanology*, **10**, 285–311.

Anderson, E. M. 1936. Dynamics of the formation of cone-sheets, ring-dykes, and cauldron-subsidences. *Proceedings of the Royal Society of Edinburgh*, **56**, 128–157.

BACHMANN, O. & BERGANTZ, G. W. 2004. On the origin of crystal-poor rhyolites: extracted from batholithic crystal mushes. *Journal of Petrology*, **45**, 1565–1582.

BAKER, S. & AMELUNG, F. 2012. Top-down inflation and deflation at the summit of Kilauea volcano, Hawai'i observed with InSAR. *Journal of Geophysical Research*, **117**, B12406, 1–14.

BAGNARDI, M., AMELUNG, F & POLAND, M. P. 2013. A new model for the growth of basaltic shields based on deformation of Fernandina volcano, Galapagos Islands. *Earth and Planetary Science Letters*, **377–378**, 358–366, http://dx.doi.org/10.1016/j.epsl.2013.07.016

BARTEL, B. A., HAMBURGER, M. W., MEERTENS, C. M., LOWRY, A. R. & CORPUZ, E. 2003. Dynamics of active magmatic and hydrothermal systems at Taal Volcano, Philippines, from continuous GPS measurements. *Journal of Geophysical Research*, **108**, 2475, http://dx.doi.org/10.1029/2002JB002194

BATTAGLIA, M., SEGALL, P., MURRAY, J., CERVELLI, P. & LANGBEIN, J. 2003. The mechanics of unrest at Long Valley caldera, California: 1. Modeling the geometry of the source using GPS leveling and two-color EDM data. *Journal of Volcanology and Geothermal Research*, **127**, 195–217.

BERRINO, G., CORRADO, G., LUONGO, G. & TORO, B. 1984. Ground deformation and gravity changes accompanying the 1982 Pozzuoli uplift. *Bulletin of Volcanology*, **47**, 187–200.

BLAKE, S. 1981. Volcanism and dynamics of open magma chambers. *Nature*, **289**, 783–785.

BONACCORSO, A. & DAVIS, P. M. 1999. Models of ground deformation from vertical volcanic conduits with application to eruptions of Mount St. Helens and Mount Etna. *Journal of Geophysical Research*, **104**, 10,531–10,542.

BONACCORSO, A., CIANETTI, S., GIUNCHI, C., TRASATTI, E., BONAFEDE, M. & BOSCHI, E. 2005. Analytical and 3-D numerical modeling of Mt. Etna (Italy) volcano inflation. *Geophysical Journal International*, **163**, 852–862.

BRACE, W. F. 1980. Permeability of crystalline and argillaceous rocks. *International Journal of Rock Mechanics and Mining Sciences and Geomechanics Abstracts*, **17**, 241–251.

BURGISSER, A. & BERGANTZ, G. W. 2011. A rapid mechanism to remobilize and homogenize highly crystalline magma bodies. *Nature*, **471**, 212–215.

CASHMAN, K. V. & SPARKS, R. S. J. 2013. How volcanoes work: a 25 year perspective. *Geological Society of America Bulletin*, **125**, 664–690.

CERVELLI, P., MURRAY, M., SEGALL, P., AOKI, Y. & KATO, T. 2001. Estimating source parameters from deformation data, with an application to the March 1997 earthquake swarm off the Izu Peninsula, Japan. *Journal of Geophysical Research*, **106**, 11,217–11,237.

CHADWICK, W. W. & DIETERICH, J. H. 1995. Mechanical modeling of circumferential and radial dike intrusion on Galapagos volcanoes. *Journal of Volcanology and Geothermal Research*, **66**, 37–52.

CHADWICK, W. W., JONSSON, S. ET AL. 2011. The May 2005 eruption of Fernandina Volcano, Galapagos; the first circumferential dike intrusion observed by GPS and InSAR. *Bulletin of Volcanology*, **73**, 679–697.

CHAUSSARD, E. & AMELUNG, F. 2012. Precursory inflation of shallow magma reservoirs at west Sundra volcanoes detected by InSAR. *Geophysical Research Letters*, **39**, L21311, http://dx.doi.org/10.1029/2012GL053817

CHESTLER, S. R & GROSFILS, E. B. 2013. Using numerical modeling to explore the origin of intrusion patterns on Fernandina volcano, Galapagos islands, Ecuador. *Geophysical Research Letters*, **40**, 4565–4569, http://dx.doi.org/10.1002/grl.50833

CHRISTENSEN, J. N. & DEPAOLO, D. J. 1993. Time scales of large volume silicic magma systems: Sr isotope systematics of phenocrysts and glass from the Bishop Tuff, Long Valley, California. *Contributions to Mineralogy and Petrology*, **113**, 100–114.

CLOUGH, C. T., MAUFE, H. B. & BAILEY, E. B. 1909. The cauldron-subsidence of Glen Coe, and the associated igneous phenomena. *Quarterly Journal of the Geological Society, London*, **65**, 611–678.

CURRENTI, G., BONACCORSO, A., DEL NEGRO, C., SCANDURA, D. & BOSCHI, E. 2010. Elasto-plastic modeling of volcano ground deformation. *Earth and Planetary Science Letters*, **296**, 311–318.

DAVIS, P. M. 1986. Surface deformation due to inflation of an arbitrarily oriented triaxial ellipsoidal cavity in an elastic half-space, with reference to Kilauea volcano, Hawaii. *Journal of Geophysical Research*, **91**, 7429–7438.

DE SILVA, S. L., ZANDT, G., TRUMBULL, R., VIRAMONTE, J. G., SALAS, G. & JIMENEZ, N. 2006. Large ignimbrite eruptions and volcano-tectonic depressions in the Central Andes: a thermomechanical perspective. *In*: TROISE, C., DE NATALE, G. & KILBURN, C. R. J. (eds) *Mechanisms of Activity and Unrest at Large Calderas*. Geological Society, London, Special Publications, **269**, 47–63.

DEL NEGRO, C., CURRENTI, G. & SCANDURA, D. 2009. Temperature-dependent viscoelastic modeling of ground deformation: application to Etna volcano during the 1993–1997 inflation period. *Physics of the Earth and Planetary Interiors*, **172**, 299–309.

DIETERICH, J. H. & DECKER, R. W. 1975. Finite-element modeling of surface deformation associated with volcanism. *Journal of Geophysical Research*, **80**, 4094–4102.

DRUITT, T. H., COSTA, F., DELOULE, E., DUNGAN, M. & SCAILLET, B. 2012. Decadal to monthly timescales of magma transfer and reservoir growth at a caldera volcano. *Nature*, **482**, 77–80.

DZURISIN, D. 2003. A comprehensive approach to monitoring volcano deformation as a window on the eruption cycle. *Reviews of Geophysics*, **41**, 1–5, http://dx.doi.org/10.1029/2001RG000107

ELLIS, S. M., WILSON, C. J. N., BANNISTER, S., BIBBY, H. M., HEISE, W., WALLACE, L. & PATTERSON, N. 2007. A future magma inflation event under the rhyolitic Taupo volcano, New Zealand: numerical models based on constraints from geochemical, geological and geophysical data. *Journal of Volcanology and Geothermal Research*, **168**, 1–27.

ENGELDER, T. & LEFTWICH, J. T. 1997. A pore-pressure limit in overpressured south Texas oil and gas fields. *In*: SURDAM, R. C. (ed.) *Seals, Traps, and the*

Petroleum System. American Association of Petroleum Geologists, Memoirs, **67**, 255–267.

ERNST, R. E., HEAD, J. W., PARFITT, E., GROSFILS, E. B. & WILSON, L. 1995. Giant radiating dyke swarms on Earth and Venus. *Earth Science Reviews*, **39**, 1–58.

ERNST, R. E., GROSFILS, E. B. & MEGE, D. 2001. Giant dike swarms: Earth, Venus and Mars. *Annual Review of Earth and Planetary Sciences*, **29**, 489–534.

FIALKO, Y., KHAZAN, Y. & SIMONS, M. 2001. Deformation due to a pressurized horizontal circular crack in an elastic half-space, with applications to volcano geodesy. *Geophysical Journal International*, **146**, 181–190.

FOLCH, A. & MARTI, J. 2004. Geometrical and mechanical constraints on the formation of ring-fault calderas. *Earth and Planetary Science Letters*, **221**, 215–225.

GALGANA, G. A., McGOVERN, P. J. & GROSFILS, E. B. 2011. Evolution of large Venusian volcanoes: insights from coupled models of lithospheric flexure and magma reservoir pressurization. *Journal of Geophysical Research*, **116**, E03009, http://dx.doi.org/10.1029/2010JE003654

GALGANA, G. A., GROSFILS, E. B. & McGOVERN, P. J. 2013. Radial dike formation on Venus: insights from models of uplift, flexure and magmatism. *Icarus*, **225**, 538–547.

GERBAULT, M. 2012. Pressure conditions for shear and tensile failure around a circular magma chamber: insight from elasto-plastic modeling. *In*: HEALY, D., BUTLER, R. W. H., SHIPTON, Z. & SIBSON, R. H. (eds) *Faulting, Fracturing and Igneous Intrusion in the Earth's Crust.* Geological Society, London, Special Publications, **367**, 111–130.

GERBAULT, M., CAPPA, F. & HASSANI, R. 2012. Elasto-plastic and hydromechanical models of failure around an infinitely long magma chamber. *Geochemistry, Geophysics, Geosystems*, **13**, Q03009, http://dx.doi.org/10.1029/2011GC003917

GERTISSER, R., SELF, S., THOMAS, L. E., HANDLEY, H. K., VAN CALSTEREN, P. & WOLFF, J. A. 2012. Processes and timescales of magma genesis and differentiation leading to the great Tambora eruption in 1815. *Journal of Petrology*, **53**, 271–297.

GESHI, N., SHIMANO, T., CHIBA, T. & NAKADA, S. 2002. Caldera collapse during the 2000 eruption of Miyake-jima Volcano, Japan. *Bulletin of Volcanology*, **64**, 55–68.

GEYER, A. & GOTTSMANN, J. 2010. The influence of mechanical stiffness on caldera deformation and implications for the 1971–1984 Rabaul uplift (Papua New Guinea). *Tectonophysics*, **483**, 399–412.

GEYER, A. & BINDEMAN, I. 2011. Glacial influence on caldera-forming eruptions. *Journal of Volcanology and Geothermal Research*, **202**, 127–142.

GOTTSMANN, J., FOLCH, A. & RYMER, H. 2006. Unrest at Campi Flegrei: a contribution to the magmatic v. hydrothermal debate from inverse and finite element modeling. *Journal of Geophysical Research*, **111**, B07203, http://dx.doi.org/10.1029/2005JB003745

GREGG, P. M., DE SILVA, S. L., GROSFILS, E. B. & PARMIGIANI, J. P. 2012. Catastrophic caldera-forming eruptions: thermomechanics and implications for eruption triggering and maximum caldera dimensions on Earth.

Journal of Volcanology and Geothermal Research, **241–242**, 1–12.

GROSFILS, E. B. 2007. Magma reservoir failure on the terrestrial planets: assessing the importance of gravitational loading in simple elastic models. *Journal of Volcanology and Geothermal Research*, **166**, 47–75.

GROSFILS, E. B. 2011. New mechanical insights into ring fault initiation and caldera formation on terrestrial planets. *In*: *42nd Lunar and Planetary Science Conference, held 7–11 March 2011 at The Woodlands, Texas.* Lunar and Planetary Institute, Houston, TX, Abstract 1170.

GROSFILS, E. B. & HEAD, J. W. 1994. The global distribution of giant radiating dike swarms on Venus: implications for the global stress state. *Geophysical Research Letters*, **21**, 701–704.

GROSSE, P., VAN WYCK DE VRIES, B., PETRINOVIC, I. A., EUILLADES, P. A. & ALVARADO, G. E. 2009. Morphometry and evolution of arc volcanoes. *Geology*, **37**, 651–654.

GUALDA, G. A. R., PAMUKCU, A. S., GHIORSO, M. S., ANDERSON, A. T., SUTTON, S. R. & RIVERS, M. L. 2012. Timescales of quartz crystallization and the longevity of the Bishop giant magma body. *PLoS ONE*, **7**, 1–12.

GUDMUNDSSON, A. 1988. Effect of tensile stress concentration around magma chambers on intrusion and extrusion frequencies. *Journal of Volcanology and Geothermal Research*, **35**, 179–194.

GUDMUNDSSON, A. 1998. Formation and development of normal-fault calderas and the initiation of large explosive eruptions. *Bulletin of Volcanology*, **60**, 160–170.

GUDMUNDSSON, A. 1990. Emplacement of dikes, sills and crustal magma chambers at divergent plate boundaries. *Tectonophysics*, **176**, 257–275.

GUDMUNDSSON, A. 2006. How local stresses control magma-chamber ruptures, dyke injections, and eruptions in composite volcanoes. *Earth Science Reviews*, **79**, 1–31.

GUDMUNDSSON, A. 2007. Conceptual and numerical models of ring-fault formation. *Journal of Volcanology and Geothermal Research*, **164**, 142–160.

GUDMUNDSSON, A. 2012. Magma chambers: formation, local stresses, excess pressures, and compartments. *Journal of Volcanology and Geothermal Research*, **237–238**, 19–41.

HEAD, J. W. & WILSON, L. 1992. Magma reservoirs and neutral buoyancy zones on Venus: implications for the formation and evolution of volcanic landforms. *Journal of Geophysical Research*, **97**, 3877–3903.

HILDRETH, W. 1981. Gradients in silicic magma chambers: implications for lithospheric magmatism. *Journal of Geophysical Research*, **86**, 10,153–10,192.

HOBBS, T. 2011. *Stress Induced Seismic Anisotropy Around Magma Chambers.* PhD thesis, University of Bristol, UK.

HOLOHAN, E. P., SCHOPFER, M. P. J. & WALSH, J. J. 2007. Mechanical and geometric controls on the structural evolution of pit crater and caldera subsidence. *Journal of Geophysical Research*, **116**, B07202, http://dx.doi.org/10.1029/2010JB008032

HUBBERT, M. K. & RUBEY, W. W. 1959. Role of fluid pressure in mechanics of overthrust faulting. *Geological Society of America Bulletin*, **70**, 115–166.

HUBER, C., BACHMANN, O. & DUFEK, J. 2012. Crystal-poor v. crystal-rich ignimbrites: a competition between stirring and reactivation. *Geology*, **40**, 115–118.

HURWITZ, D. M., LONG, S. M. & GROSFILS, E. B. 2009. The characteristics of magma reservoir failure beneath a volcanic edifice. *Journal of Volcanology and Geothermal Research*, **188**, 379–394.

JAEGER, J. C. & COOK, N. G. W. 1979. *Fundamentals of Rock Mechanics*. Chapman & Hall, New York.

JEFFREY, G. B. 1921. Plane stress and plane strain in bipolar coordinates. *Philosophical Transactions of the Royal Society of London, Series A*, **221**, 265–293.

JELLINEK, A. M. & DEPAOLO, D. J. 2003. A model for the origin of large silicic magma chambers: precursors of caldera-forming eruptions. *Bulletin of Volcanology*, **65**, 363–381.

KARLSTROM, L., RUDOLPH, M. L. & MANGA, M. 2012. Caldera size modulated by the yield stress within a crystal-rich magma reservoir. *Nature Geoscience*, **5**, 402–405.

KENDRICK, J. E., SMITH, R., SAMMONDS, P., MEREDITH, P. G., DAINTY, M. & PALLISTER, J. S. 2013. The influence of thermal and cyclic stressing on the strength of rocks from Mount St. Helens, Washington. *Bulletin of Volcanology*, **75**, 752, http://dx.doi.org/10.1007/s00445–013–0728-z

KENNEDY, B., STIX, J., VALLANCE, J. W., LAVALLEE, Y. & LONGPRE, M.-A. 2004. Controls on caldera structure: results from analogue sandbox modeling. *Geological Society of America Bulletin*, **116**, 515–524.

KOENIG, E. & POLLARD, D. D. 1998. Mapping and modeling of radial fracture patterns on Venus. *Journal of Geophysical Research*, **103**, 15183–15202.

KOHLSTEDT, D. L. & MACKWELL, S. J. 2010. Strength and deformation of planetary lithospheres. *In*: WATTERS, T. R. & SCHULTZ, R. A. (eds) *Planetary Tectonics*. Cambridge University Press, New York.

LETOURNEUR, L., PELTIER, A., STAUDACHER, T. & GUDMUNDSSON, A. 2008. The effects of rock heterogeneities on dyke paths and asymmetric ground deformation: the example of Piton de la Fournaise (Reunion Island). *Journal of Volcanology and Geothermal Research*, **173**, 289–302.

LOCKNER, D. & BYERLEE, J. D. 1977. Hydrofracture in Weber sandstone at high confining pressure and differential stress. *Journal of Geophysical Research*, **82**, 2018–2026.

LONG, S. M. & GROSFILS, E. B. 2009. Modeling the effect of layered volcanic material on magma reservoir failure and associated deformation, with application to Long Valley caldera, California. *Journal of Volcanology and Geothermal Research*, **186**, 349–360.

MACCAFERRI, F., BONAFEDE, M. & RIVALTA, E. 2011. A quantitative study of the mechanisms governing dike propagation, dike arrest and sill formation. *Journal of Volcanology and Geothermal Research*, **208**, 39–50.

MANCONI, A., WALTER, T. R. & AMELUNG, F. 2007. Effects of mechanical layering on volcano deformation. *Geophysical Journal International*, **170**, 952–958.

MANCONI, A., WALTER, T., MANZO, M., ZENI, G., TIZZANI, P., SANSOSTI, E. & LANARI, R. 2010. On the effects of 3-D mechanical heterogeneities at Campi Flegrei caldera, southern Italy. *Journal of Geophysical Research*, **115**, B08405, http://dx.doi.org/10.1029/2009JB007099

MARTI, J., GEYER, A., FOLCH, A. & GOTTSMAN, J. 2008. A review on collapse caldera modelling. *In*: GOTTSMAN, J. & MARTI, J. (eds) *Caldera Volcanism: Analysis, Modelling and Response. Developments in Volcanology*, **10**, 233–283.

MASTERLARK, T. 2007. Magma intrusion and deformation predictions: sensitivities to the Mogi assumptions. *Journal of Geophysical Research*, **112**, B06419, http://dx.doi.org/10.1029/2006JB004860

MASTIN, L. G. & POLLARD, D. D. 1988. Surface deformation and shallow dike intrusion processes at Inyo craters, Long Valley, CA. *Journal of Geophysical Research*, **93**, 13,221–13,235.

MATTHEWS, N. E., HUBER, C., PYLE, D. M. & SMITH, V. C. 2012. Timescales of magma recharge and reactivation of large silicic systems from Ti diffusion in quartz. *Journal of Petrology*, **53**, 1385–1416.

MCGARR, A. 1988. On the state of lithospheric stress in the absence of applied tectonic forces. *Journal of Geophysical Research*, **93**, 13,609–13,617.

MCGOVERN, P. J. & SOLOMON, S. C. 1993. State of stress, faulting, and eruption characteristics of large volcanoes on Mars. *Journal of Geophysical Research*, **98**, 23,553–23,579.

MCGOVERN, P. J., GROSFILS, E. B. ET AL. 2013. Lithospheric flexure: the key to the structural evolution of large volcanic edifices on the terrestrial planets. *In*: MASSIRONI, M., BYRNE, P., HIESINGER, H. & PLATZ, T. (eds) *Volcanism and Tectonism Across the Solar System*. Geological Society, London, Special Publications, **401**. First published online Month XX, 2013, http://dx.doi.org/10.1144/SP394.XX

MCLEOD, P. & TAIT, S. 1999. The growth of dykes from magma chambers. *Journal of Volcanology and Geothermal Research*, **92**, 231–245.

MCTIGUE, D. F. 1987. Elastic stress and deformation near a finite spherical magma body: resolution of a point source paradox. *Journal of Geophysical Research*, **92**, 12,931–12,940.

MERIAUX, C. & LISTER, J. R. 2002. Calculation of dike trajectories from volcanic centers. *Journal of Geophysical Research*, **107**, ETG 10-1–ETG 10-10, http://dx.doi.org/10.1029/2001JB000436

MOGI, K. 1958. Relationships between the eruptions of various volcanoes and the deformation of the ground surfaces around them. *Bulletin of the Earthquake Research Institute, University of Tokyo*, **36**, 99–134.

MORAN, S. C., NEWHALL, C. & ROMAN, D. C. 2011. Failed magmatic eruptions: Late-stage cessation of magma ascent. *Bulletin of Volcanology*, **73**, 115–122.

MULLER, O. H. & POLLARD, D. D. 1977. The stress state near Spanish Peaks, Colorado, determined from a dike pattern. *Pure and Applied Geophysics*, **115**, 69–86.

MURPHY, B. S., METCALFE, K. S. ET AL. 2012. Magma reservoir rupture beneath a Venusian edifice: When does lithospheric flexure become significant? *In*: *43rd*

Lunar and Planetary Science Conference, held 19–23 March 2012 at The Woodlands, Texas. Lunar and Planetary Institute, Houston, TX, Abstract 1060.

NEWHALL, C. G. & DZURISIN, D. 1988. *Historical Unrest at Large Calderas of the World, Volumes 1 & 2.* United States Geological Survey Bulletin, **1855**.

NEWMAN, A., DIXON, T., OFOEGBU, G. & DIXON, J. 2001. Geodetic and seismic constraints on recent activity at Long Valley Caldera, California: evidence for viscoelastic rheology. *Journal of Volcanology and Geothermal Research*, **105**, 183–206.

NEWMAN, A., DIXON, T. & GOURMELEN, N. 2006. A four-dimensional viscoelastic deformation model for Long Valley Caldera, California, between 1995 and 2000. *Journal of Volcanology and Geothermal Research*, **150**, 244–269.

NUR, A. & WALDER, J. 1990. Time-dependent hydraulics of the Earth's crust. *In: The Role of Fluids in Crustal Processes.* National Academy Press, Washington, DC, 113–127.

OWEN, S., SEGALL, P., FREYMUELLER, J., MIKIJUS, A., DENLINGER, R., ARNADOTTIR, T., SAKO, M. & BURGMANN, R. 1995. Rapid deformation of the south flank of Kilauea volcano, Hawaii. *Science*, **267**, 1328–1332.

OWEN, S., SEGALL, P., LISOWSKI, M., MIKLIUS, A., MURRAY, M., BEVIS, M. & FOSTER, J. 2000. January 30, 1997 eruptive event on Kilauea volcano, Hawaii, as monitored by continuous GPS. *Geophysical Research Letters*, **27**, 2757–2760.

PARFITT, E. A. & HEAD, J. W. 1993. Buffered and unbuffered dike emplacement on Earth and Venus: implications for magma reservoir size, depth, and rate of magma replenishment. *Earth, Moon and Planets*, **61**, 249–281.

PARFITT, E. A., WILSON, L. & HEAD, J. W. 1993. Basaltic magma reservoirs: factors controlling their rupture characteristics and evolution. *Journal of Volcanology and Geothermal Research*, **55**, 1–14.

PARKS, M. M., BIGGS, J. *ET AL.* 2012. Evolution of Santorini volcano dominated by episodic and rapid fluxes of melt from depth. *Nature Geoscience*, **5**, 749–754.

PINEL, V. & JAUPART, C. 2000. The effect of edifice load on magma ascent beneath a volcano. *Philosophical Transactions of the Royal Society of London*, **358**, 1515–1532.

PINEL, V. & JAUPART, C. 2003. Magma chamber behavior beneath a volcanic edifice. *Journal of Geophysical Research*, **108**, 2072, http://dx.doi.org/10.1029/2002JB001751

PINEL, V. & JAUPART, C. 2004. Magma storage and horizontal dyke injection beneath a volcanic edifice. *Earth and Planetary Science Letters*, **221**, 245–262.

PINEL, V. & JAUPART, C. 2005. Caldera formation by magma withdrawal from a reservoir beneath a volcanic edifice. *Earth and Planetary Science Letters*, **230**, 273–287.

POLAND, M. P., MOATS, W. P. & FINK, J. H. 2008. A model for radial dike emplacement in composite cones based on observations from Summer Coon volcano, Colorado, USA. *Bulletin of Volcanology*, **70**, 861–875.

PRITCHARD, M. E. & SIMONS, M. 2004. An InSAR-based survey of volcanic deformation in the central Andes.

Geochemistry, Geophysics, Geosystems, **5**, Q02002, http://dx.doi.org/10.1029/2003GC000610

RADEBAUGH, J., KESZTHELYI, L. P., MCEWEN, A. S., TURTLE, E. P., JAEGER, W. & MILAZZO, M. 2001. Paterae on Io: a new type of volcanic caldera? *Journal of Geophysical Research*, **106**, 33,005–33,020.

ROCHE, O., DRUITT, T. H. & MERLE, O. 2000. Experimental study of caldera formation. *Journal of Geophysical Research*, **105**, 395–416.

RONCHIN, E., MASTERLARK, T., MOLIST, J. M., SAUNDERS, S. & TAO, W. 2013. Solid modeling techniques to build 3D finite element models of volcanic systems: an example from the Rabaul caldera system, Papua New Guinea. *Computers & Geosciences*, **52**, 325–333.

RUSSO, G., GIBERTI, G. & SARTORIS, G. 1997. Numerical modeling of surface deformation and mechanical stability of Vesuvius volcano, Italy. *Journal of Geophysical Research*, **102**, 24,785–24,800.

RYMER, H. & WILLIAMS-JONES, G. 2000. Volcanic eruption prediction: magma chamber physics from gravity and deformation measurements. *Geophysical Research Letters*, **27**, 2389–2392.

SANCHEZ, L. & SHCHERBAKOV, R. 2012. Scaling properties of planetary calderas and terrestrial volcanic eruptions. *Nonlinear Processes in Geophysics*, **19**, 585–593.

SARTORIS, G., POZZI, J. P., PHILLIPE, C. & LE MOUEL, J. L. 1990. Mechanical stability of shallow magma chambers. *Journal of Geophysical Research*, **95**, 5141–5151.

SCHÖPA, A. & ANNEN, C. 2013. The effects of magma flux variations on the formation and lifetime of large silicic magma chambers. *Journal of Geophysical Research*, **118**, 926–942.

SCHULTZ, R. A. 1995. Limits on strength and deformation properties of jointed basaltic rock masses. *Rock Mechanics and Rock Engineering*, **28**, 1–15.

SEGALL, P. 2013. Volcano deformation and eruption forecasting. *In*: PYLE, D. M., MATHER, T. A. & BIGGS, J. (eds) *Remote Sensing of Volcanoes and Volcanic Processes: Integrating Observation and Modelling.* Geological Society, London, Special Publications, **380**. First published online March 20, 2013, http://dx.doi.org/10.1144/SP380.4

SHULER, A., EKSTROM, G. & NETTLES, M. 2013. Physical mechanisms for vertical-CLVD earthquakes at active volcanoes. *Journal of Geophysical Research*, **118**, 1569–1586, http://dx.doi.org/10.1002/jgrb.50131

SIGURDSSON, H., HOUGHTON, B., RYMER, H., STIX, J. & MCNUTT, S. (eds) 2000. *Encyclopedia of Volcanoes.* Academic Press, San Diego, CA.

SIMPSON, G. 2001. Influence of compression-induced fluid pressures on rock strength in the brittle crust. *Journal of Geophysical Research*, **106**, 19,465–19,478.

SIMPSON, R. W. 1997. Quantifying Anderson's fault types. *Journal of Geophysical Research*, **102**, 17,909–17,919.

SPARKS, S. R. J., SIGURDSSON, H. & WILSON, L. 1977. Magma mixing: a mechanism for triggering acid explosive eruptions. *Nature*, **267**, 315–318.

STURKELL, E., SIGMUNDSSON, F., GEIRSSON, H., ÓLAFSSON, H. & THEODÓRSSON, T. 2008. Multiple volcano deformation sources in a post-rifting period: 1989–2005 behavior of Krafla, Iceland constrained by leveling, tilt and GPS observations. *Journal of Volcanology and Geothermal Research*, **177**, 405–417.

SWARBRICK, R. E. & OSBORNE, M. J. 1998. Mechanisms that generate abnormal pressures: an overview. *In*: LAW, B. E., ULMISHEK, G. F. & SLAVIN, V. I. (eds) *Abnormal Pressures in Hydrocarbon Environments*. American Association of Petroleum Geologists, Memoirs, **70**, 13–34.

TAIT, S., JAUPART, C. & VERGNIOLLE, S. 1989. Pressure, gas content and eruption periodicity of a shallow, crystallizing magma chamber. *Earth and Planetary Science Letters*, **92**, 107–123.

TIMOSHENKO, S. & GOODIER, J. N. 1951. *Theory of Elasticity*. McGraw-Hill, New York.

TRASATTI, E., GIUNCHI, C. & BONAFEDE, M. 2003. Effects of topography and rheological layering on ground deformation in volcanic regions. *Journal of Volcanology and Geothermal Research*, **122**, 89–110.

TRASATTI, E., GIUNCHI, C. & BONAFEDE, M. 2005. Structural and rheological constraints on source depth and overpressure estimates at the Campi Flegrei caldera, Italy. *Journal of Volcanology and Geothermal Research*, **144**, 105–118.

TRASATTI, E., GIUNCHI, C. & AGOSTINETTI, N. P. 2008. Numerical inversion of deformation caused by pressure sources: application to Mount Etna (Italy). *Geophysical Journal International*, **172**, 873–884.

WATTS, A. B. 2001. *Isostasy and Flexure of the Lithosphere*. Cambridge University Press, Cambridge.

WILSON, L. & HEAD, J. W. 1994. Mars: review and analysis of volcanic eruption theory and relationships to observed landforms. *Reviews of Geophysics*, **32**, 221–263.

XU, S.-S., NIETO-SAMANIEGO, A. F. & ALANIZ-ALVAREZ, S. A. 2013. Emplacement of pyroclastic dykes in Riedel shear fraxtures: an example from Sierra de San Miguelito, central Mexico. *Journal of Volcanology and Geothermal Research*, **250**, 1–8.

YANG, X., DAVIS, P. & DIETRICH, J. 1988. Deformation from inflation of a dipping finite prolate spheroid in an elastic half-space as a model for volcanic stressing. *Journal of Geophysical Research*, **93**, 4249–4257.

ZIMBELMAN, J. R. & GREGG, T. K. P. (eds) 2000. *Environmental Effects on Volcanic Eruptions*. Kluwer Academic, New York.

ZOBACK, M. D. 2010. *Reservoir Geomechanics*. Cambridge University Press, New York.

ZUBER, M. T. & MOUGINIS-MARK, P. J. 1992. Caldera subsidence and magma chamber depth of the Olympus Mons volcano, Mars. *Journal of Geophysical Research*, **97**, 18,295–18,307.

Lateral ramps and strike-slip kinematics on Mercury

M. MASSIRONI[1,2]*, G. DI ACHILLE[3], D. A. ROTHERY[4], V. GALLUZZI[5,6],
L. GIACOMINI[1], S. FERRARI[1], M. ZUSI[3], G. CREMONESE[2] & P. PALUMBO[5]

[1]*Dipartimento di Geoscienze and CISAS, Università di Padova, Italy*

[2]*Istituto Nazionale di Astrofisica, Osservatorio Astronomico di Padova, Italy*

[3]*Istituto Nazionale di Astrofisica, Osservatorio Astronomico di Capodimonte,
Napoli, Italy*

[4]*Department of Physical Sciences, The Open University, Milton Keynes, UK*

[5]*Dipartimento di Scienze Applicate, Università di Napoli 'Parthenope', Italy*

[6]*DISTAR, Università di Napoli 'Federico II', Italy*

Corresponding author (e-mail: matteo.massironi@unipd.it)

Abstract: At a global scale, Mercury is dominated by contractional features manifested as lobate scarps, wrinkle ridges and high-relief ridges. Here, we show that some of these features are associated with strike-slip kinematic indicators, which we identified using flyby and orbital Mercury Dual Imaging System (MDIS) data and digital terrain models. We recognize oblique-shear kinematics along lobate scarps and high-relief ridges by means of (1) map geometries of fault patterns (frontal thrusts bordered by lateral ramps, strike-slip duplexes, restraining bends); (2) structural morphologies indicating lateral shearing (en echelon folding, pop-ups, pull-aparts); and (3) estimates of offsets based on displaced crater rims and differences in elevation between pop-up structures and pull-apart basins and their surroundings. Transpressional faults, documented across a wide range of latitudes, are found associated with reactivated rims of ancient buried basins and, in most cases, linked to frontal thrusts as lateral ramps hundreds of kilometres long. This latter observation suggests stable directions of tectonic transport over wide regions of Mercury's surface. In contrast, global cooling would imply an overall isotropic contraction with limited processes of lateral shearing induced by pre-existent lithospheric heterogeneities. Mantle convection therefore may have played an important role during the tectonic evolution of Mercury.

Thrust faults (on Earth and possibly other rocky planets) are often laterally confined by ramps with strikes parallel or oblique to the principal direction of tectonic transport. These lateral and oblique ramps are classically steeper than their related frontal thrusts and are associated with strike-slip kinematics. During their initial growth, all strike-slip faults typically consist of en echelon arrays of fault segments and folds (e.g. Cloos 1928; Riedel 1929; Tchalenko 1970; Wilcox *et al.* 1973). Subsequently, when displacement increases the fault segments become hard-linked, where the slip is accommodated by a double bend joining two fault segments, or soft-linked where the displacement is transferred through a rhomboidal step-over structure between two as-yet unconnected strike-slip fault segments (e.g. Wilcox *et al.* 1973; Aydin & Nur 1982) (Fig. 1a, b). Bends and step-overs insert offsets to an overall fault trace, and localize either contraction (restraining bends and step-overs) or extension (releasing bends and pull-apart basins) as a function of the relation between their geometry

(right- or left-stepping) and the fault kinematics (left- or right-lateral) (Fig. 1a, b). These features have been extensively described from field studies on Earth (e.g. 175 pull-aparts and 79 restraining bends listed in Mann 2007), analogue models (e.g. Dooley & McClay 1997; McClay & Bonora 2001; Wu *et al.* 2009; Dooley & Schreurs 2012) and numerical simulations (e.g. Segall & Pollard 1980; Du & Aydin 1995). Evidence of strike-slip faulting has also been recognized on Mars (e.g. Schultz 1989; Bistacchi *et al.* 2004; Okubo & Schultz 2006; Borraccini *et al.* 2007; Andrews-Hanna *et al.* 2008; Yin 2012), Venus (Koenig & Aydin 1998), Europa (Tufts *et al.* 1999; Hoppa *et al.* 1999; Kattenhorn 2004; Kattenhorn & Marshall 2006; Aydin 2006) and Enceladus (Patthoff & Kattenhorn 2011).

Thus far, lobate scarps and high-relief ridges (and, to some extent, wrinkle ridges) on Mercury have generally been interpreted as landforms overlying thrusts and reverse faults due to the contraction induced by secular cooling of the planet, which led to a decrease in planetary radius

From: PLATZ, T., MASSIRONI, M., BYRNE, P. K. & HIESINGER, H. (eds) 2015. *Volcanism and Tectonism
Across the Inner Solar System*. Geological Society, London, Special Publications, **401**, 269–290.
First published online June 5, 2014, http://dx.doi.org/10.1144/SP401.16
© The Geological Society of London 2015. Publishing disclaimer: www.geolsoc.org.uk/pub_ethics

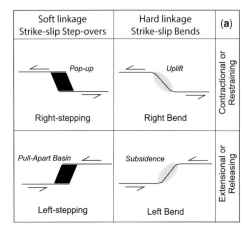

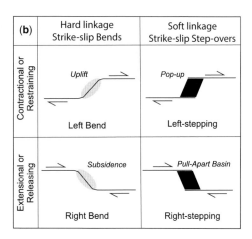

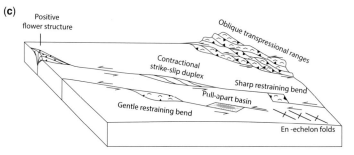

Fig. 1. Schematic sketch illustrating contractional and extensional step-overs and bends in a strike-slip fault system with (**a**) left-lateral and (**b**) right-lateral kinematics (modified after Zampieri *et al.* 2003). (**c**) Classification of structures associated with restraining and releasing step-overs and bends identified on the surface of Mercury (after Cunningham & Mann 2007).

(e.g. Strom *et al.* 1975; Solomon 1976; Watters *et al.* 1998, 2009*a*; Hauck *et al.* 2004). A notable exception to this general structural geometry has been proposed for the Beagle Rupes system, which has been interpreted as a frontal thrust with an oblique ramp on either side (Rothery & Massironi 2010).

MESSENGER (MErcury Surface, Space ENvironment, GEochemistry, and Ranging) Mercury Dual Imaging System (MDIS) data (Hawkins *et al.* 2007), acquired at high solar incidence angles (>60°) during the three Mercury flybys and from orbit, highlight geomorphological detail and hence allow better identification of landforms indicative of strike-slip motion, which are often associated with lateral or oblique ramps of frontal lobate scarps and high-relief ridges. We have integrated these data with digital terrain models (DTMs) derived from Mercury Laser Altimeter (MLA) orbital data and stereo images acquired by MDIS during the three Mercury flybys (Preusker *et al.* 2011). We find that the most prominent strike-slip features found on Mercury can be classified according to the nomenclature of Cunningham & Mann

(2007) as en echelon fold arrays, gentle and sharp restraining bends, positive flower structures, strike-slip duplexes and transpressional ranges (Fig. 1c). We have also found direct evidence of lateral displacement in the form of crater rims offset by faults. Here, we present observational evidence that supports the interpretation of several tectonic landforms on Mercury arising from transpressional tectonics (mostly along lateral ramps), and discuss the implications of these findings for the understanding of Hermean geodynamics.

Evidence of lateral ramps and strike-slip shearing on Mercury

We infer strike-slip kinematics along lobate scarps and high-relief ridges on the basis of the following different, and often concurrent, lines of evidence (Figs 1 & 2a): (1) map geometries of fault patterns suggesting an oblique or lateral displacement (i.e. frontal thrusts bordered by lateral ramps, strike-slip duplexes and restraining bends, Fig. 1);

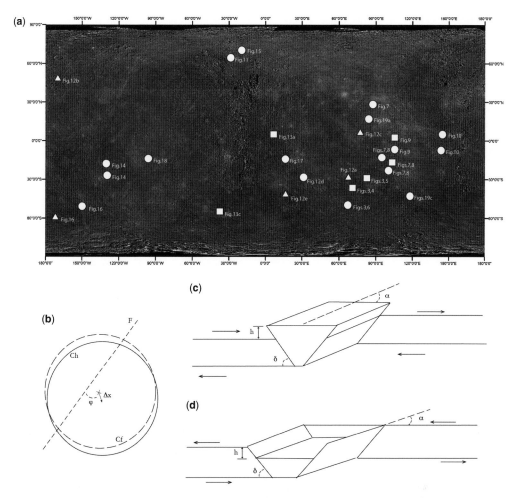

Fig. 2. (**a**) Global view in equidistant cylindrical projection (MDIS NAC_v9 orbital mosaic) with sites of structures displaying transpressional kinematics (related figures are also reported). Squares: certain oblique-slip deformation, confirmed by: (1) fault pattern geometries (i.e. frontal thrusts bordered by lateral ramps and strike-slip duplexes); (2) morphological evidence (pop-up or pull-apart structures, restraining bends, en echelon folds and/or transpressional ranges); and (3) oblique or strike-slip displacement estimates. Circles: very likely transpression, implied by (1) fault pattern geometries and (2) morphological evidence. Triangles: likely transpression, suggested by (1) fault pattern geometries. (**b**) Sketch illustrating the parameters adopted to calculate along-strike displacement d_s using crater rim offsets (see equation (1) and Table 1). F, fault trace (dashed line); Ch, crater rim at fault hanging wall (continuous circle); Cf, crater rim at fault footwall (dashed circle); Δx, horizontal displacement; φ, angle between fault strike and horizontal displacement trend. (**c**, **d**) Sketches illustrating the parameters adopted to calculate along-strike displacement d_s using pop-up and pull-apart offsets, respectively (see equation (2)). h, pop-up elevation or pull-apart depth; α, angle between the master fault strike and the pop-up or pull-apart trend; δ, dip angle of the pull-apart- or pop-up-bounding lateral faults.

(2) morphological evidence associated with strike-slip faulting (pop-up structures, restraining bends, en echelon folds and/or transpressional ranges; Fig. 1); and (3) oblique or strike-slip displacements estimated by means of offsets of crater rims (observed in six cases), the difference in elevation between a restraining bend rise and its surroundings

(found in one case) and the difference in elevation between a pull-apart floor and its surroundings (putatively identified in one case). The estimates of displacement based on crater rim offsets follow the procedure of Galluzzi *et al.* (2014). We calculate the along-strike displacement d_s of an oblique-slip movement on a fault plane from the vector connecting the centres of the two circles fitting the crater rim on the hanging wall and footwall of a fault (Δx in Fig. 2b) using the equation:

$$d_s = \Delta x \cos \varphi, \qquad (1)$$

where φ is the angle between Δx and the strike of the fault.

The elevation of a pop-up structure or the depth of a pull-apart floor with respect to the surroundings (i.e. the average elevation of the hanging wall in the case of oblique-slip thrusts) can also be used to derive the strike-slip component of displacement of the master fault. In this case,

$$d_s = \frac{2h}{\tan \delta \sin \alpha} \qquad (2)$$

where d_s is the strike-slip displacement along the master fault's strike; h is the difference in elevation between the pop-up rise or the pull-apart floor and the surroundings; α is the angle between the master fault and the lateral boundaries of the pop-up

or pull-apart structure; and δ is the dip angle of the pull-apart or pop-up bounding lateral faults (Fig. 2c, d). If an Andersonian model of faulting is assumed (Anderson 1951; Sibson 1977), the pull-apart bounding normal faults should have a dip of c. 60° whereas the pop-up bounding reverse faults most likely dip ≥60°.

Thrust systems associated with oblique or lateral ramps

The Enterprise Rupes system

The most notable evidence of lateral shearing is found along the 820 km-long Enterprise Rupes, a major tectonic landform that transects the Rembrandt basin, the second-largest preserved impact structure on Mercury (Watters *et al.* 2009*b*). Its plan-view 'bow' shape, asymmetrical relief and average difference in elevation of 2500 m between the hanging wall and the footwall are indications of a SE-verging thrust acting in response to NW–SE-oriented shortening (Fig. 3). Given this vergence and the approximate SW–NE trend of

Fig. 3. Colour-coded flyby DTM of Preusker *et al.* (2011) draped on MDIS NAC image mosaic_V8 (equidistant cylindrical projection) showing Enterprise Rupes (subdivided into Et1a and Et1b) and Belgica Rupes transecting Rembrandt Basin (a white dotted line highlights the rim). Large white arrow specifies thrust sheet vergence. White boxes show locations of Figures 4a, b, 5a, b and 6.

the scarp, its two branches (i.e. the ENE-trending western branch 'Et1a' on Figure 3 and the NE- to N-trending northern branch 'Et1b') should display opposite senses of strike-slip motion. This inference appears to be consistent with observations: Et1a shows diverse kinematic indicators that suggest dextral transpression, whereas Et1b displays several structural signs of a left-lateral component of movement. Figure 4a, b shows a strike-slip duplex (see Woodcock & Fischer 1986) consisting of two double S-shaped bends eastwards of crater A that complicate the otherwise straight Et1a, suggesting a right-lateral component of displacement (black arrow in Fig. 4a). Crater A itself has been deformed

by an oblique-slip structure (Fig. 4c) with an along-strike displacement of 2.8 km, calculated according to equation (1) (Table 1).

Two other S-shaped restraining bends are also visible (white arrows in Fig. 4a). The first (north of crater B) is barely visible but the second (SE of crater C) is very pronounced. Both structures indicate right-lateral shear along the same scarp. For the latter restraining bend, the strike-slip displacement can be calculated using the topographic section perpendicular to the bend (S–S' in Fig. 4d). This topographic profile shows that the Enterprise Rupes hanging wall is characterized by prominent relief over an elevated plateau. We interpret this relief as the expression of a pop-up structure bounded by Enterprise Rupes on one side and by a reverse fault dipping in the opposite direction on the other. Given the angle (α, c. 40°) between the orientation of this prominent relief and the main strike of Enterprise Rupes, the pop-up structure is most simply interpreted as the result of the strike-slip component along Enterprise Rupes itself. We can therefore use the elevation of the pop-up structure above the hanging-wall plateau ($h = 900$ m, Fig. 4d) to derive the strike-slip component along the Enterprise scarp d_s using equation (2), assuming a dip angle δ of c. 60° for the reverse faults enclosing the pop-up structure. This method gives a right-lateral displacement of 1.6 km. The along-strike dextral offset of crater D,

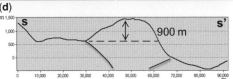

Fig. 4. (a) Et1a scarp showing a dextral strike-slip duplex east of crater A (black arrow) and two dextral sharp restraining bends (white arrows) north of crater B and SE of crater C, respectively (MDIS NAC flyby mosaic; equidistant cylindrical projection). (b) Structural interpretation of (a). Triangular teeth marks on fault traces indicate the inferred downdip direction of the fault, asymmetrical white arrows indicate strike-slip kinematics, large white arrow specifies main thrust sheet vergence, dotted line indicates S–S' geological section of (d). (c) Displaced crater A (crater '08F' in Galluzzi *et al.* 2014) (MDIS NAC_v9 orbital mosaic). Dotted line shows the strike and location of the main structure (Et1a); continuous circle fits the rim of the crater at the main structure hanging wall; dashed circle fits the rim of the crater at the main structure footwall. The horizontal displacement (Δx in Table 1) is given by the white line connecting the centre of these two circles (cross: centre of the footwall circle; point: centre of the hanging-wall circle). (d) Topographic S–S' cross-section across the sharp restraining bend SW of crater C (see (b)). The difference in elevation between the hanging-wall plateau and the sharp-bend relief feature is related to the strike-slip displacement along Et1a (see text for explanation). (e) Displaced crater D (crater '09F' in Galluzzi *et al.* 2014) (MDIS NAC_v9 orbital mosaic). Dotted line, continuous line, continuous circle and dashed circle as in (c).

Table 1. *Displaced craters*

Crater	Cross-cutting structure	Longitude (°E)	Latitude (°N)	Diameter (m)	Δx (m)	Trend (°)	Strike (°)	φ (°)	d_s (m)
A (Fig. 4a, c)	Et1a	68.1	−37.9	79 730	3070	107	263	156	2800
D (Fig. 4e)	Et1a	78.8	−35.5	59 700	2930	116	235	119	1420
E (Fig. 5a, d)	Et1b	82.9	−31.7	55 510	3960	158	216	58	2100
A (Fig. 8a, b)	Blossom thrust system	104.3	−17.0	49 690	4010	299	291	8	3970
A (Fig. 13a, b)	Unnamed scarp	6.9	5.3	64 790	1890	223	356	133	1290
Rameau (Fig. 13c, d)	Discovery Rupes	−37.1	−54.6	58 120	2480	44	204	160	2330

Δx, horizontal displacement; Trend, direction of the horizontal displacement (Δx); Strike, fault strike; φ, strike-trend angle; d_s, strike-slip component of the displacement.

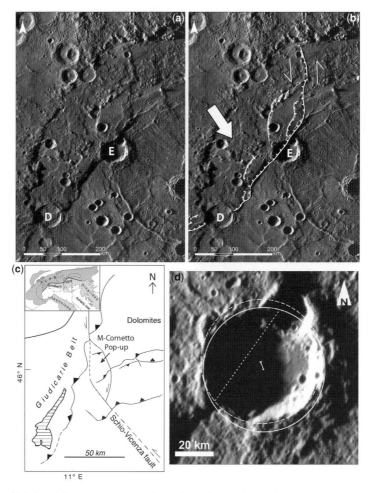

Fig. 5. (**a**) Sigmoidal-shaped flower structure associated with Et1b (MDIS NAC flyby mosaic, equidistant cylindrical projection). (**b**) Structural interpretation of (a). Asymmetrical white arrows indicate strike-slip kinematics, large white arrow specifies thrust sheet vergence. (**c**) Structural sketch of the central eastern Southern Alps (Italy) showing the M. Cornetto di Folgaria pop-up (after Zampieri *et al.* 2003). (**d**) Displaced crater E (crater '10F' in Galluzzi *et al.* 2014) (MDIS NAC flyby mosaic). Dotted line, continuous line, continuous circle and dashed circle as in Figure 4c.

located near the Et1a eastern termination, has been calculated as for crater A, returning a value of *c.* 1.4 km (Fig. 4e, Table 1). This is in agreement with the other offset estimates, considering the expected decrease of oblique slip towards the frontal part of the system.

Et1b lies within the Rembrandt basin and is characterized by a sharp thrust that grades into a set of wrinkle ridges at the interior margin of the basin. The scarp's trend changes from north–south close to its terminus to NE–SW as it begins to merge into Et1a near the Rembrandt basin rim (Fig. 5a, b). Along a thrust with a strike-slip component, such a bend would result in a local enhancement of contraction due to the coupled effect of dip-slip thrusting and lateral shearing. This is indeed suggested by the development of back-thrusts located only where the lobate scarp trend changes from north–south to NE–SW. Here the main structure has a sigmoidal-shaped positive relief (Fig. 5a, b), resembling positive flower structures and pop-ups associated with restraining bends and transpressional ranges on Earth, such as the restraining bends of the Gobi Altai intracontinental transpressional ranges (Cunningham 2007) and the Mount Cornetto di Folgaria pop-up along the Schio–Vicenza fault in the southeastern Italian Alps (Zampieri *et al.* 2003) (Fig. 5c). The flower-structure morphology is suggestive of a sinistral strike-slip component along Et1b, which is supported by the left-lateral displacement of crater E revealed in the DTM (Fig. 5d). In this case the horizontal offset is *c.* 2 km (Table 1), which is consistent with the right lateral offsets recorded along the western R1a branch.

A second scarp, parallel to Et1a and named Belgica Rupes, extends from the SW margin of the basin and trends WSW (Fig. 3) and shows a complex dextral strike-slip duplex structure (Fig. 6a, b). This structure was interpreted differently by Ruiz *et al.* (2012), who attributed modification of the otherwise straight Belgica Rupes to the effect of poorly visible extensional faulting that pre- and post-dates the inception of the scarp.

Blossom Rupes and the related thrust system

A prominent north–south-trending system, including Blossom Rupes, was observed close to the terminator between 25–32°S and 75–105°E on flyby images. Figure 7a illustrates this on the mosaic of data returned by MESSENGER's three flybys of Mercury (Becker *et al.* 2009). Care is needed to interpret these data, however, because of opposing solar illumination to either side of the seam between the flyby 1 dawn terminator and the flyby 2 sunrise terminator at *c.* 94.5°E.

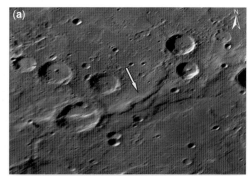

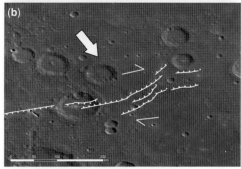

Fig. 6. (a) Part of Belgica Rupes that shows an S-shaped inflection (white arrow) where multiple thrusts suggest a focusing of contractional strain. This can be interpreted as a complex right-lateral strike-slip duplex complicating the otherwise straight Rupes (MDIS NAC flyby mosaic, equidistant cylindrical projection). (b) Structural interpretation of (a) (MDIS NAC_v8 mosaic, equidistant cylindrical projection). Asymmetrical white arrows indicate strike-slip kinematics, large white arrow specifies direction of tectonic displacement.

From this point onwards, we describe this whole system as the Blossom Rupes thrust system or belt. It consists of a 2000 km-long fold–thrust belt (cf. Byrne *et al.* 2014) bounded on either side by sharp rectilinear lobate scarps, themselves up to 150 km long and arranged in en echelon patterns (Fig. 7c, d). The array of rectilinear scarps resembles the Lebanon transpressional ranges along the Dead Sea Fault (Gomez *et al.* 2006) or the oblique deformation belts of the Eastern Tien Shan, related to the intercontinental fault systems of Central Asia (Cunningham 2007) (Fig. 7e). Unlike these examples on Earth, the structures on Mercury appear connected to arched frontal thrusts and to antiformal folds belonging to the Blossom Rupes belt (Fig. 7b, c). The en echelon arrangement of the bounding scarps suggests an opposite shear sense in the north and the south: dextral and sinistral for the northern and the southern range systems, respectively. Under this interpretation, the arrays of lobate scarps represent transpressional oblique

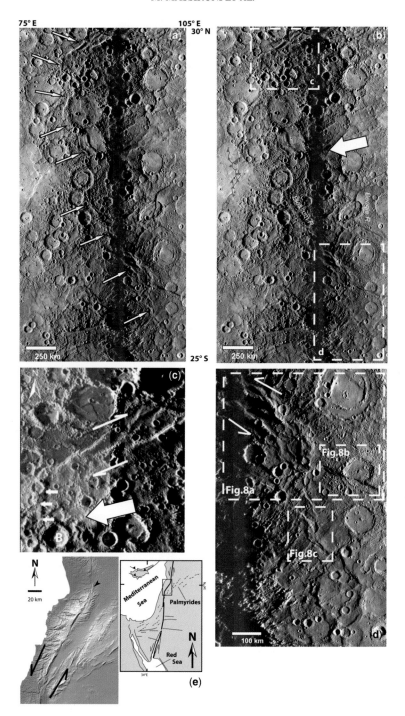

Fig. 7. (**a**) North–south orientated thrust system between 80°E and 100°E (indicated by white arrows), which includes the Blossom Rupes structure (MDIS NAC flyby mosaic, equidistant cylindrical projection). (**b**) Structural interpretation of (a). Large white arrow specifies thrust sheet vergence. White dashed boxes indicate northern and southern termination of the system ((c) and (b), respectively). (**c**) Dextral transpressional ranges (sense of movement indicated by asymmetrical white arrows) connected to the frontal thrust (small white arrows), in turn superimposed by crater B (22.1°N, 84°E). Large white arrow specifies thrust sheet vergence. Basemap is colour-coded M1 DTM of Preusker *et al.*

ramps accompanying the westwards propagation of the Blossom Rupes fold–thrust belt.

The northern system is composed of three unconnected ranges covering what is interpreted to be an ENE-striking shear zone 340 km long and 100 km wide (Fig. 7c). The WNW- to NW-trending southern system is much more complicated, with several ranges partly connected by sharp bends where thrusts and folds have developed (Figs 7d & 8a). The left-lateral kinematics along these WNW structures also seems to be recorded by a c. 4 km offset of a degraded rim of an unnamed complex crater (A in Fig. 8a, Table 1) at the very end of the en echelon ranges that limit the south of the Blossom Rupes fold–thrust belt. However, the area affected by left-lateral shear along WNW- to NW-striking structures does not seem to be limited to the abovementioned ranges but extends further to the south. Indeed, a remarkable example of a right-lateral structure is a Z-shaped high-relief feature, located at 22°S, 100°E (Fig. 8c,d). This high-relief feature, more than 100 km long and up to 22 km wide, is encompassed by what we interpret as two reverse faults defining a positive flower structure that is bounded by two WNW linear ridges with a spacing of c. 50 km. A striking similarity to classic examples of restraining structures at different scales on Earth, such as the 15 km-long Cerro de la Mica bend in Chile (McClay & Bonora 2001) (Fig. 8e) and the 80 km-long San Bernardino bend along the San Andreas fault (e.g. Jennings 1977), allows us to interpret this structural feature as a gentle double-restraining bend associated with sinistral strike-slip movement. Nevertheless, the amount of lateral slip cannot be estimated since DTMs and/or topographic profiles are not yet available at an adequate resolution for this region, and there is no unambiguous evidence for displaced features. Indeed, the NE–SW lineaments present in the area do not have an unequivocal cross-cutting relationship with the WNW transpressive structure (Fig. 8c).

The Beagle Rupes system

Beagle Rupes (Fig. 7a, b) has the same westwards vergence as the Blossom Rupes system, and consists of a frontal thrust merging into an oblique ramp on either side (Rothery & Massironi 2010) (Fig. 9a, b). The overall shortening direction is east–west as inferred from the north–south strike of the frontal thrust and the acute angle of about 60° between the

two oblique ramps; the kinematics of the oblique ramps must be transpressional, with a left-lateral component along the southern ramp and a right-lateral component along the northern ramp.

Along the northern oblique ramp is a square-shaped depression bounded by straight slopes, filled with smooth material (Fig. 9c). This feature appears to be located within a relay zone between two right-stepping transcurrent fault segments. Given its angular shape and location, we interpret it as a releasing (pull-apart) basin induced by right-lateral movement along the ramp. On the basis of a mean basin depth of 1500 m (h) derived from flyby DTM data (Preusker et al. 2011), and the dip angles of the eastern and western normal faults bounding the basin ($\delta = 60°$ if an Andersonian model of faulting is assumed), we derive a dextral strike-slip movement of c. 1.7 km (d_s) following equation (2) and assuming that the angle between the relay bend and the main fault (α) is c. 90°. This result is consistent with the 3 km of shortening observed at the Beagle Rupes frontal thrust (Rothery & Massironi 2010), given that displacement is expected to decrease with proximity to the fault tips (i.e. with increasing distance from the frontal scarp).

Lateral shearing has also been recognized transecting the Beagle Rupes frontal thrust, which consists of two arcs joined at a cusp and intersected by an en echelon fold array (Fig. 9b, d) (Rothery & Massironi 2010). The sense of stepping in the en echelon array is consistent with right-lateral movement that accounts for the more pronounced westwards propagation (nearly 13 km) of the southern thrust arc with respect to the northern portion of the frontal thrust (Fig. 9b, d).

The Paramour Rupes system

We report here a previously undescribed fold–thrust system up to 1000 km long (9°S–6°N, 140–148°E), 700 km SW of the Caloris basin rim (Fig. 10), part of which consists of Paramour Rupes. Its main topographic asymmetries and arrangement of scarps and ranges suggest an ENE direction of tectonic transport. In particular, a frontal NNW-trending portion, consisting of a series of eastwards-vergent thrusts and minor back-thrusts (white arrows in Fig. 10a, b), is bounded to the north by an irregular high-relief ridge and to the south by a series of NE ridges interpreted as a ENE-oriented transpressional structure (black arrows in

---

Fig. 7. (*Continued*) (2011), draped on MDIS NAC flyby mosaic. (**d**) Sinistral transpressional ranges (sense of movement indicated by asymmetrical arrows). White dashed boxes indicate locations of Figures 8a–c (MDIS NAC flyby mosaic, equidistant cylindrical projection). (**e**) Lebanon sinistral transpressional ranges along the Dead Sea fault (after Cowgill et al. 2004; Gomez et al. 2006).

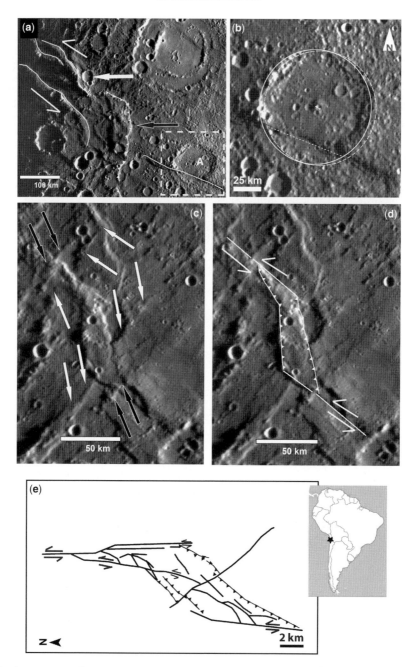

Fig. 8. Morphostructures at the southern termination of the Blossom Rupes thrust system (MDIS NAC flyby mosaic, equidistant cylindrical projection). (**a**) Sinistral transpressional ranges (white arrow) and major left-lateral restraining bend (black arrow). Asymmetrical white arrows indicate strike-slip kinematics. White dashed box indicates location of (b). (**b**) Displaced crater A in (a) (MDIS NAC flyby mosaic). Dotted line, continuous line, continuous circle and dashed circle as in Figure 4c. (**c**) Gentle double restraining bend indicating left-lateral kinematics (sense of movement indicated by asymmetrical white arrows) (MDIS NAC flyby mosaic, equidistant cylindrical projection). The cross-cutting relationship with NE–SW lineaments is equivocal since some transect the restraining bend (black arrows) and others are cut by the strike-slip feature (white arrows). (**d**) Structural interpretation of the double restraining bend of (c). (**e**) Cerro de la Mica double restraining bend (Chile) (after McClay & Bonora 2001). Note the similarity with the structure shown in (c) and (d).

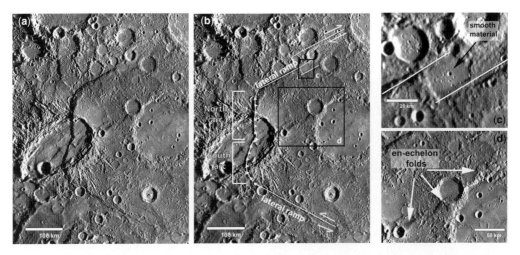

Fig. 9. (a) The Beagle Rupes fault system (MDIS NAC flyby mosaic, equidistant cylindrical projection). (b) Structural interpretation of (a). The Beagle Rupes system consists of a frontal scarp composed of two arcs, the southern arc extending further forward than the northern arc. The frontal thrust merges into an oblique lateral ramp on either side, characterized by transpressional deformation (sense of movement indicated by asymmetrical white arrows). Black boxes indicate a small pull-apart basin, filled with smooth material (c) and an en echelon fold array indicating right-lateral kinematics (d).

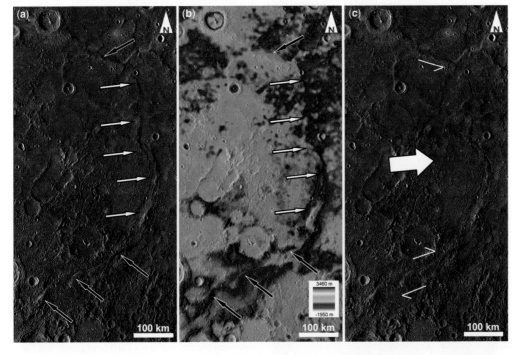

Fig. 10. Paramour Rupes thrust system. (a) MDIS NAC_v9 orbital mosaic (equidistant cylindrical projection) and (b) colour-coded flyby DTM (Preusker *et al.* 2011). The difference in elevation between the thrust sheet hanging wall and its footwall is 1150 m. White arrows indicate the frontal range composed of a thrust–back-thrust system; black arrows show the two oblique lateral ramps. (c) Structural interpretation of (a). Asymmetrical white arrows indicate strike-slip kinematics; large white arrow specifies thrust sheet vergence.

Fig. 10a, b). The components of the southern assemblage appear to define compressional splays along an overall right-lateral system (Fig. 10c). These

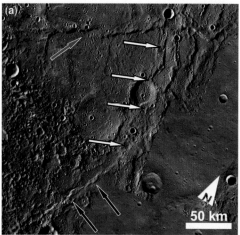

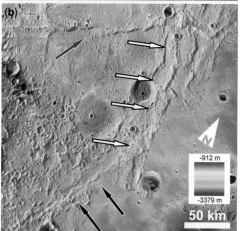

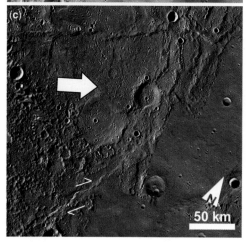

splays are consistent with the main inferred vergence of the frontal sector and imply a complementary sinistral component of movement along the northern boundary. The entire system can therefore be interpreted as a fold–thrust belt limited on both sides by oblique transpressional ramps that separate the hanging wall (at an average elevation of 150 m) from the footwall (at −1000 m elevation) (Fig. 10b).

La Dauphine Rupes system

A major fold–thrust belt system centred at 66°N, 27°W, of which La Dauphine Rupes is part, shows unequivocal evidence of eastwards tectonic vergence. Here, continuous lobate scarps with eastwards-facing steep slopes merge towards the south into a more than 100 km-long system composed of a series of NE–SW-trending sigmoidal-shaped features of positive relief arranged in en echelon pattern (Fig. 11a, b). The geometry and distribution of these topographic highs, and their relationships with the north–south-trending lobate scarps, support an interpretation of tectonic transport parallel with or slightly oblique to the system of en echelon positive reliefs, and so defining a dextral lateral ramp (Fig. 11c). In the north, the frontal system is complicated by a prominent east–west-oriented and south-verging thrust system (Fig. 11a–c). The interplay between these two systems makes identifying the location of the left-lateral ramp uncertain, although for kinematic reasons it is expected to be located to the north.

Minor thrust systems associated with oblique or lateral slip

Numerous abrupt changes in the trends of lobate scarps with an obvious vergence suggest that oblique slip on lateral or oblique ramps is common on Mercury. In particular, we note the following examples: the 390 km-long and eastwards-verging lobate scarp located between 24°S, 68°E and 31°S,

Fig. 11. La Dauphine Rupes thrust system. (**a**) MDIS NAC_v9 orbital mosaic (north polar stereographic projection) and (**b**) colour-coded Mercury Laser Altimeter (MLA) DTM (north polar stereographic projection). White arrows indicate the frontal range comprising a thrust–back-thrust system; black arrows show the southern oblique ramps with en echelon sigmoidal features indicating a left-lateral component of displacement; the red arrow indicates the east–west-orientated intersecting thrust system. (**c**) Structural interpretation of (**a**). Asymmetrical white arrows indicate strike-slip kinematics; large white arrow specifies thrust sheet vergence.

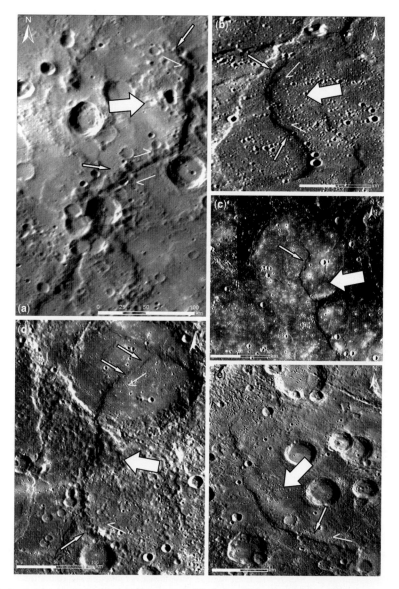

Fig. 12. Lobate scarp systems associated with segments with presumably strike-slip kinematics. White arrows show segments likely displaying a lateral component of movement; asymmetrical white arrows indicate strike-slip kinematics; large white arrow specifies thrust sheet vergence. (**a**) Lobate scarp located between 24°S, 68°E and 31°S, 62°E (MDIS NAC flyby mosaic, equidistant cylindrical projection). (**b**) West-verging lobate scarp changing its trend from NE–SW to NW–SE when it transects a 250 km-diameter double-ring basin at 49.3°N, 170°W (MDIS NAC_v9 orbital mosaic, equidistant cylindrical projection). (**c**) Highly articulated westwards-verging thrust system transecting Faulkner basin (8°N, 77°E) (MDIS NAC_v9 orbital mosaic, equidistant cylindrical projection). (**d**) Lobate scarp whose northern oblique ramp comprises en echelon ridges that cut a 160 km-diameter basin centred at 27.5°S, 31°E (MDIS NAC_v9 orbital mosaic, equidistant cylindrical projection). (**e**) SW-verging scarp SE of Debussy crater (39.9°S, 12.6°E; the rim is visible in the NW corner). The structure is associated with a wide frontal antiform and a WNW–ESE-trending straight lateral ramp (MDIS NAC_v9 orbital mosaic, equidistant cylindrical projection).

62°E whose similarities with Beagle Rupes were already highlighted by Watters *et al.* (2010) (Fig. 12a); the 375 km-long and westwards-verging lobate scarp that changes its trend from NE–SW to NW–SE at 49.3°N, 170°W, where it transects a 250 km-diameter double-ring basin (Fig. 12b); the

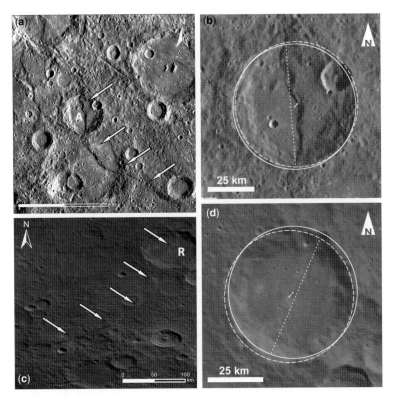

Fig. 13. (**a**) Lobate scarp (indicated by white arrows) transecting a 65-km-diameter crater at 5.3°N, 6.9°E (A) (MDIS NAC_v9 orbital mosaic, equidistant cylindrical projection). (**b**) Displaced crater A in (a). Dotted line, continuous line, continuous circle and dashed circle as in Figure 4c. (**c**) Discovery Rupes scarp (indicated by white arrows) transecting Rameau crater (R) (MDIS NAC_v9 orbital mosaic, equidistant cylindrical projection). (**d**) Displaced Rameau crater in (c). Dotted line, continuous line, continuous circle and dashed circle as in Figure 4c.

north–south-trending and westwards-verging thrust system transecting the Faulkner basin (8°N, 77°E) (Fig. 12c); the westwards-verging and 270 km-long thrust sheet whose northern oblique ramp cuts a 160 km-diameter basin centred at 27.5°S, 31°E (Fig. 12d); and the 327 km-long scarp SE of Debussy crater (centred at 39.9°S, 12.6°E) that shows a SW-vergent thrust associated with a wide frontal antiform and a WNW–ESE-oriented straight lateral ramp (Fig. 12e). Although the majority of these examples do not show measurable offsets of crater rims, a quantitative measurement has been calculated along an arc-shaped and SW-vergent lobate scarp extending for 290 km and transecting a 65-km-diameter crater at 5.3°N, 6.9°E (crater '06D' in Galluzzi *et al.* 2014) (Fig. 13a). In particular, the NW ramp of the structure, associated with an oblique slip, has a dextral strike-slip component of *c.* 1.3 km (Fig. 13b, Table 1).

All the abovementioned structures with a strike-slip component should be interpreted as lateral or oblique ramps of specific frontal thrusts. However,

we have found an unexpected oblique-slip structure along the well-known Discovery Rupes (Watters *et al.* 1998, 2001) (Fig. 13c). This structure, not yet acquired by MDIS at a solar incidence angle favourable for morphological analysis, does not display any apparent strike-slip induced topography; however, the 58 km-diameter Rameau crater appears to be displaced either by orthogonal compression or by right-lateral deformation. The strike-slip offset is *c.* 2.3 km (Fig. 13d, Table 1).

En echelon folds and restraining bends on reactivated basin walls

Several authors have recently drawn attention to lobate scarps that nucleated at the margins of ancient buried basins of diameters generally larger than 100 km (Fassett *et al.* 2012; Klimczak *et al.* 2012; Watters *et al.* 2012; Rothery & Massironi 2013; Byrne *et al.* 2014). Except for a few cases in which lobate scarps, following the basin rim, extend beyond the basin itself, most of them are limited to their margins. Rothery & Massironi (2013) highlighted that for many of these basins the deformation seems limited to portions of the interface between basin-fill and basin-rim, suggesting a preferential orientation of the main horizontal stress σ_h. We think it is unlikely that this observation is affected by illumination biases because these are large structures that are comparatively easy to see. Morphological features consisting of en echelon ridges (folds) and restraining step-overs suggest strike-slip activity at the boundary of some of the reactivated rims and support the hypothesis of a preferred orientation for σ_h. One notable feature is the Duyfken Rupes scarp along the rim of the 630 km-diameter Beethoven basin, whose western margin has been thrust westwards (Fig. 14a, b). We interpret structures on the NW margin of the basin as a series of en echelon folds, suggesting a right-lateral component of slip (Fig. 14c). The SW margin is characterized by segmented faults linked through restraining right step-overs, indicative of sinistral strike-slip kinematics (Fig. 14d).

Similar examples of this style of deformation include: a 170 km-diameter basin centred at 69.5°N, 16.5°W that shows a reactivated western margin bounded on its NW side by stepped wrinkle ridges, suggestive of a right-lateral component of slip (Fig. 15a, b); and the 690 km-diameter buried basin centred at 49.0°S, 161.5°W (Fig. 16a, b) whose SE reactivated margin merges into sharp, left-lateral restraining bends on its eastern side and into Hero Rupes to the west (the latter not displaying clear strike-slip morphologies but probably representing a right-lateral oblique ramp). Other (though less pronounced) evidence of strike-slip kinematics occurs close to the southern edge of the basin listed by Fassett *et al.* (2012) as 'b36' (730 km in diameter, 7.6°S, 21.6°E), where a sharp left-lateral step-over is visible along a prominent WNW-trending lobate scarp (Fig. 17).

Evidence of strike-slip kinematics can persist even in the case of basins whose margins have nucleated faulting all the way around their circumference, as demonstrated by the 320 km-diameter basin centred at 17.5°S, 96.6°W and listed as 'b6'

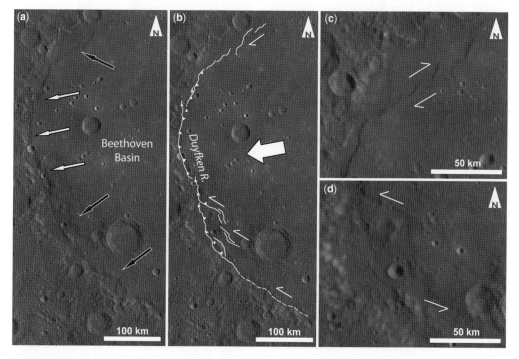

Fig. 14. (**a**) Fault nucleation around western rim of the Beethoven Basin (MDIS NAC_v9 orbital mosaic, equidistant cylindrical projection). White arrows indicate the frontal thrust (Duyfken Rupes) and black arrows the sector of reactivated rims with likely transpressional kinematics. (**b**) Structural interpretation of (a). Asymmetrical white arrows indicate strike-slip kinematics; large white arrow specifies the direction of tectonic propagation. (**c**) Series of en echelon folds at the northern termination of the Beethoven Basin reactivated rim. Asymmetrical white arrows indicate right-lateral kinematics. (**d**) Segmented faults linked through restraining right step-overs at the southern margin of the Beethoven Basin's reactivated rim. Asymmetrical white arrows indicate left-lateral strike-slip kinematics.

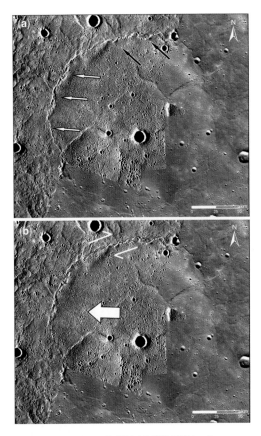

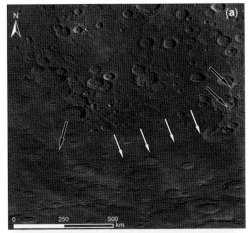

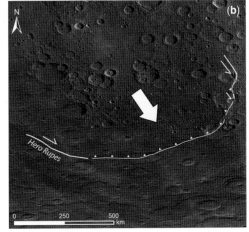

Fig. 15. (a) Basin at 69.5°N, 16.5°W (170 km in diameter) (MDIS NAC_v9 orbital mosaic, north polar stereographic projection); its western rim (white arrows) is bounded at the northern side by stepped wrinkle ridges (black arrow), suggesting right-lateral kinematics. (b) Structural interpretation of (a). Asymmetrical white arrows indicate inferred strike-slip kinematics; large white arrow specifies the direction of tectonic propagation.

Fig. 16. (a) Fault nucleation around southeastern rim of the 690 km-diameter buried basin at 49.0°S, 161.5°W (MDIS NAC_v9 orbital mosaic, equidistant cylindrical projection). The frontal thrust (white arrows) is bounded to the west by Hero Rupes (single black arrow), presumably associated with a right-lateral component of slip, and to the east by sharp, left-lateral restraining bends (black arrows). (b) Structural interpretation of (a). Asymmetrical white arrows indicate strike-slip kinematics; large white arrow specifies the direction of tectonic propagation.

by Fassett *et al.* (2012). This basin is almost entirely surrounded by outward-facing scarps, but its northern margin is bounded by what we interpret to be an ENE-trending array of folds arranged in an en echelon pattern and extending far beyond the basin itself (Fig. 18a, b). This shear structure, several hundred kilometres in length, is probably associated with dextral strike-slip kinematics and clearly merges into the western reactivated margin of basin 'b6'; it is therefore best interpreted as representing a main frontal thrust, in which case the eastern rim is a back-thrust.

Some transcurrent kinematics are also very likely to occur on either side of the prominent bulges that develop along thrusts whenever they encounter and deform rims of filled basins. For

example, this should be the case for: the Blossom Rupes thrust system where it follows the western rim of the 300 km-diameter basin centred at 15.7°N, 86.5°E (see also the 'c2' basin in Rothery & Massironi 2013) (Fig. 19a, b); and equally for the case of a NW–SE-trending and SW-verging lobate scarp that deforms the western margin of a 290 km-diameter basin centred at 43°S, 121°E (the 'c3' basin in Rothery & Massironi 2013) (Fig. 19c, d).

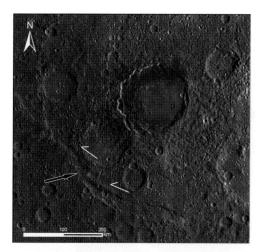

Fig. 17. Basin listed by Fassett *et al.* (2012) as 'b36' (7.6°S, 21.6°E) (MDIS NAC_v9 orbital mosaic, equidistant cylindrical projection). A sharp, left-lateral step-over is visible along a prominent WNW-trending lobate scarp (black arrow). Asymmetrical white arrows indicate inferred strike-slip kinematics

Implications for global tectonics

Three models have been invoked to explain the structural features on the surface of Mercury: global cooling (e.g. Strom *et al.* 1975; Watters *et al.* 1998); tidal despinning (e.g. Melosh 1977;

Melosh & McKinnon 1988); and mantle convection (King 2008) (Fig. 20). Although these mechanisms are not mutually exclusive (Dombard & Hauck 2008), the ubiquitous presence of shortening structures and the lack (except inside some volcanically infilled basins) of extensional features of regional significance suggests a global predominance of compressional stresses across the surface of Mercury (Watters *et al.* 2009a). According to the global cooling model (Strom *et al.* 1975, Watters *et al.* 1998), lobate scarp orientation and distribution should be isotropically arranged, and pure strike-slip kinematics would be unlikely except where induced locally by pre-existing crustal anisotropies. In particular, the constrictional prolate strain ellipsoid at Mercury's surface induced by global cooling should prevent any simple shear strain except where permitted by crustal heterogeneities. In contrast, both the other models require some lateral shearing on larger scales. Tectonic features formed in response to tidal despinning should display latitude-dependent kinematics symmetrically distributed either side of the equator and should be characterized by east–west contraction in equatorial regions, strike-slip kinematics at mid-latitudes and extension in polar regions (Melosh & McKinnon 1988). However, tidal despinning could have operated together with global contraction in an early stage of planetary evolution (up to the Late Heavy Bombardment; Dombard & Hauck 2008; Watters & Nimmo 2010), whereas global contraction should have persisted until more recent

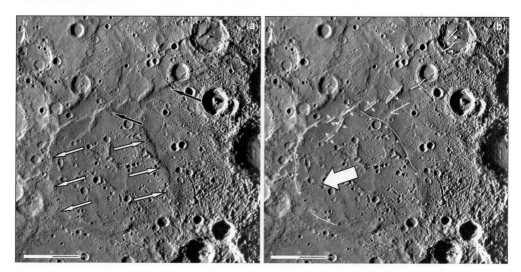

Fig. 18. (**a**) Basin listed by Fassett *et al.* (2012) as 'b6' (17.5°S, 96.6°W) (MDIS NAC flyby mosaic, equidistant cylindrical projection). The basin, almost entirely surrounded by outwards-facing scarps (white arrows), is bounded to the north by an ENE-trending fold array arranged in an en echelon pattern (black arrows). (**b**) Structural interpretation of (a). Asymmetrical white arrows indicate strike-slip kinematics; large white arrow specifies the direction of tectonic propagation.

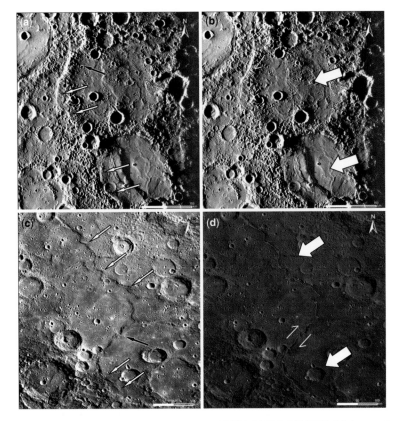

Fig. 19. (**a**) Basin listed by Rothery & Massironi (2013) as 'c2' (15.7°N, 86.5°E) (MDIS NAC flyby mosaic, equidistant cylindrical projection). The deflection (black arrow) of the main lobate scarp (white arrow) induced by the basin locally gives rise to strike-slip kinematics. (**b**) Structural interpretation of (a). Asymmetrical white arrows indicate strike-slip kinematics; large white arrows specify the direction of tectonic propagation. (**c**) Basin listed by Rothery & Massironi (2013) as 'c3' (43°S, 120.87°E) (MDIS NAC_v9 orbital mosaic, equidistant cylindrical projection). As in (a), the deflection (black arrow) of the main lobate scarp (white arrows) induced by the basin locally results in strike-slip kinematics. (**d**) Structural interpretation of (c). Asymmetrical white arrows indicate strike-slip kinematics; large white arrows specify the direction of tectonic propagation.

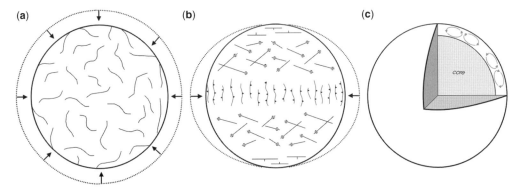

Fig. 20. Global tectonic models proposed for Mercury: (**a**) global cooling (e.g. Strom *et al.* 1975; Watters *et al.* 1998); (**b**) tidal despinning (e.g. Melosh 1977; Melosh & McKinnon 1988); and (**c**) mantle convection (King 2008).

times (Grott *et al.* 2011). Therefore, strike-slip faulting induced by tidal despinning would probably have been overprinted by the other two processes listed above, such that it remains visible (if at all) only as a remnant of an early tectonic phase.

On the other hand, crustal tectonics resulting from mantle convection should show heterogeneously distributed thrust systems with various directions of vergence and a spacing dependent on the number of convective cells, which is in turn related to mantle thickness. Our observations show contractional oblique-slip features at many latitudes (Fig. 2), which are almost always associated with lateral or oblique ramps linked to frontal thrusts, indicating various directions of tectonic transport.

Moreover, global contraction and cooling of a pre-faulted and heterogeneous lithosphere (conditions that are very likely in view of factors such as regional variations in crustal composition and structure, and infill of impact basins) might favour localized strike-slip shearing as a result of local perturbations of the regional field stress caused by pre-existing planes of weakness (e.g. Zhang *et al.* 1994; Bell 1996; Yale 2003; Zang & Stephansson 2010). Even in pure compressional environments, step-over regions between two closely spaced overlapping dip-slip faults are preferred sites for nucleation of transfer faults dominated by strike-slip kinematics, whereas tear faults usually separate compartments of different shortening within thrust systems (e.g. Boyer & Elliot 1982; Morley 1986). All these processes might account for local strike-slip kinematics along basin margins or between closely spaced lobate scarps, but not for structures hundreds of kilometres long that are associated with oblique slip and are consistently arranged as lateral or oblique ramps of major fold–thrust systems (e.g. Enterprise, Beagle, Blossom, Paramour and La Dauphine Rupes). These observations favour mantle convection (over tidal despinning) as a major dynamic contributor in shaping the surface of Mercury and are in agreement with some numerical simulations that, even when taking into account the updated estimates of the core radius (2074 km of 2440 km total planet radius, Smith *et al.* 2012), predict mantle convection during a considerable part of Mercury evolution if a mantle thickness over 300 km is considered (Michel *et al.* 2013).

The absence of transtensional features (with the exception of the northern Beagle Rupes ramp) such as releasing step-overs and pull-apart basins can be explained by the prevalence of global contraction. This process imposes a compressional stress state throughout Mercury's lithosphere, suppressing even local extension that would have been induced by any strike-slip kinematics along irregular and wavy scarps. Even the lobate scarps

and high-relief ridges with the most straightforward indicators of lateral shearing are associated with a substantial contractional component and may therefore display transpressional behaviour.

By suggesting that the strike-slip components of Mercury's compressive structures are more important than previously thought, our findings might impact future estimates of crustal shortening and planetary contraction. In fact, previous studies of Mercury's radius decrease that were based on the visible compressive structures assumed that they acted as pure dip-slip and low-angle faults (e.g. Watters *et al.* 1998; Watters & Nimmo 2010; Di Achille *et al.* 2012). This would tend to result in an overestimation of their shortening since at least a portion of the considered structures most likely had a significant strike-slip component and hence, according to an Andersonian mode of faulting, should be associated with higher fault dip angles (see also Galluzzi *et al.* 2014).

Conclusions

Numerous strike-slip indicators (en echelon fold arrays, restraining bends, positive flower structures, strike-slip duplexes and displaced crater rims) have been found associated with contractional features such as lobate scarps and high-relief ridges on Mercury. The structures with a lateral shear component are found in a wide range of latitudes and are often linked to frontal thrusts as lateral or oblique ramps; in some examples, they are associated with reactivated margins of ancient buried basins. All these examples can be interpreted as possessing strike-slip to transpressional motion along faults underlying some lobate scarps and, in some cases, indicate stable directions of tectonic transport over wide regions of Mercury's surface. This evidence is not consistent with global cooling alone, because that would imply an overall isotropic contraction with only limited processes of lateral shearing induced by local stress perturbations along pre-existing lithospheric heterogeneities or within narrow step-overs between closely spaced thrusts. It is possible, however, that mantle convection could have played an important role during the tectonic evolution of Mercury, most likely overlapping with the effects of global cooling and, secondarily, despinning to form the observed distribution of lobate scarps and high-relief ridges on the planet.

We thank C. Klimczak and N. Woodcock for their fruitful revisions, P. K. Byrne for a critical review of an early version of the paper and R. Pozzobon who provided help in data mosaicking during a preliminary stage of this work. This research was supported by the Italian Space Agency (ASI) within the SIMBIOSYS project (ASI-INAF agreement n. I/022/10/0). MM, SF and LG received financial support from Progetto di Ricerca di Ateneo CPDA112213/11, Università di Padova. DAR acknowledges support from STFC PP/E/E002412/1.

References

ANDERSON, E. M. 1951. *The Dynamics of Faulting.* Oliver and Boyd, Edinburgh, UK.

ANDREWS-HANNA, J. C., ZUBER, M. T. & HAUCK, S. A., II 2008. Strike-slip faults on Mars: Observations and implications for global tectonics and geodynamics. *Journal of Geophysical Research,* **113,** E08002, http://dx.doi.org/10.1029/2007JE002980

AYDIN, A. 2006. Failure modes of the lineaments on Jupiter's moon, Europa: implications for the evolution of its icy crust. *Journal of Structural Geology,* **28,** 2222–2236.

AYDIN, A. & NUR, A. 1982. Evolution of pull-apart basins and their scale independence. *Tectonics,* **1,** 91–105.

BECKER, K. J., ROBINSON, M. S. *ET AL.* 2009. Near Global Mosaic of Mercury. *Eos Transactions AGU,* **90** (Fall Meeting Suppl.), Abstract P21A–1189.

BELL, J. S. 1996. In situ stresses in sedimentary rocks (part 2): applications of stress measurements. *Geoscience Canada,* **23,** 135–153.

BISTACCHI, N., MASSIRONI, M. & BAGGIO, P. 2004. Large-scale fault kinematic analysis in Noctis Labyrinthus (Mars). *Planetary and Space Science,* **52,** 215–222.

BORRACCINI, F., DI ACHILLE, G., ORI, G. G. & WEZEL, F. C. 2007. Tectonic evolution of the eastern margin of the Thaumasia Plateau (Mars) as inferred from detailed structural mapping and analysis. *Journal of Geophysical Research,* **112,** E05005, http://dx.doi.org/10.1029/2006JE002866

BOYER, S. E. & ELLIOT, D. 1982. Thrust systems. *American Association of Petroleum Geologists Bulletin,* **66,** 1196–1230.

BYRNE, P. K., CELÂL ŞENGÖR, A. M., KLIMCZAK, C., SOLOMON, S. C., WATTERS, T. R. & HAUCK II, S. A. 2014. Mercury's global contraction much greater than earlier estimates. *Nature Geoscience,* **7,** 301–307, http://dx.doi.org/10.1038/ngeo2097

CLOOS, H. 1928. Experimente zur inneren Tektonic. *Centralblatt fur Mineralogie,* **1928B,** 609–621.

COWGILL, E., YIN, A., ARROWSMITH, R., FENG, W. & SHUANHONG, Z. 2004. The Akato Tagh bend along the Altyn Tagh fault, northwest Tibet, 1: smoothing by vertical-axis rotation and the effect of topographic stresses on bend-flanking faults. *Geological Society of America Bulletin,* **116,** 1423–1442.

CUNNINGHAM, W. D. 2007. Structural and topographic characteristics of restraining bend mountain ranges of the Altai, Gobai Altai and Easternmost Tien Shan. *In:* CUNNINGHAM, W. D. & MANN, P. (eds) *Tectonics of Strike-Slip Restraining and Releasing Bends.* Geological Society, London, Special Publications, **290,** 219–237.

CUNNINGHAM, W. D. & MANN, P. 2007. Tectonics of strike-slip restraining and releasing bends. *In:* CUNNINGHAM, W. D. & MANN, P. (eds) *Tectonics of Strike-Slip Restraining and Releasing Bends.* Geological Society, London, Special Publications, **290,** 1–12.

DI ACHILLE, G., POPA, C., MASSIRONI, M., MAZZOTTA EPIFANI, E., ZUSI, M., CREMONESE, G. & PALUMBO, P. 2012. Mercury's radius change estimates revisited using MESSENGER data. *Icarus,* **221,** 456–460.

DOMBARD, A. J. & HAUCK, S. A. 2008. Despinning plus global contraction and the orientation of lobate scarps on Mercury: predictions for MESSENGER. *Icarus,* **198,** 274–276.

DOOLEY, T. & MCCLAY, K. 1997. Analog modeling of pull-apart basins. *American Association of Petroleum Geologists Bulletin,* **81,** 1804–1826.

DOOLEY, T. P. & SCHREURS, G. 2012. Analogue modelling of intraplate strike-slip tectonics: A review and new experimental results. *Tectonophysics,* **574–575,** 1–71.

DU, Y. & AYDIN, A. 1995. Shear fracture patterns and connectivity at geometric complexities along strike slip faults. *Journal of Geophysical Research,* **100,** 18093–18102.

FASSETT, C. I., HEAD, J. W. *ET AL.* 2012. Large impact basins on Mercury: global distribution, characteristics, and modification history from MESSENGER orbital data. *Journal of Geophysical Research,* **117,** E00L08, http://dx.doi.org/10.1029/2012JE004154

GALLUZZI, V., DI ACHILLE, G., FERRANTI, L., POPA, C. & PALUMBO, P. 2014. Faulted craters as indicators for thrust motions on Mercury. *In:* PLATZ, T., MASSIRONI, M., BYRNE, P. K. & HIESINGER, H. (eds) *Volcanism and Tectonism Across the Inner Solar System.* Geological Society, London, Special Publications, **401.** First published online June 12, 2014, http://dx.doi.org/10.1144/SP401.17

GOMEZ, F., KHAWLIE, M., TABET, C., DARKAL, A. N., KHAIR, K. & BARAZANGI, M. 2006. Late Cenozoic uplift along the northern Dead Sea transform in Lebanon and Syria. *Earth and Planetary Science Letters,* **241,** 913–931.

GROTT, M., BREUER, D. & LANEUVILLE, M. 2011. Thermo-chemical evolution and global contraction of mercury. *Earth and Planetary Science Letters,* **307,** 135–146.

HAUCK, S. A., II, DOMBARD, A. J., PHILLIPS, R. J. & SOLOMON, S. C. 2004. Internal and tectonic evolution of Mercury. *Earth and Planetary Science Letters,* **222,** 713–728.

HAWKINS, S. E., BOLDT, J. D. *ET AL.* 2007. The Mercury dual imaging system on the MESSENGER spacecraft. *Space Science Reviews,* **131,** 247–338.

HOPPA, G., TUFTS, B. R., GREENBERG, R. & GEISSLER, P. 1999. Strike-slip faults on Europa: global shear patterns driven by tidal stress. *Icarus,* **141,** 287–298.

JENNINGS, C. W. 1977. *Geologic Map of California.* Geologic Data Map 2, scale 1:750,000. California Division of Mines and Geology, http://www.consrv.ca.gov/cgs/cgs_history/Pages/2010_geologicmap.aspx

KATTENHORN, S. A. 2004. Strike-slip fault evolution on Europa: evidence from tailcrack geometries. *Icarus,* **172,** 582–602.

KATTENHORN, S. A. & MARSHALL, S. T. 2006. Fault-induced perturbed stress fields and associated tensile and compressive deformation at fault tips in the ice shell of Europa: implications for fault mechanics. *Journal of Structural Geology,* **28,** 2204–2221.

KING, S. D. 2008. Pattern of lobate scarps on Mercury's surface reproduced by a model of mantle convection. *Nature Geoscience*, **1**, 229–232.

KLIMCZAK, C., WATTERS, T. R. ET AL. 2012. Deformation associated with ghost craters and basins in volcanic smooth plains on Mercury: strain analysis and implications for plains evolution. *Journal of Geophysical Research*, **117**, E00L03, http://dx.doi.org/10.1029/2012JE004100

KOENIG, E. & AYDIN, A. 1998. Evidence for large-scale strike-slip faulting on Venus. *Geology*, **26**, 551–554.

MANN, P. 2007. Global catalogue, classification and tectonic origins of restraining- and releasing bends on active and ancient strike-slip fault systems. *In*: CUNNINGHAM, W. D. & MANN, P. (eds) *Tectonics of Strike-Slip Restraining and Releasing Bends*. Geological Society, London, Special Publications, **290**, 13–142.

MCCLAY, K. & BONORA, M. 2001. Analog models of restraining stepovers in strike-slip fault systems. *American Association of Petroleum Geologists Bulletin*, **85**, 233–260.

MELOSH, H. J. 1977. Global tectonics of a despun planet. *Icarus*, **31**, 221–243.

MELOSH, H. J. & McKINNON, B. 1988. The tectonics of Mercury. *In*: MATTHEWS, M. S., CHAPMAN, C. & VILAS, F. (eds) *Mercury*. University of Arizona Press, Tucson, AZ, 401–428.

MICHEL, N. C., HAUCK, S. A., SOLOMON, S. C., PHILLIPS, R. J., ROBERTS, J. H. & ZUBER, M. T. 2013. Thermal evolution of Mercury as constrained by MESSANGER observations. *Journal of Geophysical Research*, **118**, 1033–1044.

MORLEY, C. K. 1986. A classification of thrust fronts. *American Association of Petroleum Geologists Bulletin*, **70**, 12–25.

OKUBO, C. H. & SCHULTZ, R. A. 2006. Variability in Early Amazonian Tharsis stress state based on wrinkle ridges and strike-slip faulting. *Journal of Structural Geology*, **28**, 2169–2181.

PATTHOFF, D. A. & KATTENHORN, S. A. 2011. A fracture history on Enceladus provides evidence for a global ocean. *Geophysical Research Letters*, **38**, http://dx.doi.org/10.1029/2011GL048387

PREUSKER, F., OBERST, J., HEAD, J. W., WATTERS, T. R., ROBINSON, M. S., ZUBER, M. T. & SOLOMON, S. C. 2011. Stereo topographic models of Mercury after three MESSENGER flybys. *Planetary and Space Science*, **59**, 1910–1917.

RIEDEL, W. 1929. Zur Mechanik geologischer Brucherscheinungen. *Zentralblatt für Mineralogie, Geologie und Paleontologie*, **1929B**, 354–368.

ROTHERY, D. A. & MASSIRONI, M. 2010. Beagle Rupes – evidence for a basal decollement of regional extent in Mercury's lithosphere. *Icarus*, **209**, 256–261.

ROTHERY, D. A. & MASSIRONI, M. 2013. A spectrum of tectonised basin edges on Mercury. Paper presented at the 44th Lunar and Planetary Science Conference, 18–22 March, 2013, http://www.lpi.usra.edu/meetings/lpsc2013/pdf/1175.pdf

RUIZ, J., LÓPEZ, V., DOHM, J. M. & FERNÁNDEZ, C. 2012. Structural control of scarps in the Rembrandt region of Mercury. *Icarus*, **219**, 511–514.

SCHULTZ, R. A. 1989. Strike-slip faulting of ridged plains near Valles Marineris, Mars. *Nature*, **341**, 424–426.

SEGALL, P. & POLLARD, D. 1980. Mechanics of discontinuous faults. *Journal of Geophysical Research*, **85**, 4337–4350.

SIBSON, R. H. 1977. Fault rocks and fault mechanisms. *Journal of the Geological Society, London*, **133**, 191–213.

SMITH, D. E., ZUBER, M. T. ET AL. 2012. Gravity field and internal structure of Mercury from MESSENGER. *Science*, **336**, 214–217.

SOLOMON, S. C. 1976. Some aspects of core formation in Mercury. *Icarus*, **28**, 509–521.

STROM, R. G., TRASK, N. J. & GUEST, J. E. 1975. Tectonism and volcanism on Mercury. *Journal of Geophysical Research*, **80**, 2478–2507.

TCHALENKO, J. S. 1970. Similarities between shear zones of different magnitudes. *Geological Society of America Bulletin*, **81**, 1625–1640.

TUFTS, B. R., GREENBERG, R., HOPPA, G. V. & GEISSLER, P. 1999. Astypalaea Linea: a San Andreas-sized strike-slip fault on Europa. *Icarus*, **141**, 53–64.

WATTERS, T. R. & NIMMO, F. 2010. Tectonism on Mercury. *In*: SCHULTZ, R. A. & WATTERS, T. R. (eds) *Planetary Tectonics*. Cambridge University Press, Cambridge, UK, 15–80.

WATTERS, T. R., ROBINSON, M. S. & COOK, A. C. 1998. Topography of lobate scarps on Mercury: New constraints on the planet's contraction. *Geology*, **26**, 991–994.

WATTERS, T. R., ROBINSON, M. S. & COOK, A. C. 2001. Large-scale lobate scarps in the southern hemisphere of Mercury. *Planetary and Space Science*, **49**, 1523–1530.

WATTERS, T. R., SOLOMON, S. C., ROBINSON, M. S., HEAD, J. W., ANDRE, S. L., HAUCK, S. A. & MURCHIE, S. L. 2009a. The tectonics of Mercury: the view after MESSENGER'S first flyby. *Earth and Planetary Science Letters*, **285**, 283–296.

WATTERS, T. R., HEAD, J. W. ET AL. 2009b. Evolution of the Rembrandt Impact Basin on Mercury. *Science*, **324**, 618–621.

WATTERS, T. R., SOLOMON, S. C., ROBINSON, M. S., OBERST, J., PREUSKER, F. THE MESSENGER TEAM 2010. Evidence of extension on Mercury unrelated to impact basin deformation. Paper presented at the 41st Lunar and Planetary Science Conference 2010, http://www.lpi.usra.edu/meetings/lpsc2010/pdf/1477.pdf

WATTERS, T. R., SOLOMON, S. C. ET AL. 2012. Extension and contraction within volcanically buried impact craters and basins on Mercury. *Geology*, **40**, 1123–1126.

WILCOX, R. E., HARDING, T. P. & SEELY, D. R. 1973. Basic wrench tectonics. *American Association of Petroleum Geologists Bulletin*, **57**, 74–96.

WOODCOCK, N. & FISCHER, M. 1986. Strike-slip duplexes. *Journal of Structural Geology*, **8**, 725–735.

WU, J. E., McCLAY, K., WHITEHOUSE, P. & DOOLEY, T. 2009. 4D analogue modelling of transtensional pull-apart basins. *Marine and Petroleum Geology*, **26**, 1608–1623.

YALE, D. P. 2003. Fault and stress magnitude controls on variations in the orientation of in situ stress. *In*: AMEEN, M. (ed.) *Fracture and in-situ Stress*

Characterization of Hydrocarbon Reservoirs. Geological Society, London, Special Publication, **209**, 55–64.

YIN, A. 2012. Structural analysis of the Valles Marineris fault zone: possible evidence of large-scale strike-slip faulting on Mars. *Lithosphere*, **4**, 286–330.

ZAMPIERI, D., MASSIRONI, M., SEDEA, R. & SPARACINO, V. 2003. Strike-slip contractional stepovers in the Southern Alps (northeastern Italy). *Eclogae Geologicale Helvetiae*, **96**, 115–123.

ZANG, A. & STEPHANSSON, O. 2010. *Stress Field of the Earth's Crust.* Springer, New York.

ZHANG, Y. Z., DUSSEAULT, M. B. & YASSIR, N. A. 1994. Effects of rock anisotropy on and heterogeneity on stress distributions at selected sites in North America. *Economic Geology*, **37**, 181–197.

Age dating of an extensive thrust system on Mercury: implications for the planet's thermal evolution

L. GIACOMINI[1,2]*, M. MASSIRONI[1,2], S. MARCHI[3], C. I. FASSETT[4],
G. DI ACHILLE[5,6] & G. CREMONESE[7]

[1]*Centro Interdipartimentale di Studi e Attività Spaziali, University of Padova,
Padova, Italy*

[2]*Department of Geosciences, University of Padova, Padova, Italy*

[3]*NASA Lunar Science Institute Center for Lunar Origin and Evolution, Southwest Research
Institute, 1050 Walnut Street, Boulder, CO 80302, USA*

[4]*Department of Astronomy, Mount Holyoke College, South Hadley, MA 01063, USA*

[5]*Istituto Nazionale di Astrofisica, Osservatorio Astronomico di Capodimonte, Napoli, Italy*

[6]*Present address: Istituto Nazionale di Astrofisica-Osservatorio Astronomico di Collurania,
Teramo, Italy*

[7]*Istituto Nazionale di Astrofisica, Osservatorio Astronomico di Padova,
Padova, Italy*

*Corresponding author (e-mail: lorenza.giacomini@unipd.it)

Abstract: The tectonic evolution of Mercury is dominated at a global scale by contractional features such as lobate scarps that are widely distributed across the planet. These structures are thought to be the consequence of the secular cooling of Mercury. Therefore, dating these features is essential to place constraints on the timing of planetary cooling, which is important for understanding the thermal evolution of Mercury. In this work, we date an extended thrust system, which we term the Blossom Thrust System, located between 80°E and 100°E, and 30°N and 15°S, and which consists of several individual lobate scarps exhibiting a north–south orientation and a westward vergence. The age of the system was determined using several different methods. Traditional stratigraphic analysis was accompanied by crater counting of units that overlap the thrust system and by using the buffered crater-counting technique, allowing us to determine an absolute model age for the tectonic feature. These complementary methods give consistent results, implying that activity on the thrust ended between 3.5 and 3.7 Ga, depending on the adopted absolute-age model. These data provide an important insight into this portion of Mercury's crust, which may have implications for models of the thermal evolution of the planet as a whole.

Mercury has a substantial number of landforms that have formed as a result of widespread deformation of the planet's crust. Contractional landforms are generally globally distributed, whereas extensional landforms are basin-localized (Watters & Nimmo 2010). Most evidence for compressive deformation resulting from crustal shortening is manifest as lobate scarps, high-relief ridges and wrinkle ridges. Lobate scarps are linear or arcuate features that cut across all major geological units and display a range of orientations. They are asymmetrical in cross-section and are probably the expression of surface-breaking thrust faults (Strom *et al.* 1975; Cordell & Strom 1977; Melosh & McKinnon 1988; Watters *et al.* 1998, 2001).

High-relief ridges, morphologically similar to lobate scarps, are generally symmetrical in cross-section and appear to be formed by high-angle reverse faults. Wrinkle ridges are generally more complex morphologically, and consist of a broad, low-relief arch with a narrow superimposed ridge. They are thought to reflect a combination of folding and thrust faulting, and are found largely in smooth plains material (Watters & Nimmo 2010; Byrne *et al.* 2014). Evidence of crustal extension on Mercury is considerably less common, and is almost entirely restricted to the interior plains of impact basins, like the Caloris (e.g. Strom *et al.* 1975), Raditladi (Prockter *et al.* 2009) and Rachmaninoff basins (Prockter *et al.* 2010), and to 'ghost'

From: PLATZ, T., MASSIRONI, M., BYRNE, P. K. & HIESINGER, H. (eds) 2015. *Volcanism and Tectonism Across the Inner Solar System*. Geological Society, London, Special Publications, **401**, 291–311.
First published online August 13, 2014, http://dx.doi.org/10.1144/SP401.21
© The Geological Society of London 2015. Publishing disclaimer: www.geolsoc.org.uk/pub_ethics

craters in the northern plains (Head *et al.* 2011; Freed *et al.* 2012; Klimczak *et al.* 2012; Watters *et al.* 2012). Extension on Mercury is manifested by complex networks of narrow linear and sinuous troughs that are interpreted as graben (Strom *et al.* 1975; Melosh & McKinnon 1988; Watters *et al.* 2005).

The array of tectonic features on Mercury provides important clues about the properties and evolution of the planet's crust and lithosphere. Moreover, understanding the deformation of Mercury's surface is important to place constraints on models of its thermal history (e.g. Melosh & McKinnon 1988; Hauck *et al.* 2004; Grott *et al.* 2011). So far, lobate scarps and high-relief ridges on Mercury have generally been interpreted as expressions of global contraction due to interior cooling (Strom *et al.* 1975; Cordell & Strom 1977; Thomas *et al.* 1988; Watters *et al.* 2009), with some contribution also from tidal despinning, mantle convection, or a combination of thermal contraction and tidal despinning (Melosh & Dzurisin 1978; Pechmann & Melosh 1979; Melosh & McKinnon 1988; Dombard & Hauck 2008; King 2008).

In this chapter we consider a recently documented Mercurian tectonic system (Di Achille *et al.* 2012) (Fig. 1) that includes a lobate scarp called Blossom Rupes and, for this reason, it will be hereafter named the 'Blossom Thrust System'. The main aim of this work is dating the end of the system's activity. Indeed, owing to its considerable spatial extent (2000 km), this system can provide insight into the stress field that affected that region of Mercury's surface. Therefore, calculating absolute age dates for this tectonic system is useful in determining its duration of deformation and to gain a better understanding of Mercury's thermal evolution.

Previous studies on the timing of linear features

Understanding the timing of tectonic activity on a planet is a key topic of interest in planetary science. Several techniques have been developed for such study. The first relies on cross-cutting relationships between structures and geological units (e.g. Lucchitta & Watkins 1978; Wise *et al.* 1979; Plescia & Saunders 1982; Dohm & Tanaka 1999). The principle underlying this method is that tectonic structures must post-date the surfaces that they deform. Therefore, if a surface hosts or is cross-cut by a tectonic structure, this structure must be younger than that surface.

In contrast, if a surface unit overlaps or embays a tectonic structure, the surface must be younger than the activity that formed that structure. This method relies on precise knowledge of unit boundaries as well as a clear understanding of the stratigraphic relationships between structures and adjacent terrains. Although these characteristics can be hard to determine, owing to limitations in image resolution or illumination, stratigraphy represents a robust method with which to understand temporal relationships between features.

Mangold *et al.* (2000) developed this method further. They determined the range of ages of faulting events at a number of sites on Mars by counting the craters located in geological units deformed by wrinkle ridges, excluding those craters intersected by the faults. This approach allowed Mangold *et al.* (2000) to determine a maximum age closer to that of the actual deformation, rather than simply the age of the pre-existing terrain. To bound the end of tectonic activity, it is therefore necessary to identify and date a unit that buries at least a portion of a tectonic structure, and which thus must post-date the structure's last phase of activity.

These techniques require that the difference in age between the older and the younger surfaces be small. Otherwise, the age range of a fault is not well constrained. Unfortunately, such an interval is often very broad. To improve the accuracy of dating linear features, Tanaka (1982) developed a technique that can be used to derive an age for landforms such as faults, ridges and channels using only their relationships with impact craters, rather than relying on stratigraphic and cross-cutting relationships between units and faults. This method, hereafter referred to as the 'buffered crater-counting technique', is based on the observation that a linear feature of limited or inconsequential surface area has a density of superimposed impact craters that depends on the area defined by the crater diameters. This technique was subsequently employed by Wichman & Schultz (1989) to study large-scale Martian extensional tectonics, and has recently been used by Fassett & Head (2008), Hoke & Hynek (2009) and Bouley *et al.* (2010) to determine the age of Martian valley networks.

Smith *et al.* (2009) proposed yet another method used to date tectonic structures on the Thaumasia Plateau on Mars. To assess the age of faulting, they counted craters on the entire faulted surface, separating the crater population into (a) craters superposed on the faults and (b) craters cut by the faults. The craters not located on the fault at all were classified into a third category of unknown age relationship. The unfaulted crater population by itself provided the younger bound on the age of the structure, whereas the corresponding older age bound was determined by considering both the population of faulted craters and the craters belonging to the third category. In this study, we date the Blossom Thrust System on Mercury

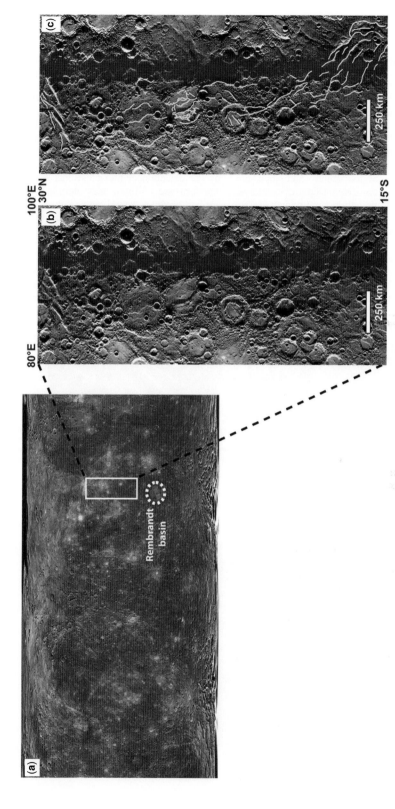

Fig. 1. (**a**) Global MESSENGER MDIS orbital image mosaic of Mercury. The white box encloses the thrust system considered in this work. The dashed white circle outlines the rim of the Rembrandt basin. (**b**) Enlarged view of the thrust system. (**c**) The identified tectonic features are mapped in yellow (online version; solid white lines in printed version). The basemap in (b) and (c) is a mosaic of MESSENGER flyby images.

using both the traditional stratigraphic methodology and these more recent dating techniques.

A very long fold and thrust belt on Mercury

During its third flyby of Mercury, the MErcury Surface, Space ENvironment, GEochemistry, and Ranging (MESSENGER) Mercury Dual Imaging System (MDIS) (Hawkins *et al.* 2007) acquired images at high solar incidence angles (i.e. with the Sun near the horizon) of the planet's surface, which contained long shadows that enhanced the appearance of surface topography and, thus, are suitable to use for observing tectonic structures. These favourable illumination conditions allow us to identify a series of lobate scarps that form an almost 2000 km-long thrust system, the Blossom Thrust System. This system may be another example of a fold-and-thrust belt on Mercury, as documented by Byrne *et al.* (2014). The system to which Blossom Rupes belongs shows a general north–south trend, and is located between 80°E

and 100°E, and 30°N and 15°S (Fig. 1). Because of the high incidence angles in the images where the thrust faults were observed, there is a possible observational bias in the detected orientations of structures as the prevalent east–west illumination of the MESSENGER flyby images may have precluded the detection of some east–west-orientated features (Di Achille *et al.* 2012). However, since all of the lobate scarps are linked together at the surface, north–south-orientated and with a westwards vergence, we interpret them as belonging to a single thrust system. This interpretation is also supported by the detection of transpressional features at the southern and northern ends of the system, which are interpreted as lateral ramps linked to the frontal thrusts (Massironi *et al.* 2014). Indeed, these lobate scarps are kinematically consistent with the westward propagation of these thrust sheets (Fig. 2) (Massironi *et al.* 2014).

The lobate scarps that make up the thrust system are all arcuate in plan view and have considerable lengths, ranging between 100 and 350 km.

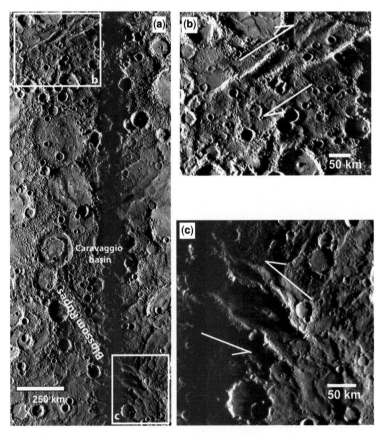

Fig. 2. (**a**) The Blossom Thrust System and its oblique lateral ramps. Dextral (**b**) and sinistral (**c**) transpressional ranges connected to frontal thrusts. The basemap is a mosaic of MESSENGER flyby images.

Moreover, most of the scarps are associated with antiforms, which are probably the result of a fault-bend folding process. At the locations of two large craters, 294 and 216 km in diameter, respectively (and located at 20°12′–12°47′N, 83°21′–90°54′E and 12°39′–7°51′N, 86°16′–91°24′E), the tectonic features have developed a more complex geometry (Fig. 3). The thrusts follow the western segment of crater rims, suggesting that the formation of these structures has been influenced by pre-existing mechanical discontinuities in the crust (see also Watters *et al.* 2001, 2004; Fassett *et al.* 2012; Rothery & Massironi 2013). The semi-circular shape could be explained by the interaction between the crater cavity and a thrust born on a shallow décollement (Mangold *et al.* 1998). The transition between the different lithologies of basin and fill material could provide this décollement (e.g. Watters 2004; Byrne *et al.* 2014).

Related to these lobate scarps are westward-verging thrusts that cross the middle of the basins, and which are interpreted as being linked to the main fault plane by a branch line (Fig. 3). In one of these craters, the lobate scarp is also associated with a back-thrust with the opposite sense of vergence (i.e. eastwards) (Fig. 3). In some cases (see

Fig. 4), lobate scarps are superposed by craters filled with smooth material that is, in turn, deformed by a number of contractional structures. This indicates that tectonic activity persisted after these basin-forming impacts. One of the clearest example of this long-lived tectonism is located near the equator (3°18′N, 87°6′E), where a peak-ring basin, Caravaggio, is partly intersected by Blossom Rupes. The area inside the basin's peak ring is filled with smooth plains material, which is, in turn, deformed (Fig. 4a). The stratigraphic relationships in this region attest to a complex interaction between peak-ring basin formation and thrust faulting, which occurred in a number of stages.

(i) First, thrusts and folds formed a sequence of isolated parallel or subparallel structures probably oriented perpendicular to the maximum horizontal stress (here, east–west-trending) (Fig. 5a). (ii) Later, these structures become progressively longer and deeper such that they overlapped to form a long thrust fault system (Fig. 5b) (e.g. Ben-Zion & Sammis 2003; Kim & Sanderson 2005). (iii) At some point thereafter, this system was superimposed by the Caravaggio peak-ring basin (Fig. 5c). The main fault continued to be active,

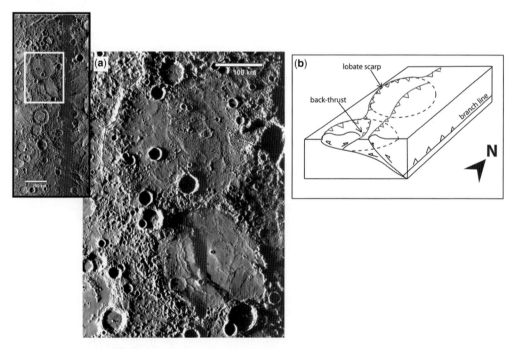

Fig. 3. (**a**) Enlarged view of the two basins intersected by thrust faults. The basemap is a mosaic of MESSENGER flyby images. (**b**) Sketch of our geological interpretation of the structures. Lobate scarps bounding the rims occur at the interface between the basement and basin filling. Thrusts going through the centre of the basins are linked to the frontal scarps by a branch line. The thrust located in the SE basin is a back-thrust and has a vergence (eastwards) opposite to that of the main thrust.

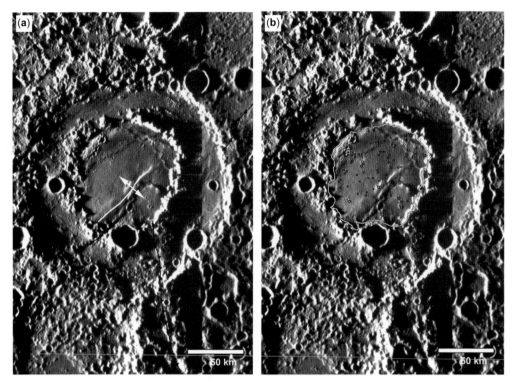

Fig. 4. (**a**) Caravaggio peak-ring basin showing smooth plains with several tectonic structures attesting to the presence of compressive stresses. These features are localized in the smooth plains, implying syndeformational formation with a 350 km-long lobate scarp. Outlined in white is a wrinkle ridge; the scarp is drawn in red (online version). (**b**) The smooth plains, encircled in white, have been dated by crater counting. Red and yellow circles (online version) indicate primary and secondary craters, respectively. The crater mapping was performed using CraterTools by Kneissl *et al.* (2011). The basemap is a mosaic of MESSENGER flyby images. Crater counting was performed on two high-resolution (190 m/pixel) MDIS NAC images.

eventually propagating to at least the basin's peak ring but even to within the inner-ring floor deposits (Fig. 5d). (iv) At this stage, the inner-ring floor was filled by smooth materials covering the main structure (Fig. 5e). Alternatively, the propagating fault system could have developed after the emplacement of the smooth plains deposits, whereupon it encountered a rheological boundary in the form of the more mechanically strong smooth plains. (v) Finally, sustained regional compression caused the formation of a thrust inside the peak-ring basin. This fault propagated through the peak ring and deformed the smooth plains within, developing a wrinkle ridge (symmetric in its transverse section), with a NNE–SSW trend that is slightly different from the primary north–south trend of the main scarp. This implies a slight reorientation of the maximum horizontal stress axis but provides evidence of ongoing regional compression, with the maximum horizontal stress axis in a relatively long-lived, stable orientation.

The regular strike of this wrinkle ridge allows us to exclude an origin solely related to subsidence of the basin floor due to the smooth material emplacement and loading, as suggested for other wrinkle ridges within Mercurian basins such as Rachmaninoff and Mozart (Watters *et al.* 2012; Blair *et al.* 2013). Indeed, in the case of these basins, wrinkle ridges orientated to follow pre-existing regional stresses have not been observed, nor would they be predicted. In addition, the discontinuous and symmetrical shape of the wrinkle ridge cross-cutting the Caravaggio smooth plains in cross-section implies that the fault beneath this structure was not able to propagate up to the surface along all of the wrinkle ridge length. This, coupled with the evidence that the wrinkle ridge did not propagate outside the rim, suggests that it is in an early stage of growth. All of these observations support the hypothesis that the deformation of the inner smooth plains is the result of a persistent compressional state of stress through time.

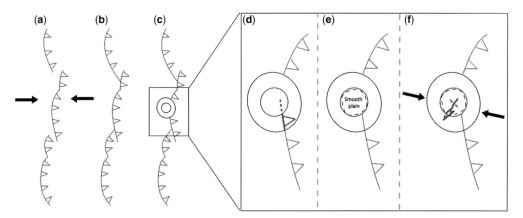

Fig. 5. Geological sketch of lobate scarp formation and its relationship with the Caravaggio peak-ring basin shown in Figure 4. Former isolated thrusts (**a**) form a single, long thrust system through multiple linkages (**b**). This major structure was superimposed by a Caravaggio peak-ring basin (**c**). The main fault was reactivated and propagated up to at least the peak ring of the basin and eventually into the inner floor (**d**). A smooth plains unit filled the area encircled by the peak ring that covers the former structure (**e**). A thrust, with a slightly different orientation with respect to the main scarp, formed inside the crater and propagated into the smooth plains, creating a wrinkle ridge that has not reached the external rim (**f**). The teeth along each scarp trace are located on the hanging-wall block and indicate the dip direction of the fault. Black arrows indicate the inferred maximum horizontal stress.

A similar scenario is shown in Figure 6, where smooth plains within the Ruysch and Savage craters were affected by several small, isolated thrusts that developed within the plains but which show the same fault strike but an opposite vergence (in these cases, eastwards-directed) with respect to the primary lobate scarp. The fact that these minor contractional features follow the same trend as the larger structures is consistent with all such structures having formed under the same stress field and so they are possibly the result of a reactivation of the main thrust. However, the lack of complete linkage at the surface between the thrust segments means that the main compression event ended before the development of the structure within the craters was complete. The formation of shorter structures with less topographical relief within the smooth-plains-filled basins, relative to the preceding structures outside the basins, indicates that the magnitude of compressional stress in this region probably decreased with time.

Data and methodology

We used MESSENGER MDIS narrow-angle camera (NAC) images available in the Planetary Data System archive. These include the images acquired during both the three MESSENGER flybys as well as during orbital operations, and range in spatial resolution from 26 to 250 m/pixel. The images were calibrated and georeferenced with USGS Integrated Software for Imagers and Spectrometers (ISIS)

software and imported into the ArcGIS environment. Tectonic structures were then digitized as vector shapefiles. Mapping the structures in this way allowed us to examine the stratigraphic relationship between linear features and geological units located in the area. Crater counting was subsequently performed to better characterize these relationships. Owing to the considerable size of the Blossom Thrust System, the buffered crater-counting technique is probably the most appropriate method for dating as it allows a substantial number of craters to be taken into account. This method takes advantage of the fact that large craters occupy more surface area than small ones, so that the effective count area is a function of the crater diameter under consideration. We employed the same approach used by Fassett & Head (2008) for Martian valleys, adapting their technique to the dating of Mercurian tectonic features. In particular, for each lobate scarp within the tectonic system, we considered a buffer on each side of the thrust, the areal extent of which was given by the relationship:

$$S_{\text{buffer}} = 1.5D + 0.5W_{\text{v}} \tag{1}$$

where W_{v} is the width of the antiform fold related to the thrust and D is the crater diameter under consideration (Fig. 7). In other words, we considered all the craters whose centres fell within a distance S_{buffer} of the lineament. The width of the buffer zone (W_{v}) was obtained by splitting the structures into 50 km-long segments and estimating the antiform width

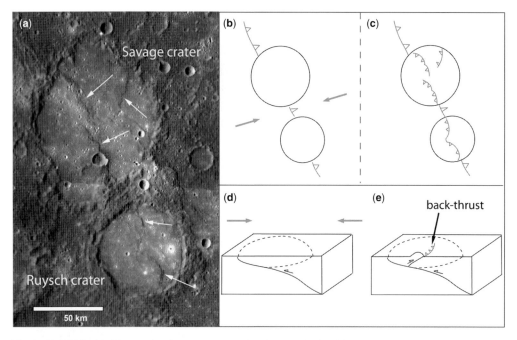

Fig. 6. (**a**) MDIS orbital images showing craters crossed by thrusts and with deformed floors. These craters were flooded by smooth material that was then clearly subjected to deformation (shown with arrows). (**b, c**) Such deformation is represented by small thrusts showing the same trend but with a different vergence with respect to the principal lobate scarp. This, coupled with the relatively small dimensions of the minor thrusts, leads to (**d**) a hypothesized scenario where, after the infilling of the craters, the compression was restored under the same stress regime to form (**e**) smaller back thrusts. Grey arrows indicate the orientation of the main horizontal stress. Since these craters formed when east–west compression was still underway, they were excluded from the counts; otherwise the age of the thrust system would have been overestimated.

at the middle of every segment to finally calculate the average value.

For every crater that satisfied the S_{buffer} condition, we calculated the buffered area as:

$$A(D) = 2LS_{\text{buffer}} = (3D + W_{\text{v}})L \qquad (2)$$

where L is the length of the thrust. Hence, for every crater, a corresponding buffered area was derived. These areas were obtained using the Buffer tool within ArcGIS, which computes a buffer by expanding the area around a mapped feature for a specified distance (S_{buffer}), and combines regions where the buffers around individual segments overlap.

As for Fassett & Head (2008), the buffer distance we considered is selected under the assumption that it is possible to determine stratigraphic relationships for craters that have rims falling within one crater diameter of the tectonic feature. This distance was chosen in accordance with Melosh (1989, p. 90), who predicted that the extent of continuous ejecta is approximately $1D$. However, in the study area, the stratigraphic relationships

between structures and craters are not obvious since the ejecta are seldom clearly discernible. Even colour MDIS wide-angle camera (WAC) images do not help for this purpose, as many of the craters close to the thrust system have ejecta that are superposed and masked by younger crater ejecta. For this reason, we also applied an alternative, more stringent, definition of the buffer. Following Tanaka (1982) and Wichman & Schultz (1989), we included in this count only craters located directly on the antiform whose rims are directly superposed on the features to be age-dated. This approach provided a conservative cross-check on our results from the straightforward buffered crater-counting technique.

As highlighted in previous work (Wichman & Schultz 1989; Fassett & Head 2008; Bouley *et al.* 2010), one of the most important constraints of the buffered crater-counting method is the limited number of crater statistics for small features. However, as the Blossom Thrust System is very long (about 2000 km), the count involved a large effective area (6.8×10^5 km^2) and, thus, we consider its crater count-derived statistics to be robust.

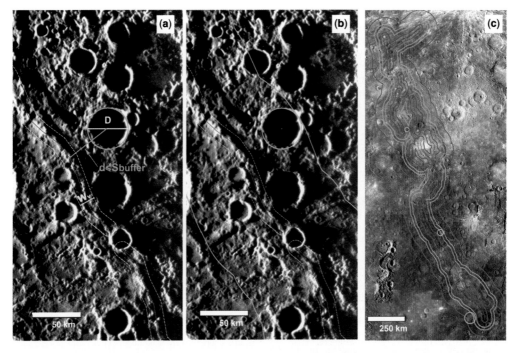

Fig. 7. Schematic sketch illustrating the buffered crater-counting technique, which adapts the method proposed by Fassett & Head (2008) to Mercurian tectonic structures. (**a**) A buffer is established for a given crater size, and craters that intersect the structures and/or have their centres in the buffer are counted. (**b**) For each crater, a count area is then calculated. (**c**) Four examples of buffered areas obtained for four distinct craters (cyan circles).

In order to confirm and/or better determine the age of the tectonic system we investigated, it can be helpful to date both the surfaces that are faulted and those that superimpose the thrust. This can be used to place constraints on the maximum and minimum age, respectively, of activity along the

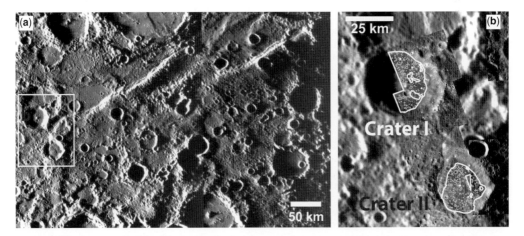

Fig. 8. MDIS flyby images (250 m/pixel) showing the northern branch of the thrust system. (**a**) Several faulted craters predating the thrust were observed. The white box outlines the two craters taken into account for dating. (**b**) Crater counting of the two crater floors was performed to establish a maximum age of the thrust activity. Areas occupied by chains or clusters of secondary craters have been excluded from the count. The basemap for crater counting is a MDIS image mosaic at high resolution (26 m/pixel).

fault system. Two faulted craters, located in the northern branch of the thrust system, were chosen as representative of crater floors where formation predates the faulting activity (Fig. 8). Unfortunately, surfaces clearly post-dating the thrust are very limited in size and had not been imaged at high resolutions prior to or during this study, preventing crater counting. For this reason we chose to count a surface only partially deformed by the thrust system. Caravaggio basin, discussed above, is ideal for this purpose (Fig. 4). Indeed, the region encircled within its peak ring is covered by smooth plains deposits that post-date the main thrust system but which were, in turn, deformed by a wrinkle ridge that we interpret to have developed during the reactivation of the fault. Therefore, dating these smooth plains provides information on the timing of the final stage of the deformation in this region.

Dating methods

The results of the crater counting are presented in the traditional manner as a crater size–frequency distribution (SFD), which describes the frequency of craters of specified size per unit area. The measurement of crater-size frequencies in individual regions allows an estimation to be made of the relative surface age, under the assumption that the higher the crater-size frequency, the longer the surface has been subject to impacts (and so the older it is).

To obtain the crater-size frequency for the buffered counting area, we first sorted the craters that satisfied the S_{buffer} from the largest to the smallest. Then we calculated the frequency. For the largest buffered area, the frequency is $N_c = 1/A_1$, where A_1 is the buffered area. The subsequent cumulative count is the sum of the individual counts. Then, for the next largest buffered area (A_2), the frequency is $1/A_1 + 1/A_2$, and so on. To summarize, the cumulative count (N_c) for every buffered area (A_j) is:

$$N_c = \sum_{j=1}^{k} 1/A_j. \qquad (3)$$

The data were subsequently analysed as a cumulative size–frequency distribution that allowed us to determine the absolute model ages, assuming an appropriate crater production function for Mercury. The production function represents the SFD of a crater population accumulated per unit time. The production function allows us to estimate the crater density on a given terrain at a reference crater size (usually 1 km), which, in turn, can be used to calculate the model age. The only available calibration for these absolute ages comes from the radiometric and exposure ages of the Apollo and Luna lunar samples, which were used to establish a lunar chronology. The crater chronology for Mercury is derived from that of the Moon. On Mercury, different production functions have been proposed (Neukum 1983; Strom & Neukum 1988; Ivanov *et al.* 2001; Neukum *et al.* 2001*a*, *b*; Marchi *et al.* 2009, 2011). In this work, we considered two of these functions. The first is that presented by Neukum *et al.* 2001*a*, *b*. Their Mercurian production function is inferred from the lunar Neukum Production Function (NPF) (Neukum 1983; Neukum & Ivanov 1994; Neukum *et al.* 2001*a*, *b*), by applying to Mercury the scaling laws relative to the impactor flux and its velocity distribution at the Moon, and by taking into account the differences in gravitational acceleration and target properties between the Moon and Mercury (Hartmann 1977; Strom & Neukum 1988; Neukum *et al.* 2001*a*, *b*). The shape of the production function is assumed to have remained stable over the history of the solar system.

The second production function we considered is the Model Production Function (MPF), proposed recently by Marchi *et al.* (2009, 2011). The MPF is based on dynamic models that describe the current Main Belt Asteroid (MBA) and Near Earth Object (NEO) impactor fluxes SFD to Mercury, which is converted into a crater SFD via the Holsapple & Housen (2007) scaling law. This scaling law depends on the physical properties of the target material (specifically its tensile strength and density). To date, two different materials have been taken into account when applying this scaling law: cohesive soil/fractured rocks; and hard rock (Massironi *et al.* 2009, 2010; Marchi *et al.* 2011). Here we adopted the terminology of Holsapple & Housen (2007), and by cohesive soil we mean a material whose strength is substantially reduced with respect to the hard rock case by, for instance, the presence of diffuse and pervasive fractures. The assumed values of tensile strength are $Y = 2 \times 10^7$ and 2×10^8 dyne cm^{-2} for fractured material and hard rock, respectively.

As for the NPF method, an absolute model age was obtained for Mercury by extrapolating from the lunar chronology. Further, the MPF model allows for the treatment of a non-constant impactor flux SFD through time, as well as the possibility of investigating variable target crustal layering (Massironi *et al.* 2009, 2010; Marchi *et al.* 2011). This latter functionality may be used to interpret kinked crater SFDs that otherwise would not fit the production function for a target of uniform material. As we discuss in the next section, the MPF model predicts that, if no other process such as superposition of later geological units has altered the crater SFD, the kink in the crater SFD may relate to rheological layering of the investigated portion of the crust (Massironi *et al.* 2009, 2010; Marchi *et al.* 2011). In particular, the inflection in a given crater SFD at smaller crater diameters could be due to the transition from a cohesive soil-fractured terrain scaling law to a hard-rock scaling law (see Holsapple & Housen 2007). If so, the diameter at which this kink occurs should vary with, and correspond to, the thickness of an upper, heavily fractured crustal layer (Massironi *et al.* 2009, 2010; Marchi *et al.* 2011). The thickness (H) of this upper layer is potentially highly variable across a given planet, depending on: (i) the rheology and variations in lithology and layering of a specific region; (ii) the age of the region itself (with a thicker fractured layer expected in older regions); and (iii) the presence of impact basins and associated processes of regional modification of the target material (Massironi *et al.* 2009, 2010).

From our measurements, we determined the age of the Blossom Thrust System by fitting the measured crater SFD with both the NPF and the MPF in order to compare the age obtained with these two different approaches. Note that the NPF-based model ages are often published with two decimal figures (e.g. Neukum *et al.* 2001a, b); here, however, we rounded them to one decimal place. Similarly, MPF-based ages are rounded to one decimal place.

Results

For the buffered crater-counting technique we identified 103 craters located within the S_{buffer} distance (see Table 1) (Fig. 9). Those craters with floor material that had been deformed by the scarps (e.g. the Ruysch and Savage craters) were excluded from this count since they attest to contractional tectonism subsequent to crater formation (Fig. 6). The measured crater SFDs for the NPF and MPF techniques gives a model age of 3.8 ± 0.1 and 3.7 ± 0.1 Ga, respectively (Fig. 9) (Table 2).

Interestingly, the crater SFD for the buffered area shows an S-slope between crater diameters of 8 and 10.5 km. The crater SFD derived using the more stringent buffer definition of Tanaka (1982) (Table 1) gives a younger age, of about 3.7 ± 0.1 Ga for NPF and 3.5 ± 0.1 Ga for MPF (Fig. 9) (Table 2), although the S-slope is still present, appearing for crater diameters of between 11 and 12.5 km.

We subsequently performed crater counting on the floor materials of the two faulted craters. Following Platz *et al.* (2013), the areas occupied by secondary crater clusters were clipped and excluded from the counting area. The best fits obtained for the crater SFD resulted in the same model age of $3.7-3.8 \pm 0.1$ Ga, both for the NPF and MPF models, with a surface having the mechanical properties of a cohesive soil (fractured terrain) for the latter model (Fig. 10).

Finally, the smooth plains unit encircled by the peak ring of the Caravaggio basin (see Fig. 11) was dated. The count area followed the morphological boundaries of the smooth plains, excluding regions occupied by clusters and chains. All craters within this region were then counted (see Table 1). The 261 craters on this surface provided a model age of about 3.8 ± 0.1 Ga using the NPF method (Table 2). The MPF method gives two different ages depending on what rheological properties are assumed for the terrain: we obtained an age of 3.7 ± 0.1 Ga for a fractured target material, and an age of 2.3 ± 0.4 Ga was calculated using the properties of hard rock (Fig. 11). The considerable age difference between a target consisting of cohesive soil-fractured rock and one of hard rock results from the fact that porous material dissipates the impactor energy more efficiently than does intact rock. Therefore, for a given impactor mass and velocity, smaller craters will form in fractured material than will in hard rock. Hence, the same measured crater size–frequency distribution will give a substantially

Table 1. *Statistics of all the craters detected on the smooth plain and the thrust system*

Count	Crater I	Crater II	Smooth plain	Thrust system (broad buffer*)	Thrust system (stringent buffer[†])
All[‡]	254	214	295	103	62
Bonafide[§]	254	214	261	103	62
Secondaries[‖]	–	–	34	–	–

*'Broad buffer' indicates the buffered crater-counting technique as used by Fassett & Head (2008).
[†]'Stringent buffer' indicates the technique proposed by Tanaka (1982) (see the text for details).
[‡]'All' indicates all crater-like features.
[§]'Bonafide', the primary craters.
[‖]'Secondaries', the secondary craters.
For the age calculation, only the bonafide craters have been considered.

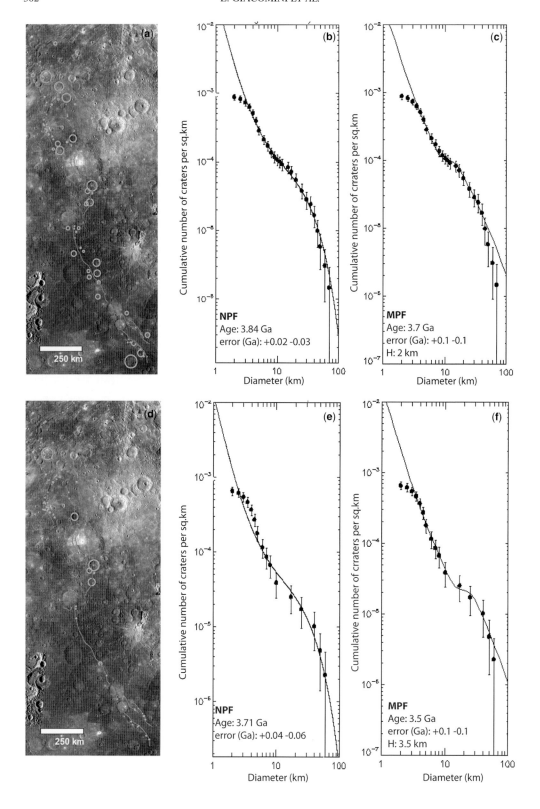

Table 2. *Age determination for the thrust system and smooth plain based on NPF and MPF*

	Age (NPF*) (Ga)	Age (MPF[†]) (Ga)
Thrust system (broad buffer[‡])	3.8 (±0.1)	3.7 (±0.1) – 2 km-thick fractured layer
Thrust system (stringent buffer[§])	3.7 (±0.1)	3.5 (±0.1) – 3.5 km-thick fractured layer
Faulted craters:		
Crater I	3.8 (±0.1)	3.7 (±0.1) – cohesive soil-fractured terrain
Crater II	3.7 (±0.1)	3.7 (±0.1) – cohesive soil-fractured terrain
Smooth plain	3.8 (±0.1)	3.7 (±0.1) – cohesive soil-fractured terrain
		2.3 (±0.4) – hard rock

*NPF indicates the Neukum Production Function.
[†]MPF indicates the Model Production Function. The MPF age for the thrust system subtends the presence of an upper fractured layer (see the text for details).
[‡]'Broad buffer' indicates the buffered crater-counting technique as used by Fassett & Head (2008).
[§]'Stringent buffer' indicates the technique proposed by Tanaka (1982).

older age if fitted by an MPF curve that assumes a fractured material than if a hard rock target is considered.

Discussion

The range of results obtained with the two different buffer approaches (broad and stringent) provides a sense of the magnitude of the uncertainty in the age of the thrust system we investigated. Indeed, we can infer that its age may vary between 3.8 ± 0.1 (broad buffer) and 3.7 ± 0.1 Ga (stringent buffer) for the NPF method, and between 3.7 ± 0.1 (broad buffer) and 3.5 ± 0.1 Ga (stringent buffer) for the MPF method.

According to the MPF technique (Massironi *et al.* 2009, 2010; Marchi *et al.* 2011), the S-slope deflection observed on the crater SFDs obtained with both the broader and the stringent buffer methods may correspond to a fractured layer on top of a more coherent rock unit. Using a broader buffer, this fractured layer has an inferred thickness of about 2 km, whereas with a more stringent buffer, a greater fractured layer thickness of 3.5 km is implied. The latter value is in better agreement with a transition depth of 4.5 km between fractured material and the underlying intact rock suggested by Schultz (1993) for volcanic basalt sequences on Mercury. Therefore, the stringent buffer method may give more geologically

reasonable results. The transition between jointed/ fractured terrain and intact rock may represent a mechanical discontinuity along which the detachment below the thrust sheet developed. Since the thrust system is quite long, this value may represent an average detachment depth along the system's entire length.

Faulted crater floor ages yield the same age of 3.7–3.8 ± 0.1 Ga for both NPF and MPF methods. These results further suggest that the stringent buffer method better characterizes the age of the most recent activity along the thrust system, since that activity must be younger than the crater floors (see Table 2).

The apparent discord in age correlation given by the broad buffer is most likely to be attributable to the uncertainty in the stratigraphic relationships between the contractional structures and craters at the resolution of the available data. Hence, it is possible that craters older than these structures have been erroneously included in our crater counting, with a consequent overestimation of the thrust system's age.

The age obtained with the NPF method for the smooth plains encircled by the Caravaggio peak ring is approximately 3.8 Ga, which therefore slightly predates the end of activity along the thrust system (dated to *c.* 3.7 Ga, according to the stringent buffer method). The two ages are geologically consistent with observations, wherein the Caravaggio smooth plains were emplaced during

Fig. 9. (**a**) Craters used to date the thrust system with the buffered crater-counting technique. The distribution of small craters appears inhomogeneous as the thrust fold has a variable width. The relative crater SFD obtained was dated both with (**b**) NPF and (**c**) MPF models. (**d**) A more conservative method with a narrower buffer, following Tanaka (1982), was also used to date the system. The crater SFD was fit with (**e**) the NPF and (**f**) the MPF models (see the text for details). A steepening in (**e**) for the crater range from 2 to 4 km was observed. Since secondaries were excluded from the crater count, this steepening may be due to target rheology. None the less, this steepening does not invalidate the age determination since the fitting was performed considering crater diameters ≥5 km. Curves in the plots indicate the isochrons shown in the legend. The basemap is a mosaic of MESSENGER MDIS orbital images.

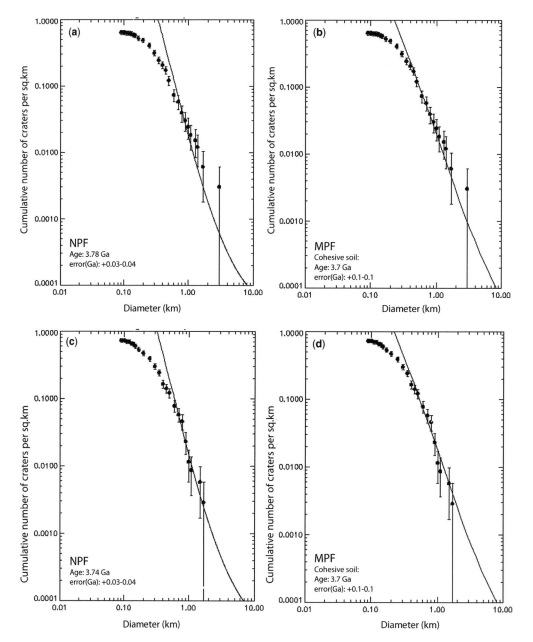

Fig. 10. MPF and NPF best fit of the primary crater SFD for the two faulted craters shown in Figure 8. (**a**) The Crater I NPF best fit according to Neukum *et al.* 2001*a, b.* (**b**) The MPF best fit for Crater I obtained when considering a fractured target. (**c**) The Crater II NPF best fit according to Neukum *et al.* 2001*a, b.* (**d**) The MPF best fit for Crater II obtained when considering a fractured target. The fitting in all of the plots was performed considering crater diameters to be ≥600 m. Curves indicate the isochrons shown in the legend.

the last phase of thrust system activity, as discussed earlier in this paper.

Using the MPF, two distinct scenarios can be outlined, based on the two ages obtained for the different rheological properties assumed for the smooth plains (i.e. cohesive soil-fractured material v. hard rock) (Fig. 12). These scenarios are described below.

- Under the first scenario (where the rheological properties of cohesive soil are assigned to the smooth plains), these units would have been emplaced at about 3.7($\pm$0.1) Ga. If we consider that the thrust system activity ended at 3.5 $\pm$ 0.1 Ga (following the results returned using the stringent buffer method), then the following sequence is possible: (i) the Blossom Thrust System formed before 3.7 Ga, since it is clearly covered by the smooth plain within the Caravaggio peak ring, but (ii) the thrust system continued its activity up to about 3.5 Ga. Similar to the NPF-generated results, these ages appear consistent with geological observations, although the time span between the smooth plains emplacement and the end of tectonic activity in this region appears larger.

- Under the second scenario (where the rheological properties of solid rock are assigned to Caravaggio's smooth plains), the deposits might have been emplaced at 2.3 $\pm$ 0.4 Ga. This is more than 1 Gyr after the main thrust activity started, and is well beyond when the Carravagio basin, which is cross-cut by the thrust, formed. A volcanic episode occurring this long after basin formation during on-going regional compression, which progressively inhibits the vertical ascent of magma, seems unlikely. A contractional state of stress is expected to exist both globally and regionally, which would tend to inhibit magma ascent and eruption (Wilson & Head 2008). Moreover, since the smooth plains are, in turn, deformed by a wrinkle ridge, activity along the thrust system would have to persist until at least 2.3 Ga, which is not consistent with the model ages obtained for the thrust system using the buffered crater-counting technique (i.e. 3.5 Ga).

All of the incongruities raised by the hard-rock scenario led us to consider fractured material properties to be more representative of the smooth plains deposits we studied. Although a cohesive soil/fractured rock material does not ideally fit our expectations for the rheological properties of a volcanic terrain, hard rocks are unlikely to be representative of smooth plains either. As highlighted by Schultz (1993, 1995), volcanic units, consisting of basaltic flows, are typically associated with cooling joints and fractures that considerably weaken the rock mass relative to an intact hand sample. This weakening causes a reduction in tensile and compressive strengths at shallower depth (such that the crust is mechanically weaker near the surface). Only at substantial depths does the crust acquire the same physical properties as an intact rock sample, since all of the inherent discontinuities gradually close as a consequence of the increasing confining pressure. On Mercury, the crustal rock mass is expected to be intact from an average depth of 4.5 km (Schultz 1993). Therefore, particularly when small craters are taken into account (as for the crater-counted crater floors and the unit within the Caravaggio peak ring), the fractured material scaling seems to be the most apt for consideration when using the MPF approach.

Conclusions

A 2000 km-long thrust system extends between 80°E and 100°E on Mercury. This thrust system, of which Blossom Rupes is part and so here named as the Blossom Thrust System, consists of several linked north–south-trending lobate scarps that have a westward vergence. We dated this thrust system using crater size–frequency distributions derived for portions of the surface that have various stratigraphic relationships with the thrust system itself, applying broad and strict buffered crater-counting techniques. Two chronology models, based on the Neukum Production Function (NPF) and Model Production Function (MPF), respectively, were considered for the dating. Our results indicate that the model ages derived with the more stringent buffer appear to be the most reliable, and which date the end of the tectonic activity along the thrust system at around 3.5 Ga using the MPF and at about 3.7 Ga with the NPF.

These ages are in broad agreement with stratigraphic observations that lobate scarps developed after the emplacement of the intercrater plains (c. 4 Ga: Marchi et al. 2013), although an earlier start of contraction cannot be excluded, and continued both during and after the emplacement of the smooth plains (Watters et al. 2004; Solomon et al. 2008). However, the ejecta of the basins surrounding the Blossom Thrust System (the Rachmaninoff, Raditladi, Steichen and Eminescu craters) obliterate almost all of the adjacent structures (Di Achille et al. 2012) (Fig. 13). As the oldest of the dated ejecta among the basins named above belong to Rachmaninoff (3.6 $\pm$ 0.1 Ga following the MPF method: Marchi et al. 2011), most of the tectonic activity in this area could have ended between 3.6 and 3.5 Ga (MPF).

These results help provide new constraints on the thermal evolution of the planet. At the beginning of its history, Mercury underwent an expansion due to core formation. Core separation must have occurred prior to the emplacement of the oldest terrains that remain visible (c. >4 Ga: Marchi et al. 2013), since there are no large-scale extensional features on heavily cratered terrain to attest to this expansion (Solomon 1977). However, current thermochemical

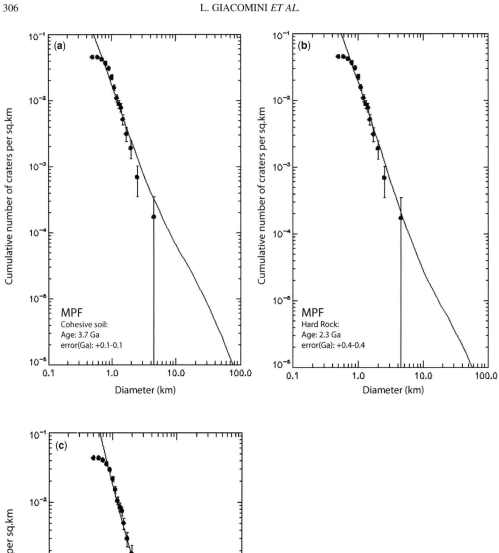

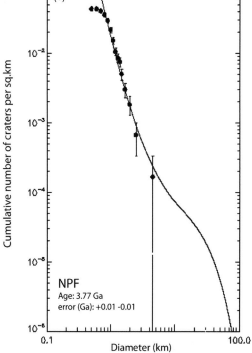

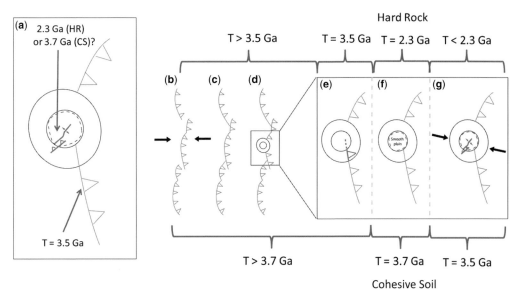

Fig. 12. (a) Schematic sketch of the tectonic scenarios outlined with the MPF method on the base of the different rheological proprieties of the Caravaggio basin smooth plains. If a cohesive soil-fractured material (CS) scaling law is applied, the age of the plains is 3.7 Ga. Since the thrust system activity ended at 3.5 Ga, then the following scenario can be proposed: (**b**) & (**c**) the Blossom Rupes formed before 3.7 Ga; (**d**) it was then superimposed by the Caravaggio basin; (**e**) it propagated to the peak ring; and (**f**) it was subsequently covered by the smooth plains. (**g**) The thrust system continued its activity up to approximately 3.5 Ga, forming the wrinkle ridge on the smooth plains. If a hard rock (HR) scaling law is applied, the age of the plains is 2.3 Ga, which implies that: (b) & (c) the thrust system activity formed before 3.5 Ga; (d) it was superposed by the Caravaggio basin; and (e) it propagated up to the peak ring until the main activity ceased, at about 3.5 Ga. (f) The smooth plains were then emplaced at 2.3 Ga; and (g) the thrust reactivated to form the wrinkle ridge.

models do not unequivocally show when global expansion ended and when global contraction began. Contraction of the planet is due to a decrease in its internal temperature causing, in turn, phase changes such as those associated with the solidification of the inner core (e.g. Solomon 1976, 1977; Schubert *et al.* 1988; Hauck *et al.* 2004; Grott *et al.* 2011). The tectonic system analysed in this work suggests that the most prominent and widespread contraction ended around 3.7 ± 0.1 Ga for the NPF method and 3.5 ± 0.1 Ga for the MPF in the sector of the planet where the system is located. Then, subsequent, localized contraction could have developed along pre-existing shear zones. This result can provide important inputs for the Mercurian thermochemical models. In particular, it may help to constrain parameters like

mantle temperature, the possible presence of an insulating layer (e.g. Grott *et al.* 2011), mantle rheology and core sulphur content (e.g. Hauck *et al.* 2004).

The crustal shortening calculated in the past from the displacements of lobate scarps indicated a modest change in radius that ranged from between 1 and 2 km (Strom *et al.* 1975; Watters *et al.* 2009) to 3.5 km (Di Achille *et al.* 2012). To satisfy a radius change constraint of around 3 km, some published models attempted to inhibit planetary cooling and suggested that global contraction began well after the end of the Late Heavy Bombardment (LHB) (e.g. Grott *et al.* 2011; Tosi *et al.* 2013) However, recently, a new estimate of the crustal shortening accommodated by wrinkle ridges and lobate scarps finds that Mercury's

Fig. 11. MPF and NPF best fit of the primary crater SFD for the inner smooth plain. (**a**) MPF best fit obtained considering a fractured target. (**b**) The MPF best fit considering a hard rock target. (**c**) The NPF best fit according to Neukum *et al.* (2001*a*, *b*). The fitting in all plots was performed considering crater diameters ≥700 m. Curves indicate the isochrons shown in the legend. A steepening is observable in the crater SFD on the MPF plots of Figure 9a–c, for a range in crater diameter from 2 to 4 km. However, if we exclude the two largest size bins that have poor statistics and therefore are less significant, then the crater SFD on the NPF and MPF plots are comparable.

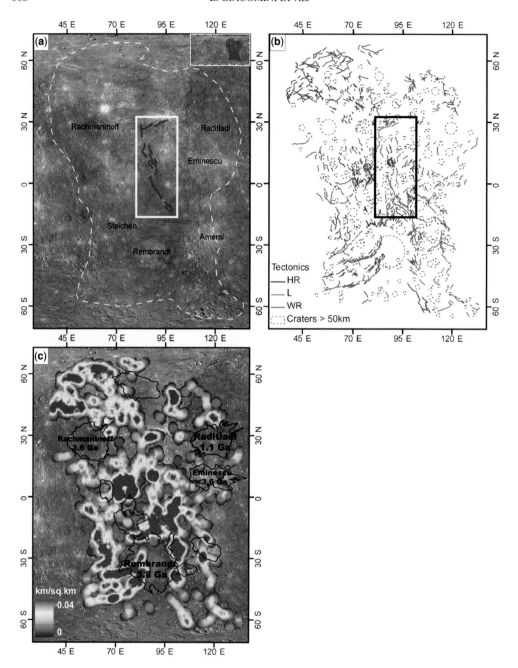

Fig.13. Thrust system and major crater ejecta stratigraphic relationship. (**a**) Thrust system (outlined in red and bounded by a white box) and the major surrounding craters are shown on the MDIS mosaic. (**b**) A geological sketch where the lineaments within the studied area (black box) and its surroundings have been indicated: lobate scarps (L; red), wrinkle ridges (WR; blue) and high-relief ridges (HR; purple) in relation to the rims (black dashed lines) of craters with a diameter larger than 50 km. (**c**) Colour-coded line density map of tectonic lineaments; black lines encompass the ejecta of major craters. Gaps occur in the surroundings of the Rachmaninoff, Raditladi, Steichen and Eminescu craters, and the eastern part of the Rembrandt basin, suggesting that they post-date large portions of the lineaments located in this sector of the planet. Crater model ages were derived from Marchi *et al.* (2011: Rachmaninoff and Raditladi crater), Ferrari *et al.* (2014: Rembrandt basin) and Schon *et al.* (2011: Eminescu crater). The Steichen crater has not yet been dated. Figure modified after Di Achille *et al.* (2012).

radius decreased by about 7 km (Byrne *et al.* 2014). Our results support contraction along a major thrust system having probably started before 3.8 Ga (Strom *et al.* 1975; Solomon 1977; Hauck *et al.* 2004; Watters *et al.* 2004; Dombard & Hauck 2008). If our age estimate can be applied to the majority of the main Mercurian contractional features (as it seems in Fegan *et al.* 2014), the larger amount of radius contraction found by Byrne *et al.* (2014) would imply a revision of Mercurian thermal evolution models to take into account an earlier onset (pre-LHB?) of planetary contraction.

We thank S. Werner, T. Platz, P. K. Byrne and an anonymous reviewer who provided detailed and helpful suggestions that improved the manuscript. We thank also E. Martellato for help with data preparation and L. Fassett for the revision of the English text. This work was supported by the Italian Space Agency (ASI) within the SIMBIOSYS Project (ASI-INAF agreement no. I/022/10/0). M. Massironi was funded by Progetto di Ricerca di Ateneo CPDA112213/11, Università di Padova.

References

BEN-ZION, Y. & SAMMIS, C. G. 2003. Characterization of fault zones. *Pure and Applied Geophysics*, **160**, 677–715.

BLAIR, D. M., FREED, A. M. ET AL. 2013. The origin of graben and ridges in Rachmaninoff, Raditladi, and Mozart basins, Mercury. *Journal of Geophysical Research*, **118**, 47–58.

BOULEY, S., CRADDOCK, R. A., MANGOLD, N. & ANSAN, V. 2010. Characterization of fluvial activity in Parana Valles using different age-dating techniques. *Icarus*, **207**, 686–698.

BYRNE, P. K., KLIMCZAK, C., ŞENGÖR, A. M. C., SOLOMON, S. C., WATTERS, T. R. & HAUCK, S. A. 2014. Mercury's global contraction much greater than earlier estimates. *Nature Geoscience*, **7**, 301–307.

CORDELL, B. M. & STROM, R. G. 1977. Global tectonics of Mercury and the Moon. *Physics of the Earth and Planetary Interiors*, **15**, 146–155.

DI ACHILLE, G., POPA, C., MASSIRONI, M., MAZZOTTA EPIFANI, E., ZUSI, M., CREMONESE, G. & PALUMBO, P. 2012. Mercury's radius change estimates revisited using MESSENGER data. *Icarus*, **221**, 456–460.

DOHM, J. M. & TANAKA, K. L. 1999. Geology of the Thaumasia region, Mars: plateau development, valley origins, and magmatic evolution. *Planetary and Space Science*, **47**, 411–431.

DOMBARD, A. J. & HAUCK, S. A. 2008. Despinning plus global contraction and the orientation of lobate scarps on Mercury. *Icarus*, **198**, 274–276.

FASSETT, C. I. & HEAD, J. W. 2008. Valley network-fed, open-basin lakes on Mars: Distribution and implications for Noachian surface and subsurface hydrology. *Icarus*, **198**, 37–56.

FASSETT, C. I., HEAD, J. W. ET AL. 2012. Large impact basins on Mercury: Global distribution, characteristics, and modification history from MESSENGER orbital data. *Journal of Geophysical Research*, **117**, E00L08, http://dx.doi.org/10.1029/2012JE004154

FEGAN, E. R., ROTHERY, D. A., CONWAY, S. J., ANAND, M. & MASSIRONI, M. 2014. Linking the timing of volcanic and tectonic features on Mercury: results from buffered crater counting. In: *Proceedings of the 45th Lunar and Planetary Science Conference, held 17–21 March, 2014, at The Woodlands, Texas*. Lunar and Planetary Institute, Houston, TX, Contribution No. 1777, 1780.

FERRARI, S., MASSIRONI, M., MARCHI, S., BYRNE, P. K., KLIMCZAK, C., MARTELLATO, E. & CREMONESE, G. 2014. Age relationships of the Rembrandt basin and Enterprise Rupes, Mercury. In: PLATZ, T., MASSIRONI, M., BYRNE, P. K. & HIESINGER, H. (eds) *Volcanism and Tectonism Across the Inner Solar System*. Geological Society, London, Special Publications, **401**. First published online July 29, 2014, http://dx.doi.org/10.1144/SP401.20

FREED, A. M., BLAIR, D. M. ET AL. 2012. On the origin of graben and ridges within and near volcanically buried craters and basins in Mercury's northern plains. *Journal of Geophysical Research*, **117**, E006L06, http://dx.doi.org/10.1029/2012JE004119

GROTT, M., BREUER, D. & LANEUVILLE, M. 2011. Thermo-chemical evolution and global contraction of mercury. *Earth and Planetary Science Letters*, **307**, 135–146.

HARTMANN, W. K. 1977. Relative crater production rates on planets. *Icarus*, **31**, 260–276.

HAUCK, S. A., DOMBARD, A. J., PHILLIPS, R. J. & SOLOMON, S. C. 2004. Internal and tectonic evolution of Mercury. *Earth and Planetary Science Letters*, **222**, 713–728.

HAWKINS, S. E., BOLDT, J. D. ET AL. 2007. The Mercury dual imaging system on the MESSENGER spacecraft. *Space Science Reviews*, **131**, 247–338.

HEAD, J. W., CHAPMAN, C. R. ET AL. 2011. Flood volcanism in the northern high latitudes of Mercury revealed by MESSENGER. *Science*, **333**, 1853–1856, http://dx.doi.org/10.1126/science.1211997

HOKE, M. R. T. & HYNEK, B. M. 2009. Roaming zones of precipitation on ancient Mars as recorded in valley networks. *Journal of Geophysical Research*, **114**, E08002, http://dx.doi.org/10.1029/2008JE003247

HOLSAPPLE, K. A. & HOUSEN, K. R. 2007. A crater and its ejecta: an interpretation of deep impact. *Icarus*, **187**, 345–356.

IVANOV, B. A., NEUKUM, G. & WAGNER, R. 2001. Size–frequency distributions of planetary impact craters and asteroids. In: MIKHAIL, YA. M. & RICKMAN, H. (eds) *Collisional Processes in the Solar System*. Astrophysics and Space Science Library, **261**. Kluwer Academic, Dordrecht, 1–34.

KIM, Y.-S. & SANDERSON, D. J. 2005. The relationship between displacement and length of faults: a review. *Earth-Science Reviews*, **68**, 317–334.

KING, S. D. 2008. Pattern of lobate scarps on Mercury's surface reproduced by a model of mantle convection. *Nature Geoscience*, **1**, 229–232.

KLIMCZAK, C., WATTERS, T. R. ET AL. 2012. Deformation associated with ghost craters and basins in volcanic smooth plains on Mercury: strain analysis and implications for plains evolution. *Journal of Geophysical Research*, **117**, E00L03, http://dx.doi.org/10.1029/2012JE004100

KNEISSL, T., VAN GASSELT, S. & NEUKUM, G. 2011. Map-projection-independent crater size-frequency determination in GIS environments – new software tool for ArcGIS. *Planetary and Space Science*, **59**, 1243–1254.

LUCCHITTA, B. K. & WATKINS, J. A. 1978. Age of graben systems on the moon. *In*: *Proceedings of the 9th Lunar and Planetary Science Conference, March 13–17, 1978, Houston, Texas, Volume* **3**. Pergamon Press, New York, 3459–3472.

MANGOLD, N., ALLEMAND, P. & THOMAS, P. G. 1998. Wrinkle ridges of Mars: structural analysis and evidence for shallow deformation controlled by ice-rich décollements. *Planetary and Space Science*, **46**, 345–356.

MANGOLD, N., ALLEMAND, P., THOMAS, P. G. & VIDAL, G. 2000. Chronology of compressional deformation on Mars: evidence for a single and global origin. *Planetary and Space Science*, **48**, 1201–1211.

MARCHI, S., MOTTOLA, S., CREMONESE, G., MASSIRONI, M. & MARTELLATO, E. 2009. A new chronology for the Moon and Mercury. *The Astronomical Journal*, **137**, 4936–4948.

MARCHI, S., MASSIRONI, M., CREMONESE, G., MARTELLATO, E., GIACOMINI, L. & PROCKTER, L. 2011. The effects of the target material properties and layering on the crater chronology: The case of Raditladi and Rachmaninoff basins on Mercury. *Planetary and Space Science*, **59**, 1968–1980.

MARCHI, S., CHAPMAN, C. R., FASSETT, C. I., HEAD, J. W., BOTTKE, W. F. & STROM, R. G. 2013. Global resurfacing of Mercury 4.0–4.1 billion years ago by heavy bombardment and volcanism. *Nature*, **499**, 59–61.

MASSIRONI, M., CREMONESE, G., MARCHI, S., MARTELLATO, E., MOTTOLA, S. & WAGNER, R. J. 2009. Mercury's geochronology revised by applying Model Production Function to Mariner 10 data: geological implications. *Geophysical Research Letters*, **36**, L21204, http://dx.doi.org/10.1029/2009GL040353

MASSIRONI, M., CREMONESE, G., MARCHI, S., MARTELLATO, E., MOTTOLA, S. & WAGNER, R. J. 2010. Correction to 'Mercury's geochronology revised by applying Model Production Function to Mariner 10 data: geological implications'. *Geophysical Research Letters*, **37**, L19203, http://dx.doi.org/10.1029/2010GL045068

MASSIRONI, M., DI ACHILLE, G. *ET AL.* 2014. Lateral ramps and strike-slip kinematics on Mercury. *In*: PLATZ, T., MASSIRONI, M., BYRNE, P. K. & HIESINGER, H. (eds) *Volcanism and Tectonism Across the Inner Solar System*. Geological Society, London, Special Publications, **401**, first published online June 5, 2014, http://dx.doi.org/10.1144/SP401.16

MELOSH, H. J. 1989. *Impact Cratering: A Geologic Process*. Oxford University Press, New York.

MELOSH, H. J. & DZURISIN, D. 1978. Mercurian global tectonics – a consequence of tidal despinning. *Icarus*, **35**, 227–236.

MELOSH, H. J. & MCKINNON, W. B. 1988. The tectonics of Mercury. *In*: VILAS, F., CHAPMAN, C. R. &

MATTHEWS, M. S. (eds) *Mercury*. University of Arizona Press, Tucson, AZ.

NEUKUM, G. 1983. *Meteoriten bombardement und Datierung planetarer Ober flachen*. Dissertation, University of Munich.

NEUKUM, G. & IVANOV, B. 1994. Crater size distributions and impact probabilities on Earth from lunar, terrestrial-planet, and asteroid cratering data. *In*: GEHRELS, T., MATTHEWS, S. M. & SCHUMANN, A. (eds) *Hazards Due to Comets and Asteroids*. Space Science Series. University of Arizona Press, Tucson, AZ, 359–416.

NEUKUM, G., IVANOV, B. A. & HARTMANN, W. K. 2001*a*. Cratering records in the Inner Solar System in relation to the Lunar Reference System. *Space and Science Reviews*, **96**, 55.

NEUKUM, G., OBERST, J., HOFFMANN, H., WAGNER, R. & IVANOV, B. A. 2001*b*. Geologic evolution and cratering history of Mercury. *Planetary and Space Science*, **49**, 1507–1521.

PECHMANN, J. B. & MELOSH, H. J. 1979. Global fracture patterns of a despin planet application to Mercury. *Icarus*, **38**, 243–250.

PLATZ, T., MICHAEL, G., TANAKA, K. L., SKINNER, J. A. & FORTEZZO, C. M. 2013. Crater-based dating of geological units on Mars: methods and application for the new global geological map. *Icarus*, **225**, 806–827.

PLESCIA, J. B. & SAUNDERS, R. S. 1982. Tectonic history of the Tharsis region, Mars. *Journal of Geophysical Research*, **87**, 9775–9791.

PROCKTER, L. M., WATTERS, T. R. *ET AL.* 2009. The curious case of Raditladi basin. *In*: *Proceedings of the 40th Lunar and Planetary Science Conference, March 23–27, 2009, The Woodlands, Texas*. Lunar and Planetary Institute, Houston, TX, abstract 1758.

PROCKTER, L. M., ERNST, C. M. *ET AL.* 2010. Evidence for young volcanism on Mercury from the third MESSENGER flyby. *Science*, **329**, 668–671.

ROTHERY, D. A. & MASSIRONI, M. 2013. A spectrum of tectonised basin edges on Mercury. *In*: *Proceedings of the 44th Lunar and Planetary Science Conference, March 18–22, 2013, The Woodlands, Texas*. Lunar and Planetary Institute, Houston, TX, abstract 1719.

SCHON, S. C., HEAD, J. W., BAKER, D. M. H., ERNST CAROLYN, M., PROCKTER, L. M., MURCHIE, S. L. & SOLOMON, S. C. 2011. Eminescu impact structure: insight into the transition from complex crater to peak-ring basin on Mercury. *Planetary Space Science*, **59**, 1949–1959.

SCHUBERT, G., ROSS, M. N., STEVENSON, D. J. & SPOHN, T. 1988. Mercury's thermal history and the generation of its magnetic field. *In*: VILAS, F., CHAPMAN, C. R. & MATTHEWS, M. S. (eds) *Mercury*. University of Arizona Press, Tucson, AZ, 429–460.

SCHULTZ, R. A. 1993. Brittle Strength of Basaltic Rock Masses with applications to Venus. *Journal of Geophysical Research*, **98**, 10 883–10 895.

SCHULTZ, R. A. 1995. Limits on strength and deformation properties of Jointed Basaltic Rock Masses. *Rock Mechanics and Rock Engineering*, **28**, 1–15.

SMITH, M. R., GILLESPIE, A. R., MONTGOMERY, D. R. & BATBAATAR, J. 2009. Crater-fault interactions: a metric for dating fault zones on planetary surfaces. *Earth and Planetary Science Letters*, **284**, 151–156.

SOLOMON, S. C. 1976. Some aspects of core formation in Mercury. *Icarus*, **28**, 509–521.

SOLOMON, S. C. 1977. The relationship between crustal tectonics and internal evolution in the Moon and Mercury. *Physics of the Earth and Planetary Interiors*, **15**, 135–145.

SOLOMON, S. C., MCNUTT, R. L. *ET AL*. 2008. Return to Mercury: a global perspective on MESSENGER's first Mercury flyby. *Science*, **321**, 59–62.

STROM, R. G. & NEUKUM, G. 1988. The cratering record on Mercury and the origin of impacting objects. *In*: VILAS, F., CHAPMAN, C. R. & MATTHEWS, M. S. (eds) *Mercury*. University of Arizona Press, Tucson, AZ, 336–373.

STROM, R. G., TRASK, N. J. & GUEST, J. E. 1975. Tectonism and volcanism on Mercury. *Journal of Geophysical Research*, **80**, 2478–2507.

TANAKA, K. L. 1982. A new time-saving crater-count technique with application to narrow features. *In*: HOLT, H. E. (ed.) *Planetary Geology and Geophysics Program Report*. NASA Technical Memorandum, **TM-85127**, 123–125.

THOMAS, P. G., MASSON, P. & FLEITOUT, L. 1988. Tectonic history of Mercury. *In*: VILAS, F., CHAPMAN, C. R. & MATTHEWS, M. S. (eds) *Mercury*. University of Arizona Press, Tucson, AZ, 401–428.

TOSI, N., GROTT, M., PLESA, A.-C. & BREUER, D. 2013. Thermochemical evolution of Mercury's interior. *Journal of Geophysical Research Planets*, **118**, 2474–2487.

WATTERS, T. R. 2004. Elastic dislocation modeling of wrinkle ridges on Mars. *Icarus*, **171**, 284–294.

WATTERS, T. R. & NIMMO, F. 2010. Tectonism on Mercury. *In*: SCHULTZ, R. A. & WATTERS, T. R. (eds) *Planetary Tectonics*. Cambridge University Press, Cambridge, 15–80.

WATTERS, T. R., ROBINSON, M. S. & COOK, A.-C. 1998. Topography of lobate scarps on Mercury: new constraints on the planet's contraction. *Geology*, **26**, 991–994.

WATTERS, T. R., ROBINSON, M. S. & COOK, A. C. 2001. Large-scale lobate scarps in the southern hemisphere of Mercury. *Planetary and Space Science*, **49**, 1523–1530.

WATTERS, T. R., ROBINSON, M. S., BINA, C. R. & SPUDIS, P. D. 2004. Thrust faults and the global contraction of Mercury. *Geophysical Research Letters*, **31**, L04701, http://dx.doi.org/10.1029/2003GL019171

WATTERS, T. R., NIMMO, F. & ROBINSON, M. S. 2005. Extensional troughs in the Caloris basin of Mercury: evidence of lateral crustal flow. *Geology*, **33**, 669–672.

WATTERS, T. R., SOLOMON, S. C., ROBINSON, M. S., HEAD, J. W., ANDRÉ, S. L., HAUCK, S. A. & MURCHIE, S. L. 2009. The tectonics of Mercury: the view after MESSENGER's first flyby. *Earth and Planetary Science Letters*, **285**, 283–296.

WATTERS, T. R., SOLOMON, S. C. *ET AL*. 2012. Extension and contraction within volcanically buried impact craters and basins on Mercury. *Geology*, **40**, 1123–1126, http://dx.doi.org/10.1130/G33725.1

WICHMAN, R. W. & SCHULTZ, P. H. 1989. Sequence and mechanisms of deformation around the Hellas and Isidis impact basins on Mars. *Journal of Geophysical Research*, **94**, 17 333–17 357.

WILSON, L. & HEAD, J. W. 2008. Volcanism on Mercury: a new model for the history of magma ascent and eruption. *Geophysical Research Letters*, **35**, L23205, http://dx.doi.org/10.1029/2008GL035620

WISE, D. U., GOLOMBEK, M. P. & MCGILL, G. E. 1979. Tharsis province of Mars: geologic sequence, geometry, and a deformation mechanism. *Icarus*, **38**, 456–472.

Faulted craters as indicators for thrust motions on Mercury

VALENTINA GALLUZZI[1,2]*, GAETANO DI ACHILLE[3,4],
LUIGI FERRANTI[1], CIPRIAN POPA[3] & PASQUALE PALUMBO[2]

[1]*DiSTAR, Università degli Studi di Napoli 'Federico II', Naples, Italy*

[2]*Dipartimento di Scienze e Tecnologie, Università degli Studi di Napoli 'Parthenope', Naples, Italy*

[3]*INAF – Osservatorio Astronomico di Capodimonte, Naples, Italy*

[4]*Present address: INAF – Osservatorio Astronomico di Collurania, Teramo, Italy*

*Corresponding author (e-mail: valentina.galluzzi@unina.it)

Abstract: Craters cross-cut by faults are used as markers to obtain fault geometric and kinematic properties. Assuming that the shape of these craters was originally circular, it is possible to measure the horizontal and vertical components of fault displacement as well as the slip trend. By applying trigonometric relations, slip plunge, displacement magnitude, fault true dip and fault rake can be derived from the observed values. An example application of this method on craters faulted by lobate scarps on Mercury shows that most of these inferred reverse faults have moderate oblique-slip trends. Moreover, the derived dips of thrusts vary over a wide range of angles. Some preliminary results in terms of fault rake compared with fault dip, strike and latitude are presented together with a pilot study to test and discriminate global tectonic models suggested for the evolution of Mercury. The possibility of estimating quantitative fault parameters through remotely sensed data provides significant assistance in the structural characterization of faults on planetary surfaces.

Despite the lack of direct observational data, several surface features have been confidently inferred as fault-related on most of the terrestrial planets and satellites since the beginning of planetary exploration (e.g. lobate scarps, Strom *et al.* 1975; wrinkle ridges, Plescia & Golombek 1986). The geometry and kinematics of these structures and their faults have previously been interpreted solely from visible imagery, radar and topographic data. On Earth, the typical parameters that are measured in the field to characterize fault kinematics are offset markers and kinematic indicators on fault surfaces such as slickenside *striae* and steps. In the case of remotely sensed structures such as Earth's seafloor faults or faults on other planets, satellites and minor bodies, however, markers of known pre-dislocation geometry are required to assess the fault kinematics.

Craters have been found to be an excellent deformation marker on planetary surfaces, both for kinematics and strain analysis (Strom *et al.* 1975; Thomas & Allemand 1993; Watters 1993; Golombek *et al.* 1996; Watters *et al.* 1998; Pappalardo & Collins 2005). Previous studies measured the change in shape of originally circular craters to estimate the strain of intensely deformed areas of Mars and Ganymede, especially in extensional provinces characterized by horst and graben systems (Thomas

& Allemand 1993; Golombek *et al.* 1996; Pappalardo & Collins 2005). At these locations, craters are often pervasively faulted, meaning that it is difficult to reconstruct their original shape. In some cases, however, craters are cross-cut by single faults and this offers an opportunity to better elucidate the fault kinematics. Previous studies have analysed craters shortened by structures such as lobate scarps or high-relief ridges on Mercury and Mars to characterize thrusts and reverse faults, respectively (e.g. Strom *et al.* 1975; Watters 1993; Watters *et al.* 1998). However, these studies relied on *a priori* assumptions on fault geometry and kinematics (i.e. the fault true dip angle and the slip vector).

The fault dip is a basic piece of information required to constrain the fault kinematics and is strongly dependent on the physical properties of the rock. For example, fault dip angle is required to estimate the amount of displacement and to calculate the displacement–length ratio (γ, Cowie & Scholz 1992) of faults on planetary surfaces (Watters *et al.* 1998, 2000; Watters 2003; Watters & Nimmo 2010; Byrne *et al.* 2014). Previously, fault dip angles of structures that accommodated shortening on Mars and Mercury have always been assumed in the range 25–35°, based on the results of mechanical models (Schultz & Watters

From: Platz, T., Massironi, M., Byrne, P. K. & Hiesinger, H. (eds) 2015. *Volcanism and Tectonism Across the Inner Solar System*. Geological Society, London, Special Publications, **401**, 313–325.
First published online June 12, 2014, http://dx.doi.org/10.1144/SP401.17

2001; Watters *et al.* 2002). These models assume a pure dip-slip motion for thrusts and normal faults. In most real cases, however, slip may have a substantial oblique component. Additionally, the slip direction may change along faults with a non-planar surface (e.g. Roberts 1996), as recurrently observed on low-angle dipping faults. All these complications require that the fault actual slip direction and true dip angle be determined.

The parameters that define fault geometry and kinematics are the attitudes of the fault plane and the slip line that lies on the fault plane. The azimuth of the line defined by the intersection between the fault plane and the horizontal plane is commonly known as the fault strike, and the azimuth of the horizontal projection of the slip

line is called the slip trend. The fault dip is measured perpendicularly to the fault strike, and the dip of the slip line is called the slip plunge (Fig. 1a). In addition to azimuths and dips the rake angle, which is the angle between the fault strike and the slip line within the fault plane, is of great importance in defining fault kinematics. The relative block motion across a fault usually lies somewhere between two end-member cases: strike-slip and dip-slip. In strike-slip motion, blocks move along the fault strike (rake is 0°), while in dip-slip motion they move along the dip direction (rake is 90°). When the motion on the fault is neither pure strike-slip nor pure dip-slip, it is generally called oblique-slip (rake is between 0° and 90°, Fig. 1a). Knowing these parameters, it is important

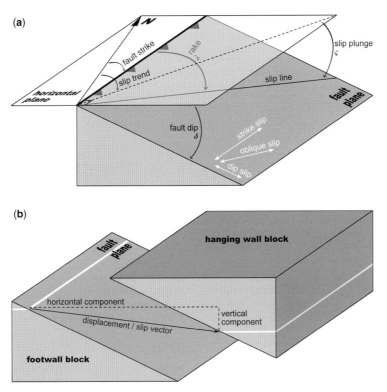

Fig. 1. Block diagram of a generic fault plane, redrawn and modified from both Allmendinger *et al.* (2011) and Twiss & Moores (2007). (**a**) Representation of the main angles defining a fault. On the fault plane, the three types of slip motion are represented by the white arrows near the bottom left-hand corner. The oblique black line represents a generic slip line as an example of oblique-slip motion. On the horizontal plane, three lines are indicated as follows: the arrowed line is the north direction, from which azimuths (i.e. strike and trend) can be measured; the bold line is the intersection of the fault plane and the horizontal plane (i.e. fault strike); and the dashed line is the horizontal projection of the slip line (i.e. slip trend). The slip plunge ς is measured as the tilt angle of the slip line with respect to the horizontal plane. Notice that the fault dip δ can be measured solely at right angles to the fault strike, while the fault rake λ can be measured only on the fault plane. (**b**) Representation of the slip vector or fault displacement. The vector orientation and tilt are the slip trend and plunge, respectively, already defined by the block diagram in (a). If a recognizable reference feature is present (white dislocated layer), the vector magnitude can be estimated and resolved into its horizontal and vertical components (dashed lines).

to represent the fault displacement as a vector (i.e. the slip vector, Fig. 1b) whose direction and dip are defined by the slip trend and plunge, respectively. Finally, the vector magnitude can be calculated from the horizontal and vertical component of displacement, which can be estimated using reference markers whose positions prior to faulting are known (Fig. 1b).

In this study we use craters as displacement markers and present a method to obtain quantitative geometric and kinematic parameters of the cross-cutting faults, such as true slip direction, horizontal and vertical components of displacement and thus the fault true dip and displacement magnitude. The parameters obtained from craters were used to estimate the fault rake and to quantitatively constrain its kinematics.

We have tested the proposed method on faulted craters found throughout the 30% of the surface of Mercury covered by the digital terrain models (DTM) of Preusker et al. (2011). The analysed craters are all cross-cut by lobate scarps interpreted as overlying reverse faults (Strom et al. 1975). The availability of the DTMs is crucial because they provide the possibility of calculating Hermean fault dips. The obtained results were plotted on several diagrams to search for possible correlations between different derived parameters. Since this method provides quantitative constraints on fault kinematics, it might be helpful to assess structural models at a regional or global scale. For Mercury, once global topography is available, this method can help to refine and validate the current hypotheses put forward to explain the tectonic evolution of the planet. Additionally, the method can be used for all types of faults, independent of their nature and of the rheology of faulted materials. If good DTM coverage is available, the method can be applied to any celestial bodies that have a brittle crust such as rocky planets, icy satellites and other minor bodies. This approach, therefore, can be very useful in planetary tectonic studies, and attempts to gather quantitative structural data from remotely sensed images.

Data and methods

We describe how craters can be used for measuring fault dislocation, assuming that the original outline of their rim was almost perfectly circular (in plan view) prior to deformation. This assumption may not always hold true; as already stated by Thomas & Allemand (1993), craters with a pristine non-circular shape may exist although an accurate analysis of the studied area can help to circumvent this problem. Specifically, it is possible to establish that the shape of a faulted crater was originally non-circular when: (a) the ellipticity is extremely high

with respect to the amount of deformation that would be reasonably expected from independent analysis of the cross-cutting fault; and (b) the shape is 'oval' with only one axis of symmetry, rather than an ellipse, which is usually the result of oblique impacts. The Sveinsdottir Crater, cross-cut by Beagle Rupes on Mercury, is an example of the latter case (Fig. 2).

The choice of a suitable spatial reference frame for the base maps is fundamental in order to obtain reliable measurements from faulted craters. We found that a stereographic projection centred on each analysed crater is the best compromise both for evaluating the crater shape and for measuring its geometry, because it does not introduce considerable distortions around the projection centre.

Since our method aims at estimating fault geometry and kinematics, we consider only those craters cross-cut by a single fault. Furthermore, we assume that a crater was rigidly deformed and its shape was extended or shortened uniquely due to the effect of the cross-cutting fault; erosion or different post-deformation processes are not taken into account.

With the exclusion of pure strike-slip (lack of vertical throw component) faults, one part of the crater will be raised and the other lowered due to

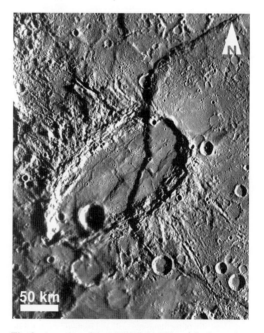

Fig. 2. A portion of the MDIS flyby 500 m/pixel mosaic (Becker et al. 2009) showing Sveinsdottir crater cross-cut by Beagle Rupes on Mercury (100.5°E; 2.5°S). The elongated shape of this crater is widely accepted as having been caused by an oblique impact.

faulting. Even if the nature of the fault is completely unknown, the hanging wall or footwall can still be identified based on the fault dip direction reconstructed from the analysis of the deformed crater.

If the rim was rigidly displaced, one should expect to see an offset along the strike of the fault between the raised and the lowered parts of the rim. The crater rim will seldom look like a perfectly cut and displaced circle, however; a method of studying the displacement effects is therefore possible using the graphical method described below.

Measuring fault slip components

The first step aims at measuring the fault horizontal and vertical components of slip, Δx and Δh, respectively (Fig. 3). To obtain Δx, a circle is drawn as the best fit of the rim portion on one side of the faulted

crater. In Figure 3a, the pink circle is assumed to represent the pre-dislocation reference shape of the originally circular crater; because of fault dislocation, it will not fit the crater rim on the opposite side of the fault (the uplifted rim). The unfitted part of the rim will consequently be either outside or inside the circle; if outside, it can be stated that the crater was extended by a normal fault, whereas, if inside, the crater was shortened by a reverse fault. In the case of a pure strike-slip fault, half the unfitted part of the rim will lie inside the circle and half will lie outside. These are end-member cases for pure dip- and strike-slip faults; however, all the intermediate cases (i.e. oblique-slip faults) are possible and their analysis is based on the following steps.

The horizontal component of slip Δx is found by shifting a copy of the previously drawn circle until it

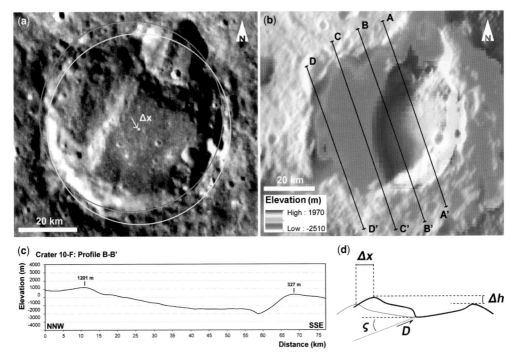

Fig. 3. Analysis of crater 10-F cross-cut by Enterprise Rupes. (**a**) A portion of the MDIS global mosaic at 250 m/pixel resolution. This specific image was chosen to show the cross-cutting fault, whereas the image mosaics with stronger contrast were used to make the fitting procedure easier. The yellow and pink circles fit the rim on the hanging wall and on the footwall of the thrust, respectively. The centre of each circle is represented by a dot with the same colour as the corresponding circle. The white arrow connecting the two centres represents Δx, and its azimuth gives the slip trend. (**b**) DTM (after Preusker *et al.* 2011) showing the vertical dislocation caused by the thrust. The black lines represent the profiles drawn parallel to the slip trend. (**c**) Vertically exaggerated ($\times 2$) cross-section diagram from line B–B′ in (b). The elevation of the hanging-wall and footwall rims is indicated. (**d**) 2D geometrical scheme representing an example of a faulted crater along a cross-section parallel to the slip direction. The reverse fault displaces the crater causing a displacement D along the fault plane. The vertical component of the displacement Δh is represented by the height difference between the two opposite rims. The horizontal component of the displacement Δx is represented by the horizontal translation of the hanging-wall rim from its original position. The angle between the fault and the horizontal plane is the slip plunge ς, the same as the fault true dip δ in the case of dip-slip faults.

fits the rim on the opposite side of the fault (yellow circle in Fig. 3a). Alternatively, a *new* circle can be built on the other side of the crater and then the two circle diameters can be compared, so as to have a better control on the predeformed crater shape. If the measured diameters are substantially different, the crater was not originally circular in shape and the method cannot be applied. Whatever graphical strategy is used for drawing the two reference circles, always using the same criterion (e.g. fitting the circles either on the highest part of the rim or on the break-in-slope just next to the crater floor) is recommended. After the two circles are drawn, the distance between their centres corresponds to the horizontal translation caused by the fault (i.e. Δx). Once Δx is measured, the rough fault kinematic parameters are already defined: the trend of the Δx segment corresponds to the slip trend (Fig. 3a).

To obtain the vertical component of slip, Δh, which is usually estimated using the fault scarp height (e.g. Watters *et al.* 1998), the use of a DTM is required. If the crater diameter is smaller than the fault length and its rim is far from the fault tips where a lesser amount of slip is expected (e.g. Kim & Sanderson 2005), we can assume that the rigid deformation caused by faulting produces an elevation difference between the two displaced crater parts that corresponds to the fault vertical displacement. To have a better control on Δh, we measure elevation at the crater floor and also at the rim. In fact, craters often present morphological irregularities at their floor such as central peaks, peak rings and topographic variations; the measurement of Δh on the rim therefore offers a more accurate estimate. This assumes that the displaced sides of the crater rim were initially at the same elevation before faulting, a condition which is met if the regional pre-impact slope was either very low or negligible across the crater area and if the crater was not tilted prior to tectonic deformation.

Under the latter assumptions, Δh is obtained by making a series of profiles across the crater based on the available DTM. Regardless of the direction in which we choose to draw the profiles, after a rigid displacement the crater would have virtually the same difference in elevation across the two faulted blocks within its perimeter. Drawing a profile perpendicular to the fault is routine in structural analysis to characterize the architecture of faulted regions; however, since we are interested in the components of the slip, we prefer to draw profiles parallel to the Δx direction (Fig. 3b) which is perpendicular to the fault trace only in the case of pure dip-slip motion. The value of Δh is obtained as the difference between the elevation of the raised rim and the lowered rim (Fig. 3c). A comparison between the values obtained at the base of the fault scarp (i.e. at the crater floor) and at the crater

rim, when both measures are available, can help to increase accuracy in the Δh estimate.

Estimating slip geometry and fault kinematics

Considering a planar fault surface as shown in Figure 1, once both the horizontal and vertical slip components are known the slip plunge ς and the amount of displacement D can be derived with simple plane trigonometry:

$$\varsigma = \tan^{-1}(\Delta h/\Delta x) \qquad (1)$$

$$D = \Delta h/\sin\varsigma \qquad (2)$$

Equation (2) was used by Watters & Nimmo (2010) to calculate the amount of thrust displacement using a hypothesized fault dip θ instead of slip plunge ς, on the assumption that all thrusts dip at 25–30° (Schultz & Watters 2001; Watters *et al.* 2002). In contrast we can calculate the slip plunge ς that, for pure dip-slip faults, matches fault true dip δ. As stated in the previous sections, most faults have moderate oblique-slip behaviour; for this reason, angle ς will in most cases be smaller than true dip δ (i.e. ς is an apparent fault dip). Despite this, the true dip of a fault can still be calculated starting from an apparent dip that is associated with a known trend, which in this case is the slip trend. Based on a commonly used relation, the true dip is:

$$\delta = \tan^{-1}(\tan\varsigma/\sin\varphi) \qquad (3)$$

where φ is the angle between the fault strike and the slip trend that was measured from the Δx segment.

At this step the fault superficial geometry is completely known, and we can proceed to investigate its kinematics. The angle between the fault strike and the slip measured within the fault plane is the fault rake, λ. In a three-dimensional space, the angle between two lines can be calculated by resolving the dot product of the two unit vectors defined by the orientations of the lines. Considering the fault strike and slip as unit vectors, the angle between them will therefore be:

$$\lambda = \cos^{-1}(f_1s_1 + f_2s_2 + f_3s_3) \qquad (4)$$

where f_{1-3} and s_{1-3} are the direction cosines of the fault strike unit vector and slip unit vector, respectively, considering the fault strike as a line with 0° plunge.

Application to Mercury

Models for the tectonic evolution of Mercury have been proposed since the 1970s, when the Mariner

10 mission revealed the first images of the planet. The surface of Mercury is characterized by the presence of widespread linear features interpreted as thrust-fault-related landforms (i.e. lobate scarps and wrinkle ridges). These tectonic features probably formed as a result of global contraction (e.g. Strom *et al.* 1975) generated by secular cooling of the planet's interior (Solomon 1976; Hauck *et al.* 2004), by tidal despinning forces possibly combined with thermal contraction (Melosh & Dzurisin 1978; Pechmann & Melosh 1979; Matsuyama & Nimmo 2009) or by mantle convection (King 2008). The MErcury Surface, Space ENvironment GEochemistry, and Ranging (MESSENGER) mission (Solomon *et al.* 2007) confirmed the predominant contractional character of Mercury's tectonics (Watters *et al.* 2009; Watters & Nimmo 2010), and lobate scarps were mapped as the major tectonic landform of the planet. Whatever the origin of these structures, their geometry and kinematics are, however, still poorly defined.

The surface of Mercury abounds in craters that although many have been moderately modified by space weathering or gravitational processes, still retain recognizable rims. We therefore applied the method described above to those craters cross-cut by lobate scarps to obtain fault dips and rakes.

Locating Mercury's faulted craters

In order to identify faulted craters on Mercury, several images and mosaics at different scales and illumination conditions were used: (a) Mercury's near-global mosaic of images acquired prior to MESSENGER's orbital operations at 500 m/pixel (including Mariner 10 images, Becker *et al.* 2009); (b) the global mosaic of MESSENGER orbital images at 250 m/pixel; and (c) different individual or partially mosaicked orbital images from the latest releases of the Mercury Dual Imaging System (MDIS) instrument on board MESSENGER spacecraft, from the Wide Angle Camera (WAC) or Narrow Angle Camera (NAC). For altimetry, the stereo-topographic models by Preusker *et al.* (2011) were used. This DTM covers 30% of Mercury's surface with a grid spacing of 1 km and allows reliable measurements of features with a horizontal extent of at least 15 km.

Within the limits imposed by data coverage, resolution and illumination, we were able to find 44 craters intersected by linear features for which topographic data were available (Fig. 4). Most of these craters had to be excluded, however, due to the following reasons and observational limits: (a) small wrinkle ridges at the crater floors, which might result from local stress fields rather than from global contraction; (b) faults too small to produce a resolvable dislocation on the rim; (c) craters too close to DTM boundaries, where there are higher uncertainties in elevation estimates; and (d) craters with complicated morphology such as palaeo-landforms inherited by older and larger underlying ghost craters, whose pre-existing slopes were not totally reset after the formation of younger craters.

Discarding all the above cases, 15 out of 44 craters were found suitable for our analyses (Figs 4 & 5).

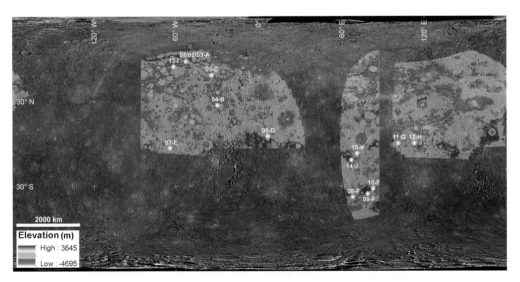

Fig. 4. Global map of Mercury in equirectangular projection (MDIS orbit mosaic *v*9 by NASA/JHUAPL/CIW, 250 m/pixel), together with the stereo-DTM by Preusker *et al.* (2011), showing the location of the studied faulted craters.

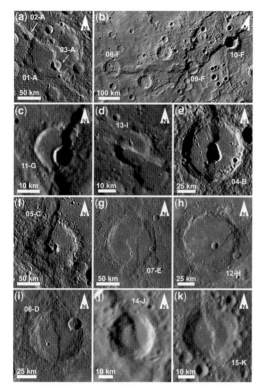

Fig. 5. MDIS images of the 15 analysed faulted craters of Mercury in stereographic projections. All the images are taken from the 250 m/pixel global mosaic, except for (f) and (j), which are taken from the 500 m/pixel mosaic by Becker *et al.* (2009). (**a**) Craters 01-A, Duccio (largest crater), 02-A and 03-A (white arrows) on Carnegie Rupes. (**b**) From west to east, craters 08-F, 09-F and 10-F on Enterprise Rupes. (**c**) Crater 11-G on Beagle Rupes. (**d**) Crater 13-I. (**e**) Crater 04-B, Geddes on Endeavour Rupes. (**f**) Crater 05-C on Victoria Rupes. (**g**) Crater 07-E on Thakur. (**h**) Crater 12-H. (**i**) Crater 06-D. (**j**) Crater 14-J. (**k**) Crater 15-K.

Results

The 15 faulted craters were numbered sequentially and accompanied by a letter corresponding to the cross-cutting fault (Table 1). This means that craters 01-A, 02-A and 03-A are cross-cut by the same lobate scarp (Carnegie Rupes, Fig. 5a), as are craters 08-F, 09-F and 10-F, which are located in the Rembrandt crater area (i.e. by Enterprise Rupes, Fig. 5b).

Table 1 shows that craters 01-A to 03-A, although cross-cut by the same fault, provide a different estimate of the true dip angle, ranging from 31° to 53°. Although some error certainly arises from the limited size of craters 02-A and 03-A when compared with crater 01-A, we believe that the

main discrepancy may result from the different age of the three craters in relation to fault activity. The two smaller craters clearly formed after the larger crater, 01-A (Duccio). Following Melosh & McKinnon (1988), who argue that lobate scarps formed after the emplacement of large craters during the Late Heavy Bombardment (LHB) period, we hypothesize that the fault formed after the imposition of crater 01-A. On the other hand, the Δx measurements show a dislocation of crater 01-A about three times larger than that recorded by craters 02-A and 03-A; on this basis we argue that the fault was already active before the formation of craters 02-A and 03-A. Because of their limited size, it is possible that during the impact the younger craters did not totally erase the pre-existing fault scarp. The hypothesized scenario would lead to a vertical offset measurement across 02-A and 03-A that is higher than the actual Δh caused by the fault motion. This would explain the observed steeper dip when compared with crater 01-A. For this reason, although craters 02-A and 03-A are still useful for evaluating the history of incremental shortening, the data obtained from Duccio crater are a more accurate estimate of the finite fault displacement. Craters 02-A and 03-A, along with craters 11-G and 13-I (Fig. 5c & d, respectively), are the smallest craters of the dataset in Table 1. While the results of the first two craters can be compared with the results of the underlying older crater 01-A, the results from craters 11-G and 13-I cannot be verified with any other feature; their reliability will therefore be treated with caution in our subsequent analysis.

Craters 08-F, 09-F and 10-F are located on Enterprise Rupes (Fig. 5b). There is no superposition relationship among the three craters and they are probably similar in age, although they in fact yield a different estimate of the fault dip δ. This is likely a function of the fault strike, which changes substantially eastwards from crater 08-F to crater 10-F (N263° to N216°; Table 1), although the slip trend remains almost the same for craters 08-F and 09-F (N107° and N116°, respectively; Table 1) and is different for crater 10-F (N158°; Table 1). The resulting rake for crater 09-F is in fact 116° (close to dip-slip), 141° (right oblique-slip) for crater 08-F and 59° (left oblique-slip) for crater 10-F. Craters 08- to 10-F confirm the existence of a consistent relationship between the fault rake and its dip, discussed in the section titled 'Relation between fault geometric and kinematic parameters'.

Craters 04-B (Geddes, Fig. 5e) and 05-C (Fig. 5f) are located at the same longitude and are cut by faults grouped into the same NNW–SSE-trending fault array. Nevertheless, since they are spaced c. 900 km from each other, we prefer to label the individual fault segments that cut them differently

Table 1. Dislocation data of Hermean faulted craters

Crater	Lon. (°)	Lat. (°)	Diameter (km)	Δx* (km)	Δh (km)	n	σΔh (km)	Trend† (°)	Plunge (°)	σ plunge (°)	Strike‡ (°)	δ (°)	σδ (°)	φ† (°)	λ (°)	σλ (°)	D (km)	σD (km)
01-A	−52.5	58.2	109.90	4.14	2.43	6	0.25	241	30	4	323	31	4	82	83	10	4.80	0.70
02-A	−55.0	58.9	18.27	1.19	1.53	1	0.14	218	52	4	297	53	4	79	83	7	1.94	0.20
03-A	−52.3	57.5	22.13	1.43	1.25	1	0.14	236	41	4	333	42	4	97	95	9	1.90	0.26
04-B	−29.6	27.1	87.30	4.52	1.14	4	0.22	104	14	3	185	14	3	81	81	14	4.66	1.29
05-C	−34.0	49.4	98.57	3.96	1.42	3	0.20	96	20	3	183	20	3	87	87	14	4.20	0.88
06-D	6.9	5.3	64.79	1.89	0.77	4	0.14	223	22	4	356	29	8	133	129	12	2.04	0.50
07-E	−64.4	−3.0	107.98	4.44	0.69	3	0.18	90	9	2	199	9	3	109	109	15	4.50	1.67
08-F	68.1	−37.9	79.73	3.07	1.89	4	0.16	107	32	3	263	57	16	156	141	7	3.60	0.46
09-F	78.8	−35.5	59.70	2.93	1.47	5	0.34	116	26	6	235	30	7	119	116	10	3.27	0.99
10-F	82.9	−31.7	55.51	3.96	0.93	6	0.22	158	13	3	216	15	5	58	59	11	4.07	1.41
11-G	101.2	0.2	16.63	1.87	0.54	1	0.14	335	16	4	12	26	10	37	40	10	1.94	0.68
12-H	113.1	0.2	85.83	4.43	0.53	5	0.22	343	7	3	55	7	3	72	72	13	4.46	2.59
13-I	−61.5	54.6	20.38	1.67	0.33	1	0.14	181	11	5	292	12	5	111	111	15	1.71	0.98
14-J	66.9	−11.6	32.14	0.82	0.79	1	0.14	85	44	6	161	45	6	76	80	10	1.13	0.23
15-K	71.3	−6.9	32.33	2.39	0.40	1	0.14	253	10	3	343	10	3	90	90	14	2.43	1.16

*The standard deviation is assumed to be 10% of the value.
†The standard deviation is 15° for all angles.
‡The standard deviation is 1° for all angles.

Average dislocation results of the 15 analysed faulted craters on Mercury. The craters are referred to numerically from 01 to 15 and associated with capital letters from A to K representing the corresponding eleven faults. Column Δx shows the measured slip horizontal component. Column Δh shows the average slip vertical component. Column n shows the number of profiles drawn on each crater to obtain the average Δh result. Fault strike was measured using the right-hand rule, and the slip trend indicates the direction of hanging-wall motion. Column δ shows the calculated true dip angles. The angle between the slip trend and the fault strike φ is also indicated. Column λ displays the rake values that were calculated on the footwall side of the fault from the strike direction to the slip vector, using the Aki & Richards (2002) convention. Each parameter is also associated with its calculated standard deviation σ.

(B and C). This is consistent with the existing nomenclature that places the two craters on separate rupēs (Geddes crater, 04-B on Endeavour Rupes and the unnamed crater 05-C on Victoria Rupes), which are evidently part of the same thrust system. The similarity between the two fault segments is marked not only by their strike (N183° and N185°; Table 1), but also by the derived dip, trend and rake. Craters 04-B and 05-C indicate that the Victoria–Endeavour Rupes system has a dip of 15–20° and near dip-slip kinematics (Table 1).

The lowest derived dips of 9° and 7° are those obtained for craters 07-E (Thakur) and 12-H, respectively (Fig. 5g, h). Such low dips are consistent with the accentuated arcuated shape in plan view of the fault scarp trace. As a comparison, crater 06-D (Fig. 5i), which is cross-cut by a more rectilinear lobate scarp than those of the above-mentioned craters, leads to a steeper dip estimate (29°; Table 1).

Uncertainties and error discussion

An analysis of the uncertainties of our derived parameters (plunge, φ, dip, rake and amount of displacement) starts with errors and uncertainties on the measured data (Δx, trend, Δh and strike, referred to as input parameters). Since the input parameters come from different sources and our approach is based on a mixture of different types of data, including images and DTMs (and their associated uncertainties), a detailed error analysis is not a simple task. For this reason, the analysis of each input parameter deserves a dedicated discussion.

The method used to measure Δx is a mixture of quantitative data derived from image features and qualitative image interpretation. Both (1) the subjectivity in crater rim fitting and shifting and (2) image resolution, linked to ground pixel size and other optical parameters, provide a source of error. Systematic errors due to subjective interpretation are difficult to estimate and depend on many factors. An empirical approach is to repeat the measurement of the same feature several times and by different users with a comparable background and experience of geological mapping, so that an average value and an associated uncertainty can be derived. Based on a test performed with three different mappers, we can associate a standard deviation of c. 10% of the measured Δx. Note that in our dataset we have already excluded craters with uncertainties due to highly irregular or unrecognizable edges.

As far as the uncertainty originating from image resolution is concerned, the ground pixel size is normally between 150 and 250 m. Despite this, we derive the measurement of crater rim displacement by fitting a feature that involves many pixels. This means that the overall position error is negligible with respect to other error sources, as it is estimated to be well below one pixel.

The slip trend is strictly connected to the Δx parameter, since they both describe the vector connecting the two circle centres (i.e. vector orientation and magnitude, respectively). It is therefore affected by similar uncertainties and, by applying the same approach as before (i.e. repeated measures), we estimate an uncertainty of c. 15°.

The DTM-derived parameter Δh is measured as a difference in DTM elevation between two points, with the measure repeated on different sections of the rim whenever the crater morphology allows it. Uncertainties in Δh are then determined by DTM errors and measurement variability from different profiles. Preusker et al. (2011) estimated a standard deviation of 135 m on the stereo-derived DTM when compared to MESSENGER Laser Altimeter (MLA) data, to which a much higher accuracy is attributed. This value was derived from MLA binning within a running 15 km-long box, which gives an indication of the 'effective resolution' of the DTM. Although we avoided taking into account craters smaller than 15 km in diameter, the remaining craters were analysed using the DTM at its higher spatial resolution (1 km). On the other hand, we are not interested in an absolute error but in the relative elevation error between points that are relatively close to each other (16–110 km), and we expect a better accuracy in this case.

In order to account for the local variability in crater rim elevations, whenever possible (e.g. for large craters, data availability) we estimate Δh along different profiles. In these cases we derive the average and the standard deviation as Δh and its associated uncertainty, respectively.

Finally, Zuber et al. (2012) found long-wavelength features in surface elevations, the origin of which still remains unclear (Byrne et al. 2014). These oscillations are up to 3 km in amplitude and have wavelengths of c. 1200 km. In the worst case, we could expect a contribution of up to c. 300 m in a 100 km-wide crater. We do not consider these oscillations to have a meaningful effect on our measurements as we can argue that, for our largest craters (>50 km), the impact completely resets the local topography and for the smallest craters (<50 km) the offset due to an underlying shallow slope is negligible. In any case, we exclude from our dataset craters embedded in complex local topography.

Based on the above discussion, we attribute to the standard deviation for each Δh a maximum value of between 135 m (from DTM uncertainty) and the standard deviation computed from the repeated measurements of each crater (Table 1).

The strike measurement is similar to the slip trend, being an angular measurement of the line linking the intersections between the fault and the crater rim. The variability of multiple measurements in this case is very low, bringing a typical standard deviation of 1°.

Taking into account the above-mentioned sources of uncertainty, we compute the uncertainties in the derived fault parameters by propagating the errors through the standard approach used for statistical errors. Most of the derived parameters (plunge, φ, dip, rake) show statistical uncertainties of between <10% and 40%. The worst cases refer to the highest relative errors on Δh associated with the smallest vertical displacement components (i.e. smallest craters). In particular, the propagated error on the displacement magnitude D is between 10% and 58% (Table 1). Moreover, the D parameter is subject to uncertainties arising from the relative age of the crater with respect to the fault activity. The value of $D \pm \sigma D$ can either correspond to the total amount of displacement accumulated by the fault or, if the fault partly acted before the emplacement of the crater, to a minimum amount of displacement recorded by the crater since the time of the impact. These considerations strongly affect the Δx parameter, which can be at its maximum in the first case or at a minimum in the second case. Finally, since the D parameter also depends on Δh (i.e. equations (1) & (2)), we stress the importance of carefully considering the crater size. Pre-existing scarps can be erased by large impacts, but this might not be the same for smaller impacts: this issue affects Δh and all its derived parameters, as already demonstrated by the example of craters 02-A and 03-A.

Discussion

Several authors have proposed general relationships between fault orientation and kinematics and their latitudinal distribution on Mercury. Melosh (1977) argued that on a despun planet, north–south-oriented thrusts at low latitudes, NW–SE and NE–SW strike-slip faults at mid-latitudes and east–west normal faults at high latitudes should be found. The tidal despinning model was refined for Mercury by Pechmann & Melosh (1979) and subsequently by Matsuyama & Nimmo (2009), who proposed a wider distribution of thrusts even at higher latitudes. These models directly correlate fault strikes and types of slip with their geographical position. Dombard & Hauck (2008) hypothesized that tidal despinning forces ceased just before the LHB period since there is a lack of latitudinal patterns in the distribution of lobate scarps, but post-LHB global contraction may have

reactivated old despinning structures, inheriting their orientation.

The formation of thrusts due to the proposed tectonic models varies from the total randomness predicted by the model of global contraction due to core solidification to the precise latitudinal variations expected by the tidal despinning model. As a consequence, it should be possible to verify and observe the same randomness or order by cross-checking fault parameters. Given the wide distribution and high number of lobate scarps present on the surface of Mercury (Byrne *et al.* 2014), the 11 analysed thrusts are too few to be representative of a global-scale tectonic interpretation. Moreover, the following plot analyses must be regarded as very preliminary given the possibility that the dataset is affected by an observational bias due to the lighting conditions in the basemaps, which could cause an overestimation of north–south-oriented structures relative to east–west-oriented structures (see also Watters *et al.* 2004; Watters & Nimmo 2010; Di Achille *et al.* 2012; Byrne *et al.* 2014). Nevertheless, these data still give important kinematic constraints and represent an example of how such parameters could be used once more data have become available.

Relation between fault geometric and kinematic parameters

The cumulative data relationships between the various geometric and kinematic fault parameters provided by the offset craters analysis are presented in Figures 6–9. Since we did not find craters cut by normal faults, the diagrams show only positive rake values (following the convention of Aki & Richards 2002) and are divided, according to the fault

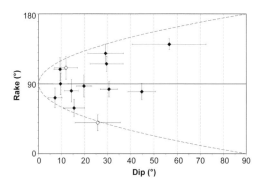

Fig. 6. Dip–rake diagram for the data in Table 1; for fault 'A', only data from crater 01-A were considered. The error bars represent the standard deviations of dip and rake in Table 1. The dashed parabola qualitatively encloses the expected trend. The hollow diamonds represent the two smaller craters 11-G and 13-I.

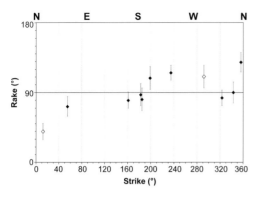

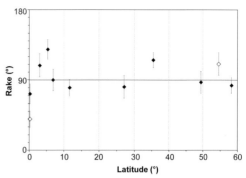

Fig. 7. Strike–rake diagram for the 11 analysed lobate scarps (data from Table 1). The error bars represent rake standard deviation in Table 1. In the upper part of the diagram, cardinal points are also indicated. Strikes are measured with the right-hand rule so that faults striking east or west lie in the same direction but dip in opposite directions (south and north, respectively); the same applies for faults striking north and south that dip east and west, respectively. The hollow diamonds represent the two smaller craters 11-G and 13-I.

Fig. 9. Latitude–rake diagram for the 11 analysed lobate scarps (data from Table 1). All latitudes are shown as positive values, independent of the hemisphere. The error bars represent rake standard deviation in Table 1. The hollow diamonds represent the two smaller craters 11-G and 13-I.

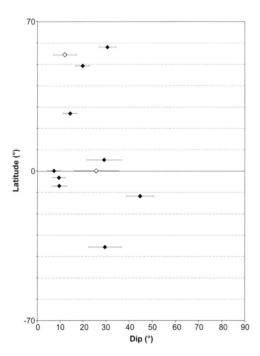

Fig. 8. Dip–latitude diagram for the 11 analysed lobate scarps (data from Table 1), showing fault true dip variations throughout the analysed latitudinal range. The error bars represent dip standard deviation in Table 1. The hollow diamonds represent the two smaller craters 11-G and 13-I.

parameters, into fields that include all the possible reverse fault types from dip-slip reverse faults to right- and left-lateral strike-slip faults, through different reverse oblique-slip faults.

Figure 6 illustrates the relation between dip on the x-axis and rake on the y-axis. To better illustrate the expected trend, we have traced a parabolic curve on the diagram that encompasses, in a qualitative manner, the plotted measurements. The curve was drawn considering two end-member cases: a horizontal pure dip-slip thrust and a vertical pure strike-slip fault, assuming homogeneous plane strain conditions. It therefore intercepts the dip axis at $(90°; 0°)$ and its vertex stands on $(0°; 90°)$, with the equation $x = (1/90) \, y^2 - 2y + 90$. This parabola qualitatively illustrates the predicted pattern for the dip–rake data, spreading from shallow dips and near dip-slip rakes to steeper dips and oblique-slip rakes. It is apparent from the diagram that faults with low dips have a near dip-slip displacement. On the other hand, the strike-slip component increases with increasing fault dips. The relationship outlined by the parabolic curve holds for both large craters and for the two smaller (and younger) craters 11-G and 13-I.

In Figure 7, fault strikes (x-axis) are plotted against fault rake (y-axis) to compare the azimuth of faults with their kinematics. Here and in the following diagrams we consider fault 'F' (i.e. Enterprise Rupes) as represented by its centremost crater 09-F and therefore avoid plotting the results from craters 08-F and 10-F (Fig. 5b). Unlike the dip–rake plot, the strike–rake diagram should not show any particular trends unless the deformation fields are associated with specific tectonic processes (e.g. tidal despinning). The resulting plot shows that some faults striking north–south to *c*. NNW–SSE cluster near dip-slip rake angles and are therefore almost pure dip-slip thrusts. On the other hand, faults striking increasingly towards the east–west direction have a progressive shift towards moderate strike-slip rake angles and are therefore oblique-reverse faults. One of the two younger and smaller craters apparently follows the trend of the older craters.

In a global-scale context, fault strike is usually analysed with respect to the fault location to understand whether a global tectonic pattern exists or not (e.g. Watters *et al.* 2004). In this case, the diagrams in Figures 8 and 9 correlate the geographical location of faults with respect to their geometry and kinematics, illustrating the latitudinal distribution of dips and rakes, respectively.

In Figure 8, fault dips (*x*-axis) are plotted against fault latitude (*y*-axis). This representation portrays how fault dips change throughout the analysed latitudinal range (from 40°S to 60°N). Although the data are limited in number, the latitudinal distribution of dip angles seems to indicate that faults with low-angle dips are preferentially located at low latitudes.

In Figure 9 fault latitude (*x*-axis) is plotted against fault rake (*y*-axis). Here we simplify the representation of fault latitude by giving only its modulus value. This provides a better interpretation of rake changes within generic latitudinal bands (i.e. low-, mid- and high-latitudes). This diagram shows that near dip-slip thrusts are present at all latitudes. The dataset is not statistically significant, however, preventing us from determining whether transpressive faults are widely distributed or latitudinally concentrated.

The database used for the proposed diagrams requires further investigation since tens of faulted craters could not be analysed because of their limited size or complicated morphology. High-resolution topography will therefore be crucial to increase the available dataset and to study trends in fault parameters.

Summary and conclusions

The use of faulted craters as kinematic markers can help in the quantitative study of planetary tectonics. Two main input parameters for the method illustrated in this paper are the horizontal (Δx) and vertical (Δh) fault displacement components that can be measured from imagery and topography, respectively. When Δx and Δh are measured, the overall fault slip geometry can be defined as well as the fault kinematics. Even if Δh is unknown (i.e. there is no available topography across the fault), Δx and Δx trend (i.e. slip trend) alone can yield information on the fault kinematics since the angle between fault strike and slip trend on the horizontal

plane φ is often very similar to the rake angle λ calculated along the fault plane.

By applying this method on 15 faulted craters found across the 30% of Mercury's surface covered by flyby stereo-derived DTM (Preusker *et al.* 2011), our results interestingly show that the derived fault dips are within a much wider range ($7° < \delta < 57°$) than the 30–35° interval predicted by Watters *et al.* (2002). This could be explained by the fact that our data are obtained from direct measurements made on multiple faulted markers, rather than from a numerical fit of the geomorphic surface of a single lobate scarp (i.e. Discovery Rupes, Watters *et al.* 2002), and therefore have the advantage of showing the natural variability inherent to fault geometry possibly related to rheological differences (e.g. faults forming within different lithologies). The possibility that the derived wide range of thrust dip angles could partly depend on stress changes at a regional scale is not, however, excluded. This is clearly indicated by the resulting rake variability ($40° < \lambda < 141°$), denoting a large variety of thrust motion directions. This emphasizes that some of the reverse faults on Mercury are not pure dip-slip thrusts, as commonly expected.

The authors gratefully acknowledge the useful comments and suggestions made by R. Herrick, M. Massironi, T. Platz, P. K. Byrne and an anonymous reviewer. We are grateful to D. Rothery for polishing the English of this manuscript. This research was supported by the Italian Space Agency (ASI) within the SIMBIOSYS project (ASI-INAF agreement no. I/022/10/0). The authors acknowledge the use of MESSENGER MDIS image mosaics processed by NASA/Johns Hopkins University Applied Physics Laboratory/Carnegie Institution of Washington.

References

AKI, K. & RICHARDS, P. G. 2002. *Quantitative Seismology.* 2nd edn. University Science Books, Sausalito, CA.

ALLMENDINGER, R. W., CARDOZO, N & FISHER, D. M. 2011. *Structural Geology Algorithms, Vectors and Tensors.* Cambridge University Press, Cambridge, UK.

BECKER, K. J., ROBINSON, M. S. ET AL. 2009. Near global mosaic of Mercury. *Eos Transactions AGU*, **90** (Fall meeting suppl.), Abstract P21A–1189.

BYRNE, P. K., KLIMCZAK, C., CELÂL ŞENGÖR, A. M., SOLOMON, S. C., WATTERS, T. R. & HAUCK II, S. A. 2014. Mercury's global contraction much greater than earlier estimates. *Nature Geoscience*, **7**, 301–307, http://dx.doi.org/10.1038/ngeo2097

COWIE, P. A. & SCHOLZ, C. H. 1992. Displacement-length scaling relationship for faults: data synthesis and discussion. *Journal of Structural Geology*, **14**, 1149–1156, http://dx.doi.org/10.1016/0191-8141(92)90066-6

DI ACHILLE, G., POPA, C., MASSIRONI, M., MAZZOTTA EPIFANI, E., ZUSI, M., CREMONESE, G. & PALUMBO,

P. 2012. Mercury's radius change estimates revisited using MESSENGER data. *Icarus*, **221**, 456–460, http://dx.doi.org/10.1016/j.icarus.2012.07.005

DOMBARD, A. J. & HAUCK, S. A. II 2008. Despinning plus global contraction and the orientation of lobate scarps on Mercury: Predictions for MESSENGER. *Icarus*, **198**, 274–276, http://dx.doi.org/10.1016/j.icarus.2008.06.008

GOLOMBEK, M. P., TANAKA, K. L. & FRANKLIN, B. J. 1996. Extension across Tempe Terra, Mars, from measurements of fault scarp widths and deformed craters. *Journal of Geophysical Research*, **101**, 26119–26130, http://dx.doi.org/10.1029/96JE02709

HAUCK, S. A. II, DOMBARD, A. J., PHILLIPS, R. J. & SOLOMON, S. C. 2004. Internal and tectonic evolution of Mercury. *Earth and Planetary Science Letters*, **222**, 713–728, http://dx.doi.org/10.1016/j.epsl.2004.03.037

KIM, Y.-S. & SANDERSON, D. J. 2005. The relationship between displacement and length of faults: a review. *Earth-Science Reviews*, **68**, 317–334, http://dx.doi.org/10.1016/j.earscirev.2004.06.003

KING, S. D. 2008. Pattern of lobate scarps on Mercury's surface reproduced by a model of mantle convection. *Nature Geoscience*, **1**, 229–232, http://dx.doi.org/10.1038/ngeo152

MATSUYAMA, I. & NIMMO, F. 2009. Gravity and tectonic patterns of Mercury: effect of tidal deformation, spin-orbit resonance, nonzero eccentricity, despinning, and reorientation. *Journal of Geophysical Research*, **114**, E01010, http://dx.doi.org/10.1029/2008JE003252

MELOSH, H. J. 1977. Global tectonics of a despun planet. *Icarus*, **31**, 221–243, http://dx.doi.org/10.1016/0019-1035(77)90035-5

MELOSH, H. J. & DZURISIN, D. 1978. Mercurian global tectonics: a consequence of tidal despinning? *Icarus*, **35**, 227–236, http://dx.doi.org/10.1016/0019-1035(78)90007-6

MELOSH, H. J. & MCKINNON, W. B. 1988. The tectonics of Mercury. *In*: VILAS, F., CHAPMAN, C. R. & MATTHEWS, M. S. (eds) *Mercury*. University of Arizona Press, Tucson, AZ, 374–400.

PAPPALARDO, R. T. & COLLINS, G. C. 2005. Strained craters on Ganymede. *Journal of Structural Geology*, **27**, 827–838, http://dx.doi.org/10.1016/j.jsg.2004.11.010

PECHMANN, J. B. & MELOSH, H. J. 1979. Global fracture patterns of a despun planet: application to Mercury. *Icarus*, **38**, 243–250, http://dx.doi.org/10.1016/0019-1035(79)90181-7

PLESCIA, J. B. & GOLOMBEK, M. P. 1986. Origin of planetary wrinkle ridges based on the study of terrestrial analogs. *Geological Society of America Bulletin*, **97**, 1289–1299, http://dx.doi.org/10.1130/0016-7606(1986)97<1289:OOPWRB>2.0.CO;2

PREUSKER, F., OBERST, J., HEAD, J. W., WATTERS, T. R., ROBINSON, M. S., ZUBER, M. T. & SOLOMON, S. C. 2011. Stereo topographic models of Mercury after three MESSENGER flybys. *Planetary and Space Science*, **59**, 1910–1917, http://dx.doi.org/10.1016/j.pss.2011.07.005

ROBERTS, G. P. 1996. Variation in fault-slip directions along active and segmented normal fault systems. *Journal of Structural Geology*, **18**, 835–845, http://dx.doi.org/10.1016/S0191-8141(96) 80016-2

SCHULTZ, R. A. & WATTERS, T. R. 2001. Forward mechanical modeling of the Amenthes Rupes thrust fault on Mars. *Geophysical Research Letters*, **28**, 4659–4662, http://dx.doi.org/10.1029/2001GL013468

SOLOMON, S. C. 1976. Some aspects of core formation in Mercury. *Icarus*, **28**, 509–521, http://dx.doi.org/10.1016/0019-1035(76)90124-X

SOLOMON, S. C., MCNUTT, R. L., GOLD, R. E. & DOMINGUE, D. L. 2007. MESSENGER mission overview. *Space Science Reviews*, **131**, 3–39, http://dx.doi.org/10.1007/s11214-007-9247-6

STROM, R. G., TRASK, N. J. & GUEST, J. E. 1975. Tectonism and volcanism on Mercury. *Journal of Geophysical Research*, **80**, 2478–2507, http://dx.doi.org/10.1029/JB080i017p02478

THOMAS, P. G. & ALLEMAND, P. 1993. Quantitative analysis of the extensional tectonics of Tharsis Bulge, Mars: geodynamic implications. *Journal of Geophysical Research*, **98**, 13 097–13 108, http://dx.doi.org/10.1029/93JE01326

TWISS, R. J. & MOORES, E. M. 2007. *Structural Geology*. 2nd edn. W. H. Freeman, New York.

WATTERS, T. R. 1993. Compressional tectonism on Mars. *Journal of Geophysical Research*, **98**, 17 049–17 060, http://dx.doi.org/10.1029/93JE01138

WATTERS, T. R. 2003. Thrust faults along the dichotomy boundary in the eastern hemisphere of Mars. *Journal of Geophysical Research*, **108**, 5054, http://dx.doi.org/10.1029/2002JE001934

WATTERS, T. R. & NIMMO, F. 2010. The tectonics of Mercury. *In*: WATTERS, T. R. & SCHULTZ, R. A. (eds) *Planetary Tectonics*. Cambridge University Press, Cambridge, 15–80.

WATTERS, T. R., ROBINSON, M. S. & COOK, A. C. 1998. Topography of lobate scarps on Mercury: new constraints on the planet's contraction. *Geology*, **26**, 991–994, http://dx.doi.org/10.1130/0091-7613(1998)026<0991:TOLSOM>2.3.CO;2

WATTERS, T. R., SCHULTZ, R. A. & ROBINSON, M. S. 2000. Displacement-length relations of thrust faults associated with lobate scarps on Mercury and Mars: comparison with terrestrial faults. *Geophysical Research Letters*, **27**, 3659–3662, http://dx.doi.org/10.1029/2000GL011554

WATTERS, T. R., SCHULTZ, R. A., ROBINSON, M. S. & COOK, A. C. 2002. The mechanical and thermal structure of Mercury's early lithosphere. *Geophysical Research Letters*, **29**, 1542, http://dx.doi.org/10.1029/2001GL014308

WATTERS, T. R., ROBINSON, M. S., BINA, C. R. & SPUDIS, P. D. 2004. Thrust faults and the global contraction of Mercury. *Geophysical Research Letters*, **31**, L04701, http://dx.doi.org/10.1029/2003GL019171

WATTERS, T. R., SOLOMON, S. C., ROBINSON, M. S., HEAD, J. W. III, ANDRÉ, S. L., HAUCK, S. A. II & MURCHIE, S. L. 2009. The tectonics of Mercury: the view after MESSENGER's first flyby. *Earth and Planetary Science Letters*, **285**, 283–296, http://dx.doi.org/10.1016/j.epsl.2009.01.025

ZUBER, M. T., SMITH, D. E. *ET AL*. 2012. Topography of the Northern Hemisphere of Mercury from MESSENGER Laser Altimetry. *Science*, **336**, 217–220, http://dx.doi.org/10.1126/science.1218805

Interactions between continent-like 'drift', rifting and mantle flow on Venus: gravity interpretations and Earth analogues

LYAL B. HARRIS[1]* & JEAN H. BÉDARD[2]

[1]Institut national de la recherche scientifique, Centre – Eau Terre Environnement (INRS-ETE), 490 de la Couronne, Québec, Canada QC G1K 9A9

[2]Geological Survey of Canada, 490 de la Couronne, Québec, Canada QC G1K 9A9

*Corresponding author (e-mail: lyal_harris@ete.inrs.ca)

Abstract: Regional shear zones are interpreted from Bouguer gravity data over northern polar to low southern latitudes of Venus. Offset and deflection of horizontal gravity gradient edges ('worms') and lineaments interpreted from displacement of Bouguer anomalies portray crustal structures, the geometry of which resembles both regional transcurrent shear zones bounding or cross-cutting cratons and fracture zones in oceanic crust on Earth. High Bouguer anomalies and thinned crust comparable to the Mid-Continent Rift in North America suggest underplating of denser, mantle-derived mafic material beneath extended crust in Sedna and Guinevere planitia on Venus. These rifts are partitioned by transfer faults and flank a zone of mantle upwelling (Eistla Regio) between colinear hot, upwelling mantle plumes. Data support the northward drift and indentation of Lakshmi Planum in western Ishtar Terra and >1000 km of transcurrent displacement between Ovda and Thetis regiones. Large displacements of areas of continent-like crust on Venus are interpreted to result from mantle tractions and pressure acting against their deep lithospheric mantle 'keels' commensurate with extension in adjacent rifts. Displacements of Lakshmi Planum and Ovda and Thetis regiones on Venus, a planet without plate tectonics, cannot be attributed to plate boundary forces (i.e. ridge push and slab pull). Results therefore suggest that a similar, subduction-free geodynamic model may explain deformation features in Archaean greenstone terrains on Earth. Continent-like 'drift' on Venus also resembles models for the late Cenozoic–Recent Earth, where westward translation of the Americas and northward displacement of India are interpreted as being driven by mantle flow tractions on the keels of their Precambrian cratons.

Supplementary material: Bouguer gravity and topographic images over a segment of the Mid-Atlantic ridge and Ross Island and surrounds in Antarctica, principal horizontal stress trajectories about mantle plumes on Earth, map and interactive 3D representations of cratonic keels beneath North America from seismic tomography, and a centrifuge simulation for comparison with Venus in support of our tectonic model are available at http://www.geolsoc.org.uk/SUP18736.

Venus, although similar to Earth in size, inferred internal composition and surface gravity, does not show any features that characterize plate tectonics on Earth, namely subduction zones, volcanic arcs, obvious seafloor-spreading ridges offset by transform faults, and large translations and rotations of distinct lithospheric plates (Anderson 1981; Kaula & Phillips 1981; Phillips *et al.* 1981; Solomon *et al.* 1991, 1992; Solomon 1993; Phillips & Hansen 1994; Simons *et al.* 1994, 1997; Grimm 1998; Hansen 2007; Smrekar *et al.* 2007; Watters & Schultz 2009; McGill *et al.* 2010; Moores *et al.* 2013). It has, nevertheless, been suggested by van Thienen *et al.* (2004) that it is theoretically possible for subduction and plate tectonics to have occurred in Venus' past, but for which no evidence remains following Venus' resurfacing (Strom *et al.* 1994). The terrestrial planets, with the notable exception of Earth, have traditionally been thought to experience a stagnant lid convection regime where a very viscous–rigid 'lid' that covers the entire planet overlies active convection in the underlying hot mantle (Solomatov & Moresi 1996, 1997; Moresi & Solomatov 1998; O'Rourke & Korenaga 2012). More recent and detailed two-dimensional (2D) numerical modelling by Armann & Tackley (2012) suggests that stagnant lid convection, alternating with episodic convection with approximately 150 million-year overturn events, best explains Venus's tectonic development. Convective cells on Venus are estimated to be around 600–900 km to (exceptionally) 2000 km wide (Solomatov & Moresi 1996). The present lithospheric thickness of Venus is estimated at <150 km by Nimmo & McKenzie (1998) and Smrekar & Parmentier (1996); van Thienen *et al.* (2004) suggested that the lithosphere may be thicker beneath continent-like planae, and to be thus similar to Earth. Orth & Solomatov (2012) proposed an average lithospheric thickness of between 300 and 500 km, and a crustal

From: PLATZ, T., MASSIRONI, M., BYRNE, P. K. & HIESINGER, H. (eds) 2015. *Volcanism and Tectonism Across the Inner Solar System*. Geological Society, London, Special Publications, **401**, 327–356.
First published online May 19, 2014, http://dx.doi.org/10.1144/SP401.9

thickness between 20 and 60 km. James *et al.* (2013) provided estimates of lithospheric thickness between 100 and 200 km, but with lower estimates of 8–25 km for the average crustal thickness, and crustal thicknesses calculated at between 41 and 60 km beneath Lakshmi Planum and surrounding fold belts (Fig. 1).

The geology and broad stratigraphic framework of Venus are portrayed in a global map by Ivanov & Head (2011, 2013). Volcanic zones, broad topographical rises, positive gravity anomalies, radiating rifts and dykes, and annular features such as coronae (Barsukov *et al.* 1986) are taken as evidence for hotspots or upwelling hot mantle plumes (Barsukov *et al.* 1986; Phillips *et al.* 1991; Bindschadler *et al.* 1992; Smrekar 1994; Smrekar & Stofan 1997; Phillips & Hansen 1998; Ernst & Desnoyers 2004; Basilevsky & Head 2007). Venus Express VIRTIS (Visible and Infrared Thermal Imaging Spectrometer) thermal emissivity

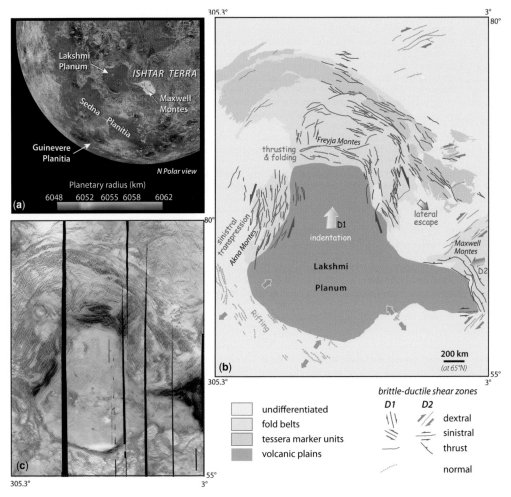

Fig. 1. (**a**) Combined Magellan radar and elevation image showing the location of areas described in Ishtar Terra, and the lowlands of Sedna and Guinevere planitia. (**b**) Simplified terrain subdivision and tectonic model for fold and shear belts surrounding Lakshmi Planum of the radar image shown in (**c**). (**c**) False-colour Magellan radar image for Lakshmi Planum and surrounds. Black bands are areas of no data. For detailed geological interpretation, see Ivanov & Head (2010*a, b*) for the Lakshmi Planum area and Marinangeli & Gilmore (2000) for the Akna Montes–Atropos Tessera region in the west of the interpreted area. The geometry of thrusts and transcurrent/transpressional shear zones on the NW, northern and NE margins is interpreted as Himalayan-style indentation and lateral escape (but without plate tectonics) resulting from the northward displacement of Lakshmi Planum in the first shortening event (D_1). Thrusts east of Lakshmi Planum and NE-striking dextral shear zones are attributed to a second event (D_2) of bulk WSW–ENE shortening. Images are from NASA/JPL; the interpretation in (**b**) is modified after Harris & Bédard (2014).

measurements were taken by Smrekar *et al.* (2010) to suggest the presence of young to possibly active hotspot-related volcanism, refuting early ideas that Venus is no longer tectonically or volcanically active (e.g. Kerr 1994). The interpretation of presently active volcanism is, however, questioned by Ivanov & Head (2010c), although they suggest volcanic eruptions have occurred in the last several decades (cf. Bondarenko *et al.* 2010). The stagnant lid, mantle plume-dominated tectonics of Venus contrast to mobile lid convection on Earth (Moresi & Solomatov 1998) where the lithosphere moves as discrete, rigid plates, and oceanic crust formed at active spreading ridges is consumed at subduction zones. In the tectonic environment presently envisaged for Venus, large, regionally coherent displacements of crustal blocks/terrains have not been considered likely.

Harris & Bédard (2014), however, documented horizontal displacements and polyphase folding and shearing on Venus interpreted from Magellan radar images of Venus' surface. The most spectacular of these examples is in western Ishtar Terra where regional folds, thrusts, and transcurrent and transpressional shear zones surrounding the craton-like Lakshmi Planum define a geometry identical to structures produced during indentation and lateral escape of the Himalayan–Indochina system on Earth (Harris & Bédard 2014; Fig. 1). Our research builds upon radar interpretations made by Kaula *et al.* (1992; who described detailed radar images of convergent fold belts, thrusts and regional transcurrent shear systems, but without any regional synthesis) and provides structural evidence for comparisons with indentation and lateral escape tectonics on Earth suggested by Crumpler *et al.* (1986; expanded upon in Moores *et al.* 2013), Markov (1986) and Head *et al.* (1990). This paper presents interpretations of enhanced gravity data on Venus to demonstrate that shear zones bounding Lakshmi Planum (interpreted from radar images) are crustal-scale structures that form part of a coherent regional system of shear zones, which is identified for the first time. Gravity data are also used to highlight regional crustal-scale rifts, the extent of which appears to have been underestimated in prior maps derived from radar images. The nature of the mechanisms causing large horizontal displacements on Venus without plate tectonics are discussed, and the interaction between rifting, transcurrent faulting, and indentation tectonics as recognized in Ishtar Terra are developed, drawing on comparisons with analogous plume-related features on Earth.

Recent geophysical and GPS data, as well as numerical modelling for Earth highlight the importance of horizontal traction on the base of deep continental roots as a force that helps to drive displacement of continents (e.g. Bokelmann 2002a, b;

Liu & Bird 2002; Eaton & Frederiksen 2007; Alvarez 2010; Faccenna *et al.* 2013; Ghosh *et al.* 2013), in addition to the action of plate boundary forces such as ridge push, slab pull, and trench suction (Forsyth & Uyeda 1975). This implies that, even on Earth, modern plate tectonics is not required for the displacement ('drift') of continents, as is frequently assumed; if there were no plate boundary forces, then continental drift and resulting orogenesis on Earth would probably still occur, albeit more slowly. There is an active debate about when plate tectonics on Earth started, and whether the formation and deformation of Archaean terrains resulted from plate tectonic processes such as subduction and arc accretion (reviewed in Bédard *et al.* 2013 and Harris & Bédard 2014). Studies of Venus thus help us better understand tectonic processes operative in an Archaean Earth (cf. Anderson 1981; McGill 1983; Morgan 1983; Markov 1986; Markov *et al.* 1989; Glukhovskiy *et al.* 1995; Sorohtin & Ushakov 2002; Stern 2004; Hansen 2007; Head *et al.* 2008; Van Kranendonk 2010).

Previous studies of faulting and brittle-ductile shearing on Venus

Despite the absence of plate tectonics, Venus displays diverse volcanic and tectonic features (folds, faults and brittle–ductile shear zones) indicative of regional lithospheric shortening and extension, many of which are similar to those developed on Earth (Crumpler *et al.* 1986; Head *et al.* 1990, 1992; Kaula *et al.* 1992; Solomon *et al.* 1992; Watters 1992; Hansen *et al.* 1997; Smrekar *et al.* 1997; Hansen 2007; Watters & Schultz 2009; McGill *et al.* 2010; Harris & Bédard 2014). Whilst detailed structural interpretations of radar images (cited above and in the following sections) have been undertaken, they are mainly concerned with relatively small regions and regional tectonic overviews have focused on the distribution of volcanic features (e.g. Pronin & Stofan 1990; Head *et al.* 1992; Stofan *et al.* 1992; Price *et al.* 1996), impact craters (Price *et al.* 1996), mafic dykes (e.g. Grosfils & Head 1994; Ernst *et al.* 1995, 2003), wrinkle-ridges and fractures (e.g. Hansen & Olive 2010), and rifts (e.g. Price *et al.* 1996; Basilevsky & Head 2000). Regional geological mapping coordinated by the USGS (United States Geological Survey) has been based almost exclusively on interpretation of geomorphological features from radar images (Tanaka *et al.* 2010; e.g. 'ridge crests', 'wrinkle ridges' or 'large ridges' are mapped instead of folds), and guidelines (Tanaka *et al.* 1994, 2011) advise curtailing the amount of structural elements included in regional maps. The global distribution of principal structural elements is portrayed by Ivanov & Head (2011); a detailed interpretation of much of the planet is also included in Hansen & Olive (2010).

Normal faults and regional rifts

The recognition of rifts on Venus has largely been based on radar-image interpretation of normal-fault-bounded graben as linear rifts radiating from volcanic centres termed novae, concentric, annular rifts rimming coronae (e.g. Solomon et al. 1992; Crumpler et al. 1993; Ernst et al. 1995; McGill et al. 2010; Studd et al. 2011), closely spaced graben in tessera terrains (Gilmore et al. 1998) and as deep (Watters & Schultz 2009) canyons/graben termed chasmata that are thousands of kilometres long (e.g. Solomon et al. 1992). Linear rifts and dyke swarms on Venus have been likened to those on Earth (Solomon 1993; Ernst et al. 1995; Foster & Nimmo 1996), but on Venus, rifts are about 3 times greater in width than plume-related rifts on Earth, which is attributed to the absence of thick sediment infill (Foster & Nimmo 1996). Low magnitudes of crustal extension due to normal faulting were calculated by Connors & Suppe (2001) from slope measurements, although they acknowledged that their calculations of crustal extension did not include the likely contribution of ductile to brittle–ductile extension and necking. Given its high mean surface temperatures (estimated at 462 °C by Seiff et al. 1985 – see NASA http://sse.jpl.nasa. gov/planets/profile.cfm?Object=Venus&Display= Facts&System=Metric (accessed June 2013) – and recently calculated as varying between c. 421 and 441 °C for Venus' southern hemisphere by Mueller et al. 2008), the brittle–ductile transition on Venus is expected to occur at a shallower depth than on Earth (McGill et al. 2010; Violay et al. 2010).

Large volcanic 'flow fields' on Venus are equated to terrestrial flood basalts or large igneous provinces (LIPs) in areas of lithospheric extension and thinning linked to mantle upwelling or mantle plumes (Lancaster et al. 1995; Magee & Head 2001; Ernst & Buchan 2003; Ernst & Desnoyers 2004; Ernst et al. 2007; Hansen 2007). Spreading centres and associated oceanic transform faults similar to those on Earth have not, however, been confirmed on Venus (Phillips et al. 1991; Watters & Schultz 2009), although transform-like faults were proposed by Crumpler et al. (1987; an interpretation we do not agree with, as discussed below). Although most models for rifting on Venus infer mantle upwelling or hot thermal plume origins, normal faulting has also been attributed to diapiric (i.e. Rayleigh–Taylor/density-driven) uplift (Hoogenboom & Houseman 2005) and gravitational collapse (e.g. Solomon 1993; Keep & Hansen 1994).

Strike-slip faulting

Regional transcurrent (strike-slip) faults with up to 450 km of displacement on individual faults, and with cumulative transcurrent displacement of up to 2000 km, were identified in early studies of Venus (e.g. Crumpler et al. 1986, 1987; Crumpler & Head 1988; Sukhanov & Pronin 1988; Kozak & Schaber 1989; Vorder Bruegge et al. 1990) and are implicit in the interpretations of Head et al. (1990). Pohn & Schaber (1992) interpreted strike-slip fault geometries that they likened to indentation-style tectonics on Earth, and conjugate strike-slip faults were interpreted by Watters (1992). The presence of broad brittle–ductile shear zones, as well as discrete faults, was also recognized early in Venus mapping. For example, Vorder Bruegge et al. (1990) drew comparisons between en échelon folds and pull-apart graben with similar structures formed during wrenching along the San Andreas fault system on Earth (cf. Wilcox et al. 1973; Sylvester 1988). Additionally, Hansen (1992) and Kaula et al. (1992) interpreted sigmoidal deflection of fold-axial traces and their angular relationship to, and truncation by, transcurrent faults as equivalent to regional-scale S/C structures (drawing comparison with structures in mylonitic rocks described by Berthé et al. 1979). Despite these early interpretations, Solomon et al. (1992, p. 13 199) and Solomon (1993, p. 50) stated that 'few large-offset strike slip faults ... have been observed' and interpreted strike-slip faults as local features developed to accommodate horizontal displacements during crustal shortening or stretching. Bindschadler et al. (1992) considered that development of long strike-slip faults on Venus is inhibited by the absence of water. The ensuing misperception for limited transcurrent displacements on Venus (as noted by Fernández et al. 2010) has been carried forward in subsequent reviews of faulting on planetary bodies (e.g. Schultz 1999; Schultz et al. 2010) and comparisons between the tectonics of Venus and Earth (e.g. Montési 2013). Numerous radar interpretation studies carried out since Solomon (1993) have, nevertheless, continued to illustrate the presence of conjugate strike-slip faults (e.g. Willis & Hansen 1996; Hansen 2007; Yin & Taylor 2011) and regional strike-slip faults and transcurrent wrench zones with displacements of up to hundreds or even thousands of kilometres along them have been mapped (e.g. Raitala 1994, 1996; Brown & Grimm 1995; Ansan et al. 1996; Hansen & Willis 1996; Koenig & Aydin 1998; Tuckwell & Ghail 2003; Kumar 2005; Romeo et al. 2005; Chetty et al. 2010; Fernández et al. 2010). Mechanisms for the formation of regional strike-slip faults and the relationship (if any) to rifting (as postulated by Markov et al. 1989) have, nevertheless, remained uncertain.

Strike-slip faults are found on other planetary bodies as well as on Earth and Venus, such as Mars (Schultz 1989; Bistacchi *et al.* 2004; Anguita *et al.* 2006; Okubo & Schultz 2006; Borraccini *et al.* 2007; Yin 2012), Mercury (Massironi *et al.* 2012, 2014), Jupiter's moons Europa (Hoppa *et al.* 1999; Tufts *et al.* 1999; Kattenhorn 2004; Aydin 2006; Kattenhorn & Marshall 2006), Io (Bunte *et al.* 2010) and Ganymede (Pappalardo & Collins 2005), and Saturn's moon Enceladus (Smith-Konter & Pappalardo 2008; Patthoff & Kattenhorn 2011). Arguments that formation of strike-slip faults implies the existence of rigid plates and, hence, modern plate tectonic processes (Pohn & Schaber 1992; Sleep 1994; Yin 2012) are clearly refuted by Fernández *et al.* (2010). Concepts for tectonic mechanisms for regional lateral displacements *without* plate tectonics are developed below.

Folds and reverse/thrust faults

Folds (including, but not limited to, 'ridge belts' and 'wrinkle ridges') deform volcanic plains, and regional folds and reverse/thrust faults have developed along the margins of plateaux, tesserae and coronae (e.g. Head *et al.* 1990; Suppe & Connors 1992). The formation of these contractional features is variably attributed to compensation of extension in adjacent rifts (Markov *et al.* 1989), local shortening and crustal thickening (Solomon 1993), flexure and underthrusting of young lithosphere against a more rigid block (Head *et al.* 1990), gravitational spreading due to elevation differences (Smrekar & Solomon 1992), thermal expansion (Solomon *et al.* 1999) or contraction (McGill *et al.* 2010), mantle downwelling (Markov *et al.* 1989; Kiefer & Hager 1991; Bindschadler *et al.* 1992), or the far-field contraction resulting from mantle flow about upwelling hot plumes (Mège & Ernst 2001). As on Earth, refolded folds on Venus (e.g. Harris & Bédard 2014) are taken as evidence for changes in the orientation of the regional stress and/or displacement field.

Bouguer gravity field and crustal thickness of Venus

A detailed map of the free-air gravity field of Venus was acquired during the 1993 Magellan mission from accelerations and decelerations calculated from line-of-sight doppler shift, where aerobraking was used to establish a low orbit with 36% ellipticity (Solomon 1993; Kaula 1996; Konopliv & Sjogren 1996) (during aerobraking, Magellan's apoapsis was lowered from 8500 to 500 km; its periapsis was 180 km). (Differences between free-air and Bouguer gravity, and their implications relevant to planetary geological interpretations, are explained in simple terms by Lakdawalla 2012.) Gravity data from the previous Venera (Barsukov *et al.* 1986) and Pioneer Venus Orbiter missions (Solomon 1993) infill many areas lacking Magellan data. As shown by comparison with early satellite gravity for Earth (e.g. Sandwell 1992), the spatial resolution of 110 km for the final, processed Magellan gravity data obtained near the equator and of 180 km at higher latitudes (Kaula 1996) makes these data suitable for regional-scale mapping of crustal- to lithospheric-scale features (especially where the total horizontal gradient is used to delineate structural features). Gravity anomalies have been used in geoid, admittance and flexural rigidity calculations (Solomon 1993 and references therein; Konopliv & Sjogren 1996; Simons *et al.* 1997; Barnett *et al.* 2000; Anderson & Smrekar 2006; Wieczorek 2007) to calculate crustal thickness (Simons *et al.* 1997; Anderson & Smrekar 2006; Wieczorek 2007; James *et al.* 2013, whose data we present below) and to create models that support the presence of mantle plumes (Bindschadler *et al.* 1992; Herrick & Phillips 1992; Solomon 1993; Kiefer & Peterson 2003; Vezolainen *et al.* 2003). Previous studies of regional free-air anomalies suggest that:

- gravity lows are due to less dense crust that extends deeper into the mantle or areas of higher temperature and less dense crust;
- there is a high correlation on Venus between the free-air gravitational field and topography, suggesting deep compensation through convective upwelling or downwelling (Solomon 1993; Rummel 2005).

The Bouguer gravitational anomaly field for Venus (Sjogren & Konopliv 2008; developed from the original 1997 gravity data updated with degree 180 harmonic coefficients by Konopliv *et al.* 1999) was calculated by these authors using a reference density of 2900 kg m^{-3} based on average basaltic crust (which contrasts to 2650 kg m^{-3} generally used on Earth, the average density of felsic/continental crust). Despite the excellent gravity coverage, Bouguer anomaly maps have not been (to our knowledge) previously used in interpreting regional, crustal-scale faults on Venus. Spectrally filtered and enhanced gravity data and edges in the horizontal gradient of Bouguer gravity (commonly termed gravity 'worms') are regularly used for mapping regional-scale faults/shear zones and lithological contacts on Earth (e.g. Blakely & Simpson 1986; Archibald *et al.* 1999; Bierlein *et al.* 2006; Vos *et al.* 2006; Austin & Blenkinsop 2008; Heath *et al.* 2009; Henson *et al.* 2010; Glen *et al.* 2013; Dufréchou *et al.* 2014; Harris & Bédard 2014), including the entire Earth (Horowitz *et al.* 2000). The same approach is applied to Venus in the following subsections.

The total Bouguer gravity field for northern polar to low southern latitudes of Venus, covering the Ishtar and western Aphrodite terrae and the Sedna, southern Snegurochka and Niobe planitae presented in Figure 2a (place names not shown in this figure are shown in Fig. 1), shows that:

- Ishtar Terra (whose geology is described by Kaula *et al.* 1992; Marinangeli & Gilmore 2000; Ivanov & Head 2008, 2010*a*, *b*) and Aphrodite Terra (including Ovda Regio mapped by Bleamaster & Hansen 2005) are marked by regional Bouguer lows down to −300 mGal.
- Subcircular Bouguer lows in the SW of the map correspond to Beta and Phoebe regionen, which are elevated volcanic centres with thick crust and radiating graben (including Devana Chasma) interpreted to overlie upwelling mantle plumes (Kiefer & Peterson 2003; Vezolainen *et al.* 2003). Beta Regio is interpreted to overlie a cluster of mantle plumes by Ernst *et al.* (2007).
- An ESE–WNW-trending broad Bouguer high (up to *c.* 130 mGal) is developed over the 1500–2000 km-wide 'lowland' volcanic plains (Stofan *et al.* 1987) of Sedna Planitia south of Ishtar Terra. An approximately 5000 km-long Bouguer gravity high bifurcates south of Lakshmi Planum. The northern branch continues for about 14 000 km towards the east and passes approximately 1500 km to the north of Aphrodite Terra, where it is locally punctuated by Bouguer lows over volcanic centres, including Bell and Tellus regionen. The southern branch, which strikes more southeasterly, corresponds to Guinevere Planitia and ends abruptly after about 7000 km to the west of Ovda Regio.
- The two linear gravity highs of Sedna and Guinevere planitias are separated by Eistla Regio, a region of lesser Bouguer anomalies dominated by tight groups of volcanos and coronae, attributed to underlying plume clusters by Ernst *et al.* (2007).
- North of Ishtar Terra, Snegurochka Planitia, a region marked by volcanic plains formed in a tensile stress field (Hurwitz & Head 2012), is also marked by a broad, diffuse gravity high. This polar region lies outside the area covered by the present research and has not been examined further.

Figure 2b portrays high-frequency (i.e. 'shallow-source') components of the Bouguer field (cf. data defined by the shallow slope of the MGNP180U power spectrum plot of Wieczorek (2007, fig. 4), equating to a spherical harmonic degree >40).

Thickness variations of Venus' crust calculated from inversion of gravity data by James *et al.* (2013) were gridded and are presented in 2D and 3D in Figure 3. From the comparison of total and 'shallow-source' Bouguer gravity and crustal thickness images, it can be seen that:

- Lakshmi Planum has a uniformly thick crust (Fig. 3a), yet can be divided into a northern shallow-source gravity high and a southern shallow-source gravity low (Fig. 2b). Similar marked lateral changes occur across all of Ishtar Terra where they correspond to strong gradients highlighted by gravity 'worms' (Fig. 2b). Jull & Arkani-Hamed (1995) and Arkani-Hamed (1996) suggested that the density perturbations over Ishtar Terra that were determined from a similar spectral analysis of gravity data required lateral variations in rock type, and that crust beneath Ishtar Terra and surrounding mountain belts may contain considerable amounts of low-density material. We concur with their conclusions and suggest that gravity lows in Lakshmi Planum may indicate either the presence of low-density, most probably felsic, rocks in the upper crust underlying the surface basaltic flows mapped by Ivanov & Head (2010*a*, *b*) or, as the contact between shallow-source gravity highs and lows coincides with their mapped textural boundaries, some flows mapped as basaltic may instead be more felsic. Although Venus has been thought to be dominated by basaltic crust, there is increasing evidence from emissivity data for felsic volcanism, and tessera/highlands material may also be felsic in composition (Nikolaeva *et al.* 1992; Hashimoto *et al.* 2008; Helbert *et al.* 2008; Mueller *et al.* 2008; Basilevsky *et al.* 2012), which is supported by petrological modelling (Shellnutt 2013). Given that felsic magma may be generated in oceanic plateaux on Earth above high-temperature mantle plumes and/or spreading ridges such as Iceland (Annen *et al.* 2006; Willbold *et al.* 2009), the presence of granites or felsic lavas in the craton-like areas of Ishtar and Aphrodite terrae is consistent with models for initial crustal thickening and plateau formation above an upwelling mantle plume. Felsic crust is not only less dense but also rheologically much weaker (more ductile) than mafic crust; this may have strong consequences for enhancing crustal deformation intensity and, especially, crustal thickening (T. Gerya, pers. comm. 2013) if felsic crust is also present on the margins of these craton-like plateaux.
- The linear gravity highs south of Ishtar Terra in the total Bouguer gravity portrayed in Figure 2a correspond to belts of thinned crust

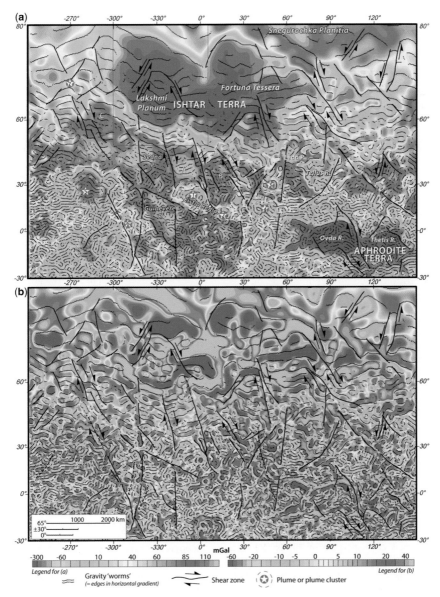

Fig. 2. (a) Bouguer gravity for the area of Venus examined in this study. Reference density is 2900 kg m^{-3}. The 'continent-like' Ishtar and Aphrodite terrae, and Beta and Phoebe regiones (interpreted to overlay mantle plumes or plume clusters) are marked by Bouguer gravity lows. Bouguer highs in Sedna and Guinevere planitias (which are not present on the short-wavelength component in b) correspond to thinned crust (Fig. 3) and mark dense mafic underplated mantle material beneath rift zones. Localized subcircular areas of lower Bouguer gravity that punctuate the linear highs lie above interpreted mantle plume centres or plume clusters from Ernst *et al.* (2007). Dots marking edges in the horizontal gradient (i.e. gravity 'worms') highlight lithological contacts and faults. Transcurrent shear and transfer zones (rift area) are interpreted from offset of gravity anomalies and worms. The rectangle shows the area in Figure 1b, c. The offset of gravity worms on either side of Lakshmi Planum confirms the indentation model based on radar-image interpretation (Harris & Bédard 2014). (b) Short wavelength (i.e. shallow-source) Bouguer gravity for the same area as in (a) overlain by worms and shear interpretation. Short-wavelength highs and lows over Ishtar and Aphrodite terrae suggest variations in lithology of the upper crust; this is supported by the correspondence between the boundary between anomalies and mapped morphological contacts in Lakshmi Planum mapped by Ivanov & Head (2010*a*), and raises the likelihood for there being more felsic flow units in addition to basalts in Venus' highlands. Names of interpreted plume clusters are given in the caption to Figure 3.

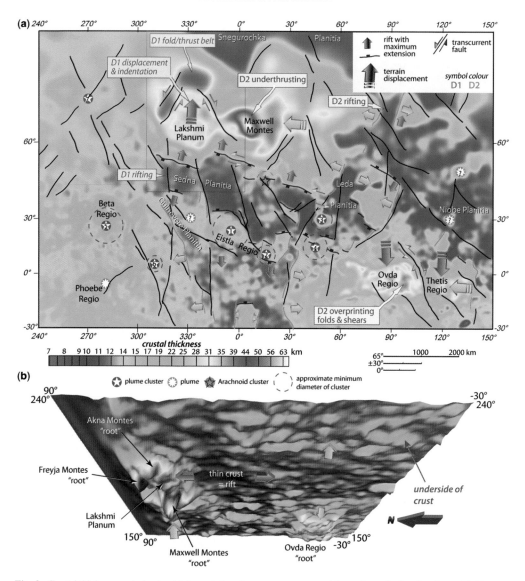

Fig. 3. Crustal thickness and structural interpretation. Images were prepared from data calculated and provided by P. James (presented in James *et al.* 2013). (**a**) Map view. The pattern of structures is explained by two generations of rift-bounding normal faults; however, the lack of any geochronological data on Venus precludes any validation that structures with the same orientation are contemporaneous. The ensuing displacements of areas of continent-like crust in terrae imply changes in regional stress field. Superposed faults are interpreted from gravity (Fig. 2). Volcanic centres: 1, western Eistla; 2, central Eistla; 3, Bell Regio; 4, eastern Eistla; 5, Beta Regio; 6, Laufey; 7, Mnemosyne, the Arachnoid cluster Bereghinya, are interpreted to overlie plume clusters, for which the minimum diameter is estimated (from Ernst *et al.* 2007). (**b**) 3D view of the same area as in (a) viewed from underneath, colour-coded for thickness. The two 'continent-like' regions of Lakshmi Planum and Ovda Regio correspond to areas of thicker crust than the surrounding plains. Thick crustal 'roots' underlie fold belts on the north and NW margins of the interpreted Lakshmi Planum D_1 indentor (Fig. 2). The thickest crust occurs beneath Maxwell Montes, where folding and underthrusting (Keep & Hansen 1994) are interpreted to have occurred in the subsequent (D_2) event. Although eastern Ovda Regio is offset by a dextral shear zone consistent with the implied D_1 approximately north–south shortening, radar interpretation (Fig. 5) presents evidence for its sinistral reactivation, consistent with the model for an orthogonal rotation of principal stresses between D_1 and D_2.

in Figure 3. The lack of a similar gravity high in the high-frequency/shallow-source gravity component in Figure 2b indicates a deep, high-density source for the Bouguer highs in Figure 2a. This can most readily be explained by the presence of denser, underplated mantle-derived mafic rocks beneath the zone of thinned crust. The interpreted rift correlates to the Guinevere, Sedna and Leda (Lednaya) planitias/volcanic lowland plains (Markov 1986). The Guinevere and Sedna planitias are interpreted as being characterized by extensional structures by Sullivan & Head (1984), which is consistent with this rift interpretation. Data from Magee & Head (2001) indicate that the interpreted rift contains over 2×10^6 km^2 of volcanic flow fields. However, only several segments, especially those bordering areas of geoid lows, are interpreted as rifts in the global map of the distribution of rift zones on Venus by Ernst *et al.* (2007) and Krassilnikov *et al.* (2012).

Interpretation of regional fault and shear zones

Interpretations of gravity and crustal thickness images. Horizontal offsets of Bouguer gravity anomalies, and offset and ductile deflection ('drag') of 'worm' edges in the total horizontal gradient, define a series of linear features (Fig. 2a). The constant horizontal displacement components along their length, the observed offset of both total Bouguer and short wavelength anomalies, and the 100 km-scale resolution of gravity data together indicate that many of these lineaments are crustal-scale transcurrent shear zones. The location and sense of horizontal displacement along shear zones interpreted from gravity images in western Ishtar Terra (Fig. 2) coincides precisely to those of transcurrent–transpressional shear zones interpreted on the margins of Lakshmi Planum from radar images (Harris & Bédard 2014; Fig. 1b). Sinistral shears on the NW margin of Lakshmi Planum are approximately 1000 km long and the curved, dextral shear zone on its NE margin is around 1800 km long. In these companion studies, Lakshmi Planum is interpreted as having acted as a rigid indentor, forming mountain belts on its northern, SE and NW margins. Gravity data thus greatly strengthen this hypothesis, and attest to the crustal scale of the structures previously interpreted only from radar images. The geometry of these structures resembles the Himalayan–Indochina system on Earth formed due to the indentation of India into Eurasia, and the length of the shear zone on the eastern margin of Lakshmi Planum is similar to the length of the Sagaing fault system on the eastern margin of the Indian indenter (Searle 2006). A similar, indenter-like geometry is defined by shear zones in SE Ishtar Terra (Fig. 2). In western Aphrodite Terra, Ovda and Thetis regiones (which correspond to Bouguer gravity lows similar to Ishtar Terra) are offset by an approximately 2500 km-long, NNW-striking dextral shear zone. Two dextral shears, approximately 1300 and 800 km long, offset the southern margin of Ovda Regio.

The linear gravity highs in Guinevere and Sedna planitias, interpreted above as broad rifts, are also cut or bounded by gravity lineaments with apparent horizontal offsets. This is especially evident in Guinevere Planitia, in the central part of the map area, where gravity highs are offset or bounded by NNW-striking linear faults. Gravity 'worms' in the centre of the map area (e.g. at locations A and B in Fig. 2a) are approximately orthogonal to the bounding faults against which they are truncated, and are oblique to the overall trend of the gravity high. This contrasts to the ductile or brittle–ductile deflection ('drag') of 'worms' along similarly orientated dextral shears that cut the same linear gravity high SW of Lakshmi Planum (e.g. location C in Fig. 2a). Blocks A and B are separated by a fault with dextral horizontal offset, but displacement sense(s) are less clear on their other margins.

Displacement senses along crustal-scale shear zones in western Aphrodite Terra. Dextral strike-slip offsets of the thick crustal blocks/Bouguer gravity lows that correspond to Ovda and Thetis regiones in western Aphrodite Terra, as previously interpreted by Crumpler *et al.* (1987) from analysis of radar and topographical images (but subsequently interpreted as oceanic fracture zone/transform faults on Earth, instead of strike-slip/transcurrent faults by Crumpler & Head 1988), is apparent on both crustal thickness and gravity images (Figs 2a & 3a). A radar image of the area immediately east of the gravity lineament in Figure 4a, however, portrays transcurrent shear zones where sinistral displacements are evident from horizontal offsets and the deflection of marker layers. Sinistral shears both parallel and step en échelon along the NW-striking feature interpreted from gravity. These sinistral shear zones, NE-striking dextral shears and east–west-trending graben are similar in geometry to subsidiary structures in transcurrent shear zones mapped on Earth and developed in analogue models (Riedel 1929; Tchalenko 1968, 1970; Wilcox *et al.* 1973; Mueller & Harris 1988; Mueller *et al.* 1988; Sylvester 1988; Richard *et al.* 1995), and define a sinistral wrench regime (Fig. 4b). In this wrench model, en échelon-stepping sinistral shears are interpreted as Riedel shears, and NE-striking dextral shears constitute secondary Riedel (R') shears (Riedel 1929; Tchalenko 1968, 1970). An

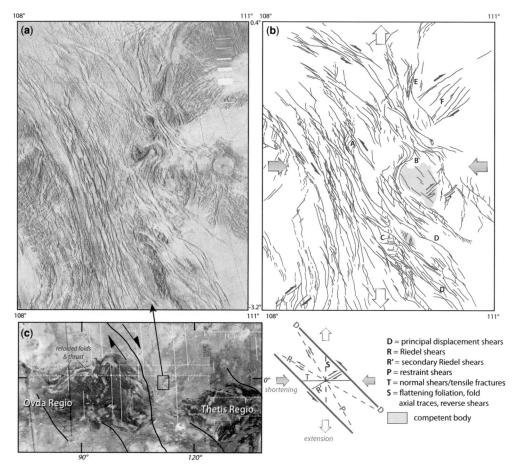

Fig. 4. Radar interpretation of western Aphrodite Terra (interpretation criteria are outlined in Harris & Bédard 2014).
(**a**) False-colour Magellan radar image. (**b**) Lineament interpretation. The geometry of interpreted transcurrent shears
and graben defines a sinistral wrench system. A, ductile deflection ('drag') of foliations indicates sinistral shear
displacement; B, foliations wrap internally fractured competent (?) body; C, localized tensile fractures/normal faults; D,
Riedel shears; E, sinistral transpressional shears; F, the 'drag' of foliations indicate that dextral shears cut older fracture
sets in tesserae. (**c**) False-colour radar image over eastern Ovda and western Thetis regions showing main shears
interpreted from Bouguer gravity in Figure 2, generated through combining left- and right-looking images to minimize
areas of no data (white lines and rectangles). The shear system interpreted in (b) is parallel to, and eastward of, the
gravity lineament (note, however, that the positions of gravity lineaments are approximate given the data resolution). As
dextral displacement along this structure is apparent on Bouguer gravity (Fig. 2), sinistral shearing in (b) is interpreted as
D_2 reactivation of an existing crustal-scale shear corridor. This interpretation is consistent with the permutations in
principal strain axes deduced from overprinting folds and thrusts in Ovda Regio west of this shear zone (c), portrayed by
Harris & Bédard (2014).

orthogonal change in the orientations of principal
strain axes is thus required, from north–south short-
ening/east–west extension during the dextral offset
of regional blocks observed on gravity images, to
bulk east–west shortening and north–south extension during sinistral wrench reactivation. These
changes in principal strain axes deduced from the
geometry and offset of transcurrent shears is corro-
borated by the same changes in principal strain axes
deduced from refolded fold interference patterns

and folding of early formed thrusts within Ovda
Regio, documented by Harris & Bédard (2014).
Our evidence for shear-zone reactivation and
change in displacement sense also reconciles previously apparent contradictions in displacement
sense (i.e. sinistral by Tuckwell & Ghail 2003
v. dextral by Kumar 2005) established for segments
of the nearby 1000 km-long, 50–200 km-wide
Thetis Boundary Shear Zone system separating the
eastern Ovda and NW Thetis regions.

Tectonics on Earth resulting from mantle plumes and global mantle flow

Before developing a tectonic model to explain our Venus observations, this section briefly reviews salient characteristics of tectonics on Earth related to mantle plumes and global mantle flow. On Earth we have a far better understanding of tectonic processes and 3D geometry through integrated field studies constrained by geochronology, stress measurements and more detailed geophysical data than for Venus. Although there has been much debate as to the existence and tectonic roles of mantle plumes on Earth (e.g. Artyushkov 1973; Forsyth & Uyeda 1975; Anderson 2000, 2013; Foulger 2010; Burke & Cannon 2014), seismic tomography (such as Lithgow-Bertelloni & Silver 1998; Montelli *et al.* 2006; Boschi *et al.* 2007; Chang & Van der Lee 2011, whose data is plotted in 2D and 3D in Fig. 5) incontestably shows the presence of deep hot mantle upwellings and cold downwellings. Recent magnetotelluric (e.g. Kelbert *et al.* 2012) and seismic tomographic data portray far more complex patterns of mantle upwelling and downwellings than early isolated plume models, including vertical walls, horizontal 'fingers' at the base of the oceanic aesthenosphere paralleling overlying plate motion (French *et al.* 2013), and spatial clusters of smaller plumes instead of single 'superplumes' (Schubert *et al.* 2014). Three-dimensional numerical modelling suggests that several plumes may rise from deeper upwelling walls/planiform structures (Hanjalić & Kenjereš 2006; Gait *et al.* 2008) in addition to forming isolated, 'mushroom-like' plumes, and that their form may change with time (Gait *et al.* 2008). Continental flood basalt/ LIPs on Earth (Pirajno 2000, 2007; Ernst & Buchan 2003; Campbell 2005) – for example, the Columbia River Province (Hooper *et al.* 2007; Camp & Hanan 2008), the Deccan traps (Cande & Stegman 2011; Sen & Chandrasekharam 2011), the Siberian traps (Saunders *et al.* 2005; Kiselev *et al.* 2012; Howarth *et al.* 2014), and the North American Mid-Continent Rift, as well as rifts in Arabia, East Africa, West Antarctica and Iceland (discussed later) – are attributed to mantle plumes. Their structural features and tomographic and gravity expressions help us to interpret the Venus gravity data, and to formulate tectonic models that can explain the interplay between rifting and lateral displacements on Venus.

Regional stress patterns about mantle plumes

Upwelling mantle plumes contribute to, and may be the prime origin of, regional stress patterns for some areas on Earth. For example, Cobbold (2008) noted that maximum horizontal principal stress axes in western Europe and Scandinavia (portrayed in the *World Stress Map* of Heidbach *et al.* 2008, 2010) converge on the Iceland mantle plume. Cobbold (2008) and Le Breton *et al.* (2012) suggested that compressive stresses generated by the Iceland plume explain thrust mechanisms of recent earthquakes in Scandinavia, post-Neogene basin inversion around Iceland, and contribute to basin inversion and onshore crustal shortening on North Atlantic margins. In an Early Tertiary tectonic reconstruction, Mège & Ernst (2001) showed that fold-axial traces in sedimentary rocks of the NW European shelf are concentrically arranged about the same mantle plume that now underlies Iceland when, at 60–50 Ma, Greenland was situated above it. Mège & Ernst (2001) attributed the implied radial shortening to plume-derived horizontal stresses. Similarly, the horizontal compressive stress trajectory in East Africa and Arabia is also controlled by upwelling mantle plumes. Mouthereau *et al.* (2012) contended that, whilst early shortening and thrusting in the Zagros is subduction and collision-related, the main driving force for Arabian plate motion since approximately 12 Ma is horizontal flow originating from upwelling mantle plumes responsible for the intracrustal extension that created the Red Sea and East African rifts (Fig. 5). Furthermore, 2D numerical modelling by Burov *et al.* (2007) and Guillou-Frottier *et al.* (2012) illustrates how plume impingement beneath shallow crust induces compressional stresses in (and, in some models, horizontal displacement of) an adjacent thicker, 'cratonic' crustal block. Zones of mantle downwelling also change the stress field in the overlying crust; for example, in the models of Behn *et al.* (2004), large compressive stresses developed in the upper crust over a zone of mantle downwelling located ahead of a continent being driven laterally by mantle flow from an upwelling plume. Horizontal projections of the instantaneous flow trajectories in numerical simulations of Hanjalić & Kenjereš (2006) at Rayleigh numbers similar to those used in other Venus simulations (e.g. Robin *et al.* 2007), although not aimed at modelling geological structures, also suggest that linear zones of focused horizontal displacement may develop between convection cells. The effect of such linear flow discontinuities on an overlying crustal 'lid' was not simulated. If such structures were to develop on Venus, could they produce the network of regional transcurrent shears we interpret from gravity, instead of conventional ideas of bulk regional shortening? Further numerical and physical models are required to test this hypothesis.

On Venus, structures formed during bulk shortening, such as wrinkle ridges and associated conjugate strike-slip faults concentric about the volcanic

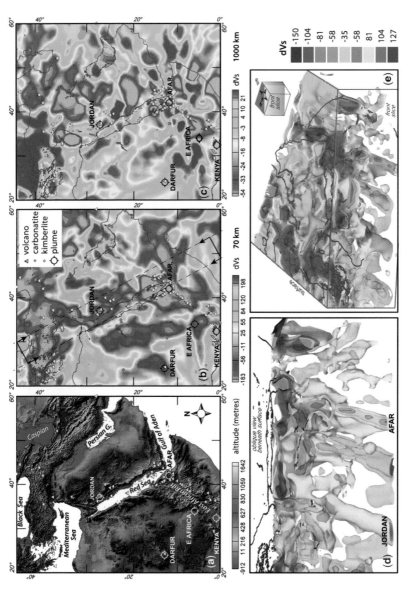

Fig. 5. S-wave seismic tomography over rifts in East Africa and Arabia showing upwelling hot and downwelling cold mantle plumes. Maps and 3D isosurfaces present velocity data from Chang & Van der Lee (2011) compared to reference model 'MEAN' of Marone *et al.* (2004). (**a**) Elevation. The majority of the interpreted mantle plumes from Chang & Van der Lee (2011) and a compiled list of plume locations (see Anderson: http://www.mantleplumes.org/CompleteHotspot.html (accessed June 2013)) correspond to areas of high topography. (**b**) & (**c**) Velocity differences at 70 and 1000 km depth, respectively. Warm colours correspond to higher temperatures that are interpreted as mantle upwelling by Chang & Van der Lee (2011). (**d**) & (**e**) 3D isosurfaces of velocity differences from 50 to 1400 km depth. Data portrayed in (d) show upwelling and downwelling plumes along a slice parallel to the Red Sea, indicated in (b), viewed from below the Earth's surface (i.e. looking upwards at the map of the Red Sea). Isosurfaces in (e), viewed looking obliquely downwards (see the inset for the orientation), show plume heads merging beneath the crust. (This region is, however, complex in detail as plumes interact with whole mantle flow, resulting in a northward entrainment of upwelling mantle material: Faccenna *et al.* 2013.)

centres in volcanic plains and folds in crustal pla-teaux, were also attributed to plumes by Mège & Ernst (2001).

Relationships between rifts and mantle upwelling

Gravity signature, topographical expression and stress patterns. The 1115–1086 Ma (Heaman et al. 2007), approximately 2500 km-long, Mid-Conti-nent Rift in the north-central USA and southern-most Canada in the Lake Superior region cuts Precambrian terrains (Fig. 6), and is marked for the most part by long linear Bouguer gravity highs (Behrendt *et al.* 1988, 1990; Hinze *et al.* 1997; Miller 2007 and references therein; Stein *et al.* 2011). The gravity signature of the Mid-Continent Rift closely resembles the Bouguer gravity pat-tern of the interpreted rift zones on Venus described above.

The Mid-Continent Rift contains an approxi-mately 30 km-thick sequence of volcanic and sedi-mentary rocks (Behrendt *et al.* 1988; Hinze *et al.* 1997), largely covered by younger, Phanerozoic sediments (Miller 2007). Behrendt *et al.* (1988, p. 81) considered that its northern part 'may con-tain the greatest thickness of intracratonic rift depos-its on Earth'. The rift is underlain by a zone of dense, underplated mafic mantle-derived rocks, which are responsible for its marked gravity high (Behrendt *et al.* 1988; Chandler *et al.* 1989; Thomas & Teskey 1994; Allen *et al.* 2006; Merino *et al.* 2013). An anomalously deep and generally flat Moho (developed syn- to post-rifting) is apparent on seismic profiles (e.g. Allen *et al.* 2006). Hinze *et al.* (1997) and Miller (2007) described successive stages in the tectonic model for rifting as resulting from impingement of an anomalously hot mantle plume at the base of the lithosphere, where the two arms of the rift radiate from the proposed plume head. This plume model, originally proposed by Burke & Dewey (1973), is supported by geologi-cal and geochemical data provided by Miller (2007 and references therein). Extension terminated before the rift–drift transition and the development of oceanic crust, presumably due to a change to regional shortening inboard of the Grenville Orogen during which the Mid-Continent Rift under-went minor compressional overprinting. Similarly, positive Bouguer anomalies also mark the 500 km-long and 100 km-wide Bransfield Rift in the Antarctic South Shetland Islands–Bransfield Strait area (Catalán *et al.* 2012), an actively extend-ing marginal basin where rifting is superposed upon an inactive continental volcanic arc (Lawyer *et al.* 1995; Fretzdorff *et al.* 2004). Forward model-ling by Catalán *et al.* (2012) again suggested that thinned continental crust of the Bransfield Rift

is underlain by a zone of anomalous, upwelled mantle.

In broad rifts that develop due to mantle flow about an active, upwelling hot plume, the area above the causative plume may correspond to a topographical high, which may or may not be cut by active rifts, and thicker crust than surrounding rifts. For example, Iceland is a topographical high cut by active normal and transform faults and frac-tures (Fig. 7a, b). Iceland and its surrounding shelf correspond to a region of lower Bouguer anomalies (Fig. 7c) that punctuate the general Bouguer highs along the Reykjanes and Kolbeinsey ridges. The interpreted mantle plume in Iceland is marked by a negative Bouguer gravity anomaly (Darbyshire *et al.* 2000); the deep source of this anomaly pro-duces a long-wavelength Bouguer low (Fig. 7d). Similarly, Ross Island in West Antarctica, which is interpreted to overlie the Erebus mantle plume (Kyle *et al.* 1992; Storey *et al.* 1999; Gupta *et al.* 2009; Mt Erebus Volcano Observatory: http:// erebus.nmt.edu/index.php/volcanology/51-vol canological-evolution (accessed June 2013)), is a topographical high flanked by intracontinental rifts (see the review by Elliot 2013). It also corresponds to an isolated negative Bouguer anomaly in a zone of rifted crust otherwise marked by Bouguer highs.

We conclude that, from comparison with these rift examples on Earth, the long, linear gravity highs of Sedna and Guinevere planitias on Venus similarly mark broad rifts of basaltic crust that are underplated by higher-density cumulates and man-tle that cause their Bouguer highs. As is the case for the Mid-Continent Rift, upper crustal faults are largely obscured by an overlying, late- to post-rift sequence (extensive lava flows on Venus), suggesting that the significance of these broad rifts may have been underestimated in previous studies based on radar mapping of surface features. Elev-ated areas with moderate Bouguer anomalies and slightly thicker crust within Eistla Regio, which are interpreted as overlying plume clusters by Ernst *et al.* (2007), show distinct similarities to Iceland and Ross Island, where upwelling plumes are postulated, supporting their plume interpret-ation. This further suggests that Eistla Regio is a topographical high that formed directly above a linear array of upwelling mantle plumes and plume clusters, and that this positive structure formed along the axis of a single rift encompassing both Sedna and Guinevere planitias.

Plume interaction in the Red Sea and East African rifts. Whilst Ernst & Buchan (2003) suggested that plumes in the geological record are characterized by areas of domal uplift, triple-point junction rift-ing and LIPs, more complex patterns also occur (e.g. Şengör & Natal'in 2001; Harris *et al.* 2004),

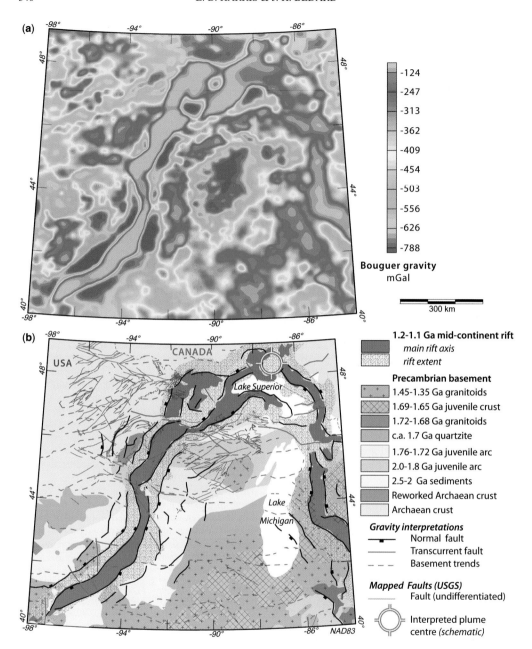

Fig. 6. Mid-Continent Rift, a plume-related rift in the NE USA and southern Canadian Great Lakes area. (**a**) Bouguer geology. The long, linear gravity high is attributed to the underplating of dense mafic rocks. A similar explanation is proposed to explain the linear Bouguer gravity highs on Venus in Figure 2a, which also correspond to thin crust (Fig. 3). (**b**) A simplified geological map, combining the interpretation of (a) and USGS geological map data (http://mrdata.usgs.gov/geology/state/).

with long linear rifts developing due to linkages between mantle plumes (cf. May 1971). Chang & Van der Lee (2011) illustrated from an S-wave seismic tomographical velocity model that rifts in

Arabia and East Africa link well-defined, distinct, upwelling hot mantle plumes in a similar manner to analogue tank experiments summarized by Harris *et al.* (2004). Three-dimensional isosurfaces

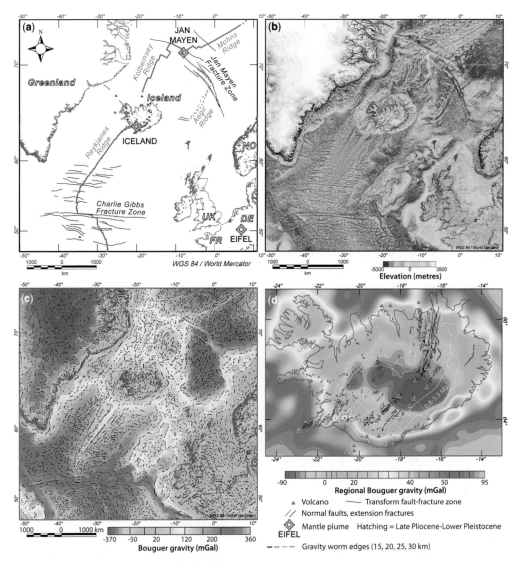

Fig. 7. Topography and gravity surrounding Iceland. (**a**) Tectonic setting. Iceland lies above a mantle plume along an active spreading ridge that is offset by rectilinear (Charlie Gibbs) and curved (Jan Mayen) fracture zones. See Le Breton *et al.* (2012) for illustrations of stages in rifting leading to this geometry. (**b**) SRTM30 Plus (Shuttle Radar Topography Mission 30 Plus) 30 arcsecond-resolution elevation image (Becker *et al.* 2009), enhanced to highlight structural features using a combination of vertical and horizontal gradients. Iceland and the surrounding shelf constitute an elliptical elevated area above the Iceland mantle plume. (**c**) EGM08 (Pavlis *et al.* 2012) Bouguer gravity anomaly with superposed edges in the total horizontal gradient upwards continued to four levels corresponding to source depths of 15, 20, 25 and 30 km (note that extreme data corrugation of variable orientation precludes worm calculation for shallower levels as too many artefacts are created). Iceland and its surrounding shelf correspond to an elliptical area with anomalies of lower magnitude than surrounding rift areas. Gravity worms highlight linear features subparallel to spreading ridges and transform faults. A similar pattern of worms and Bouguer anomalies is seen over the Sedna and Guinevere planitas on Venus (Fig. 2a). (**d**) Long-wavelength Bouguer anomaly image over Iceland showing a gravity low reflecting lower-density rocks associated with the underlying mantle plume. Volcanoes are from the Smithsonian Institution Global Volcanism Program database; geology, and faults and fissures are from the GIS version of Jóhannesson & Sæmundsson (2009). Late Pliocene–Lower Pleistocene bedrock shows a symmetry of geological units about the central, active spreading centres.

(Fig. 5d, e) show that the individual plumes coalesce in the upper mantle to produce a broad linear zone of upwelling. The alignment of interpreted upwelling mantle plumes and plume clusters in Eistla Regio (Fig. 3) is therefore analogously interpreted as a linear zone of mantle upwelling (shown schematically in Figs 7 & 8).

Lateral displacements accompanying upwelling mantle plumes and global mantle flow

Plume-related horizontal mantle flow. Upwelling mantle plumes generate a viscous force due to radial flow in the plume head that is approximately proportional to the plume's volume flux (Westaway 1993). Lithgow-Bertelloni & Silver (1998), Behn *et al.* (2004), Bobrov & Baranov (2011), van Hinsbergen *et al.* (2011) and Husson (2012) showed that

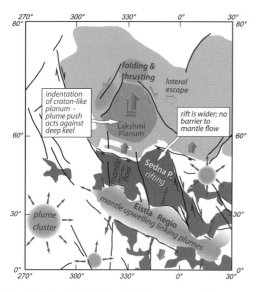

Fig. 8. Simplified relationships between upwelling mantle plumes and rifting and indentation of Lakshmi Planum. Horizontal mantle flow, directed outwards from a wall of mantle upwelling linking colinear plumes, produced the northward motion and indentation of the 'craton-like' Lakshmi Planum by pushing against, and tractions at, the base of its deep lithospheric mantle keel. Rifts are segmented and offset by transfer faults that accommodate differential extension resulting from: (i) interaction with outward, oppositely directed flow from adjacent individual plumes or plume clusters; and from (ii) implied differences in flow from where flow is impeded by the presence of Lakshmi Planum's deep keel and thickened crust beneath fold belts on its margins, in comparison with unrestricted flow east of Lakshmi Planum and the fold belts. A colour version of this figure is available in the online paper.

(on Earth) large horizontal components of global mantle flow may arise from deep mantle upwellings, and that convective mantle drag/plume-generated flow is a significant force for driving adjacent plate motions (substantiating early ideas for the role of mantle plumes in driving plate motion mooted by Morgan 1971 and Wilson 1973). For example:

- Regional mantle flow ensuing from deep mantle upwelling beneath South Africa, documented by Lithgow-Bertelloni & Silver (1998), is a major factor in driving microplate motion as far as the Mediterranean (Faccenna & Becker 2010).
- Cande & Stegman (2011) proposed that the rapid acceleration of India between about 65 and 50 Ma (i.e. prior to its collision with Eurasia) was derived from the 'plume-push' force of the Réunion mantle plume. The origin of the Deccan Traps LIP is also credited to the Réunion plume. From numerical modelling, van Hinsbergen *et al.* (2011) attributed another period of acceleration in India's displacement at around 90 Ma to be entirely related to the 'push' force from the Marion/Morondova mantle plume head, but suggest that the second approximately 65 and 50 Ma acceleration results from a combination of plume-push from the Réunion and Marion plumes and increased ridge push and slab pull forces. The ongoing northward displacement of India documented from GPS measurements is similarly connected with global mantle flow tractions at the base of the Indian continental lithosphere (Alvarez 2010).

Displacement of the Americas due to mantle flow. The westward drift of the Americas in a hotspot reference frame (Gripp & Gordon 2002; Husson *et al.* 2012) corresponds roughly to the beginning of active subduction tectonics on the west coast, with major orogenic pulses being associated with the accretion of island arc and oceanic plateau terranes (Coney *et al.* 1980; Dickinson 2006; Nelson & Colpron 2007; Ramos 2009). Can plate tectonic boundary forces account for the westward drift of the American continents? Ridge push against the eastern coast of the Americas would have been a steady westward force since the opening of the Atlantic Ocean. However, the application of a ridge push force sufficient to drive the Americas west and to create the Cordilleras requires that huge compressive stresses be transmitted through the intervening oceanic lithosphere and its junction with the continental lithosphere. Since some of the oldest, coldest, densest oceanic crust on Earth (*c.* 185 Ma old: Bryan *et al.* 1977) is located along the east coast of North America, where it is underlain by cold, oceanic mantle lithosphere and is capped by thick packages of sedimentary rocks (Steckler & Watts 1974), then the transmission of a compressive stress large enough to raise the Rocky Mountains should have triggered the initiation of subduction beneath the east coast of North America. The geological history of the North and South American west coasts are largely devoid of west-subducting episodes (Coney *et al.* 1980; Dickinson 2006; Nelson & Colpron 2007; Ramos 2009); a westward slab pull contribution therefore seems implausible. Only the slab rollback configuration of the eastward subduction of oceanic lithosphere beneath the west coast of the Americas could have contributed to westward motion. Paradoxically, terrane accretion or shallow subduction phases, which are commonly associated with phases of compression and orogenesis (Kay & Copeland 2006; Nelson & Colpron 2007; Ramos 2009), should have inhibited or even reversed westward drift, yet westward drift of the Americas was largely oblivious to these shifts in the applied force (compression v. extension). Finally, much of the western coast of North America is defined by strike-slip fault systems and should not contribute to westward motion at all. Given the complete lack of a westward slab pull force, the long strike-slip plate boundaries, and the rough balance between compressional and extensional forces above east-directed subduction zones, it seems implausible to suggest that plate boundary forces are solely responsible for the westward drift of North America. An alternative view is that large-scale convective motions in the mantle push upon the deep, stiff, mantle keels of Precambrian craton-cored continents, and that this mantle traction force plays a major role in the westward drift of the American continents (Bokelmann 2002*a*, *b*; Liu & Bird 2002; Eaton & Frederiksen 2007). This model is applied to the indentation of Lakshmi Planum, and to shear displacements between Ovda and Thetis regiones in the Discussion.

Analogue modelling of contemporaneous folding and rifting resulting from underlying flow. Contemporaneous folding, conjugate strike-slip shearing and rifting resulting from underlying flow modelled by Ramberg (1967) using a high-acceleration centrifuge showed that upwelling and downwelling of a ductile layer underlying a semi-brittle upper layer produces shear tractions that result in areas of tensile failure and separation (equivalent to rifts), and areas of localized shortening in which conjugate shear zones and folds were developed. In this model, horizontal flow (induced by a sinking weight) was symmetrical and the resulting geometry was also thus symmetrical, and there was no thickness variation of the upper, brittle–ductile 'crust'. The effects of crustal thickness variations would be expected to enhance and localize areas of folding and shearing.

Discussion

Tectonic model for linked rifting, indentation and transcurrent faulting

The comparison between observations of Venus and plume- and mantle flow-related structures on Earth outlined above leads to the schematic tectonic interpretation for the western half of the study area in 2D and 3D in Figures 8 and 9, respectively. Rifting in Sedna Planitia (Fig. 8) is attributed to crustal extension on the northern flank of a zone of mantle upwelling, linking plumes and plume clusters. Extension and rifting is produced by tractions of mantle flowing out (horizontally) away from this zone of mantle upwelling. Horizontal mantle flow pushing against the deep keel to Lakshmi Planum drives the rigid planum into the surrounding area of initially thinner crust. A fold and thrust belt is developed ahead of Lakshmi Planum (Harris & Bédard 2014), accompanied by crustal thickening. A sinistral transpressional fold belt developed on the NW planum margin also results in crustal thickening, whilst no thickening occurs in this event on its NE margin, along which dextral transcurrent displacement is interpreted by Harris & Bédard (2014). Where lateral flow away from the zone of mantle upwelling is not impeded by either the deep keel of Lakshmi Planum or the zone of mantle upwelling proposed by Mège & Ernst (2001) beneath the Bereghinya arachnoid cluster (Fig. 3a), the rift in Sedna Planitia is wider and the crust thinned more than where lateral flow is perturbed. Lateral variations in the degree of total extension are partitioned by NW-striking transfer faults.

The geometry of structures in Maxwell Montes (Keep & Hansen 1994), polyphase folding in Ovda Regio (Harris & Bédard 2014) and the change in displacement sense along the shear zone separating Ovda and Thetis regiones described above necessitate a change from north–south to approximately NE–SW to east–west shortening in the highland areas. Although the relative timing of events in these different areas remains to be established, the sequence of overprinting is the same. We suggest that this can be explained by the formation of a second set of north–south to NNE–SSW-trending rifts (e.g. Leda Planitia) linking plume centres, as shown in Figure 3a. In this second event, similar forces arising from mantle flow against the thicker crust of East Ishtar Terra formed north- to NW-trending folds during crustal underthrusting and extreme crustal thickening in Maxwell Montes on the eastern margin of Lakshmi Planum, as interpreted by Keep & Hansen (1994). The structures in Maxwell Montes and the other fold and thrust belts surrounding Lakshmi Planum imply crustal shortening, although the structural styles as described by Keep & Hansen (1994) and Harris & Bédard (2014) differ greatly. The difference reflects the initial mobility of Lakshmi Planum, leading to indentation and lateral escape about its NW, north and NE margins, whereas subsequent underthrusting against the eastern margin of Lakshmi Planum to create Maxwell Montes (as shown by Keep & Hansen 1994) reflects Lakshmi Planum's subsequent immobility.

Origin of features defined by gravity 'worms'

An intriguing outcome of this study is the identification of subparallel regional features from edges of horizontal gravity gradients in both the Sedna Planitia rift and across the whole northern part of the study area, including Ishtar Terra and for

western Aphrodite Terra (Fig. 2). 'Gravity worms' commonly correspond to the margins of short-wavelength Bouguer anomalies (Fig. 2b) and thus act as markers that help identify transcurrent faults. Gravity worms have aided mapping of regional structures on Earth (e.g. Bierlein *et al.* 2006; Vos *et al.* 2006; Austin & Blenkinsop 2008; Heath *et al.* 2009; Harris & Bédard 2014), and the correspondence between transcurrent faults established from the deflection and displacement of 'worms' on the margins of Lakshmi Planum and shear zones interpreted from radar images validates their similar use in regional structural interpretation on Venus. The linear structures defined by 'worms' in Sedna Planitia resemble parallel normal faults and patterns of gravity 'worms' about mid-oceanic spreading ridges on Earth (Fig. 7), whereas the anastomosing, more concentric pattern of 'worms' over Guinevere Planitia, also interpreted as a rift, is the

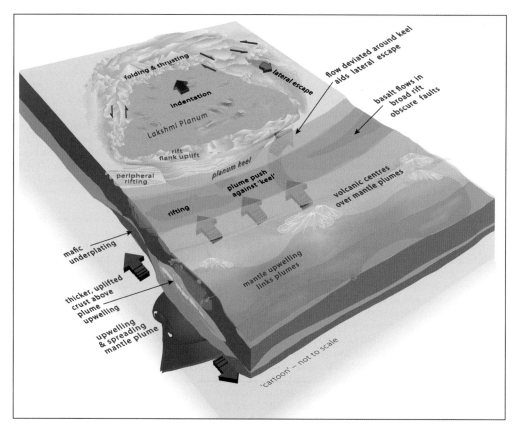

Fig. 9. Schematic, 3D 'cartoon' of indentation and lateral escape about Lakshmi Planum driven by tractions and push-force arising from horizontal mantle flow acting on its deep craton-like keel. A broad zone of mantle upwelling links mantle plumes (cf. Fig. 5 for the present-day Afar–East African rift system). Rifting on the flanks of this zone of upwelling is created through flow away from mantle upwelling. Plains volcanic material is rendered semi-transparent to reveal the underlying mantle interpretation. Bouguer gravity highs imply dense, mafic mantle underplating beneath rifts.

same as observed over mantle plumes such as at Beta and Phoebe regiones (see Fig. 3 for the location). Could gravity 'worms' in Venus' rifts act as markers of faulting about a rift axis and be used in a similar fashion to seafloor magnetic anomalies on Earth? Further detailed comparisons between geological maps, radar and emissivity images, and 'worms' are, nevertheless, required to better understand their origin(s) and to determine whether some of Venus' lowlands may have formed during symmetrical outpourings of magma about a rift axis similar to the formation of oceanic crust on Earth.

Implications of fault/shear-zone reactivation

In contrast to studies (discussed earlier) that recognize only limited strike-slip faulting on Venus, and that of McGuire *et al.* (1996), who suggested that faults are less likely to be reactivated on Venus compared with Earth, we show that large strike-slip displacements are not only widespread but that shear-zone reactivation with reversal of shear sense has occurred. Fault reactivation and inversion on Venus was also proposed by Kumar (2005) and Hansen (2006). Reactivation and reversal of displacement sense along regional transcurrent shear zones is common on Earth along crustal-scale transcurrent structures in Archaean granite–greenstone terrains (e.g. Mueller & Harris 1988; Mueller *et al.* 1988; Blewett *et al.* 2010; Leclerc *et al.* 2012; Harris & Bédard 2014). Multiple transcurrent (as well as normal and reverse) reactivation was documented along the Archaean–Cretaceous Darling Fault Zone on the western Yilgarn margin (summarized by Harris 1994*a, b* and Wilde *et al.* 1996) and in younger deformation zones (e.g. Palaeozoic–Tertiary transcurrent displacements on the Great Glen Fault in Scotland: Holgate 1969; Le Breton *et al.* 2013). Fault reactivation on Earth is attributed to fault-zone weakening (e.g. Handy *et al.* 2001; Rutter *et al.* 2001) largely due to the presence of water (e.g. Regenauer-Lieb & Yuen 2003 and references therein). Gurnis *et al.* (2000, p. 74) concluded that, on Earth, 'old weak structures are reused by the convecting system because it takes less energy to reactivate a preexisting structure than it does to create an entirely new plate margin from pristine, intact lithosphere'. Whilst Gurnis *et al.* (2000) discussed fault reactivation on Earth in the context of plate tectonics, our results show that transcurrent fault formation and reactivation during changes in principal stresses can occur due solely to changes in the mantle flow field (probably due to changes in plume activity) without plate-tectonic-related stresses. The reactivation of transcurrent faults documented in our study and inversion of normal faults (e.g. Hansen

2006) show that regional faults/shear zones on Venus may be reactivated, similar to faults on Earth, despite the absence of surface water.

Implications for mantle convection and tectonic regime of Venus

The proposed causal link between mantle flow directed horizontally from a zone of mantle upwelling and the development of rifts, indentation tectonics, and strike-slip fault zones, agrees with Kohlstedt & Mackwell (2009) who, from an analysis of a likely rheological profile for Venus, inferred that 'convection and lithospheric deformation will be strongly coupled' (p. 418) and that 'regional and planetary-scale tectonics will likely directly reflect underlying mantle processes such as convection' (p. 419). Our evidence for lateral, 'plate-like' displacement due to mantle flow does not, however, indicate that plate tectonics occurs on Venus. Even in a purely stagnant lid regime (where, by definition, convective stress is less than lithospheric yield stress: Lenardic & Crowley 2012), the lid may still move passively but does not influence convection beneath it (Solomatov & Moresi 1997). Our interpretation for horizontal translation of terrains on Venus is consistent with recent research, based largely on numerical modelling, which suggests that Venus may be better considered as exhibiting intermediate 'transient' (Robin *et al.* 2007), 'creeping stagnant lid', or transitional (Solomatov & Moresi 1996) or 'episodic' (Turcotte 1993, 1995; Loddoch *et al.* 2006; Armann & Tackley 2012; Lenardic & Crowley 2012; Papuc & Davies 2012) convective regimes, transitioning from an earlier stagnant lid convective mode during progressive cooling (O'Neill 2013); that is, the opposite evolutionary trend to that proposed by van Thienen *et al.* (2004). Our proposed conceptual model shows some similarity with the 'no subduction' regime for the hot early Earth proposed by Sizova *et al.* (2010) from 2D numerical experiments, as argued for by Bédard *et al.* (2013) and for which there is recent geochemical support (Debaille *et al.* 2013). In this regime, horizontal movements of deformable plate fragments driven by mantle flow are accommodated by coexisting zones of shortening and rifting; the latter are also associated with thick mafic crust formation. This similarity supports the proposed analogy of Venus surface dynamics with Precambrian subduction-free geodynamics. The low elevations predicted in the subductionless models of Sizova *et al.* (2010) are, however, at apparent odds with the high topography present on Venus.

In tectonic models without subduction, lithospheric shortening is accommodated by either delamination (e.g. Turcotte 1989, 1995; Pysklywec *et al.*

2010) or 'erosion' (Burov *et al.* 2007) of mantle lithosphere or through narrow zones of downwelling or 'drips' of basal lithospheric mantle into the aesthenosphere (cf. West *et al.* 2009; Pysklywec *et al.* 2010; Sizova *et al.* 2010). A similar association between mountain belts and zones of symmetrical mantle downwelling was proposed before the advent of plate tectonic theory (Wilson 1961), and analogue models of Pysklywec *et al.* (2010) show that such a mechanism is plausible to accommodate folding of the upper crust without subduction.

Further detailed integration of the displacement history along interpreted regional faults with interpreted stratigraphic relationships is required to better constrain the sequence of regional tectonic events on Venus. The structures interpreted from gravity data in our study most probably represent deformation corridors and not discrete faults, and more detailed gravity data are required to more precisely map crustal- to lithospheric-scale faults. Mapping of present-day seismicity on Venus (proposed by Balint *et al.* 2009; Hunter *et al.* 2012), through wireless seismometers (Ponchak *et al.* 2012) or remote detection of seismic waves in the atmosphere (Garcia *et al.* 2005) or ionosphere (Lognonné *et al.* 2006; Lognonné 2009), is required to provide crucial missing information on whether regional faults interpreted in our study are still active.

Conclusions

Bouguer gravity, enhanced and visualized using modern geophysical software, combined with crustal thickness estimates, constitute important datasets to map regional crustal rift zones and faults on Venus. The high Bouguer gravity signature of the interpreted rifts on Venus is comparable to that of the North American Mid-Continent Rift interpreted to result from mantle underplating. The surface expression of faulting does not provide a true indication of the scale of rifting, and rift-related extension is greater than previously proposed. Aerially extensive lava flows may mask early formed faults, such that visible faults record only the last stages of extension. Criteria for rifts on Venus must be revised to include the presence of long, straight Bouguer gravity highs; the full extent of rifts may thus not be portrayed in current maps that are based solely on structural interpretation of radar imagery.

In additional to local hotspot-like upwellings that may have a direct correlation with discrete volcanic centres and coronae, gravity data suggest the likelihood for an approximately 8000 km-long zone of sheet-like mantle upwelling formed by the coalescence of aligned, upwelling mantle plumes and plume clusters in Eistla Regio, similar to the coalescing and lateral flow of upwelling mantle plume in the Red Sea (but 3.5 times longer). This interpreted active volcanic zone above plumes is flanked by the Sedna and Guinevere rifts, which are characterized by crustal thinning and mafic underplating. Transcurrent shear displacements on Venus are concomitant with, and most probably compensate for, regional extension and rifting. Faults with transcurrent components of horizontal displacement that offset rifts differ from transcurrent faults formed during bulk shortening and indentation in Lakshmi Planum. Their origin compensates for differential extension between blocks where flow is hindered by the presence of either: (i) the deep keel of the adjacent Lakshmi Planum; or (ii) another zone of mantle upwelling (where narrower rifts develop) and blocks where underlying horizontal flow away from the zone of mantle upwelling is unhampered (and wider rifts develop). They show some similarities with transform faults on Earth but do not necessarily imply a central spreading axis. Changes in the locii of active rifting or variations in the amount of extension on orthogonal rifts are interpreted to have produced approximately 90° changes in principal horizontal stress orientations, leading to the reversal of shear sense on transcurrent shear zones separating Ovda and Thetis regiones, refolded folds and underthrusting to produce Maxwell Montes on the eastern margin of the newly immobilized Lakshmi Planum.

On Earth, regional mantle flow and/or plume push acting against deep cratonic keels (along with 'pull' from the related downwelling) contributes to the horizontal displacements of continents in addition to plate boundary forces. The Americas illustrate that continental blocks with the thickest keels are displaced further than terrains with no cratonic keel (e.g. the Caribbean plate). Similarly, the displacement of Ishtar Terra and the driving force for Himalayan–Indochina-like indentation and lateral escape, and lateral displacements between Ovda and Thetis regiones, are attributed to mantle flow acting on their deep crustal keels, whereas terrains with thin crust are not displaced. We concur with previous studies that there is no evidence for subduction and Venusian plate tectonics similar to what is seen on the present-day Earth. The pattern of structures derived from gravity data, especially in the Sedna Planitia rift, however, raises the possibility for features similar to those formed in oceanic crust on Earth. We provide an updated view of Venusian tectonics where large, coherent displacements of its constituent terrains occur without true seafloor spreading or subduction. We surmise that this new perspective of Venus provides an analogue for the tectonics of the Archaean Earth.

T. Gerya, L. Marinangeli, and editors M. Massironi, T. Platz and P. Byrne are thanked for their suggestions and additions to the text. Venus gravity and radar data were provided by NASA, the USGS Astrogeology Science Center, and the Jet Propulsion Laboratory (JPL) at the California Institute of Technology. P. James (MIT) kindly supplied his crustal thickness data, calculated from NASA gravity and presented in James *et al.* (2013). S.-J. Chang is thanked for providing her seismic velocity data and MEAN velocity model from which seismic tomography figures were generated. The GIS versions of Iceland geological maps were provided by the Icelandic Institute of Natural History. EGM08 and WGM2012 Bouguer gravity data were provided by the Bureau Gravimétrique International. The Holocene Volcanoes of the World database was downloaded from Orr and Associates (Australia). Geosoft's Oasis Montaj™ and software by P. Keating (NRCAN) were used for processing geophysical and crustal thickness data. Radar images were enhanced using the public domain software ImageJ and Adobe Photoshop™. B. Giroux (INRS-ETE) is thanked for converting NASA gravity data to a readable format. Some features in Figure 9 are modified after symbols courtesy of the Integration and Application Network, University of Maryland Center for Environmental Science (ian.umces.edu/symbols/). This is NRCAN/ESS/GSC contribution number 20130152.

References

ALLEN, D. J., BRAILE, L. W., HINZE, W. J. & MARIANO, J. 2006. Chapter 10: The Midcontinent rift system, U.S.A: A major Proterozoic continental rift. *In*: OLSEN, K. H. (ed.) *Continental Rifts: Evolution, Structure, Tectonics*. Developments in Geotectonics, **25**. Elsevier, Amsterdam, 375–407.

ALVAREZ, W. 2010. Protracted continental collisions argue for continental plates driven by basal traction. *Earth and Planetary Science Letters*, **296**, 434–442.

ANDERSON, D. L. 1981. Plate tectonics on Venus. *Geophysical Research Letters*, **8**, 309–311.

ANDERSON, D. L. 2000. The thermal state of the upper mantle; no role for mantle plumes. *Geophysical Research Letters*, **27**, 3623–3626.

ANDERSON, D. L. 2013. The persistent mantle plume myth. *Australian Journal of Earth Sciences*, **60**, 657–673.

ANDERSON, F. S. & SMREKAR, S. E. 2006. Global mapping of crustal and lithospheric thickness on Venus. *Journal of Geophysical Research: Planets*, **111**, E08006, http://dx.doi.org/10.1029/2004JE002395

ANGUITA, F., FERNANDEZ, C. *ET AL.* 2006. Evidences for a Noachian-Hesperian orogeny in Mars. *Icarus*, **185**, 331–357.

ANNEN, C., BLUNDY, J. D. & SPARKS, R. S. J. 2006. The genesis of intermediate and silicic magmas in deep crustal hot zones. *Journal of Petrology*, **47**, 505–539.

ANSAN, V., VERGELY, P. & MASSON, P. 1996. Model of formation of Ishtar Terra, Venus. *Planetary and Space Sciences*, **44**, 817–831.

ARCHIBALD, N. J., BOSCHETTI, F. & HOLDEN, D. J. 1999. Visualizing the geological 'edges' of US Gulf of Mexico structures. *Offshore*, **59**(6), 74–76.

ARKANI-HAMED, J. 1996. Analysis and interpretation of high-resolution topography and gravity of Ishtar Terra, Venus. *Journal of Geophysical Research*, **101**, 4691–4710.

ARMANN, M. & TACKLEY, P. J. 2012. Simulating the thermochemical magmatic and tectonic evolution of Venus's mantle and lithosphere: Two-dimensional models. *Journal of Geophysical Research*, **117**, E12003, http://dx.doi.org/10.1029/2012JE004231

ARTYUSHKOV, E. V. 1973. Stresses in the lithosphere caused by crustal thickness inhomogeneities. *Journal of Geophysical Research*, **78**, 7675–7708.

AUSTIN, J. R. & BLENKINSOP, T. G. 2008. The Cloncurry lineament: Geophysical and geological evidence for a deep crustal structure in the Eastern Succession of the Mount Isa Inlier. *Precambrian Research*, **163**, 50–68.

AYDIN, A. 2006. Failure modes of the lineaments on Jupiter's moon, Europa: implications for the evolution of its icy crust. *Journal of Structural Geology*, **28**, 2222–2236.

BALINT, T., CUTTS, J. *ET AL.* 2009. *Technologies for Future Venus Exploration*. White paper to the NRC Decadal Survey Inner Planets Sub-Panel, Venus Exploration Analysis Group, http://www.lpi.usra.edu/vexag/resources/decadal/TiborSBalint.pdf (accessed June 2013).

BARNETT, D. N., NIMMO, F. & MCKENZIE, D. 2000. Elastic thickness estimates for Venus using line of sight accelerations from Magellan Cycle 5. *Icarus*, **146**, 404–419.

BARSUKOV, V. L., BASILEVSKY, A. T. *ET AL.* 1986. The geology and geomorphology of the Venus surface as revealed by the radar images obtained by Veneras 15 and 16. *Journal of Geophysical Research: Solid Earth*, **91**, 378–398.

BASILEVSKY, A. T. & HEAD, J. W. III. 2000. Rifts and large volcanoes on Venus: global assessment of their age relations with regional plains. *Journal of Geophysical Research: Planets*, **105**, 24 583–24 611.

BASILEVSKY, A. T. & HEAD, J. W. III. 2007. Beta Regio, Venus: evidence for uplift, rifting, and volcanism due to a mantle plume. *Icarus*, **192**, 167–186.

BASILEVSKY, A. T., SHALYGIN, E. V. *ET AL.* 2012. Geologic interpretation of the near-infrared images of the surface taken by the Venus Monitoring Camera, Venus Express. *Icarus*, **217**, 434–450.

BECKER, J. J., SANDWELL, D. T. *ET AL.* 2009. Global bathymetry and elevation data at 30 arc seconds resolution: SRTM30_PLUS. *Marine Geodesy*, **32**, 355–371.

BÉDARD, J. H., HARRIS, L. B. & THURSTON, P. 2013. The hunting of the snArc. *Precambrian Research*, **229**, 20–48.

BEHN, M. D., CONRAD, C. P. & SILVER, P. G. 2004. Detection of upper mantle flow associated with the African Superplume. *Earth and Planetary Science Letters*, **224**, 259–274.

BEHRENDT, J. C., GREEN, A. G. *ET AL.* 1988. Crustal structure of the Midcontinent rift system: results from GLIMPCE deep seismic reflection profiles. *Geology*, **16**, 81–85.

BEHRENDT, J. C., HUTCHINSON, D. R., LEE, M. W., THORNBER, C. R., TREHU, A., CANNON, W. F. & GREEN, A. G. 1990. Seismic reflection (GLIMPCE) evidence of deep crustal and upper mantle intrusions and magmatic underplating associated with the Mid-continent Rift System of North America. *Tectonophysics*, **173**, 617–626.

BERTHÉ, D., CHOUKROUNE, P. & JEGOUZO, P. 1979. Orthogneiss, mylonite and noncoaxial deformation of granite: the example of the South Amorican shear zone. *Journal of Structural Geology*, **1**, 31–42.

BIERLEIN, F. P., MURPHY, F. C., WEINBERG, R. F. & LEES, T. 2006. Distribution of orogenic gold deposits in relation to fault zones and gravity gradients: targeting tools applied to the Eastern Goldfields, Yilgarn Craton, Western Australia. *Mineralium Deposita*, **41**, 107–126.

BINDSCHADLER, D. L., SCHUBERT, G. & KAULA, W. M. 1992. Coldspots and hotspots: Global tectonics and mantle dynamics of Venus. *Journal of Geophysical Research*, **97**, 13 495–13 532.

BISTACCHI, N., MASSIRONI, M. & BAGGIO, P. 2004. Large scale fault kinematic analysis in Noctis Labyrinthus (Mars). *Planetary and Space Science*, **52**, 215–222.

BLAKELY, R. J. & SIMPSON, R. W. 1986. Approximating edges of source bodies from magnetic or gravity anomalies. *Geophysics*, **51**, 1494–1498.

BLEAMASTER, L. F., III & HANSEN, V. L. 2005. *Geologic Map of the Ovda Regio Quadrangle (V-35), Venus*. United States Geological Survey, Geologic Investigations Series, **I-2808**.

BLEWETT, R. S., CZARNOTA, K. & HENSON, P. A. 2010. Structural-event framework for the eastern Yilgarn Craton, Western Australia, and its implications for orogenic gold. *Precambrian Research*, **183**, 203–229.

BOBROV, A. M. & BARANOV, A. A. 2011. Horizontal stresses in the mantle and in the moving continent for the model of two dimensional convection with varying viscosity. *Physics of the Solid Earth*, **47**, 801–815.

BOKELMANN, G. H. R. 2002a. Convection-driven motion of the North American craton: evidence from P-wave anisotropy. *Geophysical Journal International*, **148**, 278–287.

BOKELMANN, G. H. R. 2002b. Which forces drive North America? *Geology*, **30**, 1027–1030.

BONDARENKO, N. V., HEAD, J. W. & IVANOV, M. A. 2010. Present-day volcanism on Venus: Evidence from microwave radiometry. *Geophysical Research Letters*, **37**, L23202, http://dx.doi.org/10.1029/2010GL045233

BORRACCINI, F., DI ACHILLE, G., ORI, G. G. & WEZEL, F. C. 2007. Tectonic evolution of the eastern margin of the Thaumasia Plateau (Mars) as inferred from detailed structural mapping and analysis. *Journal of Geophysical Research*, **112**, E05005, http://dx.doi.org/10.1029/2006JE002866

BOSCHI, L., BECKER, T. W. & STEINBERGER, B. 2007. Mantle plumes: Dynamic models and seismic images. *Geochemistry, Geophysics, Geosystems*, **8**, Q10006, http://dx.doi.org/10.1029/2007GC001733

BROWN, C. D. & GRIMM, R. E. 1995. Tectonics of Artemis Chasma: a Venusian "plate" boundary. *Icarus*, **117**, 219–249.

BRYAN, W. B., FREY, F. A. & THOMPSON, G. 1977. Oldest Atlantic seafloor: mesozoic basalts from western North Atlantic margin and eastern North America. *Contributions to Mineralogy and Petrology*, **64**, 223–242.

BUNTE, M. K., WILLIAMS, D. A., GREELEY, R. & JAEGER, W. L. 2010. Geologic mapping of the Hi'iaka and Shamshu regions of Io. *Icarus*, **207**, 868–886.

BURKE, K. & CANNON, J. M. 2014. Plume–plate interaction. *Canadian Journal of Earth Sciences*, **51**, 208–221.

BURKE, K. & DEWEY, K. L. 1973. Plume generated triple junctions: key indicators in applying plate tectonics to old rocks. *Journal of Geology*, **81**, 406–433.

BUROV, E., GUILLOU-FROTTIER, L., D'ACREMONT, E., LE POURHIET, L. & CLOETINGH, S. 2007. Plume head–lithosphere interactions near intra-continental plate boundaries. *Tectonophysics*, **434**, 15–38.

CAMP, V. E. & HANAN, B. B. 2008. A plume-triggered delamination origin for the Columbia River Basalt Group. *Geosphere*, **4**, 480–495.

CAMPBELL, I. H. 2005. Large igneous provinces and the mantle plume hypothesis. *Elements*, **1**, 265–269.

CANDE, S. C. & STEGMAN, D. R. 2011. Indian and African plate motions driven by the push force of the Réunion plume head. *Nature*, **475**, 47–52.

CATALÁN, M., GALINDO-ZALDIVAR, J., DAVILA, J., MARTOS, Y. M., MALDONADO, A., GAMBÔA, L. & SCHREIDER, A. A. 2012. Initial stages of oceanic spreading in the Bransfield Rift from magnetic and gravity data analysis. *Tectonophysics*, **585**, 102–112.

CHANDLER, V. W., MCSWIGGEN, P. L., MOREY, G. B., HINZE, W. J. & ANDERSON, R. R. 1989. Interpretation of seismic reflection, gravity, and magnetic data across Middle Proterozoic Mid-Continent Rift System, north-western Wisconsin, eastern Minnesota, and central Iowa. *American Association of Petroleum Geologists Bulletin*, **73**, 261–275.

CHANG, S.-J. & VAN DER LEE, S. 2011. Mantle plumes and associated flow beneath Arabia and East Africa. *Earth and Planetary Science Letters*, **302**, 448–454.

CHETTY, T. R. K., VENKATRAYUDU, M. & VENKATASI-VAPPA, V. 2010. Structural architecture and a new tectonic perspective of Ovda Regio, Venus. *Planetary and Space Science*, **58**, 1286–1297.

COBBOLD, P. R. 2008. Horizontal compression and stress concentration at passive margins: causes, consequences, and episodicity. Paper presented at the 33rd International Geological Congress, Oslo, Norway, 6–14 August 2008, http://hal-insu.archives-ouvertes.fr/insu-00376344

CONEY, P. J., JONES, D. L. & MONGER, J. W. H. 1980. Cordilleran suspect terranes. *Nature*, **188**, 329–333.

CONNORS, C. & SUPPE, J. 2001. Constraints on magnitudes of extension on Venus from slope measurements. *Journal of Geophysical Research*, **106**, 3237–3260.

CRUMPLER, L. S. & HEAD, J. W. III. 1988. Bilateral topographic symmetry patterns across Aphrodite Terra, Venus. *Journal of Geophysical Research*, **93**, 301–312.

CRUMPLER, L. S., HEAD, J. W., III & CAMPBELL, D. B. 1986. Orogenic belts on Venus. *Geology*, **14**, 1031–1034.

CRUMPLER, L. S., HEAD, J. W., III & HARMON, J. K. 1987. Regional linear cross-strike discontinuities in western Aphrodite Terra, Venus. *Geophysical Research Letters*, **14**, 607–610.

CRUMPLER, L. S., HEAD, J. W., III & AUBELE, J. C. 1993. Relation of major volcanic center concentration on

Venus to global tectonic patterns. *Science*, **261**, 591–595.

DARBYSHIRE, F. A., WHITE, R. S. & PRIESTLEY, K. F. 2000. Structure of the crust and uppermost mantle of Iceland from a combined seismic and gravity study. *Earth and Planetary Science Letters*, **181**, 409–428.

DEBAILLE, V., O'NEILL, C., BRANDON, A. D., HAENE-COUR, P., YIN, Q.-Z., MATTIELLI, N. & TREIMAN, A. H. 2013. Stagnant-lid tectonics in early Earth revealed by ^{142}Nd variations in late Archean rocks. *Earth and Planetary Science Letters*, **373**, 83–92.

DICKINSON, W. R. 2006. Geotectonic evolution of the Great Basin. *Geosphere*, **2**, 353–368, http://dx.doi.org/10.1130/GES00054.1

DUFRÉCHOU, G., HARRIS, L. B. & CORRIVEAU, L. 2014. Tectonic reactivation of transverse basement structures in the Grenville orogen of SW Quebec, Canada: insights from potential field data. *Precambrian Research*, **241**, 61–84.

EATON, D. W. & FREDERIKSEN, A. 2007. Seismic evidence for convection-driven motion of the North American plate. *Nature*, **446**, 428–431.

ELLIOT, D. H. 2013. The geological and tectonic evolution of the Transantarctic Mountains: a review. *In*: HAMBREY, M. J., BARKER, P. F., BARRETT, P. J., BOWMAN, V., DAVIES, B., SMELLIE, J. L. & TRANTER, M. (eds) *Antarctic Palaeoenvironments and Earth-Surface Processes*. Geological Society, London, Special Publications, **381**, 7–35.

ERNST, R. E. & BUCHAN, K. L. 2003. Recognizing mantle plumes in the geological record. *Annual Review of Earth and Planetary Sciences*, **31**, 469–523.

ERNST, R. E. & DESNOYERS, D. W. 2004. Lessons from Venus for understanding mantle plumes on Earth. *Physics of the Earth and Planetary Interiors*, **146**, 195–229.

ERNST, R. E., HEAD, J. W., PARFITT, E., GROSFILS, E. & WILSON, L. 1995. Giant radiating dyke swarms on Earth and Venus. *Earth Science Reviews*, **39**, 1–58.

ERNST, R. E., DESNOYERS, D. W., HEAD, J. W. & GROSFILS, E. B. 2003. Graben–fissure systems in Guinevere Planitia and Beta Regio (264°–312°E, 24°–60°N), Venus, and implications for regional stratigraphy and mantle plumes. *Icarus*, **164**, 282–316.

ERNST, R. E., BUCHAN, K. L. & DESNOYERS, D. W. 2007. Plumes and plume clusters on earth and venus: evidence from Large Igneous Provinces (LIPs). *In*: YUEN, D. A., MARUYAMA, S., KARATO, S.-I. & WINDLEY, B. F. (eds) *Superplumes: Beyond Plate Tectonics*. Springer, Dordrecht, 537–562.

FACCENNA, C. & BECKER, T. W. 2010. Shaping mobile belts by small-scale convection. *Nature*, **465**, 602–605.

FACCENNA, C., BECKER, T. W., CONRAD, C. P. & HUSSON, L. 2013. Mountain building and mantle dynamics. *Tectonics*, **32**, 80–93, http://dx.doi.org/10.1029/2012TC003176

FERNÁNDEZ, C., ANGUITA, F., RUIZ, J., ROMEO, I., MARTÍN-HERRERO, À. I., RODRÍGUE, A. & PIMENTEL, C. 2010. Structural evolution of Lavinia Planitia, Venus: implications for the tectonics of the lowland plains. *Icarus*, **206**, 210–228.

FORSYTH, D. & UYEDA, S. 1975. On the relative importance of the driving forces of plate motion. *Geophysical Journal International*, **43**, 163–200.

FOSTER, A. & NIMMO, F. 1996. Comparisons between the rift systems of East Africa, Earth and Beta Regio, Venus. *Earth and Planetary Science Letters*, **143**, 183–195.

FOULGER, G. L. 2010. *Plates vs Plumes: A Geological Controversy*. Wiley-Blackwell, Chichester.

FRENCH, S., LEKIC, V. & ROMANOWICZ, B. 2013. Waveform tomography reveals channeled flow at the base of the oceanic asthenosphere. *Science*, **342**, 227–230.

FRETZDORFF, S., WORTHINGTON, T. J., HAASE, K. M., HÉKINIAN, R., FRANZ, L., KELLER, R. A. & STOFFERS, P. 2004. Magmatism in the Bransfield Basin: Rifting of the South Shetland Arc? *Journal of Geophysical Research*, **109**, B12208, http://dx.doi.org/10.1029/2004JB003046

GAIT, A. D., LOWMAN, J. P. & GABLE, C. W. 2008. Time-dependence in 3D mantle convection models featuring evolving plates: the effect of lower mantle viscosity. *Journal of Geophysical Research*, **113**, B08409, http://dx.doi.org/10.1029/2007JB005538

GARCIA, R., LOGNONNÉ, P. & BONNIN, X. 2005. Detecting atmospheric perturbations produced by Venus quakes. *Geophysical Research Letters*, **32**, L16205, http://dx.doi.org/10.1029/2005GL023558

GHOSH, A., BECKER, T. W. & HUMPHREYS, E. D. 2013. Dynamics of the North American continent. *Geophysical Journal International*, **194**, 651–669, http://dx.doi.org/10.1093/gji/ggt151

GILMORE, M. S., COLLINS, G. C., IVANOV, M. A., MARI-NANGELI, L. & HEAD, J. W. III. 1998. Style and sequence of extensional structures in tessera terrain, Venus. *Journal of Geophysical Research: Planets*, **103**, 16 813–16 840.

GLEN, R. A., KORSCH, R. J., HEGARTY, R., SAEED, A., POUDJOM DJOMANI, Y., COSTELLOE, R. D. & BELOU-SOVA, E. 2013. Geodynamic significance of the boundary between the Thomson Orogen and the Lachlan Orogen, northwestern New South Wales and implications for Tasmanide tectonics. *Australian Journal of Earth Sciences*, **60**, 371–412.

GLUKHOVSKIY, M. Z., VLORALEV, V. M. & KUZ'MIN, M. I. 1995. Hot belt in the early Earth and its evolution. *Geotectonics*, **28**, 367–379.

GRIMM, R. E. 1998. What do we really know about the heat flow of Venus (or anyplace else we can't stick with probes?). *The Leading Edge*, **1998**, 1544–1546.

GRIPP, A. E. & GORDON, R. G. 2002. Young tracks of hotspots and current plate velocities. *Geophysical Journal International*, **150**, 321–361.

GROSFILS, E. B. & HEAD, J. W. 1994. The global distribution of giant radiating dike swarms on Venus: implications for the global stress state. *Geophysical Research Letters*, **21**, 701–704.

GUILLOU-FROTTIER, L., BUROV, E., CLOETINGH, S., LE GOFF, E., DESCHAMPS, Y., HUET, B. & BOUCHOT, V. 2012. Plume-induced dynamic instabilities near cratonic blocks: Implications for P–T–t paths and metallogeny. *Global and Planetary Change*, **90–91**, 37–50.

GUPTA, S., ZHAO, D. & RAI, S. S. 2009. Seismic imaging of the upper mantle under the Erebus hotspot in Antarctica. *Gondwana Research*, **16**, 109–118.

GURNIS, M., ZHONG, S. & TOTH, J. 2000. On the competing roles of fault reactivation and brittle failure in generating plate tectonics from mantle convection.

In: RICHARDS, M. A., GORDON, R. G. & VAN DER HILST,
R. D. (eds) *The History and Dynamics of Global Plate
Motions*. American Geophysical Union, Geophysical
Monograph Series, **121**, 73–94.

HANDY, M. R., MULCH, R., ROSENAU, M. & ROSENBERG,
C. R. 2001. The role of transcurrent shear zones as melt
conduits and reactors and as agents of weakening in the
continental crust. *In*: HOLDSWORTH, R. E., STRACHAN,
R. A., MAGLOUGHLIN, J. F. & KNIPE, R. J. (eds) *The
Nature and Significance of Fault Zone Weakening*.
Geological Society, London, Special Publications,
186, 303–330.

HANJALIĆ, K. & KENJEREŠ, S. 2006. RANS-based very
large eddy simulation of thermal and magnetic convection at extreme conditions. *Journal of Applied Mechanics*, **73**, 430–440.

HANSEN, V. L. 1992. Non-coaxial deformation on Venus.
In: *Proceedings of the 23rd Lunar and Planetary
Science Conference*, March 16–20, 1992, Houston,
Texas. Lunar and Planetary Institute, Houston, TX,
479–480.

HANSEN, V. L. 2006. Geologic constraints on crustal
plateau surface histories, Venus: the lava pond and
bolide impact hypotheses. *Journal of Geophysical
Research*, **111**, E11010, http://dx.doi.org/10.1029/
2006JE002714

HANSEN, V. L. 2007. LIPs on Venus. *Chemical Geology*,
241, 354–374.

HANSEN, V. L. & OLIVE, A. 2010. Artemis, Venus: the
largest tectonomagmatic feature in the solar system?
Geology, **38**, 467–470.

HANSEN, V. L. & WILLIS, J. J. 1996. Structural analysis of
a sampling of tesserae: implications for Venus geodynamics. *Icarus*, **123**, 296–312.

HANSEN, V. L., WILLIS, J. J. & BANERDT, W. B. 1997. Tec-
tonic overview and synthesis. *In*: BOUGHER, S. W.,
HUNTEN, D. M. & PHILLIPS, R. J. (eds) *Venus II:
Geology, Geophysics, Atmosphere, and Solar Wind
Environment*. University of Arizona Press, Tucson,
AZ, 797–844.

HARRIS, L. B. 1994*a*. Neoproterozoic sinistral displacement along the Darling Mobile Belt, Western Australia. *Journal of the Geological Society, London*, **151**,
901–904.

HARRIS, L. B. 1994*b*. Structural and tectonic synthesis for
the Perth Basin, Western Australia. *Journal of Petroleum Geology*, **17**, 129–156.

HARRIS, L. B. & BÉDARD, J. H. 2014. Chapter 9. Crustal
evolution and deformation in a non-plate tectonic
Archaean Earth: Comparisons with Venus. *In*: DILEK,
Y. & FURNES, H. (eds) *Evolution of Archean Crust
and Early Life*. Modern Approaches in Solid Earth
Sciences, **7**. Springer, Berlin, 215–288.

HARRIS, L. B., BYRNE, D. R., WETHERLY, S. & BEESON, J.
2004. Analogue modelling of structures developed
above single and multiple mantle plumes: applications
to brittle crustal deformation on Earth and Venus. *In*:
BERTOTTI, G., BUITER, S., RUFFO, P. & SCHREURS,
G. (eds) *GeoMod 2004 – From Mountains to Sedimentary Basins: Modelling and Testing Geological
Processes*. Bolletino di Geofisica teorica ed applicata,
45(Suppl. 1), 301–303.

HASHIMOTO, G. L., ROOS-SEROTE, M., SUGITA, S.,
GILMORE, M. S., KAMP, L. W., CARLSON, R. W. &

BAINES, K. H. 2008. Felsic highland crust on Venus
suggested by Galileo near-infrared mapping spectrometer data. *Journal of Geophysical Research*, **113**,
E00B24, http://dx.doi.org/10.1029/2008JE003134

HEAD, J. W., VORDER BRUEGGE, R. W. & CRUMPLER, L. S.
1990. Venus orogenic belt environments: architecture
and origin. *Geophysical Research Letters*, **17**,
1337–1340.

HEAD, J. W., CRUMPLER, L. S., AUBELE, J. C., GUEST, J.
E. & SAUNDERS, R. S. 1992. Venus volcanism:
Classification of volcanic features and structures,
associations, and global distribution from Magellan Data. *Journal of Geophysical Research*, **97**,
13 153–13 198.

HEAD, J. W., HURWITZ, D. M., IVANOV, M. A., BASILEVSKY,
A. T. & KUMAR, P. S. 2008. Geological mapping of
Fortuna Tessera (V–2): Venus and Earth's Archean
process comparisons. *In*: *Abstracts of the Annual
Meeting of Planetary Geologic Mappers, Flagstaff,
AZ*, June 2008. NASA, Washington, DC, 463, http://
ntrs.nasa.gov/archive/nasa/casi.ntrs.nasa.gov/20080
040988.pdf

HEAMAN, L. M., EASTON, R. M., HART, T. R., HOLLINGS,
P., MACDONALD, C. A. & SMYK, M. 2007. Further
refinement to the timing of Mesoproterozoic magmatism, Lake Nipigon region, Ontario. *Canadian
Journal of Earth Sciences*, **44**, 1055–1086.

HEATH, P., DHU, T., REED, G. & FAIRCLOUGH, M. 2009.
Geophysical modelling of the Gawler Province, SA
– interpreting geophysics with geology. *Exploration
Geophysics*, **40**, 342–351.

HEIDBACH, O., TINGAY, M., BARTH, A., REINECKER, J.,
KURFESS, D. & MÜLLER, B. 2008. *The World Stress
Map Database Release 2008*, http://dx.doi.org/10.
1594/GFZ.WSM.Rel2008

HEIDBACH, O., TINGAY, M., BARTH, A., REINECKER, J.,
KURFESS, D. & MÜLLER, B. 2010. Global crustal
stress pattern based on the World Stress Map database
release 2008. *Tectonophysics*, **482**, 3–15.

HELBERT, J., MÜLLER, N., KOSTAMA, P., MARINANGELI,
L., PICCIONI, G. & DROSSART, P. 2008. Surface
brightness variations seen by VIRTIS on Venus
Express and implications for the evolution of the
Lada Terra region, Venus. *Geophysical Research
Letters*, **35**, L11201.

HENSON, P. A., BLEWETT, R. S., ROY, I. G., MILLER, J.
McL. & CZARNOTA, K. 2010. 4D architecture and tectonic evolution of the Laverton region, eastern Yilgarn
Craton, Western Australia. *Precambrian Research*,
183, 338–355.

HERRICK, R. R. & PHILLIPS, R. J. 1992. Geological
correlations with the interior density structure of
Venus. *Journal of Geophysical Research*, **97**,
16 017–16 034.

HINZE, W. J., ALLEN, D. J., BRAILE, L. W. & MARIANO,
J. 1997. The Midcontinent Rift System: a major Proterozoic continental rift. *In*: OJAKANGAS, R. W.,
DICKAS, A. B. & GREEN, J. C. (eds) *Middle Proterozoic to Cambrian Rifting, Central North America*.
Geological Society of America Special Papers, **312**,
7–35.

HOLGATE, N. 1969. Palaeozoic and Tertiary transcurrent
movements on the Great Glen Fault. *Scottish Journal
of Geology*, **5**, 97–139.

HOOGENBOOM, T. & HOUSEMAN, G. A. 2005. Rayleigh Taylor instability as a mechanism for corona formation on Venus. *Icarus*, **180**, 292–307.

HOOPER, P. R., CAMP, V. E., REIDEL, S. P. & ROSS, M. E. 2007. The origin of the Columbia River flood basalt province: Plume versus nonplume models. *In*: FOULGER, G. R. & JURDY, D. M. (eds) *Plates, Plumes and Planetary Processes*. Geological Society of America Special Papers, **430**, 635–668.

HOPPA, G., TUFTS, B. R., GREENBERG, R. & GEISSLER, P. 1999. Strike-slip faults on Europa: global shear patterns driven by tidal stress. *Icarus*, **141**, 287–298.

HOROWITZ, F. G., STRYKOWSKI, G. *ET AL.* 2000. Earthworms; 'multiscale' edges in the EGM96 global gravity field. *SEG Technical Program Expanded Abstracts*, **19**, 414–417, http://dx.doi.org/10.1190/1.1816081

HOWARTH, G. H., BARRY, P. H. *ET AL.* 2014. Superplume metasomatism: evidence from Siberian mantle xenoliths. *Lithos*, **184–187**, 209–224.

HUNTER, G. W., PONCHAK, G. E. *ET AL.* 2012. Development of a high temperature Venus seismometer and extreme environment testing chamber. Paper presented at the International Workshop on Instrumentation for Planetary Missions, http://www.lpi.usra.edu/meetings/ipm2012/pdf/1133.pdf

HURWITZ, D. M. & HEAD, J. W. 2012. *Geologic Map of the Snegurochka Planitia Quadrangle (V–1), Venus*. Pamphlet to accompany United States Geological Survey, Geologic Investigations Map, 3178, http://pubs.usgs.gov/sim/3178/sim3178_pamphlet.pdf (accessed June 2013).

HUSSON, L. 2012. Trench migration and upper plate strain over a convecting mantle. *Physics of the Earth and Planetary Interiors*, **212–213**, 32–43.

HUSSON, L., CONRAD, C. P. & FACCENNA, C. 2012. Plate motions, Andean orogeny, and volcanism above the South Atlantic convection cell. *Earth and Planetary Science Letters*, **317**, 126–135.

IVANOV, M. A. & HEAD, J. W. 2008. Formation and evolution of Lakshmi Planum, Venus: assessment of models using observations from geological mapping. *Planetary and Space Science*, **56**, 1949–1966.

IVANOV, M. A. & HEAD, J. W. 2010a. *Geologic Map of the Lakshmi Planum Quadrangle (V-7), Venus*. United States Geological Survey, Geologic Investigations Map, **3116**.

IVANOV, M. A. & HEAD, J. W. 2010b. *Geologic Map of the Lakshmi Planum Quadrangle (V-7), Venus*. Scientific Investigations Map 3116, Atlas of Venus: Lakshmi Planum Quadrangle (V-7). Pamphlet to accompany United States Geological Survey, Geologic Investigations Map, 3116.

IVANOV, M. A. & HEAD, J. W. 2010c. The Lada Terra rise and Quetzalpetlatl Corona: a region of long-lived mantle upwelling and recent volcanic activity on Venus. *Planetary and Space Science*, **58**, 1880–1894.

IVANOV, M. A. & HEAD, J. W. 2011. Global geological map of Venus. *Planetary and Space Science*, **59**, 1559–1600.

IVANOV, M. A. & HEAD, J. W. 2013. The history of volcanism on Venus. *Planetary and Space Science*, **84**, 66–92.

JAMES, P. B., ZUBER, M. T. & PHILLIPS, R. J. 2013. Crustal thickness and support of topography on Venus. *Journal of Geophysical Research: Planets*, **118**, 859–875, http://dx.doi.org/10.1029/2012JE004237

JÓHANNESSON, H. & SÆMUNDSSON, K. 2009. *Geological Map of Iceland. Tectonics, Revised edition, scale: 1:600 000*. Icelandic Institute of Natural History, Gardabaer.

JULL, M. G. & ARKANI-HAMED, J. 1995. The implications of basalt in the formation and evolution of mountains on Venus. *Physics of Earth and Planetary Interiors*, **89**, 163–175.

KATTENHORN, S. A. 2004. Strike-slip fault evolution on Europa: evidence from tailcrack geometries. *Icarus*, **172**, 582–602.

KATTENHORN, S. A. & MARSHALL, S. T. 2006. Fault-induced perturbed stress fields and associated tensile and compressive deformation at fault tips in the ice shell of Europa: implications for fault mechanics. *Journal of Structural Geology*, **28**, 2204–2221.

KAULA, W. M. 1996. Regional gravity fields on Venus from tracking of Magellan cycles 5 and 6. *Journal of Geophysical Research*, **101**, 4683–4690.

KAULA, W. M. & PHILLIPS, R. J. 1981. Quantitative tests for plate tectonics on Venus. *Geophysical Research Letters*, **8**, 1187–1190.

KAULA, W. M., BINDSCHADLER, D. L., GRIMM, R. E., HANSEN, V. L., ROBERTS, K. M. & SMREKAR, S. E. 1992. Styles of deformation in Ishtar Terra and their implications. *Journal of Geophysical Research: Planets*, **97**, 16 085–16 120.

KAY, S. M. & COPELAND, P. 2006. Early to middle Miocene backarc magmas of the Neuquén Basin: geochemical consequences of slab shallowing and the westward drift of South America. *In*: KAY, S. M. & RAMOS, V. A. (eds) *Evolution of an Andean Margin: A Tectonic and Magmatic View from the Andes to the Neuquén Basin (35°–39°S Lat.)*. Geological Society of America Special Papers, **407**, 185–213.

KEEP, M. & HANSEN, V. L. 1994. Structural history of Maxwell Montes, Venus: implications for Venusian mountain belt formation. *Journal of Geophysical Research*, **99**, 26 015–26 028.

KELBERT, A., EGBERT, G. D. & deGROOT-HEDLIN, C. 2012. Crust and upper mantle electrical conductivity beneath the Yellowstone hotspot track. *Geology*, **40**, 447–450.

KERR, R. A. 1994. A new portrait of Venus: thick-skinned and decrepit. *Science*, **263**, 59–760.

KIEFER, W. S. & HAGER, B. H. 1991. Mantle downwelling and crustal convergence. A model for Ishtar Terra, Venus. *Journal of Geophysical Research*, **96**, 20 967–20 980.

KIEFER, W. S. & PETERSON, K. 2003. Mantle and crustal structure in Phoebe Regio and Devana Chasma, Venus. *Geophysical Research Letters*, **30**, 1005, http://dx.doi.org/10.1029/2002GL015762

KISELEV, A. I., ERNST, R. E., YARMOLYUK, V. V. & EGOROV, K. N. 2012. Radiating rifts and dyke swarms of the middle Paleozoic Yakutsk plume of eastern Siberian craton. *Journal of Asian Earth Sciences*, **45**, 1–16.

Koenig, E. & Aydin, A. 1998. Evidence for large-scale strike-slip faulting on Venus. *Geology*, **26**, 551–554.

Kohlstedt, D. L. & Mackwell, S. J. 2009. Chapter 9. Strength and deformation of planetary lithospheres. *In*: Watters, T. R. & Schultz, R. A. (eds) *Planetary Tectonics*. Cambridge University Press, Cambridge.

Konopliv, A. S. & Sjogren, W. L. 1996. *Venus Gravity Handbook*. Technical Report **96–2**. Jet Propulsion Laboratory, California Institute of Technology, Pasadena, CA.

Konopliv, A. S., Banerdt, W. B. & Sjogren, W. L. 1999. Venus gravity: 180th degree and order model. *Icarus*, **139**, 3–18.

Kozak, R. C. & Schaber, G. G. 1989. New evidence for global tectonic zones on Venus. *Geophysical Research Letters*, **16**, 175–178.

Krassilnikov, A. S., Kostama, V.-P., Aittola, M., Guseva, E. N. & Cherkashina, O. S. 2012. Relationship of coronae, regional plains and rift zones on Venus. *Planetary and Space Science*, **68**, 56–75.

Kumar, P. S. 2005. An alternative kinematic interpretation of Thetis Boundary Shear Zone, Venus: evidence for strike-slip ductile duplexes. *Geophysical Research*, **110**, E07001, http://dx.doi.org/10.1029/2004JE002387

Kyle, P. R., Moore, J. A. & Thirlwall, M. F. 1992. Petrologic evolution of anorthoclase phonolite lavas at Mount Erebus, Ross Island, Antarctica. *Journal of Petrology*, **33**, 849–875.

Lakdawalla, E. 2012. Isostasy, gravity, and the Moon: an explainer of the first results of the GRAIL mission. *The Planetary Society*, http://www.planetary.org/blogs/emily-lakdawalla/2012/12110923-grail-results.html (accessed May 2013).

Lancaster, M. G., Guest, J. E. & Magee, K. P. 1995. Great lava flow fields on Venus. *Icarus*, **118**, 69–86.

Lawyer, L. A., Keller, R. A., Fisk, M. R. & Strelin, J. A. 1995. Bransfield Strait; Antarctic Peninsula. Active extension behind a dead arc. *In*: Taylor, B. (ed.) *Backarc Basins: Tectonics and Magmatism*. Plenum Press, New York, 315–343.

Le Breton, E., Cobbold, P. R., Dauteuil, O. & Lewis, G. 2012. Variation in amount and direction of sea-floor spreading along the North East Atlantic Ocean and resulting deformation of the continental margin of North West Europe. *Tectonics*, **31**, TC5006, http://dx.doi.org/10.1029/2011TC003087

Le Breton, E., Cobbold, P. R. & Zanella, A. 2013. Cenozoic reactivation of the Great Glen Fault, Scotland: additional evidence and possible causes. *Journal of the Geological Society, London*, **170**, 403–415.

Leclerc, F., Harris, L. B., Bédard, J. H., van Breeman, O. & Goulet, N. 2012. Structural and stratigraphic controls on magmatic, volcanogenic, and syn-tectonic mineralization in the Chapais-Chibougamau mining camp, northeastern Abitibi, Canada. *Economic Geology*, **107**, 963–989.

Lenardic, A. & Crowley, J. W. 2012. On the notion of well-defined tectonic regimes for terrestrial planets in this solar system and others. *The Astrophysical Journal*, **755**(2), 132.

Lithgow-Bertelloni, C. & Silver, P. G. 1998. Dynamic topography, plate driving forces and the African superswell. *Nature*, **395**, 269–272.

Liu, Z. & Bird, P. 2002. North America plate is driven westward by lower mantle flow. *Geophysical Research Letters*, **29**(24), 17-1–17-4.

Loddoch, A., Stein, C. & Hansen, U. 2006. Temporal variations in the convective style of planetary mantles. *Earth and Planetary Science Letters*, **251**, 79–89.

Lognonné, P. 2009. Seismic waves from atmospheric sources and atmospheric/ionospheric signatures of seismic waves. *Infrasound Monitoring for Atmospheric Studies*, **2009**, 281–304.

Lognonné, P., Garcia, R., Crespon, F., Occhipinti, G., Kherani, A. & Artru-Lambin, J. 2006. Seismic waves in the ionosphere. *Europhysicsnews*, **37**, 11–14.

Magee, K. P. & Head, J. W. 2001. Large flow fields on Venus: Implications for plumes, rift associations, and resurfacing. *In*: Ernst, R. E. & Buchan, K. L. (eds) *Mantle Plumes: Their Identification Through Time*. Geological Society of America Special Papers, **352**, 81–101.

Marinangeli, L. & Gilmore, M. S. 2000. Geologic evolution of the Akna Montes–Atropos Tessera region, Venus. *Journal of Geophysical Research: Planets*, **105**, 12 053–12 075.

Markov, M. S. 1986. *Structural Ensembles of the North Belt of Venus Deformations and Possible Mechanisms of their Formation*. NASA TM–88511, http://ntrs.nasa.gov/archive/nasa/casi.ntrs.nasa.gov/19870005708_1987005708.pdf. (Translation of Sreukturnyye Ansambl i Severnogo Poyasa Deformatsiy na Veneryei. Vosmozhnyye Mekhanizmy Ikh Obrazovaniya. Geotektonika, **21**, 77–87, 1986.)

Markov, M. S., Smirnov, Ya. B. & Dobrzhinetskaya, L. F. 1989. Tectonics of the Venus and the early Precambrian. *Earth, Moon, and Planets*, **45**, 101–113.

Marone, F., van der Lee, S. & Giardini, D. 2004. Three-dimensional upper-mantle S-velocity model for the Eurasia–Africa plate boundary region. *Geophysical Journal International*, **158**, 109–130.

Massironi, M., Di Achille, G. et al. 2012. Strike-slip kinematics on Mercury: Evidences and implications. *In*: *Proceedings of the 43rd Lunar and Planetary Science Conference, March 19–23, 2012, The Woodlands, Texas*. Lunar and Planetary Institute, Houston, TX, 1924, http://www.lpi.usra.edu/meetings/lpsc2012/pdf/1924.pdf

Massironi, M., Di Achille, G. et al. 2014. Lateral ramps and strike-slip kinematics on Mercury. *In*: Platz, T., Massironi, M., Byrne, P. K. & Hiesinger, H. (eds) *Volcanism and Tectonism Across the Inner Solar System*. Geological Society, London, Special Publications, **401**. First published online June 5, 2014, http://dx.doi.org/10.1144/SP401.16

May, P. R. 1971. Pattern of Triassic-Jurassic diabase dikes around the North Atlantic in the context of predrift positions of the continents. *Geological Society of America Bulletin*, **82**, 1285–7292.

McGill, G. E. 1983. The Geology of Venus. *Episodes*, **1983**, 10–17.

McGill, G. E., Stofan, E. R. & Smrekar, S. E. 2010. Venus tectonics. *In*: Watters, T. R. & Schultz,

R. A. (eds) *Planetary Tectonics*. Cambridge University Press, Cambridge, 81–120.

McGuire, J. C., Davis, D. M. & Consolmagno, G. J. 1996. Crossing fractures and the strength of Venus crustal rocks. *Lunar and Planetary Science*, **27**, 847–848.

Mège, D. & Ernst, R. E. 2001. Contractional effects of mantle plumes on Earth, Mars, and Venus. *In*: Ernst, R. E. & Buchan, K. L. (eds) *Mantle Plumes: Their Identification Through Time*. Geological Society of America Special Papers, **352**, 103–140.

Merino, M., Keller, G. R., Stein, S. & Stein, C. 2013. Variations in Mid-Continent Rift magma volumes consistent with microplate evolution. *Geophysical Research Letters*, **40**, http://dx.doi.org/10.1002/grl.50295

Miller, J. D. Jr. 2007. The Midcontinent Rift in the Lake Superior region: A 1.1 Ga large igneous province. *In*: *LIP of the Month*. Large Igneous Provinces Commission, International Association of Volcanology and Chemistry of the Earth's Interior, http://www.largeigneousprovinces.org/07nov (accessed May 2013).

Montelli, R., Nolet, G., Dahlen, F. A. & Masters, G. 2006. A catalogue of deep mantle plumes: new results from finite-frequency tomography. *Geochemistry, Geophysics, Geosystems*, **7**, Q11007, http://dx.doi.org/10.1029/2006GC001248

Montési, L. G. J. 2013. Fabric development as the key for forming ductile shear zones and enabling plate tectonics. *Journal of Structural Geology*, **50**, 254–266.

Moores, E. M., Yikilmaz, M. B. & Kellogg, L. H. 2013. Tectonics: 50 years after the revolution. *In*: Bickford, M. E. (ed.) *The Web of Geological Sciences: Advances, Impacts, and Interactions*. Geological Society of America Special Papers, **500**, 321–369.

Moresi, L. & Solomatov, V. 1998. Mantle convection with a brittle lithosphere: thoughts on the global tectonic styles of the Earth and Venus. *Geophysical Journal International*, **133**, 669–682.

Morgan, P. 1983. Hot spot heat loss and tectonic style on Venus and in the Earth's Archean. *American Journal of Science*, **14**, 515–516.

Morgan, W. J. 1971. Convection plumes in the lower mantle. *Nature*, **230**, 42–43.

Mouthereau, F., Lacombe, O. & Vergés, J. 2012. Building the Zagros collisional orogen: timing, strain distribution and the dynamics of Arabia/Eurasia plate convergence. *Tectonophysics*, **532–535**, 27–60.

Mueller, A. & Harris, L. B. 1988. Application of wrench tectonic models to mineralized structures in the Golden Mile, Kalgoorlie. *In*: Ho, S. E. & Groves, D. I. (eds) *Recent Advances in the Understanding of Precambrian Gold Deposits*. University of Western Australia Geology Department and University Extension, University of Western Australia Publications, **11**, 97–108.

Mueller, A., Harris, L. B. & Lungan, A. 1988. Structural control of greenstone-hosted gold mineralization by transcurrent shearing: a new interpretation of the Golden Mile district, Kalgoorlie, Western Australia. *Ore Geology Reviews*, **3**, 359–387.

Mueller, N., Helbert, J., Hashimoto, G. L., Tsang, C. C. C., Erard, S., Piccioni, G. & Drossart, P. 2008. Venus surface thermal emission at 1 μm in VIRTIS imaging observations: evidence for variation of crust and mantle differentiation conditions. *Journal of Geophysical Research*, **113**, E00B17.

Nelson, J. & Colpron, M. 2007. Tectonics and metallogeny of the British Columbia, Yukon and Alaskan Cordillera, 1.8 Ga to the present. *In*: Goodfellow, W. D. (ed.) *Mineral Deposits of Canada: A Synthesis of Major Deposit-Types, District Metallogeny, the Evolution of Geological Provinces, and Exploration Methods*. Geological Association of Canada, Special Publications, **5**, 755–791.

Nikolaeva, O., Ivanov, M. & Borozdin, V. 1992. Evidence on the crustal dichotomy (of Venus). *In*: *Venus Geology, Geochemistry, and Geophysics–Research Results from the USSR (A 92-39726 16-91)*. University of Arizona Press, Tucson, AZ, 129–139.

Nimmo, F. & McKenzie, D. 1998. Volcanism and tectonics on Venus. *Annual Review of Earth and Planetary Sciences*, **26**, 23–51.

Okubo, C. H. & Schultz, R. A. 2006. Variability in Early Amazonian Tharsis stress state based on wrinkle ridges and strike-slip faulting. *Journal of Structural Geology*, **28**, 2169–2181.

O'Neill, C. 2013. Tectonothermal evolution of solid bodies: terrestrial planets, exoplanets and moons. *Australian Journal of Earth Sciences*, **59**, 189–198.

O'Rourke, J. G. & Korenaga, J. 2012. Terrestrial planet evolution in the stagnant-lid regime: size effects and the formation of self-destabilizing crust. *Icarus*, **221**, 1043–1060.

Orth, C. P. & Solomatov, V. S. 2012. Constraints on the Venusian crustal thickness variations in the isostatic stagnant lid approximation. *Geochemistry, Geophysics, Geosystems*, **13**, Q11012, http://dx.doi.org/10.1029/2012GC004377.

Pappalardo, R. T. & Collins, G. C. 2005. Extensional tectonics on Ganymede as recorded by strained craters. *Journal of Structural Geology*, **27**, 827–838.

Papuc, A. M. & Davies, G. F. 2012. Transient mantle layering and the episodic behaviour of Venus due to the 'basalt barrier' mechanism. *Icarus*, **217**, 499–509.

Patthoff, D. A. & Kattenhorn, S. A. 2011. A fracture history on Enceladus provides evidence for a global ocean. *Geophysical Research Letters*, **38**, L18201, http://dx.doi.org/10.1029/2011GL048387.

Pavlis, N. K., Holmes, S. A., Kenyon, S. C. & Factor, J. K. 2012. The development and evaluation of the Earth Gravitational Model 2008 (EGM2008). *Journal of Geophysical Research: Solid Earth*, **117**, B04406, http://dx.doi.org/10.1029/2011JB008916

Phillips, R. J. & Hansen, V. L. 1994. Tectonic and magmatic evolution of Venus. *Annual Review of Earth and Planetary Sciences*, **22**, 597–654.

Phillips, R. J. & Hansen, V. L. 1998. Geological evolution of Venus: rises, plains, plumes, and plateaus. *Science*, **279**, 1492–1497.

Phillips, R. J., Kaula, W. M., McGill, G. E. & Malin, M. C. 1981. Tectonics and evolution of Venus. *Science*, **212**, 879–887.

Phillips, R. J., Grimm, R. E. & Malin, M. C. 1991. Hot-spot evolution and the global tectonics of Venus. *Science*, **252**, 651–658.

Pirajno, F. 2000. *Ore Deposits and Mantle Plumes*. Kluwer Academic, Dordrecht.

PIRAJNO, F. 2007. Mantle plumes, associated intraplate tectono-magmatic processes and ore systems. *Episodes*, **30**, 6–19.

POHN, H. A. & SCHABER, G. G. 1992. Indenter type deformation on Venus as evidence for large-scale tectonic slip, and multiple strike-slip events as a mechanism for producing tesselated terrain. *In: Proceedings of the 23rd Lunar and Planetary Science Conference, March 16–20, 1992, Houston, Texas*. Lunar and Planetary Institute, Houston, TX, 1095, www.lpi.usra.edu/meetings/lpsc1992/pdf/1539.pdf

PONCHAK, G. E., SCARDELLETTI, M. C. ET AL. 2012. High temperature, wireless seismometer sensor for Venus. *In: Wireless Sensors and Sensor Networks (WiSNet), 2012 IEEE Topical Conference*, http://ntrs.nasa.gov/archive/nasa/casi.ntrs.nasa.gov/20120004170_2012004292.pdf

PRONIN, A. A. & STOFAN, E. R. 1990. Coronae on Venus: morphology, classification, and distribution. *Icarus*, **87**, 452–474.

PRICE, M. H., WATSON, G., SUPPE, J. & BRANKMAN, C. 1996. Dating volcanism and rifting on Venus using impact crater densities. *Journal of Geophysical Research: Planets*, **101**, 4657–4671.

PYSKLYWEC, R. N., GOGUS, O., PERCIVAL, J., CRUDEN, A. R. & BEAUMONT, C. 2010. Insights from geodynamical modeling on possible fates of continental mantle lithosphere: collision, removal, and overturn. *Canadian Journal of Earth Sciences*, **47**, 541–563.

RAITALA, J. 1994. Main fault tectonics of Meshkenet Tessera on Venus. *Earth, Moon and Planets*, **65**, 55–70.

RAITALA, J. 1996. Chocolate tablet aspects of tectonics of Meshkenet Tessera on Venus. *Earth, Moon and Planets*, **74**, 191–214.

RAMBERG, H. 1967. *Gravity, Deformation and the Earth's Crust as Studied by Centrifuge Models*. Academic Press, London.

RAMOS, V. A. 2009. Anatomy and global context of the Andes: main geologic features and the Andean orogenic cycle. *In*: KAY, S. M., RAMOS, V. A. & DICKINSON, W. R. (eds) *Backbone of the Americas: Shallow Subduction, Plateau Uplift, and Ridge and Terrane Collision*. Geological Society of America, Memoirs, **204**, 31–65.

REGENAUER-LIEB, K. & YUEN, D. A. 2003. Modeling shear zones in geological and planetary sciences: solid and fluid- thermal- mechanical approaches. *Earth Science Reviews*, **63**, 295–349.

RICHARD, P. D., NAYLOR, M. A. & KOOPMAN, A. 1995. Experimental models of strike-slip tectonics. *Petroleum Geoscience*, **1**, 71–80.

RIEDEL, W. 1929. Zur mechanik geologischer brucherscheinungen. *Zentral-blatt fur Mineralogie, Geologie und Paleontologie B*, 354–368.

ROBIN, C. M. I., JELLINEK, M., THAYALAN, V. & LENARDIC, A. 2007. Transient mantle convection on Venus: the paradoxical coexistence of highlands and coronae in the BAT region. *Earth and Planetary Science Letters*, **256**, 100–119.

ROMEO, I., CAPOTE, R. & ANGUITA, F. 2005. Tectonic and kinematic study of a strike–slip zone along the southern margin of Central Ovda Regio, Venus: geodynamical implications for crustal plateaux formation and evolution. *Icarus*, **175**, 320–334.

RUMMEL, R. 2005. Gravity and topography of moon and planets. *Earth, Moon, and Planets*, **94**, 103–111.

RUTTER, E. H., HOLDSWORTH, R. E. & KNIPE, R. J. 2001. The nature and tectonic significance of fault-zone weakening: an introduction. *In*: HOLDSWORTH, R. E., STRACHAN, R. A., MAGLOUGHLIN, J. F. & KNIPE, R. J. (eds) *The Nature and Tectonic Significance of Fault Zone Weakening*. Geological Society, London, Special Publications, **186**, 1–11.

SANDWELL, D. T. 1992. Antarctic marine gravity field from high-density satellite altimetry. *Geophysical Journal International*, **109**, 437–448.

SAUNDERS, A. D., ENGLAND, R. W., REICHOW, M. K. & WHITE, R. V. 2005. A mantle plume origin for the Siberian traps: uplift and extension in the West Siberian Basin, Russia. *Lithos*, **79**, 407–424.

SCHUBERT, G., MASTERS, G., OLSON, P. & TACKLEY, P. 2014. Superplumes or plume clusters? *Physics of the Earth and Planetary Interiors*, **146**, 147–162.

SCHULTZ, R. A. 1989. Strike-slip faulting of ridged plains near Valles Marineris, Mars. *Nature*, **341**, 424–426.

SCHULTZ, R. A. 1999. Understanding the process of faulting: selected challenges and opportunities at the edge of the 21st century. *Journal of Structural Geology*, **21**, 985–993.

SCHULTZ, R. A., HAUBER, E., KATTENHORN, S. A., OKUBO, C. H. & WATTERS, T. R. 2010. Interpretation and analysis of planetary structures. *Journal of Structural Geology*, **32**, 855–875.

SEARLE, M. P. 2006. Role of the Red River Shear zone, Yunnan and Vietnam, in the continental extrusion of SE Asia. *Journal of the Geological Society, London*, **163**, 1025–1036.

SEIFF, A., SCHOFIELD, J. T., KLIORE, A. J., TAYLOR, F. W. & LIMAYE, S. S. 1985. Models of the structure of the atmosphere of Venus from the surface to 100 kilometers altitude. *Advances in Space Research*, **5**, 3–58.

SEN, G. & CHANDRASEKHARAM, D. 2011. Chapter 2. Deccan Traps flood basalt province: An evaluation of the thermochemical plume model. *In*: RAY, J., SEN, G. & GHOSH, B. (eds) *Topics in Igneous Petrology*. Springer, London, 29–53.

ŞENGÖR, A. M. C. & NATAL'IN, B. A. 2001. Rifts of the world. *In*: ERNST, R. E. & BUCHAN, K. L. (eds) *Mantle Plumes: Their Identification Through Time*. Geological Society of America Special Papers, **352**, 389–482.

SHELLNUTT, J. G. 2013. Petrological modeling of basaltic rocks from Venus: a case for the presence of silicic rocks. *Journal of Geophysical Research: Planets*, **118**, 1350–1364.

SIMONS, M., HAGER, B. H. & SOLOMON, S. C. 1994. Global Variations in the geoid/topography admittance of Venus. *Science*, **264**, 798–803.

SIMONS, M., SOLOMON, S. C. & HAGER, B. H. 1997. Localization of gravity and topography: constraints on the tectonics and mantle dynamics of Venus. *Geophysical Journal International*, **131**, 24–44.

SIZOVA, E., GERYA, T., BROWN, M. & PERCHUK, L. L. 2010. Subduction styles in the Precambrian: insight from numerical experiments. *Lithos*, **116**, 209–229.

SLEEP, N. H. 1994. Martian plate tectonics. *Journal of Geophysical Research*, **99**, 5639–5655.

SJOGREN, W. L. & KONOPLIV, A. S. 2008. *Magellan Spherical Harmonic and Gravity Map Data*, ftp:// pds-geosciences.wustl.edu/mgn/mgn-v-rss-5-gravity-l2-v1

SMITH-KONTER, B. & PAPPALARDO, R. T. 2008. Tidally driven stress accumulation and shear failure of Enceladus's tiger stripes. *Icarus*, **198**, 435–451.

SMREKAR, S. E. 1994. Evidence for active hotspots on Venus from analysis of Magellan gravity data. *Icarus*, **112**, 2–26.

SMREKAR, S. E. & PARMENTIER, E. M. 1996. The interaction of mantle plumes with surface thermal and chemical boundary layers: applications to hotspots on Venus. *Journal of Geophysical Research*, **101**, 5397–5410, http://dx.doi.org/10.1029/95JB02877

SMREKAR, S. E. & SOLOMON, S. C. 1992. Gravitational spreading of high terrain in Ishtar Terra, Venus. *Journal of Geophysical Research: Planets*, **97**, 16 121–16 148.

SMREKAR, S. E. & STOFAN, E. R. 1997. Corona formation and heat loss on Venus by coupled upwelling and delamination. *Science*, **277**, 1289–1294.

SMREKAR, S. E., KIEFER, W. S. & STOFAN, E. R. 1997. Large volcanic rises on Venus. *In*: BOUGHER, S. W., HUNTEN, D. M. & PHILLIPS, R. J. (eds) *Venus II: Geology, Geophysics, Atmosphere, and Solar Wind Environment*. University of Arizona Press, Tucson, AZ, 845–879.

SMREKAR, S. E., ELKINS–TANTON, L. *ET AL.* 2007. Tectonic and thermal evolution of Venus and the role of volatiles: Implications for understanding the terrestrial planets. *In*: ESPOSITO, L. W., STOFAN, E. R. & CRAVENS, T. E. (eds) *Exploring Venus as a Terrestrial Planet*. American Geophysical Union, Geophysical Monograph Series, **176**, 45–71.

SMREKAR, S. E., STOFAN, E. R. *ET AL.* 2010. Recent hotspot volcanism on Venus from VIRTIS emissivity data. *Science*, **328**, 605–608.

SOLOMATOV, V. S. & MORESI, L.-N. 1996. Stagnant lid convection on Venus. *Journal of Geophysical Research*, **101**, 4737–4753.

SOLOMATOV, V. S. & MORESI, L.-N. 1997. Three regimes of mantle convection with non-Newtonian viscosity and stagnant lid convection on the terrestrial planets. *Geophysical Research Letters*, **24**, 1907–1910.

SOLOMON, S. C. 1993. The geophysics of Venus. *Physics Today*, **46**, 48–55.

SOLOMON, S. C., HEAD, J. W. *ET AL.* 1991. Venus tectonics: initial analysis from Magellan. *Science*, **252**, 297–312.

SOLOMON, S. C., SMREKAR, S. E. *ET AL.* 1992. Venus tectonics: an overview of Magellan observations. *Journal of Geophysical Research: Planets*, **97**, 13 199–13 255.

SOLOMON, S. C., BULLOCK, M. A. & GRINSPOON, D. H. 1999. Climate change as a regulator of tectonics on Venus. *Science*, **286**, 87–90.

SOROHTIN, O. G. & USHAKOV, S. A. 2002. Chapter 6. Tectonic activity nature of the Earth. *In*: *Development of the Earth*. Moscow State University Press, Moscow, 144–199 (in Russian, original title: Глава 6. Развитие. В: Земли *ПРИРОДА ТЕКТОНі́ ИЧЕСКОЙ АКТИВНОСТИ ЗЕМЛИ*), http:// evolbiol.ru/sorohtin.htm

STECKLER, M. & WATTS, A. B. 1974. Subsidence history and tectonic evolution of Atlantic-type continental margins. *In*: SCRUTTON, R. A. (ed.) *Dynamics of Passive Margins*. American Geophysical Union, Geophysical Monograph Series, **6**, 184–196.

STEIN, S., VAN DER LEE, S. *ET AL.* 2011. Learning from failure: the SPREE Mid-Continent Rift Experiment. *GSA Today*, **21**, 5–7.

STERN, R. J. 2004. Subduction initiation: spontaneous and induced. *Earth and Planetary Science Letters*, **226**, 275–292.

STOFAN, E. R., HEAD, J. W. & CAMPBELL, D. B. 1987. Geology of the southern Ishtar Terra/Guinevere and Sedna Planitae region on Venus. *Earth, Moon, and Planets*, **38**, 183–207.

STOFAN, E. R., SHARPTON, V. L., SCHUBERT, G., BAER, G., BINDSCHADLER, D. L., JANES, D. M. & SQUYRES, S. W. 1992. Global distribution and characteristics of coronae and related features on Venus: implications for origin and relation to mantle processes. *Journal of Geophysical Research: Planets*, **97**, 13 347–13 378.

STOREY, B. C., LEAT, P. T., WEAVER, S. D., PANKHURST, R. J., BRADSHAW, J. D. & KELLEY, S. 1999. Mantle plumes and Antarctica-New Zealand rifting: evidence from mid-Cretaceous mafic dykes. *Journal of the Geological Society, London*, **156**, 659–671.

STROM, R. G., SCHABER, G. G. & DAWSON, D. D. 1994. The global resurfacing of Venus. *Journal of Geophysical Research*, **99**, 10 899–10 926.

STUDD, D., ERNST, R. E. & SAMSON, C. 2011. Radiating graben–fissure systems in the Ulfrun Regio area, Venus. *Icarus*, **215**, 279–291.

SUKHANOV, A. L. & PRONIN, A. A. 1988. Spreading features on Venus (abstract). *In*: *Proceedings of the 19th Lunar and Planetary Science Conference, March 14–18, 1988, Houston, Texas*. Lunar and Planetary Institute, Houston, TX, 1147–1148.

SULLIVAN, K. & HEAD, J. W. 1984. Geology of the Venus Lowlands: Guinevere and Sedna Planitia. *In*: *Proceedings of the 15th Lunar and Planetary Science Conference, March 12–16, 1984, Houston, Texas*. Lunar and Planetary Institute, Houston, TX, 836–837.

SUPPE, J. & CONNORS, C. 1992. Critical Taper Wedge Mechanics of Fold-and-Thrust Belts on Venus: initial Results From Magellan. *Journal of Geophysical Research*, **97**, 13 545–13 561.

SYLVESTER, A. G. 1988. Strike-slip faults. *Geological Society of America Bulletin*, **100**, 1666–1703.

TANAKA, K. L., MOORE, H. J. *ET AL.* 1994. *The Venus Geologic Mappers Handbook*, 2nd edn. United States Geological Survey, Open–File Report, **94–438**.

TANAKA, K. L., ANDERSON, R. *ET AL.* 2010. Planetary structural mapping. *In*: WATTERS, T. A. & SCHULTZ, R. A. (eds) *Planetary Tectonics*. Cambridge Planetary Science, **11**. Cambridge University Press, Cambridge, 351–396.

TANAKA, K. L., SKINNER, J. A. JR. & HARE, T. M. 2011. *Planetary Geologic Mapping Handbook–2011*. United States Geological Survey, Astrogeology Science Center, Flagstaff, AZ.

TCHALENKO, J. S. 1968. The evolution of kink-bands and the development of compression textures in sheared clays. *Tectonophysics*, **6**, 159–174.

TCHALENKO, J. S. 1970. Similarities between shear-zones of different magnitudes. *Geological Society of America Bulletin*, **81**, 1625–1640.

THOMAS, M. D. & TESKEY, D. J. 1994. An interpretation of gravity anomalies over the Midcontinent rift, Lake Superior, constrained by GLIMPCE seismic and aeromagnetic data. *Canadian Journal of Earth Sciences*, **31**, 682–697.

TUCKWELL, G. W. & GHAIL, R. C. 2003. A 400-km-scale strike-slip zone near the boundary of Thetis Regio, Venus. *Earth and Planetary Science Letters*, **211**, 45–45.

TUFTS, B. R., GREENBERG, R., HOPPA, G. V. & GEISSLER, P. 1999. Astypalaea Linea: a San Andreas-sized strike-slip fault on Europa. *Icarus*, **141**, 53–64.

TURCOTTE, D. L. 1989. A heat pipe mechanism for volcanism and tectonics on Venus. *Journal of Geophysical Research: Solid Earth*, **94**, 2779–2785.

TURCOTTE, D. L. 1993. An episodic hypothesis for Venusian tectonics. *Journal of Geophysical Research: Planets*, **98**, 17061–17068.

TURCOTTE, D. L. 1995. How does Venus lose heat? *Journal of Geophysical Research*, **100**, 16 961–16 940.

VAN HINSBERGEN, D. J. J., STEINBERGER, B., DOUBROVINE, P. & GASSMÖLLER, R. 2011. Acceleration and deceleration of India – Asia convergence since the Cretaceous: roles of mantle plumes and continental collision. *Journal of Geophysical Research*, **116**, B06101.

VAN KRANENDONK, M. J. 2010. Two types of Archean continental crust: plume and plate tectonics on early Earth. *American Journal of Science*, **310**, 1187–1209.

VAN THIENEN, P., VLAAR, N. J. & VAN DEN BERG, A. P. 2004. Plate tectonics on the terrestrial planets. *Physics of the Earth and Planetary Interiors*, **142**, 61–74.

VEZOLAINEN, A. V., SOLOMATOV, V. S., HEAD, J. W., BASILEVSKY, A. T. & MORESI, L.-N. 2003. Timing of formation of Beta Regio and its geodynamical implications. *Journal of Geophysical Research*, **108**, 5002, http://dx.doi.org/10.1029/2002JE001889

VIOLAY, M. E. S., GIBERT, B., MAINPRICE, D., EVANS, B., PEZARD, P. A., FLOVENZ, O. G. & ASMUNDSSON, R. 2010. The brittle ductile transition in experimentally deformed basalt under oceanic crust conditions: evidence for presence of permeable reservoirs at supercritical temperatures and pressures in the Icelandic crust. Paper presented at the World Geothermal Congress 2010, Bali, Indonesia, 25–29 April 2010.

VORDER BRUEGGE, R. W., HEAD, J. W. & CAMPBELL, D. B. 1990. Orogeny and large-scale strike-slip faulting on Venus: tectonic evolution of Maxwell Montes. *Journal of Geophysical Research*, **95**, 8357–8381.

VOS, I. M. A., BIERLEIN, F. P., BARLOW, M. A. & BETTS, P. G. 2006. Resolving the nature and geometry of major fault systems from geophysical and structural analysis: the Palmerville Fault in NE Queensland, Australia. *Journal of Structural Geology*, **28**, 2097–2108.

WATTERS, T. R. 1992. A system of tectonic features common to Earth, Mars, and Venus. *Geology*, **20**, 609–612.

WATTERS, T. R. & SCHULTZ, R. A. 2009. Planetary tectonics: introduction. *In*: WATTERS, T. R. & SCHULTZ, R. A. (eds) *Planetary Tectonics*. Cambridge University Press, Cambridge.

WEST, J. D., FOUCH, M. J., ROTH, J. B. & ELKINS-TANTON, L. T. 2009. Vertical mantle flow associated with a lithospheric drip beneath the Great Basin. *Nature Geoscience*, **2**, 439–444.

WESTAWAY, R. 1993. Forces associated with mantle plumes. *Earth and Planetary Science Letters*, **119**, 331–348.

WIECZOREK, M. A. 2007. The gravity and topography of the terrestrial planets. *Treatise on Geophysics*, **10**, 165–206.

WILCOX, R. E., HARDING, T. P. & SEELY, D. R. 1973. Basic wrench tectonics. *American Association of Petroleum Geologists Bulletin*, **57**, 74–96.

WILDE, S. A., MIDDLETON, M. F. & EVANS, B. J. 1996. Terrane accretion in the southwestern Yilgarn Craton: evidence from a deep seismic crustal profile. *Precambrian Research*, **78**, 179–196.

WILLBOLD, M., HEGNER, E., STRACKE, A. & ROCHOLL, A. 2009. Continental geochemical signatures in dacites from Iceland and implications for models of early Archaean crust formation. *Earth and Planetary Science Letters*, **279**, 44–52.

WILLIS, J. J. & HANSEN, V. L. 1996. Conjugate shear fractures at "Ki Corona", southeast Parga Chasma, Venus. *American Journal of Science*, **27**, 1443–1444.

WILSON, J. T. 1961. Continental and oceanic differentiation. *Nature*, **192**, 125–128.

WILSON, J. T. 1973. Mantle plumes and plate motions. *In*: IRVING, E. (ed.) *Mechanisms of Plate Tectonics. Tectonophysics*, **19**, 149–164.

YIN, A. 2012. Structural analysis of the Valles Marineris fault zone: Possible evidence for large-scale strike-slip faulting on Mars. *Lithosphere*, **4**, 286–330.

YIN, A. & TAYLOR, M. H. 2011. Mechanics of V-shaped conjugate strike-slip faults and the corresponding continuum mode of continental deformation. *Geological Society of America Bulletin*, **123**, 1798–1821.

Automatic detection of wrinkle ridges in Venusian Magellan imagery

M. T. BARATA[1]*, F. C. LOPES[1,2], P. PINA[3], E. I. ALVES[1,2] & J. SARAIVA[3]

[1]*Centre for Geophysics of the University of Coimbra (CGUC), Geophysical and Astronomical Observatory, University of Coimbra, Santa Clara, 3040-004 Coimbra, Portugal*

[2]*Department of Earth Sciences (DCT), University of Coimbra, 3000-272 Coimbra, Portugal*

[3]*CERENA, Instituto Superior Técnico, University of Lisbon, Avenida Rovisco Pais, 1049-001 Lisbon, Portugal*

*Corresponding author (e-mail: mtbarata@gmail.com)

Abstract: Wrinkle ridges constitute one of the most abundant tectonic features on terrestrial planetary surfaces. On Venus, evidence suggests a connection between wrinkle ridges and the climatic evolution of the planet. However, like other planets and moons that experience more active surface geological processes, such as Earth, Mars, Europa, Io and Titan, visible impact craters on the Venusian surface are less common because they are eroded, buried or transformed by tectonics or other geological processes over time. It is of great importance to identify and understand some characteristics of those surface morphologies, such as orientation, length, spacing, original dimension and topography. Nevertheless, these parameters can only be computed on remotely sensed images after their segmentation. Until now, the manual identification of these features has been focused on those of major geological significance, leaving many more to be identified, mapped and studied. The main aim of this paper is to provide a method for automatic detection of wrinkle ridges from Magellan Synthetic Aperture Radar (SAR) imagery at different scales. The proposed algorithm, based on a combination of fractal dimension and morphological operators, identifies regions of interest to this study, namely those of anisotropic behaviour, but also impact craters and their ejecta blankets. The high performances achieved in a variety of situations demonstrate that its robustness can be applied to an automated procedure.

The most common tectonic features on terrestrial planets are wrinkle ridges (Watters 2007). First found in lunar *maria*, wrinkle ridges are described as linear to sinuous features, with a considerable complex morphology, which, however, is similar on all the planets (Plescia & Glombeck 1986). Wrinkle ridges represent contractional deformation of the surface, and are usually the result of anti-cline folding, thrust faulting and fault propagation folds (Bilotti & Suppe 1999).

The Venusian surface is essentially constituted by plains (about 80%) generally of low relief, and the large majority of them are characterized by abundant wrinkle ridges (Banerdt *et al.* 1997). Those already analysed are long, narrow and exhibit sinuous features (Banerdt *et al.* 1997; Anderson & Smrekar 1999), and they occur in sets of evenly spaced, parallel ridges (Banerdt *et al.* 1997). The Venusian wrinkle ridges are simpler than the wrinkle ridges observed on other planetary surfaces, and they do not have the complex morphology of broad rises and gentle arches as other planets do (Mars, Moon, Earth). The origin of those wrinkle ridges has been under discussion for at least three decades (Banerdt *et al.* 1997; Kreslavky & Basilevsky 1998), since they are features that can

relevantly contribute to understanding the complex thermal evolution of Venus. Explicitly, as pointed out by Anderson & Smrekar (1999), Bullock & Grinspoon (1999, 2001) and Solomon *et al.* (1999), they can provide important clues to help understand the global climate change on Venus. These authors suggest that temporal variations in climate and tectonic deformations of the crust are connected. Solomon *et al.* (1998) put forward, based on an analytical model, that wrinkle ridges may be formed due to variations in atmospheric temperature related to climate changes. Also, according to Solomon *et al.* (1998), the atmospheric cooling during the initial phases of climate change would be sufficient to cause extension, followed by compression, of the volcanic plains, causing the formation of wrinkle ridges. Anderson & Smrekar (1999) suggested that climate changes on Venus have the potential to cause small-strain global surface deformation, and the authors also suggest that these climate changes can also create wrinkle ridges.

About 960 impact craters have been identified on the surface of Venus, with diameters ranging from 1.5 to 270 km, morphologically similar to impact craters found on other planets (Basilevsky & Head

From: PLATZ, T., MASSIRONI, M., BYRNE, P. K. & HIESINGER, H. (eds) 2015. *Volcanism and Tectonism Across the Inner Solar System*. Geological Society, London, Special Publications, **401**, 357–376.
First published online January 9, 2014, http://dx.doi.org/10.1144/SP401.5

2003). The importance of these structures is related to the determination of the absolute age of the geological units of the planets, which is a significant contribution towards the understanding of the planets' geological evolution (Basilevsky & Head 2002). However, owing to the sparse number of preserved impact craters on Venus, their density can only be used for the determination of Venus' mean surface age.

The present investigation is a novel method for the automatic detection of wrinkle ridges on Venus, based on fractal dimension and morphological transforms using Synthetic Aperture Radar (SAR) images obtained by the Magellan mission.

Our motivation for this work comes from the global distribution map of the wrinkle ridges on Venus presented by Bilotti & Suppe (1999). As explained by these authors, the global map was produced based on on-screen digital mapping of wrinkle ridges, from 'browse' images of C1-MIDRs (SAR images of Magellan, with 225 m per pixel), to avoid the time-consuming process of digitalization. Bilotti & Suppe (1999) also pointed out the possibility of obtaining other results in their research, had they used images with better spatial resolution, such as the F-MIDRs (Full Mosaicked Image Data Record). Each F-MIDR mosaic covers an area of approximately 5° square, with 75 m per pixel resolution. The C1-MIDR images (Once-Compressed Mosaicked Image Data Record) are obtained by compression of F-MIDR images.

First, during the development of the automatic detection of wrinkle ridges, it was necessary to identify impact craters and their ejecta blankets to define masks where it is unnecessary to look for wrinkle ridges. The maturity of some methodologies for detecting morphological patterns on planetary surfaces, namely craters (Bandeira *et al.* 2007; Martins *at al.* 2009; Salamunicar *et al.* 2011), polygonal terrains (Smrekar *et al.* 2002; Moreels & Smrekar 2003; Pina *et al.* 2008; Saraiva *et al.* 2009), dunes and dust devils (Silvestro *et al.* 2010; Bandeira *et al.* 2011, 2013; Statella *et al.* 2012), tectonic structures (Vaz *et al.* 2012), and their importance to characterize and quantify tectonic features, also contributed to this work's motivation. It must be noted that, although all Venusian craters are already identified and catalogued, information on the morphological characteristics of ejecta blankets is still missing.

Identification and characterization of Venusian wrinkle ridges

Wrinkle ridges are defined as morphologically complex linear to sinuous structures, which result from compressive stress and crustal shortening associated with folding, thrust faulting and fault propagation folds. Observed in radar images, the Venusian wrinkle ridges occur in sets of approximately evenly spaced, parallel ridges (Banerdt *et al.* 1997), and appear on radar imagery as bright, thin, elongated structures when compared to other constituents of the surfaces upon which they occur, as a consequence of their roughness.

Figure 1a shows the study area, Atla Regio, located in the Guinevere Planitia region, between 0–20°N, and 180–210°E. The image in Figure 1b (located in the white square of Fig. 1a), is used to illustrate several steps of our methodology, which clearly exhibits these structures. In particular, this image, used by Banerdt *et al.* (1997) with a geographical region common to the image used by Solomon *at al.* (1992, 1999), shows a Magellan SAR image of typical wrinkle ridges on a Venusian plain, located in Rusalka Planitia. The black dashed lines and vertical lines in Figure 1a, as in other images throughout this paper, are artefacts of the Magellan imaging system. They represent gaps where no data have been acquired.

The methodology that was developed for automatic identification is explained in the flowchart depicted in Figure 2, and consists of integrating two different algorithms with different aims: the first is to perform the automatic segmentation of wrinkle ridges; and the second intends to eliminate all the structures that do not correspond to wrinkle ridges.

The detailed explanation of the proposed methodology will be described in the next sections.

Segmentation

The great sensitivity of SAR images to the structures' textural characteristics is vital if we want to distinguish different features on a surface, and it helps to discriminate wrinkle ridges from the Venusian background (Fig. 3). Unfortunately, the SAR images are also sensitive to speckle, an interference phenomenon, and consequently these images are very noisy. Speckle increases the variability of the spatial and radiometric information, and influences SAR image interpretation. Although it is difficult to remove the entire speckle present in an image, it is possible to apply methods (e.g. Lee and Lee-Sigma filters, Gabor filter, Frost filter and Wavelet transform) that can significantly reduce noise (Mansourpour *et al.* 2006). However, the above-mentioned aspects make it difficult to define the texture on SAR images and, consequently, several approaches have been developed to deal with texture on this type of images (Reed & Dubuf 1993; Tuceryan & Jain 1998; Oliver & Quegan 2004). Accordingly, fractal dimension is used in this work to discriminate wrinkle ridges on Magellan

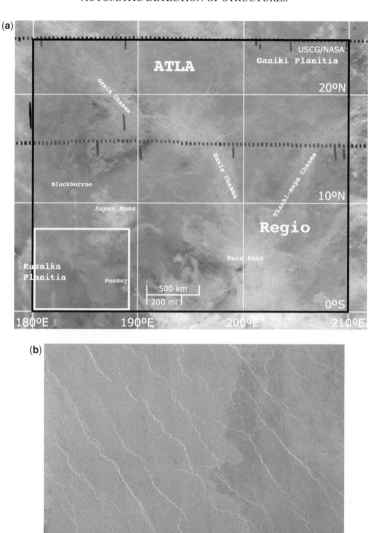

Fig. 1. (**a**) Atla Regio, in Guinevere Planitia region. The black square represents the whole study area; the white square represents the area used to illustrate the algorithm for wrinkle ridge detection. Image taken from http://planetarymapping.wr.usgs.gov/Project/viewjsessionid=CF425FC18B3D2137859B208B7E238FE3. (**b**) Image (C1-MIDR.00N180, framelet 19) in Rusalka Planitia. The image shows a characteristic plain with a radar-bright response to wrinkle ridges.

SAR images, since these images are sensitive to the structures' textural characteristics and, as stated above, wrinkle ridges are rough structures. The ability of fractal analysis to determine the texture heterogeneity of images is the reason for the fractal dimension choice in this algorithm. Also,

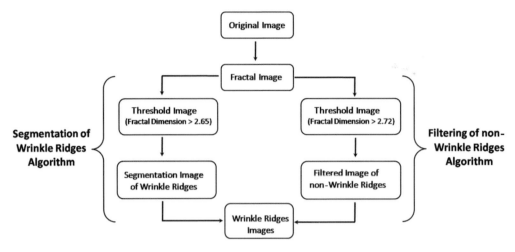

Fig. 2. Flowchart of the methodology applied in this work.

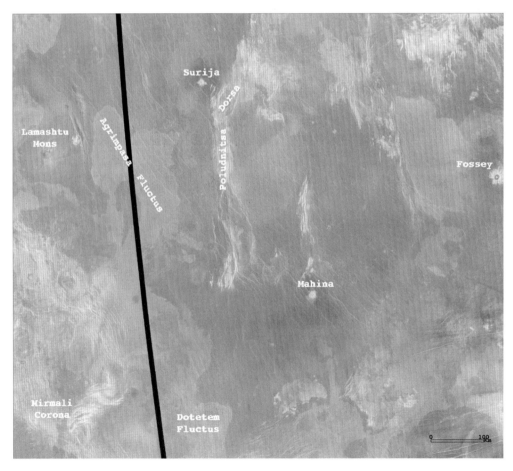

Fig. 3. Image of Atla Regio.

conventional methods based on topology cannot distinguish rough features from smooth ones and, therefore, important information can be lost. In this sense, fractal analysis offers an important tool for characterizing complex objects and land surface patterns in remotely sensed images, and several examples of this can be found in literature (Sun *et al.* 2006; Di Martino *et al.* 2010, 2012; Santos *et al.* 2012; Rani & Aggarwal 2013). One important application of this analysis consists of the use of fractal analysis to describe and determine the texture of complex surface patterns in remote sensed images (Sun *et al.* 2006), based on the fact that natural features are discontinuous and fragmented.

A fractal set is defined according to:

$$N_n = \frac{C}{r_n^D} \tag{1}$$

where N_n is the number of fragments with a characteristic linear dimension, r_n, C is a constant of proportionality and D is fractal dimension (Mandelbrot 1967). So, to determine D, equation (1) can be written as:

$$D = \frac{\log(N_{n+1}/N_n)}{\log(r_n/r_{n+1})}. \tag{2}$$

Several methods are used to compute fractal dimension on remote sensing images but, as has been pointed out by Sun *et al.* (2006), they are similar in the sense that they all establish a statistical relationship between the measured quantities of an object and the scale or resolution to obtain the fractal dimension (D).

The fractal dimension was calculated for the image of Figure 1b, and the result is shown in Figure 4a. Owing to the textural characteristics of wrinkle ridges, they present higher digital levels, and thus higher fractal dimension values are expected. The wrinkle ridges, by a visual analysis of fractal image (Fig. 4a), show a higher fractal dimension ($D \geq 2.65$). This value was chosen based on a careful visual analysis of fractal images. The compromise was to determine the best threshold value and the best results. Note that the fractal value found to detect wrinkle ridges allows these structures to be enhanced, but it does not automatically extract them. However, fractal dimension has proven to be an excellent filter, since speckle was considerably reduced, and it is now easier to perform the automatic segmentation of wrinkle ridges. Based on values higher than 2.65, a threshold transform was performed and the binary image obtained can be seen in Figure 4b.

Wrinkle ridges are correctly identified, including their irregular shape and orientation, since they are anisotropic structures; but the image is very noisy owing to the speckle present in the original image. In order to distinguish the wrinkle ridges and to correctly preserve their shape, it is necessary to eliminate most of the noise, and this can be done by using morphological operators. The methodology presented in this work is based on mathematical morphology for binary images (Soille 2003).

The algorithm starts with a filtering of the fractal dimension image to eliminate excess noise, by applying an opening area of size 16, followed by reconstruction. This operation has the advantage of filtering the image without altering the shape of the significant structures, as can be observed in Figure 4c, and simultaneously filling small holes. The algorithm continues with the goal of performing the correct wrinkle ridge segmentation. This can be achieved by finding appropriate markers for the wrinkle ridges; one way to do this is through the distance function (Fig. 4d) of the filtered image (Fig. 4c). The distance function allows the image's regional minima to be identified in order to be used as the wrinkle ridge markers. Figure 4e shows the result obtained from this operation. As it can be observed, the markers' contours are very irregular and, sometimes, even disconnected. In order to minimize these effects, the next step of the algorithm consists of performing the markers' thickening, followed by dilation. As a result of these operations (Fig. 4f), the markers' contours are more regular and connected. However, the markers are too wide and they do not correctly express the wrinkle ridges, so it is necessary to reduce the markers' thickness to 1 pixel, by applying a thinning. The direct application of thinning leads to undesirable artefacts; so, before this operation, the image of Figure 4f was previously subjected to a hole-filling procedure (Fig. 4g). Figure 4h shows the final result of this operation and, in Figure 4i, the wrinkle ridges are superimposed on the original image. This algorithm finishes with the automatic extraction of the wrinkle ridges in different directions, by applying an ASF (alternate sequential filter) starting with a closing (a morphological operator). Since wrinkle ridges are anisotropic structures, the application of ASF uses an anisotropic structuring element; in this case, a straight line in different directions. The direction of the structuring element varies according to the main wrinkle ridges' directions, and thus the algorithm allows wrinkle ridges to be detected in different directions. The results are shown in Figure 5 for, respectively, the 135° (Fig. 5a), 45° (Fig. 5b), 0° (Fig. 5c) and 90° (Fig. 5d) orientations, superimposed on the original image. An important aspect that must be mentioned and taken into account, is the fact that the entire surface of Venus was not covered by Magellan SAR imagery.

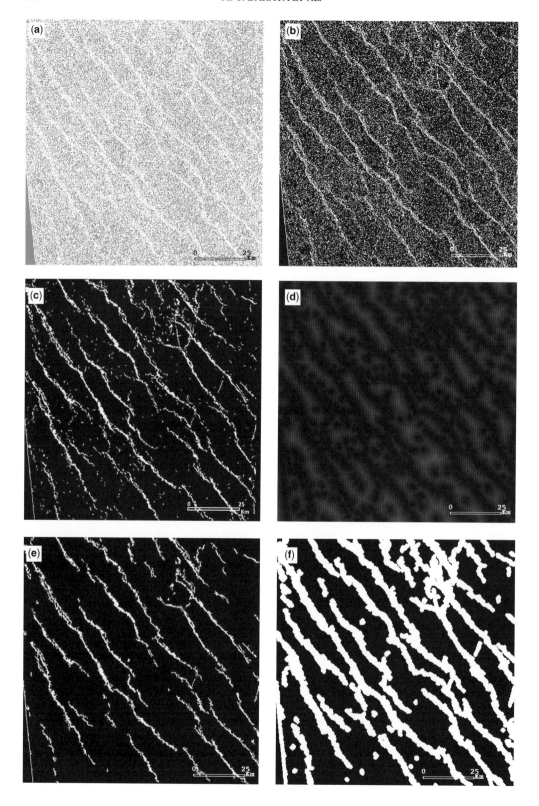

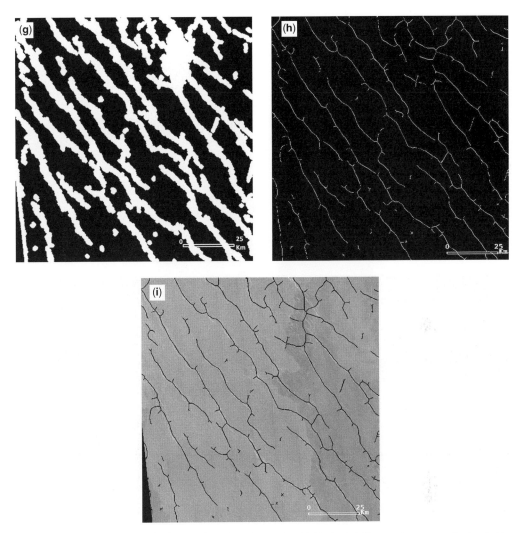

Fig. 4. Identification of wrinkle ridges of Rusalka Planitia: (**a**) fractal dimension of the image in Figure 1; (**b**) threshold of (a) based on the fractal value of wrinkle ridges ($D \geq 2.65$); (**c**) area opening followed by reconstruction; (**d**) distance function of (c); (**e**) regional minima of distance function (markers of wrinkle ridges); (**f**) thickness of the markers, followed by dilation; (**g**) hole fill operation; (**h**) thinning operation; and (**i**) superposition of wrinkle ridges on the original image.

The lower left-hand corner of the image in Figure 1b illustrates this aspect, and the application of the algorithm detects the contour of the 'non-data' as a wrinkle ridge. This can be easily eliminated by performing a threshold of the original image – in this case, the image of Figure 1 – to create a mask. The complementation of this image followed by erosion reduces the strip contour, and the next step consists of performing a multiplication of this mask by the wrinkle ridges image, and thus the wrinkle ridges that are wrongly detected are eliminated.

This method was tested in other regions of the planet, with wrinkle ridges similar to the wrinkle ridges in Figure 1. Figure 6 shows the results obtained by the application of this algorithm to the Mahuela Tholus region, in the Aphrodite Terra region (C1-MIDR.030S171, framelets 45 (Fig. 6a), 46 (Fig. 6b) and 47 (Fig. 6c)), and a visual qualitative evaluation indicates that the wrinkle ridges are well identified.

Other examples of wrinkle ridges are shown in Figure 7, where other types of structures – such as the impact craters in Figure 7a, b, namely the

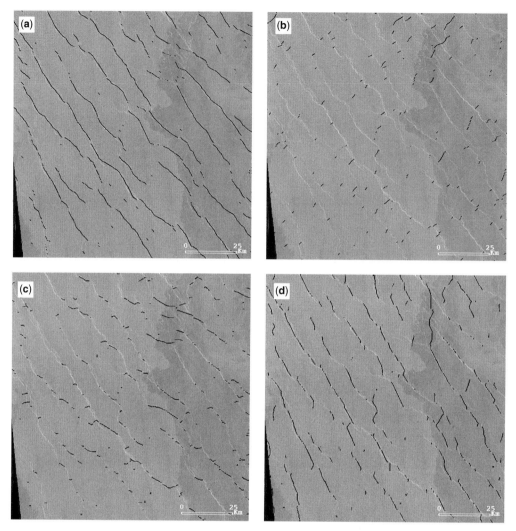

Fig. 5. Wrinkle ridges in different orientations overlapping the original image: (**a**) 135° direction; (**b**) 45° direction; (**c**) 0° direction; and (**d**) 90° direction.

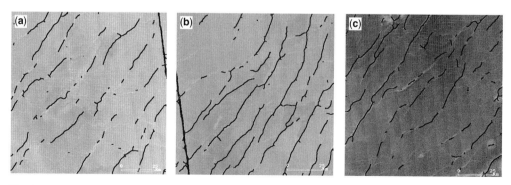

Fig. 6. Automatic extractions of wrinkle ridges in Mahuela Tholus, Aphrodite Terra (C1-MIDR.030S171): (**a**) framelet 45; (**b**) framelet 46; and (**c**) framelet 47.

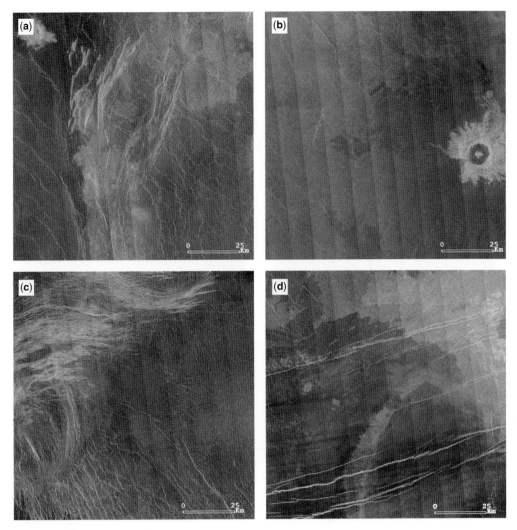

Fig. 7. Images of plains of wrinkle ridges, with different structures: (**a**) the Surija crater and deformation belt of Poludnista Dorsa (C1-MIDR.00N180, framelet 12) in Rusalka Planitia, Niobe Planita; (**b**) the Fossey crater (C1-MIDR.00N180, framelet 24) in Atla Regio, Guinevere Planitia; (**c**) ridge belt (C1-MIDR.00N180, framelet 50) in Atla Regio, Guinevere Planitia; and (**d**) fractures (C1-MIDR.00N180, framelet 56) in Atla Regio, Guinevere Planitia.

Surija (Rusalka Planitia) and Fossey craters (Atla Regio) – can also be observed. Some structures exhibit a certain directionality. For instance, the SAR image in Figure 7a shows part of the deformation belt of Poludnista Dorsa, near the Surgija crater. Other examples of directional structures can be observed in the images of Figures 7c, d, respectively, ridge belts and fractures in Atla Regio.

Filtering of non-wrinkle ridges

Although the wrinkle ridges' segmentation algorithm appears to show good results, due to its

directional character, it can also detect other aniso-tropic structures (such as densely fractured terrain, rifts, ridge belts and fractures belts) or structures that can exhibit some directionality (such as tes-serae, coronae, craters and channels). This problem is avoided by the application of an algorithm that allows the automatic identification of these struc-tures. Subsequently, they will be filtered in the wrinkle ridges image.

Once again, due to the high roughness of those structures, higher digital levels (white pixels) in the fractal image are expected as they have a strong signal response in the SAR images. Despite

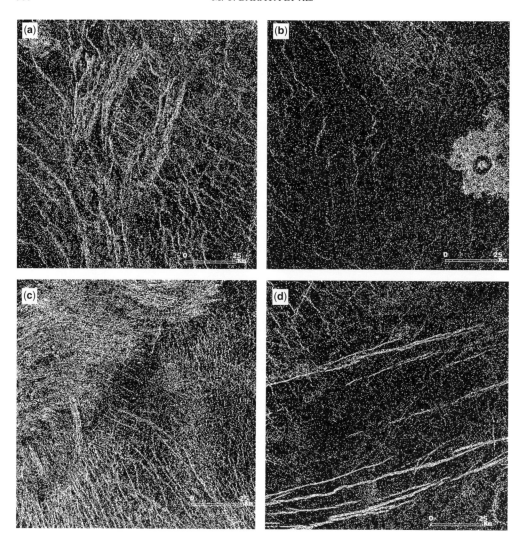

Fig. 8. Threshold of the original images in Figure 7, based on fractal value ($D \geq 2.72$), for different structures: (**a**) the Surija crater and deformation belt of Poludnista Dorsa; (**b**) the Fossey crater; (**c**) ridge belt; and (**d**) fractures.

the fact that different structures may have different fractal values, it is fundamental to find a value that can characterize those structures and, at the same time, it must be different from the value found for wrinkle ridges. According to these requirements, the fractal dimension was calculated for several SAR Magellan images, with several types of structures, and the results obtained show that these structures are characterized by the highest values ($D \geq 2.72$) when compared to wrinkle ridges ($D \geq 2.68$). Based on values higher than 2.72, a threshold transform was performed and the binary images obtained of the original images of Figure 7a–d can be seen in Figure 8a–d. This operation allows these structures to be enhanced but it does not automatically extract them. Besides the images' high noise (Fig. 8a–d), some wrinkle ridges prevail after the threshold operation and

Fig. 9. Filtering of non-wrinkle ridges: (**a**) erosion followed by reconstruction of the image of Figure 6a; (**b**) hole fill operation: marker image; (**c**) hole fill of the image in Figure 7a; (**d**) mask image: erosion followed by reconstruction and then submitted to an opening; and (**e**) reconstruction from the marker and the mask image.

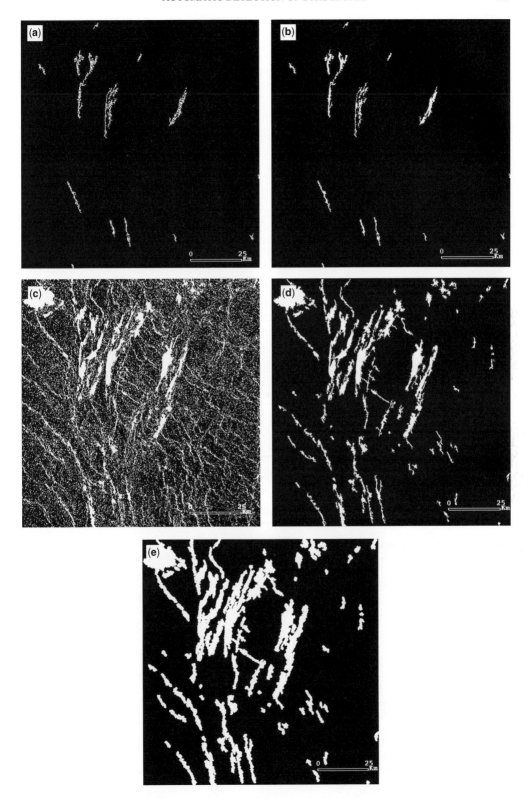

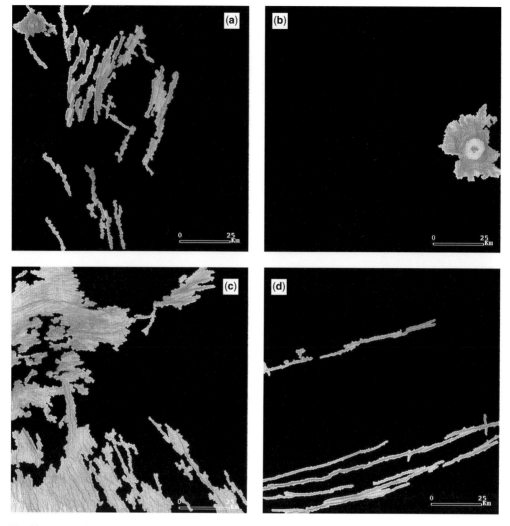

Fig. 10. Images of the structures obtained from the algorithm of filtering non-wrinkle ridges: (**a**) the Surija crater and deformation belt of Poludnista Dorsa; (**b**) the Fossey crater; (**c**) ridge belt; and (**d**) fractures.

they must be removed. This can be done through filtering operations, based on mathematical morphology operators, as already mentioned. The algorithm for filtering non-wrinkle ridge structures will be explained and exemplified in the image of Figure 8a, which shows the Surija crater and part of the deformation belt of Poludnista Dorsa.

The first step consists of filtering the image to remove the excessive noise, by applying an erosion with an isotropic structuring element (a disk), followed by reconstruction. The goal of the erosion consists of removing all isolated pixels (noise), but this transformation also destroys or modifies small structures of interest. The reconstruction can smooth the effect of the erosion and thus preserve some smaller structures. Figure 9a shows the final result of the application of both operations. Compared to the image of Figure 8a, this image does not present noise, and it does not maintain all of the structures either. To avoid this outcome, a hole-filling operation was performed to obtain simply connected structures. This result can be observed in the image of Figure 9b.

Although these transforms do not exactly preserve the structures, the final result contains all their components (Fig. 9b). This image is a marker image and it contains all of the structures of the Surija crater and part of Poludnista Dorsa to be later reconstructed. Now it is only necessary to create a mask image of these structures, again from

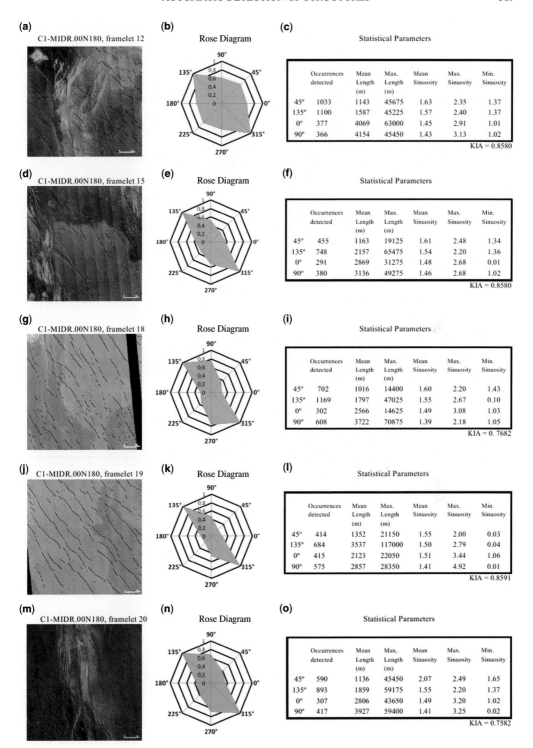

Fig. 11. Automatic detection of wrinkle ridges on some framelets of Atla Regio: (**a, d, g, j, m**) examples of superposition of wrinkle ridges in the 135° orientation on the original image; (**b, e, h, k, n**) rose diagrams; and (**c, f, i, l, o**) statistical parameters.

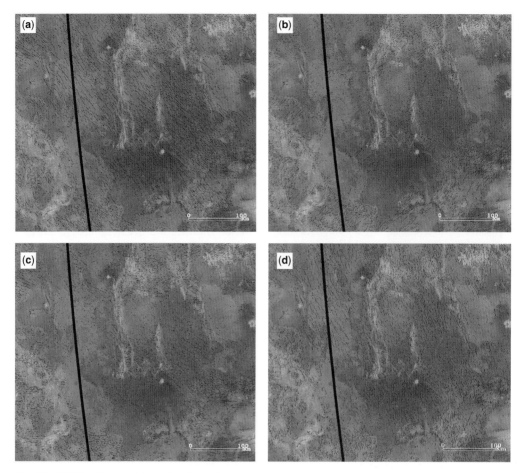

Fig. 12. Wrinkle ridges for Atla Regio (0°N180°), principal directions: (**a**) 135° direction; (**b**) 45° direction; (**c**) 0° direction; and (**d**) 90° direction.

the thresholded image (Fig. 8a), where the marker can be reconstructed. Thus, this image is subjected, again, to a hole-filling operation and the result is presented in Figure 9c. As expected, all of the components of the crater and the deformation belt are simply connected, but the image is still noisy. It is necessary to eliminate all of the unwanted structures, but without destroying the structures' masks. Once again, this can be done by performing an erosion by an isotropic structuring element (a disk), followed by a reconstruction. The erosion removes small structures, and the reconstruction recovers part of what had been lost after the erosion. However, the structures' outer limit in the original image (Fig. 8a) is very irregular, and small disconnected parts can disappear in this operation. Thus, the reconstructed image was submitted to an opening (the dual transform of the closing). The result of these operations has allowed a strong

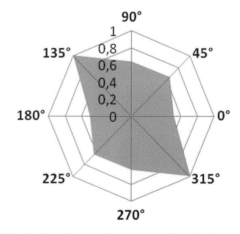

Fig. 13. Rose diagram of orientations of Atla Regio (00°N180°).

marker of the crater and the deformation belt to be obtained, and is presented in Figure 9d.

Now it is necessary to reconstruct the Surija crater and part of Poludnista Dorsa, using both images: the marker image (Fig. 9b) and the mask image (Fig. 9d). The result of this operation is shown in Figure 9e. Since some small holes and unwanted concavities are still present, this image was submitted to a dilation, where all of the components of these structures are now connected and their shape preserved. Figure 10a shows the Surija crater and part of Poludnista Dorsa of the original image extracted by this algorithm.

The images in Figure 10a–d are the final results obtained from the application of this algorithm to the images of Figure 8a–d. The final aim of this algorithm is to eliminate these structures from the images, since only the wrinkle ridges need to be obtained. This can be done by multiplying the

structures' mask image (where the structures take the value of 0) and the wrinkle ridges images. At the end of this operation, only the wrinkle ridges, for the different directions, are presented, as can be seen in the images of Atla Regio (Figs 11 & 12; the white square in Fig. 1a).

Characterization

In order to evaluate the results achieved through this methodology, the automatically identified wrinkle ridges are compared to 'ground-truth' images, in which all of the wrinkle ridges have been manually digitized. There is an important aspect that must be taken into account in this evaluation, since it can influence the final results: the quality of some images is strongly affected by their acquisition and by the processing made in the construction of the C1-MIDRs images. This aspect manifests itself

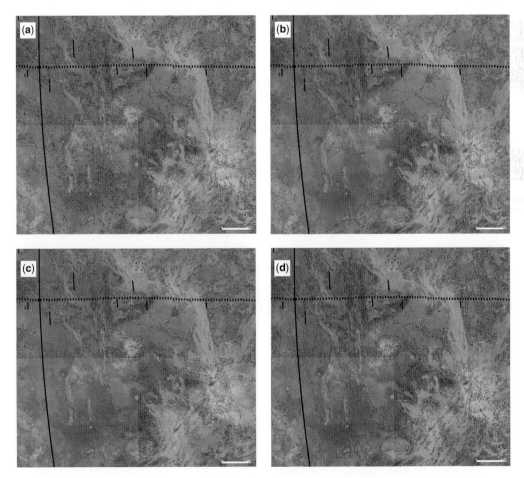

Fig. 14. Wrinkle ridges for Atla Regio (0–20°N, 180–210°E), principal directions: (**a**) 135° direction; (**b**) 45° direction; (**c**) 0° direction; and (**d**) 90° direction (the black dashed lines and vertical lines are strips of missing data).

in certain images by almost vertical lines, such as the image in Figure 4b, d, and, in some cases, they can be wrongly detected as wrinkle ridges. To evaluate the sensitivity of the method, the Kappa Index of Agreement (KIA) was calculated. In general, and for most images, the values of KIA are above 75% (in many images the values are approximately 90%), with very low levels of false positives. These values are especially obtained for the images with better possible quality and when there are no complex geological structures; that is, when the images mainly cover plains of wrinkle ridges. Figure 11 shows some images of Atla Regio and the values obtained for the KIA. Except for one image, that of framelet 15 (Fig. 11b), all of the values of KIA are higher than 0.75, meaning that the method of automatic detection of wrinkle ridges allows good results to be achieved.

One advantage of performing the automatic detection of geological structures, such as wrinkle ridges, is the possibility to easily extract other types of information. Figure 11 also shows the rose diagrams displaying the orientations of identified wrinkle ridges/structures (Fig. 11b, e, h, k, n) and some statistical parameters obtained for each of the images. The analysis of these rose diagrams shows that the wrinkle ridges are predominant in the 135° direction, except for the image of framelet 18 (Fig. 11g), where the 90° direction predominates.

The analysis of the statistical parameters reveals that the maximum length of wrinkle ridges occurs in the 135° direction, except for the images of framelets 12 (Fig. 11a) and 18 (Fig. 11g), where the maximum length appears in the 0° and 90° direction, respectively. Notice that the minimum length detected by the algorithm is equal or superior to the dimension of the pixel (225 m). Another parameter calculated in this work was the sinuosity, which relates the true length of the wrinkle ridges with their length along a straight line. Values close to 1 mean that wrinkles tend to be rectilinear, in contrast to values greater than 2, which show high sinuosity. Intermediate values suggest transitional forms, both regular and irregular. In the examples of Figure 11, both the maximum and minimum sinuosity occur mainly in the 0° and 90° directions. The wrinkle ridges, for the 45° and 135° directions, tend to have transitional forms.

The described methodology was applied to a small region (the region covers the white square of Fig. 1a). This region covers 56 C1-MIDRs images, and all of the images were processed as above and subsequently integrated into a mosaic that covers the entire region. The final result obtained for the four principal directions are presented in Figure 12a–d; respectively, the 135°, 45°, horizontal and vertical directions. The corresponding rose diagram was calculated and it is plotted in Figure 13. As expected, the main orientation trend for this region is 135°, and it is in accordance with the work presented by Bilotti & Suppe (1999) and with the results obtained for individual images – Figure 11 shows some examples. Figure 14 shows the images obtained for the entire region (i.e. for Atla Regio: black square of Fig. 1a) for the four principal directions.

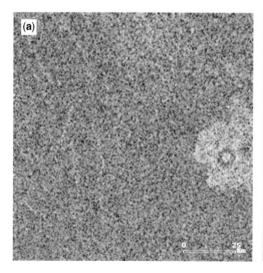 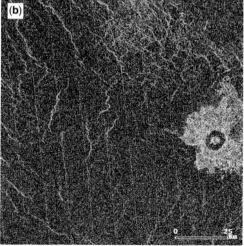

Fig. 15. The Fossey crater: (**a**) fractal dimension of the image in Figure 7b; (**b**) threshold based on fractal values of $D \geq 2.65$.

Identification of impact craters and their ejecta blanket

Impact craters and associated ejecta form radar-bright regions when compared to the surrounding terrains due to their topographical roughness, with ejecta deposits characterized by irregular shapes. A good example of this is the Fossey crater, which can be seen in Figure 7b, located in Atla Regio. Based on the appearance displayed by this impact crater, the fractal dimension was once again used, but this time with the aim of discriminating impact craters.

Impact craters' segmentation

The fractal dimension was calculated for each pixel of this image; the result is shown in Figure 15a. Since the ejecta are a rougher terrain, the fractal values of the corresponding pixels are high ($D \geq 2.65$). Based on this value, a threshold was applied and the binary image obtained can be seen in Figure 15b. The crater and the ejecta surrounding stand out, but the image is rather noisy due to the speckle present on the original image. Other structures are also still visible, such as wrinkle ridges, in the upper left-hand corner of the image. In

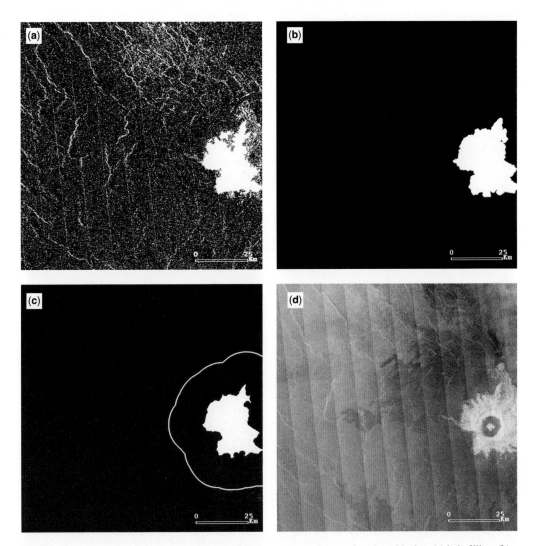

Fig. 16. Sequence of morphological transforms to delineate craters and respective ejecta blanket: (**a**) hole filling; (**b**) erosion reconstruction; (**c**) internal and external markers; (**d**) watershed transform (final contour in red superimposed on the initial image).

order to extract the crater and the ejecta deposits, it is thus necessary to eliminate all of the other objects while preserving and enhancing the area of interest.

First, it is necessary to reinforce contiguous components that roughly correspond to the target which it is necessary to isolate. Again, mathematical morphology for binary images was applied, by performing a hole-filling operation (Fig. 16a). Then the image was filtered, by applying an erosion reconstruction sequence; this removes all isolated pixels (corresponding to noise), but it also destroys or modifies some parts of the crater and ejecta. These are then recovered with the help of dilation, followed by another hole filling, resulting in a much cleaner image (Fig. 16b).

This image is free of noise, but it does not show the complete extent of the ejecta. To recover it, the watershed transform is applied. This requires a pair of markers: one is internal, to define the starting region of the application of the transform; the other is external, to delimitate the area of its application. The first one is obtained from the erosion of Figure 16b, and the second one from the dilation of the same image. The union of the internal marker with the contour of the external one is shown in Figure 16c. The watershed is then applied to the gradient of the initial grey-level image (Fig. 7b) using this pair of markers and results in the red contour line seen in Figure 16d, which marks the outline of the ejecta blanket surrounding the crater.

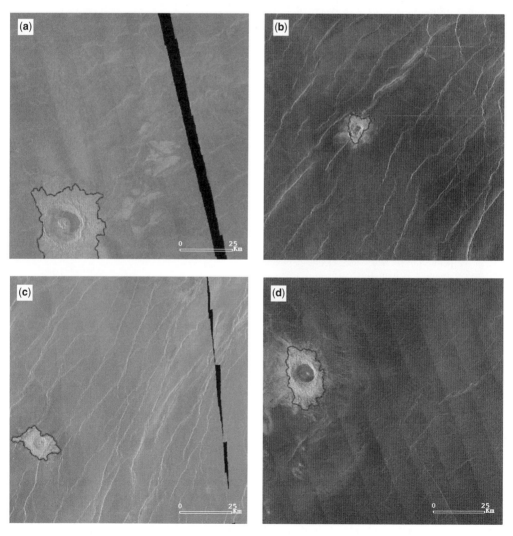

Fig. 17. Examples of the application of the methodology to other SAR images of craters and ejecta blankets on Venus.

Other examples of the application of the method are shown in Figure 17a–d. The shapes and areas of craters and ejecta blankets visible in these are different, but the set of parameters used when running the algorithm was the same for all examples. This highlights the robustness of the methodology.

Conclusions

Over recent years, digital image processing has begun to be applied to planetary imagery in such a way that it is now possible to identify and characterize many surface morphological features of diverse nature.

The first aim of this work has been the development of a methodology for automatically extracting wrinkle ridges on Venusian imagery, based on fractal dimension and morphological transforms, and the results obtained are very promising. Fractal dimension has allowed wrinkle ridges and impact craters to be enhanced and, simultaneously, to eliminate most of the intrinsic noise of SAR images. Through the application of morphological transforms, the automatic detection of wrinkle ridges has been achieved with good results. The main advantages of this method are its applicability to different scales and different orientations, and the future possibility of extracting other important characteristics of wrinkle ridges, such as geometric and, even, rheological parameters. Also, by using this method, there is no need to perform on-screen digitization of wrinkle ridges and, therefore, the results are less subjective and also less time-consuming.

A second goal has been the automatic extraction of impact craters and their ejecta deposits from radar images. This method is useful in determining some parameters of shape in an automatic way. The determination of these parameters will allow ghost craters to be identified with greater precision.

Future work involves the improvement of these algorithms, especially applying them to higher-resolution images, to perform a better detection of wrinkle ridges. The textural segmentation, based on fractal dimension, must be optimized; this is independent of any visual analysis. Morphological transforms and other operators that allow extracting measures (length, size and spacing) will be incorporated into the algorithm. Thus, the physiographical characterization of wrinkle ridges can be performed, such as the characteristics of arches and ridges, and the ratio between width and height. Also, the performance of the algorithm will be tested and, if necessary, adapted, to make the applicability to other kinds of images, such as optical images, possible. Thus, these methods could be applied to detect wrinkle ridges on other planetary surfaces (e.g. Mercury, Mars and the Moon) because they are important morphological structures in understanding the tectonic evolution of planets and moons.

We are very grateful for the constructive criticism of the anonymous referees and the editor, Thomas Platz.

This research is funded by FCT, the Portuguese Science Foundation, under the contract PEst-OE/CTE/UI0611/2012 – Centre for Geophysics.

References

ANDERSON, F. S. & SMREKAR, S. E. 1999. Tectonics effects on climate change on Venus. *Journal of Geophysical Research*, **104**, 30 743–30 756.

BANDEIRA, L., SARAIVA, J. & PINA, P. 2007. Impact crater recognition on Mars based on a probability volume created by template matching. *Geoscience and Remote Sensing Letters, IEEE*, **45**, 4008–4015.

BANDEIRA, L., MARQUES, J. S., SARAIVA, J. & PINA, P. 2011. Automated detection of Martian dune fields. *Geoscience and Remote Sensing Letters, IEEE*, **8**, 626–630.

BANDEIRA, L., MARQUES, J. S., SARAIVA, J. & PINA, P. 2013. Advances in automated detection of sand dunes on Mars. *Earth Surface Processes and Landforms*, **38**, 275–283.

BANERDT, W. B., McGILL, G. E. & ZUBER, M. T. 1997. Plains tectonics on Venus. *In*: BOUGHER, S. W., HUNTEN, D. M. & PHILLIPS, R. J. (eds) *Venus II – Geology, Geophysics, Atmosphere, and Solar Wind Environment*. Arizona University Press, Tucson, AZ, 901–930.

BASILEVSKY, A. & HEAD, J. 2002. Venus: analysis of the degree of impact crater deposit degradation and assessment of its use for dating geological units and features. *Journal of Geophysical Research*, **107**, 5-01–5-38.

BASILEVSKY, A. & HEAD, J. 2003. The surface of Venus. *Reports on Progress in Physics*, **66**, 1699–1734.

BILOTTI, F. & SUPPE, J. 1999. The global distribution of wrinkle ridges on Venus. *Icarus*, **139**, 137–157.

BULLOCK, M. A. & GRINSPOON, D. H. 1999. Global climate change on Venus. *Scientific American*, **March**, 50–57.

BULLOCK, M. A. & GRINSPOON, D. H. 2001. The recent evolution of climate on Venus. *Icarus*, **150**, 19–37.

DI MARTINO, G., IODICE, A., RICCIO, D., RUELLO, G. & ZINNO, I. 2010. Fractal based filtering of SAR images. *In*: *Proceedings of the IEEE International Geoscience & Remote Sensing Symposium, IGARSS 2010, July 25–30, Honolulu, Hawaii, USA*. Institute of Electrical and Electronics Engineers, New York, 2984–2987.

DI MARTINO, G., IODICE, A., RICCIO, D., RUELLO, G. & ZINNO, I. 2012. On the fractal nature of volcano morphology detected via SAR image analysis: the case of Somma–Vesuvius Volcanic Complex. *European Journal of Remote Sensing*, **45**, 177–187.

KRESLAVKY, M. A. & BASILEVSKY, A. T. 1998. Morphometry of wrinkle ridges on Venus. Comparison with other planets. *Journal of Geophysical Research*, **103**, 11 103–11 111.

MANDELBROT, B. 1967. How long is the coast of Britain? Statistical self-similarity and fractional dimension. *Science*, **156**, 636–638.

MANSOURPOUR, M., RAJABI, M. A. & BLAIS, J. A. R. 2006. Effects and performance of speckle noise reduction filters on active radar and SAR images. *In: ISPRS International Society for Photogrammetry and Remote Sensing Conference*, 14–16 February, Ankara, Turkey, **XXXV1-1/w41**, http://www.isprs.org/proceedings/ XXXVI/1-W41/makaleler/Rajabi_Specle_Noise.pdf

MARTINS, R., PINA, P., MARQUES, J. S. & SILVEIRA, M. 2009. Crater detection by a boosting approach. *Geoscience and Remote Sensing Letters, IEEE*, **6**, 127–131.

MOREELS, P. & SMREKAR, S. E. 2003. Watershed identification of polygonal patterns in noisy SAR images. *IEEE Transactions on Image Processing*, **12**, 740–750.

OLIVER, C. & QUEGAN, S. 2004. *Understanding Synthetic Aperture Radar Images*. SciTech Publishing, Herndon, VA.

PINA, P., SARAIVA, J., BANDEIRA, L. & ANTUNES, J. 2008. Polygonal terrains on Mars: a contribution to their geometric and topological characterization. *Planetary and Space Science*, **56**, 1919–1924.

PLESCIA, J. B. & GLOMBECK, M. P. 1986. Origin of planetary wrinkle ridges based on the study of terrestrial analogs. *Geological Society of America Bulletin*, **97**, 1289–1299.

RANI, M. & AGGARWAL, S. 2013. Fractal texture: a survey. *Advances in Computational Research*, **5**, 149–152.

REED, T. R. & DUBUF, J. M. H. 1993. A review of recent texture segmentation and feature extraction techniques. *CVGIP: Image Understanding*, **57**, 359–372.

SALAMUNICAR, G., LONCARIC, S., PINA, P., BANDEIRA, L. & SARAIVA, J. 2011. MA130301GT catalogue of Martian impact craters and advanced evaluation of crater detection algorithms using diverse topography and image datasets. *Planetary and Space Science*, **59**, 111–131.

SANTOS, J. A., GOSSELIN, P., PHILIPP-FOLIGUET, S., TORRES, R. S. & FALCÃO, A. X. 2012. Multiscale classification of remote sensing images. *IEEE Transactions on Geoscience and Remote Sensing*, **50**, 3764–3775.

SARAIVA, J., PINA, P., BANDEIRA, L. & ANTUNES, J. 2009. Polygonal networks on the surface of Mars; applicability of Lewis, Desch and Aboav–Weaire laws. *Philosophical Magazine Letters*, **89**, 185–193.

SILVESTRO, S., FENTON, L. K., VAZ, D. A., BRIDGES, N. T. & ORI, G. G. 2010. Ripple migration and dune activity on Mars: evidence for dynamic wind processes. *Geophysical Research Letters*, **37**, L20203, http://dx.doi. org/10.1029/2010GL044743

SMREKAR, S. E., MOREELS, P. & FRANKLIN, B. J. 2002. Characterization and formation of polygonal fractures on Venus. *Journal of Geophysical Research – Planets*, **107**, 8–1–8–17.

SOILLE, P. 2003. *Morphological Image Analysis – Principles and Applications*, 2nd edn. Springer, Berlin.

SOLOMON, S. C., SMREKAR, S. E. *ET AL.* 1992. Venus tectonics: an overview of Magellan observations. *Journal of Geophysical Research – Planets*, **97**, 13 199–13 255.

SOLOMON, S. C., BULLOCK, M. A. & GRINSPOON, D. H. 1998. Climate change as a regulator of global tectonics on Venus. *Lunar and Planetary Science*, **XXIX**, abstract 1624, http://www.lpi.usra.edu/meetings/ LPSC98/pdf/1624.pdf

SOLOMON, S. C., BULLOCK, M. A. & GRINSPOON, D. H. 1999. Climate change as a regulator of tectonics on Venus. *Science*, **286**, 87–90.

SUN, W., XU, G., GONG, P. & LIANG, S. 2006. Fractal analysis of remotely sensed images: a review of methods and applications. *International Journal of Remote Sensing*, **27**, 4963–4990.

STATELLA, T., PINA, P. & DA SILVA, E. A. 2012. Image processing algorithm for the identification of Martian dust devil tracks in MOC and HiRISE Images. *Planetary and Space Science*, **70**, 46–58.

TUCERYAN, M. & JAIN, A. 1998. Texture analysis. *In:* CHEN, C. H., PAU, L. F. & WANG, P. S. P. (eds) *The Handbook of Pattern Recognition and Computer Vision*, 2nd edn. World Scientific, Washington, DC, 207–248.

VAZ, D., ACHILLE, G., BARATA, T. & ALVES, E. I. 2012. Tectonic lineament mapping of the Thaumasia Plateau, Mars: comparing results from photointerpretation and a semi-automatic approach. *Computers & Geosciences*, **48**, 162–172.

WATTERS, T. R. 2007. Planetary wrinkle ridges: a tale of scale. *In: 2007 GSA Denver Annual Meeting*, 28–31 October. Geological Society of America, Boulder, CO.

Rupes Recta and the geological history of the Mare Nubium region of the Moon: insights from forward mechanical modelling of the 'Straight Wall'

AMANDA L. NAHM[1-4]* & RICHARD A. SCHULTZ[5]

[1]*Department of Geological Sciences, University of Texas at El Paso, 500 West University Avenue, El Paso, TX 79968, USA*

[2]*Center for Lunar Science and Exploration, USRA – Lunar and Planetary Institute, 3600 Bay Area Boulevard, Houston, TX 77058, USA*

[3]*NASA Lunar Science Institute*

[4]*Department of Geological Sciences, University of Idaho, 875 Perimeter Drive, MS 3022, Moscow, ID 83844-3022, USA*

[5]*ConocoPhillips, 600 North Dairy Ashford, PR-2010, Houston, TX 77079, USA*

Corresponding author (e-mail: nahm@uidaho.edu)

Abstract: Rupes Recta, also known as the 'Straight Wall', is an individual normal fault located in eastern Mare Nubium on the nearside of the Moon. Age and cross-cutting relationships suggest that the maximum age of Rupes Recta is 3.2 Ga, which may make it the youngest large-scale normal fault on the Moon. Based on detailed structural mapping and throw distribution analysis, fault nucleation is interpreted to have occurred near the fault centre, and the fault has propagated bi-directionally, growing northwards and southwards by segment linkage. Forward mechanical modelling of fault topography gives a best-fitting fault dip of approximately 85°, and suggests that Rupes Recta accommodated approximately 400 m of maximum displacement and extends to a depth of around 42 km. The cumulative driving stresses required to form Rupes Recta are similar in magnitude to those that formed normal faults in Tempe Terra, Mars. The spatial and temporal association with Rima Birt, a sinuous rille to the west of Rupes Recta, suggests a genetic relationship between both structures and implies regional extension at the time of formation.

In contrast to graben, individual normal faults are rare on the Moon, although a few examples exist (Watters & Johnson 2010). The best-known example is Rupes Recta (Ashbrook 1960; Watters & Johnson 2010). Rupes Recta, or the 'Straight Wall', is a 120 km-long normal fault scarp located in eastern Mare Nubium on the lunar near side (Fig. 1). Age and cross-cutting relationships between mare and crater materials suggest that the maximum age of Rupes Recta is 3.2 Ga, approximately 400 Ma younger than when large-scale normal faulting may have ceased on the Moon (3.6 Ga: Boyce 1976; Lucchitta & Watkins 1978; Watters & Johnson 2010).

Several hypotheses for the formation of Rupes Recta have been suggested, varying from fault reactivation to differential mare subsidence. Discovered in 1686 by Christiaan Huygens, Rupes Recta did not become generally known until it was independently rediscovered and mapped by Johann Schröter in 1791 (Whitaker 1999). Although this fault has been known for centuries, little is known about how it formed. Understanding the origin of this structure can give insight into the geological history of Mare Nubium and the Nubium Basin, and to recent normal fault formation mechanisms on the Moon.

Regional and geological setting

Rupes Recta (RR) is a generally straight, NW-trending normal fault scarp located in eastern Mare Nubium that formed within an approximately 200 km-diameter pre-Nectarian (4.52–3.92 Ga: Stöffler *et al.* 2006) crater (Baldwin 1963; Fielder 1963; Holt 1974; DeHon 1979; Wood 2003; Watters & Johnson 2010), which has informally been called 'Ancient Thebit' by Wood (2003). The crater rim is visible in the NE, eastern and SE portions of 'Ancient Thebit' (Fig. 1), while only the western portion is buried by basalt (Holt 1974). The inferred location of the western portion of the crater rim is outlined by wrinkle ridges

From: PLATZ, T., MASSIRONI, M., BYRNE, P. K. & HIESINGER, H. (eds) 2015. *Volcanism and Tectonism Across the Inner Solar System*. Geological Society, London, Special Publications, **401**, 377–394.
First published online December 16, 2013, http://dx.doi.org/10.1144/SP401.4

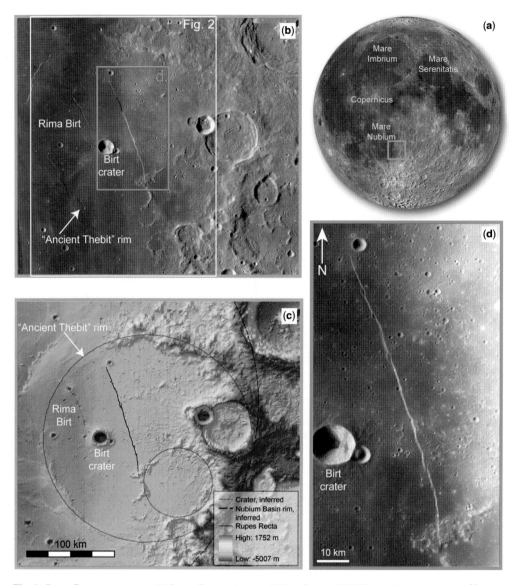

Fig. 1. Rupes Recta context map. (**a**) Lunar Reconnaissance Orbiter Camera (LROC) nearside mosaic (http://wms. lroc.asu.edu/lroc_browse/view/wac_nearside (NASA/GSFC/Arizona State University): 145 m/px) with key locations labelled for reference. The orange box shows the extent of (b) and (c). (**b**) Regional view showing Rupes Recta, Rima Birt, rim of 'Ancient Thebit' and the Birt crater. Portion of LROC nearside mosaic, corner coordinates: 17.5°S, 348.4°E and 26.9°S, 358.5°E. The white box shows the extent of Figure 2. The blue box shows the location of (d). (**c**) LRO Lunar Orbiter Laser Altimeter (LOLA) topography overlain on the shaded relief map. Resolution: 512 pixels per degree (ppd). 'Ancient Thebit's' rim terminates abruptly along the northern and southern extrapolation of the Rupes Recta scarp (Holt 1974). (**d**) Close-up view of Rupes Recta and the crater Birt. LROC Wide Angle Camera (WAC) image: M119903078ME. Original image resolution: 59 m/px.

(Holt 1974), which are probably the surface expression of crater rim localization of contractional deformation at depth (e.g. Allemand & Thomas 1995; Watters & Johnson 2010). The rim of 'Ancient Thebit' coincides approximately with the

northern and southern extrapolation of the younger Rupes Recta scarp (Holt 1974) (Fig. 1c).

Approximately 40–45 km to the west of Rupes Recta sits Rima Birt, a 55 km-long sinuous rille that is roughly parallel to Rupes Recta (Figs 1 & 2).

Based on the geological maps of both Holt (1974) and Bugiolacchi *et al.* (2006), Rima Birt is likely to be Eratosthenian in age (3.2–0.8 Ga: Stöffler *et al.* 2006). The surface expression of Rima Birt shows right-stepping en echelon segments with a generally straight trend, except where the southern portion curves to the west (mirroring the behaviour of RR). In addition, Rima Birt is characterized by relatively shallow pits of various sizes along its length, suggestive of collapse into the subsurface. Recent work by Gustafson *et al.* (2012) has identified a dark-mantle deposit at the northern edge of Rima Birt, which has been interpreted as evidence for late-stage pyroclastic eruptions. No obvious lava flows originate at Rima

Birt. Traditionally, sinuous rilles on the Moon are interpreted to be the surface manifestation of lava flowing either on the surface (where rilles form through mechanical or thermal erosion) (e.g. Hurwitz *et al.* 2012 and references therein) or in the subsurface (where the rilles form by collapse of these tubes, leaving observable depressions) (e.g. Oberbeck *et al.* 1969; Greeley 1971; Cruik-shank & Wood 1972). However, the segmented nature of Rima Birt suggests that flowing lava, either at the surface or in the subsurface, cannot be responsible for the morphology of the structure. Thus, we interpret these observations as Rima Birt probably being the surface manifestation of a sub-surface dyke that erupted pyroclastic materials.

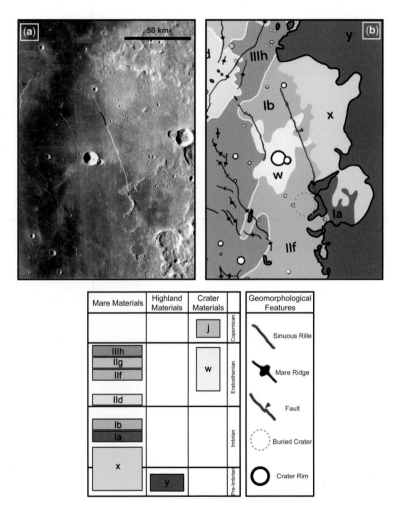

Fig. 2. (**a**) WAC regional mosaic (portion of LROC nearside mosaic) and (**b**) simplified geological and geomorphological map (modified from Bugiolacchi *et al.* 2006) of the Rupes Recta region. Rupes Recta is shown as the purple line and Rima Birt as the red line.

Rupes Recta cuts several geological units within Mare Nubium (Fig. 2). The northern portion of the fault crosses Imbrian mare (unit Ib), the centre of the fault crosses Eratosthenian ejecta from Birt crater (unit w), and the southern portion traverses both Eratosthenian mare material (unit IIf) and pre-Imbrian highlands material (unit y) (Bugiolacchi *et al.* 2006) (Fig. 2). These map relations generally support the observations previously made by Holt (1974) that the northern tip of RR crosses the unit Ip (Imbrian plains), the majority of RR cuts across Imbrian mare material (Im), and the southern portion of the fault cuts through both Eratosthenian mare (Em) and Imbrian and pre-Imbrian undivided terra (IpIt) material.

Mare Nubium is contained in the pre-Nectarian Nubium Basin, which is approximately 700 km in diameter (Stuart-Alexander & Howard 1970; Wilhems 1987; Bugiolacchi *et al.* 2006), although the margins of the basin are degraded and not easily defined (Stuart-Alexander & Howard 1970; DeHon 1977; Wood 2003). The original depth of the Nubium Basin is not known, but has been inferred to be 5 km or less (e.g. Tompkins *et al.* 1994) based on the topography of younger basins such as Orientale (e.g. Head 1974), and gravitational modelling. The latter approach suggests that older basins are shallower than younger ones (Bratt *et al.* 1985) and may be related to viscous relaxation of older basin relief (e.g. Solomon *et al.* 1982). The Nubium Basin does not display a mascon (Stuart-Alexander & Howard 1970; DeHon 1977; Zuber *et al.* 2013), implying that isostatic adjustment has occurred (DeHon 1977). The ambient crustal thickness for the Nubium Basin has been calculated to be approximately 60 km (Bratt *et al.* 1985), while the average crustal thickness near Rupes Recta is around 20–35 km (Wieczorek *et al.* 2013).

In order to understand which materials Rupes Recta cuts through, we examine the broad-scale crustal stratigraphy of the Mare Nubium region. Lava thicknesses in Mare Nubium have been estimated from the height of exposed crater rims of partially flooded craters (Marshall 1963; DeHon 1977, 1979), by spectroscopic studies of craters on Mare Nubium (Tompkins *et al.* 1994; Bugiolacchi *et al.* 2006) and by gravity modelling (Bratt *et al.* 1985), and estimates vary significantly. Bugiolacchi *et al.* (2006) found the mare thickness in the vicinity of Rupes Recta to be 101 m based on the composition of ejecta blankets surrounding small (<14 km diameter) craters, although this value is likely to be a lower bound given that these small craters may have impacted into the youngest surface and the uncertainty in measuring the crater floor compositions. Gravity modelling indicates that 1 km is a reasonable minimum thickness of the basalt

fill within the basin (Bratt *et al.* 1985). Basalt thickness was estimated, from measurements of lava embayment of craters, to be approximately 1500 m, while the average thickness for basalts in the Nubium Basin is around 500 m (DeHon 1977). Tompkins *et al.* (1994) estimated the thickness of basalt units to be as much as 6 km based on spectrographic analysis of the crater Bullialdus, located in western Mare Nubium. Therefore, comparing these values suggests that the thickness of the lava in Mare Nubium is between 100 m and 6 km.

The regional-scale crustal stratigraphy of western Mare Nubium (Fig. 3) was deduced based on spectrographical studies of the 61 km-diameter Eratosthenian-aged crater Bullialdus (Tompkins *et al.* 1994), as well as from basalt thickness estimates described above. Materials present in the central peak of a crater of this size are thought to have originated from a depth of about 6 km (e.g. Melosh 1989; Cintala & Grieve 1998), and are thus expected to be representative of the materials within, and perhaps below the floor of, the entire Nubium Basin (Tompkins *et al.* 1994). The stratigraphic sequence of the lunar crust at Nubium is thus interpreted to be (Fig. 3) basalt from the surface to a maximum depth of approximately 6 km and anorthositic norite–noritic anorthosite to an unknown depth (Tompkins *et al.* 1994), assumed here to be represented by anorthosite to a depth of 23 km and norite to the base of the crust, determined to be about 35–40 km (Ishihara *et al.* 2009; Wieczorek *et al.* 2013).

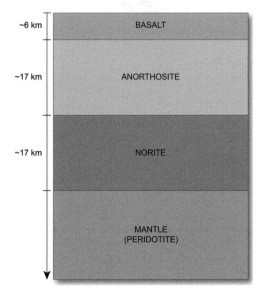

Fig. 3. Schematic crustal stratigraphy for the Mare Nubium region.

Rupes Recta age determination

Basalts in Mare Nubium display subtle colour (e.g. Whitaker 1972; McCord et al. 1976; Pieters 1978), compositional (e.g. Elphic et al. 2000; Lawrence et al. 2000; Lucey et al. 2000; Hiesinger et al. 2003; Bugiolacchi et al. 2006) and age differences (Hiesinger et al. 2003; Bugiolacchi et al. 2006). The ages of Mare Nubium basalt flows have been estimated by crater counting from Lunar Orbiter images (Hiesinger et al. 2003; Bugiolacchi et al. 2006). Volcanic activity in Mare Nubium occurred over a period of approximately 1.08 Gyr, between 2.77 and 3.85 Ga. The mean model age of basalts in the Mare Nubium–Cognitum region is 3.3 (Bugiolacchi et al. 2006) or 3.4 Ga (Hiesinger et al. 2003). Rupes Recta is contained within a unit with a model age of 3.32 Ga (Hiesinger et al. 2003). However, Bugiolacchi et al. (2006) found that Rupes Recta cuts several units, delineated on the basis of both composition and model age (Fig. 2); the basalts are approximately 3.5 Ga in the north and 3.1 Ga in the south (Bugiolacchi et al. 2006). Based on their classification, Bugiolacchi et al. (2006) mapped a remnant of old (3.8 Ga) basalt (unit x; Fig. 2) located between Rupes Recta and the highlands to the east. These results are similar to the map made by Holt (1974), where he distinguished an older Imbrian unit in the north and younger Imbrian unit in the south that are both cut by the fault. The material that makes up the fault scarp was mapped as Eratosthenian in age (Holt 1974).

Based on the geological interpretation of Bugiolacchi et al. (2006), Rupes Recta cuts Birt crater ejecta. Birt crater, and therefore its ejecta, has been estimated to be Eratosthenian in age based on the freshness of the rim (Bugiolacchi et al. 2006) and a well-preserved ejecta blanket. Faulting at Rupes Recta must, therefore, post-date both crater Birt and the mare basalts that RR cross-cuts. Watters & Johnson (2010) suggested that Rupes Recta is relatively young based on the observation that the fault offsets two approximately 2 km-diameter craters and that no craters of this size (or larger) are superposed on the fault scarp. The generally crisp appearance of RR is also an indicator of its young age.

Based on these age relationships, we take the maximum age of RR to be Eratosthenian (0.8–3.2 Ga: Stöffler et al. 2006) and the minimum age to be Copernican (0.8 Ga–present: Stöffler et al. 2006). Ages derived from crater degradation estimates suggest that large lunar normal faults formed before about 3.6 ± 0.2 Ga (Boyce 1976; Lucchitta & Watkins 1978; Watters & Johnson 2010), thought to be primarily in response to mascon loading of craters and basins.

Errors in ages derived from crater counting on Eratosthenian terrains are suggested to be a minimum of 100 myr (Kirchoff et al. 2013). Statistical errors associated with the crater counts produced by Bugiolacchi et al. (2006) are considered to be small, so the upper limit of the ages given for the units used for the determination of relative and absolute ages for Rupes Recta are reliable (M. R. Kirchoff pers. comm. 2013). However, the ages derived by Boyce (1976) and Lucchitta & Watkins (1978) are less reliable. The method used to obtain the ages presented in their work is based on the degradation states of counted craters. While this method was reasonable given the information available at the time, crater degradation rates continue to be poorly understood and, thus, the ages and associated errors presented by Boyce (1976) and Lucchitta & Watkins (1978) (i.e. 3.6 ± 0.2 Ga) should be used with caution (M. R. Kirchoff pers. comm. 2013). Because the ages from Boyce (1976) and Lucchitta & Watkins (1978), and from Bugiolacchi et al. (2006), were determined with two different techniques, direct comparison between these ages is not recommended. Nonetheless, as the ages from Boyce (1976) and Lucchitta & Watkins (1978) are the only available data for the timing of widespread normal faulting on the lunar near side, we propose that the Eratosthenian age for Rupes Recta is significantly younger than the timing of the end of widespread normal faulting suggested by current model age data. However, derivation of the absolute ages of nearside normal faulting using modern methods is needed for the relationship between Rupes Recta and normal faulting elsewhere on the nearside to be better reconciled.

Formation hypotheses

Several hypotheses for the formation of Rupes Recta have been suggested, including reactivation as an Imbrium-radial fault (Wilhelms 1972; Mason et al. 1976) after mare materials were emplaced (Holt 1974) or as a result of mantle uplift beneath Oceanus Procellarum (Wilhelms 1987), by differential settling of mare basalts over a crater rim (DeHon 1977), due to subsidence of the Nubium Basin due to infilling by mare basalts (Baldwin 1963; Wood 2003), or because the region surrounding Rupes Recta formed as an uplift that faulted first in the north–south direction, presumably along Rupes Recta (Fielder 1963).

The motivation behind this work is to gain a better understanding of the range of normal fault geometries on the Moon, which gives constraints on mechanisms for lunar normal fault formation. To evaluate these formation hypotheses, the fault was investigated in detail. First, a detailed structural map was produced using high-resolution imagery and topography, permitting determination of the distribution of fault throw along strike. Next, forward mechanical modelling was used to infer pertinent characteristics of the fault at depth and to provide estimates the cumulative amount of stress required to form Rupes Recta.

Data and methods

Data

The primary image data used in this analysis were from the Lunar Reconnaissance Orbiter (LRO) Wide Angle Camera (WAC) and Narrow Angle Camera (NAC: Robinson *et al.* 2010). Regional views were obtained from the nearside WAC mosaic created by the LRO camera team (http://wms.lroc.asu.edu/lroc_browse/view/wac_nearside, 145 m/px: Fig. 1). Individual WAC image numbers are given in their respective figure captions. Higher-resolution views of Rupes Recta (NAC imagery, *c.* 0.5–2 m/px) used in the course of structural mapping were obtained using the lunar ACT-REACT-QuickMap (http://target.lroc.asu.edu/q3/). The NAC images were used primarily to clarify relationships between fault segments and to corroborate the conclusions from the WAC imagery.

Topographical profiles used in this analysis were derived from gridded LRO Lunar Orbiter Laser Altimeter (LOLA) data (512 ppd; *c.* 57 m/px at the equator: Smith *et al.* 2010). For mechanical modelling, seven topographical profiles (25 km long) were taken near the centre of Rupes Recta, perpendicular to the fault scarp and approximately 500 m apart. These profiles were stacked and averaged to obtain the mean topography (Fig. 4a). Long (*c.* 200 km in length) regional profiles indicate the presence of a regional slope (Fig. 5a), which was removed from all topographical profiles prior to mechanical modelling (Figs 4b & 5b).

Throw distribution

Relay ramps along the length of Rupes Recta are indicative of the extensional kinematics of the fault system (e.g. Peacock 2002). Under such an extensional scenario, the distribution of displacement and/or relief (throw) along fault strike can give independent constraints on the history of fault growth and evolution (e.g. Watterson 1986; Walsh & Watterson 1987, 1988; Marrett & Allmendinger 1991; Cowie & Scholz 1992; Gillespie *et al.* 1992; Dawers *et al.* 1993, Bürgmann *et al.* 1994; Peacock & Sanderson 1996). The throw along strike can be mapped by measuring the change in elevation across the scarp and how it varies along

the fault length (e.g. Dawers *et al.* 1993). Detailed structural mapping of Rupes Recta has revealed the presence of four distinct fault segments (Fig. 6). Analysis of throw distributions can clarify which segments are mechanically linked and which segments are separated by relay ramps (Peacock & Sanderson 1991). To fully characterize the throw distribution along the strike of Rupes Recta, 70 profiles perpendicular to the fault scarp were taken, and the throw was measured from both original and detrended profiles in order to observe any potential effects on throw by removing the regional slope from the profiles.

Mechanical modelling

To determine Rupes Recta's important fault characteristics, we adopt the standard technique of inversion of fault-related topography (e.g. Cohen 1999; Schultz & Lin 2001). Models of this type are used to calculate surface displacements due to underlying faults with prescribed geometries and displacement magnitudes (Schultz & Lin 2001). Forward mechanical modelling has been used successfully to model the surface displacements from faults on Mercury (Watters *et al.* 2002), Earth (e.g. King *et al.* 1988; Toda *et al.* 1998; Cohen 1999; Muller & Aydin 2005; Resor 2008), asteroids (Watters *et al.* 2011), Mars (e.g. Schultz 2000; Schultz & Lin 2001; Schultz & Watters 2001; Grott *et al.* 2007) and the Moon (Nahm 2012; Nahm *et al.* 2013; Williams *et al.* 2013). This approach provides remarkably good fits to the structural topography above a fault for a relatively narrow range of parameters (e.g. Cohen 1999; Schultz & Lin 2001). Correspondence between the output model displacements and observed LOLA (Lunar Orbiter Laser Altimeter) topography would suggest that the fault parameters obtained from modelling are representative of the characteristics of Rupes Recta (e.g. Schultz & Lin 2001; Nahm *et al.* 2013).

We use the forward mechanical dislocation modelling program *Coulomb* (available from http://earthquake.usgs.gov/research/software/#Coulomb) (Lin & Stein 2004; Toda *et al.* 2005) to model surface displacements induced by normal faulting at depth. Stress and material displacement calculations are made in a half-space with uniform isotropic elastic properties following the equations derived by Okada (1992). In these models, a fault is idealized as a rectangular plane with a specified sense of slip (i.e. normal, thrust, strike-slip or oblique), magnitude of average displacement (D), fault dip angle (δ), depth of faulting (T) and fault length (L) (Fig. 7). For this work, we consider pure dip-slip displacement (no strike-slip component) (Schultz & Lin 2001) along Rupes Recta, consistent with its

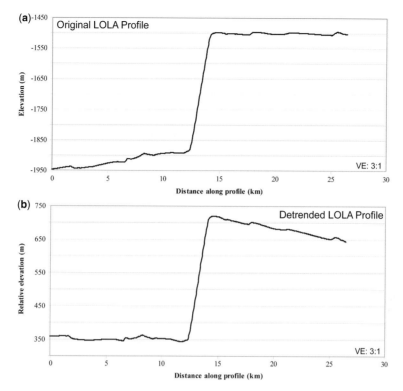

Fig. 4. Local (short) topographical profiles perpendicular to central Rupes Recta derived from LRO LOLA DEM (512 ppd). Average of seven profiles 500 m apart. West is to the left. West end point: 22.0°S, 351.9°W; east end point: 21.8°S, 352.8°W. (**a**) Original profile showing elevations referenced to a sphere of 1734 km radius. (**b**) Detrended profile after removal of the regional slope. Note the steeper slope on the east and the flatter mare surface on the west of the scarp compared to the original profile.

morphology including relay ramps and segmentation (Fig. 6) (e.g. Peacock 2002). Rupes Recta is considered here to be approximately planar at depth as observational evidence argues against listric faulting (Golombek & McGill 1983; Forslund & Gudmundsson 1992).

The initial displacement magnitude is estimated from the scarp relief and subsequently iteratively adjusted based on the model output. For the unfaulted plains material, a Young's modulus (E) of 100 GPa, a Poisson's ratio (v) of 0.3 and a coefficient of friction (μ) of 0.85 are assumed for basalt (Schultz 1995, 1996; Schultz & Lin 2001; Turcotte & Schubert 2002; Fossen 2010). Modest variations in these parameters, including an order-of-magnitude decrease in Young's modulus, have negligible effects on the modelled fault-related topography on the Moon (e.g. Nahm *et al.* 2013) and at Rupes Recta. Good fits to the LOLA topography are obtained by iteratively adjusting the values of fault dip angle, displacement, and depth of faulting in the model. Final model parameters were determined based initially on visual fits to

the shape of the footwall between observed (LOLA) and predicted (modelled) topography (Watters *et al.* 2002), as the shape of the footwall uplift is characteristic of normal fault topography (e.g. Jackson & McKenzie 1983). Once several possible models are identified visually, root mean square (RMS) errors and residuals were used to identify the best-fitting model. The modelling approach is not sensitive to the removal of regional slopes (i.e. detrending) carried out during the LOLA topographical data processing (e.g. Grott *et al.* 2007; Nahm *et al.* 2013).

Stress

In order to investigate formation mechanisms for Rupes Recta, we calculate the required cumulative driving stress for its formation. The cumulative driving stress required to form a fault or a population of faults can be determined by relating the elastic properties of the material, such as E and v, to the fault characteristics, namely the relationship between displacement and fault length (γ) and the

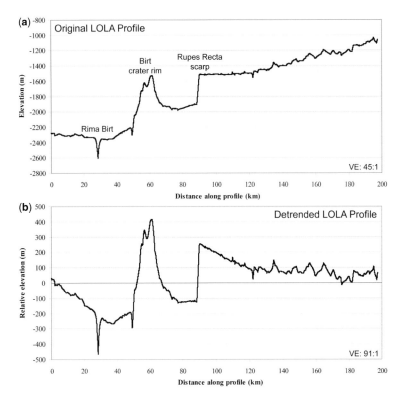

Fig. 5. Regional (long) topographical profiles perpendicular to central Rupes Recta derived from LRO LOLA DEM (512 ppd). West is to the left. West end point: 22.7°S, 349.4°W; east end point: 20.6°S, 355.8°W. (**a**) Original profile showing elevations referenced to a sphere of 1734 km radius. Note the regional slope beginning at about 40 km along the profile. (**b**) Detrended profile after removal of the regional slope. Note the steeper slope on the east and the flatter mare surface on the west of the scarp compared to the original profile. Also, note the difference in vertical exaggeration between (a) and (b).

fault dip angle (δ) (Schultz 2003), determined from modelling.

The incremental driving stress (i.e. stress drop) (σ_d) for one episode of fault slip (i.e. one moonquake event along the growing fault) is given by:

$$\sigma_d = \frac{\gamma_{quake}E}{2C(1-v^2)} \quad (1)$$

where γ_{quake} is the displacement:length ratio for one slip event ($10^{-5}-10^{-6}$: e.g. Scholz 1997), E is the Young's modulus of the basaltic rock mass (taken to be between 10 and 40 GPa in the near-surface: Schultz 1995), C is the effective driving stress factor which varies for typical rock yield stress values of between 0.4 and 0.6 (Schultz 2003), and v is the Poisson's ratio.

The cumulative driving stress represents the sum of the incremental driving stresses that led to the geological offset along a fault (Schultz 2003). It is defined as the ratio of the displacement:length

ratio for faults (γ_{fault}) to the displacement:length ratio for individual seismic events (γ_{quake}) times the incremental driving stress (σ_d) calculated using equation (1) (Cowie & Scholz 1992; Schultz 2003):

$$\text{total } \sigma_d = \left(\frac{\gamma_{fault}}{\gamma_{quake}}\right)\sigma_d. \quad (2)$$

The cumulative driving stress, calculated by equations (1) and (2), typically represents an average over the depth of faulting (T). The value of driving stress at any depth within the faulted volume (σ_{dz}) can be calculated for a linear increase in σ_{dz} with depth, following stress gradients well documented on the Earth and plausible for the Moon and other planetary bodies, by Schultz (2003):

$$\sigma_{dz} = \frac{3}{2}\frac{\text{total } \sigma_d}{T}. \quad (3)$$

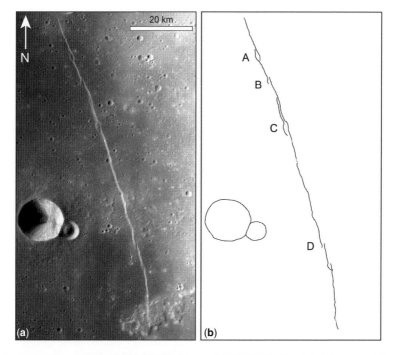

Fig. 6. Structural map of Rupes Recta. (**a**) LROC WAC image, M119903078ME; original image resolution 59 m/px. (**b**) Line map showing the complex morphology of Rupes Recta; outline of Birt crater is shown for reference. Letters A–D identify the segment boundaries referred to in the text and Figure 8.

Results and discussion

Throw distribution

Typical along-strike profiles of throw for faults or individual fault segments increase gradually from the fault or segment tips, reaching a maximum near the centre (Dawers *et al.* 1993; Kim & Sanderson 2005), creating what is referred to as a peaked profile. Deviations from this typical shape can give insight into the growth and evolution of individual faults and fault arrays (e.g. Peacock & Sanderson 1991).

Several throw maxima occur along Rupes Recta's strike. Although the maximum throw occurs at approximately 115 km along strike (Fig. 8), this is interpreted as an indicator of material strength rather than fault characteristics (see below). The second largest throw along Rupes Recta is found to occur slightly north of the fault centre and, in general, fault throw decreases from this point to a minimum at the fault tips (Fig. 8). This maximum is interpreted to be indicative of fault characteristics rather than a function of material properties. The location of the maximum displacement or throw is considered to indicate the approximate location of fault nucleation, with

symmetrical (or peaked) profiles indicating equidimensional propagation from the nucleation point (Peacock & Sanderson 1991; Dawers & Anders 1995). However, asymmetrical throw v. length profiles, such as that obtained for Rupes Recta, indicate unequal propagation tendency (Peacock & Sanderson 1991; Dawers & Anders 1995) with the direction of faster growth (in the absence of lateral tip restriction) indicated by the location of the offset throw maximum and steeper throw gradient. The asymmetry in the throw distribution (Fig. 8) implies that Rupes Recta grew generally south to north, while propagating southwards and northwards from the point of maximum throw. The curved ends of the segments (A–D, Fig. 6) indicate that the segments were propagating towards each other, affecting the local stress field, which resulted in a change in segment end orientation, thus supporting the hypothesis that the fault segments were propagating both north and south while the entire array generally grew from south to north.

Mechanical linkage of smaller fault segments is an important mechanism for fault growth (e.g. Peacock 1991; Peacock & Sanderson 1991; Cartwright *et al.* 1995; Willemse *et al.* 1996; Wilkins & Gross 2002). Faults grow through this mechanism

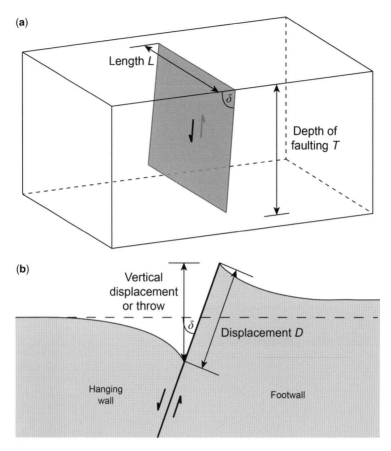

Fig. 7. Schematic representation of model parameters and definition of terms used in this study. (**a**) Block diagram of single rectangular normal fault showing slip direction and geometry parameters: fault length (L), depth of faulting (T) and fault dip angle (δ). (**b**) Cross-sectional view showing the relationship between average fault displacement (D), dip angle (δ) and vertical displacement or throw for a single normal fault. Modified after Schultz & Lin (2001), Schultz *et al.* (2010) and Nahm *et al.* (2013).

in three stages (Peacock & Sanderson 1991): isolated faults propagate towards each other and eventually interact when their tips are close and/or overlapping. The fault segments may continue to grow without direct linkage of segments (soft linkage), forming relay ramps, or link by breaching relay ramps and mechanically linking (hard linkage). Interactions of this type can be seen in both the morphology of the fault or fault zone and the throw v. length profiles. Throw v. length profiles for multisegmented faults may show throw minima at soft-linked segment boundaries (e.g. Peacock & Sanderson 1991, 1994; Cartwright *et al.* 1995; Dawers & Anders 1995; Kim & Sanderson 2005). Summing the displacements or throw in the relay regions and segment boundaries reduces the deficits, resulting in a more continuous profile shape (e.g. Dawers & Anders 1995) and enabling estimates of cumulative stress for the entire fault.

Local deficits in fault throw are observed at segment boundaries A and D (Fig. 8). If the throws for the eastern and western fault segments at locations A and D are summed (dashed lines in Fig. 8a), the deficits are removed and the shape of the throw v. length distribution is uninterrupted. This then implies that the fault segments are mechanically interacting and efficiently transferring throw between them at points A and D. Thus, Rupes Recta is a series of normal fault segments that grew into a continuous fault trace by segment linkage. Systematically higher throw in basin impact ejecta (unit y: Bugiolacchi *et al.* 2006) at the southern end of the fault implies that this material is less stiff (i.e. lower Young's modulus) than the basin mare fill units that are also cut by Rupes Recta. Differences in mechanical properties and contacts between lithological units can inhibit fault propagation (e.g. Rippon 1985;

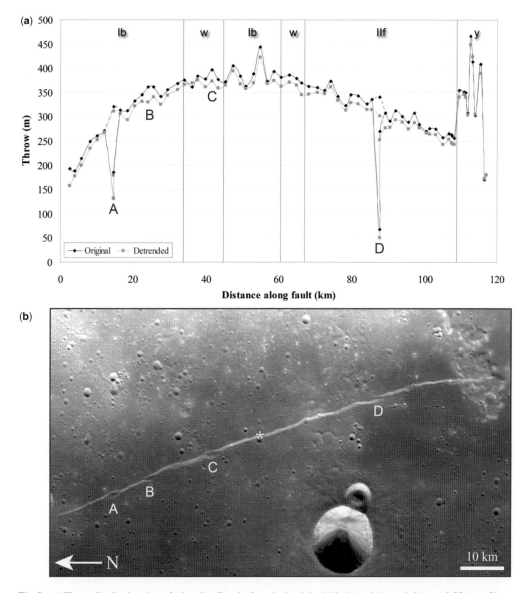

Fig. 8. (a) Throw distribution along fault strike. Results from both original (black) and detrended (grey) LOLA profiles shown. Dashed lines in (a) indicate cumulative throw across relay ramps at A and D. Letters A–D correspond to segment boundaries as defined in Figure 6 and shown in (b). Letters at the top of the panel refer to geological units from Bugiolacchi *et al.* (2006) traversed by Rupes Recta (Fig. 2). Note the systematically higher throw in the Pre-Imbrian highlands material (unit y). (b) LROC WAC image M119903078ME (original image resolution 59 m/px) rotated to show locations of throw measurements along strike. North is to the left. Asterisk indicates the inferred position of fault nucleation. See the text for details.

Bürgmann *et al.* 1994; Wilkins & Gross 2002). This may explain the low throw magnitude near the contact between unit IIf on the north and unit y on the south (Fig. 8), but it does not easily explain the asymmetric throw distribution of the entire fault.

For Rupes Recta, a developmental sequence can be inferred based on photogeological evidence (Fig. 6) and analysis of the throw v. length distribution (Fig. 8). Initial fault nucleation occurred near the fault centre (*c.* 55–60 km south from the northern tip: asterisk in Fig. 8b) and propagation progressed both northwards and southwards, as indicated by the curved segment boundaries. At the same time, smaller fault segments nucleated north and south of the main central segment, and propagated south and north, respectively, towards the central segment. The growth of segments within a narrow linear zone implies that Rupes Recta formed at depth and propagated up to the lunar surface. As the segments grew towards each other, relay ramps formed at boundaries A–D, visible in imagery (Fig. 6). Displacement continued to accumulate, and relay ramps at B and C were breached, hard-linking the segments together. The lack of throw minima at these segment boundaries (Fig. 8) suggests that linkage was essentially complete, forming a kinematically coherent central fault segment with a length of approximately 60 km. The remaining northernmost and southernmost segments did not breach the relay ramps at A and D, meaning that these segments are primarily soft linked to the central fault segment, indicated in the throw v. length diagram as local throw minima.

Modelling

The parameters varied during the modelling were the fault dip angle (δ), average fault displacement (D) and the depth of faulting (T). Based on results from previous modelling of normal faults on the Moon (Nahm 2012; Nahm et al. 2013) and throw measurements, the dip angle was varied between 60° and 89°, displacement varied from 150 m to 1 km, and the depth of faulting varied between 20 and 50 km. Based on conclusions obtained from segment linkage and interaction analyses, the main portion of the fault was taken to be 63 km in length (from B to D in Figs 6 & 8).

Best-fit models were chosen based initially on visual fits to the footwall uplift alone (e.g. Nahm et al. 2013). The shape of the footwall uplift is more characteristic of normal fault topography than the shape of the hanging wall (e.g. Jackson & McKenzie 1983), and the model predicts material displacements relative to an original flat, horizontal datum to which displacements are referenced. The best-fit parameters from the models suggest that the fault dips at approximately 85°, accommodates around 400 m of displacement and extends to a depth of about 42 km (Fig. 9). An RMS error for the footwall of 14.7 m (c. 7% of the maximum relief) was obtained for this model, with the average residuals being 9.2 m (c. 4%). Visible as roughness in the regional topographical profile (Fig. 5) at approximately 120 km is ejecta from the three craters to the NE of RR (Thebit, Thebit A and Thebit L). RMS errors and residuals increase markedly along the profile past this distance. Ignoring the topographical variation resulting from the

ejecta emplaced on top of the originally flat surface, the RMS error is reduced to 6.8 m and the average residual is 5.3 m. Given the inherent uncertainty in the approach of forward modelling techniques, these results represent the approximate best-fitting solution to the topography data. However, the fit of the model output matches the averaged topography well (Fig. 9) and, thus, the values for the best-fit parameters are significant. Very shallow fault dips (i.e. less than 50°) are not consistent with the footwall topography.

Steeply-dipping normal faults (i.e. $\delta > 70°$) have been observed in Iceland (e.g. Forslund & Gudmundsson 1992) and Volcanic Tableland, California, USA (Dawers et al. 1993). These steeply dipping and near-vertical faults are probably influenced by the dip of joint surfaces in jointed near-surface rocks such as fractured basalt seen in Iceland and inferred to be present in Mare Nubium.

Vertical joints extending from the surface to several hundreds of metres in depth can influence the surface manifestation of faulting (e.g. Gudmundsson 1992; Schultz 1996; Ferrill & Morris 2003; Grant & Kattenhorn 2004). As oblique displacement accumulates along a dipping fault propagating upwards towards a jointed rock mass at the surface (i.e. columnar basalt), vertical en echelon fractures may form above the fault's upper tip (Grant & Kattenhorn 2004; Martel & Langley 2006). These en echelon fractures would connect the dipping fault to the surface by utilizing preexisting cooling joints; thus, surface bending would occur preferentially at the relay ramps between the en echelon fractures (Grant & Kattenhorn 2004). The segments that make up Rupes Recta are en echelon, and relay ramps exist between the segments (e.g. Fig. 6). These observations may suggest a similar formation mechanism for parts of Rupes Recta, but testing this hypothesis is outside the scope of this paper.

Recent forward mechanical modelling results (Nahm et al. 2013) have shown that lunar normal faults can be quite deep (c. 30–40 km). While the faults modelled in that study were created as a result of the stresses during the Orientale basin-forming event, those results show that fractures in the lunar crust may extend down to 40 km. It has been hypothesized that deep moonquakes may be from faults that extend to depths of 50 km (Nakamura et al. 1979). In addition, the crust in the area of Rupes Recta has been estimated, based on gravity modelling, to be up to 35 km thick (Wieczorek et al. 2013). Thus, faults to depths of 35–42 km on the Moon are plausible.

The depth of 'Ancient Thebit', calculated using available lunar crater scaling laws (Pike 1977), is likely to be about 5 km. This, together with a basalt thickness of up to 6 km (Fig. 3), implies

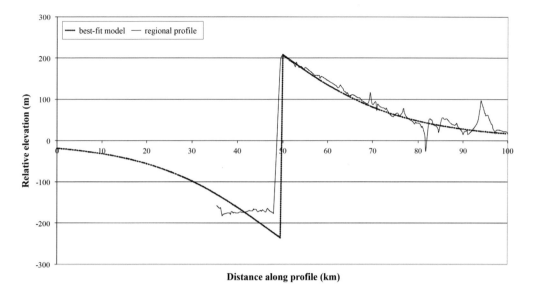

Fig. 9. Detrended regional topographical profile (solid black line) derived from LRO LOLA DEM (512 ppd). Best-fit forward mechanical modelling results shown as a dotted heavy black line. West is to the left. Best-fit model parameter values: fault dip angle (δ) is 85°, average fault displacement (D) is 400 m and the depth of faulting (T) is 42 km.

that the Rupes Recta fault probably cuts through the basalt fill, through the floor of 'Ancient Thebit', into the underlying pre-impact material, through the floor of the Nubium Basin and towards the base of the lunar crust (Fig. 3). The inferred depth of faulting suggests that the stresses responsible for forming the fault probably did not arise locally, such as from crater flexure or subsidence, but that an origin related to larger-scale (hundreds of km) features, such as long-wavelength topographical slopes, is implied.

Stress

The maximum displacement along faults has been shown to scale linearly with the planimetric fault length (e.g. Cowie & Scholz 1992; Clark & Cox 1996) regardless of fault kinematics, tectonic setting or on which planetary body they occur (e.g. Schultz *et al.* 2006). In its simplest form, the relationship between displacement D and fault length L is given by $D = \gamma L$. The D/L ratio γ_{fault} is controlled by three parameters: the modulus of the deformed rock; shear driving stress; and yield strength. All of these are influenced, to some extent, by planetary gravity (Schultz *et al.* 2006). Based on analysis of the effect of gravity on the D/L ratio, γ for lunar faults should be approximately 4% of that for terrestrial faults forming in the same material and under the same (anhydrous) pore-fluid conditions (Schultz *et al.* 2006).

Measurements of maximum relief and fault lengths of lunar thrust faults have yielded estimates of γ_{fault} of $\approx 10^{-2}$ (Watters & Johnson 2010; Watters *et al.* 2010; Banks *et al.* 2012), which is smaller than determined for terrestrial faults ($\gamma_{\text{fault}} \approx 1 \times 10^{-2}–5 \times 10^{-2}$) (Watters *et al.* 2000; Schultz *et al.* 2006 and references therein) as expected. However, for Rupes Recta, the ratio of displacement (400 m) to fault length (63 km) yields a γ_{fault} of 6×10^{-3}, which is of the same order of magnitude as predicted for lunar faults by Schultz *et al.* (2006).

Bounding values for incremental driving stress for one moonquake along Rupes Recta have been calculated using equation (1). For $v = 0.3$, $C = 0.5$ and the parameter ranges noted here, the minimum σ_d is calculated to be 44 kPa ($\gamma_{\text{quake}} = 10^{-6}$, $E = 10$ GPa) and the maximum is 1.75 MPa ($\gamma_{\text{quake}} = 10^{-5}$, $E = 40$ GPa). Values for cumulative driving stress (total σ_d) are calculated by equation (2), with $\gamma_{\text{fault}} = 6 \times 10^{-3}$, to be between 26.3 ($\gamma_{\text{quake}} = 10^{-6}$, $E = 10$ GPa) and 105 MPa ($\gamma_{\text{quake}} = 10^{-5}$; $E = 40$ GPa). In Iceland, joints commonly start to develop into normal faults in jointed columnar basalt at a depth of 1 km (Forslund & Gudmundsson 1992) and this would scale to approximately 6 km for the Moon given the difference in surface gravities. Accounting for the difference in stress gradient with depth between Earth and the Moon determined by differences in gravitational acceleration, we calculate the cumulative differential stresses along Rupes Recta averaged linearly over the 0–6 km depth range of interest. Cumulative differential stresses at 6 km, calculated from equation (3), are between 6.6 ($\gamma_{quake} = 10^{-6}$, $E = 10$ GPa) and 26.3 MPa ($\gamma_{quake} = 10^{-5}$, $E = 40$ GPa).

These values for cumulative driving stress are about one half of those calculated for the Tempe Terra normal fault population on Mars (Schultz 2003), to date the only other planetary normal fault population that has been analysed using this method. The driving stresses required for faulting on planetary bodies are proportional to their gravitational acceleration (Schultz et al. 2006) and the gravity on the Moon is about half of that for Mars (1.62 v. 3.71 m s^{-2}). Thus, accounting for differences in gravitational acceleration between the Moon and Mars, and factors such as tectonic setting and fault density, these values are remarkably similar. This suggests that the cumulative driving stress required to form Rupes Recta and, perhaps, other lunar normal faults and graben of comparable scale, may have arisen from stresses similar in magnitude to those that formed the faults in Tempe Terra, which were also inferred to have formed by regional rather than local tectonic processes.

The presence of Rima Birt and associated pyroclastic deposits implies that local lateral extension was sufficient to open conduits that allowed magma to ascend and erupt on the lunar surface (cf. Solomon 1978; Solomon & Head 1979). The observation that Rima Birt and Rupes Recta are parallel, segmented, curve at their southern ends, may be coeval and are extensional in origin suggests that both Rima Birt and Rupes Recta formed in the same stress field. Normal faults and dilational cracks (such as dykes or joints) form perpendicular to the horizontal least-compressive stress. Thus, we interpret the least-compressive stress at the time of formation to be oriented in the NE–SW direction, perpendicular to the strike of the structures. The stress field may have rotated clockwise to ENE–WSW during along-strike propagation of the rille and fault, as evidenced by the westerly curves of Rima Birt and RR at their southern ends.

Rupes Recta is distinct from lunar graben by lacking an antithetic fault. However, many attributes including throw magnitude and length are comparable for lunar graben and Rupes Recta (e.g. Watters & Johnson 2010). Furthermore, many lunar graben occur as members of sets having approximately regular spacings. Both graben widths and regular graben spacings have been associated with extensional deformation of lithospheres having several characteristic length-scales (e.g. Golombek 1979; Polit et al. 2009). For example, the widths of planetary graben have been related to the thickness of a faulted stratigraphic sequence such as basalt or megaregolith, with graben depth mapping out the depth to a significant change in strength or stiffness in the lithosphere (e.g. Golombek 1979; Golombek & McGill 1983). Similarly, regularly spaced graben are consistent with faulting of a deeper, regionally consistent mechanical stratigraphy (Polit et al. 2009). The solitary nature of Rupes Recta may be related to faulting of a lunar stratigraphic sequence that lacked sufficient or regular contrasts in material strength or stiffness to nucleate antithetic or regularly spaced faults within the 'Ancient Thebit' basin.

Conclusions

Although Rupes Recta has been known for centuries, systematic analyses have not been carried out with the goal of understanding how this fault formed. In this study, we produced a detailed structural map, which permitted determination of the fault throw distribution along strike. Forward mechanical modelling was used to determine characteristics of the fault at depth, which allowed estimates of the cumulative amount of stress required to form Rupes Recta to be determined.

The throw distribution along Rupes Recta reveals that it grew by segment linkage and is comprised of four distinct fault segments. The asymmetry in the distribution of throw implies Rupes Recta nucleated near the centre of the array and grew bi-directionally, as evidenced by the curved segment ends. Systematically higher throw in basin impact ejecta implies this material is weaker than basin mare fill also cut by Rupes Recta, as is expected. Forward mechanical modelling of fault-related topography suggests that the fault dips at approximately 85°, accommodates about 400 m of displacement and extends to a depth of around 42 km. These values can provide insight into the formation of Rupes Recta.

On Earth, normal faults in jointed basalt flows are suggested to form through either of two primary mechanisms (Forslund & Gudmundsson 1992): coalescence of large-scale tension fractures into shear fractures (normal faults); or development of a shear fracture from a set of joints in a lava pile. In the case of Rupes Recta and Mare Nubium, as subsidence of the basin interior progressed, the columnar cooling joints in the basalt became slightly tilted (with shallower dip) towards the basin interior. During tilting, the state of stress along the joint faces would become oblique to the orientation of the principal stresses (σ_1 vertical, σ_3 horizontal), and shear displacement along joints may have been initiated (e.g. Forslund & Gudmundsson 1992; Schultz & Zuber 1994).

It has been previously proposed that Rupes Recta formed as a reactivated Imbrium-radial fault (e.g. Wilhelms 1972; Holt 1974; Mason *et al.* 1976). If fracturing and/or faulting from the Imbrium basin-forming impact occurred in the location of present-day Rupes Recta, those fractures may have influenced the location of fault formation. However, no other individual normal faults orientated radial to Imbrium have yet been recognized, rendering the hypothesis that Rupes Recta formed through reactivation of an Imbrium-radial structure unlikely.

The authors thank M. Weller (Rice University) for discussions about fault modelling, T. Öhman and P. McGovern (LPI) for fruitful discussions, G. Kramer (LPI) for help with Clementine analysis, M. Kirchoff (SwRI) and S. Robbins (CU-Boulder) for discussions regarding errors in model ages derived from crater counting, and S. Kattenhorn (University of Idaho) for discussion about faults in basalt. The authors also thank C. Okubo, S. Tavani and M. Massironi for comments that significantly improved the paper. This research was partially funded by NASA Lunar Science Institute contract NNA09DB33A (PI David A. Kring) and NASA Outer Planets Research grant NNX09AP33G (PI B. R. Smith-Konter). ConocoPhillips is thanked for granting permission to publish this paper. This is LPI Contribution 1767.

References

ALLEMAND, P. & THOMAS, P. G. 1995. Localization of Martian ridges by impact craters: mechanical and chronological implications. *Journal of Geophysical Research*, **100**, 3251–3262.

ASHBROOK, J. 1960. The lunar straight wall. *Publications of the Astronomical Society of the Pacific*, **72**, 55–58.

BALDWIN, R. B. 1963. *The Measure of the Moon*. University of Chicago Press, Chicago, IL.

BANKS, M. E., WATTERS, T. R., ROBINSON, M. S. & WILLIAMS, N. R. 2012. Updating the displacement-length relationship of thrust faults associated with lobate scarps on the Moon: preliminary results. *In: American Geophysical Union Fall Meeting, San Francisco, CA, 3–7 December 3–7*. American Geophysical Union, Washington, DC, Abstract P53A-2040.

BOYCE, J. M. 1976. *Interagency Report: Astrogeology 82; A Summary of Relative Ages of the Lunar Nearside Maria Based on Lunar Orbiter IV Photography*. United States Geological Survey, Open-File Report, 77–68.

BRATT, S. R., SOLOMON, S. C., HEAD, J. W. & THURBER, C. H. 1985. The deep structure of lunar basins: implications for basin formation and modification. *Journal of Geophysical Research*, **90**, (B4), 3049–3064.

BUGIOLACCHI, R., SPUDIS, P. D. & GUEST, J. E. 2006. Stratigraphy and composition of lava flows in Mare Nubium and Mare Cognitum. *Meteoritics and Planetary Science*, **41**, 285–304.

BÜRGMANN, R., POLLARD, D. D. & MARTEL, S. J. 1994. Slip distributions on faults: effects of stress gradients, inelastic deformation, heterogeneous host-rock stiffness, and fault interaction. *Journal of Structural Geology*, **16**, 1675–1690.

CARTWRIGHT, J. A., TRUDGILL, B. D. & MANSFIELD, C. S. 1995. Fault growth by segment linkage: an explanation for scatter in maximum displacement and trace length data from the Canyonlands Grabens of SE Utah. *Journal of Structural Geology*, **17**, 1319–1326.

CINTALA, M. J. & GRIEVE, R. A. F. 1998. Scaling impact melting and crater dimensions: implications for the lunar cratering record. *Meteoritics and Planetary Science*, **33**, 889–912.

CLARK, R. M. & COX, S. J. D. 1996. A modern regression approach to determining fault displacement-length relationships. *Journal of Structural Geology*, **18**, 147–152.

COHEN, S. C. 1999. Numerical models of crustal deformation in seismic zones. *Advances in Geophysics*, **41**, 133–231.

COWIE, P. A. & SCHOLZ, C. H. 1992. Physical explanation for the displacement-length relationship of faults using a post-yield fracture mechanics model. *Journal of Structural Geology*, **14**, 1133–1148.

CRUIKSHANK, D. P. & WOOD, C. A. 1972. Lunar rilles and Hawaiian volcanic features: possible analogues. *The Moon*, **3**, 412–447, http://dx.doi.org/10.1007/BF00562463

DAWERS, N. H. & ANDERS, M. H. 1995. Displacement-length scaling and fault linkage. *Journal of Structural Geology*, **17**, 607–614.

DAWERS, N. H., ANDERS, M. H. & SCHOLZ, C. H. 1993. Growth of normal faults: displacement-length scaling. *Geology*, **21**, 1107–1110.

DEHON, R. A. 1977. Mare Humorum and Mare Nubium: Basalt thickness and basin-forming history. *Proceedings of the Lunar Science Conference*, **8**, 633–641.

DEHON, R. A. 1979. Thickness of the western mare basalts. *Proceedings of the Lunar and Planetary Science Conference*, **10**, 2935–2955.

ELPHIC, R. C., LAWRENCE, D. J., FELDMAN, W. C., BARRACLOUGH, B. L., MAURICE, S., BINDER, A. B. & LUCEY, P. G. 2000. Lunar rare earth element distribution and ramifications for FeO and TiO_2: Lunar Prospector neutron observations. *Journal of Geophysical Research*, **105**, 20 333–20 345.

FERRILL, D. A. & MORRIS, A. P. 2003. Dilational normal faults. *Journal of Structural Geology*, **25**, 183–196.

FIELDER, G. 1963. Topography and tectonics of the lunar straight wall. *Planetary and Space Science*, **11**, 23–30.

FORSLUND, T. & GUDMUNDSSON, A. 1992. Structure of Tertiary and Pleistocene normal faults in Iceland. *Tectonics*, **11**, 57–68.

FOSSEN, H. 2010. *Structural Geology*. Cambridge University Press, Cambridge.

GILLESPIE, P. A., WALSH, J. J. & WATTERSON, J. 1992. Limitations of dimensions of displacement data from single faults and consequences for data analysis and interpretation. *Journal of Structural Geology*, **14**, 1157–1172.

GOLOMBEK, M. P. 1979. Structural analysis of lunar grabens and the shallow crustal structure of the Moon.

Journal of Geophysical Research, **84**, 4657–4666, http://dx.doi.org/10.1029/JB084iB09p04657

GOLOMBEK, M. P. & McGILL, G. E. 1983. Grabens, basin tectonics, and the maximum total expansion of the Moon. *Journal of Geophysical Research*, **88**, 3563–3578.

GRANT, J. V. & KATTENHORN, S. A. 2004. Evolution of vertical faults at an extensional plate boundary, southwest Iceland. *Journal of Structural Geology*, **26**, 537–557, http://dx.doi.org/10.1016/j.jsg.2003.07.003

GREELEY, R. 1971. Lava tubes and channels in the lunar Marius Hills. *The Moon*, **3**, 289–314, http://dx.doi.org/10.1007/BF00561842

GROTT, M., HAUBER, E., WERNER, S. C., KRONBERG, P. & NEUKUM, G. 2007. Mechanical modeling of thrust faults in the Thaumasia region, Mars, and implications for the Noachian heat flux. *Icarus*, **186**, 517–526, http://dx.doi.org/10.1016/j.icarus.2006.10.001

GUDMUNDSSON, A. 1992. Formation and growth of normal faults at the divergent plate boundary in Iceland. *Terra Nova*, **4**, 464–471.

GUSTAFSON, J. O., BELL, J. F., III, GADDIS, L. R., HAWKE, B. R. & GIGUERE, T. A. 2012. Characterization of previously unidentified lunar pyroclastic deposits using Lunar Reconnaissance Orbiter Camera data. *Journal of Geophysical Research*, **117**, E00H25, http://dx.doi.org/10.1029/2011JE003893

HEAD, J. W. 1974. Orientale multi-ringed basin interior and implications for the petrogenesis of lunar highland samples. *Moon*, **11**, 327–356

HIESINGER, H., HEAD, J. W., III, WOLF, U., JAUMANN, R. & NEUKUM, G. 2003. Ages and stratigraphy of mare basalts in Oceanus Procellarum, Mare Nubium, Mare Cognitum, and Mare Insularum. *Journal of Geophysical Research*, **108**, 5065, http://dx.doi.org/10.1029/2002JE001985

HOLT, H. E. 1974. *Geologic Map of the Purbach Quadrangle of the Moon (1:1,000,000 Scale)*. United States Geological Survey, Miscellaneous Investigations Series, **Map I-485**.

HURWITZ, D. M., HEAD, J. W., WILSON, L. & HIESINGER, H. 2012. Origin of lunar sinuous rilles: modeling effects of gravity, surface slope, and lava composition on erosion rates during the formation of Rima Prinz. *Journal of Geophysical Research*, **117**, E00H14, http://dx.doi.org/10.1029/2011JE004000

ISHIHARA, Y., GOOSSENS, S. ET AL. 2009. Crustal thickness of the Moon, Implications for farside basin structures. *Geophysical Research Letters*, **36**, L19202, http://dx.doi.org/10.1029/2009GL039708

JACKSON, J. & McKENZIE, D. 1983. The geometrical evolution of normal fault systems. *Journal of Structural Geology*, **5**, 471–482, http://dx.doi.org/10.1016/0191-8141(83)90053-6

KIM, Y.-S. & SANDERSON, D. J. 2005. The relationship between displacement and length of faults: a review. *Earth Science Review*, **68**, 317–334, http://dx.doi.org/10.1016/j.earscirev.2004.06.003

KING, G. C. P., STEIN, R. S. & RUNDLE, J. 1988. The growth of geological structures by repeated earthquakes 1. Conceptual framework. *Journal of Geophysical Research*, **93**, 13,307–13,318, http://dx.doi.org/10.1029/JB093iB11p13307

KIRCHOFF, M. R., CHAPMAN, C. R., MARCHI, S., CURTIS, K. M., ENKE, B. & BOTTKE, W. F. 2013. Ages of large lunar impact craters and implications for bombardment during the Moon's middle age. *Icarus*, **225**, 325–341, http://dx.doi.org/10.1016/j.icarus.2013.03.018

LAWRENCE, D. J., FELDMAN, W. C. ET AL. 2000. Thorium abundances on the lunar surface. *Journal of Geophysical Research*, **105**, 20,307–20,331.

LIN, J. & STEIN, R. S. 2004. Stress triggering in thrust and subduction earthquakes and stress interaction between the southern San Andreas and nearby thrust and strike-slip faults. *Journal of Geophysical Research*, **109**, B02303, http://dx.doi.org/10.1029/2003JB002607

LUCCHITTA, B. K. & WATKINS, J. A. 1978. Age of graben systems on the Moon. *Proceedings of the Lunar and Planetary Science Conference*, **9**, 3459–3472.

LUCEY, P. G., BLEWETT, D. T. & HAWKE, B. R. 2000. Lunar iron and titanium abundance algorithms based on final processing of Clementine ultraviolet–visible images. *Journal of Geophysical Research*, **105**, 20,297–20,305.

MARRETT, R. & ALLMENDINGER, R. W. 1991. Estimates of strain due to brittle faulting: sampling of fault populations. *Journal of Structural Geology*, **13**, 735–738.

MARSHALL, C. H. 1963. Thickness and structure of the Procellarum system in the Lansberg Region of the moon. *In: Astrogeologic Studies Annual Progress Report, August 25, 1961 to August 24, 1962*. United States Geological Survey, Open-File Report, **63-123**, 19–31.

MARTEL, S. J. & LANGLEY, J. S. 2006. Propagation of normal faults to the surface in basalt, Koae fault system, Hawaii. *Journal of Structural Geology*, **28**, 2123–2143.

MASON, R., GUEST, J. E. & COOKE, G. N. 1976. An Imbrium pattern of graben on the Moon. *Proceedings of the Geologists' Association*, **87**, 161–168.

McCORD, T. B., PIETERS, C. M. & FEIERBERG, M. A. 1976. Multi-spectral mapping of the lunar surface using groundbased telescopes. *Icarus*, **29**, 1–34.

MELOSH, H. J. 1989. *Impact Cratering*. Oxford University Press, New York.

MULLER, J. R. & AYDIN, A. 2005. Using mechanical modeling to constrain fault geometries proposed for the northern Marmara Sea. *Journal of Geophysical Research*, **110**, B03407, http://dx.doi.org/10.1029/2004JB003226

NAHM, A. L. 2012. Forward mechanical modeling of Rima Ariadaeus, the Moon. *In: Abstracts, European Geosciences Union General Assembly 2012, Vienna, Austria*, 22–27 April 2012. EUG, Munich.

NAHM, A. L., ÖHMAN, T. & KRING, D. A. 2013. Normal faulting origin for the Cordillera and Outer Rook Rings of Orientale Basin, the Moon. *Journal of Geophysical Research: Planets*, **118**, 190–205, http://dx.doi.org/10.1002/jgre.20045

NAKAMURA, Y., LATHAM, G. V., DORMAN, H. J., IBRAHIM, A.-B. K., KOYAMA, J. & HORVATH, P. 1979. Shallow moonquakes – Depth, distribution and implications as to the present state of the lunar interior. *Proceedings of the Lunar and Planetary Science Conference*, **10**, 2299–2309.

OBERBECK, V. R., WILLAM, L. Q. & GREELEY, R. 1969. On the origin of lunar sinuous rilles. *Modern Geology*, **1**, 75–80.

OKADA, Y. 1992. Internal deformation due to shear and tensile faults in a half-space. *Bulletin of the Seismological Society of America*, **82**, 1018–1040.

PEACOCK, D. C. P. 1991. Displacement and segment linkage in strike-slip fault zones. *Journal of Structural Geology*, **13**, 1025–1035.

PEACOCK, D. C. P. 2002. Propagation, interaction and linkage in normal fault systems. *Earth Science Reviews*, **58**, 121–142.

PEACOCK, D. C. P. & SANDERSON, D. J. 1991. Displacement and segment linkage and relay ramps in normal fault zones. *Journal of Structural Geology*, **13**, 721–733.

PEACOCK, D. C. P. & SANDERSON, D. J. 1994. Geometry and development of relay ramps in normal fault systems. *American Association of Petroleum Geologists Bulletin*, **78**, 147–165.

PEACOCK, D. C. P. & SANDERSON, D. J. 1996. Effects of propagation rate on displacement variations along faults. *Journal of Structural Geology*, **18**, 311–320.

PIETERS, C. M. 1978. Mare basalts types on the front side of the Moon. *Proceedings of the Lunar and Planetary Science Conference*, **9**, 2825–2849.

PIKE, R. J. 1977. Size-dependence in the shape of fresh impact craters on the moon. *In*: RODDY, D. J., PEPIN, R. O. & MERRILL, R. B. (eds) *Impact, Explosion Cratering: Planetary, terrestrial implications. Proceedings of the Symposium on Planetary Cratering Mechanics [held at] Flagstaff, Arizona, September 13–17, 1976*. Pergamon, New York, 489–509.

POLIT, A. T., SCHULTZ, R. A. & SOLIVA, R. 2009. Geometry, displacement-length scaling, and extensional strain of normal faults on Mars with inferences on mechanical stratigraphy of the Martian crust. *Journal of Structural Geology*, **31**, 662–673.

ROBINSON, M. S., BRYLOW, S. M. ET AL. 2010. Lunar Reconnaissance Orbiter Camera (LROC) Instrument Overview. *Space Science Reviews*, **150**, 81–124.

RESOR, P. G. 2008. Deformation associated with a continental normal fault system, western Grand Canyon, Arizona. *Geological Society of America Bulletin*, **120**, 414–430.

RIPPON, J. H. 1985. Contoured patterns of throw and shade of normal faults in the Coal Measures (Westphalian) of north-east Derbyshire. *Proceedings of the Yorkshire Geological Society*, **45**, 147–161.

SCHOLZ, C. H. 1997. Earthquake and fault populations and the calculation of brittle strain. *Geowissenschaften*, **15**, 124–130.

SCHULTZ, R. A. 1995. Limits on strength and deformation properties of jointed basaltic rock masses. *Rock Mechanics and Rock Engineering*, **28**, 1–15.

SCHULTZ, R. A. 1996. Relative scale and the strength and deformability of rock masses. *Journal of Structural Geology*, **18**, 1139–1149, http://dx.doi.org/10.1016/0191-8141(96)00045-4

SCHULTZ, R. A. 2000. Localization of bedding plane slip and backthrust faults above blind thrust faults: keys to wrinkle ridge structure. *Journal of Geophysical Research*, **105**, 12,035–12,052, http://dx.doi.org/10.1029/1999JE001212

SCHULTZ, R. A. 2003. A method to related initial elastic stress to fault population strains. *Geophysical Research Letters*, **30**, 1593, http://dx.doi.org/10.1029/2002GL016681

SCHULTZ, R. A. & LIN, J. 2001. Three-dimensional normal faulting models of the Valles Marineris, Mars, and geodynamic implications. *Journal of Geophysical Research*, **106**, 16,549–16,566, http://dx.doi.org/10.1029/2001JB000378

SCHULTZ, R. A. & WATTERS, T. R. 2001. Forward mechanical modeling of the Amenthes Rupes thrust fault on Mars. *Geophysical Research Letters*, **28**, 4659–4662.

SCHULTZ, R. A. & ZUBER, M. T. 1994. Observations, models, and mechanisms of failure of surface rocks surrounding planetary surface loads. *Journal of Geophysical Research*, **99**, 14 691–14 702.

SCHULTZ, R. A., OKUBO, C. H. & WILKINS, S. J. 2006. Displacement–length scaling relations for faults on the terrestrial planets. *Journal of Structural Geology*, **28**, 2182–2193.

SCHULTZ, R. A., SOLIVA, R., OKUBO, C. H. & MÈGE, D. 2010. Fault populations. *In*: WATTERS, T. R. & SCHULTZ, R. A. (eds) *Planetary Tectonics*. Cambridge University Press, Cambridge, 457–510.

SMITH, D. E., ZUBER, M. T. ET AL. 2010. The Lunar Orbiter Laser Altimeter investigation on the Lunar Reconnaissance Orbiter mission. *Space Science Reviews*, **150**, 209–241.

SOLOMON, S. C. 1978. On volcanism and thermal tectonics on one-plate planets. *Geophysical Research Letters*, **5**, 461–464.

SOLOMON, S. C. & HEAD, J. W. 1979. Vertical movement in mare basins: relation to mare emplacement, basin tectonics, and lunar thermal history. *Journal of Geophysical Research*, **84**, 1667–1682.

SOLOMON, S. C., COMER, R. P. & HEAD, J. W. 1982. The evolution of impact basins: viscous relaxation of topographic relief. *Journal of Geophysical Research*, **87**, 3975–3992.

STÖFFLER, D., RYDER, G., IVANOV, B. A., ARTEMIEVA, N. A., CINTALA, M. J. & GRIEVE, R. A. F. 2006. Cratering history and lunar chronology. *In*: JOLLIFF, B. L., WIECZOREK, M. A., SHEARER, C. K. & NEAL, C. R. (eds) *New Views of the Moon*. Reviews in Mineralogy & Geochemistry, **60**, 519–596, http://dx.doi.org/10.2138/rmg.2006.60.05

STUART-ALEXANDER, D. E. & HOWARD, K. A. 1970. Lunar maria and circular basins – a review. *Icarus*, **12**, 440–446.

TODA, S., STEIN, R. S., REASENBERG, P., DIETERICH, J. & YOSHIDA, A. 1998. Stress transferred by the 1995 $M_\omega = 6.9$ Kobe, Japan, shock: effect on aftershocks and future earthquake probabilities. *Journal of Geophysical Research*, **103**, 24 543–24 565, http://dx.doi.org/10.1029/98JB00765

TODA, S., STEIN, R. S., RICHARDS-DINGER, K. & BOZKURT, S. B. 2005. Forecasting the evolution of seismicity in southern California: animations built on earthquake stress transfer. *Journal of Geophysical Research*, **110**, B05S16, http://dx.doi.org/10.1029/2004JB003415

TOMPKINS, S., PIETERS, C. M., MUSTARD, J. F., PINET, P. & CHEVREL, S. D. 1994. Distribution of materials excavated by the lunar crater Bullialdus and implications for the geologic history of the Nubium region. *Icarus*, **110**, 261–274.

TURCOTTE, D. L. & SCHUBERT, G. 2002. *Geodynamics: Application of Continuum Physics to Geological Problems*, 2nd edn. Cambridge University Press, Cambridge.

WALSH, J. J. & WATTERSON, J. 1987. Distribution of cumulative displacement and seismic slip on a single normal fault surface. *Journal of Structural Geology*, **9**, 205–226.

WALSH, J. J. & WATTERSON, J. 1988. Analysis of the relationship between displacements and dimensions of faults. *Journal of Structural Geology*, **10**, 239–247.

WATTERS, T. R. & JOHNSON, C. L. 2010. Lunar tectonics. *In*: WATTERS, T. R. & SCHULTZ, R. A. (eds) *Planetary Tectonics*. Cambridge University Press, Cambridge, 121–182.

WATTERS, T. R., ROBINSON, M. S. & SCHULTZ, R. A. 2000. Displacement–length relations of thrust faults associated with lobate scarps on Mercury and Mars: comparison with terrestrial faults. *Geophysical Research Letters*, **27**, 3659–3662.

WATTERS, T. R., SCHULTZ, R. A., ROBINSON, M. S. & COOK, A. C. 2002. The mechanical and thermal structure of Mercury's early lithosphere. *Geophysical Research Letters*, **29**, 37-1–37-4, http://dx.doi.org/10.1029/2001GL014308

WATTERS, T. R., ROBINSON, M. S. ET AL. 2010. Evidence of recent thrust faulting on the Moon revealed by the Lunar Reconnaissance Orbiter Camera. *Science*, **329**, http://dx.doi.org/10.1126/science.1189590

WATTERS, T. R., THOMAS, P. C. & ROBINSON, M. S. 2011. Thrust faults and the near-surface strength of asteroid 433 Eros. *Geophysical Research Letters*, **38**, L02202, http://dx.doi.org/10.1029/2010GL045302

WATTERSON, J. 1986. Fault dimensions, displacements and growth. *Pure and Applied Geophysics*, **124**, 365–373.

WHITAKER, E. A. 1972. Lunar color boundaries and their relationship to topographic features: a preliminary survey. *Moon*, **4**, 348–355.

WHITAKER, E. A. 1999. *Mapping and Naming the Moon*. Cambridge University Press, Cambridge.

WIECZOREK, M. A., NEUMANN, G. A. ET AL. 2013. The crust of the Moon as seen by GRAIL. *Science*, **339**, 671–675.

WILHELMS, D. E. 1972. *Geologic Map of the Taruntius Quadrangle of the Moon (Scale 1:1,000,000)*. United States Geological Survey, Miscellaneous Investigations Series, **Map I-722**.

WILHELMS, D. E. 1987. *Geologic History of the Moon*. United States Geological Survey, Professional Papers, **1348**.

WILKINS, S. J. & GROSS, M. R. 2002. Normal fault growth in layered rocks at Split Mountain, UT: influence of mechanical stratigraphy on dip linkage, fault restriction and fault scaling. *Journal of Structural Geology*, **24**, 1413–1429.

WILLEMSE, E. J. M., POLLARD, D. D. & AYDIN, A. 1996. Three-dimensional analyses of slip distributions on normal fault arrays with consequences for fault scaling. *Journal of Structural Geology*, **18**, 295–309.

WILLIAMS, N. R., WATTERS, T. R., PRITCHARD, M. E., BANKS, M. E. & BELL III, J. F. 2013. Fault dislocation modeled structure of lobate scarps from Lunar Reconnaissance Orbiter Camera digital terrain models. *Journal of Geophysical Research: Planets*, **118**, 224–233, http://dx.doi.org/10.1002/jgre.20051

WOOD, C. A. 2003. Mare Nubium. *In*: *The Modern Moon: A Personal View*. Sky Publishing, Cambridge, 144–151.

ZUBER, M. T., SMITH, D. E. ET AL. 2013. Gravity field of the Moon from the Gravity Recovery and Interior Laboratory (GRAIL) Mission. *Science*, **339**, 668–671.

Physical analogue modelling of Martian dyke-induced deformation

DANIELLE Y. WYRICK*, ALAN P. MORRIS, MARY K. TODT &
MORGAN J WATSON-MORRIS

Southwest Research Institute, 6220 Culebra Road, San Antonio, TX, USA 78247

Corresponding author (e-mail: dwyrick@swri.org)

Abstract: The Tharsis region of Mars is characterized by large volcanic and tectonic centres that have been active throughout Martian geological history, including distinct sets of graben that extend radially for distances of hundreds to thousands of kilometres. Formation of these graben has been attributed to crustal extension and/or dyke propagation. Physical analogue models using layered sand and liquid paraffin wax were constructed to test the magnitude and style of deformation in the host rock associated with dyke injection. A variety of igneous morphologies was produced, including dykes and plugs. Results suggest that, in the absence of pre-existing faults, vertical dykes do not produce significant deformation in the surrounding rock. Deformation associated with other magmatic intrusions produced primarily contractional features rather than extensional features, similar to previous numerical studies and terrestrial field investigations.

The Tharsis region of Mars is characterized by large volcanic and tectonic centres that have been active throughout Martian geological history. Tectonic patterns (Fig. 1) correlate with at least five main episodes of activity concentrated around distinctive centres, dominated in the later stages by large volcanic provinces (Anderson *et al.* 2001). Many of these tectonic complexes exhibit distinct sets of graben that extend radially for distances of hundreds to thousands of kilometres (Carr 1974; Plescia & Saunders 1982; Scott & Tanaka 1986; Tanaka *et al.* 1991; Mège & Masson 1996; Anderson *et al.* 2001). Formation of these graben has been attributed to crustal extension (Plescia & Saunders 1982; Banerdt *et al.* 1992; Phillips *et al.* 2001) and/or dyke propagation (Mège & Masson 1996; Scott *et al.* 2002; Wilson & Head 2002; Schultz *et al.* 2004).

The dyke-induced graben formation hypothesis stems from both numerical (Rubin & Pollard 1998; Rubin 1992) and analogue modelling (Mastin & Pollard 1988) studies of terrestrial dyke intrusion. These models assumed slip along pre-existing faults to determine the extent of graben subsidence due to dyke intrusion. In contrast to these models, many dyke-intrusion models on Mars do not incorporate pre-existing faults and graben; rather, they rely upon dyke injection as a graben formation mechanism (Scott *et al.* 2002; Wilson & Head 2002; Head *et al.* 2003). The fundamental assumption of this interpretation is that the internal pressures within a dyke cause significant structural deformation in the surrounding host rock, specifically that a vertical dyke will allow a graben to form above the tip of the dyke.

In this study, we have constructed and analysed physical analogue models of magmatic injection as a primary mechanism for the production of graben on Mars. In particular, our models were designed to explore the extent to which magmatic injection, under varying emplacement conditions, will induce near-subsurface and surface deformation.

Methodology

A series of physical analogue models using layered sand and paraffin were constructed to test the magnitude and style of deformation associated with dyke injection. Physical analogue modelling techniques have been proven to provide robust simulations of a wide variety of tectonic and geological features and processes from global- or regional-scale crustal studies to outcrop and sub-metre-scale processes (e.g. Cadell 1888; Hubbert 1937; Ramberg 1963; Mulugeta 1988; Koyi 1997; Rahe *et al.* 1998; Wyrick *et al.* 2011; Byrne *et al.* 2013; Sims *et al.* 2013). Among the most important strengths of physical analogue modelling are: (1) the capability to model complex three-dimensional configurations; (2) the capacity to simulate discontinuous processes such as fracturing and faulting; and (3) the ability to document evolving deformation over time. Analogue materials such as sand are selected to reproduce, at a small scale, the geometric and kinematic features of natural brittle deformation structures.

In the experiments described here, dry Oklahoma #1 sand sieved to 500 μm was used as an analogue for brittle upper crust. Dry sand behaves in a

From: PLATZ, T., MASSIRONI, M., BYRNE, P. K. & HIESINGER, H. (eds) 2015. *Volcanism and Tectonism Across the Inner Solar System*. Geological Society, London, Special Publications, **401**, 395–403.
First published online June 23, 2014, http://dx.doi.org/10.1144/SP401.15
© The Authors 2015. Publishing disclaimer: www.geolsoc.org.uk/pub_ethics

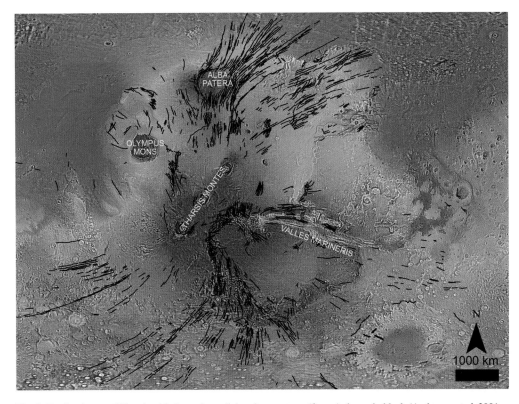

Fig. 1. Regional map of Tharsis with the major radial graben systems (fossae) shown in black (Anderson *et al.* 2001; Skinner *et al.* 2006). Viking MDIM2.1 overlain with Mars Orbiter Laser Altimeter (MOLA) topography with reds representing high elevations and blues representing low elevations.

time-independent manner at the strain rates used in the experiments and its material properties and behaviour have been well documented (Cloos 1968; Weijermars 1986; Withjack & Jamison 1986; Krantz 1991). Shear testing of the sand yielded average cohesion values of 209 Pa and an average friction angle of 34°. These values scale to natural rock cohesion values of *c.* 27 MPa (Hubbert 1937; Schellart 2000). We chose a model/nature length ratio of $L^* = 5 \times 10^{-6}$, such that one centimetre in the model represents 2 km in nature. Faults produced in these analogue materials are geometrically and kinematically similar to those observed in terrestrial fault systems (Rahe *et al.* 1998; Wyrick *et al.* 2011; Sims *et al.* 2013). Although the absolute values of the principal stresses are not measurable, the experiments were set up so that the maximum principal compressive stress σ_1 was vertical and the relative magnitudes of the initial minimum principal stress σ_3 were controlled by varying boundary constraints.

These models were specifically designed to simulate deformation in the surrounding host rock in response to the injection of magma. Paraffin wax was used in these experiments as a magma analogue as it is capable of preserving the three-dimensional structure of the intrusive body relative to the surrounding host rock. The paraffin was kept liquid at 66 °C until injected into the overlying sand pack. An average of 300 cm^3 of wax was injected in the experiments over 1–3 s before cooling and hardening. Rubber tubing (6.35 mm interior diameter) with a 20 cm-long slit parallel to the length of the tube along its uppermost edge was secured to the model base by aluminium plates such that its axis was parallel to the *y*-direction of the model rig (Fig. 2a). This configuration was designed to produce a linear injection into the overlying model. Layers of 1 cm-thick coloured sand were placed on top of the dyke model set-up. The total overlying sand thickness ranged from 6 to 10 cm in the models. The sand pack was cooled prior to injection of the liquid paraffin with solid carbon dioxide (dry ice) placed *c.* 2 cm above the sand pack for at least one hour. Once the sand was cooled, liquid paraffin was injected under pressure

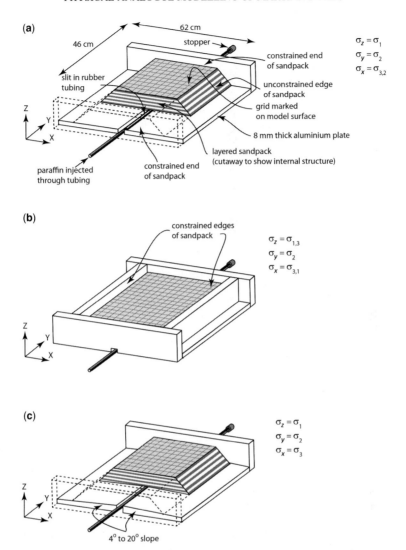

Fig. 2. General model set-up. Absolute principal stresses were not measured, but are shown schematically in relation to the model symmetry axes x, y and z; differential stress is expected to be very low, and injection pressures are close to lithostatic at the point of exit from the slit in the rubber tubing. (**a**) Sand pack with unconstrained sides parallel to the axis of dyke injection. σ_x is shown as either σ_3 or σ_2 because during dyke injection the pressure of the injecting paraffin against the sand pack is expected to create a back pressure that may locally cause the magnitude of σ_x to approach (or exceed) that of σ_y. (**b**) Configuration as for (a) but with all sides of the sand pack constrained. The pressure of the injecting paraffin is expected to create a back pressure that may locally cause the magnitude of σ_x to approach (or exceed) that of σ_z. (**c**) Model configuration as in (a) but with a basal slope away from the axis of dyke injection to reduce back pressure and decrease the magnitude of σ_x.

(1.0–1.8 bar or 15–26 psi) into the rubber tubing, which caused the paraffin to intrude upwards into the overlying sand layers where it cooled and solidified.

Three model configurations were used (Fig. 2) with some variation of parameters within each configuration. Figure 2a illustrates model configuration

A with partially unconstrained sand pack. In this set-up, the sides of the sand pack parallel to the y-direction (in the model reference framework) were unconstrained and had a slope equal to the angle of repose of the sand used (c. $30°$). The sand pack was composed of horizontal layers of coloured sand without mechanical differences between

Table 1. *Model names, configurations and principal variables*

Model	Configuration	Outward dip of base plates (°)	Sand pack thickness above injection site (cm)	Injection pressure (psi)	Sand pack width at top (cm)
05JAN12	B	0	6	15	n/a
24JAN12	B	0	10	15	n/a
08FEB12	A	0	6	15	22
13FEB12	A	0	8	15	10
15FEB12	A	0	8	15	25
20FEB12	A	0	8	15	18
22FEB12	A	0	8	18	18
21AUG12	A	0	8	15	18
27AUG12	A	0	8	15	18
30NOV12	A	0	8	15	18
12MAR13a	A	0	6	20	40
12MAR13b	A	0	8	20	28
13MAR13a	A	0	6	22	38
13MAR13b	A	0	8	22	28.5
15MAR13	C	4	8	23	38.5
28MAR13a	C	11	8	23	34
28MAR13b	C	11	6	23	29
02APR13a	C	11	8	26	16
02APR13b	C	22	8	23	14
03APR13	C	23.5	8	23	14
04APR13	C	10	8	26	17
31MAY13	A	0	8	22	45

layers; the base of the sand pack was horizontal. Experiments were run with different sand pack thicknesses (6–10 cm) and with different widths in the x-direction (10–48 cm measured on the top of the sand pack). A second model configuration B, illustrated in Figure 2b, was used in a small number of runs; in this set-up, the sand pack was constrained in all four horizontal directions with no free faces except the top. Figure 2c illustrates the third model configuration C, which is the same as in Figure 2a but with slopes at the base of the sand pack directed away from the injection slit at 4–23.5°. A total of 22 models were run (Table 1).

Although the dynamic stress conditions within the sand pack during injection of the paraffin were not measured, the different model configurations were intended to influence the stress state into which the dykes were injected. Configurations A and C were intended to simulate either a near-hydrostatic stress state, or a weakly normal faulting stress state with the least principal compressive stress σ_3 directed in the x-direction within the model reference framework. Increasing sand pack thickness increased the vertical stress (maximum principal compressive stress σ_1) and varying the (y-direction) width of the sand pack varied the magnitude of σ_3. Because of the fully enclosed sand pack, configuration B likely generated a

reactive stress in the y-direction, resulting in a weak reverse-faulting stress state with σ_1 directed parallel to the y-axis.

After dyke injection, the models were wetted to allow for dissection. Models were typically sliced perpendicular to the dyke injection (x–z-plane in the reference framework of Fig. 2) at c. 1 cm intervals to determine the style and magnitude of deformation in the sand layers surrounding the paraffin injection (Fig. 2a). In some models, the paraffin structures were dissected at the same intervals as the sand pack; in others, the sand pack was carefully dissected around the wax to preserve the full 3D structure of the intact dyke. All models were photographed during experimentation and dissection to analyse deformation patterns.

Results

Various dyke and plug geometries were produced during experimentation, similar to previously modelled magmatic intrusions (Galland *et al.* 2006, 2009; Mathieu *et al.* 2008). The paraffin wax injections preserved the detailed morphology of these intrusions. Although the purpose of these models was to characterize deformation associated with vertical dykes, the deformation styles associated

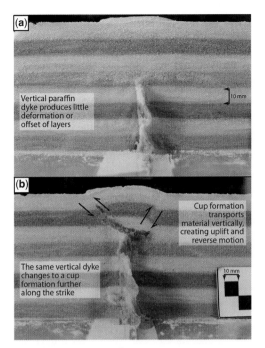

Fig. 3. (a) A vertical paraffin dyke produced no discernible deformation or offset of sand layers, either ahead of or above the intrusion. Model configuration as in Figure 2a, with paraffin injected at 1.5 bar (22 psi) into 8 cm of sand overburden. (b) The same dyke further along-strike produced a cup formation near the surface, transporting the sand layers above it both vertically and laterally, creating uplift and reverse faulting directly above the cup.

with other intrusive geometries are presented here to inform future interpretations of surface structure.

Vertical paraffin dykes produced no discernible deformation or offset of sand layers, either ahead of or above the intrusion (Fig. 3a). Vertical dykes appeared more often with higher injection pressures of 1.5–1.8 bar (22–26 psi) and under model configurations A and C. Several vertical dykes produced cup formations along their strike. These cups formed near the surface, transporting the sand layers above both vertically and laterally, creating uplift and reverse faulting directly above the cup (Fig. 3b). Non-vertical dykes typically occurred in models with lower injection pressures of 1.0–1.4 bar (15–20 psi) under varying overburden and model configurations A, B and C. These dykes induced reverse motion in overlying layers, expressed either as reverse faults or folds (Fig. 4).

Plug formation at the model base led to the translation of overlying material vertically, producing reverse faulting at the plug margins (Fig. 5a). A few complex morphologies were formed, such as a plug that developed vertical dykes at its margins. Dissection of this model suggested that reverse faulting developed above the plug first; feeder dykes at the plug margin then created steep anticlines on either side of the intrusion (Fig. 5b). The first appearance of a reverse fault occurred c. 1 cm ahead of the leading edge of the rising plug. Immediately above the leading edge of the plug, a topographic high formed and the deformed subsurface layers indicate the development of an anticline. Further along the dyke, reverse faults developed nearly to the surface and became slightly asymmetrical as the plug became more cup-like in geometry along strike. The main paraffin plug developed two distinct smaller vertical dykes at the margins that produced anticlines above the dyke tips. The uppermost layers developed folds above the width of the dyke that translated to the surface as symmetrical topographic highs of c. 0.4 cm on the sides of the dykes. It should be noted however that the region above the plug and dyke tips did not drop below the regional model elevation.

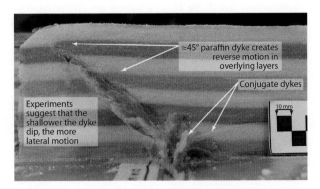

Fig. 4. Non-vertical dykes induced reverse motion on the overlying layers, expressed as either reverse faults or folds. Model configuration as in Figure 2a with paraffin injected at 1.2 bar (18 psi) into 8 cm of sand overburden.

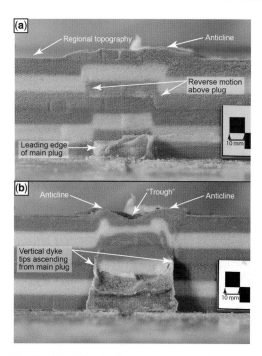

Fig. 5. (**a**) Paraffin plug formation translates overlying material vertically, producing reverse faulting at the plug margins. Model configuration as in Figure 2b, with paraffin injection at 1.0 bar (15 psi) into 6 cm of sand overburden. (**b**) Further dissection of this model suggested that reverse faulting developed above the plug first, and then feeder dykes at the plug margin created steep anticlines on either sides of the intrusion.

Many of the paraffin dykes in the models developed en echelon branching segments at or near their tip lines (Fig. 6). These branches typically formed *c*. 30° rotated from the dyke plane. In one model where the horizontal stresses were inferred

to be close in magnitude (i.e. $\sigma_2 \approx \sigma_3$; model configuration C), a fairly symmetrical spiral plug was formed with branching segments rotated *c*. 30° from the spiralling edge of the plug.

Discussion

Vertical paraffin dykes in our models – without a significant dip or base plug – produced no discernible deformation, either at the surface or in the subsurface. This result is consistent with previous numerical studies and field investigations of arrested dykes (Gudmundsson 2003; Geshi *et al.* 2010). Dykes in the models that transitioned to cup formation in the near surface may be reflections of near-surface effects similar to terrestrial non-feeder dykes (Geshi *et al.* 2010).

Non-vertical dykes produced reverse faulting and folds. In our models, surface deformation above and around the paraffin dykes primarily took the form of anticlines and reverse faults (i.e. contractional features) rather than graben and normal faults (i.e. extensional features). Our models suggest that, for cases of a non-vertical dyke, the shallower the dip the more likely reverse motion above and along the dyke injection surface is to occur. Previous numerically modelling efforts have examined the role of a widening subsurface dyke in the absence of regional extension and under various configurations of mechanically layered stratigraphy (Gudmundsson & Loetveit 2005; Wyrick & Smart 2009). Their results suggested that deformation was accommodated primarily by contractional fold development or reverse faulting on existing graben. The results of our models suggest dyke injection would produce reverse motion on pre-existing faults, similar to terrestrial examples of dyke-induced reverse faulting in graben (Gudmundsson *et al.* 2008)

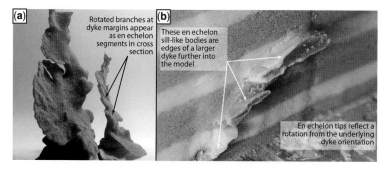

Fig. 6. (**a**) Paraffin model of dyke intrusion showing rotated en echelon branching segments. These branches typically formed *c*. 30° from the dyke plane. (**b**) In cross-section, these branching segments appeared as en echelon sill-like bodies.

Various intrusion types were produced during experimentation, including dykes, cups and plugs. Many of the paraffin dykes produced bladed 'branches' that formed at *c.* 30° to the main dyke plane. However, in cross-section these branches often appeared as separate horizontal en echelon intrusion bodies rather than the edges of a much larger subvertical igneous structure. At Martian scales, these en echelon segments would likely be mapped as separate dykes in satellite imagery rather than belonging to a larger parent dyke. Estimates of the stress conditions of dyke emplacement based on the orientation of these en echelon branches may be missing the larger *in situ* stress state for the parent dyke.

The origins of Tharsis-radial graben have been debated for many years (e.g. Plescia & Saunders 1982; Banerdt *et al.* 1992; Mège & Masson 1996; Wilson & Head 2002). The Tharsis radial graben systems are characterized by the 'simple graben' morphology: long narrow graben bounded by normal faults, with down-dropped flat floors unbroken by antithetic faults. To gain a better understanding of the role and extent of dyke-induced deformation on Mars, physical analogue experiments were performed to document deformation in response to magmatic intrusion. Modelling of magmatic dyke intrusion did not produce the hypothesized extensional graben formation; instead, magma intrusion was primarily characterized by contractional strain patterns. The primary result was surface deformation in the form of contractional folds producing uplift at the surface (i.e. bounding anticlines with a synclinal trough) rather than extensional graben over the dyke tip producing subsidence (i.e. bounded by normal faults with a down-dropped floor).

It should be noted that although the models reported here were performed with homogeneous sand layers, Mars probably has a layered, heterogeneous crust. The role of mechanical stratigraphy in fault and dyke development has been investigated by several researchers (e.g. Gudmundsson 2003; Schöpfer *et al.* 2007a, b; Wyrick & Smart 2009) and likely plays a role in the development of the Tharsis radial graben. Our homogeneous analogue model results produced deformation patterns similar to numerical model results that do incorporate mechanical stratigraphy, however (Gudmundsson 2003; Wyrick & Smart 2009).

Both dyke emplacement and normal faulting occur in extensional environments (Anderson 1951), so distinguishing the relative roles of each in deforming the near-surface crust is difficult. This study suggests that a vertical dyke intrusion alone does not produce significant deformation in the near surface; vertical dykes in the analogue models produced no resolvable deformation above

or ahead of their propagation. This result is similar to terrestrial field evidence of arrested dykes without graben and normal faults ahead of their tips (Gudmundsson 2003). If the dyke dip is not vertical, folding and contractional faulting develop adjacent to and above the dyke. The type of contractional deformation found in the models is not associated with the Tharsis radial graben systems, suggesting that dyke intrusion was not the primary deformation driver.

An alternative interpretation is that the dykes have intruded into pre-existing faults and fractures that provide paths of least resistance through the host rock. These pre-existing faults would also be subject to the same regional stress environment that would favour dyke formation (Anderson 1951). Many of these faults and fault segments would likely have a high slip tendency, with the potential to be more dilatant (Morris *et al.* 1996; Ferrill & Morris 2003). This condition favours fluid flow (Ferrill & Morris 2003) and likely influences the emplacement pathways of an intruding dyke. Further injection of magma likely produces displacement along these faults in the form of a contractional reverse-fault slip (Gudmundsson *et al.* 2008). Together, these investigations suggest a more passive role in dyke emplacement rather than the active graben-producing hypothesis.

The dyke-induced graben hypothesis has been widely used to interpret underlying dykes and dyke swarms and to help understand the magmatic history of the Tharsis region. This study suggests that the Tharsis-radial graben were not formed primarily in response to magmatic dyke intrusion. Instead, the graben (and their fault segments) likely predate dyke emplacement. These pre-existing faults would have responded to the same regional extensional stress environment predicted for dyke emplacement, with some fault segments becoming dilatant and influencing the emplacement pathway of the magma. It is likely that many of the graben in the Tharsis region are underlain (filled) as dykes. The graben alone are not evidence of an underlying dyke, however; additional evidence such as lava flows, cinder cones and gravity anomalies are required to accurately interpret subsurface magmatic intrusions. Understanding the dynamic interaction between magmatic activity and the structural response of the host rock is crucial for understanding the volcanic and tectonic history of Mars and has implications for astrobiological research at past and present geothermally active sites. Our results from this project provide constraints on the relative contributions of dyke- and fault-related deformation in the Tharsis region and contribute to the evolving debate regarding the timing and sources of stress for Tharsis (cf. Mège & Masson 1996; Dimitrova *et al.* 2006).

These results are consistent with terrestrial field examples and are probably applicable to other planetary bodies, such as the Moon (Head & Wilson 1993) and Venus (Ernst *et al.* 2001), where dyke-induced graben formation has been inferred.

This work was supported by NASA's Mars Fundamental Research programme (NASA Grant # NNX009AK18G). The authors thank K. Smart, D. Ferrill, A. Gudmundsson, G. Corti and M. Massironi for their constructive reviews.

References

ANDERSON, E. M. 1951. *The Dynamics of Faulting and Dyke Formation.* 2nd edn. Oliver & Boyd, Edinburgh.

ANDERSON, R. C., DOHM, J. M. ET AL. 2001. Primary centers and secondary concentrations of tectonic activity through time in the western hemisphere of Mars. *Journal of Geophysical Research*, **106**, 20563–20585.

BANERDT, W. B., GOLOMBEK, M. P. & TANAKA, K. L. 1992. Stress and tectonics on mars. *In*: KIEFFER, H. H., JAKOSKY, B. M., SNYDER, C. W. & MATTHEWS, M. S. (eds) *Mars*. The University of Arizona Press, Tucson, 249–297.

BYRNE, P. K., HOLOHAN, E. P., KERVYN, M., VAN WYK DE VRIES, B., TROLL, V. R. & MURRAY, J. B. 2013. A sagging-spreading continuum of large volcano structure. *Geology*, **41**, 339–342.

CADELL, H. M. 1888. Experimental researches in mountain-building. *Transactions of the Royal Society of Edinburgh*, **35**, 337–357.

CARR, M. H. 1974. Tectonism and volcanism of the Tharsis region of mars. *Journal of Geophysical Research*, **79**, 3943–3949.

CLOOS, E. 1968. Experimental analysis of Gulf Coast fracture patterns. *American Association of Petroleum Geologists Bulletin*, **52**, 420–444.

DIMITROVA, L. L., HOLT, W. E., HAINES, A. J. & SCHULTZ, R. A. 2006. Toward understanding the history and mechanisms of Martian faulting: the contribution of gravitational potential energy. *Geophysical Research Letters*, **33**, L08202, http://dx.doi.org/10.1029/2005GL025307

ERNST, R. E., GROSFILS, E. B. & MÈGE, D. 2001. Giant mafic dike swarms on Earth, Venus and Mars. *Annual Reviews of Earth and Planetary Science*, **29**, 489–534.

FERRILL, D. A. & MORRIS, A. P. 2003. Dilational normal faults. *Journal of Structural Geology*, **25**, 183–196.

GALLAND, O., COBBOLD, P. R., HOLLOT, E., DE BREMOND D'ARS, J. & DELAVAUD, G. 2006. Use of vegetable oil and silica powder for scale modeling of magmatic intrusion in a deforming brittle crust. *Earth and Planetary Science Letters*, **243**, 786–804.

GALLAND, O., PLANKE, S., NEUMANN, E. & MALTHE-SØRESSEN, A. 2009. Experimental modeling of shallow magma emplacement: Application to saucer-shaped intrusions. *Earth and Planetary Science Letters*, **277**, 373–383.

GESHI, N., KUSUMOTO, S. & GUDMUNDSSON, A. 2010. Geometric difference between non-feeder and feeder dikes. *Geology*, **38**, 195–198.

GUDMUNDSSON, A. 2003. Surface stresses associated with arrested dykes in rift zones. *Bulletin of Volcanology*, **65**, 606–619.

GUDMUNDSSON, A. & LOETVEIT, I. F. 2005. Dyke emplacement in a layered and faulted rift zone. *Journal of Volcanology and Geothermal Research*, **144**, 311–327.

GUDMUNDSSON, A., FRIESE, N., GALINDO, I. & PHILIPP, S. L. 2008. Dike-induced reverse faulting in a graben. *Geology*, **36**, 123–126.

HEAD, J. W., III & WILSON, L. 1993. Lunar graben formation due to near-surface deformation accompanying dike emplacement. *Planetary Space Science*, **41**, 719–727.

HEAD, J. W., WILSON, L. & MITCHELL, K. L. 2003. Generation of recent massive water floods at Cerberus Fossae, Mars by dike emplacement, cryospheric cracking, and confined aquifer groundwater release. *Geophysical Research Letters*, **30**, 1577, http://dx.doi.org/10.1029/2003GL017135

HUBBERT, M. K. 1937. Theory of scale models as applied to the study of geological structures. *Geological Society of America Bulletin*, **48**, 1459–1520.

KOYI, H. 1997. Analogue modeling: from a qualitative to a quantitative technique, a historical outline. *Journal of Petroleum Geology*, **20**, 223–238.

KRANTZ, R. W. 1991. Measurement of friction coefficients and cohesion for faulting and fault reactivation in laboratory models using sand and sand mixtures. *Tectonophysics*, **188**, 203–207.

MASTIN, L. & POLLARD, D. 1988. Surface deformation and shallow dike intrusion processes at Inyo Craters, Long Valley, California. *Journal of Geophysical Research*, **93**, 13,221–13,235.

MATHIEU, L., VAN WYK DE VRIES, B., HOLOHAN, E. P. & TROLL, V. R. 2008. Dikes, cups, saucers, and sills: analogue experiments on magma intrusion into brittle rocks. *Earth and Planetary Science Letters*, **271**, 1–13.

MÈGE, D. & MASSON, P. 1996. A plume tectonics model for the Tharsis province, Mars. *Planetary Space Science*, **44**, 1499–1546.

MORRIS, A. P., FERRILL, D. A. & HENDERSON, D. B. 1996. Slip-tendency analysis and fault reactivation. *Geology*, **24**, 275–278.

MULUGETA, G. 1988. Modeling the geometry of Coulomb thrust wedges. *Journal of Structural Geology*, **10**, 847–859.

PHILLIPS, R. J., ZUBER, M. T. ET AL. 2001. Ancient geodynamics and global-scale hydrology on Mars. *Science*, **291**, 2587–2591.

PLESCIA, J. B. & SAUNDERS, R. S. 1982. Tectonic history of the Tharsis region, Mars. *Journal of Geophysical Research*, **87**, 9775–9791.

RAHE, B., FERRILL, D. A. & MORRIS, A. P. 1998. Physical analog modeling of pull-apart basin evolution. *Tectonophysics*, **285**, 21–40.

RAMBERG, H. 1963. Experimental study of gravity tectonics by means of centrifuge models. *Bulletin of the Geological Institute of the University of Uppsala*, **42**, 1–97.

RUBIN, A. 1992. Dike-induced faulting and graben subsidence in volcanic rift zones. *Journal of Geophysical Research*, **97**, 1839–1858.

RUBIN, A. M. & POLLARD, D. D. 1988. Dike-induced faulting in rift zones of Iceland and Afar. *Geology*, **16**, 413–417.

SCHELLART, W. P. 2000. Shear test results for cohesion and friction coefficients for different granular materials: scaling implications for their usage in analogue modeling. *Tectonophysics*, **324**, 1–16.

SCHÖPFER, M. P. J., CHILDS, C. & WALSH, J. J. 2007a. Two-dimensional distinct element modeling of the structure and growth of normal faults in multilayer sequences: 1. Model calibration, boundary conditions, and selected results. *Journal of Geophysical Research*, **112**, B10401, http://dx.doi.org/10.1029/2006JB004902

SCHÖPFER, M. P. J., CHILDS, C. & WALSH, J. J. 2007b. Two-dimensional distinct element modeling of the structure and growth of normal faults in multilayer sequences: 2. Impact of confining pressure and strength contrast on fault zone geometry and growth. *Journal of Geophysical Research*, **112**, B10404, http://dx.doi.org/10.1029/2006JB004903

SCHULTZ, R. A., OKUBO, C. H., GOUDY, C. L. & WILKINS, S. J. 2004. Igneous dikes on Mars revealed by Mars Orbiter Laser Altimeter topography. *Geology*, **32**, 889–892.

SCOTT, D. H. & TANAKA, K. L. 1986. *Atlas of Mars 1:15 000 000 Scale Geologic Map Western Equatorial Region*. US Geological Survey I-1802-A, 1:15M scale, http://astrogeology.usgs.gov/PlanetaryMappingOld/DIGGEOL/mars/marswest/mw.pdf

SCOTT, E. D., WILSON, L. & HEAD, J. W. 2002. Emplacement of giant radial dikes in the northern Tharsis region of Mars. *Journal of Geophysical Research*, **107**, http://dx.doi.org/10.1029/2000JE001431

SIMS, D. W., MORRIS, A. P. ET AL. 2013. Analog modeling of normal faulting above Middle East domes during regional extension. *American Association of Petroleum Geology Bulletin*, **97**, 877–898.

SKINNER, J. A., HARE, T. M. & TANAKA, K. L. 2006. Digital renovation of the Atlas of Mars, 1:15 000 000-scale Global Geologic Series Maps, LPSC XXXVII, abstract #2331, http://www.lpi.usra.edu/meetings/lpsc2006/pdf/2331.pdf

TANAKA, K. L., GOLOMBEK, M. P. & BANDERDT, W. B. 1991. Reconciliation of stress and structural histories of the Tharsis region of Mars. *Journal of Geophysical Research*, **96**, 15617–15633.

WEIJERMARS, R. 1986. Flow behaviour and physical chemistry of bouncing putties and related polymers in view of tectonic laboratory applications. *Tectonophysics*, **124**, 325–358.

WILSON, L. & HEAD, J. 2002. Tharsis-radial graben systems as the surface manifestation of plume-related dike intrusion complexes: models and implications. *Journal of Geophysical Research*, **107**, http://dx.doi.org/10.1029/2001JE001593

WITHJACK, M. O. & JAMISON, W. R. 1986. Deformation produced by oblique rifting. *Tectonophysics*, **126**, 99–124.

WYRICK, D. & SMART, K. J. 2009. Dike-induced deformation and Martian graben systems. *Journal of Volcanology and Geothermal Research*, **185**, 1–11, http://dx.doi.org/10.1016/j.jvolgeores.2008.11.022

WYRICK, D. Y., MORRIS, A. P. & FERRILL, D. A. 2011. Normal fault growth in analog models and on Mars. *Icarus*, **212**, 559–567, http://dx.doi.org/10.1016/j.icarus.2011.01.011

Physical modelling of large-scale deformational systems in the South Polar Layered Deposits (Promethei Lingula, Mars): new geological constraints and climatic implications

LUCA GUALLINI[1,2]*, CRISTINA PAUSELLI[3], FRANCESCO BROZZETTI[4] & LUCIA MARINANGELI[1,2]

[1]*Dipartimento di Scienze Psicologiche, Umanistiche e del Territorio, Università d'Annunzio, Via dei Vestini, 31, 66100 Chieti Scalo, Italy*

[2]*International Research School of Planetary Sciences, Viale Pindaro, 42, 65127 Pescara, Italy*

[3]*Dipartimento di Scienze della Terra, Sezione di Geologia Strutturale e Geofisica, Università degli Studi di Perugia, Piazza dell'Università, 1, 06123 Perugia, Italy*

[4]*Geodynamics and Seismogenesis Laboratory, Dipartimento di Scienze Psicologiche, Umanistiche e del Territorio, Via dei Vestini, 31, 66100 Chieti Scalo, Italy*

*Corresponding author (e-mail: guallini@irsps.unich.it)

Abstract: Deformation systems (DSs) locally affect the South Polar Layered Deposits (SPLDs) along the margins of the Promethei Lingula ice sheet (part of the southern Martian ice-dome). One example is the 'S$_2$' deformation system, characterized by a complex pattern of brittle and brittle–ductile structures related to kilometre-scale shear zones that deform the sequence. Moreover, soft-sediment structures affect one layer located at the base of the S$_2$. An earlier structural analysis suggested that: (1) two deformation stages (D$_1$, in which the shear zones developed, and D$_2$, in which the D$_1$ structures were reactivated by deep-seated gravitational slope deformation) occurred, driven by gravity; and (2) there are variations in the bulk composition of the SPLD (which is inferred to be mainly composed of water ice plus basaltic dust). This work supports these structural results through thermal and mechanical modelling of the S$_2$ sequence. Our modelling results suggest that several layers within the S$_2$ system are probably composed of, or are mixed with, CO$_2$ ice, and that the development of the observed deformation is inconsistent with present-day physical conditions. Soft-sediment structures probably formed under warmer surface temperatures during the past, with those warmer temperatures favouring or even triggering ice flow/basal sliding of the Promethei Lingula.

Supplementary material: The complete description of the stratigraphy of the analysed S$_2$ sequence (in table format) is available at http://www.geolsoc.org.uk/SUP18746.

Regional context and geological setting

The South Polar Layered Deposits (SPLDs) form part of the Planum Australe region on Mars and consist of kilometre-thick sequences of porous bedding planes (Apl Unit: Kolb & Tanaka 2001; Tanaka & Kolb 2001). The surface of these bedding planes has been dated as Late Amazonian (10–100 Ma: Plaut *et al.* 1988; Herkenhoff & Plaut 2000; Koutnik *et al.* 2002). An impure (dust-rich) water ice composition (e.g. Cutts 1973; Kieffer *et al.* 1976; Thomas *et al.* 1992; Mellon 1996; Clifford *et al.* 2000) has been indirectly inferred by Martian subsurface radar sounders (Nunes & Phillips 2006; Plaut *et al.* 2007) and by density assessments, derived from the analysis of gravity anomalies associated with Planum Australe (Zuber *et al.* 2007; Wieczorek 2008). However, other secondary components have not been excluded (e.g. Mellon 1996; Kolb & Tanaka 2001; Wieczorek 2008; Phillips *et al.* 2011).

In an earlier study, Guallini *et al.* (2012) presented a semi-quantitative structural and kinematic analysis of the large-scale deformation systems observed in the SPLDs along the margins of the Promethei Lingula ice sheet (Fig. 1a, c). By systematically collecting structural data of the surveyed deformation features (particularly normal and transtensional faults), and by characterizing the cross-cutting relationships of the structures, these authors proposed a deformation history consisting of two superimposed, well-defined tectonic stages, and argued for soft-sediment and deep-seated gravitational slope deformation (DSGSD)

From: PLATT, T., MASSIRONI, M., BYRNE, P. K. & HIESINGER, H. (eds) 2015. *Volcanism and Tectonism Across the Inner Solar System*. Geological Society, London, Special Publications, **401**, 405–421.
First published online April 25, 2014, http://dx.doi.org/10.1144/SP401.13

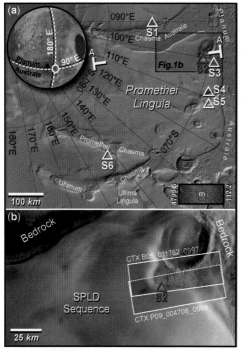

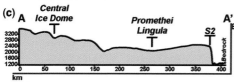

Fig. 1. (**a**) Location map of the observed deformation systems in Promethei Lingula (white triangles; the red triangle is the location of the S_2 system) and topography of the region. (**b**) Detailed location of the S_2 system. (**c**) Surface topographical section along transect A–A′ of the Promethei Lingula ice sheet. Topographical data are derived from the MOLA 512 ppd base map (images modified after Guallini *et al.* 2012).

mechanisms within the SPLD. Here, we focus on the representative 'S_2' deformation system analysed by Guallini *et al.* (2012) (Fig. 1a, b).

Structural characterization of the deformation systems

The first tectonic stage (D_1) of the S_2 system (Guallini *et al.* 2012) is characterized mainly by the development of transtensional faults that bound high-angle brittle–ductile shear zones (Figs 2 & 3a–c). Some of the fault planes are associated with clear kinematic indicators (mainly

S- and Z-shaped drag folds), outlining their left- and right-lateral senses of shear (see the 'LyD' marker layers in Fig. 3b). Some of these faults are associated with an extensional dip-slip component, as shown by the downwards displacement of certain layers, which were observed using Digital Terrain Models (DTMs) from High Resolution Imaging Science Experiment (HiRISE) stereo-pairs and from the Mars Orbiter Altimeter (MOLA) dataset. The major faults are arranged in en echelon sets. Other notable deformation structures occur within the major shear zones as boudins (which entail the extension of layers between fault planes and contrasting competences within the SPLD; see the 'LyB' marker layers in Fig. 3a) and S–C-like bands. Other typical structures related to the D_1 event include the true convolute folds and 'ball-and-pillow' and 'flame' structures (Fig. 3c) that affect one particular marker bed (labelled 'LyP' in Fig. 3c) at the base of the S_2 system. Although developed within an ice sheet (with a high sediment content), this deformation is geometrically similar to soft-sediment tectonic structures observed in certain terrestrial sequences, such as the deposits created by gravitational flows during the first stages of sediment consolidation in the presence of water (e.g. Rossetti 1999).

The second tectonic stage (D_2) of the S_2 system (Guallini *et al.* 2012) affects the weaker and, in some cases, pre-deformed, stratigraphic levels of the SPLD succession. This phase is mainly characterized by extensional deformation due to gravity. The D_2 deformation may be related to large DSGSD in the S_2, the onset of which should be subsequent to, and influenced by, the fault systems and shear bedding planes of the D_1 stage. The deformation along the weaker layers may have reactivated the high-angle faults as extensional or transtensional tears, leading to the formation of morphostructures (e.g. Agliardi *et al.* 2001, 2009a, b) such as trenches and topographical scarps (Fig. 2). The mechanisms and amount of deformation identified here closely resemble those observed on mountain slopes affected by DSGSD on Earth (Guallini *et al.* 2012).

Rationale for this study

On the basis of structural observations of ice sheets on Earth, the presence of complex deformation in the SPLD implies a large-scale movement of the Promethei Lingula. Ice flow and/or basal/internal sliding are the most plausible mechanisms of movement. This means that at a certain point in time, the forces within the Promethei Lingula ice sheet locally exceeded the strength of the SPLD layers, leading to their deformation. On the basis

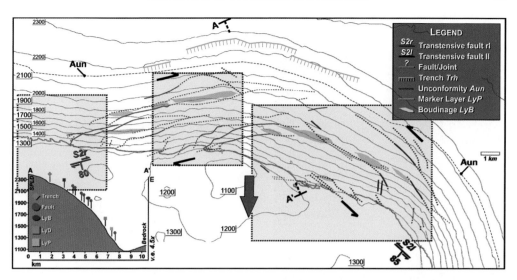

Fig. 2. Structural map and topographical section A–A' (lower left) of the S_2 deformational system (modified after Guallini *et al.* 2012). The grey insets are, from right to left, the locations of Figure 3a–c. The red arrow represents the possible direction of motion of the Promethei Lingula ice sheet during deformation of the S_2 section. The black arrows represent the main transtensive sense of motion (right- and left-lateral) of the two main D_1 shear zones that characterize the S_2. Trenches are related to the D_2 phase.

of a structural analysis, Guallini *et al.* (2012) suggested that the combined action of four main factors predisposed and/or triggered deformation of the SPLD:

- *Gravity (i.e. lithostatic stress)*. By analogy to ice sheets on Earth, it is assumed that the outwards movement and associated deformation of the Promethei Lingula ice sheet was driven only by gravity (Guallini *et al.* 2012).
- *Soft-sediment deformation*. Soft-sediment plastic deformation mainly affects the 'LyP' layer in the S_2 system, implying that local physical and rheological conditions were conducive to softening/melting, which reduced the value of the internal friction along this bedding plane. This observation suggests that this stratum worked as a preferential shear/detachment plane, locally enabling basal/internal sliding of the Promethei Lingula ice sheet.
- *Climate change*. Large-scale motion of Promethei Lingula is negligible at present (e.g. Pathare & Paige 2005; Fishbaugh & Hvidberg 2006; Guallini *et al.* 2012). Moreover, the estimated current viscosity/rheology of the ice would not result in the development of the observed regional deformation. However, conditions amenable to SPLD deformation were more likely during the past. As suggested by previous studies (e.g. Touma & Wisdom 1993; Head 2001; Laskar *et al.* 2002; Costard *et al.* 2002),

periods of high obliquity of Mars periodically caused major increases in the insolation over the SPLD. Under such a scenario, we propose that the heating of the SPLD softened/melted some inner layers (such as the 'LyP' layer in Fig. 3c), triggering or accelerating the ice sheet's outwards movement and resulting deformation.
- *Rheological behaviour and compositional variations*. Beyond the strain rate, the deformation of layers within the SPLD depends on two primary factors. (i) The layers' stratigraphic depths and thus the ambient temperature and pressure (T and P) conditions are important. At shallower levels (with commensurately lower T and P values), any deformation should be brittle in nature. At deeper levels, the principal deformation would occur in the brittle–ductile and ductile regimes (e.g. Durham & Stern 2001). The observed deformation appears to represent these different rheological behaviours, displaying predominantly brittle faults and fractures at shallower elevations, and grading into mainly brittle–ductile, ductile and soft-sediment structures towards the base of the SPLD. (ii) The relative consistency/density of the S_2 layers is a controlling factor. The layers display a marked disharmonic behaviour under stress, as evidenced by almost undisturbed layers/stratigraphic intervals that alternate with bedding planes/layers deformed by brittle and ductile structures. This is consistent with the damage

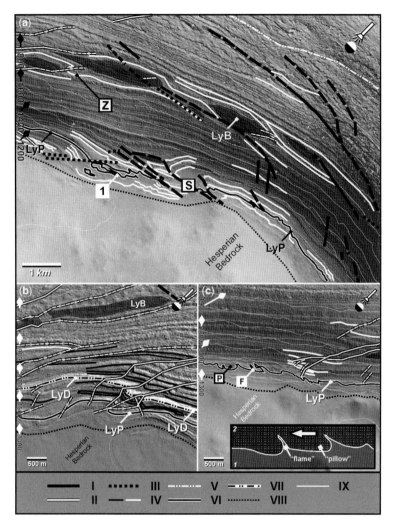

Fig. 3. Main deformation structures observed in the S_2 system. (**a**) Boudins (layers 'LyB') and drag-folds (S and Z) related to fault kinematics and soft-sediment structures affecting the 'LyP'. (**b**) Drag-folds/kink-folds affecting the layer 'LyD' and the sequence above, and 'en-echelon' fault planes. (**c**) 'Ball-and-pillows' (P)/flames (F) deforming the layer 'LyP'. Through comparison with terrestrial analogues (e.g. the inset), the top layer (2) is denser than, and sank into, the bottom layer (1). The assemblage of the observed structures (cf. Anketell *et al.* 1970) possibly indicates shear motion of the 'LyP' towards the left (white arrow in the inset) under melting conditions. In all of the figures, the symbols are as follows: I, transtensional left-lateral faults; II, transtensional right-lateral faults; III, inferred fault planes (with unknown sense of displacement) or fractures; IV, mapped layers (black or white depending on the background image); V, unconformity surface; VI, the 'LyP' layer; VII, the 'LyD' layer; VIII, topographical contours (100 m apart); IX, the boundary between the SPLD scarp and the bedrock. Images are modified after Guallini *et al.* (2012).

zones observed in layered sequences on Earth, in which deformation preferentially occurs in interbedded high- and low-competency materials (e.g. Moran 1971; Brodzikowski & Van Loon 1980; Visser *et al.* 1984). These observations imply marked contrasts in the competency of the deformed and undeformed bedding planes

in the S_2 sequence. In addition, differences in the relative density between bounding layers is suggested by the soft-sediment deformation in the 'LyP' layer. To account for the changes in the mechanical and rheological behaviour of the S_2 layers, we suggest that differences in lithology between layers accounts for the contrasting levels of deformation observed (cf. the next section on 'Stratigraphic model').

The deformational behaviour of the S_2 system could be a useful tool with which to investigate the internal properties of its constituent layers and the physical conditions present during deformation. The purpose of this study is therefore to investigate the validity of the above-mentioned four sets of observations suggested by the structural analysis of Guallini et al. (2012). In particular, this study aims to: (1) physically characterize the surveyed deformation; (2) understand the rheology of the Promethei Lingula ice sheet; (3) better identify the global-scale triggering mechanisms of the observed types of deformation; and (4) determine what effect (if any) past climate changes had on the South Pole of Mars and on the compositional variations within the SPLD. These aims have been pursued through the thermal and mechanical modelling of the SPLD, and through characterizing the S_2 system under former and present-day physical conditions.

Stratigraphic model

To parameterize thermal and mechanical modelling using the detailed geological and structural observations described above, a stratigraphic model ('SM', which included layers L_1–L_{23} in Fig. 4a, b) was used. The model was derived from the analysis and mapping of the S_2 log (Guallini et al. 2012) using high-resolution georeferenced images (Context Camera (CTX) images, with resolutions of 6.0 m/px: Malin et al. 2007) and topography (the MOLA topographical base map, sampled on a 512 px/degree (ppd) grid with a resolution of c. 115 m/px: Smith et al. 2001).

The S_2 stratigraphy is characterized by a sequence (approximately 1000 m thick) of strata (from St1 to St65) that vary in thickness from the metre scale to the decametre scale, and which are distinguished on the basis of their relative albedos (i.e. surface reflectance), surface morphology and apparent deformation. However, to identify the most important forms of deformational behaviour of the S_2 bedding planes through modelling, it was necessary to use as simple a stratigraphy as possible. This requirement was achieved by grouping together contiguous S_2 strata that show similar types of deformation and/or that are assumed to have the same composition (i.e. they are physically homogeneous). Thus, each layer of the model represented multiple bedding planes in the S_2 sequence. The SM is therefore an 'end-member' model that helps in constraining and inferring the true composition of the ice stack (see the 'Discussion' section), following the results of the thermal and mechanical models. On the basis of the observations reported in the previous section on 'Regional context and geological setting', the model assumed the following:

- The main constituent layers are composed of water ice with 10% by volume (on average) of dust particles.
- Only ice layers with a dust content greater than approximately 10% by volume can lead to lower flow and internal deformational rates (e.g. Durham et al. 1992). In addition, dust content between 0 and 30% did not change the results of thermal modelling in a substantial way (Wieczorek 2008). Thus, the inferred dust content in our model (see the previous point above), and its possible range of variation in the water-ice layers, did not meaningfully affect the development of deformational features within the model. In addition, the CO_2 layers were assumed to be pure, with their density and mechanical parameters already sufficient to justify their deformation with respect to the dirty water-ice layers (Wieczorek 2008).
- The marked changes in style of deformation of different layers within the S_2 sequence, and the observed resultant deformation structures, are consistent with local interbedding of layers with different ice–matrix compositions (Wieczorek 2008; Guallini et al. 2012). According to previous studies (e.g. Longhi 2005, 2006; Wieczorek 2008; Phillips et al. 2011), CO_2 and/or CO_2 clathrate hydrate are possible alternatives to water ice. However, in this study, only CO_2 was considered for modelling because the presence of CO_2 clathrate hydrates slows the nucleation rate of ice crystals (Longhi 2006), and so, even if CO_2 clathrate hydrate did contribute to the composition of SPLD (forming at the contacts between solid CO_2 and H_2O), it would be volumetrically insignificant, particularly in the presence of the massive layers we investigate here. Moreover, large uncertainties exist regarding the mechanical behaviour of CO_2 clathrate hydrate under planetary conditions, given the lack of theoretical/experimental parameters as inputs for physical modelling.
- In the model, the possible bedding planes of CO_2 corresponded to: (i) the previously described marker layer 'LyP', which appears affected by soft-sediment deformation and drag folds, and has a distinctly lighter-toned albedo (layer L_2 in Fig. 4, and 'LyP' in Fig. 4b); (ii) the marker layer 'LyD', which has distinct drag folds associated with faulting and a light-toned albedo (L_6 in Fig. 4, and 'LyD' in Fig. 4b); and (iii) the marker layers 'LyB', characterized by well-defined trains of boudins (layers L_{10}, L_{13}, L_{16},

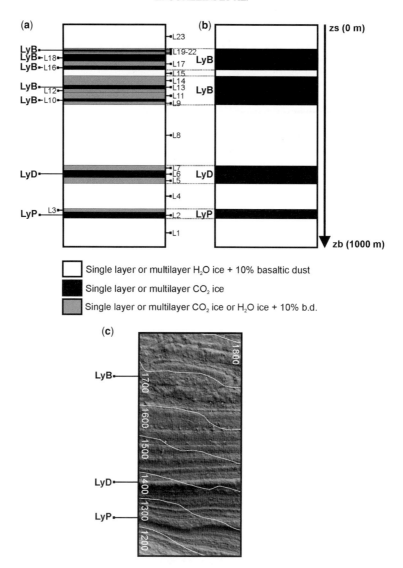

Fig. 4. Stratigraphic models (SMs), as obtained from the simple structural analysis of the S_2 sequence (Guallini *et al.* 2012). (**a**) & (**b**) Models SMa and SMb, respectively, representing simplified versions of the S_2 system for use in the thermal and mechanical models. (**c**) An example of a stratigraphic log of the S_2 system used to parameterize the stratigraphic model, in which 'LyB', 'LyD' and 'LyP' layers are indicated. Illumination of the image is from the top.

L_{18} and L_{21} in Fig. 4a, and 'LyB' in Fig. 4b), observed by Guallini *et al.* (2012).

- For the purposes of structural modelling, the layers labelled 'LyB' were assigned a high competency relative to their surrounding strata (e.g. Ramsay & Huber 1987). In fact, the boudinaged layers, representing a zone of stretching, generally occur in competent beds bounded by incompetent ones. A typical example in

sedimentary sequences is boudins developed in siltstones, which are in turn bounded by silty claystones. Further, soft-sediment deformation can result in an inverse density gradient (Visser *et al.* 1984; Rossetti 1999), causing the upper layer to founder into the lower, less dense one. Thus, the layer labelled 'LyP' should be denser than the strata at its base during the softening/melting conditions.

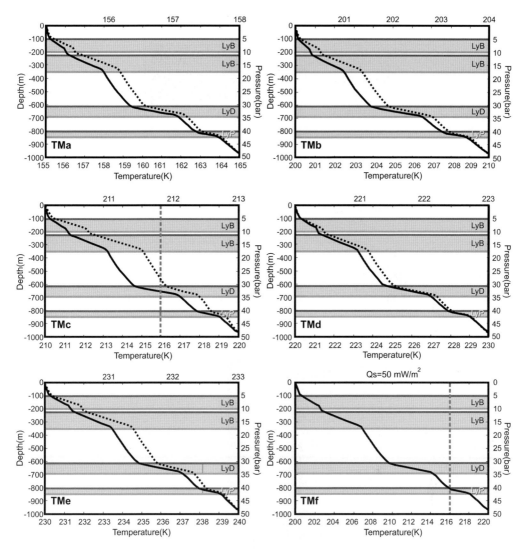

Fig. 5. Thermal models (TMa–TMaf; see Tables 1 & 2) representing (**a**) present-day physical conditions and (**b**)–(**f**) possible earlier conditions of the SPLD in Promethei Lingula. Models TMa–TMe are calculated for $Q_s = 8$ mW m^{-2} (dotted lines; see the upper x-axis for temperatures) and $Q_s = 25$ mW m^{-2} (solid lines; see the bottom x-axis for temperatures). Model TMf is calculated for $Q_s = 50$ mW m^{-2}. In each model, the dark and light grey horizontal lines represent, respectively, the top and the bottom of each stratum inferred to be composed of CO_2 (i.e. layers 'LyB', 'LyD' and 'LyP'). The vertical grey dashed lines indicate the melting point of CO_2 ice at considered depths, according to the H_2O–CO_2 phase diagram (see Fig. 6).

Thermal analysis

Methods

To understand whether present-day and ancient physical conditions (in terms of the boundary temperatures, pressures and surface heat fluxes) are consistent with the observed rheological deformation of the S$_2$ sequence, as well as with the assumed

compositions of the layers of the stratigraphic model, we performed several thermal model runs (Fig. 5). These models resolved the heat equation for conduction, discretized the heat conservation equation with finite difference and investigated the effects of different initial conditions. We obtained the solution of the 2D Eulerian temperature equation by modifying a code proposed by Gerya (2010) for a moving medium with variable

Table 1. *Adopted physical parameters of the stratigraphic model layers for the thermal and mechanical models*

SPLD composition	K (W m^{-1} K^{-1})	ρ (kg m^{-3})	C_p (J kg^{-1} K^{-1})	E (MPa)	ν
Solid CO_2	0.65[*]	1560[*]	1000[†]	1.0×10^{10}[‡]	0.28[‡]
Solid H_2O (plus c. 10% dust[§])	3.50[¶]	1250[**]	2100 (pure ice)[††]	9.4×10^{9}[‡] (pure ice)[††]	0.33[‡] (pure ice)[††]

K is the thermal conductivity, ρ is the density, C_p is the specific heat, E is Young's modulus and ν is Poisson's ratio.
[*]After Wieczorek (2008).
[†]Washburn *et al.* (1929).
[‡]After Kieffer (2007).
[§]By volume, hypothesizing pure basaltic insoluble dust.
[¶]After Clifford (1993).
[**]After Zuber *et al.* (2007) and Wieczorek (2008).
[††]Omitting dust content does not affect modelling results.

conductivity and resolved with the implicit method. The input parameters (i.e. the relevant lithologies and layers thicknesses) were obtained from the stratigraphic model. The physical parameters necessary for modelling (i.e. the thermal conductivity, density and specific heat) were obtained from published values (Table 1). The values chosen fall within the range of those estimated for the geological formations on Mars; small changes in these values lead to changes in the results within experimental errors (estimated to be of the order of 10%).

For each thermal model, the upper surface temperature ($T_{surface}$, i.e. at a depth (z) of 0 m) and the lower surface temperature (T_{basal}, at a depth (z) of c. 1000 m) were left constant. As the boundary conditions of the right-hand and left-hand side of the model were also constant, the horizontal heat flow was assumed to be zero. The simulations were performed for a period sufficiently long enough to reach a steady state. The current surface temperature between 70°S and 90°S is approximately 165–170 K (as averaged over one solar year), and has been derived on the basis of the polar albedo and thermal inertia maps (Paige & Keegan 1994; Paige *et al.* 1994).

It has been assumed that, during the Late Amazonian, the surface thermal heat flux (and thus the thermal gradient) was constant and equal to present-day conditions (Clifford 1993; Clifford *et al.* 2010). In contrast, lithostatic pressure and/or surface temperatures in the past may have exceeded present-day values. However, the addition of approximately 500 m of material to the stratigraphic model (equal to the supposed portion removed by erosion: e.g. Kolb & Tanaka 2006) does not substantially change our modelling results (Fig. 6). Thus, the lithostatic pressure (i.e. the sequence thickness in the stratigraphic model) was also assumed to remain constant through time. Conversely, as previously assumed (see the earlier subsection on

'Rationale for this study'), warmer temperatures at the Martian poles would have played a key role in influencing the rheology of the SPLD. Based on a likely scenario under which the planet's orbital axis is inclined by 45°, the poles would be tilted towards the Sun during the summer solstice, with daily average temperatures at the surface reaching as high as 223–293 K at 70°S (Costard *et al.* 2002).

The value of surface heat flow (Q_s) for the Late Amazonian was estimated to be 25–40 mW m^{-2} by Schubert *et al.* (1992), Clifford (1993) and Zuber *et al.* (2000). In the last decade, the range of estimated heat flux has been revised to 8–25 mW m^{-2} (Clifford *et al.* 2010). For that reason, and to evaluate a more complete scenario for our thermal models, we used the minimum and the maximum values of heat flux to obtain two different thermal gradients (representing the two end-members), calculated from a chosen surface temperature and from the thermal conductivity assigned at each layer (as reported in Table 1).

The first model (TMa: Fig. 5) featured present-day surface temperatures for both the minimum and maximum estimated heat fluxes. The subsequent five models represent possible past thermal conditions. The models TMb–TMe were run with incremental increases of 10 K at the surface for each, and the last model, TMf, had a Q_s of 50 mW m^{-2} and a thermal gradient of 0.02 K m^{-1} (see Table 2). In our thermal modelling, the influence of strain or frictional heating generated by basal and/or internal sliding, a previously discussed possible mechanism in the SPLD, was not taken into account. The reason for this choice is due to the difficulty in quantifying the shear heating intensity (W m^{-3}) (especially at planetary conditions), which is equal to the differential stress multiplied by the strain rate (Stuwe 2002), under the basic assumption that all mechanical energy is dissipated as heat (Joule 1850). However, considering that

Table 2. *Calculated T_{basal} for different initial $T_{surface}$* and Q_s*

Model	$T_{surface}$[†] (K)	T_{basal}[‡] (K)	Q_s[§] (mW m^{-2})
TMa	155	165	25
	155	158	8
TMb	200	210	25
	200	203	8
TMc	210	220	25
	210	213	8
TMd	220	230	25
	220	223	8
TMe	230	240	25
	230	233	8
TMf	200	220	50

*The value of conductivity has been chosen from Table 1 following the stratification of the stratigraphic model of Figure 4 (see the section on 'Thermal analysis').
[†]At a depth (z) of 0 m.
[‡]At a depth (z) of c. 1000 m.
[§]Q_s is the heat flux.

shear heating is generally faster than the conduction of heat towards the surface, the presence of shear heating could cause a rise in surface heat flow (Hartz & Podladchikov 2008); such a rise is taken into account in the thermal model TMf.

Using these input parameters, the thermal models were used to determine the temperature field from the surface to the base of the stratigraphic model until the first softening/melting was observed. Our attention was focused mainly on the inferred CO_2 marker layers (see the earlier 'Stratigraphic model' section) and, in particular, on the layer labelled 'LyP'.

Results from thermal model

The results obtained from the thermal models were compared with the H_2O–CO_2–CO_2 clathrate hydrate phase diagram (Miller 1974; Wagner *et al.* 1994; Span & Wagner 1996) (see Fig. 6). For the SPLD, at $z = 0$ m (c. 8×10^{-3} bar) and $z = 1000$ m pressures (c. 50 bar: Fig. 6), the melting temperature of water ice is approximately 273 K. For depths equivalent to pressures between about 5 ($z = 100$ m) and 50 bar ($z = 1000$ m), the melting temperature of dry ice is around 216 K.

Present-day conditions. Under the assumption that $T_{surface}$ is 155 K, and given the constraints provided by the stratigraphic and thermal models, the calculated T_{basal} is equal to 165 K (for $Q_s = 25$ mW m^{-2}) and 158 K (for $Q_s = 8$ mW m^{-2}) at $z = 1000$ m (TMa: Fig. 5). At the lithostatic pressures we considered, this value is far from the conditions required for softening/melting of both CO_2 and H_2O ice. Thus, all layers within the stratigraphic models are in the solid state. This suggests that: (1) in general, the brittle–ductile and ductile deformation that characterizes the 'LyB' and 'LyD' bedding planes is less likely to occur under

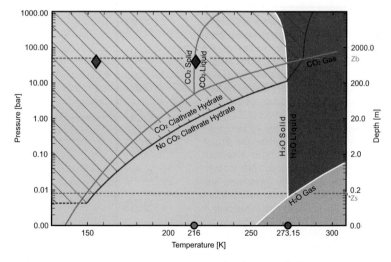

Fig. 6. Phase diagram of H_2O (blue areas bounded by white lines), CO_2 (light blue lines) and CO_2 clathrate hydrate (dark blue lines; the pinstriped area represents its stability as solid). The green rhomb is located at the calculated P–T boundary conditions of the marker layer 'LyP' under present-day physical conditions (i.e. model TMa; see Fig. 4). The red rhomb is located at the calculated P–T boundary conditions of the marker layer 'LyP' under possible earlier physical conditions (i.e. models TMc, TMd and TMf; see Fig. 4). The top and bottom grey dashed lines are, respectively, the bottom (c. 1000 m) and top (0 m) depths of the SM. Adapted from Mellon (1996).

present-day thermal conditions; and (2) the soft-sediment structures observed within the 'LyP' layer, in particular, are not due to the current surface temperatures on Mars.

Past conditions. Assuming a minimum $T_{surface}$ of 200 K (model TMb: Fig. 5b), the stratigraphic model is still in the solid state from top to bottom ($T_{basal} \leq 210$ K for the two chosen surface heat flows). However, by hypothesizing a thermal gradient higher than that at present, the resulting temperature at the depth of the 'LyP' layer is equal to about 215 K (model TMf: Fig. 5). This implies that the depth of the 'LyP' bedding plane (from 820 to 850 m) coincides more or less with the CO_2 softening/melting isotherm for surface conditions warmer than present, although these temperatures are still largely below the freezing temperature of water ice. In addition, all layers below the 'LyP' stratum (i.e. the base of the Promethei Lingula ice sheet) composed of CO_2 should be completely melted. This is consistent with the soft-sediment structures we observe within the 'LyP' layer. Conversely, all CO_2 layers above the 'LyP' bedding plane should still be in a solid state. This is also true for water-ice layers above and below 'LyP', given that the T_{basal} value (approximately 220 K) is very far from the melting point of water ice.

For a $T_{surface}$ of 210 K and $Q_s = 25$ mW m^{-2}, the calculated temperature at the depth of the 'LyP' stratum is equal to approximately 218 K (model TMc; Fig. 5c). This implies that the CO_2 layers can melt at depths equal to and greater than the 'LyP' bedding plane during times when the $T_{surface}$ value is still substantially below the freezing temperature of water ice. This, again, is consistent with the soft-sediment structures observed in the 'LyP' bedding plane. In addition, the temperature at the depth of the 'LyD' layer (from 620 to 680 m) is approximately 215 K, which supports ductile deformation, as is observed in this bedding plane. All layers above the 'LyD' plane composed of CO_2 ice should remain in the solid state. This is also true for all water-ice layers as, once again, the T_{basal} (220 K) is still far from the conditions required to melt water ice. For a $T_{surface}$ of 210 K and $Q_s = 8$ mW m^{-2}, at the lithostatic pressures we considered, the calculated temperature is also far from the conditions necessary for softening/melting of both CO_2 and H_2O ice.

Finally, for $T_{surface} \geq 220$ K (models TMd and TMe: Fig. 5d, e), all CO_2 layers in the SM should be melted for both of the chosen Q_s values. In contrast, because $T_{basal} = 230$ K for $Q_s = 25$ mW m^{-2} and $T_{basal} = 223$ K for $Q_s = 8$ mW m^{-2} (for $T_{surface} = 220$ K), water remains in its solid state. To melt water, it is necessary that $T_{surface}$ be very close to 0 °C ($T_{surface} > 273.15$ K), the occurrence

of which is more unlikely than the surface temperatures investigated by the thermal models.

Mechanical analysis

Methods

We used an engineering, design and finite-element analysis commercial code (COMSOL Multiphysics®) to test the mechanical behaviour of the stratigraphic layer, under the assumption that all the constituent layers are in their solid state. The mechanical model (Fig. 7) was applied to a two-dimensional (2D) geometry that represents the S_2 sequence. The finite-element method was used to solve the elastic equation of Cauchy–Navier. The stresses were obtained from the model strains using Hooke's law, assuming plane strain conditions. This means that the out-of-plane principal stress was the intermediate one, and that no appreciable strain occurred in that direction. Moreover, the out-of-plane stress is always a principal stress and is perpendicular to the model. The other two

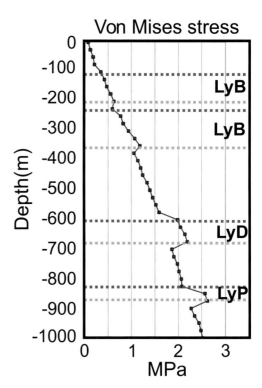

Fig. 7. von Mises' stress distribution across the stratigraphic model. The marker layer 'LyP' is affected by the strongest shear stress rate, followed by the 'LyD'. Both can preferentially act as detachment planes.

principal stresses must, therefore, lie in the plane of the model.

To solve the Cauchy–Navier equation, we defined the geometry of the S_2 system and the appropriate boundary conditions. The geometry of the model was derived from the stratigraphic model (Fig. 4). However, the dimensions of the model were greater than those reported in Figure 4 to avoid the influence of the boundary conditions. The physical parameters used for each layer are given in Table 1. The upper surface was free to deform horizontally and vertically, whereas the lower boundary and the left and right sides were fixed. The only pressure acting in the model was the gravity force per unit area (i.e. the out-of-plane lithostatic stress). This value is the Martian gravitational acceleration (3.71 m s^{-2}: Smith et al. 2001) multiplied by the density of each material and the depth at which the material is located plus the atmospheric pressure at polar surface (8 × 10^{-3} bar: Leovy 2001). In all our simulations, the layers were parallel and their inclination did not affect the final results.

Results are represented through the von Mises scalar quantity:

$$V_M = \sqrt{\frac{(\sigma_1 - \sigma_2)^2 + (\sigma_2 - \sigma_3)^2 + (\sigma_3 - \sigma_1)^2}{2}}$$

(1)

where σ_1, σ_2 and σ_3 are the maximum, intermediate and minimum principal stresses, respectively. The value of σ_3 was obtained using the relationship:

$$\sigma_3 = \nu(\sigma_1 + \sigma_2)$$

(2)

derived from the equation of linear elasticity in the case of plane strain (where ν is Poisson's ratio). Equation (1) expresses the difference between the principal components of stress and gives an indication of the amount of shear stress. This parameter is proportional to the octahedral shear stress (Jaeger & Cook 1971) by a constant factor and, thus, can be directly compared with the yield strength of the materials under consideration to produce an estimate of the possibility of failure.

It is important to emphasize that the mechanical model used here considered the elastic behaviour of the analysed materials until failure. In other words, the main purpose of our mechanical model was to identify the most probable sites of brittle failure across layer-parallel surfaces under the force of gravity, in order to define the eventual detachment planes. The approach we used was unable to define the rheological and physical conditions associated with the observed viscous–plastic deformation (such as the drag folds, boudins, etc.). The modelling of such structures is beyond the scope of the present work, in particular given the large uncertainties regarding the parameters needed to simulate such deformation.

Results from mechanical model

The mechanical model clearly shows that the maximum shear stresses in the von Mises distribution are focused on the marker layer 'LyP' and, to a lesser degree, in the marker layer 'LyD'. As predicted by the structural analysis of Guallini et al. (2012), this is consistent with the possibility of a detachment surface at the base of the 'LyP' layer and with shear structures along the 'LyD' bedding plane.

Conversely, the mechanical model does not indicate the likely presence of large shear stresses across the 'LyB' layers, excluding those strata as possible detachment planes. Nevertheless, according to the structural constraints, the presence of boudins in these layers is evidence that they experienced extension due to layer-parallel shear forces under brittle–ductile conditions. 'LyB' layers, therefore, represent shear zones but probably did not experience layer-parallel failures.

Discussion

Geological constraints from thermal modelling

The thermal modelling clearly shows that brittle–ductile and soft-sediment structures can develop in the interbedding layers 'LyB', 'LyD' and 'LyP', which are inferred to be composed of CO_2 ice and which have a lower thermal conduction value than water ice. In particular, while the brittle–ductile/ductile behaviour observed in the 'LyB' and 'LyD' layers is explainable through those strata having temperatures just below the melting point of dry ice, the development of the observed soft-sediment structures in the 'LyP' layer first requires that its physical state must change from solid-like to liquid-like (e.g. Allen 1986), at least along its bottom surface.

The occurrence of liquefaction in solid ice layers can be in response to various triggering forces, such as temperature variations, seismic or impact shocks or critical loading (e.g. Anketell et al. 1970; Rossetti 1999). In the case of the S_2 system, most of these triggers can be excluded on the basis of geological observations (Guallini 2012; Guallini et al. 2012), yet the thermal models support the temperature variation hypothesis. In other words, both our structural analysis and the thermal modelling suggest that the deformation within the 'LyP' layer was probably triggered by an increase in surface temperature that led to the melting of the CO_2 layers.

As previously discussed, the conditions required for the melting of CO_2 are not satisfied with the present-day surface temperatures, which are too low to cause CO_2 liquefaction at the depth of the 'LyP' layer. However, the 'LyP' plane would melt under temperatures higher than present (i.e. those in the TMf model, with a Q_s of 50 mW m^{-2}, or those in the TMc, with a Q_s of 25 mW m^{-2}). The TMd and TMe models represent less likely scenarios because they would also imply the melting of the 'LyD' and 'LyB' layers, which is not supported by the observations. The findings from the thermal modelling are consistent with the temperature modelling of Costard *et al.* (2002), which was related to variations in the obliquity of Mars' axis. In particular, at the considered latitudes, our results suggest an axial obliquity above about 26°, which is higher than that at present. According to Laskar *et al.* (2002), the higher axial obliquity implies that the DS could be older than at least 500 kyr, when the average obliquity ranged from 15° to 35°.

Under melting conditions, the observed deformation in the 'LyP' layer was driven by gravity. In particular, and similar to soft-sediment deformation on Earth (e.g. Lowe 1975; Rossetti 1999), the 'LyP' was probably part of a sequence of layers characterized by reverse density stratification due to differential liquefaction of the stacked layers. Under this scenario, gravity acted on an unstable system characterized by more- and less-fluidized sediments, such that the upper layer sank down into the less dense bottom layer and vice versa: the less dense layer rose towards the less liquefied upper layer (e.g. Mills 1983; Nichols *et al.* 1994; Owen 1996). If so, then more bedding planes than just the single 'LyP' layer must have been melted, such as the one immediately below 'LyP'. This suggests that a portion of the layers below the 'LyP' and the 'LyP' layer itself (i.e. the 'LyP' sequence) are composed of CO_2 ice that experienced melting, with an increasing degree of liquefaction from top to bottom.

The different degrees of liquefaction of such CO_2 layers can be explained through one or more interacting factors. (1) The thermal gradient may have caused an increase in temperature with depth, such that the 'LyP' layer was more viscous/plastic and denser (i.e. less liquefied or more compacted) than the bottom bedding plane(s). The thermal models TMc (with $Q_s = 25$ mW m^{-2}) and TMf (with $Q_s = 50$ mW m^{-2}) follow this condition. (2) The differences in dust content could have altered the density of the constituent layers. Therefore, 'LyP' would have a higher dust content (which increased its density) than the bottom bedding plane(s). (3) The differences in the void ratios of

the layers in succession may have influenced their original density. In this case, the 'LyP' layer had fewer voids (i.e. it was denser) than the bottom bedding plane(s).

The absence of soft-sediment deformation above the 'LyP' layer implies that the transmission of the liquefaction process was probably interrupted by the overlying solid ice layers, coupled with a temperature decrease towards shallower depths. Under the temperatures modelled with TMc–TMf, this implies that the overlying layer(s) is (are) probably composed of water ice, at temperatures far from the melting point. Alternatively, this could also suggest that the overlying layer(s) is (are) composed of CO_2 ice and that the 'LyP' layer occurs at roughly the same depth as the isotherm of melting of the CO_2 ice (see thermal model TMc).

However, yet another explanation for the differential liquefaction of the 'LyP' sequence is the difference in the bulk composition of the icy layers within the sequence (i.e. the sequence is characterized by alternating CO_2 and non-CO_2 layers). Based on the possible alternatives following the reverse-density stratigraphy condition (see Table 1), we can assume, for example, that the 'LyP' ice is composed of CO_2 and that the bottom bedding plane(s) involved in the differential liquefaction is (are) composed of H_2O. However, if the water ice layers are melted, then the surface temperatures must be higher ($T_{surface} \geq 265$ K) than those considered by the TMc and TMf models in order to cause the melting of water ice. Although this is not impossible, this requirement seems improbable.

A further possibility is that in the lower portion of the 'LyP' sequence, dry ice and water ice could coexist in the bulk composition of some bedding planes (individually or in the form of CO_2 clathrate hydrate), leading to different thermal behaviours, consistencies and densities relative to neighbouring layers as a function of water/CO_2 proportions.

More generally, beyond these mechanical and structural considerations (see the previous subsections on 'Rationale for this study' and 'Geological constraints from mechanical modelling'), the thermal models allow us to rule out a water-ice-only composition for all of the S_2 layers. Such a composition would be in contrast with the general rheology of the S_2 system. Moreover, if all layers were composed of water ice, $T_{surface}$ values ≥ 265 K would be needed to cause the melting of the 'LyP' sequence. However, this condition would have also driven the melting of all of the water ice layers at the bottom of the 'LyP' sequence (equal to up to 100 m in thickness), which forms the base of the S_2 system upon the bedrock below. This situation is not supported by structural analysis (Guallini *et al.* 2012) and, in any case, would have caused the complete collapse of the S_2 sequence. Lastly, if the

SPLD were composed of only water ice layers, then $T_{surface}$ values ≥ 265 K would have resulted in the (by no means localized, as we observe) occurrence of widespread gravitational collapse of the Promethei Lingula ice sheet, which is not supported by our geological mapping.

Finally, it should be noted that if the Promethei Lingula ice sheet were actively flowing along a detachment between the bottom portion of the 'LyP' sequence and the underlying layers, this could have enhanced the heat flow across the detachment surface itself (Wieczorek 2008). The use of the base as a shear plane seems to be reinforced by the type of soft-sediment structures themselves, which are consistent with horizontal motions (see the following subsection). However, taking into account the possible shear heat flow (see TMf in Fig. 5), the temperatures at the surface would still need to be higher than those today to cause the melting of the 'LyP' layer.

Geological constraints from mechanical modelling

Similar to the results of the thermal models, and as predicted by the structural analysis of Guallini et al. (2012), the mechanical model supports the hypothesis that some bedding planes in the stratigraphic model are composed of CO_2. If so, then shear strain was focused on the dry-ice layers, which worked as preferential planes of failure under those conditions that cause the lithostatic stress to exceed the layers' strength. Moreover, under stress, CO_2 layers would have recorded diverse deformation structures as a function of the relative competency of the neighbouring layers and their intrinsic rheological/physical conditions (i.e. those parameters as a function of the layers' depths in the sequence). Conversely, where all of the stratigraphic model's constituent layers were assumed to be water ice (either pure or impure), the mechanical model did not indicate the presence of preferential failure planes.

The mechanical model has some specific implications on the composition and mechanical behaviour of the 'LyB', 'LyD' and 'LyP' layers, in particular those listed below:

- The mechanical model does not indicate brittle failure due to gravity across the 'LyB' layers. However, according to the structural observations, the trains of boudin can be restricted to discrete layer-parallel shear zones (Visser et al. 1984). This observation implies that, although the 'LyB' layers experienced some degree of extension and were deformed by forces that exceeded their mechanical strength under brittle–ductile physical conditions (which is at

least consistent with the past conditions investigated by the TMc thermal model with $Q_s = 25$ mW m^{-2}), they did not experience failure. Nevertheless, on the basis of the experimental tests at planetary conditions, the layers composed of CO_2 ice (which we infer the 'LyB' strata to be) should be less competent (i.e. they should have lower mechanical strength) than water ice layers (Durham et al. 1996; Kargel & Ross 1998). Thus, assuming that neighbouring layers are composed of water ice, the CO_2 composition of the 'LyB' layers would have contrasting mechanical constraints, such that the boudinaged layers must be more competent than the bounding layers (see the 'Stratigraphic model' section of this paper). Unfortunately, the available data do not allow for an unequivocal solution, and thus, we must hypothesize that if the 'LyB' layers are characterized by dry ice (as the thermal models suggest) or by water ice, then the neighbouring layers are similarly composed of dry or water ice but have lower insoluble dust content because the insoluble dust content increases the mechanical strength of icy material (e.g. Durham et al. 1996), and/or the neighbouring layers have more voids. However, cases in which the 'LyB' bedding planes are thicker than the adjacent layers could result from a contrast in competency that is already sufficient to lead to different styles of deformation. Alternatively, the 'LyB' layers may be composed of water ice, whereas neighbouring layers are composed, or at least contain a certain amount, of CO_2. None the less, this condition does not appear particularly consistent with the constraints suggested by the thermal models that investigated hypothetical past conditions (i.e. TMc–TMf). Yet another alternative may be that the 'LyB' layers are composed of CO_2 clathrate hydrate, which has the highest mechanical strength of the icy materials discussed here. Even if it is less probable as a major component in the SPLD bulk composition, such a material cannot be excluded a priori. We note, however, that even under this scenario, this composition is less consistent with the findings of thermal models TMc–TMf than a CO_2 composition.
- The mechanical model is consistent with a CO_2 composition for the 'LyD' layer. Under present-day conditions (as investigated by the thermal model TMa), dry ice at the depth of the 'LyD' stratum could already be weak enough to act as a preferential plane of failure. However, the observed geological structures across the 'LyD' layer (e.g. drag folds) suggest that primarily plastic deformation occurred. A synthesis of the results from the structural analysis of Guallini

et al. (2012), and the thermal and mechanical modelling, suggests, again, that the observed deformation probably occurred in the past instead of under current conditions.

- The mechanical model also indicates that the 'LyP' layer has a CO_2 composition and supports the strong possibility of the presence of detachments through it. Moreover, the soft-sediment structures within the 'LyP' layer are inconsistent with its state being solid under stress, indicating that such deformation occurred during past conditions. Thus, our mechanical model supports the results from the structural analysis and is strengthened by the thermal models TMc ($Q_s = 25$ mW m^{-2}) and TMf (with $Q_s = 50$ mW m^{-2}), in which the plastic/ liquid behaviour of the 'LyP' stratum further favoured layer-parallel slip of the S_2 sequence under gravity.

It is therefore possible that the 'LyP' layer initially deformed due to viscous–plastic ice-flow (i.e. ductile deformation) in the absence of horizontal motion. During this phase (the D_1 deformation stage: Guallini *et al.* 2012), the base of the 'LyP' layer acted as a plastic shear plane along which deformation was gravity-driven. With a progressive increase in surface temperatures, up to at least those conditions investigated by the thermal models TMc (with $Q_s = 25$ mW m^{-2}) and TMf (with $Q_s = 50$ mW m^{-2}), liquefaction of the 'LyP' layer formed the soft-sediment structures, with only vertical movements at first. This deformation was followed by a horizontal displacement (Anketell *et al.* 1970), such that the unstable part of the S_2 sequence at the top of the 'LyP' layer slipped on the 'LyP' layer itself, which acted as a detachment plane (the D_2 deformation stage: Guallini *et al.* 2012). As previously mentioned, this hypothesis seems to be confirmed by the morphology of the observed soft-sediment structures: the 'ball-and-pillows'/flame structures display an en echelon pattern with a trend parallel to the inferred layer-parallel movement (Fig. 3c). It is also possible that, under melting temperatures, the 'LyP' layer experienced a 'positive-feedback' loop, in which gravity (i.e. the lithostatic pressure) drove the deformation and horizontal shear of the layer, leading to an increase in liquefaction that in turn reduced the mechanical strength, and accelerated the deformation and the horizontal motion, of the S_2 sequence (e.g. Ashby 1965).

Conclusions

The SPLDs on Mars are thought to be composed mainly of a mixture of water ice and basaltic dust, but the observed ductile and brittle–ductile

deformation of the SPLD in certain places (e.g. the S_2 system of Promethei Lingula) require a marked contrast in competency (probably due to differences in bulk composition) between hosting layers. The complex pattern of deformation affecting the S_2 system, in particular the soft-sediment structures, indicate deformation mechanisms that were probably active when surface temperatures at Mars' South Pole were warmer than today. Under these conditions, broad shear/detachment surfaces developed along weaker bedding planes under gravity, which in turn resulted in the deformation we observe.

In this work, the structural constraints have been integrated with thermal and mechanical modelling to develop an overall qualitative and quantitative overview of the S_2 deformation system. Our conclusions are based upon the synthesis of physical and geological data through the comparison between theoretical modelling results and the observations of real structures. Using this approach, both the thermal and mechanical modelling results are consistent with the structural assessment of Guallini *et al.* (2012). In particular, we find that the bedding planes 'LyB', 'LyD' and 'LyP': (1) have been affected by distinctive deformational structures; (2) are probably composed (in part or entirely) of CO_2 ice; (3) were melted (in the case of the 'LyP' layer only) or were deformed when surface temperatures were higher than present day; and (4) acted, in different ways, as preferential shear and/ or detachment planes, driving the deformation of the S_2 system. Our results also exclude CO_2 clathrate ice (as hypothesized by Guallini *et al.* 2012) as an alternative to water ice (except for some possible bulk inclusions and/or interfaces between water- and dry-ice layers) because this material does not meet the competency and thermal requirements for ductile and soft-sediment deformation under the present-day and past physical conditions we investigated.

Our modelling results are consistent with Wieczorek (2008), who argued that basal melting in the SPLD occurred during earlier conditions, with higher surface temperatures, higher heat flows and thicker sequences than today, together with a possible active flow of the polar cap. According to this study, at present-day conditions and in the presence of interbedded layers of water ice and solid CO_2, at least 10–25% carbon dioxide by volume would be needed to cause dry ice to melt in the thickest part (>3000 m) of the ice cap. However, given that the average CO_2 concentrations by volume are approximately 55% (Wieczorek 2008), sequences thicker than about 2000 m (but thinner than *c.* 3000 m to prevent the collapse of the sequence: Nye *et al.* 2000) are sufficient to achieve melting conditions. Thus, dry ice cannot melt under conditions similar to those of the present Promethei Lingula ice sheet, where the maximum thickness is approximately 1500 m (Plaut *et al.* 2007) and, in particular, not within the S$_2$ sequence, which has a thickness of approximately 1000 m. None the less, the results from our study reinforce the modelling work of Wieczorek (2008) by supplying more evidence in support of the presence of CO$_2$ as massive interbedding layers in the SPLD. Our work also provides clear structural evidence of basal melting of at least part of the Promethei Lingula ice sheet in the past, which was (at a certain point) actively flowing. Our results provide important support to the various deformation structures observed by Murray *et al.* (2001) and Grima *et al.* (2011) in the neighbouring region of Ultima Lingula. In particular, we propose that the interbedded layers of CO$_2$ ice could serve as detachment planes, compatible with their model requirements, to develop slumps and listric faults.

Whether the entirety or only part of the Promethei Lingula ice sheet was actively moving, and why only some sectors of Promethei Lingula experienced deformation, are still open issues. As suggested by the several deformation systems observed in Promethei Lingula (Guallini *et al.* 2012), it is most probable that, similar to terrestrial ice sheets, ice flow/basal sliding processes affected Promethei Lingula at a regional scale when temperatures of the South Pole were warmer during high-obliquity conditions on Mars. However, the conditions required for failure were achieved only in places where local factors favoured the triggering of deformation. The variation in the ice sheet thickness (which causes imbalances between internal forces), the roughness of the basement topography (which affects internal tensions), the presence of dust mantling the ice sheet surface (which can alter the heat conduction within the ice due to its low thermal inertia: Mellon *et al.* 2008) and other factors may have been instrumental in the inhomogeneous distribution of stresses within the ice sheet. None the less, the presence of interbedded CO$_2$ layers at the base of the sequence seems necessary (and sufficient) to trigger deformation within the SPLD. It is likely, in fact, that these layers aided development of shear/detachment planes that may have accelerated the basal sliding of Promethei Lingula under softening/melting conditions.

We are grateful to P. K. Byrne (the associate editor for this manuscript), to T. Platz (the editor-in-chief) and to another anonymous reviewer for their constructive comments and suggestions that substantially improved the manuscript. Our research was funded by the Italian Space Agency (ASI) and the Italian Ministry of University and Research (MIUR).

References

AGLIARDI, F., CROSTA, G. & ZANCHI, A. 2001. Structural constraints on deep-seated deformation kinematics. *Engineering Geology*, **59**, 83–102.

AGLIARDI, F., CROSTA, G., ZANCHI, A. & RAVAZZI, C. 2009*a*. Onset and timing of deep-seated gravitational slope deformations in the Eastern Alps, Italy. *Geomorphology*, **103**, 113–129.

AGLIARDI, F., CROSTA, G. & ZANCHI, A. 2009*b*. Tectonic v. gravitational morphostructures in the central Eastern Alps (Italy): constraints on the recent evolution of the mountain range. *Tectonophysics*, **474**, 250–270.

ALLEN, J. R. L. 1986. Earthquake magnitude-frequency, epicentral distance, and soft sediment deformation in sedimentary basins. *Sedimentary Geology*, **46**, 67–75.

ANKETELL, J. M., CEGLA, J. & DZULYNSKI, S. 1970. On the deformational structures in systems with reverse density gradients. *Annales de la Société Géologique de Pologne*, **XL**, 1.

ASHBY, W. R. 1965. *An Introduction to Cybernetics*. University Paperbacks, Methuen, London.

BRODZIKOWSKI, K. & VAN LOON, A. J. 1980. Sedimentary deformations in Saalian glaciolimnic deposits near Wlostów (Zary area, western Poland). *Geologie en Mijnbouw*, **59**, 251–272.

CLIFFORD, S. 1993. A model for the hydrologic and climatic behavior of water on Mars. *Journal of Geophysical Research*, **98**, 10 973–11 016.

CLIFFORD, S. M., CRISP, D. ET AL. 2000. The state and future of Mars polar science and exploration. *Icarus*, **144**, 210–242.

CLIFFORD, S. M., LASUE, J., HEGGY, E., BOISSON, J., MCGOVERN, P. J. & MAX, M. D. 2010. Depth of the Martian cryosphere: revised estimates and implications for the existence and detection of subpermafrost groundwater. *Journal of Geophysical Research*, **115**, 1–17.

COSTARD, F., FORGET, F., MANGOLD, N. & PEULVAST, J. P. 2002. Formation of recent Martian debris flows by melting of near-surface ground ice at high obliquity. *Science*, **295**, 110–113.

CUTTS, J. A. 1973. Nature and origin of layered deposits in the Martian polar regions. *Journal of Geophysical Research*, **78**, 4231–4249.

DURHAM, W. B. & STERN, L. A. 2001. Rheological properties of water ice – applications to satellites of the outer planets. *Annual Review of Earth and Planetary Sciences*, **29**, 295–330.

DURHAM, W. B., KIRBY, S. H. & STERN, L. A. 1992. Effects of dispersed particulates on the rheology of water ice at planetary conditions. *Journal of Geophysical Research*, **97**, 20 883–20 897.

DURHAM, W. B., STERN, L. A. & KIRBY, S. H. 1996. Rheology of water ices V and VI. *Journal of Geophysical Research*, **101**, 2989–3001.

FISHBAUGH, K. E. & HVIDBERG, C. 2006. Martian north polar layered deposits stratigraphy: implications for accumulation rates and flow. *Journal of Geophysical Research*, **111**, E06012, http://dx.doi.org/10.1029/2005JE002571

GERYA, T. (ed.) 2010. *Introduction to Numerical Geodynamic Modelling*. Cambridge University Press, Cambridge.

GRIMA, C., COSTARD, F. *ET AL*. 2011. Large asymmetric polar scarps on Planum Australe, Mars: characterization and evolution. *Icarus*, **212**, 96–109, http://dx. doi.org/10.1016/j.icarus.2010.12.017

GUALLINI, L. 2012. *New Geologic Constraints on South Polar Layered Deposits (Promethei Lingula region) and Light-Toned Layered Deposits (Iani Chaos Region): Two Different Facets of an Ancient Water Activity and Climate Synamic on Planet Mars*. PhD thesis, Università d'Annunzio, Pescara, Italy.

GUALLINI, L., BROZZETTI, F. & MARINANGELI, L. 2012. Large-scale deformational systems in the South Polar Layered Deposits (Promethei Lingula, Mars): 'soft-sediment' and deep-seated gravitational slope deformations mechanisms. *Icarus*, **220**, 821–843, http://dx.doi.org/10.1016/j.icarus.2012.06.023.

HARTZ, E. H. & PODLADCHIKOV, Y. Y. 2008. Toasting the jelly sandwich: the effect of shear heating on lithospheric geotherms and strength. *Geology*, **36** 331–334.

HEAD, J. W. 2001. Mars: evidence for geologically recent advance of the south polar cap. *Journal of Geophysical Research*, **106**, 10 075–10 085.

HERKENHOFF, K. E. & PLAUT, J. J. 2000. Surface ages and resurfacing rates of the polar layered deposits on Mars. *Icarus*, **144**, 243–255.

JAEGER, J. C. & COOK, N. G. W. 1971. *Fundamentals of Rocks Mechanics*. Chapman & Hall, London.

JOULE, J. P. 1850. On the mechanical equivalent of heat. *Philosophical Transactions of the Royal Society of London*, **140**, 61–82.

KARGEL, J. S. & ROSS, R. G. 1998. Thermal conductivity of solar system ices, with special reference to Martian polar caps. *In*: SCHMITT, D. (ed.) *Solar System Ices*. Kluwer Academic, Dordrecht.

KIEFFER, H. H. 2007. Cold jets in the Martian polar caps. *Journal of Geophysical Research*, **112**, E08005, http://dx.doi.org/10.1029/2006JE002816

KIEFFER, H. H., CHASE, S. C., MARTIN, T. Z., MINER, E. D. & PALLUCONI, F. D. 1976. Martian north pole summer temperatures: dirty water ice. *Science*, **194**, 1341–1344.

KOLB, E. J. & TANAKA, K. L. 2001. Geologic history of the polar regions of Mars based on Mars global surveyor data. II. Amazonian period. *Icarus*, **154**, 22–39.

KOLB, E. J. & TANAKA, K. L. 2006. Accumulation and erosion of South Polar Layered Deposits in the Promethei Lingula region, Planum Australe, Mars. *Mars*, **2**, 1–9.

KOUTNIK, M., BYRNE, S. & MURRAY, S. 2002. South Polar Layered Deposits of Mars: the cratering record. *Journal of Geophysical Research*, **107**, 1029.

LASKAR, J., LEVRARD, B. & MUSTARD, J. F. 2002. Orbital forcing of the Martian polar layered deposits. *Nature*, **419**, 375–377.

LEOVY, C. 2001. Weather and climate on Mars. *Nature*, **412**, 245–249.

LONGHI, J. 2005. Phase equilibria in the system CO_2–H_2O. I. New equilibrium relations at low temperatures. *Geochimca et Cosmochimica Acta*, **69**, 529–539.

LONGHI, J. 2006. Phase equilibrium in the system CO_2–H_2O: application to Mars. *Journal of Geophysical Research*, **111**, E06011, http://dx.doi.org/10.1029/2005JE002552

LOWE, D. R. 1975. Water escape structures in coarse-grained sediments. *Sedimentology*, **22**, 157–204.

MALIN, M. C., BELL, J. F. III. *ET AL*. 2007. Context Camera Investigation on board the Mars Reconnaissance Orbiter. *Journal of Geophysical Research*, **112**, E05S04, http://dx.doi.org/10.1029/2006JE002808

MELLON, M. T. 1996. Limits of the CO_2 content of the Martian polar deposits. *Icarus*, **124**, 268–279.

MELLON, M. T., FERGASON, R. L. & PUTZIG, N. E. 2008. The thermal inertia of the surface of Mars. *In*: BELL, J. F. (ed.) *The Martian Surface: Composition, Mineralogy and Physical Properties*. Cambridge University Press, Cambridge.

MILLER, S. L. 1974. The nature and occurrence of clathrate hydrates. *In*: KAPLAN, I. R. (ed.) *Natural Gases in Marine Sediments*. Marine Science, Plenum Press, New York, **3**, 151–177.

MILLS, P. C. 1983. Genesis and diagnostic value of 'soft-sediment' deformation structures – a review. *Sedimentary Geology*, **35**, 83–104.

MORAN, S. R. 1971. Glaciotectonic structures in drift. *In*: GOLDTHWAIT, R. P. (ed.) *Till – A Symposium*. Ohio State University Press, Columbus, OH, 127–148.

MURRAY, B., KOUTNIK, M., BYRNE, S., SODERBLOM, L., HERKENHOFF, K. & TANAKA, K. L. 2001. Preliminary geological assessment of the Northern Edge of Ultimi Lobe, Mars south polar layered deposits. *Icarus*, **154**, 80–97, http://dx.doi.org/10.1006/icar.2001.6657

NICHOLS, R. J., SPARKS, R. S. J. & WILSON, C. J. N. 1994. Experimental studies of the fluidization of layered sediments and the formation of fluid escape structures. *Sedimentology*, **41**, 233–253.

NUNES, D. C. & PHILLIPS, R. J. 2006. Radar subsurface mapping of the polar layered deposits on Mars. *Journal of Geophysical Research*, **111**, 6.

NYE, J., DURHAM, W. B., SCHENK, P. M. & MOORE, J. M. 2000. The instability of a South Polar Cap on Mars composed of carbon dioxide. *Icarus*, **144**, 449–455, http://dx.doi.org/10.1006/icar.1999.6306

OWEN, G. 1996. Experimental 'soft-sediment' deformation: structures formed by the liquefaction of unconsolidated sands and some ancient examples. *Sedimentology*, **43**, 279–294.

PAIGE, D. A. & KEEGAN, K. E. 1994. Thermal and albedo mapping of the polar regions of Mars using Viking thermal mapper observations. 2. South polar region. *Journal of Geophysical Research*, **99**, 25 993–26 031.

PAIGE, D. A., BACHMAN, J. E. & KEEGAN, K. E. 1994. Thermal and albedo mapping of the polar regions of Mars using Viking thermal mapper observations. 1. North polar region. *Journal of Geophysical Research*, **99**, 25 959–25 991.

PATHARE, A. V. & PAIGE, D. A. 2005. The effects of Martian orbital variations upon the sublimation and relaxation of north polar troughs and scarps. *Icarus*, **174**, 419–443.

PHILLIPS, R. J., DAVIS, J. B. *ET AL*. 2011. Massive CO_2 ice deposits sequestered in the South Polar Layered Deposits of Mars. *Science*, **332**, 838–841, http://dx. doi.org/10.1126/science.1203091

PLAUT, J., KAHN, R., GUINNESS, E. & ARVIDSON, R. 1988. Accumulation of sedimentary debris in the south polar region of Mars and implications for climate history. *Icarus*, **76**, 357–377.

PLAUT, J. J., PICARDI, G. ET AL. 2007. Subsurface radar sounding of the south polar layered deposits of Mars. *Science*, **316**, 92–96.

RAMSAY, J. G. & HUBER, M. I. (eds) 1987. *The Techniques of Modern Structural Geology*. Vol. 2: *Folds and Fractures*. Academic Press, London.

ROSSETTI, D. D. F. 1999. 'soft-sediment' deformation structures in late Albian to Cenomanian deposits, São Luís Basin, northern Brazil: evidence for paleoseismicity. *Sedimentology*, **46**, 1065–1081.

SCHUBERT, G., SOLOMON, S. C., TURCOTTE, D. L., DRAKE, M. J. & SLEEP, N. H. 1992. Origin and thermal evolution of Mars. *In*: KIEFFER, H. H., JAKOSKY, B. M., SNYDER, C. W. & MATTHEWS, M. S. (eds) *Mars*. University of Arizona Press, Tucson, AZ, 147–183.

SMITH, D. E., PHILLIPS, R. J. ET AL. 2001. Mars Orbiter Laser Altimeter: experiment summary after the first year of global mapping of Mars. *Journal of Geophysical Research*, **106**, 23 689–23 722, http://dx.doi.org/10.1029/2000JE001364

SPAN, R. & WAGNER, W. 1996. A new equation of state for carbon dioxide covering the fluid region from the triple point temperature to 1100 K at pressures up to 800 MPa. *Journal of Physical and Chemical Reference Data*, **25**, 1509–1596.

STUWE, K. 2002. *Geodynamics of the Lithosphere: An Introduction*. Springer, Berlin.

TANAKA, K. L. & KOLB, E. J. 2001. Geologic history of the polar regions of Mars based on Mars Global Surveyor data. I. Noachian and Hesperian periods. *Icarus*, **154**, 3–21, http://dx.doi.org/10.1006/icar.2001.6675

THOMAS, P., SQUYRES, S., HERKENHOFF, K., HOWARD, A. & MURRAY, B. 1992. Polar deposits of Mars. *In*: KIEFFER, H. H., JAKOSKY, B. M., SNYDER, C. W. & MATTHEWS, M. S. (eds) *Mars*. University of Arizona Press, Tucson, AZ, 767–795.

TOUMA, J. & WISDOM, J. 1993. The chaotic obliquity of Mars. *Science*, **259**, 1294–1297.

VISSER, J. N. J., COLLISTON, W. P. & TERBLANCHE, J. C. 1984. The origin of 'soft-sediment' deformation structures in Permo-Carboniferous glacial and proglacial beds, South Africa. *Journal of Sedimentary Petrology*, **54**, 1183–1196.

WAGNER, W., SAUL, A. & PRUSS, A. 1994. International equations for the pressure along the melting and along the sublimation curve for ordinary water substance. *Journal of Physical and Chemical Reference Data*, **23**, 515–527.

WASHBURN, E. W., WEST, C. J. & DORSEY, N. E. (eds) 1929. *International Critical Tables*. McGraw-Hill, New York.

WIECZOREK, M. A. 2008. Constraints on the composition of the Martian south polar cap from gravity and topography. *Icarus*, **196**, 506–517.

ZUBER, M. T., SOLOMON, S. C. ET AL. 2000. Internal structure and early thermal evolution of Mars from Mars Global Surveyor topography and gravity. *Science*, **287**, 1788–1793.

ZUBER, M. T., PHILLIPS, R. J. ET AL. 2007. Density of Mars South Polar Layered Deposits. *Science*, **317**, 1718–1719.

Tectonism and magmatism identified on asteroids

DEBRA L. BUCZKOWSKI[1]* & DANIELLE Y. WYRICK[2]

[1]*Johns Hopkins University Applied Physics Laboratory, Laurel, Maryland, USA*

[2]*Southwest Research Institute, San Antonio, Texas, USA*

*Corresponding author (e-mail: debra.buczkowski@jhuapl.edu)

Abstract: Linear features generally accepted as tectonic structures have been observed on several asteroids and their presence has implications for the internal structure, strength and evolution of these various bodies. Lineaments observed on the Martian moon Phobos led to the prediction that other cratered small bodies would be similarly lineated. Observations of several small bodies, including Gaspra, Ida, Mathilde, Eros, Itokawa, Steins, Lutetia and Vesta, have identified different physical mechanisms by which linear features can be formed. Analysis shows that asteroid lineaments appear to have different origins, including impact, and that lineament orientations and magnitudes provide important constraints on the interior structure of these bodies. Being a differentiated proto-planet, Vesta is a unique body with which to study the role that internal rheologies and structures play on surface tectonic expression. Vesta presents fractures and grooves similar to the other observed asteroids, but also large-scale graben and trough structures more characteristic of terrestrial planet tectonics. Unlike many terrestrial planets, Vesta's main stressors have, however, been primarily exogenic (e.g. impacts) rather than internally driven. Nevertheless, while the search for volcanic features on Vesta has not been successful, the evaluation of the Brumalia Tholus feature suggests that the geological history of Vesta may have included magmatism.

Asteroids are small rocky bodies, sometimes referred to as minor planets, which orbit the Sun. Orbits vary widely from within Earth's orbit to beyond that of Saturn. A large number of asteroids are found between the orbits of Mars and Jupiter; this region is commonly referred to as the 'main belt', the 'asteroid belt' or the 'main asteroid belt'.

In addition to these main-belt asteroids there are also near-Earth asteroids (NEAs), defined as those asteroids whose closest distance to the Sun (i.e. perihelion distance) is less than 1.3 AU. NEAs can be divided into subgroups (apollos, atens and amors) based on the semi-major axes of their orbits, as well as their perihelion and aphelion distances. Apollos cross Earth's orbit, with orbital semi-major axes within the range 1–1.0167 AU and perihelion at $\leq$1.017 AU; they account for 62% of all NEAs (e.g. McFadden *et al.* 1989; Morbidelli *et al.* 2002). Atens also cross Earth's orbit; they have orbital semi-major axes that are <1 AU and aphelion within the range 0.983–1.0167 AU, and include 6% of all NEAs (e.g. McFadden *et al.* 1989; Morbidelli *et al.* 2002). Amors, which make up 32% of all NEAs, have orbital semi-major axes of >1 AU and perihelion of 1.017–1.3 AU (e.g. McFadden *et al.* 1989; Morbidelli *et al.* 2002); they can cross Mars' orbit but never cross that of Earth. Asteroids can be sub-spherical (e.g. Magnusson *et al.* 1989; Parker *et al.* 2002) to elongate (e.g. Magnusson *et al.* 1989; Lagerros 1997) in shape and appear to have a wide range of compositions (e.g. Chapman *et al.* 1975; Bus & Binzel 2002). It has been suggested that these bodies are remnant planetesimals that became gravitationally perturbed into tilted, eccentric orbits and then began to strike each other catastrophically instead of accumulating into larger bodies (e.g. Petit *et al.* 2002; Bottke *et al.* 2005). Asteroids are also thought to be the source bodies of the majority of meteorites on Earth (e.g. Greenberg & Chapman 1983; Burbine *et al.* 2002). Meteorites are among the oldest rocks in the solar system, with radiometric dates of up to 4.6 billion years. We can therefore assume that they, and thus their source bodies, are leftovers from the formation of the solar system (e.g. Petit *et al.* 2002; Bottke *et al.* 2005).

There are three types of meteorites: stones, irons and stony-irons. The stone meteorites can be further divided into ordinary chondrites, carbonaceous chondrites and achondrites (e.g. Weisberg *et al.* 2006; Krot *et al.* 2007). Ordinary chondrites are the most common type of meteorite (e.g. Bischoff & Geiger 1995; Krot *et al.* 2007). However, reflectance spectroscopy of asteroids indicates that *c.* 75% of all known asteroids are the carbon-rich C-type asteroids (Gradie *et al.* 1989), which most closely resemble carbonaceous chondrites spectroscopically. S-type or silica-rich asteroids comprise less than 17% of known asteroids (Gradie *et al.* 1989), but may be the source of the ordinary chondrites.

From: PLATZ, T., MASSIRONI, M., BYRNE, P. K. & HIESINGER, H. (eds) 2015. *Volcanism and Tectonism Across the Inner Solar System.* Geological Society, London, Special Publications, **401**, 423–441. First published online July 28, 2014, http://dx.doi.org/10.1144/SP401.18

Theoretically, the density of an asteroid can be used to determine its composition. An asteroid density close to 5 g cm^{-3} should be indicative of a stony-iron composition (Britt & Consolmagno 2003), while a density close to 3.3 g cm^{-3} should be more consistent with an ordinary chondrite (Wilkison & Robinson 2000; Britt & Consolmagno 2003). However, the densities of S-type asteroids are less than 3.3 g cm^{-3} (e.g. Belton et al. 1995; Yeomans et al. 2000).

This density discrepancy may be due to the internal structure of the asteroids. Assuming that an asteroid has the same density as a corresponding meteorite, the asteroid must be a solid body. However, there are four states of asteroid structural modification (Wilkison et al. 2002): (1) completely coherent; (2) coherent but fractured; (3) heavily fractured (e.g. Chapman 1978; Davis et al. 1979); and (4) rubble pile (e.g. Hartmann 1979; Asphaug et al. 1998; Wilson et al. 1999). The density of a rubble pile would be considerably less than that of a solid body with a similar chemical composition (as low as 3.5 g cm^{-3} for a stony-iron and perhaps 2.5 g cm^{-3} for an ordinary chondrite).

The density of an asteroid can therefore provide insight into its internal structure. If the bulk density of an S-type asteroid is lower than the measured density of comparable ordinary chondrite meteorites (c. 3.3 g cm^{-3}), the asteroid probably has a high porosity inconsistent with a completely coherent asteroid (Yeomans et al. 2000). However, the presence of long structural features on the surface of an asteroid is indicative of a significant internal strength, inconsistent with a rubble pile. Since small solar system bodies (<200 km radius) do not have sufficient internal heat energy to drive terrestrial-style tectonics (Thomas & Prockter 2010), determining how these features formed yields important information about the nature and geological history of such lineated asteroids.

Several different types of linear structural features – including grooves, troughs, pit crater chains and ridges (Fig. 1) – have been observed on a number of asteroids (e.g. Veverka et al. 1994; Sullivan et al. 1996; Prockter et al. 2002). Grooves are characterized as a shallow V-shaped gash in the asteroid; they are most likely the result of simple fracturing of the surface or perhaps the surface representation of larger fractures whose distinct edges have been muted by burial under regolith and crater ejecta. Troughs are wider than grooves and have distinct walls and floors; they may be the result of reactivation of pre-existing grooves or fractures, perhaps by later impacts causing further widening of existing cracks. Pit crater chains are linear assemblages of small depressions, theorized to be grooves or troughs that were covered by regolith that may now be draining into the

underlying structure (e.g. Thomas et al. 1979; Horstman & Melosh 1989; Wyrick et al. 2004). Ridges are linear topographic highs, as determined by image illumination and/or topographic data, and are probably the surface representation of thrust faulting due to a compressive state of stress.

In addition, it has been hypothesized that some asteroids may have experienced magmatism and potentially volcanism. For example, volcanic activity has long been predicted on the asteroid 4 Vesta. Spectroscopic studies of Vesta (e.g. McCord et al. 1970; Gaffey 1997) compared with laboratory studies of the howardite eucrite–diogcnite (IIED) meteorites (Drake 1979) indicate that the HEDs are Vestan fragments (e.g. Consolmagno & Drake 1977; Takeda 1997; Drake 2001). Since the HEDs are all igneous in nature, this implies that Vesta may have undergone volcanic and/or magmatic activity at some point in its history (Keil 2002). Finding evidence for intrusive magmatism on Vesta would imply that even a small planetary body can sustain melting of sufficient volume to erupt into a pre-existing crust.

Here we present a review of the solar system asteroids that have been visited by spacecraft – 951 Gaspra, 243 Ida, 253 Mathilde, 433 Eros, 25143 Itokawa, 2867 Steins, 5535 Annefrank, 21 Lutetia and 4 Vesta – and discuss how analyses of linear structures observed on these small terrestrial bodies have implications for tectonism and magmatism. The focus is primarily on the tectonics of the asteroids, models of linear structure formation and implications for the internal structure of the asteroids. We then review Vesta in the context of a unique planetoid in our solar system, sharing characteristics of asteroid tectonics as well as planetary-scale tectonics. We discuss the evidence for magmatism on Vesta and the implications for other asteroids.

Models of tectonism on asteroids

Decades ago, linear structural features were identified on the Martian moon Phobos in Viking orbiter imagery (Fig. 2a). Thomas & Veverka (1979) suggested that the grooves on Phobos are most likely the result of the large impact that formed Stickney crater, with which the majority of the grooves are associated. They also predicted that similar lineaments would be observed on other small cratered bodies, a reasonable prediction given the natural tendency for craters to possess radial fractures (e.g. Fujiwara & Asada 1983; Reimold et al. 1998). However, the subsequent imaging of a variety of asteroids led to new models being proposed for the formation of asteroid lineaments. These formation mechanisms are described below.

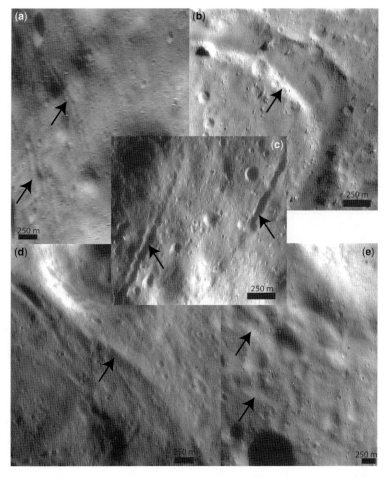

Fig. 1. Examples of different lineament morphologies on Eros, from Buczkowski *et al.* (2008). Arrows point to features. (**a**) Grooves (image 143673751, 9.48 m/pixel). (**b**) Flat-floored trough (image 134011958, 4.76 m/pixel). (**c**) Pit crater chains (image 135344864, 4.91 m/pixel). (**d**) Ridge (image 131011232, 9.91 m/pixel). (**e**) Shallow troughs (image 131034292, 9.18 m/pixel).

Formation by impact

Numerical calculations indicate that impacts into asteroids could be responsible for the formation of fractures. Axi-symmetrical calculations of an impact that would generate a Stickney-sized impact in a Phobos-like ellipsoid predict sizes of spall that compare favourably with the spacing of grooves and fractures seen on Phobos (Asphaug & Melosh 1993). In other numerical calculations where impacts into main-belt asteroid Ida (Fig. 2b) are considered, Asphaug *et al.* (1996) indicated that fractures can be generated far from the impact. Indeed, a 3D simulation of the formation of a large crater at one elongate end of Ida shows fracturing as far away as its antipode, where grooves have been observed on the asteroid (Veverka *et al.*

1994). In this study, calculations also indicate that impacts into the flat portion of an elongated ellipsoid generate circumferential fractures around the edge of the asteroid perpendicular to the impact normal; impacts on the curved ends of the asteroid result in fracturing mainly at the antipode. These calculations assume extremely simplified asteroid shapes and that the modelled 'asteroids' are physically homogeneous.

Fabric inherited from a parent body

Another hypothesis suggests that the lineated small bodies are in fact fragments of larger parent bodies, and it is on these precursor planetary bodies that the lineaments actually formed. Two large-scale

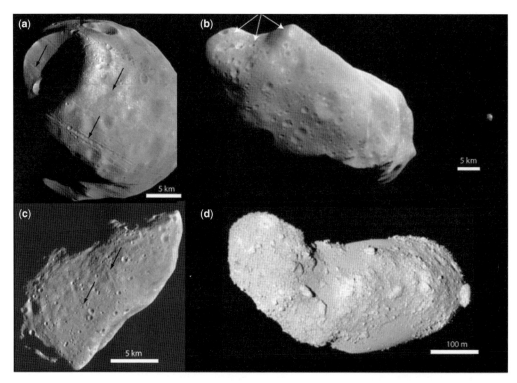

Fig. 2. (**a**) Mosaic of three images of Phobos taken by the Viking 1 orbiter at Mars on 19 October 1978. Large crater in upper left is Stickney. Arrows points to lineaments. (**b**) Mosaic of images of Ida and its satellite Dactyl, taken by the Galileo spacecraft on 28 August 1993 while on its way to Jupiter. White arrow points to Vienna Regio. (**c**) Mosaic of highest resolution (54 m/pixel) images of Gaspra taken by the Galileo spacecraft on 29 October 1991 while on its way to Jupiter. Arrows points to lineaments. (**d**) Image of Itokawa, taken by the Hayabusa spacecraft on 4 October 2005. Note that the scale is markedly different from the other asteroids. Release 051101-2 ISAS/JAXA.

lineaments on Eros – the Rahe Dorsum ridge and the shallow troughs of Calisto Fossae (Fig. 3) – were found by Thomas *et al.* (2002) to be coplanar with a large flat region (the southern 'facet') on one end of the asteroid (Fig. 3b). Thomas *et al.* (2002) determined the unit normal or pole of a plane described by a combination of Rahe Dorsum and Calisto Fossae and compared it with the pole of the plane described by the southern 'facet' of Eros. The two poles are approximately the same and so Thomas *et al.* (2002) suggested that the three features represent parallel planes indicative of a pre-existing structure throughout most of the asteroid, consistent with a fabric inherited from a parent body.

Downslope scouring

An alternate hypothesis proposed for the formation of grooves on Phobos is scouring by rolling boulders (e.g. Head & Cintala 1979; Wilson & Head 1989). Prockter *et al.* (2002) determined that downslope scouring could not be the primary cause of the globally distributed lineaments on Eros. Boulders have

been identified in association with the lineaments on Itokawa, however, and are thought to be the cause of their formation (Sasaki *et al.* 2006).

Thermal stresses

It has also been suggested that some lineaments could be the result of thermal stresses (Dombard & Freed 2002). The thermal stress model for lineament formation invokes long-term secular changes in the daily/yearly average surface temperature of a near-Earth asteroid as it first moves from the asteroid belt into the inner solar system, and then wanders around the near-Earth region (Dombard & Freed 2002). Expected expansion and subsequent contraction of the asteroid could lead to observed lineaments, the orientations of which would strongly depend on the shape of the asteroid.

Tectonic structures on asteroids

Several asteroids have been visited by spacecraft in recent years. The Galileo spacecraft flew by and

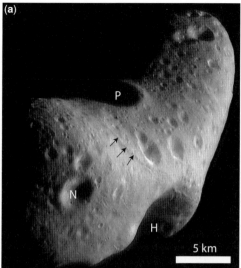

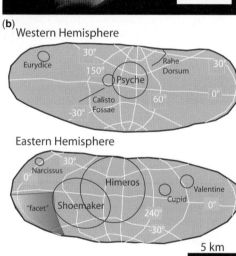

Fig. 3. (**a**) Mosaic of the northern hemisphere of Eros, taken by the NEAR-Shoemaker spacecraft (images 127275100, 127275164, 127275246, 127275310, 127275374, 127275456, 127275520). Resolution *c.* 27 m/pixel. Arrows point to the Rahe Dorsum ridge. P, Psyche crater; H, Himeros crater; N, Narcissus crater. (**b**) Cartoon showing location of certain features on Eros (after Buczkowski *et al.* 2008).

imaged main-belt asteroids 951 Gaspra in 1991 and 243 Ida in 1993, while on its way to the Jupiter system. The Near-Earth Asteroid Rendezvous (NEAR)–Shoemaker spacecraft orbited the S-type near-Earth asteroid 433 Eros for a year during 2000–2001. On its way to Eros, NEAR flew by 253 Mathilde, a C-type main-belt asteroid, in 1997 and successfully imaged one hemisphere. In 2002,

the Stardust spacecraft flew by the main-belt asteroid 5535 Annefrank. The Hayabusa spacecraft orbited the near-Earth asteroid 25143 Itokawa in 2005 before landing on the asteroid and then returning to Earth in 2010. The Rosetta probe flew by and imaged 2867 Steins in 2008 and 21 Lutetia in 2010, both main-belt asteroids. The Dawn spacecraft orbited and extensively imaged 4 Vesta in the main belt from 11 July 2011 to 5 September 2012.

Phobos has been imaged by multiple missions, including Viking, Mars Global Surveyor and Mars Reconnaissance Orbiter. As discussed in the previous section, tectonic features on Phobos have been both observed and analysed (e.g. Head & Cintala 1979; Thomas & Veverka 1979; Wilson & Head 1989; Asphaug & Melosh 1993). However, since Phobos is a moon rather than an asteroid, further discussion of Phobos is not included in this paper, although it has been hypothesized to be a captured asteroid (e.g. Singer 1968; Lambeck 1979; Cazenave *et al.* 1980).

Gaspra

The first asteroid flyby was of the S-type 951 Gaspra (Fig. 2c) on 29 October 1991 by NASA's Galileo spacecraft (Belton *et al.* 1992); about 80% of the asteroid was imaged (Veverka *et al.* 1994). Gaspra is an extremely angular and non-spherical asteroid, with an average radius of 6.1 km (Thomas *et al.* 1994). There are planar surfaces and evidence of surficial cracks, both perhaps due to impacts. Strangely, Gaspra is covered with mostly small craters (<1 km diameter); there are only five medium or large craters (1.2–2.8 km diameter), which is unexpected given the asteroid belt's high cratering rate (Belton *et al.* 1992). Either Gaspra's surface is very strong (perhaps made of metal) or very young, although it is also possible that the lack of large craters is an observational bias (Thomas & Prockter 2010).

Grooves are visible in the highest resolution (54 m/pixel) images of Gaspra (Fig. 2c). The linear depressions are commonly pitted in appearance, can extend to 2.5 km in length and can be up to 400 m wide (Veverka *et al.* 1994). The pitted nature of the grooves is theorized to indicate the presence of pre-existing fractures or troughs that were covered by a regolith of impact-pulverized rock that later drained into the underlying structure (e.g. Thomas *et al.* 1979; Horstman & Melosh 1989; Wyrick *et al.* 2004). The orientation of the rooves is consistent with the orientation of 'facets' in the asteroid's shape identified by Thomas *et al.* (1994). The combined groove data led Veverka *et al.* (1994) to suggest that Gaspra is most likely a coherent fragment of precursor body rather than a rubble pile.

Ida

The Galileo spacecraft also flew by 243 Ida (Fig. 2b) on 28 August 1993 (Belton *et al.* 1994). Ida, like Gaspra, is an S-type asteroid. Unlike Gaspra, Ida has a satellite named Dactyl. The existence of Dactyl has allowed scientists to determine Ida's mass, using Dactyl's orbital period and Kepler's Third Law of Motion. Ida's bulk density was thus constrained to between 2.0 and 3.1 g cm^{-3}, which strongly favours an ordinary chondrite composition for the asteroid with some internal void space. Colour images of Ida indicate that areas of small, deep craters are less red than the older undisturbed surfaces, and show prominent absorption bands of pyroxene and olivine. These characteristics are compatible with ordinary chondrites.

Grooves on Ida are generally continuous linear features that extend up to 4 km in length but are commonly <100 m in width (Sullivan *et al.* 1996). Morphologically, the grooves are similar to those on Gaspra and Phobos (Sullivan *et al.* 1996). There is no obvious orientation of the grooves relative to craters on the asteroid. However, in a numerical simulation where impacts into Ida were considered, Asphaug *et al.* (1996) indicated that the formation of a large impact crater (Vienna Regio) at one elongate end of Ida could cause fracturing as far away as its antipode, where grooves have been observed on the asteroid (Veverka *et al.* 1994). Sullivan *et al.* (1996) concluded that the morphology of the grooves on Ida is consistent with the surface expression of fractures in a more coherent internal body overlain by a surface regolith.

Mathilde

Although the NEAR-Shoemaker spacecraft was designed to visit Eros, it flew by 253 Mathilde (Fig. 4a) on 27 June 1997 and was able to image approximately half of the asteroid (Veverka *et al.* 1997*a*). Mathilde was found to be irregularly shaped and heavily cratered, with a dark surface that shows no albedo or colour variations (Veverka *et al.* 1997*a*). No strong evidence of a pervasive structural fabric was identified on Mathilde (Veverka *et al.* 1999), although the slightly polygonal shape of two named and several unnamed craters suggested that there might be some subsurface structural jointing (Thomas *et al.* 1999). However, only one definite linear structure was identified. A 20 km-long ridge with a step-like topography of *c.* 200 m was observed and likened to a fault scarp, as its eastern side was noted to be higher than the western side (Thomas *et al.* 1999).

Eros

Because of its repeated close encounters with Earth, there has been more than a century of ground-based study of 433 Eros (Fig. 3), the first near-Earth asteroid to be discovered. The NEAR-Shoemaker spacecraft orbited 433 Eros from 2000–2001. The

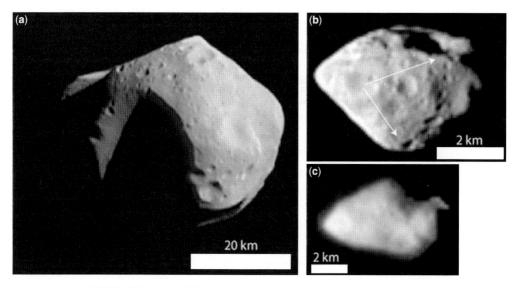

Fig. 4. (**a**) Asteroid 253 Mathilde seen by NEAR Shoemaker from a distance of 2400 km (380 m/pixel). (**b**) Asteroid 2867 Steins seen from a distance of 800 km (80 m/pixel), taken by the OSIRIS imaging system of the Rosetta space probe in 2008. White arrows point to a putative pit crater chain (Keller *et al.* 2010; Besse *et al.* 2012). (**c**) Asteroid 5535 Annefrank seen by the Stardust probe from a distance of 3300 km.

NEAR Multi-Spectral Imager (MSI) (Veverka *et al.* 1997*b*) collected tens of thousands of high-resolution (*c.* 3–10 m/pixel) images (Murchie *et al.* 2002; Prockter *et al.* 2002) and, as a result, Eros is one of the most thoroughly studied asteroids in the solar system.

This global high-resolution imagery allowed for a far more comprehensive study of linear structures than had been possible on any previously visited asteroid. Buczkowski *et al.* (2008) completed a global database of all of the linear structures on 433 Eros, to better determine potential formation mechanisms. Using images and laser-ranging data from the NEAR-Shoemaker mission, they identified several sets of linear structures whose formation is clearly related to impact craters. However, although there are some regions of Eros where there are hardly any impact craters visible (Thomas & Robinson 2005), lineaments at a number of different scales and orientations were found to be ubiquitous on the surface of the asteroid (Buczkowski *et al.* 2008). The fact that not all of the linear structures are associated with impact craters suggests that different parts of the asteroid may have undergone different stress histories; some of these sets infer internal structure, at least on a local level (Buczkowski *et al.* 2008). This internal structure may derive from Eros' parent body and suggests that, although largely coherent, Eros' interior may have portions that have not undergone a common history (Buczkowski *et al.* 2008).

Lineament types present on Eros include shallow grooves, flat-floored troughs, pit crater chains and ridges (Fig. 1). Whether the linear structures observed on Eros are due to impact events or are the surface expression of a pre-existing internal structural fabric, the presence of large-scale structures on Eros seems to indicate that the asteroid is, for the most part, a coherent body. The density of Eros is, however, 2.6 g cm^{-3} (Yeomans *et al.* 2000), suggesting substantial porosity (Wilkison *et al.* 2002).

Buczkowski *et al.* (2008) classified the mapped lineaments into sets on the basis of location and orientation. Some lineament sets are clearly related to specific impact craters. Other lineament sets have no obvious relationship to impact events but are instead found globally and may be related to interior structures. It is not obvious on a non-spherical body whether lineaments are associated with each other in a systematic way indicative of internal structure (Buczkowski *et al.* 2008). However, because the Eros lineaments were mapped directly on to the Eros shape model, they have a 3D component and can be fit to planes that cut through the asteroid. The unit normal of these planes yields a pole that can be compared with the poles of the entire lineament dataset. Buczkowski *et al.* (2008)

compared the pattern of these lineament sets with impact crater location and to models of: (1) interior configuration and structure; and (2) cratering mechanics.

Linear structures were identified radial to 13 craters on Eros (Buczkowski *et al.* 2008). Although none of these craters has lineaments arranged 360° around their centres, each has lineaments radiating for at least 45° of arc (Buczkowski *et al.* 2008, fig. 6). Given their orientations relative to these craters, it seems likely that these lineaments were formed as a direct result of an impact event. There is no obvious correlation between crater diameter or location and the occurrence of radial lineaments on Eros. Most of the craters with radial lineaments are among the largest of all the Eros impact craters, but there are multiple craters in the largest 30% that do not have radial lineaments, whereas several of the smallest craters do (Buczkowski *et al.* 2008). Eight of the 13 craters with associated radial lineaments are located at the ends of the elongate asteroid (Buczkowski *et al.* 2008). There is some correlation between the presence of radial lineaments and the volume of large ejecta blocks per unit area, a proxy for the thickness of the low-velocity ejecta from the Shoemaker impact (IAU designation: Charlois Regio) (Thomas *et al.* 2001). A majority of the craters with radial lineaments occur in regions of thin Shoemaker ejecta, implying that other Eros impact craters could have formed radial lineament sets that are simply buried. However, some of the radial lineaments occur in areas of high ejecta-block volume; this observation may therefore indicate that these particular impacts are younger than Shoemaker (Buczkowski *et al.* 2008).

Buczkowski *et al.* (2008) found that the primary set of lineaments on Eros is consistent with the pattern expected from fragmentation due to impact on the long side of an ellipsoid target (Asphaug *et al.* 1996). Buczkowski *et al.* (2008) inferred that these structures were formed as a result of the Psyche and/or Himeros impacts (Fig. 3b), although they did not discount the possibility that the Shoemaker impact played a role in their formation. Several of the structures in the set are extremely long, up to tens of kilometres in length (Buczkowski *et al.* 2008, fig. 7). It was suggested that the 24 troughs in the set could be the result of reactivation by a second impact, which caused further widening of existing cracks. The five pit chains in the set were thought to indicate that some of the grooves or troughs were covered by regolith that may now be draining into the underlying structure (e.g. Thomas *et al.* 1979; Horstman & Melosh 1989; Wyrick *et al.* 2004). When compared with maps of Shoemaker ejecta volume (Thomas & Robinson 2005, fig. 1d), the pit chains are only located in regions of moderate ejecta coverage (green regions); there are no pit chains in regions of greatest coverage or in regions of no ejecta coverage (Buczkowski *et al.* 2008).

Another set of lineaments, identified as 'set 2' by Buczkowski *et al.* (2008, fig. 8), was found encircling Eros in the restricted longitude range of *c.* 170–240° (Fig. 3b). The pit chains in this set also correspond to a region of moderate Shoemaker ejecta volume (Thomas & Robinson 2005). A planar analysis of these structures did not show any correlation to predictions of models of lineament formation by impact, and Buczkowski *et al.* (2008) suspected that these lineaments represent a pre-existing internal structure.

Thermal stresses from the expected expansion and subsequent contraction of the asteroid could lead to observed features whose orientations strongly depend on the shape of Eros, but in general are predicted to trend east–west at the poles and north–south at the lower latitudes (Dombard & Freed 2002). Some lineaments on Eros are observed to be consistent with formation by thermal stresses (Buczkowski *et al.* 2008), but an in-depth investigation has yet to be performed. Downslope scouring is not thought to be a primary cause of the globally distributed lineaments on Eros (Buczkowski *et al.* 2008). Prockter *et al.* (2002) found only a few grooves on Eros that are associated with boulders, mostly in the interior of craters.

In addition to the observation and analysis of linear structures on Eros (e.g. Prockter *et al.* 2002; Buczkowski *et al.* 2008), an evaluation of certain craters supports the identification of a structural fabric within the asteroid (Thomas *et al.* 2002). Some craters on Eros, such as Valentine and Tutanekai, have orthogonal shapes that suggest an underlying structural control on their formation (Prockter *et al.* 2002; Robinson *et al.* 2002). However, it has not yet been determined if this pre-existing structural fabric is a remnant of a putative Eros parent body or is due to fragmentation of Eros during an earlier impact event.

Itokawa

From Earth-based observations, 25143 Itokawa (Fig. 2d) is interpreted to be an S-type asteroid in composition (Binzel *et al.* 2001). Radar imaging by Goldstone revealed a somewhat elongated shape and a 12.5 h rotation period (Ostro *et al.* 2004). The Hayabusa spacecraft, which went into orbit around Itokawa on 12 September 2005, confirmed these findings. Results from that mission also indicate that Itokawa is both a rubble pile asteroid and a contact binary, formed by two or more small asteroids that gravitated towards each other and stuck together (Fujiwara *et al.* 2006). The asteroid was

found to have a bulk density of 1.9 g cm^{-3} and a bulk porosity of 40% (Abe *et al.* 2006).

There is a surprising lack of impact craters on Itokawa, which has a very rough surface studded with boulders (Fujiwara *et al.* 2006). Although the density of Ida could indicate that it is not a completely coherent asteroid, Itokawa is the first visited asteroid to actually look like a rubble pile. Despite its rubble pile appearance, linear structures have been identified on the asteroid. These lineaments have, however, been interpreted to be the result of boulder movement on its surface (Sasaki *et al.* 2006), and their presence does not require a coherent interior.

Steins

While the Rosetta mission was launched by the European Space Agency (ESA) to rendezvous with the comet 67P/Churyumov–Gerasimenko, it achieved a close (803 km) flyby of the main-belt asteroid 2867 Steins (Fig. 4b) on 5 September 2008 (Keller *et al.* 2010). The surface of Steins is mostly covered by shallow craters, but a few linear structures were observed. A series of craters extending from a 2.1 km-diameter crater (Keller *et al.* 2010, fig. 1) was interpreted as a pit crater chain (Keller *et al.* 2010; Besse *et al.* 2012) and, on the opposite side of the asteroid, a linear depression was identified (Keller *et al.* 2010). It was suggested that these features together indicate a partial drainage of loose surface material into a stronger, deeper material (Keller *et al.* 2010; Besse *et al.* 2012), and that this may mark pre-existing physical inhomogeneities in the asteroid (Keller *et al.* 2010).

Lutetia

After its flyby of Steins, the Rosetta spacecraft flew by the main-belt asteroid 21 Lutetia (Fig. 5) on 10 July 2010. The surface of Lutetia is highly diverse with craters, lineaments, landslides, scarps and boulders, indicating a complex history (Thomas *et al.* 2012). There is a high density of linear structures on Lutetia, most of which show region-dependent organization and structure (Thomas *et al.* 2012). Several categories of lineaments have been described, including irregular troughs, larger faults, pit chains, intra-crater trenches, ridges and scarps (Thomas *et al.* 2012).

Approximately 50% of Lutetia was imaged by Rosetta, and this observed half was divided into eight geological regions (Sierks *et al.* 2011a; Massironi *et al.* 2012): Achaia, Noricum, Narbonesis, Etruria, Baetica, Pannonia, Raetia and Goldsmith (Fig. 5b). Three of these regions (Pannonia, Raetia and Goldsmith) did not have illumination

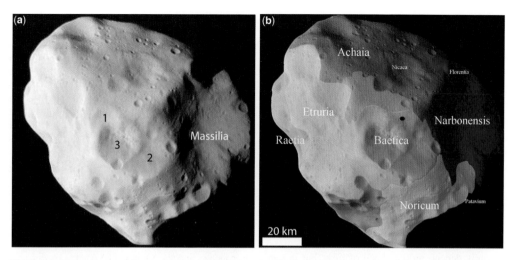

Fig. 5. (**a**) Rosetta image of the north pole of 21 Lutetia, taken at closest approach on 11 July 2010. The North Pole Crater Cluster (NPCC) is centred in the image, and the largest component craters are labelled 1, 2 and 3 (Massironi *et al.* 2012). The Massilia depression is also labelled. (**b**) Map of the eight Lutetia regions, from Thomas *et al.* (2012, fig. 2).

conditions favourable for geological mapping, and are recognized as regions because of only a single morphological boundary (Massironi *et al.* 2012). The Etruria region is distinguished by its clear morphological boundaries with Raetia, Noricum, Achaia and Baetica. The Noricum and Achaia regions are the most heavily cratered (and thus represent some of the oldest surfaces on Lutetia) and are divided by the Narbonesis region, which is defined by the *c.* 55 km-wide Massilia depression (Massironi *et al.* 2012). Meanwhile, the north polar Baetica region is dominated by the North Polar Crater Cluster (NPCC) and its associated ejecta (Sierks *et al.* 2011*a*; Massironi *et al.* 2012; Thomas *et al.* 2012). The NPCC comprises at least three craters: a 21 km-diameter crater (NPCC-3; Gades) superimposed on a 14 km-diameter crater (NPCC-1; Hispalis), both of which are in turn superimposed on a 34 km-diameter crater (NPCC-2; Corduba) (Massironi *et al.* 2012; Thomas *et al.* 2012).

Many of the linear structures on Lutetia (in the Noricum, Eturia and Achaia regions) are concentric to the NPCC. Although no obvious relationship with any particular NPCC crater was observed by Thomas *et al.* (2012), the concentric orientation of the structures around the NPCC as a whole does suggest that one or more of its component craters was responsible for their formation. However, structures in the Narbonensis region have a completely different orientation, suggesting that either: (1) these particular features were formed by a different impact; (2) there was a different *in situ* stress environment in the Narbonensis region (and thus different lineament orientations developed) during

the impact-generated seismicity that formed the other lineaments (Thomas *et al.* 2012); or (3) the lineaments represent structures inherited from a parent body. In addition, the presence of pit crater chains in Narbonensis was thought to indicate that this region has been heavily resurfaced since its formation, more so than the rest of Lutetia, in which pit crater chains were scarce to nonexistent (Thomas *et al.* 2012). No linear structures on Lutetia were observed in an obvious radial orientation to any impact crater (Thomas *et al.* 2012).

In addition to these observed lineaments, Massironi *et al.* (2012) suggested that there may also be a pre-existing planar structure within Lutetia such as that suggested for Eros (Thomas *et al.* 2002). They determined that the shape of individual craters within the NPCC, specifically the asymmetrical topographic profiles of NPCC2 and NPCC3, indicates that there was a structural control on the formation of these craters. Similarly, Thomas *et al.* (2012) suggested that if the Massilia depression did indeed form due to an impact event, its irregular shape may have been controlled by a pre-existing structural fabric. This is not dissimilar to what has been observed on Eros (Prockter *et al.* 2002; Robinson *et al.* 2002).

Vesta

As the Dawn spacecraft (Russell & Raymond 2011) approached Vesta in July of 2011, large-scale linear troughs encircling the asteroid became immediately evident in Framing Camera (FC) (Sierks *et al.* 2011*b*) images (Fig. 6). Buczkowski *et al.* (2012)

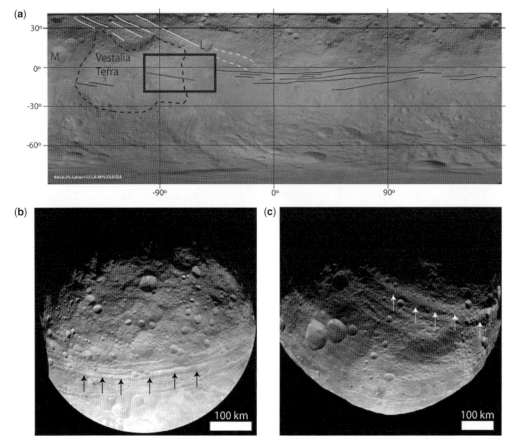

Fig. 6. (a) Simple cylindrical global mosaic image of Vesta, showing location of Vestalia Terra (dashed line), the equatorial troughs (black lines), the northern troughs (white lines) and Figure 8 (black box). Marcia and Lepida craters are labelled M and L, respectively. (b) Framing Camera clear-filter image of Vesta from Dawn's late approach phase (RC3), showing equatorial troughs. Black arrows point to Divalia Fossa A. (c) Clear-filter FC image of Vesta showing northern troughs. White arrows point to Saturnalia Fossa A. Adapted from Buczkowski *et al.* (2012).

evaluated the morphology of the Vestan structures and determined that their dimensions and shape suggest that they are graben, more similar to those observed on terrestrial planets as opposed to the fractures or grooves found on smaller asteroids.

The most prominent set of large linear structures encircle Vesta, roughly aligned with the equator (Fig. 6b). Although most of these structures are wide troughs bounded by steep scarps, extending in a similar orientation are muted troughs, grooves and pit crater chains. A fault plane analysis (such as that performed on Eros by Buczkowski *et al.* 2008) suggests that the formation of the equatorial troughs was triggered by the impact event that formed the Rheasilvia basin (Jaumann *et al.* 2012). Divalia Fossa A (Fig. 6), the largest of the equatorial troughs, is *c.* 465 km in length and has a width within the range 14.5–21.8 km. Its morphology

varies along strike; the southern scarp is clearly higher and steeper than the northern scarp along much of the trough, but there are locations along strike where Divalia Fossa A resembles a classic flat-floored graben. The large equatorial troughs encircle *c.* 60% of Vesta, but do not cut the topographically high plateau known as Vestalia Terra (Fig. 6a) (Buczkowski *et al.* 2012).

A second set of large-scale linear structures extend to the NW from the equatorial troughs (Buczkowski *et al.* 2012). The primary structure in this group is a trough named Saturnalia Fossa A (Fig. 6c), which is up to 39.2 km wide and extends north for 366 km into Vesta's north polar region. Saturnalia Fossa A has generally shallower scarps than Divalia, with rounded edges and infilling on the trough floor; both the trough scarps and floor appear to have been more heavily cratered than the equatorial troughs. Together, these observations suggest that Saturnalia is older than Divalia (Buczkowski *et al.* 2012) and the other equatorial structures, an inference that is consistent with the fault plane analysis that ties its formation to the Veneneia basin (Jaumann *et al.* 2012).

Displacement profiles across Divalia and Saturnalia Fossae indicate that these graben are not expressed as a single long continuous fault (Buczkowski *et al.* 2012). The southern Divalia fault effectively tips to zero displacement along its length and then increases again, suggesting that the southern Divalia fault comprises at least two large, linked faults (Buczkowski *et al.* 2012). It is also possible that a drop in displacement along the northern Divalia fault represents the linkage site of two faults that comprise the northern bounding fault (Buczkowski *et al.* 2012). Similarly, a small downturn in displacement on the Saturnalia southern fault might suggest that two faults link at this location, with strain being instead primarily accommodated on the northern fault here (Buczkowski *et al.* 2012). Meanwhile, the large vertical displacement on the graben south of Lepida crater suggests that these faults were reactivated, perhaps by the event that formed the equatorial troughs (Buczkowski *et al.* 2012).

As mentioned above, linear structures have been identified on several asteroids in a concentric orientation around impact craters, including Ida, Eros and Lutetia. Although Vesta's equatorial and northern troughs are in a similar orientation around a basin (Jaumann *et al.* 2012), their geomorphology and displacement profiles are far more similar to planetary faults than to the features on Ida, Eros and Lutetia. Although the fault plane analysis (Jaumann *et al.* 2012) indicates that impact may have been responsible for triggering the formation of these features as on other asteroids, the morphological and structural differences in the resulting features implies that there must be inherent differences between Vesta and the other asteroids.

Buczkowski *et al.* (2012) discussed how this difference could be due to asteroid composition. Vesta is composed of a range of rock types (e.g. De Sanctis *et al.* 2012; Reddy *et al.* 2012) while Ida (Sullivan *et al.* 1996) and Eros (Trombka *et al.* 2000) are both compositionally homogeneous. In contrast to other asteroids, Buczkowski *et al.* (2012) suggested that Vesta may behave more akin to planetary bodies that have mechanical stratigraphy (i.e. alternating layers of mechanically strong and weak rock), allowing for larger-scale features such as graben to form. Vesta is now known to be a differentiated body with a mantle and core (Russell *et al.* 2012), and Buczkowski *et al.* (2012) suggest that it is not unlikely that the strain rate due to a giant impact on Vesta would be high enough to

result in the instantaneous release of strain in its ductile interior, with concurrent large-scale brittle strain on its surface. Many researchers have, however, run numerical models that show that a differentiated interior could also be responsible for the formation of graben on Vesta by amplifying and reorienting the stresses resultant from impacts (e.g. Buczkowski *et al.* 2012; Bowling *et al.* 2013; Ivanov & Melosh 2013; Stickle *et al.* 2013), as compared with impacts on an undifferentiated body (Buczkowski *et al.* 2012). Some of these models also show that impact stresses unaffected by a core would more likely yield the fractures and smaller flat-floored trough morphologies observed on undifferentiated asteroids such Eros and Ida (Buczkowski *et al.* 2012). Although Lutetia is partially differentiated (Sierks *et al.* 2011*a*), it does not display features on the scale of Vesta's troughs; this is possibly because it does not have a fully differentiated core and therefore has lower density contrasts throughout its interior.

Annefrank

On its way to the comet 81P/Wild 2, the Stardust spacecraft flew by the main-belt asteroid 5535 Annefrank (Fig. 4c). Less than 40% of the asteroid was imaged at resolutions of 185–300 m/pixel (Duxbury *et al.* 2004). To date, Annefrank is the only asteroid on which no linear structures have been observed. A dark line running approximately north–south on Annefrank was suggested to be either an albedo difference or a surface slope discontinuity at the contact between the prism-shaped main body and some smaller rounded bodies attached to its base (Duxbury *et al.* 2004). However, the lack of observed linear structures may simply be a function of the asteroid not being imaged at a sufficiently high resolution to observe such features.

Models of volcanic activity on asteroids

Of the asteroids that have been visited to date, only one was expected to show signs of volcanism. It has long been suspected that Vesta may have undergone volcanic and/or magmatic activity at some point in its history (e.g. Wilson & Keil 1996; Keil 2002); spectroscopic studies of Vesta (e.g. McCord *et al.* 1970; Gaffey 1997) show that it has similar spectral signatures to the howardite–eucrite–diogenite (HED) meteorites (e.g. Consolmagno & Drake 1977; Drake 1979, 2001; Takeda 1997). This similarity indicates that the HEDs may be Vestan fragments (e.g. Binzel & Xu 1993; Drake 2001) and, since the HEDs are all igneous in nature, this in turn suggests that Vesta might have experienced volcanism. For example, eucrites are basaltic rocks composed of calcium-poor pyroxene, pigeonite and calcium-rich plagioclase (Takeda 1997; Keil 2002). Diogenites are also igneous rocks, dominated by magnesium-rich orthopyroxene with small amounts of plagioclase and olivine (Takeda 1997; Keil 2002; Beck & McSween 2010); the crystal size is generally larger than in basaltic eucrites, suggesting that diogenites experienced slower cooling consistent with a plutonic origin. Howardites are impact breccias comprising fragments of eucrites and diogenites (Takeda 1997; Keil 2002).

Accordingly, there had been hypotheses of igneous intrusion on Vesta prior to Dawn's arrival at the asteroid. Wilson & Keil (1996) suggested that igneous intrusions in the form of dykes could occur on Vesta, based on mathematical and petrological modelling. Their models indicated that both shallow and deep dykes were possible, with volumes of 3–10 000 km^3. Specifically, they predicted shallow dykes with widths of c. 1 m and vertical extents of <10 km and/or deep dykes 3 m thick and 30 km in lateral extent. Barrat et $al.$ (2010) performed a detailed study of the trace-element chemistry of diogenites. Results suggested that their petrogenesis is more complex than that of simple early crystallization products, and Barrat et $al.$ (2010) proposed that they possibly formed as later-stage plutons injected into the eucritic Vestan crust. Wilson & Keil (2012) suggested that there may be many sill-like intrusions at the base of Vesta's lithosphere.

The search for volcanic and magmatic features and the description of their spatial distribution was therefore a primary focus of Dawn's mission at Vesta. However, a comprehensive evaluation of lobate flows on Vesta yielded no unequivocal morphological evidence of ancient volcanic activity (Williams et $al.$ 2013). Instead, Williams et $al.$ (2013) found that all lobate flow materials were associated with impact craters or steep slopes, indicating that their formation was either due to impact or erosional mass-wasting processes.

Magmatic activity on Vesta

Some indication of magmatic processes has been identified on Vesta. Raymond et $al.$ (2013) identified a high Bouger gravity anomaly associated with the topographically elevated region named Vestalia Terra (Fig. 6a), indicating that the plateau is composed of a denser material than the rest of Vesta's crust. They suggested that the density difference is due to an ancient mantle plume underlying the plateau. Another similar possibility is that Vestalia Terra is the location of one of the sill-like intrusions theorized to be at the base of Vesta's lithosphere (Wilson & Keil 2012). Vestalia Terra is therefore a logical region to continue the search for signs of past magmatic activity on Vesta.

The identification of Brumalia Tholus

The equatorial region of Vesta displays numerous wide, flat-floored troughs whose formation has been tied to the Rheasilvia impact event (Jaumann et $al.$ 2012); these troughs do not cut Vestalia Terra however (Fig. 6a) (Buczkowski et $al.$ 2012), supporting the conclusion that the plateau is composed of stronger and denser material than the surrounding region. Nevertheless, there are three long pit crater chains observed on the surface of the plateau (Fig. 6a, dotted lines) (Buczkowski et $al.$ 2012, 2014). Pit crater chains are hypothesized to form when dilational motion on buried normal faults causes overlying material to collapse into the opening portions of the buried fault, and a strong correlation between pit crater chains and fault-bounded graben has been observed on other

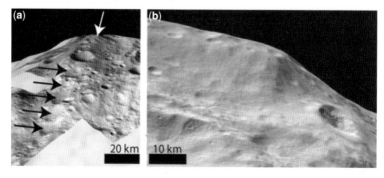

Fig. 7. (a) Framing Camera LAMO mosaic of Albalonga Catena draped over the Vesta shape model. Image looks west over the pit crater chain. Merged pits (black arrows) transition into Brumalia Tholus (white arrow). (b) LAMO image draped over Vesta topography looks south towards the northern face of Brumalia Tholus. High-albedo deposits are ejecta from Teia crater.

planetary bodies (Wyrick *et al.* 2004). Consistent with this hypothesis, the merged pits of the Vestalia Terra pit crater chains show signs of collapse, but distinct fault faces can also be observed (Buczkowski *et al.* 2014). It has therefore been suggested that the pit crater chains on Vestalia Terra are representative of subsurface faulting of the plateau (Buczkowski *et al.* 2012).

As the pit crater chain Albalonga Catena progresses westwards, it transitions from being a topographically low feature consisting of merged pits (Fig. 7a, black arrows) into the topographically high Brumalia Tholus (Figs 7 & 8). Brumalia Tholus is an elongate hill that is evident in both photographic and topographic data of Vesta (Fig. 8a, b). Westwards of the hill, merged pits are again visible in the slope data (Fig. 8c). If Albalonga Catena does represent a buried normal fault, as the work of Wyrick *et al.* (2004) suggests, then the topographic high that emerges along its length would most likely have been formed as some form of magmatic intrusion. Under this scenario, molten material utilized the pre-existing subsurface fault as a conduit to travel towards the surface, intruding into and deforming the rock above it. If this hypothesis is correct, then the core of Brumalia Tholus is composed of a more plutonic rock than the basaltic eucrites and brecciated howardites that have been observed on the majority of the equatorial region of Vesta's surface (De Sanctis *et al.* 2012).

Brumalia Tholus is 36 km wide and 68 km long; this is considerably larger than any dyke predicted by the Wilson & Keil (1996) model. Also, topographic profiles indicate that Brumalia Tholus is dome-shaped along both its long and short axes. Together with its size, this morphology suggests that Brumalia Tholus could be the surface representation of a laccolith forming over a dyke. Laccoliths develop when a magmatic intrusion does not reach the surface but magma pressure is high enough to dome the overlying material. Since cooling below the surface takes places slowly, larger crystals such as those observed in diogenites (or cumulate eucrites) have time to form.

Teia crater impacts the northern face of Brumalia Tholus (Fig. 7b), and its ejecta likely provide samples of Brumalia's core material (Buczkowski *et al.* 2014). Framing Camera (FC) colour data indicate that ejecta from Teia have a distinct composition (Fig. 8d). The false-colour orange and reds correspond to the ratio of 749/438 nm, and are observed to be directly related to ejecta material with a smeared and flow-like texture in the highest resolution (20 m/pixel) FC images from Dawn's Low Altitude Mapping Orbit (LAMO) (Buczkowski *et al.* 2014). Spectral analysis by the VIR instrument has shown that while the background Vestalia Terra material is howarditic (De Sanctis *et al.* 2012),

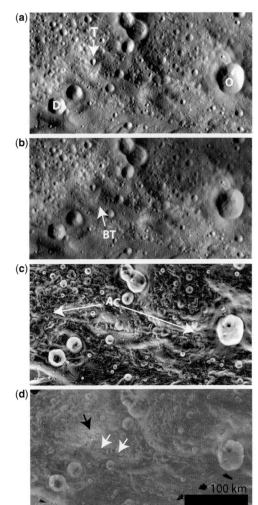

Fig. 8. Region of eastern Vestalia Terra at 22.5°S–10.4°N, 254–313.5°E, adapted from Buczkowski *et al.* (2014). (**a**) Framing camera LAMO mosaic of region. O, Oppia crater; D, Drusilla crater; T, Teia crater. (**b**) Topographic data overlying LAMO mosaic. BT, Brumalia Tholus. (**c**) Slope map of region. White arrows point to Albalonga Catena (AC), which is distinctly visible in this dataset. (**d**) FC colour data of region. Black arrow points to Teia ejecta; white arrows point to false-red deposits on the top of Brumalia Tholus. Red = 749/438 nm; blue = 749/917 nm; green = 438/749 nm.

consistent with the plateau being covered with ejecta from the large impacts surrounding it, Teia's ejecta are more diogenitic in composition (De Sanctis *et al.* 2014). The identification of diogenite in the ejecta of Teia crater is consistent with the hill being the surface representation of a magmatic intrusion.

The formation of Brumalia Tholus

The source of the molten material that putatively formed Brumalia Tholus requires further consideration. Simple models of the thermal evolution of asteroidal bodies, and comparison with HED meteorites, suggest volcanism on Vesta ceased by 10–100 Ma after formation, that is, >4.4 Ga (e.g. Schiller *et al.* 2010; McSween *et al.* 2011). Albalonga Catena is roughly aligned with the Divalia Fossae (Buczkowski *et al.* 2012, 2014), suggesting that its underlying fault may also have formed during the Rheasilvia impact event. However, Rheasilvia occurred either 1 Ga (Schenk *et al.* 2012) or 3.6 Ga (Schmedemann *et al.* 2014) ago, and therefore volcanism should have ended long before the Rheasilvia impact.

It is possible that the Albalonga fault predates the end of Vestan volcanism, and thus the Rheasilvia impact (Buczkowski *et al.* 2014). For example, it could be the result of an earlier impact such as that which formed the basin at the northern boundary of Vestalia Terra (Buczkowski *et al.* 2014). The apparent alignment of Albalonga with the equatorial troughs could then be due to small reorientation of the underlying fault after the Rheasilvia impact, similar to that proposed for the southernmost Saturnalia Fossae (Buczkowski *et al.* 2012). In such a case, the magmatic material injected into the Albalonga fault could have been sourced by: (1) the magma plume theorized by Raymond *et al.* (2013); (2) the sill-like intrusions hypothesized by Wilson & Keil (2012); or (3) from one of the diogenitic intrusions into the eucritic crust predicted by Barrat *et al.* (2010).

It was also suggested that the putative molten material might be diogenitic impact melt associated with Rheasilvia's formation (Buczkowski *et al.* 2014), similar to impact melt dykes identified around the Sudbury (Riller 2005) and Vredfort

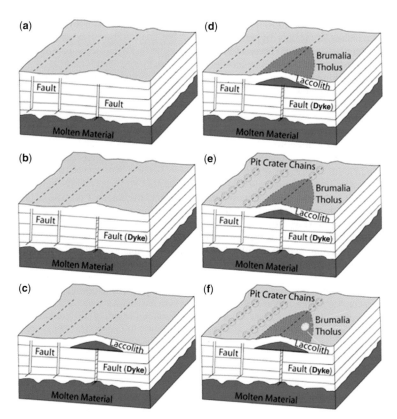

Fig. 9. Schematic diagram representing the three Vestalia Terra faults, the pit crater chains forming over them and the creation of Brumalia Tholus due to laccolith formation (from Buczkowski *et al.* 2014). Although the three faults are orientated in a way that suggests a common formation mechanism, only one extends to sufficient depth to sample sub-plateau partial melt (red). This fault serves as a conduit for the molten material to rise towards the surface (red hatch marks), deforming the surface material above it and creating Brumalia Tholus.

(Reimold & Gibson 2006) impact basins on Earth. In this hypothesis, Rheasilvia impact melt drained through fractures in the basin floor and was injected into the Albalonga fault, forming Brumalia Tholus (Buczkowski *et al.* 2014). However, modelling indicates that melt production on asteroids would be limited due to the lower impact velocities of main-belt impactors (Keil *et al.* 1997); to date, only very little such melt has been tentatively identified (Schenk *et al.* 2012; McSween *et al.* 2013; Williams *et al.* 2013). Impact melt is, therefore, a less likely source of the molten material (Buczkowski *et al.* 2014).

Whatever the magma source, Buczkowski *et al.* (2014) determined that the Albalonga fault most likely formed prior to magma injection and not as a result of the magma intrusion itself. Although dyke injection has been associated with graben formation on planetary bodies (e.g. Ernst *et al.* 2001; Wilson & Head 2002), in the case of the Vestalia Terra only the Albalonga fault/catena is associated with a magmatic intrusion. The similar orientation of the three catenae relative to each other suggests that they share a common formation mechanism (Buczkowski *et al.* 2014), strongly implying that the faults predate intrusive activity. The putatively greater depth of the Albalonga fault likely allowed it to sample the deep molten material, whereas the other faults were not deep enough to do the same (Buczkowski *et al.* 2014).

Buczkowski *et al.* (2014) suggest that the following sequence of events may have occurred on Vesta. Ancient fracturing and faulting occurred in the Vestalia Terra subsurface, forming the Albalonga, Robigalia and unnamed subsurface faults in the same orientation (Fig. 9a). The Albalonga fault then sampled a region of partial melt and served as a conduit for this mantle or lower crustal material to move upwards (Fig. 9b) and deform the surface (Fig. 9c). Brumalia Tholus then formed due to magmatic injection and laccolith doming (Fig. 9d). The molten material cooled slowly at depth, forming diogenite. Some time later the Rheasilvia impact occurred, reactivating (and perhaps reorienting) the Albalonga and other Vestalia Terra faults. The surface of Vestalia Terra was later covered by loose regolith material (i.e. ejecta), which collapsed into dilational openings along the same steep subsurface faults to form the pit crater chains (Fig. 9e). Finally, the Teia impact event occurred and incorporated the diogenitic Brumalia core material into its ejecta (Fig. 9f).

Conclusions

Asteroid lineaments observed on those small bodies that have been visited by spacecraft appear to have several different origins and are indicative of variable interior structures. Many of the linear structures, such as those on Ida, Eros, Lutetia and Vesta, appear to be due to impact, but some lineaments have no obvious relationship to impact craters. For example, pitted grooves on Gaspra are indicative of a fabric in a coherent asteroid inherited from a parent body, as are some of the linear structures on Eros. These results are consistent with previous suggestions that Gaspra and Eros are fragments of larger parent bodies. Pervasive subsurface fracturing can also be distinguished by the polygonal shapes of some craters on Mathilde, Eros and Lutetia. The presence of long structural features on the surfaces of some asteroids is indicative of substantial internal strength, despite low-density values that indicate high porosity. Meanwhile, lineaments on Itokawa have been associated with boulders and are consistent with the excavation of regolith by boulder movement on a 'rubble pile' asteroid. It is therefore clear that determining how linear features formed on these asteroids yields important information about their internal structure and strength, as well as on the nature and history of the asteroid itself.

Vesta presents an intermediate style of tectonic deformation, with fractures and grooves similar to those observed on other asteroids as well as large-scale graben and trough structures more characteristic of tectonics on a terrestrial planet. Being a differentiated proto-planet, Vesta is a unique body with which to study the roles played by internal rheologies and structures on the surface expressions of tectonism. Unlike many terrestrial planets, Vesta's main stressors have been primarily exogenic (i.e. impacts) rather than internally driven; however, the inferred formation path of Brumalia Tholus suggests that Vesta's geological history may have included endogenic magmatism.

As a group, asteroids represent some of the earliest remnants of the early solar system. Deciphering the tectonic histories of these bodies provides insight into the complex dynamic and geological history of the inner solar system. Although currently limited by available observations, our understanding of asteroid composition and structure has grown exponentially in the last few decades, leading to improved recognition and classification of asteroid characteristics based on strength and cohesion (i.e. solid bodies v. rubble piles). Impact processes dominate the tectonic styles observed on many asteroids, but there is also evidence for structure inherited from parent bodies. Once more, Vesta represents a transitional form of tectonics that reflects its internal differentiation and impact history.

The authors would like to thank M. Massironi, S. Besse, P. Byrne and T. Platz for their reviews and comments.

The writing of this review was funded in part by the Dawn at Vesta Participating Science Program, grant number NNX10AR58G S06.

References

ABE, S., MUKAI, T. ET AL. 2006. Density, porosity and interior structure of asteroid Itokawa by the Hayabusa mission. Abstract A-00493, presented at the European Planetary Science Congress, Berlin, Germany.

ASPHAUG, E. & MELOSH, H. J. 1993. The Stickney impact of Phobos: a dynamical model. *Icarus*, **101**, 144–164, http://dx.doi.org/10.1006/icar.1993.1012

ASPHAUG, E., MOORE, J. M., MORRISON, D., BENZ, W., NOLAN, M. C. & SULLIVAN, R. J. 1996. Mechanical and geological effects of impact cratering on Ida. *Icarus*, **120**, 158–184, http://dx.doi.org/10.1006/icar.1996.0043

ASPHAUG, E., OSTRO, S. J., HUDSON, R. S., SCHEERES, D. J. & BENZ, W. 1998. Disruption of kilometer-sized asteroids by energetic collisions. *Nature*, **393**, 437–440.

BARRAT, J. A., YAMAGUCHI, A., ZANDA, B., BOLLINGER, C. & BOHN, M. 2010. Relative chronology of crust formation on asteroid Vesta: insights from the geochemistry of diogenites. *Geochimica et Cosmochimica Acta*, **74**, 6218–6231.

BECK, A. W. & MCSWEEN, H. Y. 2010. Diogenites as polymict breccias composed of orthopyroxenite and harzburgite. *Meteoritics & Planetary Science*, **45**, 850–872, http://dx.doi.org/10.1111/j.1945-5100.2010.01061.x

BELTON, M. J., VEVERKA, S. J. ET AL. 1992. Galileo encounter with 951 Gaspra: first pictures of an asteroid. *Science*, **257**, 1647–1652, http://dx.doi.org/10.1126/science.257.5077.1647

BELTON, M. J. S., CHAPMAN, C. R. ET AL. 1994. First images of 243 Ida. *Science*, **265**, 1543–1547, http://dx.doi.org/10.1126/science.265.5178.1543

BELTON, M. J. S., CHAPMAN, C. R. ET AL. 1995. Bulk density of asteroid 243 Ida from the orbit of its satellite Dactyl. *Nature*, **374**, 785–788.

BESSE, S., LAMY, P., JORDA, L., MARCHI, S. & BARBIERI, C. 2012. Identification and physical properties of craters on Asteroid (2867) Steins. *Icarus*, **221**, 1119–1129, http://dx.doi.org/10.1016/j.icarus.2012.08.008

BINZEL, R. P. & XU, S. 1993. Chips off of asteroid 4 Vesta: evidence for the parent body of basaltic achondrite meteorites. *Science*, **260**, 186–191, http://dx.doi.org/10.1126/science.260.5105.186

BINZEL, R. P., RIVKIN, A. S., BUS, S. J., SUNSHINE, J. M. & BURBINE, T. H. 2001. MUSES-C target asteroid (25143) 1998 SF36: a reddened ordinary chondrite. *Meteoritics & Planetary Science*, **36**, 1167–1172, http://dx.doi.org/10.1111/j.1945-5100.2001.tb01950.x

BISCHOFF, A. & GEIGER, T. 1995. Meteorites for the Sahara: find locations, shock classification, degree of weathering and pairing. *Meteoritics*, **30**, 113–122, http://dx.doi.org/10.1111/j.1945-5100.1995.tb01219.x

BOTTKE, W. F., DURDA, D. D., NESVORNY, D., JEDICKE, R., MORBIDELLI, A., VOKROUHLICKY, D. & LEVISON, H. 2005. The fossilized size distribution of the main asteroid belt. *Icarus*, **175**, 111–140, http://dx.doi.org/10.1016/j.icarus.2004.10.026

BOWLING, T. J., JOHNSON, B. C. & MELOSH, H. J. 2013. Formation of equatorial graben following the Rheasilvia impact on asteroid 4 Vesta. Abstract 1673, presented at the 44th Lunar & Planetary Science Conference. Lunar & Planetary Institute, Houston, TX.

BRITT, D. T. & CONSOLMAGNO, G. J. 2003. Stony meteorite porosities and densities: a review of the data through 2001. *Meteoritics & Planetary Science*, **38**, 1161–1180, http://dx.doi.org/10.1111/j.1945-5100.2003.tb00305.x

BUCZKOWSKI, D. L., BARNOUIN-JHA, O. S. & PROCKTER, L. M. 2008. 433 Eros lineaments: global mapping and analysis. *Icarus*, **193**(1), 39–52, http://dx.doi.org/10.1016/j.icarus.2007.06.028

BUCZKOWSKI, D. L., WYRICK, D. Y. ET AL. 2012. Large-scale troughs on Vesta: a signature of planetary tectonics. *Geophysical Research Letters*, **39**, L18205, http://dx.doi.org/10.1029/2012GL052959

BUCZKOWSKI, D. L., WYRICK, D. Y. ET AL. 2014. The unique geomorphology and physical properties of the Vestalia Terra plateau. *Icarus*, http://dx.doi.org/10.1016/j.icarus.2014.03.035

BURBINE, T. H., MCCOY, T. J., MEIBOM, A., GLADMAN, B. & KEIL, K. 2002. Meteoritic parent bodies: their number and identification. *In*: BOTTKE, W. F., CELLINO, A., PAOLICCHI, P. & BINZEL, R. P. (eds) *Asteroids III*. University of Arizona Press, Tucson, 653–667.

BUS, S. J. & BINZEL, R. P. 2002. Phase II of the small main-belt asteroid spectroscopy survey: a feature-based taxonomy. *Icarus*, **158**, 146–177, http://dx.doi.org/10.1006/icar.2002.6856

CAZENAVE, A., DOBROVOLSKIS, A. & LAGO, B. 1980. Orbital history of the Martian satellites with inferences on their origin. *Icarus*, **44**, 730–744, http://dx.doi.org/10.1016/0019-1035(80)90140-2

CHAPMAN, C. R. 1978. Asteroid collisions, craters, regoliths and lifetimes. *In*: MORRISON, D. & WELLS, W. C. (eds) *Asteroids: An Exploration Assessment*. NASA Conference Publication, **2053**, National Technical Information Service, Springfield, VA, 145–160.

CHAPMAN, C. R., MORRISON, D. & ZELLNER, B. 1975. Surface properties of asteroids: a synthesis of polarimetry, radiometry, and spectrophotometry. *Icarus*, **25**(1), 104–130, http://dx.doi.org/10.1016/0019-1035(75)90191-8

CONSOLMAGNO, G. J. & DRAKE, M. J. 1977. Composition of the eucrite parent body: evidence from rare Earth elements. *Geochimica et Cosmochimica Acta*, **41**, 1271–1282.

DAVIS, D. R., CHAPMAN, C. R., GREENBERG, R., WEIDENSCHILLING, S. J. & HARRIS, A. W. 1979. Collisional evolution of asteroids: populations, rotations and velocities. *In*: GEHRELS, T. (ed.) *Asteroids*. University of Arizona Press, Tucson, 528–557.

DE SANCTIS, M. C., AMMANNITO, E. ET AL. 2012. Spectroscopic characterization of mineralogy and its diversity across Vesta. *Science*, **336**, 697–700, http://dx.doi.org/10.1126/science.1219270

DE SANCTIS, M. C., AMMANNITO, E. ET AL. 2014. Compositional evidence of magmatic activity on Vesta. *Geophysical Research Letters*, http://dx.doi.org/10.1002/2014GL059646

DOMBARD, A. J. & FREED, A. M. 2002. Thermally induced lineaments on the asteroid Eros: evidence of orbit transfer. *Geophysical Research Letters*, **29**, 65-1–65-4, http://dx.doi.org/10.1029/2002GL015181

DRAKE, M. J. 1979. Geochemical evolution of the eucrite parent body: possible evolution of Asteroid 4 Vesta? *In*: GEHRELS, T. (ed.) *Asteroids*. University of Arizona Press, Tucson, 765–782.

DRAKE, M. J. 2001. Presidential address: the eucrite/Vesta story. *Meteoritics & Planetary Science*, **36**, 501–513, http://dx.doi.org/10.1111/j.1945-5100.2001.tb01892.x

DUXBURY, T. C., NEWBURN, R. L. *ET AL*. 2004. Asteroid 5535 Annefrank size, shape and orientation: stardust first results. *Journal of Geophysical Research*, **109**, E02002, http://dx.doi.org/10.1029/2003JE002108

ERNST, R. E., GROSFILS, E. B. & MEGE, D. 2001. Giant dike swarms: Earth, Venus and Mars. *Annual Review of Earth & Planetary Sciences*, **29**, 489–534.

FUJIWARA, A. & ASADA, N. 1983. Impact fracture patterns on Phobos ellipsoids. *Icarus*, **56**, 590–602, http://dx.doi.org/10.1016/0019-1035(83)90176-8

FUJIWARA, A., KAWAGUCHI, J. *ET AL*. 2006. The rubble-pile asteroid Itokawa as observed by Hayabusa. *Science*, **312**, 1330–1334, http://dx.doi.org/10.1126/science.1125841

GAFFEY, M. J. 1997. Surface lithologic heterogeneity of asteroid 4 Vesta. *Icarus*, **127**, 130–157, http://dx.doi.org/10.1006/icar.1997.5680

GRADIE, J. C., CHAPMAN, C. R. & TEDESCO, E. F. 1989. Distribution of taxonomic classes and the compositional structure of the asteroid belt. *In*: BINZEL, R. P., GEHRELS, T. & MATTHEWS, M. S. (eds) *Asteroids II*. University of Arizona Press, Tucson, 316–335.

GREENBERG, R. & CHAPMAN, C. R. 1983. Asteroids and meteorites: parent bodies and delivered samples. *Icarus*, **55**, 455–481, http://dx.doi.org/10.1016/0019-1035(83)90116-1

HARTMANN, W. K. 1979. A special class of planetary collisions: theory and evidence. *Proceedings of the 10th Lunar and Planetary Sciences Conference*, 1897–1916.

HEAD, J. W. & CINTALA, M. J. 1979. *Grooves on Phobos: evidence for Possible Secondary Cratering Origin*. Reports of the Planetary Geology Program, 1978–1979, NASA Technical Memorandum 80339.

HORSTMAN, K. C. & MELOSH, H. J. 1989. Drainage pits in cohesionless materials: implications for the surface of Phobos. *Journal of Geophysical Research*, **94**, 12 433–12 441.

IVANOV, B. A. & MELOSH, H. J. 2013. Two-dimensional numerical modelling of the Rheasilvia impact formation. *Journal of Geophysical Research*, **118**, 1545–1557, http://dx.doi.org/10.1002/jgre.20108

JAUMANN, R., WILLIAMS, D. A. *ET AL*. 2012. Vesta's shape and morphology. *Science*, **336**, 687–690, http://dx.doi.org/10.1126/science.1219122

KEIL, K. 2002. Geologial history of asteroid 4 Vesta: the 'smallest terrestrial planet'. *In*: BOTTKE, W. F., CELLINO, A., PAOLICCHI, P. & BINZEL, R. P. (eds) *Asteroids III*. University of Arizona Press, Tucson, 573–584.

KEIL, K., STOFFLER, D., LOVE, S. G. & SCOTT, E. R. D. 1997. Constraints on the role of impact heating and melting in asteroids. *Meteoritics & Planetary Science*, **32**, 349–363, http://dx.doi.org/10.1111/j.1945-5100.1997.tb01278.x

KELLER, H. U., BARBIERI, C. *ET AL*. 2010. E-type asteroid (2867) Steins as imaged by OSIRIS on board Rosetta. *Science*, **327**, 190–193, http://dx.doi.org/10.1126/science.1179559

KROT, A. N., KEIL, K., SCOTT, E. R. D., GOODRICH, C. A. & WEISBERG, M. K. 2007. 1.05-Classification of Meteorites. *In*: HOLLAND, H. D. & TUREKIAN, K. K. (eds) *Treatise on Geochemistry 1*. Pergamon, Oxford, 1–52, http://dx.doi.org/10.1016/B0-08-043751-6/01062-8

LAGERROS, J. S. V. 1997. Thermal physics of asteroids III: irregular shapes and albedo variegations. *Astronomy & Astrophysics*, **325**, 1226–1236.

LAMBECK, K. 1979. On the orbital evolution of the Martian satellites. *Journal of Geophysical Research*, **84**, 5651–5657.

MAGNUSSON, P., BARUCCI, M. A. *ET AL*. 1989. Determination of pole orientations and shapes of asteroids. *In*: BINZEL, R. P., GEHRELS, T. & MATTHEWS, M. S. (eds) *Asteroids II*. University of Arizona Press, Tucson, 67–97.

MASSIRONI, M., MARCHI, S. *ET AL*. 2012. Geological map and stratigraphy of asteroid 21 Lutetia. *Planetary & Space Science*, **66**, 125–136, http://dx.doi.org/10.1016/j.pss.2011.12.024

McCORD, T. B., ADAMS, J. B. & JOHNSON, T. V. 1970. Asteroid Vesta: spectral reflectivity and compositional implications. *Science*, **168**, 1445–1447, http://dx.doi.org/10.1126/science.168.3938.1445

McFADDEN, L. A., THOLEN, D. J. & VEEDER, G. J. 1989. Physical properties of Aten, Apollo and Amor asteroids. *In*: BINZEL, R. P., GEHRELS, T. & MATTHEWS, M. S. (eds) *Asteroids II*. University of Arizona Press, Tucson, 442–467.

McSWEEN, H. Y. J., MITTLEDFEHLDT, D. W., BECK, A. W., MAYNE, R. G. & McCOY, T. J. 2011. HED meteorites and their relationship to the geology of Vesta and the Dawn Mission. *Space Science Reviews*, **163**, 141–174, http://dx.doi.org/10.1007/s11214-010-9637-z

McSWEEN, H. Y., AMMANNITO, E. *ET AL*. 2013. Composition of the Rheasilvia basin, a window into Vesta's interior. *Journal of Geophysical Research*, **118**, 335–346, http://dx.doi.org/10.1002/jgre.20057

MORBIDELLI, A., BOTTKE, W. F., JR., FROESCHLE, Ch. & MICHEL, P. 2002. Origin and evolution of near-Earth objects. *In*: BOTTKE, W. F., CELLINO, A., PAOLICCHI, P. & BINZEL, R. P. (eds) *Asteroids III*. University of Arizona Press, Tucson, 409–422.

MURCHIE, S. L., ROBINSON, M. *ET AL*. 2002. In flight calibration of the NEAR multispectral imager: II. Results from Eros approach and orbit. *Icarus*, **155**, 229–243, http://dx.doi.org/10.1006/icar.2001.6746

OSTRO, S. J., BENNER, L. A. M. *ET AL*. 2004. Radar observations of Asteroid 25143 Itokawa (1998 SF36). *Meteoritics and Planetary Science*, **39**, 407–424.

PARKER, J. W., STERN, S. A. *ET AL*. 2002. Analysis of the first disk-resolved images of Ceres from ultraviolet observations with the Hubble Space Telescope. *The Astronomical Journal*, **123**, 549–557, http://dx.doi.org/10.1086/338093

PETIT, J.-M., CHAMBERS, J., FRANKLIN, F. & NAGASAWA, M. 2002. Primordial excitation and depletion of the main belt. *In*: BOTTKE, W. F., CELLINO, A., PAOLICCHI,

P. & BINZEL, R. P. (eds) *Asteroids III*. University of Arizona Press, Tucson, 711–723.

PROCKTER, L., THOMAS, P. *ET AL*. 2002. Surface expressions of structural features on Eros. *Icarus*, **155**, 75–93, http://dx.doi.org/10.1006/icar.2001.6770

RAYMOND, C. A., PARK, R. S. *ET AL*. 2013. Vestalia Terra: an ancient mascon in the southern hemisphere of Vesta. Abstract 2882, presented at the 44th Lunar & Planetary Science Conference. Lunar & Planetary Institute, Houston, TX, http://www.lpi.usra.edu/meetings/lpsc2013/pdf/2882.pdf

REDDY, V., NATHUES, A. *ET AL*. 2012. Color and albedo heterogeneity of Vesta from Dawn. *Science*, **336**, 700–704, http://dx.doi.org/10.1126/science.1219088

REIMOLD, W. U. & GIBSON, R. L. 2006. The melt rocks of the Vredefort impact structure – Vredefort Granophyre and pseudotachylitic breccias: implications for impact cratering and the evolution of the Witwatersrand Basin. *Chemie der Erde*, **66**, 1–35, http://dx.doi.org/10.1016/j.chemer.2005.07.003

REIMOLD, W. U., BRANDT, D. & KOEBERL, C. 1998. Detailed structural analysis of the rim of a large complex impact crater: Bosumtwi Crater, Ghana. *Geology*, **26**, 543–546.

RILLER, U. 2005. Structural characteristics of the Sudbury impact structure, Canada: impact-induced v. orogenic deformation – a review. *Meteoritics & Planetary Science*, **40**, 1723–1740.

ROBINSON, M. S., THOMAS, P. C., VEVERKA, J., MURCHIE, S. L. & WILCOX, B. B. 2002. Invited review: the geology of Eros. *Meteoritics and Planetary Science*, **37**, 1651–1684.

RUSSELL, C. T. & RAYMOND, C. A. 2011. The Dawn mission to Vesta and Ceres. *Space Science Reviews*, **163**, 3–23, http://dx.doi.org/10.1007/s11214-011-9836-2

RUSSELL, C. T., RAYMOND, C. A. *ET AL*. 2012. Dawn at Vesta: testing the protoplanetary paradigm. *Science*, **336**, 684–686, http://dx.doi.org/10.1126/science.1219381

SASAKI, S., Saito, J. *ET AL*. 2006. Observations of 25143 Itokawa by the Asteroid Multiband Imaging Camera (AMICA) of Hayabusa: morphology of brighter and darker areas. Abstract 1671, presented at the 37th Lunar & Planetary Science Conference. Lunar & Planetary Institute, Houston, TX, http://www.lpi.usra.edu/meetings/lpsc2006/pdf/1671.pdf

SCHENK, P., O'BRIEN, D. P. *ET AL*. 2012. The geologically recent giant impact basins at Vesta's south pole. *Science*, **336**, 694–697, http://dx.doi.org/10.1126/science.1223272

SCHILLER, M., BAKER, J. A., BIZZARO, M., CREECH, J. & IRVING, A. J. 2010. Timing and mechanisms of the evolution of the magma ocean on the HED parent body. Abstract 5042, presented at the 73rd Annual Meeting of the Meteoritical Society. Lunar & Planetary Institute, New York.

SCHMEDEMANN, N., KNEISSL, T. *ET AL*. 2014. The cratering record, chronology and surface ages of (4) Vesta in comparison to smaller asteroids and ages of HED meteorites. *Planetary & Space Science*, http://dx.doi.org/10.1016/j.pss.2014.04.004

SIERKS, H., LAMY, P. *ET AL*. 2011a. Images of Asteroid 21 Lutetia: a remnant planetisimal from the early solar system. *Science*, **334**, 487–490, http://dx.doi.org/10.1126/science.1207325

SIERKS, H., KELLER, H. U. *ET AL*. 2011b. The Dawn Framing Camera. *Space Science Reviews*, **163**, 263–327, http://dx.doi.org/10.1007/s11214-011-9745-4

SINGER, S. F. 1968. The origin of the moon and its geophysical consequences. *Geophysical Journal of the Royal Astronomical Society*, **15**, 205–226.

STICKLE, A. M., SCHULTZ, P. H. & CRAWFORD, D. A. 2013. Subsurface shear failure in spherical bodies: a possible formation mechanism for the surface troughs on 4 Vesta. Abstract 2417, presented at the 44th Lunar & Planetary Science Conference. Lunar & Planetary Institute, Houston TX.

SULLIVAN, R., GREELEY, R. *ET AL*. 1996. Geology of 243 Ida. *Icarus*, **120**, 119–139, http://dx.doi.org/10.1006/icar.1996.0041

TAKEDA, H. 1997. Mineralogical records of early planetary processes on the howardite, eucrite, diogenite parent body with reference to Vesta. *Meteoritics and Planetary Science*, **32**, 841–853, http://dx.doi.org/10.1111/j.1945-5100.1997.tb01574.x

THOMAS, P. C. & PROCKTER, L. M. 2010. Tectonics of small bodies. *In*: WATTERS, T. R. & SCHULTZ, R. A. (eds) *Planetary Tectonics*. Cambridge University Press, Cambridge, 233–263.

THOMAS, P. C. & ROBINSON, M. S. 2005. Seismic resurfacing by a single impact on the asteroid 433 Eros. *Nature*, **436**, 366–369.

THOMAS, P. & VEVERKA, J. 1979. Grooves on asteroids: a prediction. *Icarus*, **40**, 394–405, http://dx.doi.org/10.1016/0019-1035(79)90032-0

THOMAS, P. C., VEVERKA, J., BLOOM, A. & DUXBURY, T. 1979. Grooves on Phobos: their distribution, morphology and possible origin. *Journal of Geophysical Research*, **84**, 8457–8477.

THOMAS, P. C., VEVERKA, J. *ET AL*. 1994. The shape of Gaspra. *Icarus*, **107**, 23–36, http://dx.doi.org/10.1006/icar.1994.1004

THOMAS, P. C., VEVERKA, J. *ET AL*. 1999. Mathilde: size, shape and geology. *Icarus*, **140**, 17–27, http://dx.doi.org/10.1006/icar.1999.6121

THOMAS, P. C., VEVERKA, J., ROBINSON, M. S. & MURCHIE, S. 2001. Shoemaker crater as the source of most ejecta blocks on the asteroid 433 Eros. *Nature*, **413**, 394–396.

THOMAS, P. C., PROCKTER, L., ROBINSON, M., JOSEPH, J. & VEVERKA, J. 2002. Global structure of asteroid 433 Eros. *Geophysical Research Letters*, **29**, 46-1–46-4, http://dx.doi.org/10.1029/2001GL014599

THOMAS, N., BARBIERI, C. *ET AL*. 2012. The geomorphology of (21) Lutetia: results from the OSIRIS imaging system onboard ESA's Rosetta spacecraft. *Planetary and Space Science*, **66**, 96–124, http://dx.doi.org/10.1016/j.pss.2011.10.003

TROMBKA, J. I., SQUYRES, S. W. *ET AL*. 2000. The elemental composition of asteroid 433 Eros: Results of the NEAR-Shoemaker X-ray Spectrometer. *Science*, **289**, 2101–2105, http://dx.doi.org/10.1126/science.289.5487.2101

VEVERKA, J., THOMAS, P. *ET AL*. 1994. Discovery of grooves on Gaspra. *Icarus*, **107**, 399–411, http://dx.doi.org/10.1006/icar.1994.1007

VEVERKA, J., THOMAS, P. *ET AL*. 1997a. NEAR's flyby of 253 Mathilde: images of a C asteroid. *Science*, **278**, 2109–2114, http://dx.doi.org/10.1126/science.278.5346.2109

VEVERKA, J., BELL, J. F., III *ET AL.* 1997*b*. An overview of the NEAR multispectral imager – near-infrared spectrometer investigation. *Journal of Geophysical Research*, **102**, 23,709–23,727, http://dx.doi.org/10.1029/97JE01742

VEVERKA, J., THOMAS, P. *ET AL.* 1999. NEAR encounter with asteroid 253 Mathilde: overview. *Icarus*, **140**, 3–16, http://dx.doi.org/10.1006/icar.1999.6120

WEISBERG, M. K., MCCOY, T. J. & KROT, A. N. 2006. Systematics and evaluation of meteorite classification. *In*: LAURETTA, D. S. & MCSWEEN, H. Y. (eds) *Meteorites and the Early Solar System II*. University of Arizona Press, Tucson, 19–52.

WILKISON, S. L. & ROBINSON, M. S. 2000. Bulk density of ordinary chondrite meteorites and implications for asteroidal internal structure. *Meteoritics & Planetary Science*, **35**, 1203–1213, http://dx.doi.org/10.1111/j.1945-5100.2000.tb01509.x

WILKISON, S. L., ROBINSON, M. S. *ET AL.* 2002. An estimate of Eros's porosity and implications for internal structure. *Icarus*, **155**, 94–103, http://dx.doi.org/10.1006/icar.2001.6751

WILLIAMS, D. A. *ET AL.* 2013. Lobate and flow-like features on asteroid Vesta. *Planetary & Space Science*, http://dx.doi.org/10.1016/j.pss.2013.06.017

WILSON, L. & HEAD, J. W. 1989. Dynamics of groove formation on Phobos by ejecta from Stickney. *In*: *Lunar &* *Planetary Science 20*. Lunar & Planetary Institute, Houston, 1211–1212.

WILSON, L. & HEAD, J. W. 2002. Tharsis-radial graben systems as the surface manifestations of plume-related dike intrusion complexes: models and implications. *Journal of Geophysical Research*, **107**, 1–24, http://dx.doi.org/10.1029/2001JE001593

WILSON, L. & KEIL, K. 1996. Volcanic eruptions and intrusions on the asteroid 4 Vesta. *Journal of Geophysical Research*, **101**, 18 927–18 940, http://dx.doi.org/10.1029/96JE01390

WILSON, L. & KEIL, K. 2012. Volcanic activity on differentiated asteroids: a review and analysis. *Chemie der Erde – Geochemistry*, **72**, 289–322, http://dx.doi.org/10.1016/j.chemer.2012.09.002

WILSON, L., KEIL, K. & LOVE, S. J. 1999. The internal structures and densities of asteroids. *Meteoritics & Planetary Science*, **34**, 479–483, http://dx.doi.org/10.1111/j.1945-5100.1999.tb01355.x

WYRICK, D., FERRILL, D. A., MORRIS, A. P., COLTON, S. L. & SIMS, D. W. 2004. Distribution, morphology and origins of Martian pit crater chains. *Journal of Geophysical Research*, **109**, E06005, http://dx.doi.org/10.1029/2004JE002240

YEOMANS, D. K., ANTREASIAN, P. G. *ET AL.* 2000. Radio science results during the NEAR-Shoemaker spacecraft rendezvous with Eros. *Science*, **289**, 2085–2088, http://dx.doi.org/10.1126/science.289.5487.2085

Index

Page numbers in *italic* denote Figures. Page numbers in **bold** denote Tables.